INTERNATIONAL ENCYCLOPAEDIA OF FOOD AND NUTRITION

VOLUME 17

NUTRITION OF ANIMALS OF AGRICULTURAL IMPORTANCE

VOLUME EDITOR

SIR DAVID CUTHBERTSON, C.B.E., F.R.S.E., *Glasgow*

PART 2

INTERNATIONAL ENCYCLOPAEDIA OF FOOD AND NUTRITION

NUTRITION OF ANIMALS
OF AGRICULTURAL IMPORTANCE

PART 2

ASSESSMENT OF AND FACTORS AFFECTING REQUIREMENTS OF FARM LIVESTOCK

VOLUME EDITOR

SIR DAVID CUTHBERTSON, C.B.E., F.R.S.E.

Department of Pathological Biochemistry, Glasgow University and Royal Infirmary, Glasgow C.4

1966
THE QUEEN'S AWARD TO INDUSTRY 1966

PERGAMON PRESS

OXFORD · LONDON · EDINBURGH · NEW YORK
TORONTO · SYDNEY · PARIS · BRAUNSCHWEIG

Pergamon Press Ltd., Headington Hill Hall, Oxford
4 & 5 Fitzroy Square, London W.1
Pergamon Press (Scotland) Ltd., 2 & 3 Teviot Place, Edinburgh 1
Pergamon Press Inc., Maxwell House, Fairview Park, Elmsford, New York 10523
Pergamon of Canada Ltd., 207 Queen's Quay West, Toronto 1
Pergamon Press (Aust.) Pty. Ltd., 19A Boundary Street, Rushcutters Bay, N.S.W. 2011, Australia
Pergamon Press S.A.R.L., 24 rue des Écoles, Paris 5e
Vieweg & Sohn GmbH, Burgplatz 1, Braunschweig

First edition 1969

Library of Congress Catalog Card No. 69-14229

PRINTED IN GREAT BRITAIN BY THE ANCHOR PRESS LTD., TIPTREE, ESSEX

08 003584 1

Contents

PART 2

*These chapters are reprinted from Agricultural Research Council Publications, published by H.M.S.O.

Contents

PART 1

List of Contributors

BOLTON, W., *Agricultural Research Council, Poultry Research Centre, Edinburgh*

CARE, A. D., *The Rowett Research Institute, Bucksburn, Aberdeen*

COATES, M. E., *National Institute for Research in Dairying, Shinfield, Reading*

CORBETT, J. L., *C.S.I.R.O., Pastoral Research Laboratory, Armidale, New South Wales*

CUTHBERTSON, D. P., *Department of Pathological Biochemistry, Royal Infirmary, Glasgow C.4*

DODSWORTH, T. L., *The North of Scotland College of Agriculture, Aberdeen*

FELL, B. F., *The Rowett Research Institute, Bucksburn, Aberdeen*

FLATT, W., *United States Department of Agriculture, Beltsville, Maryland*

GARTON, G. A., *The Rowett Research Institute, Bucksburn, Aberdeen*

GREENHALGH, J. F. D., *The Rowett Research Institute, Bucksburn, Aberdeen*

HILL, F. W., *Department of Nutrition, University of California, Davis, California*

HOBSON, P. N., *The Rowett Research Institute, Bucksburn, Aberdeen*

JAKOBSEN, P. E., *Department of Animal Physiology, Rolighedsvej 25, Copenhagen*

JAYNE-WILLIAMS, D. J., *National Institute for Research in Dairying, Shinfield, Reading*

JONES, A. S., *The Rowett Research Institute, Bucksburn, Aberdeen*

LAMMING, G. E., *University of Nottingham, School of Agriculture, Sutton Bonington, Loughborough*

LODGE, G. A., *University of Nottingham, School of Agriculture, Sutton Bonington, Loughborough*

McDONALD, I. W., *Division of Animal Physiology, Ian Clunies Ross Animal Research Laboratory, C.S.I.R.O., Parramatta, New South Wales*

MILLS, C. F., *The Rowett Research Institute, Bucksburn, Aberdeen*

OLSSON, N., *Verberöd, Sweden*

PAYNE, W. J. A., *F.A.O./U.N.D.P., Dairy Training and Research Institute, Philippines*

PHILLIPSON, A. T., *Department of Veterinary Clinical Studies, The University of Cambridge*

PRESTON, T. R., *Instituto de Ciencia Animal, Catalina de Güines, Habana, Cuba*

PULLAR, J. D., *The Rowett Research Institute, Bucksburn, Aberdeen*

RIIS, P. M., *Department of Animal Physiology, Rolighedsvej 25, Copenhagen*

ROBINSON, K. L., *Agricultural Chemistry Department, The Queen's University of Belfast*

ROLFE, E., *National College of Food Technology, Weybridge*

ROY, J. H. B., *National Institute for Research in Dairying, Shinfield, Reading*

SCOTT, M. L., *Department of Poultry Science and Graduate School of Nutrition, Cornell University, Ithaca, New York*

SLEN, S. B., *Canada Department of Agriculture, Research Station, Lethbridge, Alberta*

VAN SOEST, P. J., *United States Department of Agriculture, Beltsville, Maryland*

WOODHAM, A. A., *The Rowett Research Institute, Bucksburn, Aberdeen*

WRAGG, S. R., *Department of Economics (Agricultural Economics), the University, Bristol*

Preface

THE purpose of Volume 17 is to provide first a general background to the science of nutrition of animals of agricultural importance and then to assess their requirements in the light of existing knowledge. Thus the contents are presented in two parts.

Part 1 provides first an introduction to the study of nutrition of animals of agricultural importance and there is a chapter on the newer knowledge of the composition and methods of analysis of feeding stuffs. Then follow chapters on the physiology, biochemistry and microbiology of digestion and absorption in the main classes of livestock. These are followed by details of energy metabolism and the metabolism of specific groups of nutrients and on the relation of nutrition to reproduction and the endocrine control of growth and metabolism. The final chapter in this first part is on the preservation of animal products for human consumption.

Part 2 is concerned with the special nutrition of the different classes of farm livestock, including the horse, and the assessment of their requirements. 'Requirement' has been defined as the amount of a nutrient that must be supplied in the diet to meet the needs of the normal healthy animal given a completely adequate diet in an environment compatible with good health. In general, the requirements include no margins of safety and are estimated requirements, but some discussion on allowances is provided. The methods of estimating these requirements have differed for the different nutrients and are further expounded in the Technical Reviews of the Nutrient Requirements of Farm Livestock published for 'Poultry', 'Ruminants' and 'Pigs' by the U.K. Agricultural Research Council, London. By kind permission of the Agricultural Research Council, Summaries of these Technical Reviews are included in these volumes as separate chapters, and the comparable publications on Nutrient Requirements of Domestic Animals by the Committee on Animal Nutrition of the U.S. National Academy of Sciences National Research Council also provide much valuable knowledge.

While most of the observations relate to temperate climates, attention has also been paid to the problems of the nutrition of ruminants in the tropics. There is also a chapter on nutritional disorders and the so-called 'metabolic' diseases.

Recent developments in our basic understanding of ruminant nutrition, in particular, and in the evolution of a new system of measuring nutrient requirements and feeding values have a very special significance in agricultural economics. Economists will regard current progress in this field of animal science as a means of making the direct econometric approach more effective through a better understanding of the underlying principles of the input/output relationships which this new knowledge provide. Some of the information on metabolizable energy requirements and feed values may allow them to predict, for example, the economic allowances of concentrates for various milk/concentrate price ratios. A number of the basic assumptions in linear programming may be upset by the fact that nutrient affects are not additive. These matters of import are dealt with in the penultimate chapter of Part 2. The final chapter surveys world needs.

As Editor of this volume I should like to say how fortunate I have been in securing as authors many of the leading investigators in the science of nutrition and its application to the animal production industry, and I am deeply conscious of the honour they do us in participating in the writing of this work. I am greatly indebted to my very efficient secretaries Miss R. A. Watson and Mrs. E. Coulter. We should also like to take this opportunity of thanking the publishers for their great help and care.

General Introduction

THE three primitive instincts of all higher animals including man are nutrition, self-preservation and sex. Of these nutrition is the earliest since all animals and plants ingest and metabolize food. Hence the science of nutrition and the practical application of that science interest those engaged in a wide variety of disciplines. For instance, the chemist studies the composition of foodstuffs, the biochemist their transformations in the body with consequent growth and energy-production, the physiologist the normal processes of metabolism and the pathologist the changes that result from abnormalities of those processes. The clinician must recognize the disorders produced by faulty or insufficient food and the dietist helps to correct these; agriculturalists and economists are concerned with the production and distribution of food whether animal or vegetable, sociologists and politicians with the national and international aspects of that production and distribution.

There are two factors that have made desirable the collection and integration of existing knowledge of food and nutrition at the present time. First, this knowledge has been advanced very rapidly in the past few years largely by the great discoveries in biochemistry. Secondly, there is no more urgent problem in the world today than the nutritional disasters that are now present amongst us and will increase with the explosive rise in the population of the world. Hundreds of thousands of years have been needed to reach our present population of 3400 million, but another 3400 million persons will probably be produced in the next 35 years. And it is alleged that two-thirds of the present population is underfed. Whatever the proportion, we have a food deficit that is certain to increase and must be met by increased production of food whether by more efficient methods, bringing new lands into cultivation or by the use of new types of foodstuffs, for instance from oil.

Nor need the other third be complacent about its present food and nutrition, for the rapid developments in food technology have created new problems as foods are increasingly processed and sophisticated. We now treat foods with a variety of chemicals. Such additives are essential for two reasons First, the housewife increasingly demands partially cooked foods in plastic wrappers which she merely warms at home; she no longer prepares and cooks what she gathered from her garden. Secondly, we have to produce the maximum amount of food from the land and then store it and transport it over long distances. Therefore we very properly use chemical fertilizers, pesticides, hormones, antibiotics and improvers (such as vitamins A and D in margarine). All these activities create problems.

An encyclopaedia is a work that contains exhaustive information on a branch of knowledge, arranged systematically and not necessarily alphabetically. The aims, scope and purpose of this one were discussed with Mr. Robert Maxwell, the publisher, at the International Congresses of Nutrition in Washington, Edinburgh and Hamburg in 1960, 1963 and 1966. The plan is to provide an authoritative and comprehensive presentation of knowledge in the whole field of nutrition of man and lower animals, with brief reference to the nutrition of plants and microorganisms. Each volume is written primarily for nutritional scientists, but is of value to others such as chemists, biochemists, medical practitioners, dietists, veterinary surgeons, agriculturalists and economists. Experimental details of an essentially practical nature are not emphasized in the Encyclopaedia; but a thorough background of the theoretical basis of techniques is included so that these may be understood and applied at the fullest extent. The Encyclopaedia provides for investigations in the fundamental biological sciences—biophysics, biochemistry, physiology, pathology, psychology, experimental therapeutics, experimental zoology and related fields—data which will help each in his own field. Further, the advanced student in both the basic medical sciences and clinical specialities will find these volumes an invaluable guide and an authoritative source of reference to literature.

As Editor-in-Chief I have been supported by five Regional Editors and by an Honorary Editorial Advisory Board composed of the world's leading nutritional scientists and under the Chairmanship of the Publisher, Mr. Robert Maxwell. We regret that the authorship of the articles has not been as widely international as is desirable, but while in some volumes there is a preponderance of articles from the English-speaking countries, greater emphasis on international authorship has been achieved in many parts of the Encyclopaedia. Perhaps it is inevitable that in multi-author works delays in one or two chapters should affect a whole volume, but we believe that this has not detracted seriously from the usefulness of the work as a whole. Indeed, the rapid advances in the science of nutrition create a problem. To meet it an annual review will be published entitled 'Progress in the Science of Nutrition and Food Technology' which will report advances in the respective fields covered by the Encyclopaedia and give full bibliographical details of recent works for easy reference.

H. M. SINCLAIR
Oxford

CHAPTER 17

The Nutritional Value of Grassland Herbage

J. L. CORBETT

C.S.I.R.O., Pastoral Research Laboratory, Armidale, New South Wales

INTRODUCTION

It has been estimated that about one-quarter of the earth's land surface is covered by grasslands.[1] In a natural state they vary from the treeless prairies, steppes and pampas of the temperate regions to the savannahs of the tropics and sub-tropics with their varying proportions of trees and shrubs. The vegetation and the innumerable wild animals of many species they sustain have presumably acted as reciprocal factors in natural selection.[2]

Man has domesticated only a few of these species of animals and he has so far brought into cultivation only a few grasses; perhaps 40 of about 10,000 known species account for over 90 per cent of sown pastures[3] but in the main their chemical composition, which is frequently used as an index of nutritional value, is not especially distinctive. When compared with many of the species they are intended to replace, the cultivated grasses do usually show a greater ability to respond by increased growth to an increase in soil fertility. They often provide qualitative as well as quantitative benefits to the animal in that maturation, seed-shed, senescence and decay may proceed less rapidly and may be delayed more readily by management.

Grassland communities are naturally dynamic. The balance of the species is determined by many factors—climatic, topographic, edaphic and biotic—and can be changed markedly by the type of management imposed.[4] The persistence of the more productive grasses depends to a large extent upon a continued supply of plant nutrients. Legumes, as a result of their symbiotic relation with *Rhizobia* play a major role in improving the nitrogen status of soils.[5,6] They have therefore come to be an important constituent of grasslands and of the diet of grazing animals. Efforts to obtain legumes that can be established satisfactorily in tropical and sub-tropical pastures have only recently begun to meet with success.[7]

Many other plant species contribute to grassland communities. Dicotyledonous herbaceous plants, known collectively as forbs, often contain higher concentrations of minerals than associated grasses and legumes[8] but produce lower yields of dry matter and may be less digestible.[9] These species and the many woody 'browse' plants that make an important contribution to the feed supply of grazing animals in many areas will not be considered in this review. In addition, toxicoses and other disorders in livestock due to plant constituents or contaminants, such as alkaloids in *Phalaris*,[10] oestrogenic substances,[11] and photosensitization caused by the fungus *Pithomyces chartarum*[12] will not be discussed.

The botanical, chemical and morphological diversity of grassland vegetation is matched by the diversity of methods of utilization. These vary from highly capitalized systems of mechanical harvesting for the feeding of fresh or conserved herbage to confined animals, to pastoral systems where animals are wholly dependent on grazing throughout the year. Even in Britain and the U.S.A., where the grazing season may be short and there are large areas of arable crops, grasslands remain for economic reasons the chief source of feed for sheep and cattle and provide at least 60 per cent of their total annual feed requirements.[13,14] At the other extreme Australia's 160 million sheep (about one-sixth of the world's total) and 19 million cattle obtain virtually all their feed from grazing.

NON-RUMINANTS

Some non-ruminant animals that utilize grassland herbage will be mentioned briefly here. Pigs and poultry are commonly held on pasture though sometimes the consumption of herbage, which may be as much as 2 kg of dry matter daily by sows[15] and 20 g daily by laying hens,[16] is incidental to other intended benefits such as the acquisition from the soil of iron and other nutrients. Dried herbage is used in large amounts in rations compounded for these animals and here the paramount considerations are low levels of structural carbohydrate, the concentration and the quality of the protein (see Chapter 8), and the concentrations of vitamins and of minerals. The horse is able to utilize cellulosic material as a result of microbial digestion in its caecum and colon[17] which may account for half the total capacity of the alimentary tract.[18] The end products of this digestion include volatile fatty acids and newly synthesized water-soluble vitamins (cf. Chapters 3 and 4—Digestion in the ruminant). Microbial digestion is preceded by gastric digestion and the stomach of the horse accounts for only about one-eighth of the capacity of the alimentary tract. If the digestibility of the fodder eaten is low, so that the volume of indigestible residues is large, working animals especially may be in negative energy balance. The grazing behaviour of the horse tends to reduce the effective nutritional value of an area of pasture for this species.[19]

Where they defaecate they do not graze, and mature herbage accumulates; conversely where they graze they do not defaecate, and may overgraze.

CRITERIA OF NUTRITIONAL VALUE

The nutritional value of feeds for domestic livestock is expressed by the level of animal production of the desired type that is achieved and, unless the animal is killed at an early age for meat, by the length of productive life. It is a function not only of the proportion of the feed made available to the animal by digestion, and the suitability of the digested nutrients for the various metabolic processes but also, and most importantly, of the quantity consumed. Such information is obtained from biological measurements and data from chemical and other types of analyses can be interpreted only in their light.

The nutritional value of a feed is generally given in terms of the value per unit weight but that of grassland herbage has also to be viewed in a wider context. A far larger amount of herbage is harvested directly by animals themselves in grazing than is fed to them by hand. In grazing, the animal is directly subject to the climate and it gains a certain weight of herbage with more or less effort, and so with smaller or greater net benefit, according to the abundance, the pattern of distribution, and the quality of the herbage available. These characteristics of the pasture themselves reflect the climate, the previous activity of the animal, the net growth rate of the plant cover (actual growth less wastage by decay, etc.), the condition of the soil and the complex interactions that ensue.

The nutritional value of grassland herbage, both fresh and conserved and whether grazed or hand-fed, is determined primarily by its ability to meet the requirements of the animal for energy and protein.[20] The caloric need can be regarded as the primary need to which the requirements for all nutrients are linked[21] and in fact the grassland feeds of temperate zones do not usually have protein contents lower than those required by the animal at whatever level of production it attains in response to the amount of energy these feeds supply. Problems of protein supply are largely confined to the tropics and sub-tropics though even here the primary need is often for an increase in energy supply.[267] There are no universal inadequacies (or excesses) of minerals, or deficiencies of those vitamins that the ruminant does not derive from microbial syntheses in its digestive tract; such problems are of a relatively local or temporary nature.

Energy value especially, and protein value, will be discussed in the following sections on voluntary feed intake, digestibility, the utilization of absorbed nutrients and feed evaluation. Specific nutrients, herbage conservation and various methods of feed preparation will then be considered, and a final section will deal with nutritional consequences of grazing.

VOLUNTARY FEED INTAKE

The intake of feed in the long form only will be discussed in this section; processed feeds will be considered later.

The eating behaviour of higher animals, including the regulation of intake, appears ultimately to be subject to central nervous control,[22,23] but of greater immediate importance is the nature of the stimuli that provoke the neural responses. A number of homeostatic mechanisms for quantitative regulation have been proposed, [22,24,25] e.g. the stimulation of receptors sensitive to the concentrations of certain metabolites in blood, and in non-ruminant animals physical mechanisms appear to be of relatively less importance. In contrast the regulation of intake in ruminant animals can be described to a considerable extent, though not exclusively, in physical terms. Much evidence has accumulated that cattle and sheep tend to eat so that a certain degree of distention of the digestive tract—in particular of the reticulo-rumen—with feed or with indigestible residues is maintained or restored.[26,27,28] At this point appropriate physiological responses are presumably evoked by receptors sensitive to mechanical stimuli such as stretch.[29]

The reticulo-rumen normally contains about three-quarters of the total contents of the gut.[18] The more rapidly the breakdown and digestion of feed proceeds in this organ the faster can be the rate of passage of digesta onward through the small orifice leading to the omasum and the rest of the alimentary canal and, ultimately, the excretion of residues in faeces; hence the greater is the quantity of feed that must be eaten in a given period of time if a certain degree of 'fill' is to be maintained. The amount of feed eaten is thus a function of the rates of digestion in and of passage from the reticulo-rumen but these processes are the major steps in the degradation of feed in the alimentary tract of ruminants[30] and so it is found that they can be related to the extent of digestion as measured in digestibility trials.[27,28,31] The general form of the relationship between the digestibility* and the voluntary intake of feed by cattle and sheep is shown diagrammatically in Fig. 1.

In order to compare the feed intakes by animals differing widely in liveweight (W), they are often expressed in terms of metabolic body size ($W^{0.73}$ or $W^{0.75}$).[31,32,33,125] $W^{0.75}$ has recently been recommended[204] for general adoption. Intakes have also been related to $W^{1.00}$ [34,35,36] and some of the published exponents,[37,38,39,40] or those that can be calculated from the raw data given by some workers,[32,35,41] have fiducial limits that do not exclude either 0·73 or 1·00. Because of its frequent usage hitherto, $W^{0.73}$ will be used here.

When intakes are adjusted for differences in liveweight the values found for a feed of a given digestibility vary with the age, physiological state and

* $\frac{\text{Feed} - \text{Faeces}}{\text{Feed}} \times 100$. Known as the apparent digestibility; the faeces includes matter of endogenous origin as well as feed residues.

other characteristics of the animal.[42,76] For example, a lactating dairy cow might eat 120 g dry matter/kg $W^{0.73}$/day, whereas a steer or wether sheep might eat no more than 80 g/kg $W^{0.73}$. Differences in intake are found between animals of a similar type; coefficients of variation ranging from 7 to 14 per cent have been reported.[31,32,33,37,43]

FIG 1—Diagrammatic representation of relationships between feed digestibility and the intake and excretion of dry matter (D.M.) per unit of live weight ($W^{0.73}$).

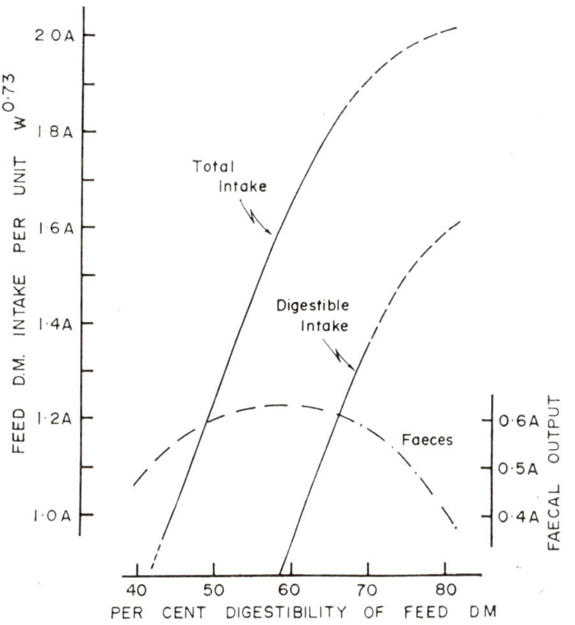

For the purposes of this diagram the dry matter intake, A, was taken to be 40 g/kg $W^{0.73}$ at a digestibility of 45 per cent; the intake was assumed to increase by 1.5 g/kg $W^{0.73}$ per unit change in digestibility over the range <45 to about 65 per cent (see text and Table 1).

The abscissa in Fig. 1 may be scaled in terms of the apparent digestibility of the feed dry matter, of the organic matter, or of the energy, and the actual values will range from a minimum of about 30 per cent to a maximum of about 85 per cent. In a certain range of digestibility, approximately 60 to 70 per cent, it has been found[31,34,44,45] that there is an inflexion in the feed intake line. If the complement of this line is drawn, namely that describing the change in faecal output, it is seen that as digestibility falls from the higher values the quantity of faeces excreted first increases[37,45] and then becomes approximately constant,[45] but ultimately it will decrease. The finding that in the higher range of digestibility ruminant animals do not eat to the extent that they have to deal with feed residues as large as they evidently can excrete, implies that regulation of intake in this range is primarily physiological in type. Thermostatic or chemostatic mechanisms may then supervene at threshold levels determined by the metabolic activity of the animal, e.g. the level of milk production, so that intake will be adjusted to meet the physiological demand for energy.[28,31,34,37,44,45,62] As feed quality falls physiological regulation becomes overshadowed, but certainly not superseded, by primarily physical effects to which most attention has been given in work done up to the present time.

In this discussion the chief interest lies in how the slope of the feed intake line, that is the change in intake per unit change in digestibility, and the position of the line, that is intake at a given digestibility, varies between feeds. Up to the present time (1966), only a few publications have either given this information directly or provided data that are adequate in form and extent for re-calculation by regression analysis; available information is given in Table 1.

The values for the regression coefficients will be considered first. Digestibilities of 70 per cent or more were, with few exceptions, found only with the temperate species. These species were examined to determine if the regressions of intake on digestibility departed significantly from linearity in the manner illustrated in Fig. 1, but the observations were too few in number, or showed too much scatter, and no significant quadratic or higher terms were detected. The significant regression coefficients given in Table 1 show that for a change in digestibility of one unit in both temperate and tropical feeds, the corresponding change in intake varied between about 1·0 and 1·7g DM/kg $W^{0.73}$. The effects that such a change in nutritional value will have upon animal production may be large. Blaxter[31] has cited the example of two hays of 50 and 55 per cent digestibility respectively. If a sheep eats 62 g/kg $W^{0.73}$ of the former, and there is an increase in intake per unit of digestibility of 2 g/kg $W^{0.73}$, it will gain 28 per cent more energy in digested nutrients when offered the latter. This represents nearly a two-fold increase in the amount of energy available above maintenance needs, and so a 10 per cent increase in digestibility may promote nearly 100 per cent increase in liveweight gain. The particular importance of voluntary intake as a measure of the nutritional value of tropical roughages has been stressed by Milford.[41] For a large part of the year the intake of many of these feeds may be below that required to meet the animals' needs.

The failure to obtain significant regression relationships for some of the feeds listed may reflect the variability between animals in intake performance and a paucity of data, but it is also possible that in these cases the rates of breakdown in the rumen were not adequately described by the digestibility data. Measurements of rates of breakdown or of rumen fill under steady state conditions[48] might resolve such anomalies unless a mechanism of intake

TABLE 1

Relationships between the Digestibility and the Voluntary Intake of Feed by Sheep. Intakes (I) expressed as g dry matter per kg liveweight$^{0.73}$

Plant species	No. of observations	Range in digestibility %(*D*)	Range in *I*	Change in *I* per unit change in *D*	Intakes at *D* = 70	Intakes at *D* = 55
Temperate						
1. *Lolium multiflorum* var. S22	10	60·0 – 84·6	62·5 – 90·9	1·12	74·2	
2. *L. multiflorum x L. perenne* var. H1	10	67·3 – 82·9	54·0 – 93·8	1·66	67·0	
3. *Festuca arundinacea* var. S170 (year 1)	6	65·7 – 79·5	51·8 – 81·4	1·33 (N.S.)[a]	68·9	
4. *Phleum pratense* var. S48 (year 1)	7	66·8 – 75·4	56·4 – 68·8	– 0·55 (N.S.)	64·8	
5. *Festuca pratensis* var. S215	8	60·9 – 81·2	52·1 – 91·8	1·55	71·8	
6. *Dactylis glomerata* var. Germinal	11	58·0 – 77·9	39·4 – 90·2	1·64	71·2	
Species 1 – 6, combined data	52	58·0 – 84·6	39·4 – 93·8	1·36	69·6	
F. arundinacea and *P. pratense* (year 2), combined data	11	62·5 – 83·3	39·9 – 64·8	0·77	51·9	
Tropical						
1. *Setaria sphacelata* var. Nandi	11	46·7 – 63·0	28·3 – 70·0	1·26 (N.S.)		36·8
2. *Digitaria decumbens*	15	40·7 – 68·9	25·7 – 65·5	0·72 (N.S.)		37·7
3. *Chloris gayana* (CPI16144)	7	54·0 – 60·0	34·3 – 46·1	0·57 (N.S.)		39·7
4. *C. gayana* var. Callide	9	48·7 – 62·5	51·3 – 64·3	0·53 (N.S.)		58·1
5. *Sorghum almum*	33	35·6 – 67·8	37·8 – 86·0	0·74		57·7
6. *Pennisetum clandestinum*	16	52·9 – 68·7	21·4 – 75·1	1·18 (N.S.)		51·5
7. *Cenchrus ciliaris* var. Molopo	22	34·1 – 60·6	21·6 – 78·9	1·20		63·2
8. *Phaseolus atropurpureus* var. Siratro (Legume)	12	40·3 – 56·8	50·5 – 82·2	1·05		69·2
9. *Glycine javanica* var. Cooper (Legume)	5	46·1 – 60·7	68·7 – 91·5	1·38		82·0
10. Various	90	28·0 – 76·0	9·9 – 82·3	1·22		49·5
11. Various (fresh herbage and dried fodders)	21	46·3 – 68·4	30·0 – 69·3	0·42 (N.S.)		43·2
12. Various	20	55·0 – 71·3	42·6 – 103·3	2·05		54·3

[a]. Regression not significant.

Sources: Temperate species ref. 33; Tropical species nos. 1–9 ref. 47; no. 10 ref. 41; no. 11 ref. 35; no. 12 ref. 350. Values of *I* for nos. 10, 11 and 12 recalculated from the published data.

regulation was involved other than through gut distension. This mechanism might involve humoral factors such as high concentrations of acetic acid in peripheral blood, [49,50,51] or the nitrogen status of the animal.[58,59]

The drastic reduction in intake and the low faecal excretion when feed digestibility approaches the bottom of the range can still be explained fairly well in physical terms but also reflects the nutritional status of the animal and the state of the symbiotic 'functional field'[52] of the rumen. Breakdown and digestion of feed in the rumen, though aided by the mechanical effects of mastication[53] and rumination,[52,54] is mainly achieved by the activity of the resident micro-flora and -fauna. This activity profoundly affects the nutrition of the host (Chapter 3) but can also be regarded from another standpoint, that of the nutrition of the microbial population. Growth and activity of the population will be limited by, amongst other things, the supply of appropriate nutrients and it can be expected that variation in the supply will be reflected in the rates of processes in the rumen.[55] Roughages of low digestibility invariably contain low proportions of available nitrogen and there is evidence that a supply of nitrogen insufficient for the maintenance of an actively functional ruminal population is a major factor limiting the utilization of these feeds by ruminants.[55,58,59] Shortages of some other nutrients for bacteria such as certain inorganic ions and branched chain fatty acids [29,56] may also be associated with the limitation of intake when poor quality feeds are given.[67,59]

There is a limited amount of direct information on the minimum level of dietary nitrogen necessary. Recycling of nitrogen, particularly of urea in saliva and by diffusion through the rumen epithelium, plays an important role in the nitrogen economy of the animal[29,347] so that the effects of a change from a higher to a lower nitrogen ration may not immediately become apparent. In a long term trial with sheep given steadily decreasing amounts of nitrogen in the feed (800 g dry matter per day), Moir and Harris[55] found that intakes started to fall when the nitrogen intake was below 7 g per day, that is about 0·8 per cent (5 per cent crude protein) in the feed that was used. Large changes were found within the rumen: the concentration of bacteria was greatly reduced as was the rate of digestion of cotton thread. In many experiments where the nitrogen content of poor quality roughage diets has been raised by the addition of urea (see Chapter 4), the intake of these feeds has been increased and concomitant increases in the rates of digestion and passage have been observed.[61] There have, however, been conflicting reports[57,60] and recent work[59] suggests that increased nitrogen intakes, apart from any effects upon ruminal bacteria and the rates of digestion and passage of feed, may promote increases in voluntary intake because of an improvement in the nitrogen status of the animal. When casein was infused into the duodenum of sheep, their intakes of wheaten straw supplemented with urea increased although there were no measurable changes in the rates of cotton thread digestion in the rumen, in dry matter digestibility, mean retention time of the feed in the alimentary tract, and the extent of rumination activity. Improvements in acetate and propionate tolerances were found suggesting that when dietary nitrogen levels and the nitrogen status of animals are low, intake is, at least in part, regulated chemostatically by the ability of the animal to cope with these (and probably other) metabolites.

Some other studies on the amounts of protein in feed associated with satisfactory consumption and efficient utilization have given results in general agreement with those of Moir and Harris.[55] Elliott and Topps[63] found that dry matter intake was positively related to crude protein content over the range 2·6 to 10 per cent; the data suggested that the effect was greater below rather than above 8 per cent. Blaxter and Wilson[64] suggested that a minimum level of 8·5 per cent was required and Milford and Minson[47] found that the intake of two sub-tropical herbage species declined significantly only when their crude protein content was 7 per cent or less. From a statistical study of a large volume of published data, Glover and Dougall[65] found that the digestibility of crude fibre plus nitrogen free extractive began to fall markedly at crude protein levels below about 6 per cent.

All these reports suggest that improvement in the crude protein content of tropical and sub-tropical feeds, which is often well below even 6 per cent for long periods of the year, may be one of the important goals of the plant breeder in these areas. In temperate zones, conserved herbage and particularly hay may contain as little as 1 per cent of nitrogen because the material has been cut at a mature stage to obtain high yields per unit area, and may also have been badly processed. In these cases protein insufficiency is generally of less importance than the energy value, particularly if the feed forms only a part of the ration or is given for only part of the year when other feeds of higher protein content, such as fresh grass, are not available. If the herbage is cut at an earlier stage of growth in order to gain a conserved product of higher energy value then the protein content will also be higher.

The other feature of the data in Table 1 to be discussed is the variation in intake at a given digestibility which, with feeds that are not deficient in cobalt[193] or some other nutrient, reflects the subjective evaluation by the animal of its feed. For example, the intake by sheep of *Setaria sphacelata* at a digestibility of 55 per cent was barely two-thirds the intake of *Sorghum almum* at the same digestibility by similar animals.[47] There appears to be greater variability of this type with tropical than with temperate feeds. In any case, significant relationships between digestibility and intake are more likely to be found within rather than between fodder species or varieties, and legumes, including lucerne (*Medicago sativa*) which is not very highly digestible (see Fig. 2), tend to be eaten in greater amounts than are grasses.[351]

Any understanding of the animal's subjective evaluation is necessarily limited and must stem from empirical observations on feed acceptability.

FIG. 2—**Changes in the digestibility of temperate and tropical grasses and legumes with increasing age and maturity of the herbage.**

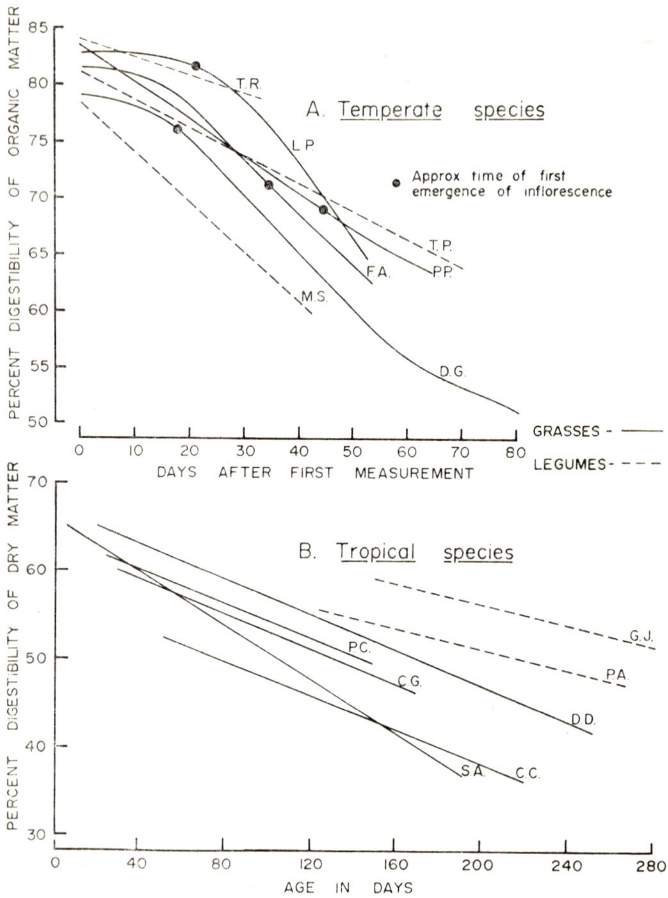

Temperate species. Spring growths; dates of first measurements varied between species and between localities (U.K., refs. 33, 96, 99; Netherlands, ref. 260; Sweden, ref. 95).

L.P. *Lolium perenne* var.S23 (ryegrass)
P.P. *Phleum pratense* var.S48 (timothy)
F.A. *Festuca arundinacea* var.S170 (tall fescue)
D.G. *Dactylis glomerata* var.S37 (cocksfoot or orchard grass)
T.R. *Trifolium repens* var.S100 (white clover)
T.P. *Trifolium pratense* var. Ultuna (red clover)
M.S. *Medicago sativa* var. Dupuits (lucerne or alfalfa)

Tropical species. (ref. 47)

S.A. *Sorghum almum*
C.C. *Cenchrus ciliaris* var. Molopo (Buffel grass)
C.G. *Chloris gayana* var. Callide (Rhodes grass)
P.C. *Pennisetum clandestinum* (Kikuyu grass)
D.D. *Digitaria decumbens* (Pangola grass)
P.A. *Phaseolus atropurpureus* var. Siratro
G.J. *Glycine javanica* var. Cooper

These have been made in many trials where free choice of a number of different cut grasses or types of pasture[66,67,68] has been allowed, but attempts to link the observations with objective measurements have so far met with little success. At the present state of knowledge animals appear to be fickle in their choice; apart from differences due to their species, physiological state, experience and individuality the preferences of the same animal may vary at different times.[69,70,71] The nature of the choice generally has little if any significance in the sense of self-regulatory nutritional 'wisdom', i.e. the deliberate selection of a particular feed in an attempt to meet the need for a particular nutrient, though there are some exceptions as in the case of induced sodium deficiency.[126] The results of the choice may well be of importance from their effect on the amount of feed eaten and the resulting productivity of the animal. A striking illustration is provided by the syndrome of 'ill thrift' in grazing lambs in New Zealand where it has been found that a major cause of poor growth during autumn is an abnormally low intake of herbage at this time.[72] A similar phenomenon, though smaller in magnitude, occurs elsewhere in the world.[37,73] In wholly pastoral areas, relative unacceptability of some plants in a grazed community could be advantageous for these plants would then form a reserve of feed for use during periods of feed scarcity.[261]

The senses of taste, smell, touch and sight are probably all involved in the determination of feed preferences.[74] Many characters of the plant that might perhaps affect the animal's choice such as hairiness or scent,[69] could be measured only with difficulty, if at all; the characters that were measured, say the degree of contamination with excreta, soil or fungi,[68] might be scaled in inappropriate units or their importance unduly weighted, whilst others contributing to the net assessment by the animal might be totally overlooked.

If part of the animal's response to a feed results from events occurring at the cellular level, e.g. in taste buds in the tongue, it is unlikely that chemical data expressed in such terms as crude fibre or crude protein can be very useful indexes. Discrimination may be determined by the presence of an apparently novel constituent such as the unidentified substance(s) extracted from a cultivar of *Phalaris arundinacea*.[75] Of other plant constituents there is conflicting evidence on the importance of soluble carbohydrate level.[67,77,78] This may be an imperfect index of acceptability because it is only one of a large number of associated variables in the feed. The same comment applies to observations made on the effects of the water content of a feed, both intra-cellular and on the surface as rain, etc.[76,79] The report by Arnold[80] that intake is directly related to water content in the range 14 to 28 per cent was based on measurements made on grazing animals during a period of active plant growth. The direct introduction of large volumes of water into the rumen, unless retained in a rubber bladder that physically reduces rumen volume, does not reduce intake by the animal.[81]

Much of the discussion on the effect of water stems from observations that the intake of silage increases with increasing dryness[82] but it is clear that water *per se* is not the causal factor.[83] Addition of water to or its removal from a silage does not alter voluntary intake which remains lower than the intake of hay made at the same time from the same type of herbage. Variation in the water content of silage is only one consequence of variation in the original material or in the ensiling processes; other constituents such as acetic acid[49,50,51] and possibly some nitrogenous compounds,[83] but not histamine,[84] appear to have a metabolic effect that results in reduced intake.

It is well known that phosphorus deficient animals have depraved appetites[190] and that low feed intakes are characteristic in cobalt deficiency.[193] Some effects of major mineral fertilizers will be discussed in a later section.

DIGESTIBILITY

Probably the most useful single measure of the nutritional value of a grassland feed is its apparent digestibility. Apart from the effect on feed intake already discussed, the digestibility value is a measure of the largest and most variable single loss to the animal, by excretion, of energy and nutrients. For temperate grasses the minimum loss of energy in faeces represents about 15 per cent of intake and the maximum rarely exceeds 50 to 55 per cent; corresponding values for tropical materials are about 25 to 70 per cent respectively. In contrast the losses of energy in urine or as methane do not, in either case, usually exceed 8 to 10 per cent of gross energy intake.[21]

The digestibility values obtained for a feed using healthy ruminant animals have shown only small variation due to differences in species, breed, physiological state, age, activity and physical environment, provided that the feed is given to the animals at a standard level.[86,87,88,92] Level of feeding can be assessed by reference to the quantity required to achieve zero energy retention (maintenance), and the effects of variation in level are illustrated by the results of a large number of trials reported by Blaxter.[89] An increase from a maintenance to a twice maintenance level in the intake of dried grasses and hays given in long form was accompanied by a fall in digestibility of 0·119 times the indigestibility at the lower level, the fall consequently being greater for feeds of lower rather than higher quality. An effect due to level of feeding was also found with fresh herbage that was preserved by freezing;[90] that it was not detected in another study[91] possibly reflects the difficulties of measuring dry matter intakes accurately when green material, which shows short-term changes in moisture content, is given.

The greatest differences in digestibility between feeds are those due to differences in stage of growth, and as the plants mature their composition shows large changes. Some examples of the effect of age on the digestibility and composition of temperate herbage species are shown in Fig. 2A and Table 2.

The general pattern of change is much the same for all species but there are some characteristic differences. Legumes as a class, with the notable exception of lucerne (*Medicago sativa*), maintain a place amongst the most digestible components of pasture herbage and this is also true of tropical legumes (Fig. 2B). The digestibility of some of the temperate grasses such as ryegrass (*Lolium perenne*) and cocksfoot (*Dactylis glomerata*) remains roughly constant for some time after active growth has commenced in the spring.[33] At about the time of ear emergence digestibility starts to fall at a rate of 0·3 to 0·5 units daily, a change that occurs in timothy (*Phleum pratense*) from the start of growth. The digestibility of the cocksfoot varieties now in use is consistently lower than that of many other cultivated species though it may prove possible to breed improved varieties.[94]

Chemical and botanical changes are exemplified by the detailed analyses of a ryegrass variety (Table 2). As with digestibility, the changes in temperate species conform to one general pattern though the absolute values for composition will vary. For example, the concentration of soluble carbohydrate tends to be lower in cocksfoot and timothy than in ryegrass,[208] and legumes generally contain larger amounts of nitrogen and minerals than grasses. The earliest growth increments in spring are due to increases in leaf lamina and leaf sheath. These parts of the plant contain proportionately more nitrogen, lipid, carotene and minerals and less lignin, and are more digestible, than the stem which begins to elongate rapidly once the flowering head emerges.[100,112] The leaf lamina then becomes a steadily smaller proportion of the whole plant.

To the animal the most important change is the progressive increase in structural constituents and especially lignin which, with a variety of other substances, contribute to the crude fibre fraction that is commonly determined (see Chapter 2). Even in the young plant, cellulose and hemicellulose comprise at least one-third of the dry matter, but are then highly digestible.[97, 101] As their concentration in the plant increases to 50 per cent or more, so their digestibility decreases. This may be due in part to physico-chemical changes, such as an increasing crystallinity[97,102,103] or degree of polymerization of the cellulose, but is mainly due to increasing lignification. Lignin is virtually indigestible—the disappearance of small amounts in passage through the gut may be an artifact arising from difficulties in exact chemical determination[104, 105]—and it reduces the extent and the rate of breakdown of the structural carbohydrates by combining chemically with them to form materials that are resistant to enzymic degradation, as well as by forming physical barriers to digestion.[97,102,105,106,348]

The increase in structural components is accompanied by a decrease in the crude protein (nitrogen × 6·25) content of the plant and lignification of cell

TABLE 2
Composition and Nutritional Value of a Grass, Lolium perenne *var. S23, at Four Stages of Growth*
(Sources: Refs. Nos. 97, 98, 99)

	CUT 1	CUT 2	CUT 3	CUT 4
Date cut	11 May	25 May	1 June	29 June
Stage of growth[(a)]	Young leafy	Late leafy	Head emergence	Seed setting
%Leaf lamina	73	58	48	25
Yield of dry matter (DM), kg/ha	2490	3814	4071	6089
Chemical composition (% in DM)				
1. Ether solubles	9·1	7·6	6·5	4·7
2. Water-soluble carbohydrates[(b)]	13·8	11·8	11·3	10·6
3. Organic acids	4·2	4·9	4·6	2·9
4. Protein nitrogen × 6·25	14·6	11·6	10·6	6·4
5. Non-protein nitrogen × 6·25	3·9	3·6	3·2	3·2
6. Cellulose	21·3	22·1	23·9	26·7
7. Hemicelluloses[(c)]	15·8	18·9	19·4	25·7
8. Pectin	2·4	2·1	2·2	2·2
9. Lignin	2·7	3·6	4·3	7·3
10. Ash	8·1	8·5	7·8	5·7
Total 1–10	95·9	94·7	93·8	95·4
Crude protein (4 + 5)	18·5	15·2	13·8	9·6
Crude fibre	21·2	24·8	25·8	31·2
Gross energy, kcal/gDM	4·634	4·479	4·480	4·565
Apparent digestibilities %[(d)]				
Energy	83·0	77·6	77·0	63·7
Organic matter	86	83	79	62
Protein nitrogen	82	78	71	50
Non-protein nitrogen	75	78	79	76
Cellulose	92	89	87	73
Hemicelluloses[(c)]	93	84	79	56
Apparently digestible energy (ADE), kcal/gDM	3·85	3·48	3·45	2·91
Molar proportions in ruminal liquor				
Acetic acid	62·7	64·8	68·6	68·6
Propionic acid	22·7	21·9	22·2	20·8
Butyric acid	14·6	13·3	9·2	10·6
%gross energy intake lost[(d)]—in urine	8·2	7·1	6·1	4·1
—in methane	8·7	9·2	8·9	7·6
Metabolizable energy (ME), kcal/gDM[(d)]	3·06	2·75	2·78	2·37
M.E. as % of ADE[(d)]	79·6	79·0	80·7	81·7
Efficiency of utilization of M.E.				
—for maintenance	78·1	75·5	74·5	73·8
—for fattening	52·5	53·5	46·7	33·6
Voluntary intake by sheep per kg liveweight $^{0·73}$				
—g DM	69	79	78	60
—kcal ADE	266	275	269	175
Animal production/ha				
—no. of sheep-days	2085	2789	3013	5728
—liveweight gain (kg)	450	662	543	240

(a) 75 per cent heads emerged by 17 June.
(b) Hexoses, sucrose and fructosan.
(c) Xylan, araban, glucan, galactan and aldobiouronics.
(d) At the maintenance level of feeding.

walls will impede the digestion of this and other constituents. Values obtained for the apparent digestibility of the crude protein, however, reflect the large amounts of metabolic faecal nitrogen (MFN) excreted by the animal.[349] This excretion includes cellular detritus from the gut mucosa, glandular secretions such as mucoproteins and digestive enzymes and microbial protein, and can be related to the amount of feed eaten, about 0·4 g nitrogen being excreted per 100 g feed dry matter.[107,108] As dietary crude protein levels fall from initially high values the MFN becomes an increasingly large fraction of total faecal N. Apparent digestibility consequently declines with increasing rapidity from a maximum of 80 to 85 per cent until at a crude protein concentration in the dry matter of from 2 to 4 per cent, zero and negative values are found.[109,110] Most of the nitrogen in the plant is organically combined though as much as 30 per cent may be in non-protein form. This fraction is soluble in water and normally includes some 5 to 10 per cent of inorganic nitrogenous compounds.[111,112]

The water-soluble carbohydrates can constitute from as little as 1 to more than 30 per cent of the plant dry matter and are wholly digestible. The concentration does not follow a readily definable pattern but reflects the balance between the rate of production by photosynthesis and the rates of utilization in energy metabolism, for the elaboration of polysaccharides and protein for growth, and for storage in root or seed, as determined by the physiological state of the plant.[100,112]

Under many systems of management, grassland herbage may be utilized almost as it grows, allowing little opportunity for large changes to occur in digestibility and composition due to maturation.When successive month-old regrowths of temperate grasses are examined it is generally found that they are not as digestible as the first growths in spring although in late summer the plants may consist almost entirely of leaf lamina and contain considerably more crude protein.[99,113,114] Later growths of leaves have been found to contain greater amounts of structural carbohydrates than first growths[115] and water soluble carbohydrate contents are usually low.[116] If herbage left standing through the winter suffers frost damage and leaching the material remaining green can be of good quality but the digestibility of the entire plant falls appreciably.[117]

Tropical herbage plants are almost all adapted to a short growing season before flowering. Seasonality of rainfall and of adequate soil moisture often restricts growth to short period(s) of the year and though the first, vegetative, growth can be 75 per cent digestible and contain from 15 to 20 per cent of crude protein the rapid onset of flowering causes a rapid decline in quality.[118] The crude protein content of the grasses becomes so low that negative apparent digestibilities and negative nitrogen balances in animals maintained on these feeds are common.[41] If frost occurs in subtropical areas it kills the aerial growth of many of the legumes and of indigenous grasses; dry mature growths of cultivated grasses here and in tropical regions often decline in quality rather slowly during the dry season (0·05 to 0·15 units of digestibility per day—Fig. 2B) as do surviving legumes.[47] Analyses as detailed as those given in Table 2 have not been reported for tropical grasses but these plants do show similar changes in composition with age. Even young growths, however, rarely contain less than 30 per cent of crude fibre in the dry matter, later increasing to over 40 per cent when crude protein contents can fall to as low as 1 or 2 per cent.[41,119,120]

UTILIZATION BY THE ANIMAL

Not all the energy of the apparently digested nutrients is available to the animal for cellular metabolism. The microbial digestion of cellulosic feeds results in the production of considerable volumes of methane and its removal by eructation, and the loss of energy from the rumen as heat of fermentation which is of no consequence to the animal in a thermo-neutral environment. By convention, the apparently digestible energy minus the methane energy, but not the heat of fermentation, and minus the gross energy of the urine is referred to as metabolizable energy (ME).

At maintenance levels of feeding and above, the urine energy does not usually amount to more than 8 per cent of the gross energy of the feed.[89] The loss tends to decrease with increasing intake[98] but shows little relationship to feed digestibility. It is better described by the amount of carbon in the urine[89] than by the amount of nitrogen[123,124] which reflects the nature of the diet. Herbages of high digestibility generally have a high nitrogen content and in excreting more nitrogen in the urine than from feeds of lower digestibility and nitrogen content, the animals lose a larger proportion of energy intake.[73,98] Only part of the nitrogen is excreted as urea with a calorific value of 5·45 kcal/g nitrogen. Large amounts of hippuric acid, the purine derivatives uric acid and allantoin, and detoxication products are excreted as well as xylose and other non-nitrogenous compounds, as in the special case of essential oils from some desert range plants.[21,121] The excretion of these substances is in part a consequence of ruminal digestion and the metabolic activity of the microorganisms,[122] and may result in a heat of combustion as high as 35 kcal/g urinary nitrogen.[21]

In general the methane produced in the digestion of herbage represents about 8 per cent of total energy intake. The loss tends to decrease with decreasing apparent digestibility of the feed; again as a percentage of total intake, it tends to decrease as the level of feed intake rises.[89,98] These effects reflect changes in the extent of breakdown of holocellulose which chiefly determines the digestibility of the herbage diet as a whole.[97] Several workers have reported that around 4·7 g of methane are produced per 100 g digested carbohydrate.[125] These relationships are not entirely satisfactory perhaps because 'digested carbohydrate' is determined only indirectly by the Weende system of summative analysis (see Chapter 2) and is the digested crude fibre plus digested nitrogen-free extractive. A better relationship would be expected with directly determined chemical entities as has been found in one case where methane and digested cellulose were examined.[73] Wolin[127] has proposed a theoretical rumen fermentation balance which allows the estimation of methane production from the molar proportions of acetic, propionic and butyric acids in ruminal liquor; this approach has received some support from experimental observations.[128] Calculation from the balance equation of the heat production associated with cellulose fermentation gives a value of about 7 kcal/100 kcal cellulose digested which is in good agreement with that of 6 kcal obtained by Marston.[129] If methane is used as an index of fermentation heat, Marston's data yield the relationship of 0·8 kcal/kcal methane which is the most generally accepted of the few estimates available.[130,131]

The loss of up to 15 per cent of the gross energy of roughages as methane plus heat can be regarded as a price to be paid for the microbial digestion that permits the utilization of these feeds by the animals. Quantitatively, the most important end products of the digestion are the steam-volatile fatty acids (VFA) acetic, propionic and *n*-butyric and together they probably provide well over half the total energy supply of ruminants.[132,133,134,262] The efficiency with which the energy of VFA is utilized by the animals appears to vary with the molar proportions of the acids in the mixture and the purpose for which they are used. The efficiency of lipogenesis increased as the proportion of acetic acid fell[135] though conflicting results have recently been reported.[352] The efficiency of milk synthesis changes in a similar way unless the fall is very large, and the propionic acid levels are high; milk fat production then decreases markedly due to a change in the supply of acetate and β-hydroxybutyrate precursors.[136] A marked decrease, to 2 per cent or less of fat in milk rarely occurs when cows are at pasture[137] unless the feed contains exceptionally low amounts of structural carbohydrate. It is also uncommon when rations, even those including large amounts of readily fermentable carbohydrate feeds such as cooked maize, contain more than about 25 per cent of long roughage.[138] The efficiency of utilization of VFA for maintenance also appears to vary inversely with the proportion of acetic acid but the range of the variation is much smaller than that shown when fat or other materials are synthesized.[73,98,139]

Of the VFA produced from herbage feeds, acetic acid normally constitutes from 50 to over 70 per cent of the molar total.[134] Many studies have now shown that a change from lower to higher values, and a corresponding reduction in the proportion of propionic plus butyric acids, accompanies increasing maturity and fibre content and decreasing digestibility of the feed.[140,141] This change is also reflected in the performance of animals; the percentage utilization of the ME of the diet falls (see Table 2). Variation in the proportion of acetic acid is also related, inversely, to water-soluble carbohydrate which promotes the formation of propionic acid.[134,140,142] Relationships between the composition or digestibility of the feed and the proportions of propionic or butyric acids have generally not been so precise as those with acetic acid. There appears to be a fine balance in the rumen between conditions that favour the relatively greater production of either the C_3 or the C_4 acid;[140] the outcome may depend in part upon the extent of acetic-butyric acid interconversion.[143]

As with digestibility values, VFA proportions show some variation between herbage species. They show rather small changes with increasing maturity of pasture legumes and the acetic acid level tends to be higher when cocksfoot, rather than other temperate grass species, is eaten.[140] In addition, tropical herbages yield higher proportions of acetic acid than temperate feeds;[144] this would be expected from the difference in crude fibre contents.

There has been considerable discussion on the nutritional significance of the water-soluble carbohydrates[114] which quite often constitute as much as one-quarter of the digestible dry matter of temperate herbages. Substances so reactive, and in such amounts, must have a considerable effect on the course of microbial digestion. It is known that ammonia concentration in the rumen falls when a readily available carbohydrate, starch, is added and there are probably effects on the nitrogen economy of the animal[145] (see Chapter 3). In addition, the production of VFA containing a high proportion of propionic acid could be expected to improve animal performance but there is, as yet, little supporting evidence. Workers in New Zealand[146,147] reported that carcass-weight gains by sheep at pasture could be related to the carbohydrate contents of the herbages but no measurements were made of the quantities eaten and these effects could well have been confounded with those due to differences in feed intake.[276] More certain evidence was obtained in a calorimetric trial[73] where studies on two herbages of very similar chemical and botanical composition, apart from a difference in soluble carbohydrate content, and of the same digestibility, yielded a consistent series of observations. The cellulose of the herbage with the higher sugar content was digested to a significantly smaller extent and less methane was produced; the VFA produced in the rumen included a significantly higher proportion of propionic acid, and the metabolizable energy of the feed was used by the animal more efficiently.

ASSESSMENT OF NUTRITIONAL VALUE

It is desirable that any feed can be described in terms of its nutritional worth in a particular set of circumstances so that the performance of individual animals can be predicted, or determined, with some precision.[21] When confronted with the immense range of feeds produced by grasslands the problem is to select one or more common characteristics that can be determined readily and will give reliable estimates of feeding value.[148]

BIOLOGICAL METHODS

The ultimate test of a feed is animal performance and if this is measured in terms of energy and nitrogen retentions, uncertainties about the caloric and protein contents of liveweight gains, etc. are largely resolved. Nitrogen retention can usually be measured at the cost of little extra analytical work when digestibility trials are made. On the other hand the measurement of energy retention involves much skilled but laborious and time-consuming work[149] (see Chapters 14 and 15).

Feeding systems based on the net energy concept, which stems from the fundamental calorimetric work of Kellner and Armsby have, however, been used in practice in Europe for many years. The evaluation of feeds for these systems—starch equivalent (SE), Scandinavian Food Unit, etc.—is based on measurements of digestibility as is the Total Digestible Nutrient (TDN) system of North America. It is done by summing the quantities in the feed of the digestible nutrients determined as true or crude protein, crude fibre, nitrogen-free extractives and crude fat with weightings that approximately represent their relative energy values. To calculate the SE of roughages the values obtained are then reduced by between 0·29 and 0·58 times the crude fibre content of the feed as used, depending upon the amount present. This has the apparently odd result that if the lower factor (0·29) is appropriate to a fresh herbage the higher factor will apply to the same material after it has been dried.[150,151]

That a single feed may be assigned values which differ widely according to the method of assessment is due to inconsistent assumptions about the efficiency with which the energy of digested nutrients is used by the animal; it also reflects the inherent but mistaken view that a unit of SE or TDN, etc., has constant value to the animal regardless of the purpose for which it is used.[21] Nevertheless, measurements of digestibility remain as a firm basis for the assessment of energy value. As Dijkstra[151] has pointed out, the calculation of SE is essentially the application of the crude fibre 'correction' to the digestible organic matter (DOM) content of the feed. The situation with TDN is similar. Swift[15] has described its determination as a laborious, cumbersome, indirect and inaccurate effort to determine digestible energy. From a study of many roughages he found that the caloric value of the TDN was $4{\cdot}41 \pm 0{\cdot}13$ kcal/g. Very similar values, showing slightly larger variation, are commonly found for the energy value per g of the digestible dry matter (DDM) or DOM of grassland feeds. This result reflects the large contributions of cellulose (4·2 kcal/g) and other polysaccharides with varying, but smaller, amounts of protein (5·7 kcal/g) to the energy they supply. Crude fat is a relatively small part of this supply and a significant fraction of the ether-extractable material of herbage comprises plant pigments, fatty acids, nitrogenous substances and even some ash;[112] consequently the standard value for the heat of combustion of fat, 9·5 kcal/g, is usually inappropriately high.

Measurements of the amounts of DDM or DOM consumed by cattle and sheep therefore provide good estimates of apparently digestible energy intake and may be obtained fairly readily. Measurements in terms of OM are often preferred to those on a DM basis because mineral matter which, as such, has no energy value to the animal, is excluded and because the ash content of the feed, which is normally some 5 to 10 per cent of the dry matter, may be increased massively by contamination with soil. The calorific value of DOM, however, tends to be a little more variable than that of DDM[153] probably because variation in the amount of protein has a proportionately greater effect.

Moir[154] has shown that there is almost a 1 : 1 relationship between the per cent apparently digestible energy (Y) and per cent dry matter digestibility (X) over a range of X from 30 to 73. The regression equation that he obtained: $Y = 1{\cdot}006\ X - 2{\cdot}013$ accounted for 99·6 per cent of the variation in Y. For a large number of feeds of different types (but apparently excluding oil-seeds which have high heats of combustion), of digestibilities (X) from 30 to 83, and given at various levels to both cattle and sheep, he found that 96·2 per cent of the variation in digestible energy kcal/g DM (Y^1) was accounted for in the equation $Y^1 = 0{\cdot}0462\ X - 0{\cdot}158$. The range in Y^1 was from 1·24 to 3·65 kcal/g. A similar and precise relationship for sub-tropical herbages has been reported.[153]

The determination of per cent apparently digestible energy (ADE), or of DM digestibility if a bomb calorimeter is not available, can also lead directly towards the evaluation of feeds for the feeding system proposed by Blaxter[21,155] In this system rations are devised so as to predict animal performance on the bases of the amount and concentration of metabolizable energy (ME) in the feed(s) to be supplied (Chapter 24). The determination of ME requires the measurement of methane production as well as of energy losses in faeces and urine but only small errors accrue if ME is assumed to be 80 per cent of ADE[21,98,152].

The apparent digestibility (D) of the energy, DM or OM of a feed times its voluntary intake (I) gives the total digestible intake (DI). As has been seen (Fig. 1), I and D are correlated and the question arises: which of the two is the more important in determining DI and thence the level of animal production? The data of Miller,[35] Milford,[41] Minson,[33] and Conrad[36] were examined by the method of Henderson and Hayman[156] who showed that when a regression analysis of the type $\log_{10} DI = \log_{10} I + \log_{10} D$ is made, the regression coefficient $b_{x_i y} = (\text{Cov}\,(y,x_i)/\text{var}\,y)$ represents the proportion of the variation in the dependent variable y that is attributable to variation in x_i.

It was found with both temperate and tropical herbage that from 60 to 75 per cent of the changes in *DI* were accounted for by changes in intake as such. Voluntary intake therefore appears to be a better measure of the value of a feed than its digestibility determined under the usual conditions of restricted, constant, daily intake. Measurements of digestibility, and thence of *DI*, can, of course, be obtained whilst voluntary intake is being studied but there is a practical difficulty. The variation in digestive efficiency between animals, expressed as a coefficient of variation, is generally found to be of the order of $\pm$ 2 per cent but the variation in intake is some five times greater. Thus if groups of three animals are used there would be a 90 per cent chance of showing a digestibility difference between two feeds of about 5 percentage units, or more, significant at $P = 0{\cdot}05$ but there would be a similar chance of showing a significant difference in intakes only if the difference was as large as approximately 40 per cent. There may well not be the facilities available for groups of ten animals which are required if differences in intake of from 15 to 20 per cent are to be detected.[33,157]

If the voluntary intake of ADE can be measured it does predict quite closely the performances of similar animals given the feed *ad libitum*. Crampton and his colleagues[32] have devised a 'nutritive value index' (NVI) which is essentially the ADE (kcal) consumed per kg liveweight$^{0{\cdot}75}$; their index could more simply be expressed in these terms than by taking the further step of dividing by the arbitrary constant 80 (the 'expected intake of Standard Forage' in g/kg$^{0{\cdot}75}$) to obtain the NVI.[153] Correlations between NVI and liveweight change of from + 0·88 to + 0·94 have been reported when a change of 0·5 kg in liveweight over a period of 11 days was found to reflect a change in NVI of about 8 units.[32]

Digestibility values themselves certainly provide a good general index of the nutritional value of feeds but how meaningful is this index when differences in digestibility are small? For example, are the differences of about five percentage units between ryegrass and cocksfoot at similar stages of growth (Fig. 2) reflected in animal performance? It has been found in practice[158] and in controlled experiments[159,160] that cocksfoot pastures do not sustain milk production by cows at as high a level as that concurrently obtained from ryegrass; a corresponding difference in lamb growth-rates has also been reported.[161] These effects, however, are probably not due simply to the lower digestibility or intake of cocksfoot; it is likely that they also reflect differences between the grasses in the metabolites they yield on digestion. When growing lambs were given equal amounts of digestible dry matter as either cocksfoot or ryegrass it was found[162] that significantly more of the digestible energy of the ryegrass was retained as a significantly

the increasing amounts of MFN (p. 607) excreted relative to the amounts of undigested feed N. This is illustrated by Mitchell's[172] equation $Y = 42{\cdot}64\ (X - 5)^{0.215}$ which he derived from a large body of data. Glover, Duthie and French[109] suggested that their equation $Y = 70\ (\log_{10} X) - 15$, which was also derived from a large amount of published information, was more satisfactory at the lower values of X. Other rather similar equations to describe relatively restricted ranges of temperate[114,173] and sub-tropical[110] forages have been reported. The small differences between values of Y predicted from the various equations for a given value of X probably reflect differences between genera of animals (sheep vs. cattle) and between species (*Bos indicus* vs. *B. taurus*)[174] as well as between feeds. Whereas Glover *et al.* found that Y became zero at a crude protein content of 1·8 per cent, Milford and Haydock[110] observed a value of 3·9 per cent and also found that zero N retention in the animal was attained at a digestibility of 49·3 (95 per cent confidence limits approximately $\pm$ 2·5).

The amount of DCP (Y^1) in the feed may be of more interest than the digestibility of the crude protein when animals are hand-fed. Rectilinear regression equations of Y^1 on the amount of crude protein in the feed have been reported [173,175] that had low residual standard deviations in the range $\pm$ 0·46 to 0·85 per cent.

Other Methods

Digestibility in vitro. A quasi-biological method, the *in vitro* digestibility technique, is now being used increasingly. In one two-stage technique[176,177] that is widely employed, small (0·5 g) samples of dried, ground, forage are incubated anaerobically at 38°C in glass tubes containing buffered ruminal liquor. After 48 hr the tubes are centrifuged, or their contents are filtered, and the solid matter is incubated with pepsin solution for a further 48 hr. From the weights of dry matter or organic matter remaining, which are adjusted by applying the results of determinations on blank and on standard samples, the *in vitro* digestibility is calculated. In another technique[178] the proportion of the cellulose in a feed that has disappeared after incubation for 12 hr with buffered ruminal liquor is related to the Nutritive Value Index. A further method[179] is to insert into the rumen of an animal by way of a fistula, a nylon mesh bag containing a sample of feed and to measure the rate of loss from the bag.

Criticism of these techniques on the grounds that they do not—and cannot—exactly match *in vivo* digestion is misplaced. Their purpose is to obtain a useful and repeatable measurement, and the use of ruminal liquor as a source of a wide range of cellulolytic and other enzymes is primarily a matter of convenience rather than of imitation. It may become possible to replace the liquor by standardized enzyme preparations.[189] Unfortunately, the relationships of the results to digestibility values obtained for the same feeds from animals are often reported as correlation coefficients (r). The information required is not simply that the coefficient of determination ($r^2 \times 100$) is, say, 85 per cent, which may in any case be unduly affected by a few of the results obtained, but it is the precision with which a value can be predicted.

Tilley and Terry[176] applied their two-stage technique to measure the *in vitro* digestibility (X) of 148 different samples of grasses and legumes of known *in vivo* digestibility (Y). The residual standard deviation of the rectilinear regression equation of Y on X was $\pm 2 \cdot 31$ units of digestibility. This value, which is of the same order as those found from other, similar experiments,[177] still means that estimated values of Y that differ by less than 6 digestibility units will not be significantly different. Nevertheless this is a greater degree of precision than can be achieved with any other general relationship for the prediction of digestibility, such as those based on chemical composition. The technique can be applied to problems which purely chemical methods of analysis have been unable to solve, and which could not readily be studied *in vivo*. For example, the separate digestibilities of the leaf and stem fractions of growing herbage can be studied and the information obtained sheds light on the spring 'plateau' of digestibility (Fig. 2).[176,180] It is also of great value in plant breeding work where very large numbers of individual plants, each yielding rather small amounts of herbage, have to be examined; plants showing significant differences can be chosen for further study. For this work the index for the initial selection can be *in vitro* digestibility as such. If care is taken to maintain anaerobic conditions during the fermentation stage, and well defined standard samples of high and low digestibility are included in each run for comparison, the standard error of the mean of duplicate measurements is usually less than one digestibility unit; between runs it is a little over one unit.[176,177] If the intention is to predict *in vivo* digestibility or metabolizable and net energy values,[270] then a regression equation must be obtained by the individual worker for his own situation.

Date of cutting. Reid[181] reported that the dry matter digestibility (Y) of first growths of herbage species grown in the north-east regions of the U.S.A. was related to time of cutting in days elapsing after 30 April (X) by the equation $Y = 85 \cdot 0 - 0 \cdot 48\ X$; the residual standard deviation was $\pm\ 1 \cdot 65$ per cent. This is a general description of digestibility: time relationships appropriate to the latitude where the data were obtained; other starting dates would apply in other latitudes.[95] The rate of decline in digestibility, 0·48 units daily, approximately corresponds to the rates illustrated in Fig. 2 but would not take account of the early 'plateau' in digestibility of ryegrass and cocksfoot found in the U.K. Cocksfoot varieties grown in Europe have not been found to be as digestible as 85 per cent, the theoretical maximum from the equation which does not distinguish between this and other species or between early and late growing varieties. Minson *et al.*[99] pointed out that on one day the digestibilities of first growths of cocksfoot (var. S37), ryegrass (S24) and ryegrass (S23) were 59, 65 and 75 respectively.

This type of equation is not readily derived for second and later growths of herbage but is of value in practice when the time to harvest a first cut for conservation has to be decided.

Plant morphology. Although a leafy herbage can be expected to be more digestible than one containing a high proportion of stem the relationship between digestibility and the per cent of dry matter yield as leaf lamina is by no means precise. When monthly regrowths of ryegrass and cocksfoot were examined[99] it was found that the second growth was more digestible than the first although it had the lowest leaf percentage in the series and contained some flowering stems. In subsequent regrowths the leaf percentages increased but so also did the contents of hemi-cellulose and lignin[115] and the digestibility of the whole plants fell. Data on the *in vitro* digestibility of leaf, leaf sheath and stem fractions, and on the *in vivo* digestibility of feeds made up to contain these fractions in different proportions, show that stem, particularly if of recent growth, can be at least as digestible as leaf.[33]

Entries in tables of feed composition are frequently in such terms as 'meadow hay, good average'.[182] Morphological characters are also the basis of the official hay grading system in the U.S.A.[183] When forages so identified are given to animals the results are frequently a salutory reminder of human fallibility.

Miscellaneous. Attempts to substitute small laboratory animals for the larger ruminants to determine digestibility have not been very successful.[184] Other methods of evaluation proposed have included measurements of the weight gains of cricket nymphs given ground herbage,[185] microscopic examination,[186] the rate of breakdown during mechanical maceration,[187] leaf strength[188] and various enzymic and chemical procedures.[171,189]

General

The nearest approach to a single, comprehensive measure of nutritional value is probably the voluntary intake by animals of digestible dry matter but if low values are found they may be due to any of a number of causes that will not be directly measured. The criteria of nutritional value chosen will depend on the facilities available, the precision required and any special circumstances. Biological methods are to be preferred to chemical if the number of feeds to be examined is not too large, but it may be wise to determine mineral content(s) when deficiencies are liable to occur. In temperate areas energy value is of prime importance; in sub-tropical areas Milford[41] suggests that voluntary intake, the digestibility of dry matter and crude protein, and nitrogen balance should be measured. If prediction equations are used their validity should be checked and notice should be taken of their precision as indicated by the residual standard deviations rather than by correlation coefficients. Digestibility determinations *in vitro* are generally more precise, and with proper care are more accurate, than values calculated from the chemical composition of the feed.

MINERALS, FERTILIZERS

An extensive review of world literature on deficiencies and excesses of minerals in pasture herbage was published in 1956.[190] Since that time it has been recognized that selenium, already known to be present in excess amounts in the vegetation of considerable areas of the North American continent where chronic selenium poisoning occurs, is an essential nutrient and that deficiency states in grazing animals are of widespread occurrence.[191,192,193,265] The major clinical manifestation is a muscular dystrophy in lambs and calves often referred to, descriptively, as white muscle disease; infertility in ewes has also been observed. Growth is retarded when available selenium levels in herbage are below about 0·1 ppm; there have been many reports of increased rates of liveweight gain after the subcutaneous injection or oral administration of 1 to 5 mg (lambs) or 10 mg (calves) of the element, usually as selenate or selenite. The very narrow margin between sufficiency and toxicity makes the direct application of seleniferous fertilizer to pasture too hazardous a means for correcting Se deficiency.

A second major advance made in recent years has been the development of the cobalt 'bullet', now very widely and successfully used for the prevention and correction of Co deficiency in grazing animals.[194] The bullet, about 2 cm^3 in volume, contains cobalt oxide and in order to reduce chance regurgitation[264] is so made that it has a high specific gravity. It remains in the reticulo-rumen after oral administration and yields a steady supply of Co to the ruminal liquor. A steel grub-screw of similar size is also usually administered so that the surface of the bullet will be abraded continually, preventing the deposition of an impermeable coat of calcium phosphates.[195]

The concentrations of minerals in plants reflect the inter-relations between (1) genus, species or strain of plant; (2) the nature of the soil; (3) climate and season; and (4) the stage of maturity of the plant.[193] It has to be remembered, however, that a change in the amount of a mineral in plant matter does not necessarily represent a change in the amount available to the animal[196] and that a dietary level adequate in one situation may be insufficient in another (as, for example, in the case of the copper-molybdenum-sulphur synergism, Chapter 11).

Apart from large genetic differences between plants, as between legumes and grasses where legumes almost invariably contain higher concentrations of calcium and phosphorus and of many but not all trace elements, variation in mineral content has also been observed where the differences are relatively small. For example, ten-fold differences in iodine content[197] were found among strains of ryegrass growing together and differences between species and varieties of grasses in the uptakes of major minerals (i.e. those measured as per cent rather than ppm of dry matter) have been reported.[198,199,200] Changes in the mineral contents of plants as they mature do not conform to a single pattern and are generally of less importance in temperate than in tropical areas where, for example, the progressive fall in phosphorus and sodium levels will aggravate any incipient deficiencies of these nutrients.[119]

The mineral composition of plants reflects to some degree the nature of the soil and responds to the alterations brought about by applications of fertilizers and trace elements or other treatments, e.g. changes in pH due to liming.[201,202] Thus an increase in the relative concentration in the soil of any one cation will increase its concentration in the plant and reduce that of the others but fertilizers have additional effects of wider significance, namely the promotion of increased plant growth and of changes in the botanical composition of grasslands. Discussion of the effects of fertilizers will be confined to those that supply the major plant nutrients nitrogen, phosphorus and potassium; it will be assumed that there are no gross deficiencies of these or other minerals essential for plant growth.

Plants normally need more nitrogen than any other nutrient and a shortage of nitrogen commonly limits herbage production.[203] Unless applications in fertilizer so enhance the growth of grasses that the uptake of nitrogen does not keep pace with the large increases in dry matter yield that can be obtained,[5] the nitrogen content of the grasses will be raised. When light intensity and photosynthetic activity are low but there is abundant soil moisture and a high temperature this increase may, rather uncommonly, be due in part to the accumulation of nitrate with consequent risk to animal health.[111,205] Nitrogenous fertilizers applied to temperate grasslands have often been shown to reduce the clover population.[206] Unless the amount applied is large enough, say 40 kg N/ha/year or more, to compensate for the reduced contribution of N from the rhizobia associated with the clovers, the crude protein content of the whole herbage and even its dry matter yield may be reduced. Nitrogen does not appear to have any large direct effects on the digestibility of herbage plants[245] but their energy value may be changed because of the marked reduction in fructosan content that it causes.[100,207,208] The ruminal liquor of sheep grazing pasture heavily fertilized with N was found[209] to contain higher concentrations of soluble non-protein N, ammonia and nitrate, a lower total VFA concentration and proportion of butyric acid, and a higher proportion of acetic acid than that of sheep grazing lightly fertilized pasture. Nitrogenous fertilizers increase the carotene content of grasses[100] but only minor changes have been found in trace element concentrations even with applications as large as 390 kg N/ha/year for several years.[206,210]

It is most improbable that the needs of animals for potassium would not be met from grassland herbage, even that grown where the amounts available for the plants were grossly insufficient. Potassic fertilizers applied for the maintenance of a high level of herbage production, which is particularly necessary if the growth is cut and removed for conservation, will raise the K content of the feed. It has been suggested, notably by Dutch workers,[211] that excessive applications will, by altering mineral composition, promote the development of hypomagnesaemic tetany in lactating dairy cows and sheep. Experimental evidence generally does not give very much support to this view. Hypomagnesaemia, if found, may be of little significance unless it is also found that there is an increase in the incidence of clinical tetany; the number of animals used in experiments has frequently been too small, and the frequency of tetany too low, to yield unequivocal results[212] (see Chapter 33).

Phosphatic fertilizers increase the P content of herbage but by far their most important use is to raise the phosphorus status of the soils of the pastoral areas of the world, thence to promote the growth of legumes and set in train events that lead to enormous increases in pasture productivity. As an example the indigenous pastures of the northern tablelands of New South Wales, Australia, can with difficulty support two wether sheep per hectare. The application of a few hundred kg of superphosphate and oversowing with clover rapidly allows a safe stocking rate of 10 breeding ewes per hectare to be achieved. Whether from a higher requirement for P or an inability to compete successfully with grasses for this nutrient,[213] many legumes do not persist when the soil P status is low. One noteworthy exception is Townsville lucerne (*Stylosanthes humilis*) which is becoming an increasingly important pasture plant in northern Queensland.[214]

There are some reports on the effects of the major mineral fertilizers on the acceptability of the herbage to animals. Reid and Jung[215] found that sheep offered the free choice of a number of herbages tended to prefer those containing the higher levels of phosphorus and to eat less of those from plots heavily fertilized with nitrogen. There was, however, a large variation between individual animals in their preferences and when the feeds were given singly, without choice, there were no differences in the intakes. Milford[47] reported that the intake of three sub-tropical feeds was reduced by nitrogenous fertilizer but no such effect was found[79] in an experiment with temperate pasture species.

VITAMINS

Vitamin B_{12}–cobalt, and vitamin E–selenium relationships in ruminant animals are discussed in Chapter 11.

Fresh herbage, silage made from fresh rather than wilted herbage[100] and dried grass normally contain abundant β-carotene, the principal precursor of vitamin A, for the immediate needs of cattle and sheep and provide a surplus for storage in the liver. Carotene levels are low in mature dry herbage, such as that available during drought or persisting in cold winters, and decrease in hay and dried grass during storage due to oxidation.[216] The liver stores of animals normally maintain a supply of vitamin A for several months to animals given such feeds[217] but reproductive efficiency and the viability of offspring may ultimately deteriorate. If young are born they may show signs of vitamin A deficiency, particularly night blindness, in the first few months of life because lambs and calves as a rule have only small stores of this vitamin in the liver at birth.[218]

The vitamin D content of fresh green herbage is not as low as was once thought[219] and deficiency states in grazing animals are rare. They have been reported in lambs when a rachitogenic factor in the feed, possibly carotene, was implicated.[220,223,330] Ultraviolet irradiation converts ergosterol in the skin of animals to ergocalciferol (vitamin D_2) and a similar change occurs in cut grass during haying. If the herbage is cut for ensiling there may be only a short period of exposure to the sun and the gain in vitamin D activity may be small. The gain will also vary between different latitudes because of variation in the period of time that the elevation of the sun exceeds 35° and thence in the amount of incident ultraviolet radiation. The variation that has been reported in the anti-rachitic activity of American and European herbages,[219,221,222,223] both fresh and preserved in various ways, has not been shown to have practical importance in animal feeding though effects on the vitamin D content of milk could be of significance in human nutrition.

CONSERVATION

Drying and ensiling are the two main methods employed in practice for the preservation of cut herbage so that losses in store due to fungal and microbial spoilage and to enzyme activity are minimized. There are innumerable variations in technique.[224] Drying may be done entirely in the field, as in haymaking, or partly or wholly indoors where air, either at ambient temperature or heated, is blown through the material; the final moisture content should be 15 per cent or less. In the ensiling process the pH of the herbage is reduced to a level where bacterial and enzymic activity is inhibited either as a result of natural fermentation, sometimes aided by the addition of fermentable material such as molasses, or by the addition of inorganic or organic acids or other chemicals; some of the latter substances may act primarily as sterilants.

The digestibility of well conserved fodder dried or ensiled is similar to, and reflects the stage of growth of, the herbage when cut. Modern techniques and notably the introduction of the flail type of harvester allow much freedom of choice in the time of cutting and the decision that is made can be determined by the requirements of the animals that will be given the product. If they are simply to be maintained for a period of time then harvesting may be delayed until the weight of crop approaches the maximum. If the feeds are to be used for the production of meat or milk then earlier harvesting is necessary so that the digestibility is higher (Fig. 2). The yields of dry matter will then be reduced, though there will be some compensatory production from aftermath; the yields per unit area of metabolizable energy and production net energy may also be lower. On the other hand the actual output of animal products can be two to four times greater because voluntary intakes by the animals, and thence the quantities of energy available for production after maintenance needs are met, will be much higher than with late-cut fodders.[64,82,98]

The method of conservation employed may be determined by economic and technical considerations such as the implements and storage facilities that are available, the greater dependence of haymaking than ensiling upon weather conditions, and the difficulty of handling short growths of herbage dried in the field. Some nutritional consequences of various methods of conservation and the nature and extent of the losses of nutrients incurred are discussed below.

Hay

Herbage continues to respire after it has been cut until its moisture content is reduced to below about 35 per cent.[225] The major changes involved are the oxidation of readily available carbohydrates and the hydrolysis of protein. There will be further losses due to leaching and perhaps to mould growth if rain falls on cut grass in the field, and even with careful handling there will be mechanical losses which will be greater from the leaf than from the stem fraction. The final yield of dry matter is often found to be less than 80 per cent of that in the original material[82] but the reduction in feeding value will usually be greater because soluble components and leaf have suffered proportionately greater losses than structural materials and stem. In an experiment where small quantities of hay were made carefully, it was found[226] that the digestibility of the organic matter was 2 percentage units lower than for the fresh material; the difference was due entirely to a decrease in the digestibility of protein. The net availabilities of the metabolizable energy of the fresh and dried feeds, measured in a calorimeter, were identical although an increase in methane production when the hay was given indicated that there was some change in the pattern of rumen fermentation.

Unless the moisture content of the hay in the stack remains so high that it ignites spontaneously the quantitative losses in store will be much smaller than those usually incurred during the making. It was found in the Netherlands[227] that when hays were stacked at initial moisture contents of 15, 20 and 25 per cent the losses of dry matter were 1·7, 3·3 and 5·0 per cent respectively. The moisture content usually falls in store but if it does not the losses of dry matter increase sharply above about 15 per cent moisture.[228] Sugars and fructosans are broken down, and moulds that are likely to grow reduce the feeding value because animals will reject hay that is affected.

Qualitative losses due to the heating that will occur increasingly with rising moisture content may be more serious.[227,229] The digestibility of the crude protein is affected first and may be reduced to zero if heating is severe. The digestibility of the organic matter also falls markedly due to the production of polymeric materials from xylan and other substances.[101,105] Dutch workers[23] have devised a colorimetric method for estimating the degree to which hay has been heated and the extent of the fall in nutritional value.

Dried Grass

In a well organized grass-drying enterprise the fresh herbage is placed in the dryer soon after it is cut and respiration losses will be small. The air blown through the material is in some machines heated to over 200°C but provided that the drying time is measured in minutes rather than in hours, and the outlet temperature is low, the changes in nutritional value will be small. Heat damage is insignificant below about 50°C, but the apparent lignin content, of young grass especially, may be increased three-fold by ordinary drying at 80 to 100°C.[105] Prolonged drying at these or higher temperatures, particularly in the presence of moisture which catalyses damage reactions, will cause a reduction in the digestibility of protein and organic matter. In one experiment[231] the ME content of grass dried in the laboratory at 100°C was reduced by 4 per cent compared with the fresh material but the efficiency of utilization for fattening was unaccountably increased by 9 per cent. Evidence that rumen fermentation patterns differ between fresh and dried grass has been reported.[232]

Silage

In silage made from freshly cut herbage by fermentation processes the object is to reduce the pH of the material as quickly as possible to a value below 4·2; at this point the undesirable activity of *Clostridium butyricum* in particular is inhibited.[233,234] Lactobacilli continue to ferment soluble carbohydrate when conditions are more acid and at pH 4·0 the silage dry matter contains about 6 per cent of lactic acid. Some hemicellulose is hydrolysed[235] but rapid transfer of the cut grass to the silo and compaction to promote anaerobic conditions will minimize the loss by respiration of the more readily available carbohydrate substrates. If the sugar content of the herbage is low the addition of molasses, whey, or similar materials will be advantageous.

There is considerable breakdown of protein to a wide variety of substances of low molecular weight so that, in temperate herbages, a 1 : 4 ratio of water-soluble:-insoluble nitrogen compounds may be reversed; nitrogen equilibrium is reached at about pH 4·3[237,238]. Undesirable types of fermentation are characterized by the formation of large amounts of butyric acid and volatile bases including ammonia. The pH of silage is not a certain index of its quality. Satisfactory silage is frequently made from herbage wilted in the field to about 40 per cent dry matter even though high concentrations of lactic acid are not obtained and the pH is about 5·0;[236] it appears that the high osmotic pressure retards the development of undesirable types of bacteria.[234] The activity of *Cl. butyricum* is also inhibited above a temperature of about 50°C but the nutritional value of the resulting feed is then reduced by heat-damage. Many workers advocate that during fermentation the temperature of the silage should not rise above 20 to 30°C;[234] usually, little butyric acid is formed unless the herbage is ensiled when very wet.[239]

Losses of dry matter are incurred by plant respiration, bacterial fermentation, seepage from the silo and surface wastage and in practice they rarely total less than 20 per cent of the weight in the fresh herbage;[224,236] surface wastage may be very large during storage unless air and rain are excluded.[240] The losses may be overestimated if in the determination of dry matter content account is not taken of the volatile substances in silage that are lost during oven-drying.[241]

Provided that the silage has not overheated, its digestibility generally differs little from that of the original herbage however it is made[242] but its energy value may be lower because of the breakdown of sugars to lactic acid.[100] The breakdown of protein to a wide variety of less complex nitrogenous compounds may not be of great importance to ruminant animals but poor silage containing large amounts of volatile bases and butyric said is not eaten in as large amounts as is hay made at the same time from the came raw materials.[83] This phenomenon, discussed earlier (p. 604), is perhaps the major problem in the nutrition of animals given silage, but appears to be overcome if the herbage is wilted to about 50 per cent dry matter before ensiling.[243] If inorganic acids are used to preserve herbage the acid-base balance of animals given the resulting silage may be disturbed; the addition of a neutralizing mineral mixture to the feed is recommended.[234]

Cold Storage

Cold storage is a laboratory technique that is useful if, for example, the feeding value of herbage at a particular date and stage of growth is to be studied.[244,266] Quantities sufficient for a digestibility trial, which with the preliminary feeding period will normally extend over a 3-week period, can be frozen quickly and then stored satisfactorily at −18°C. The product is readily eaten by animals which soon become accustomed to it, and the process causes only minor changes in chemical and physical characteristics, in digestibility and in energy value.[231]

PROCESSING

Chopping, wafering, grinding, and the pelletting of ground feed facilitate the handling, storage and rationing of dried fodders but such treatments may have nutritional consequences.

Chopping, usually into lengths of 1 to 2 cm, and wafering, where chopped or long fodder is compressed into blocks 5 to 12 cm in diameter or square and 1 to 5 cm thick, have little effect upon nutritional value.[246,247] The rate of eating (g/min) might be increased, and wastage may be reduced because it will be more difficult for the animals to eat selectively, and to reject any less acceptable (or noxious) components, than when the same feed is given in long form.

Grinding has a number of important consequences. Feed passes through the alimentary tract of ruminants more rapidly when it is given in ground rather than in long form and the amount eaten, the digestibility, the products of digestion, and production by the animal can all be affected.[246,247]

As described earlier, voluntary feed intake is related to the rate of emptying of the reticulo-rumen, and to leave this organ the ingesta must be in a finely divided state. Intake of a feed may not be altered by grinding if it is of very poor quality or is given as a dry meal, perhaps because animals dislike the dustiness, but it is generally increased if the meal is moistened[248] or pelletted. The extent of the increase shows an inverse relationship with the quality of the feed as illustrated by data[249] on the voluntary intake of dried lucerne, cocksfoot and timothy, each harvested at three stages of growth at dry matter digestibilities in the range 71 to 50 per cent. At the highest digestibilities, intakes were increased by some 40 per cent by pelletting and the actual intakes of pellets at the various growth stages were not significantly different; in contrast, the intakes of the long forages decreased markedly with advancing maturity and differences between the two physical forms became more pronounced so that at the lowest digestibilities only about half as much of the long as of the pelletted feeds were eaten.

The increase in the intake of digestible energy that occurs is smaller than the increase in gross energy intake because grinding reduces digestibility,[249,250,251] and in particular that of the structural carbohydrate components. This effect is probably a consequence of the increased rate of passage;[24,251,252] crude fibre is exposed to microbial attack for a shorter period of time and there is evidence that the rate of cellulose digestion decreases and that changes occur in the microbial population.[54,253] These changes may stem from the alterations in eating behaviour that follow the substitution of long feed by pellets. Pellets are eaten more rapidly, rumination times fall, and it appears that the volume of saliva entering the rumen decreases which, because of reduced buffering, would account for the lower ruminal pH observed.[247,254,255] The changes in digestibility are reflected in the volatile fatty acids produced; the proportion of propionic acid tends to be higher and that of acetic acid to be lower.[259] This effect can cause a reduction in the fat percentage in milk secreted by dairy cows.[136] Liveweight gains of fattening animals can be much increased when a feed is pelletted and given *ad lib.* but have been found to be similar to those obtained from long roughage when dry matter intakes of the feed in its two physical forms are equalized.[246] This is a surprising result because digestible dry matter intakes will differ but it has been confirmed in a calorimetric study.[250] It was found that an increase in the faecal loss of energy (from 19·4 to 26·9 per cent of gross energy intakes) due to pelletting a dried grass was compensated for by a reduction in methane production and an increase in the efficiency of utilization of the ME by the fattening sheep that were used. The latter effect could have been due to a higher proportion of propionic acid in the ruminal VFA. Similar results were obtained in an experiment where energy gains by sheep were estimated from carcass analyses.[256] There could be a lower expenditure of energy in the prehension, mastication and rumination of pellets; energy costs of the order of 0·5 kcal/hr/kg liveweight for eating and 0·2 kcal/hr/kg for rumination have been reported[257] and a diminution in these activities, which is a typical result of pelletting,[254] could appreciably reduce the energy costs per g of feed.[258]

GRAZING

Information on the nutritional value of cut herbage is not directly applicable to grazing conditions, and if combined with information gained from separate studies on the effects of cutting on the growth of ungrazed swards, will not necessarily lead to improved systems of grazing management. Very many experiments on grass plots, notably those of Woodman and his colleagues at Cambridge some 30 to 40 years ago, have shown that the desirable characteristics of leafiness and digestibility are positively related to the frequency and severity of defoliation while the annual yield of herbage dry matter is negatively related. It came to be accepted, but usually without reference to the grazing animal, that a satisfactory compromise between quantity and quality in the herbage grown and thence maximum animal production from a pasture could be achieved only by intermittent grazing. The length of the grazing and intermittent rest periods, and the area of pasture allotted per animal, would vary with the current rate of herbage growth. When actually put to the test, these expectations have not been realized.

It has been found that intermittent grazing can assist management for the needs of particular classes of livestock, e.g. lactating *vs.* non-lactating cattle, and for the conservation of herbage surplus to immediate requirements but the anticipated major benefit, a large increase in animal production, has generally not been obtained.[271,272,273,254] Provided that comparisons are made using similar numbers of animals on similar areas of pasture, animal productivity under intermittent grazing is usually very little different from that obtained under a system that was thought to be harmful to both pasture and animal outputs, namely the continuous grazing of a pasture with a more or less fixed number of animals. A detailed discussion of this topic is outside the scope of this chapter but when it is stated that animal production from pasture is determined by:

(1) the amount, the quality and the seasonal pattern of the herbage growth,

(2) the proportion of the growth that is harvested by the animals, and

(3) the efficiency with which the animals utilize the herbage they consume —it has to be remembered that all these factors are interdependent. Grazed pastures yield an ever-varying supply of feed to the animal from a heterogeneous plant population in which substantial and progressive changes, botanical, morphological and chemical, are continually occurring. All the changes reflect the mutual interactions of climate, soil moisture and fertility, plant species, the species, type, number and activities of the grazing animals, and the dynamic populations of parasites and other organisms.[268,269,275] Pastoral enterprises are in fact very complicated exercises in applied ecology.

Significant progress has been made only during the past two decades in the development of techniques that are essential if the understanding of the nutrition of grazing animals is to be increased; even now, the measurement of the quantity and quality of herbage grazed by livestock is beset with difficulties. Without this information, claims such as that the higher liveweight gains by lambs grazing on one type of pasture rather than another are due to differences between the herbages in their carbohydrate contents[146,147] are doubtful. Nor, lacking data on herbage intakes, is it sufficient to state that there were no apparent differences in the amounts of feed available to the animals.[277] What herbage is left on a pasture at any one time is a function not only of the amounts removed by grazing but also of the rates of growth by and the decay of the plants.[285] Visual appraisals of quality can also be misleading; the apparent predominance of one species in a mixed pasture may be the result of rejection by the animals which in fact are hard-grazing some other species that is, in consequence, not readily seen to be present.

Estimation of Feed Intake

Estimates of intakes gained from measurements of the amounts of herbage on a pasture before and after periods of grazing are subject to many errors[278] and are too imprecise to be of any real value in nutritional studies. There has been intensive work on another method[278,279,280,281] that depends on measurements of faecal output (F), which is the apparently indigestible fraction of the herbage grazed, and of the apparent digestibility (D) of that herbage. Intakes (I) can then be calculated from the expression $I = A \times F$, where $A = (100/100 - D) =$ (feed/faeces) and is termed the Intake Factor. The intake of digested feed is given by the term $(F \times (A - 1))$.

Faecal outputs can be measured directly by collection in bags attached to harness worn by the animals, but is more often estimated from the concentration in the faeces of a totally inert material that is administered regularly in standard amounts. The substance that is most frequently used for this purpose is chromium sesquioxide (Cr_2O_3). In a typical application, sheep are dosed by mouth twice daily with a capsule or impregnated paper[282] containing 1 g Cr_2O_3. The mean daily excretion in the faeces is similar to the amount (2 g) given and, provided that an adequate sampling procedure is adopted,[283] a concentration of 4 mg Cr_2O_3/g faecal organic matter indicates that the daily organic matter output is 500 g. The same concentration in the faeces of cattle given two 10 g Cr_2O_3 doses daily would indicate an output of 5 kg organic matter/day. This method is usually more practical than harnessing when a large number of animals is to be studied, and it generally causes less disturbance to them.

Herbage plants or parts of plants at many different stages of growth, varying widely and continually changing in digestibility, are normally present in a pasture. The mowing machine removes this material unselectively down to a certain height and does not simulate the selective harvesting by the animal. In consequence, digestibility values obtained directly from cut herbage cannot validly be applied to the herbage that is grazed. The values can be related to the concentrations of various index substances in the faeces. From such relationships, established in digestibility trials with stalled animals,[284] the digestibility of herbage eaten by animals at pasture can be estimated from the concentrations of the indices in their faeces. The nitrogen concentration is most widely employed. Faecal nitrogen comprises undigested feed nitrogen, which is 10 per cent or less of that in the fresh herbage eaten (see p. 615) and large amounts of metabolic faecal nitrogen (MFN, see p. 607). Since the MFN excretion is positively related to feed dry matter intake, the greater the digestibility of the dry matter the higher is the total nitrogen concentration in the faeces. Among other indices that are used, chromogen is derived from the chlorophylls and related pigments of the plants.[278] The chromogen concentration in an acetone extract of faeces is positively related to digestibility because herbage of high digestibility generally has a high concentration of the pigments.

In recent years techniques for the permanent fistulation of ruminants at the oesophagus have been developed[286] so that animals can now be quite simply prepared and maintained, and they are widely used. Herbage can thus be collected as it is ingested; measurements can be made of botanical[93] and chemical composition, and of digestibility by *in vitro* procedures (see p. 616).

All these techniques demand great care in their application lest the results obtained should include serious bias. Chromium sesquioxide is not excreted in a uniform manner, and because the calculation of intake requires the *in*digestibility value, an estimated digestibility that is in error by 5 units will give a bias of 20 per cent or more in the estimated intake. Such an error can readily occur if, for example, a faecal index : digestibility relationship derived from herbage cut from one pasture is applied to another, or even to the same pasture at a different time of year.[280,284,318] Measurements cannot be made on herbage samples collected from oesophageally fistulated animals without difficulties. The material is contaminated with saliva,[289] is damaged to some extent during prehension and mastication,[53] and may deteriorate due to fermentation or during drying while being prepared for analysis. In addition sampling errors arise from variability between animals, and between times of day for individual animals.[287,288,289]

With care and in favourable circumstances feed intakes by individual animals on temperate pastures might be estimated with a standard error as low as ± 6 per cent,[37,278] but one may have to be content with ± 20 per cent.[290] There have been rather few reports of measurements of intakes on tropical and sub-tropical pastures;[344,345,346] significant faecal index: digestibility relationships have not always been obtained.[85]

Selective Grazing

Measurements of intakes by the techniques referred to above have yielded information on the extent of the selectivity exercised in grazing by animals, a feature that has long been recognized. Selectivity[70,319] is partly mechanical—the consumption of herbage nearest the muzzle. Since this will often lead to the consumption of more leaf than stem the digestibility of the feed eaten and its content of nitrogen and sugars will tend to be higher, and of fibre lower, than for the pasture herbage as a whole. These effects will be enhanced by the deliberate selection of leaf, and of green rather than dry or dead material; choices by the animal among the usually large number of species present in a pasture further distinguish grazing from mechanical cutting. There do not seem to have been any reports that grazed feed was less digestible than the material from which it was selected. It may be only slightly more digestible when the pasture includes a large proportion of new growth[291] but where the range and opportunity for selection is large the difference can be 15 units or more.[292]

Selection by the animals of the higher quality fractions of the feed offsets the decline that occurs in the digestibility of pasture herbage as it grows and matures so that a fall in the digestibility of the grazed herbage will tend to occur more slowly. Perhaps for this reason, and because of the difficulties in measurement, there is uncertainty about the importance under grazing conditions of relationships between digestibility and intake (Fig. 1) though these have been observed.[37,45,294]

Probably the greatest opportunity for selection occurs when plant growth is most rapid, that is in spring or at the start of the wet season; in temperate areas the rate of increase in herbage dry matter can approach a maximum of about 200 kg/ha/day, but in winter or in drought it will be virtually nil. If no attempt is made to restrict the area available for grazing the animals may at times consume much less than 10 per cent of the current herbage production[295] though the discrepancy can be reduced by management, as in seasonal dairying, so that the patterns of animal demand and herbage growth roughly correspond.[296] Selective grazing can be turned to advantage if lactating dairy cows or, by suitable fencing, sucking lambs but not their mothers are allowed the first pick at a growth of pasture and the residual herbage is subsequently grazed by animals in a less productive state.[269,297,319]

Availability of Herbage

The term 'availability' refers not only to the actual amount present on the area of pasture allotted to an animal but also to the manner of its distribution. Sheep would have considerable difficulty in gathering sufficient feed from a pasture yielding a uniform 50 g dry matter/m^2 (i.e. 500 kg/ha) but not if the same amount of material were disposed in a few large clumps. There are undoubtedly critical limits in height/density patterns of growth below which intakes by animals, and thence their productivity, will be reduced. Comprehensive measurements of the yield and structure of pasture are difficult to make[269,293,298,299,300,317] and must take account of the fact that animals, especially sheep, can graze herbage only just above ground level and below the height at which it can be cut by many types of machine. Sheep will in fact dig in very bare pastures to gain underground parts of plants. It has been reported[301] that the liveweight gains of grazing lambs increased with an increasing supply of green herbage up to a level of 1570 kg DM/ha but gains did not continue to rise at higher yields. Because of the immense range of pasture types, of differences in grazing behaviour and in the acquisition of herbage, as between sheep and cattle,[302,303,304,319] and the difficulty of making appropriate measurements, any such relationships, if found,[305] are likely to be of rather local applicability.

Grazing Behaviour

The total time spent in grazing by cattle and sheep, often about 8 hr/day,[19] varies with the availability of feed and may have a positive but imprecise relationship with feed intake.[70,306] When the herbage is very sparse, grazing times increase[307] but it does not seem that animals will work at feed collection for more than half of the 24 hr even under the most difficult conditions. The remainder of their time is spent in resting, and in ruminating which generally occupies some 6–8 hr/day; the longer rumination periods occur when the herbage is of rather poor quality.[307] Adaptation of grazing habits to new conditions occurs very rapidly; thus the total grazing time is made up of a number of periods of activity which are most intensive in the hour or two after sunrise and again towards sunset,[269] but if the day is very hot the main grazing activity may be deferred until the cooler night hours.[308,309]

The effects of a 'little and often' regime of feeding have been studied with housed animals.[310] It has been found on a number of occasions that liveweight gains were increased by giving a ration as a number of small feeds at intervals throughout the day rather than as one or two larger feeds. This result could be due to greater uniformity in the supply of metabolites to the tissues of the animal.[311] A frequent-feeding regime also tends to increase *ad libitum* intakes, and a herbage is generally consumed in greater amounts when grazed than when given freshly cut to confined animals, as in the 'zero-grazing' system of management.[312]

Advocates of this system point to the wastage of herbage by trampling[313] and fouling with excreta that occurs during grazing, and the increased expenditure of energy by animals at pasture compared with those that are penned (see p. 633), but when it is applied to dairy cows the milk production per unit area of grassland is generally similar to that obtained by grazing.[312] In the latter case the animals have greater opportunities for dietary selection and may benefit from more frequent feeding. Zero-grazing might give the higher yields of animal products per unit area if instead of utilizing swards that were developed primarily for grazing, herbage plants were bred specifically for cutting. A growth habit that intercepted the maximum amounts of solar energy could then be selected.[314] It can also be noted here that although large quantities of plant nutrients are returned to pastures in animal excreta they are often distributed very unevenly.[315] With animals that are gregarious during rest a large proportion of a pasture may even be depleted of nutrients due to their transfer in excreta on to the small habitual resting areas.[316]

Large amounts of nitrogenous substances are returned by animals, especially when they graze intensively-managed pastures yielding herbage of high crude protein content. Concentrates or other feeds might be given in these circumstances, the intention being to increase the supply of non-protein energy and perhaps to improve the utilization of dietary nitrogen, or in other circumstances to add to the energy supplied by herbage of rather low digestibility. In practice such supplements have neither large nor economically favourable effects on the production of animals provided that there is sufficient herbage readily available to meet their needs.[273,309,320,321] Unless the herbage is scanty, supplementary feeds are more correctly described as substitutes because, when they are consumed, herbage intakes fall. This effect is illustrated by the results of a number of experiments with dairy cows which yielded[322] the expression $I = 1{\cdot}27 - 0{\cdot}015\ D$, where I is the increase in total feed intake in kg per kg concentrate consumed, and D is the digestibility of the herbage organic matter. It will be seen that the increase in the total amount of feed eaten is likely to be small; unless the concentrate is of much higher nutritional value than the herbage it replaces, the net effect on nutrient intake will also be small.

Stocking Rate

The stocking rate (the number of animals per unit area) is the major factor determining the output of animal products from any given pasture and certainly has very much larger effects than changes in grazing management as such.[272,323,324] Doubling the number of animals on an understocked pasture might almost double the production. An increase in stocking rate from a very low level may initially increase the output per head although there is little change in the amount of high quality feed available for selection.[294] With sheep this effect was thought to be due to a low intake per bite from understocked erect open growths of herbage compared with a shorter and more dense growth.[70] With further increases output per head falls, at first slowly but then with increasing rapidity. The increasing number of animals generally gives increasing output per unit area beyond the point where the productivity of individual animals starts to fall quite sharply.[323,324] Ultimately the production on either basis must fail entirely.

The continual problem facing the grazier is the attainment of optimum outputs per head and/or per pasture; in milk or meat production there might be greater emphasis on output per head than, for example, in wool production. In addition the stocking rate has to be adjusted with an eye on the maintenance of a productive sward. Continued hard grazing and severe defoliation will reduce leaf area and so reduce the amount of solar energy that is intercepted and used in photosynthetic activity and regrowth.[22,75,325]

Energy Costs

Although increasing the stocking rate reduces the amount of herbage available per animal, it has been reported[294, 326] that feed intake increases as individual productivity declines, indicating a progressively inefficient utilization of feed by the grazing animal.[327] Quite a large number of estimates of efficiency have now been made from measurements of feed intakes (I), usually as digestible organic matter, and of the liveweights (W) and production of the animals as milk (M), gain in weight (G), etc. Values for the constants in equations of the type $I = aW^k + bG + cM$, or $I = aW^k \cdot e^{bG}$, are obtained by least-square procedures.

When sheep or cattle have been confined to small areas of pasture yielding abundant high quality herbage it has been found that the value for the constant a, which is taken to be the amount of energy required by the animal for maintenance, is about 20 per cent greater than that obtained from similar animals given cut herbage in stalls.[328,329] As the area available for grazing and perhaps the activity of the animal and the opportunity for the consumption of feed in excess of immediate requirements increase, so the maintenance costs appear to rise.[322] The greatest increases, however, have been observed where there was a deficit rather than an excess of pasture herbage. An increase to three times or more the indoor maintenance requirement was found[326] on a very sparse pasture where sheep had the utmost difficulty in obtaining feed, and they were able to maintain themselves only at a liveweight very much lower than the normal for these animals. It appeared that the less herbage stock have on offer, the more they need.

These input/output studies are subject to several criticisms. Firstly, of all the variables only milk yield can readily be measured with accuracy and precision. Liveweights often show large short-term changes that reflect only changes in the weight of gut contents.[330,331] In addition the estimated intakes could include large biasses. Secondly, the mathematical model used may be biologically inappropriate; for example, it is usually assumed that G and M are directly related in a rectilinear manner to feed intake and no allowance is made for diminishing returns effects. Thirdly, the statistical method used might be inappropriate in a situation where the variables are often correlated and where the data may be heterogeneous. Regression analysis may be convenient where the object is to predict intakes from a number of independent variables (W, G, M, etc.) even though these are not measured without error, but a functional type of analysis that examines the underlying relationships between all variables will probably give rather different, and perhaps more valid results.[329]

The apparent large increases in maintenance requirements observed on a number of occasions[332,333,334,335,336,337] in addition to those already mentioned have been shown[21,338] to have some plausibility by calculations based on separate measurements made indoors of the energy costs of locomotion[88], standing, grazing, rumination etc., and the effects of climatic factors such as wind on the rate of heat loss by the animal[340,341]. These calculations are, however, simply summative in type whereas the mathematical model truly representing the biological situation is likely to be very much more complicated. Calorimetric techniques that allow estimation of the energy expenditure by grazing animals have recently been developed[342,343,353]. Estimates of the energy requirements for maintenance of sheep determined from their oxygen consumption and carbon dioxide production[343], and those derived from measurements of the rate of entry of carbon dioxide in their bodies[353], showed that even under extremely difficult grazing conditions the requirements were directly proportional to liveweight (W). The value obtained[354], approximately 132 kcal/kg$W^{0.75}$/24h, is some 50 per cent greater than would be expected for similar animals kept indoors but is not inexplicably high, and does cast doubt on those derived from measurements of feed intakes.

The application to grazing animals of calorimetric and other techniques[263] that have so far been used almost exclusively under necessarily stereotyped laboratory conditions is long overdue. They must be applied if there is to be real understanding of the relationships between animals and pastures in all their fascinating complexity.

REFERENCES

1. Schantz, H. L. *Ecology*, **35**, 143 (1954).
2. Barnard, C. (Ed.) *Grasses and Grasslands*. MacMillan, London, 1964.
3. Hartley, W. and Williams, R. J. *Proc. 7th Int. Grassl. Congr., New Zealand*, p. 190 (1956).
4. Symposium on Grassland Botany. *Neth. J. Agric. Sci.* **13**, No. 2 (1965).
5. Russell, E. W. *Soil Conditions and Plant Growth*. 9th edn. Longmans, London, 1961.
6. Hallsworth, E. G. (Ed.) *Nutrition of the Legumes*. Butterworths, London, 1958.
7. Hutton, E. M. *J. Aust. Inst. Agric. Sci.* **31**, 102 (1965).
8. Thomas, B., Thomson, A., Oyenuga, V. A. and Armstrong, R. H. *Emp. J. Exp. Agric.* **20**, 10 (1952).
9. Thomas, B., Rogerson, A. and Armstrong, R. H. *J. Brit. Grassl. Soc.* **11**, 10 and 82 (1956).
10. Gallagher, C. H., Koch, J. H., Moore, R. M. and Steel, J. D. *Nature, Lond.* **204**, 542 (1964).
11. Moule, G. R., Braden, A. W. H. and Lamond, D. R. *Anim. Breed. Abstr.* **31**, 139 (1963).
12. Mortimer, P. H. *et al. Res. Vet. Sci.* **3**, 147, 161, 369 (1962); **4**, 166 (1963).
13. Jennings, R. D. *U.S. Dept. Agric. Stat. Bull.*, No. 215 (1957).
14. *Report of the Committee on Grassland Utilization*. Cmd. 547. H.M.S.O., London, 1958.
15. Eyles, D. E. and Alder, F. E. *Proc. Brit. Soc. Anim. Prod.*, p. 77 (1955).

*16. Cowlishaw, S. J. and Eyles, D. E. *Nutr. Abstr. Rev.* **26**, 983 (1957).

*17. Alexander, F. In *Progress in Nutrition and Allied Sciences*, p. 259 (D. P. Cuthbertson, Ed.), Oliver & Boyd, Edinburgh, 1963.

*18. Makela, A. *Acta Agral. Fenn.* **85** (1956).

*Review articles

19. HAFEZ, E. S. E. (Ed.) *The Behaviour of Domestic Animals.* Balliere, London, 1962.
20. CRAMPTON, E. W. *J. Anim. Sci.* **16**, 546 (1947).
21. BLAXTER, K. L. *The Energy Metabolism of Ruminants.* Hutchinson, London, 1962.
*22. ANAND, B. K. *Physiol. Rev.* **41**, 677 (1961).
*23. ANDERSSON, B. and LARSSON, S. *Pharm. Rev.* **13**, 1 (1961).
*24. BALCH, C. C. and CAMPLING, R. C. *Nutr. Abstr. Rev.* **32**, 669 (1962).
*25. MAYER, J. In *Nutrition : A Comprehensive Treatise*, Vol. 1, p. 1 (A. H. Beaton and E. W. McHenry, Eds.) Academic Press, New York 1964.
*26. BLAXTER, K. L. In *Scientific Principles of Feeding Farm Livestock.* Farmer & Stockbreeder Publications Ltd., London, 1958.
27. CAMPLING, R. C., FREER, M. and BALCH, C. C. *Brit. J. Nutr.* **15**, 531 (1961).
28. FREER, M. and CAMPLING, R. C. *Brit. J. Nutr.* **17**, 79 (1963).
*29. KAY, R. N. B. and HOBSON, P. N. *J. Dairy Res.* **30**, 261 (1963).
30. HARRIS, L. E. and PHILLIPSON, A. T. *Anim. Prod.* **4**, 97 (1962).
31. BLAXTER, K. L., WAINMAN, F. W. and WILSON, R. S. *Anim. Prod.* **3**, 51 (1961).
32. CRAMPTON, E. W., DONEFER, E. and LLOYD, L. E. *J. Anim. Sci.* **19**, 538 (1960).
33. MINSON, D. J., HARRIS, C. E., RAYMOND, W. F. and MILFORD, R. *J. Brit. Grassl. Soc.* **19**, 298 (1964).
34. CONRAD, H. R., PRATT, A. D. and HIBBS, J. W. *J. Dairy Sci.* **47**, 54 (1964).
35. MILLER, T. B., RAINS, A. B., and THORPE, R. J. *J. Brit. Grassl. Soc.* **18**, 158, 223 (1963) **19**, 77 (1964).
36. CONRAD, H. R., PRATT, A. D., HIBBS, J. W. and DAVIS, R. R. *Ohio Agric. Exper. Sta. Res. Bull.* 914 (1962).
37. CORBETT, J. L., LANGLANDS, J. P. and REID, G. W. *Anim. Prod.* **5**, 119 (1963).
38. HUTCHINSON, K. J. *Aust. J. Agric. Res.* **12**, 696 (1961).
39. COOP, I. E. and HILL, M. K. *J. Agric. Sci.* **58**, 187 (1962).
40. HOLMES, W., JONES, J. G. W. and DRAKE-BROCKMAN, R. M. *Anim. Prod.* **3**, 251 (1961).
41. MILFORD, R. *Austr. J. Agric. Res.* **11**, 121, 138 (1960).
*42. CORBETT, J. L. *J. Roy. Agric. Soc. Engl.* **122**, 175 (1961).
43. BLAXTER, K. L. and WILSON, R. S. *Anim. Prod.* **4**, 351 (1962).
44. MONTGOMERY, M. J. and BAUMGARDT, B. R. *J. Dairy Sci.* **48**, 569 (1965).
45. HOLMES, W. and JONES, J. G. W. *Anim. Prod.* **7**, 39 (1965).
46. ELLIOTT, R. C., FOKKEMA, K. and FRENCH, C. H. *Rhodesia Agric. J.* **58**, 124 (1961).
47. MILFORD, R. and MINSON, D. J. *Proc. 9th Int. Grassl. Congr., Brazil* (1964) and personal communication.
48. COMMONWEALTH SCI. INDUSTR. RES. ORG. AUST. Divn. Trop. Pastures. *Ann. Rept.* 1964-65, p. 78.
MINSON, D. J. *Brit. J. Nutr.* **20**, 757 and 765 (1966).
49. DOWDEN, D. R. and JACOBSON, D. R. *Nature, Lond.* **188**, 148 (1960).
50. MONTGOMERY, M. J., SCHULTZ, L. H. and BAUMGARDT, B. R. *J. Dairy Sci.* **46**, 1380 (1963).
51. ULYATT, M. J. *N.Z. J. Agric. Res.* **8**, 397 (1965).
*52. BAKER, F. *Proc. Roy. Soc. B.* **139**, 204 (1952); *Nutr. Abstr. Rev.* **17**, 1 (1947).
53. HOGAN, J. P. *Aust. J. Agric. Res.* **16**, 855 (1965); REID, C. S. W., LYTTLETON, J. W., and MANGAN, J. L. *N.Z. J. Agric. Res.* **5**, 237 (1962).
54. PEARCE, G. R. and MOIR, R. J. *Aust. J. Agric. Res.* **15**, 635 (1964).
55. MOIR, R. J. and HARRIS, L. E. *J. Nutr.* **77**, 285 (1962).
*56. CUTHBERTSON, D. P. and HOBSON, P. N. In *World Review of Nutrition and Dietetics* **3**, 69 (G. H. Bourne, Ed.) (1960). Pitman, London.
57. HEMSLEY, J. A. and MOIR, R. J. *Aust. J. Agric. Res.* **14**, 509 (1963).
58. EGAN, A. R. and MOIR, R. J. *Aust. J. Agric. Res.* **16**, 437 (1965).
59. EGAN, A. R. *Aust. J. Agric. Res.* **16**, 451, 463, 473 (1965).
60. HEMSLEY, J. A. *Proc. Aust. Soc. Anim. Prod.* **5**, 321 (1964).
61. CAMPLING, R. C., FREER, M. and BALCH, C. C. *Brit. J. Nutr.* **16**, 115 (1962).
62. DONEFER, E., LLOYD, L. E. and CRAMPTON, E. W. *J. Anim. Sci.* **22**, 425 (1963).

*Review articles

63. ELLIOTT, R. C. and TOPPS, J. H. *Anim. Prod.* **5**, 269 (1963).
64. BLAXTER, K. L. and WILSON, R. S. *Anim. Prod.* **5**, 27 (1963).
65. GLOVER, J. and DOUGALL, H. W. *J. Agric. Sci.* **55**, 391 (1960).
*66. IVINS, J. D. *Herbage Abstr.* **25**, 75 (1955).
67. BLAND, B. F. and DENT, J. W. *J. Brit. Grassl. Soc.* **19**, 306 (1964).
68. COWLISHAW, S. J. and ALDER, F. E. *J. Agric. Sci.* **54**, 257 (1960).
*69. TRIBE, D. E. *J. Brit. Grassl. Soc.* **5**, 81, 209 (1950).
*70. ARNOLD, G. W. In *Grazing in Terrestrial and Marine Environments*, p. 133 (D. J. Crisp, Ed.) Blackwell, Oxford, 1964.
*71. ARNOLD, G. W. *Proc. Aust. Soc. Anim. Prod.* **5**, 258 (1964).
72. CLARKE, E. A. *Proc. N.Z. Soc. Anim. Prod.* **19**, 91 (1959).
73. CORBETT, J. L., LANGLANDS, J. P., MCDONALD, I. and PULLAR, J. D. *Anim. Prod.* **8**, (1966).
74. BELL, F. R. *Vet. Record*, **71**, 1071 (1959).
75. ROE, R. and MOTTERSHEAD, B. E. *Nature, Lond.*, **193**, 255 (1962).
*76. CAMPLING, R. C. *Proc. Nutr. Soc.* **23**, 80 (1964).
77. PLICE, M. J. *J. Range Mgmt.* **5**, 69 (1952).
78. MARTIN, G. C. and DONKER, J. D. *J. Dairy Sci.* **47**, 773, 871 (1964).
79. HOLMES, J. C. and LANG, R. W. *Anim. Prod.* **5**, 17 (1963).
80. ARNOLD, G. W. *Aust. J. Agric. Res.* **13**, 701 (1962).
81. CAMPLING, R. C. and BALCH, C. C. *Brit. J. Nutr.* **15**, 523 (1961).
*82. MURDOCH, J. C. *Proc. Nutr. Soc.* **23**, 99 (1964).
83. THOMAS, J. W., MOORE, L. A., OKAMOTO, M. and SYKES, J. F. *J. Dairy Sci.* **44**, 1471 (1961).
84. OKAMOTO, M., WALDO, D. R. MILLER, R. W. and MOORE, L. A. *J. Dairy Sci.* **47**, 1231 (1964).
*85. MILFORD, R. In: Some Concepts and Methods in Sub-tropical Pasture Research. *Bull. Commonw. Bur. Past. Fld. Crops*, No. 47, p. 144. Commonw. Agric. Bur., Farnham Royal, Bucks., 1964.
86. RAYMOND, W. F., HARRIS, C. E. and KEMP, C. D. *J. Brit. Grassl. Soc.* **9**, 209 (1954).
87. DIJKSTRA, N. D., WEIDE, H. J. and VAN ADRICHEM, P. W. M. *Verslag. landbouw. Ondersoek.* **68**, No. 9 (1962).
88. CLAPPERTON, J. L. *Brit. J. Nutr.* **18**, 39, 47 (1964).
89. BLAXTER, K. L. In *Proc. 2nd Symp. on Energy Metabolism*, p. 211; Europ. Assoc. Anim. Prod. Publ. No. 10. Rome, 1961.
BLAXTER, K. L., CLAPPERTON, J. L. and MARTIN, A. K. *Brit. J. Nutr.* **20**, 449 (1966).
90. RAYMOND, W. F., MINSON, D. J. and HARRIS, C. E. *J. Brit. Grassl. Soc.* **14**, 75 (1959).
91. ANDERSON, P. E., REID, J. T., ANDERSON, M. J. and STROUD, J. W. *J. Anim. Sci.* **18**, 1299 (1959).
92. BLAXTER, K. L. and WAINMAN, F. W. *J. Agric. Sci.* **57**, 419 (1961).
93. HARKER, K. W., TORRELL, D. T. and VAN DYNE, G. M. *J. Anim. Sci.* **23**, 465 (1964).
94. COOPER, J. P., TILLEY, J. M. A., RAYMOND, W. F. and TERRY, R. A. *Nature, Lond.* **195**, 1276 (1963).
95. KIVIMAE, A. *Acta. Agric. Scand.* Suppl. 5 (1959).
96. HARKESS, R. D. *J. Brit. Grassl. Soc.* **18**, 62 (1963).
97. WAITE, R., JOHNSTON, M. J. and ARMSTRONG, D. G. *J. Agric. Sci.* **62**, 391 (1964).
98. ARMSTRONG, D. G. *J. Agric. Sci.* **62**, 399 (1964).
BLAXTER, K. L. and CLAPPERTON, J. L. *Brit. J. Nutr.* **19**, 511 (1965).
99. MINSON, D. J., RAYMOND, W. F. and HARRIS, C. E. *J. Brit. Grassl. Soc.* **15**, 174 (1960).
*100. WAITE, R. *Proc. Nutr. Soc.* **24**, 38 (1965).
101. GAILLARD, B. D. E. *J. Agric. Sci.* **59**, 369 (1962).
*102. HALLIWELL, G. In *Digestive Physiology and Nutrition of the Ruminant*, p. 119 (D. Lewis, Ed.) Butterworths, London, 1961; and HEAD, M. J. ibid., p. 263.
*103. HALLIWELL, G. *Nutr. Abstr. Rev.* **29**, 747 (1959).
104. BALCH, D. A., BALCH, C. C. and ROWLANDS, S. J. *J. Sci. Fd. Agric.* **5**, 584 (1954).

*Review articles

*105. VAN SOEST, P. J. *J. Anim. Sci.* **23**, 838 (1964).
106. JOHNSTON, M. J. and WAITE, R. *J. Agric. Sci.* **64**, 211 (1965).
107. BLAXTER, K. L. and MITCHELL, H. H. *J. Anim. Sci.* **7**, 351 (1948).
*108. EL-SAMMAN, S. *The Biological Value of Proteins in Mixed Grass Hays.* Thesis, Agricultural University of Wageningen, 1961.
109. GLOVER, J., DUTHIE, D. W. and FRENCH, M. H. *J. Agric. Sci.* **48**, 373 (1957).
110. MILFORD, R. and HAYDOCK, K. P. *Aust. J. Exp. Agric. and Anim. Husb.* **5**, 13 (1965).
111. FERGUSON, W. S. and TERRY, R. A. *J. Sci. Fd. Agric.* **5**, 515 (1954); *J. Agric. Sci.* **48**, 149 (1957).
112. WAITE, R. and GORROD, A. R. N. *J. Sci. Fd. Agric.* **10**, 308 and 317 (1959).
113. MINSON, D. J., RAYMOND, W. F. and HARRIS, C. E. *Proc. 8th Int. Grassl. Congr., Reading*, p. 470 (1960).
114. DEMARQUILLY, C. and JARRIGE, R. *Ann. Zootech.* **13**, 301 (1964).
115. JARRIGE, R. and MINSON, D. J. *Ann. Zootech.* **13**, 117 (1964).
116. WAITE, R. and BOYD, J. *J. Sci. Fd. Agric.* **4**, 197, 257 (1953).
117. MILES, D. G., GRIFFITH, G. ap. and WALTERS, R. J. K. *J. Brit. Grassl. Soc.* **19**, 75 (1964).
118. DAVIES, J. G. and SHAW, N. H. In: Some Concepts and Methods in Sub-tropical Pasture Research, *Bull. Commw. Bur. Past. Fld. Crops*, No. 47, p. 1. Commw. Agric. Bur., Farnham Royal, Bucks., 1964.
*119. FRENCH, M. H. *Herbage Abstr.* **25**, 1 (1957).
*120. OYENUGA, V. A. *Nutr. Abstr. Rev.* **28**, 985 (1958).
122. TOPPS, J. H. and ELLIOTT, R. C. *Nature, Lond.* **208**, 498 (1965).
121. COOK, C. W., STODDART, L. A. and HARRIS, L. E. *J. Anim. Sci.* **11**, 578 (1952).
123. PALADINES, O. L., REID, J. T., VAN NIEKERK, B. D. H. and BENSADOUN, A. *J. Anim. Sci.* **23**, 528 (1964).
124. STREET, J. C., BUTCHER, J. E., and HARRIS, L. E. *J. Anim. Sci.* **23**, 1039 (1964).
125. KLEIBER, M. *The Fire of Life.* Wiley, New York, 1961.
126. DENTON, D. A. and SABINE, J. R. *Behaviour*, **20**, 364 (1963).
127. WOLIN, M. J. *J. Dairy Sci.* **43**, 1452 (1960).
128. COPPOCK, C. E., FLATT, W. P., MOORE, L. A. and STEWART, W. E. *J. Dairy Sci.* **47**, 1330, 1359 (1964).
129. MARSTON, H. R. *Biochem. J.* **42**, 564 (1948).
130. FRIES, J. A. *Proc. Amer. Soc. Anim. Prod.*, p. 33. 1922.
131. KROGH, A., SCHMIT-JENSEN, H. O. *Biochem. J.* **14**, 686 (1920).
*132. WARNER, A. C. I. *Nutr. Abstr. Rev.* **34**, 399 (1964).
133. LENG, R. A. and LEONARD, G. J. *Brit. J. Nutr.* **19**, 469 (1965).
*134. ROOK, J. A. F. *Proc. Nutr. Soc.* **23**, 71 (1964).
135. ARMSTRONG, D. G., BLAXTER, K. L., GRAHAM, N. McC. and WAINMAN, F. W. *Brit. J. Nutr.* **12**, 177 (1958).
*136. ROOK, J. A. F. *Dairy Sci. Abstr.* **23**, 251, 303 (1961).
STORRY, J. E., and ROOK, J. A. F. *Brit. J. Nutr.* **20**, 217 (1966).
137. MCCLYMONT, G. L. *Aust. Vet. J.* **26**, 111 (1950).
*138. SHAW, J. C. *Proc. 8th Int. Congr. Anim. Prod., Hamburg, Gen. Rept.*, p. 29. 1961.
139. BROUWER, E., VAN ES., A. J. H. and NIJKAMP, H. J. *Proc. 2nd Symp. on Energy Metabolism*, Eur. Assoc. Anim. Prod. Publ. No. 10. Rome, 1961.
140. BATH, I. H. and ROOK, J. A. F. *J. Agric. Sci.* **64**, 67 (1965).
141. PARKS, P. F., RIEWE, M. E., LYMAN, C. M. and KUNKEL, H. O. *J. Anim. Sci.* **23**, 344 (1964).
142. TILLEY, J. M. A., DERIAZ, R. E. and TERRY, R. A. *Proc. 8th Int. Grassl. Congr., Reading*, p. 533. 1960.
143. LENG, R. A. and BRETT, D. J. *Brit. J. Nutr.* **20**, 541 (1966).
144. TOPPS, J. H., REID, W. D. C. and ELLIOTT, R. C. *J. Agric. Sci.* **64**, 397 (1965).
145. LEWIS, D. In *Digestive Physiology and Nutrition of the Ruminant*, p. 127 (D. Lewis, Ed.) Butterworths, London, 1961.
146. JOHNS, A. T. *Proc. N.Z. Soc. Anim. Prod.* **22**, 88 (1962).

*Review articles

147. JOHNS, A. T., ULYATT, M. J. and GLENDAY, A. C. *J. Agric. Sci.* **61**, 201 (1963); BAILEY, R. W. *N.Z. J. Agric. Res.* **7**, 496 (1964).

*148. RAYMOND, W. F., TILLEY, J. M. A., DERIAZ, R. E. and MINSON, D. J. In *Chemical Aspects of the Production and Use of Grass*, p. 181. Monograph No. 9, Society of Chemical Industry, London, 1960.

*149. BLAXTER, K. L. *J. Dairy Sci.* **39**, 1396 (1956).

*150. WATSON, S. J. *Nutr. Abstr. Rev.* **18**, 1 (1948).

*151. DIJKSTRA, N. D. *Neth. J. Agric. Sci.* **2**, 273 (1954).

*152. SWIFT, R. W. *Penn. State Univ. Bull.* 615. N.E. Regional Publ. 27, 1957.

153. MINSON, D. J. and MILFORD, R. *Aust. J. Agric. Res.* **17**, 411 (1966).

154. MOIR, R. J. *Aust. J. Exp. Agric. Anim. Husb.* **1**, 24 (1961).

155. BLAXTER, K. L. *J. Roy. Agric. Soc. Engl.* **123**, 7 (1962).

156. HENDERSON, A. E. and HAYMAN, B. I. *Aust. J. Agric. Res.* **11**, 851 (1960).

157. RAYMOND, W. F., HARRIS, C. E. and HARKER, V. G. *J. Brit. Grassl. Soc.* **8**, 301 (1953).

158. PATERSON, R. In *The Measurement of Grassland Productivity*, p. 194 (J. D. Ivins, Ed.) Butterworths, London, 1959; *J. Brit. Grassl. Soc.* **11**, 93 (1956).

159. BOYD, M. and JENNINGS, J. W. *Res. Exp. Record Min. Agric. N. Ireland*, **9**, 135 (1960); CASTLE, M. E., DRYSDALE, A. D. and WATSON, J. N. *J. Dairy Res.* **29**, 199 (1962).

160. DEMARQUILLY, C. *Ann. Zootech.* **12**, 69 (1963).

161. MILFORD, R. and MINSON, D. J. *J. Brit. Grassl. Soc.* **20**, 177 (1965).

162. MILFORD, R. and MINSON, D. J. *Brit. J. Nutr.* **19**, 373 (1965).

163. DOWNES, A. M. and MCDONALD, I. W. *Brit. J. Nutr.* **18**, 153 (1964).

164. THOMSON, D. J. and TERRY, R. A. *Proc. Nutr. Soc.* **24**, XXV (1965).

165. KIVIMAE, A. *Proc. 8th Int. Grassl. Congr., Reading*, p. 466. 1960. *Acta Agric. Scand.* **14**, 165 (1964).

166. WALKER, D. M. and HEPBURN, W. R. *Agric. Progr.* **30**, 118 (1955).

167. ARMSTRONG, D. G. *Agric. Progr.* **39**, 109 (1964).

*168. RAYMOND, W. F. In *The Measurement of Grassland Productivity*, p. 156 (J. D. Ivins, Ed.) Butterworths, London, 1959.

169. WATSON, S. J. and HORTON, E. A. *J. Agric. Sci.* **26**, 142 (1936).

*170. KEFFORD, N. P. *J. Aust. Inst. Agric. Sci.* **24**, 297 (1958).

171. DONEFER, E., NIEMANN, P. J., CRAMPTON, E. W. and LLOYD, L. E. *J. Dairy Sci.* **46**, 965 (1963).

*172. MITCHELL, H. H. *J. Anim. Sci.* **1**, 159 (1942).

173. HOLTER, J. A. and REID, J. T. *J. Anim. Sci.* **18**, 1339 (1959).

174. ASHTON, G. C. *J. Agric. Sci.* **58**, 333 (1962).

175. MILFORD, R. and MINSON, D. J. *J. Brit. Grassl. Soc.* **20** (1965).

176. TILLEY, J. M. A. and TERRY, R. A. *J. Brit. Grassl. Soc.* **18**, 105 (1963).

177. ALEXANDER, R. H. and MCGOWAN, M. *J. Brit. Grassl. Soc.* **16**, 275 (1961).

178. DONEFER, E., CRAMPTON, E. W. and LLOYD, L. E. *J. Anim. Sci.* **19**, 545 (1960).

179. VAN DYNE, G. M. and WEIR, W. C. *J. Range Mgmt.* **17**, 327 (1964). *J. Anim. Sci.* **23**, 1116 (1964).

180. TERRY, R. A. and TILLEY, J. M. A. *J. Brit. Grassl. Soc.* **19**, 363 (1964).

181. REID, J. T., KENNEDY, W. K., TURK, K. L., SLACK, S. T., TRIMBERGER, G. W. and MURPHY, R. P. *J. Dairy Sci.* **42**, 567 (1959); *Agron. J.* **51**, 213 (1959).

182. RATIONS FOR LIVESTOCK. *Bull. Min. Agric.* No. 48. H.M.S.O., London, 1960.

183. U.S.D.A. GRAIN BRANCH, PRODUCTION AND MARKETING ADMINISTRATION. *Handbook of Official Hay and Straw Standards*, 1949.

184. RICHARDS, C. R., HAENBIN, G. F. W., CONOLLY, J. D. and CALHOUN, M. C. *J. Anim. Sci.* **21**, 73 (1962); INGALLS, J. R., THOMAS, J. W. and TESAR, M. B. *J. Anim. Sci.* **24**, 1165 (1965).

185. PFANDER, W. H., HARGUS, W. A., TYREE, W. and STONE, P. C. *Fed. Proc.* **23**, 135 (1964).

186. REGAL, V. *Proc. 8th Int. Grassl. Congr. Reading*, p. 522, 1960; MARTIN, D. J. In *Grazing in Terrestrial and Marine Environments* (D. J. Crisp, Ed.). Blackwell, Oxford, 1964.

*Review articles

187. TROELSEN, J. E. and BIGSBY, F. W. *J. Anim. Sci.* **23**, 1139 (1964); CHENOST, M., *Proc. 10th Int. Grassl. Congr.*, Helsinki, 1966.
188. EVANS, P. S. *N.Z. J. Agric. Res.* **7**, 508 (1964); WILSON, D. *J. Agric. Sci.* **65**, 285 (1965).
189. JOHNSON, R. R., DEHORITY, B. A., MCCLURE, K. E. and PARSONS, J. L. *J. Anim. Sci.* **23**, 112 (1964).
*190. RUSSELL, F. C. and DUNCAN, D. L. *Minerals in Pasture: Deficiencies and Excesses in Relation to Animal Health, Tech. Comm. Commonw. Bur. Anim. Nutr.* No. 15 (2nd edn.). Aberdeen, 1956.
191. BLAXTER, K. L. *Brit. J. Nutr.* **17**, 105 (1963).
192. ALLAWAY, W. H. and HODGSON, J. F. *J. Anim. Sci.* **23**, 271 (1964).
193. UNDERWOOD, E. J. *Trace Elements in Human and Animal Nutrition* (2nd edn.). Academic Press, New York, 1962.
194. DEWEY, D. W., LEE, H. J. and MARSTON, H. R. *Nature, Lond.* **181**, 1367 (1958).
195. OWEN, E. C., ELLIS, S. E., VOSS, R. C., WILSON, A. L. and ROBERTSON, J. *Proc. Nutr. Soc.* **19**, XXI (1960).
*196. SYMP. PROC. AVAILABILITY OF MINERALS IN FOODS OF PLANT ORIGIN. *Proc. Nutr. Soc.* **24**, 81 (1965).
197. BUTLER, G. W. and GLENDAY, A. C. *Aust. J. Biol. Sci.* **15**, 183 (1962).
198. SAID, I. M. *Verslag. landbouw. Onderzoek.* **65**, No. 16 (1959).
*199. DIJKSHOORN, W. *Proc. 1st Reg. Conf. Int. Potash Inst., Wexford*, p. 43. Int. Potash Inst., Berne, 1963.
200. GRIFFITH, G. ap., JONES, D. I. H. and WALTERS, J. J. K. *J. Sci. Fd. Agric.* **16**, 94 (1965).
*201. WARD, G. M. *J. Dairy Sci.* **42**, 277 (1959).
*202. WILLIAMS, R. D. *Minor Elements and their Effects on the Growth and Chemical Composition of Herbage Plants*, Mimeo, Commonw. Bur. Past. Fld. Crops, No. 1, revised edn., 1963.
203. WALKER, T. W., *Proc. 7th Int. Grassl. Congr., New Zealand*, p. 157. 1956.
204. BLAXTER, K. L. (Ed.) *Energy Metabolism* (Proc. 3rd Conf. Eur. Assoc. Anim. Prod.). Academic Press, New York, 1965.
205. GRIFFITH, G. ap. and JOHNSTON, T. D. *J. Sci. Fd. Agric.* **11**, 622, 626 (1960).
206. REITH, J. W. S. *et al. J. Agric. Sci.* **56**, 17 (1961); **63**, 209 (1964).
207. NOWAKOWSKI, T. Z. *J. Agric. Sci.* **59**, 387 (1962).
208. JONES, D. I. H., GRIFFITH, G. ap. and WALTERS, R. J. K. *J. Agric. Sci.* **64**, 323 (1965).
209. BRYANT, A. M. and ULYATT, M. J. *N.Z. J. Agric. Res.* **8**, 109 (1965).
210. STEWART, A. B. and HOLMES, W. *J. Sci. Fd. Agric.* **4**, 401 (1953).
*211. BROUWER, E. In *Digestive Physiology and Nutrition of the Ruminant*, p. 154 (D. Lewis, Ed.) Butterworths, London, 1961.
212. HEMINGWAY, R. G. and RITCHIE, N. S. *Proc. 1st Reg. Conf. Int. Potash Inst., Wexford*, p. 161. Int. Potash Inst., Berne, 1963.
213. MOUAT, M. C. H. and WALKER, T. W. *Plant and Soil*, **11**, 30, 41 (1959).
214. ANDREW, C. S. *J. Aust. Inst. Agric. Sci.* **31**, 3 (1965).
215. REID, R. L. and JUNG, G. A. *J. Anim. Sci.* **24**, 615 (1965).
216. KREYGER, J., HUISMAN, M. H. *Verslag. landbouw. Onderzoek.* **65**, No. 10 (1959).
217. GUILBERT, H. R. and HART, G. H. *J. Nutr.* **8**, 24 (1934).
218. BARRON, N. S. *Vet. Rec.* **54**, 29 (1942); and BAKER, F. H., POPE, L. S. and MACVICAR, R. *J. Anim. Sci.* **13**, 802 (1954).
219. HENRY, K. M., KON, S. K., THOMPSON, S. Y., MCCALLUM, J. W. and STEWART, J. *Brit. J. Nutr.* **12**, 462 (1958).
220. GRANT, A. B. and O'HARA, P. B. *N.Z. J. Sci. Tech.* **38A**, 548 (1957).
221. MOORE, L. A., THOMAS, J. W., JACOBSON, W. C., MELIN, C. G. and SHEPHERD, J. B. *J. Dairy Sci.* **31**, 489 (1948).
222. KEENER, H. A. *J. Dairy Sci.* **37**, 1337 (1954).

*Review articles

223. WEITS, J. *Meded. Landbouwhogesch. Wageningen*, **56**, No. 1 (1956). (in Dutch; abstr. 3225, *Nutr. Abstr. Rev.* **27**, 708 (1957)); *Int. Z. Vitaminforsch.* **30**, 399 (1960).
224. WATSON, S. J. and NASH, M. J. *The Conservation of Grass and Fodder Crops.* Oliver & Boyd, Edinburgh, 1960.
225. GREENHILL, W. L. *J. Sci. Fd. Agric.* **9**, 495 (1959).
226. GRAHAM, N. McC. *Aust. J. Agric. Res.* **15**, 974 (1964).
*227. DIJKSTRA, N. D. *Neth. J. Agric. Sci.* **5**, 271 (1957).
228. GREENHILL, W. L., COUCHMAN, J. F. and DE FREITAS, J. *J. Sci. Fd. Agric.* **12**, 293 (1961).
229. DIJKSTRA, N. D. and VAN DER SCHAAF, D. *Verslag. landbouw. Onderzoek.* **61**, No. 15 (1955); **63**, No. 9 (1957).
230. *Manual for the Calculation of the Nutritive Value of Roughages.* Lab. Soil and Crop Testing, Oosterbeek, Netherlands, 1958.
231. EKERN, A., BLAXTER, K. L. and SAWERS, D. In *Energy Metabolism*, p. 217 (K. L. Blaxter, Ed.) Academic Press, New York, 1965.
232. CHRISTIAN, K. R. and WILLIAMS, V. J. *N.Z. J. Sci. Tech.* **38A**, 1003 (1957).
233. BARNETT, A. J. G. *Silage Fermentation.* Butterworths, London, 1954.
234. BREIREM, K. and ULVESLI, O. *Herb. Abstr.* **30**, 1 (1960).
235. McDONALD, P., *et al.* Studies on Ensilage. *Tech. Bull. Edin. Uni. Sch. Agric.* No. 24, 1960.
236. MURDOCH, J. C. *J. Brit. Grassl. Soc.* **16**, 253 (1961).
237. MACPHERSON, H. T. *J. Sci. Fd. Agric.* **3**, 362, 365 (1952).
238. BRADY, C. J. *J. Sci. Fd. Agric.* **11**, 276 (1960).
239. McDONALD, P., STIRLING, A. C., HENDERSON, A. R. and WHITTENBURY, R. *J. Sci. Fd. Agric.* **13**, 581 (1962).
240. MINSON, D. J. and LANCASTER, R. J. *N.Z. J. Agric. Res.* **8**, 542 (1965).
241. WILSON, R. F., TILLEY, J. M. A. and STEEMERS, M. A.-TH. *J. Sci. Fd. Agric.* **15**, 197 (1964).
242. HARRIS, C. E. and RAYMOND, W. F. *J. Brit. Grassl. Soc.* **18**, 204 (1963).
243. GORDON, C. H., DERBYSHIRE, J. C., JACOBSON, W. C. and MOORE, L. A. *Proc. 9th Int. Grassl. Congr., Brazil*, 1964.
244. RAYMOND, W. F., HARRIS, C. E. and HARKER, V. G. *J. Brit. Grassl. Soc.* **8**, 315 (1953).
*245. BLASER, R. E. *J. Anim. Sci.* **23**, 246 (1964).
*246. MINSON, D. J. *J. Brit. Grassl. Soc.* **18**, 39 (1963).
*247. MOORE, L. A. *J. Anim. Sci.* **23**, 230 (1964); BEARDSLEY, D. W. ibid., p. 239.
248. MEYER, J. H., WEIR, W. C., DOBIE, J. B. and HULL, J. L. *J. Anim. Sci.* **18**, 976 (1959).
249. HEANEY, D. P., PIGDEN, W. J., MINSON, D. J. and PRITCHARD, G. I. *J. Anim. Sci.* **22**, 752 (1963).
250. BLAXTER, K. L. and GRAHAM, N. McC. *J. Agric. Sci.* **47**, 207 (1956).
251. RODRIGUE, C. B. and ALLEN, N. N. *Canad. J. Anim. Sci.* **40**, 23 (1960).
252. BLAXTER, K. L., GRAHAM, N. McC. and WAINMAN, F. W. *Brit. J. Nutr.* **10**, 69 (1956).
253. CAMPLING, R. C., FREER, M. and BALCH, C. C. *Brit. J. Nutr.* **17**, 263 (1963).
254. FREER, M. and CAMPLING, R. C. *Brit. J. Nutr.* **19**, 195 (1965).
255. BALCH, D. A. and ROWLAND, S. J. *Brit. J. Nutr.* **11**, 288 (1957).
256. PALADINES, D. L., REID, J. T., VAN NIEKERK, B. D. H. and BENSADOUN, A. *J. Nutr.* **83**, 49 (1964).
257. GRAHAM, N. McC. *Aust. J. Agric. Res.* **15**, 969 (1964).
258. YOUNG, B. A. *Aust. J. Agric. Res.* **17**, 355 (1966).
259. BALCH, C. C., BROSTER, W. H., ROOK, J. A. F. and TUCK, V. J. *J. Dairy Res.* **32**, 1 (1965).
260. DIJKSTRA, N. D. and BRANDSMA, S. *Verslag. landbouw. Onderzoek.* **61**, No. 5 (1955).
261. LEIGH, J. H. and MULHAM, W. E. *Proc. Aust. Soc. Anim. Prod.* **5**, 251 (1964).
262. BERGMAN, E. N., REID, R. S., MURRAY, M. G., BROCKWAY, J. M. and WHITELAW, F. G. *Biochem. J.* **97**, 53 (1965).

*Review articles

263. LENG, R. A., CORBETT, J. L. and BRETT, D. J. *Brit. J. Nutr.* **22**, 57 (1968).
264. SKERMAN, K. D., O'HALLORAN, M. W. and MUNDAY, B. L. *Aust. Vet. J.* **37**, 181 (1961); MILLAR, K. R. and ANDREWS, E. D. *N.Z. Vet. J.* **12**, 9 (1964).
265. ROSENFELD, I. and BEATH, O. A. *Selenium : Geobotany, Biochemistry, Toxicity and Nutrition.* Academic Press, New York, 1964.
266. PIGDEN, W. J., PRITCHARD, G. T., WINTER, K. A. and LOGAN, V. S. *J. Anim. Sci.* **20**, 796 (1961).
267. HARDISON, W. A. *J. Anim. Sci.* **24**, 909 (1965).
*268. CROFTON, H. D. Nematode Parasite Population in Sheep and on Pasture, *Commonw. Bur. Helminth. Tech. Comm.*, No. 35 Commonw. Agric. Bur., Farnham Royal, Bucks., 1963.
269. SPEDDING, C. R. W. *Sheep Production and Grazing Management.* Bailliere, London, 1965.
270. ARMSTRONG, D. G., ALEXANDER, R. H., MCGOWAN, M. *Proc. Nutr. Soc.* **23**, XXVI (1964).
*271. MCMEEKAN, C. P. *Proc. 8th Int. Grassl. Congr., Reading*, p. 21. 1960.
*272. WHEELER, J. L. *Divn. Plant Ind. Rept.* No. 20. Commonw. Sci. Ind. Res. Org., Canberra, 1960.
*273. HOLMES, W. *J. Brit. Grassl. Soc.* **17**, 30 (1962).
274. MCMEEKAN, C. P. and WALSHE, J. M. *J. Agric. Sci.* **61**, 147 (1963).
*275. SPEDDING, C. R. W. *J. Brit. Grassl. Soc.* **20**, 7 (1965).
276. MCLEAN, J. W., THOMSON, G. G., JAGUSCH, K. T. and LAWSON, B. M. *Proc. Ruakura Fmrs. Conf. Wk.* (N.Z. Dept. Agric.), p. 34. 1965.
277. HIGHT, G. K. and SINCLAIR, D. P. *N.Z. J. Agric. Res.* **8**, 1079 (1965).
*278. CORBETT, J. L. and GREENHALGH, J. F. D. In *Chemical Aspects of the Production and Use of Grass*, p. 167. Monogr. No. 9, Soc. Chem. Ind., London, 1960.
*279. CORBETT, J. L. *Proc. 8th Int. Grassl. Congr., Reading*, p. 438. 1960.
280. Research Techniques in use at the Grassland Research Institute, Hurley. *Commonw. Bur. Pastures Fld. Crops* No. 45. Commonw. Agric. Bur., Farnham Royal, 1961.
281. MOULE, G. R. (Ed.), *Field Investigations with Sheep. A Manual of Techniques.* Commonw. Sci. Ind. Res. Org., Melbourne, 1965.
282. CORBETT, J. L., GREENHALGH, J. F. D., MCDONALD, I. and FLORENCE, E. *Brit. J. Nutr.* **14**, 289 (1960).
283. LANGLANDS, J. P., CORBETT, J. L., MCDONALD, I. and REID, G. W. *Brit. J. Nutr.* **17**, 211 and 219 (1963).
284. GREENHALGH, J. F. D. and CORBETT, J. L. *J. Agric. Sci.* **55**, 371 and 377 (1960); **61**, 221 (1963).
285. CAMPBELL, A. G. *Proc. N.Z. Soc. Anim. Prod.* **24**, 17 (1964).
*286. VAN DYNE, G. M. and TORELL, D. T. *J. Range Mgmt.* **17**, 7 (1964).
287. ARNOLD, G. W., MCMANUS, W. R., BUSH, I. G. and BALL, J. *Aust. J. Exper. Agric. Anim. Husb.* **4**, 71 (1964).
288. VAN DYNE, G. M. and LOFGREEN, G. P. *J. Anim. Sci.* **23**, 823 (1964).
289. LANGLANDS, J. P. *Nature, Lond.* **207**, 666 (1965); *Anim. Prod.* **8**, 253 (1966).
290. LAMBOURNE, L. J. and REARDON, T. F. *Aust. J. Agric. Res.* **14**, 257 (1963).
291. FELS, H. E., MOIR, R. J. and ROSSITER, R. C. *Aust. J. Agric. Res.* **10**, 237 (1959).
292. TAYLOR, J. C. and DERIAZ, R. E. *J. Brit. Grassl. Soc.* **18**, 29 (1963).
293. CAMPBELL, A. G., PHILLIPS, D. S. M. and O'RIELLY, E. D. *J. Brit. Grassl. Soc.* **17**, 89 (1962); HYDE, F. J. and LAWRENCE, J. T. *Electron. Engng.* **36**, 666 (1964); ALCOCK, M. B. *Nature, Lond.* **203**, 1309 (1964); JOHNS, G. C., NICOL, G. R. and WATKIN, B. R. *J. Brit. Grassl. Soc.* **20**, 212 (1965).
294. ARNOLD, G. W., MCMANUS, W. R. and BUSH, I. G. *Aust. J. Exper. Agric. Anim. Husb.* **4**, 392 (1964).
295. Grazing Management : Continuous and Rotational Grazing by Merino Sheep. *Bull. Council Sci. Ind. Res. Aust.*, No. 201, Melbourne, 1946.
*296. MCMEEKAN, C. P. *Proc. 6th Int. Grassl. Congr.*, p. 149. Pennsylvania, 1952.

*Review articles

297. BLASER, R. E., HAMMES, R. C., BRYANT, H. T., HARDISON, W. A., FONTENOT, J. P. and ENGEL, R. W. *Proc. 8th Int. Grassl. Congr., Reading*, p. 601. 1960.
*298. BROWN, D. Methods of Surveying and Measuring Vegetation. *Bull. Commonw. Bur. Past. Fld. Crops*, No. 42, Commonwealth Agric. Bureaux, Farnham Royal, 1954; MANNETJE, L. T. and HAYDOCK, K. P. *J. Brit. Grassl. Soc.* **18**, 268 (1963); WILSON, J. W. *Aust. J. Bot.* **11**, 95 (1963); TEARE, D. and MOTT, G. O. *Crop Science*, **5**, 311 (1965).
299. *Pasture and Range Research Techniques* (various authors). Comstock, Ithaca, New York, 1962.
300. Basic Problems and Techniques in Range Research. *Publ. Nat. Res. Council* No. 890. National Academy of Sciences, Washington, 1962.
301. WILLOUGHBY, W. M. *Aust. J. Agric. Res.* **10**, 248 (1959).
302. VAN DYNE, G. M. and HEADY, H. F. *J. Range Mgmt.* **18**, 78 (1965).
303. COWLISHAW, S. J. and ALDER, F. E. *J. Agric. Sci.* **54**, 257 (1960).
304. COOK, C. W., BLAKE, J. T. and CALL, J. W. *J. Anim. Sci.* **22**, 579 (1963).
305. WHEELER, J. L., REARDON, T. F. and LAMBOURNE, L. J. *Aust. J. Agric. Res.* **14**, 364 (1963).
306. BRUMBY, P. J. *N.Z. J. Agric. Res.* **2**, 797 (1959).
307. HANCOCK, J. *J. Agric. Sci.* **44**, 420 (1954).
308. PAYNE, W. J. A., LAING, W. I. and RAIVOKA, E. N. *Nature, Lond.* **167**, 610 (1951).
309. CORBETT, J. L. and BOYNE, A. W. *J. Agric. Sci.* **51**, 95 (1958); CORBETT, J. L. *Proc. Brit. Soc. Anim. Prod.* p. 3. 1958.
*310. MOCHRIE, R. D. *Fed. Proc.* **23**, 85 (1964).
311. KNOX, K. L. and WARD, G. M. *J. Dairy Sci.* **44**, 1550 (1961).
312. GREENHALGH, J. F. D. and RUNCIE, K. V. *J. Agric. Sci.* **59**, 95 (1962).
313. EDMOND, D. B. *N.Z. J. Agric. Res.* **7**, 1 (1964).
314. MITCHELL, K. G. *Proc. N.Z. Soc. Anim. Prod.* **20**, 82 (1960).
315. PETERSEN, R. G., LUCAS, H. L. and WOODHOUSE, W. W. *Agron. J.* **48**, 440 (1956).
316. HILDER, E. J. *Proc. Aust. Soc. Anim. Prod.* **5**, 241 (1964).
317. ARNOLD, G. W. *Proc. Aust. Soc. Anim. Prod.* **5**, 275 (1964).
318. LAMBOURNE, L. J. and REARDON, T. F. *Nature, Lond.* **196**, 961 (1962).
*319. FONTENOT, J. P. and BLASER, R. E. *J. Anim. Sci.* **24**, 1202 (1965).
*320. HUFFMAN, C. F. *J. Dairy Sci.* **42**, 1495 (1959).
321. ALLDEN, W. G. and JENNINGS, A. C. *Proc. Aust. Soc. Anim. Prod.* **4**, 145 (1962); HOLDER, J. M. ibid., p. 152.
*322. HOLMES, W. and JONES, J. G. W. *Proc. Nutr. Soc.* **23**, 88 (1964).
*323. MOTT, G. O. *Proc. 8th Int. Grassl. Congr., Reading*, p. 606. 1960.
324. RIEWE, M. E. *Agron. J.* **53**, 309 (1961).
*325. ALCOCK, M. B. In *Grazing in Terrestrial and Marine Environments*, p. 25 (D. J. Crisp, Ed.) Blackwell, Oxford, 1964.
326. LAMBOURNE, L. J. and REARDON, T. F. *Aust. J. Agric. Res.* **14**, 272 (1963).
327. ARNOLD, G. L., MCMANUS, W. R. and DUDZINSKI, M. L. *Aust. J. Exper. Agric. Anim. Husb.* **5**, 396 (1965).
328. CORBETT, J. L., LANGLANDS, J. P. and BOYNE, A. W. *Proc. 8th Int. Congr. Anim. Prod. Hamburg*, Vol. III, p. 245. 1961.
329. LANGLANDS, J. P., CORBETT, J. L., MCDONALD, I. and PULLAR, J. D. *Anim. Prod.* **5**, 1 and 11 (1963).
330. TAYLER, J. C. *Proc. Brit. Soc. Anim. Prod.* p. 3 (1954).
331. BALCH, C. C. and LINE, C. *J. Dairy Res.* **24**, 11 (1957).
*332. WALLACE, L. R. *Proc. 7th Int. Grassl. Congr. Reading*, p. 134. 1956.
333. HUTTON, J. B. *Proc. N.Z. Soc. Anim. Prod.* **22**, 12 (1962).
334. JONES, J. G. W., DRAKE-BROCKMAN, R. M. and HOLMES, W. *Anim. Prod.* **7**, 141 (1965).
335. COOP, I. E. *J. Agric. Sci.* **58**, 179 (1962); COOP, I. E. and HILL, M. K. ibid., p. 187.
336. COOP, I. E. and DREW, K. R. *Proc. N.Z. Soc. Anim. Prod.* **23**, 53 (1963).
337. HADJIPIERIS, G., JONES, J. G. W. and HOLMES, W. *Anim. Prod.* **7**, 309 (1965).

*Review articles

338. GRAHAM, N. McC. *Proc. Aust. Soc. Anim. Prod.* 5, 272 (1964); *Aust. J. Agric. Res.* **15**, 969 (1964).
339. Nat. Inst. Res. Dairying, Reading, *Ann. Repts.* 1958, p. 105; 1959, p. 106; 1960, p. 114.
340. JOYCE, J. P. and BLAXTER, K. L. *Brit. J. Nutr.* **18**, **5** (1964); *Res. Vet. Sci.* **5**, 506 (1964).
341. BLAXTER, K. L., JOYCE, J. P. and WAINMAN, F. W. *Nature, Lond.* **198**, 1115 (1963).
342. FLATT, W. P., WALDO, D. R., SYKES, J. F. and MOORE, L. A. *Proc. 1st Symp. Energy Metabolism, Copenhagen*, p. 101. Publ. Eur. Assoc. Anim. Prod., No. 8. Rome, 1958.
343. CORBETT, J. L., LENG, R. A. and YOUNG, B. A. *Proc. 4th Symp. Energy Metabolism*, Warsaw, 1967 (in press).
344. ELLIOTT, R. C. and FOKKEMA, K. *Rhodesia Agric. J.* **57**, 439 and 446 (1960); **58**,49 and 124 (1961).
345. SCAUT, A. *Publ. Inst. Nat. Etude Agron. du Congo Belge* (*I.N.E.A.C.*) No. 81 (1959) and No. 91 (1961).
346. PHILLIPS, G. D. and LAMPKIN, G. H. *Emp. J. Exper. Agric.* **32**, 60 (1964).
*347. PHILLIPSON, A. T. In *Mammalian Protein Metabolism*, Vol. I, p. 71 (H.N. Munro and J. B. Allison, Eds.). Academic Press, New York, 1964.
348. BOLKER, H. I. *Nature, Lond.* **197**, 489 (1963).
349. BLAXTER, K. L. In *The Role of the Gastro-Intestinal Tract in Protein Metabolism*, p. 143 (H. N. Munro, Ed.), Blackwell, Oxford, 1964.
350. GRIEVE, C. M. and OSBOURN, D. F. *J. Agric. Sci.* **65**, 411 (1965).
351. OSBOURN, D. F., THOMSON, D. J., and TERRY, R. A. *Proc. 10th Int. Grassl. Congr.*, Helsinki, 1966.
352. ØRSKOV, E. R. and ALLEN, D. M. *Brit. J. Nutr.* **20**, 295, 307, 509 and 519 (1966).
353. YOUNG, B. A., LENG, R. A., WHITE, R. G., MCCLYMONT, G. L. and CORBETT, J.L. *Proc. 4th Symp. Energy Metabolism*, Warsaw, 1967 (in press).
354. YOUNG, B. A. and CORBETT, J. L. *Proc. Aust. Soc. Anim., Prod.* **7**, 327 (1968)

*Review articles

CHAPTER 18

The Nutrition of the Dairy Calf

J. H. B. ROY

National Institute for Research in Dairying, Shinfield, Reading

THE INFLUENCE OF PLANE OF NUTRITION

The plane of nutrition of the calf can vary from that which will just maintain body weight to that which will allow the maximum rate of gain, and thus of fat and protein deposition. The maximum rate of weight gain attainable is limited by the voluntary productive energy intake of the animal. The optimum rate of weight gain is mainly an economic decision, determined by a combination of such factors as the end product required, be it meat or milk or a newborn calf, the relative cost of the productive energy from different sources of nutrients, and the desired rate of turnover of capital.

The higher the plane of nutrition, the earlier is the onset of puberty and thus the quicker the return on capital whether the animal is to be used for veal or beef production, or for the production of a calf and milk. The earliest age at first oestrus of Friesian calves reared on a very high plane of nutrition was 6½–8 months when their weights varied from 250 to 325 kg (Roy, Stobo and Gaston, 1965). Friesian calves that were reared on a plane of nutrition, which resulted in their first oestrus at 11 months, and were mated at their second oestrus gave similar milk yields to those reared on a lower plane of nutrition and mated 11 months later (Loosli, 1963). Similar results of successful intensive rearing of young heifers have been reported from Israel by Volcani (1963). However, Ayrshire heifers calving at 20 months when weighing only 330 kg gave only half the yield of less intensively reared animals, although no troubles were experienced in getting the intensively reared animals in calf (Anon, 1965).

Until these more recent experiments, there was evidence from a number of countries that a moderate to low plane of nutrition resulted in higher yields of milk and butterfat and increased longevity compared with a high plane of nutrition. At Cornell University (Reid *et al.*, 1964), Friesian heifers given a low plane calved at a weight of 440 kg, those given a medium plane at 539 kg, and those given a high plane at 614 kg. The average age at first oestrus was 9 months for the high, 11 months for the medium, and 20 months for the low plane groups respectively. Thus, difficulty was experienced in getting the low plane animals in calf at the prescribed age of 18 months and their calving was also more difficult. Milk and butterfat yields in the first five lactations were similar in all three groups, but the milk yield per unit of metabolic body weight was lower for the high plane group in each lactation. Crichton, Aitken and Boyne (1959, 1960 a, b) reared Friesian and Ayrshire twin calves on high (H) and low (L) planes of nutrition to 44 weeks, and then subdivided both groups onto high (H) and low (L) planes. All calves showed their first oestrus at a weight of 239–258 kg, which they reached on the HH, LH, LL and HL treatments at 372, 440, 474 and 552 days respectively. All the heifers were mated at 18 months of age; those given the HL plane of nutrition had a significantly lower milk yield in the first three lactations than those reared on the other three treatments. Similar results with heifers reared on different planes of nutrition but calving at the same age have been obtained in Sweden (Hansson, 1956) and in New Zealand (Anon., 1952, 1958). In Denmark, no relationship was discernible between level of rearing and the length of a cow's productive life, although conception rates were higher in animals reared on a moderate or low plane of nutrition compared with those reared on a high plane (Larsen and Larsen, 1956).

At present there is insufficient evidence to advocate a very high plane of nutrition for animals intended for milk production, but clearly much higher planes than those suggested in the earlier experiments can be used, for these experiments nearly all suffered from the shortcoming that all heifers were calved at the same age irrespective of the level of rearing.

THE DIGESTIVE SYSTEM OF THE CALF

The young calf at birth, like the adult ruminant, has a stomach of four parts, although only the abomasum or fourth stomach, with a capacity about twice that of the other three compartments, is functional (Grossman, 1949). The reticulo-rumen, although non-functional, has an inherent capacity of about 2 l. at birth (Warner, Flatt and Loosli, 1956). (See Chapter 4.)

In the young calf, liquid food can by-pass the reticulo-rumen, when the oesophageal groove, which begins at the cardia and extends to the reticulo-omasal orifice, is closed. Wester (1930) originally showed that the soluble proteins and salts of milk caused the oesophageal groove to close and Riek (1954) found that, of the solutions tested, 60 ml of a 10 per cent solution of $NaHCO_3$ were the most effective in causing closure. However, with some animals glucose invariably passed into the abomasum. Watson (1944) suggested that the reflex closure of this groove is probably evoked as part of the sucking pattern of behaviour, and not as part of the thirst pattern, for water does not elicit as strong an effect as milk. A great variation, however, exists between calves in the response to various stimuli and the reflex action becomes weaker as the calf becomes older. Until the calf is about eight weeks old the oesophageal groove reflex is stimulated by water, as well as by milk and milk by-products, but after this time water is less effective. Up to the age of 12 weeks, milk is effective whether given by nipple feeder or open bucket, but after this age the nipple feeder is superior. Capsules given to calves between liquid meals pass into the rumen, whereas those given at the same time as the calf is drinking pass into the abomasum (Hegland *et al.*, 1957). Comline and Titchen (1951) showed that the closure of the groove could be brought about by stimulation of the glossopharyngeal nerve, and that the efferent nerve supply was through the vagus.

The longer the period that the calf has access to a plentiful supply of milk, the less will be its urge to supplement its diet with other foods. The calf may be made dependent on the products of monogastric digestion throughout its productive life, as in veal production, by withholding dry food. Given a limited quantity of liquid diet, calves will begin eating dry food and grass at a young age. These dry foods pass into the rumen, where bacteria and other microorganisms become established and convert fibrous and starchy foods into a form of energy available to the young calf, synthesize B vitamins, and create protein from simpler nitrogenous compounds.

Calves having a restricted concentrate intake, but access to large amounts of roughage have a considerably increased capacity of the rumen. Blaxter *et al.* (1952) found the increase resulted from the stretching of the tissues but no increase in weight of the rumen walls occurred. Warner, Flatt and Loosli (1956) and Stobo, Roy and Gaston (1966a), however, found that inclusion of dry food, besides increasing the capacity of the reticulo-rumen, also increased the weight of tissue. Only a relatively small increase occurred in the thickness of the muscular wall, but there was a considerable increase in the thickness of the mucosa, due to the development of rumen papillae, which was much greater in concentrate-fed calves (Brownlee, 1956). Papillary development was found to be stimulated by the end-products of rumen fermentation, rather than by the coarse nature of the food (Warner *et al.*, 1953; Warner *et al.*, 1955; Warner, Flatt and Loosli, 1956; Flatt, Warner and Loosli, 1958). Both Sander *et al.* (1959) and Tamate *et al.* (1962) found that solutions of sodium butyrate and to a less extent sodium propionate caused extensive papillary growth, whereas sodium acetate had decidedly less effect.

The extent of the papillary development has no demonstrable effect on the calf in its ability to digest diets of concentrate and hay. A period as short as 3 weeks is sufficient for increased papillary development to occur in calves transferred from a high roughage diet to one of all-concentrates, and for retrogression of papillae if transferred from a high-concentrate diet to one of all-hay (Stobo, Roy and Gaston, 1966b). Similarly, Balch *et al.* (1960) compared high- or low-roughage diets in the rearing period from 4 months to 2 years of age, and found that the method of rearing had no effect on the ability of their heifers to digest hay or mixed rations of hay and concentrates.

The mode of digestion, whether monogastric or ruminant, has a marked effect on the weight of contents of the alimentary tract and thus on the relationship between live weight and empty body weight. With pre-ruminant calves, the total weight of contents has been shown to be 4·1 per cent, 8 hr after the last feed, at 7 days of age (Kesler, Ronning and Knodt, 1951), 6·2 per cent, 3 hr after the last feed, at 21 days of age (Stobo, Roy and Gaston, 1966a) and 3·6 per cent, 17 hr after the last feed, for veal calves at 104 days of age (Roy *et al.*, 1964). With ruminant calves at 12 weeks of age, the values are very much higher, varying from 27 per cent on a diet of *ad lib.* hay to 11 per cent on a diet of *ad lib.* concentrates. Stobo, Roy and Gaston (1966a) attempted to predict the weight of the alimentary tract contents by the following equation:

$$y = 0{\cdot}20x_1 - 3{\cdot}33x_2 + 0{\cdot}40x_3 + 2{\cdot}90$$

where y = wt. of contents (kg)
x_1 = live weight (kg)
x_2 = daily concentrate intake (kg)
x_3 = daily hay intake (kg)

In addition, the weight of the alimentary tract itself is much greater in the ruminant calf, the total weight of contents and alimentary tract being as high as 36 per cent of live weight for all-roughage diets and 19 per cent for all-concentrate diets, compared with 11 per cent for the monogastric calf.

MAXIMUM WEIGHT GAIN

The maximum potential weight gain of calves from birth to 22 per cent of mature weight can be represented by the exponential growth curve

$$W = Be^{0{\cdot}0150t}$$

where B = birth weight (kg) and t = age in days from birth.

No difference in maximum potential growth rate, other than that associated with differences in birth weight, was found between the Jersey and Ayrshire breeds up to the same stage of development, but the Friesian breed had a greater maximum potential growth rate in relation to its birth weight. Birth weight of calves is about 6·6 per cent of mature weight and maximum daily growth rate to 22 per cent of mature weight approaches 3·0 per cent of birth weight. It should thus be possible to estimate the maximum potential weight gain of a particular breed or crossbreed from a knowledge of the mean birth weight (Roy, Stobo and Gaston, 1966).

In practice, this maximum potential is rarely realized except in veal production, where very high energy diets are used.

VOLUNTARY INTAKE

DRY MATTER

The maximum voluntary dry matter intake differs between pre-ruminant and ruminant calves. Figure 1 shows the maximum voluntary intake of Friesian bull calves given liquid diets containing 2·8 per cent fat in two feeds per day, and that of similar calves given *ad lib.* concentrates containing 16–20 per cent crude protein together with hay *ad lib.* Up to 70 kg live weight, a greater amount of dry matter is ingested by the pre-ruminant calf, but at an empty body weight of 100 kg, equivalent to 112 kg live weight for the ruminant and 104 kg live weight for the pre-ruminant calf, the voluntary intake is 1·95 kg and 3·33 kg dry matter for the pre-ruminant and ruminant calf respectively, but the net energy intake is virtually the same for both forms of digestion (Roy and Stobo, 1965).

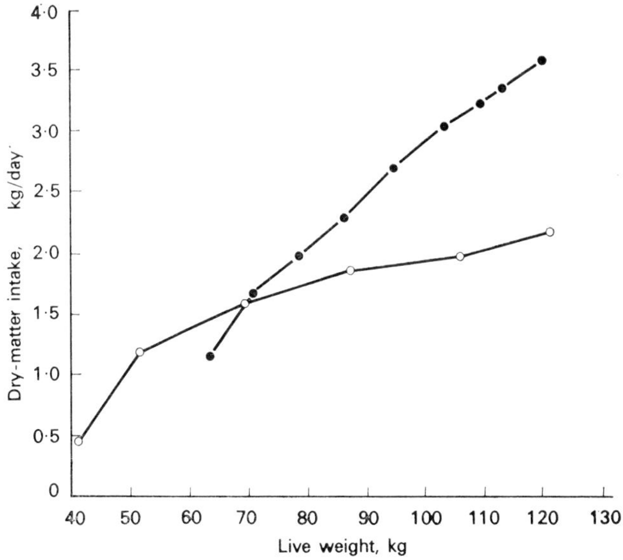

FIG. 1.—Maximum voluntary intake of dry matter in relation to live weight.
o——o Calves given liquid diet containing 2·8 per cent fat, *ad lib.* in two feeds daily.
•——• Calves given dry diet consisting of concentrates containing 16-20 per cent crude protein offered *ad lib* together with hay *ad lib.*

The maximum dry matter intake given in Figure 1 increases from 2·2 kg/100 kg live weight at 50 kg live weight to 3·0 kg/100 kg at 120 kg live weight, the latter value being the generally accepted maximum dry matter intake for adult ruminants. It represents a value of 110 g/kg$^{0·73}$, which may be compared with a maximum intake of 100 g/kg$^{0·73}$ found in sheep when the apparent digestibility of energy was 82 per cent (Blaxter, 1962).

With the pre-ruminant calf, no difference in dry matter intake/kg live weight was found between Friesian, Ayrshire and Jersey calves in the weight range 60–110 kg (Roy, Stobo and Gaston, 1966).

With liquid diets, the voluntary intake appears to be higher when the energy content of the dry matter is low, whereas dry matter intake by calves on solid food tends to rise with increase in the digestible energy content of the diet. However, diets of high energy content are usually in the form of concentrates and tend to pass rapidly from the rumen and through the alimentary tract (Blaxter, Graham and Wainman, 1956; Paloheimo and Mäkelä, 1959; Freer and Campling, 1963).

WATER

Water intake of calves is affected not only by the dry matter intake and environmental temperature but also by the proportion of salt (Riggs, Colby and Sells, 1953) and protein (Ritzman and Benedict, 1924; Sykes, 1955; Stobo, 1964) in the diet. At a given live weight, low levels of dry matter intake are generally associated with high levels of water intake/unit of dry matter (Leitch and Thomson, 1944-5).

Atkeson, Warren and Anderson (1934) found with calves, whose milk intake was restricted, that water intake/unit of dry matter was high during the first 6 weeks of life, averaging about 6·5 kg water/kg dry matter consumed. With calves given a milk substitute diet *ad lib.*, the total amount of water ingested is similar to that of the ruminant calf, since at 100 kg live weight, Friesian calves were ingesting 11·6 kg water or 6 kg water/kg dry matter at an environmental temperature of 13°C, whereas similar calves of 100 kg live weight given concentrates *ad lib.* were ingesting 10 kg water or 3·3 kg/kg dry matter (Roy and Stobo, 1965). Pettyjohn, Everitt and Mochrie (1963) have found that the maximum utilization of a milk substitute diet occurs at a reconstitution rate of 15 per cent or 6·6 kg water/kg dry matter. At higher concentrations, nutrient utilization decreases.

Up to an environmental temperature of 10°C, water intake in ruminant calves remains constant at about 3·1–3.5 kg/kg dry matter but at a temperature of 21·1°C the value is about 4·4 kg/kg dry matter ingested and at 29·4°C about 5·9 kg/kg dry matter (Winchester and Morris, 1956; MacDonald and Bell, 1958).

THE NUTRIENT REQUIREMENTS OF CALVES

The nutrient requirements of calves have recently been reviewed in Great Britain, and the recommendations have been published in Agricultural Research Council (1965). (See Chapter 24.)

ENERGY

The fasting metabolism of young calves is considerably greater per unit of body weight or per unit of surface area than that of adult ruminants. With Ayrshire calves of 1 month of age, Blaxter obtained a value of 122 kcal/kg$^{0 \cdot 73}$ using a spirometer, and 120 kcal/kg$^{0 \cdot 73}$ from 24 hr determinations in a respiration chamber (Blaxter and Howells, 1951; Blaxter and Wood, 1951a, b; Blaxter, 1962).

Fasting metabolism has been shown to decline with age. A fall of 1 per cent per day was observed by Gonzalez-Jimenez and Blaxter (1962), and Roy, Huffman and Reineke (1957) observed an increase during the first 3 days of life followed by an exponential decline over the next 18 days, which could be represented by the regression equation

$$y = 105 \cdot 9 - 48 \cdot 03\,(x - 0 \cdot 9685)$$

where y = heat production (kcal/kg$^{0 \cdot 73}$/24 hr)
x = age (log days)

Gonzalez-Jimenez and Blaxter (1962) suggested that the rise in metabolism found by Roy, Huffman and Reineke (1957) was a response to the cold environment in which the observations were made. However, Settlemire, Hibbs and Conrad (1964) in comparing the basal metabolism of normal and anaemic calves found very similar trends to those of Roy, Huffman and Reineke (1957). Values for fasting metabolism of 186 kcal/kg$^{0 \cdot 73}$ for the 8-day old Friesian heifer falling to 133 kcal/kg$^{0 \cdot 73}$ at 2 months of age were reported by Ritzman and Colovos (1943).

Few critical studies of differences between breeds have been made. Prediction equations for the basal metabolism of different breeds were given by Brody (1945) as follows:

Jersey females $Q/m = 29 \cdot 5e^{-0 \cdot 096t} + 13 \cdot 1$
Jersey males $Q/m = 28 \cdot 1e^{-0 \cdot 105t} + 13 \cdot 3$
Holstein females $Q/m = 24 \cdot 7e^{-0 \cdot 084t} + 10 \cdot 8$
Hereford females $Q/m = 14e^{-0 \cdot 091t} + 14 \cdot 0$

where Q = basal metabolism in kcal/day
m = weight in kg
^{t} = age in months

Tkacev and Taranenko (1963) compared the basal metabolism of beef and dairy calves and showed that the results for the former were 10 per cent lower at 2 months of age and 19 per cent lower at 12 months. Brody (1945) found that, at a given age after 6 months, Holsteins had a higher resting metabolism/unit of surface area than Jerseys, which he attributed to the more rapid growth of the former breed. At a given weight up to 100 kg live weight, metabolism/unit of surface area appeared to be greater for the Jersey than for the Holstein.

Both Blaxter and Wood (1951b) and Roy, Huffman and Reineke (1957) observed a close relationship between pulse rate and heat production. Blaxter (1962) has suggested that the extra energy expenditure from body movement of calves grouped in a pen and given a high plane of nutrition is not likely to exceed 10 per cent of the fasting metabolism. An increase in the play activity of calves with increasing plane of nutrition was noted by Brownlee (1954).

The critical temperature of Ayrshire calves given 4 l. milk daily was studied by Gonzalez-Jimenez and Blaxter (1962) who found a value of 12·8°C on the third day of life declining to 8·2°C on the 20th day. For those given 6 l. milk, the critical temperature was not below 7°–8°C in spite of the higher heat production.

For maintenance of live weight of the pre-ruminant calf, Tomme and Taranenko (1939) obtained a value equivalent to 44·7 kcal net energy/kg empty body weight/24 hr whilst Blaxter and Wood (1951c) found a value of 52·4 kcal digested energy/kg live weight/24 hr, equivalent to 43 kcal net energy/kg live weight/24 hr or 45 kcal/kg empty body weight/24 hr. Similarly, Roy *et al.* (1958) estimated the maintenance requirement as 42·9 kcal net energy/kg live weight/24 hr or 44·7 kcal net energy/kg empty body weight/24 hr. For the first 10 days of life it is clear that the values will be considerably higher (Roy, Huffman and Reineke, 1957).

For ruminant calves, Preston (1963) obtained a value for maintenance of 22·6 kcal net energy for fattening/kg live weight/24 hr with calves weaned at 3 weeks of age. Preston obtained this value from the estimated starch equivalent intake; it is equivalent to 31·1 kcal net energy/kg live weight/24 hr. Stobo and Roy (1963) on the other hand found a value for maintenance of 55·5 kcal digested energy/kg/24 hr which on conversion by the factors of Blaxter (1962) gave a value of 33·9 kcal net energy/kg live weight/24 hr, equivalent to 42·5 kcal net energy/kg empty body weight/24 hr, a value similar to that for the pre-ruminant calf.

The digestibility of the energy of whole milk and milk substitutes, containing fats of similar digestibility to that of butterfat, was found to be 95 per cent (Tomme and Taranenko, 1939) or 97 per cent (Roy *et al.*, 1964). Even at *ad lib.* intakes of liquid diets, there is only a slight fall in digestibility (Roy, Stobo and Gaston, 1965). The highest digestibility of energy of concentrate diets given *ad lib.* was found by Stobo (1964) to be 76·5 per cent, whereas on all hay diets the value was 48 per cent.

With pre-ruminant calves, Blaxter (1952) and Gonzalez-Jimenez and Blaxter (1962) found that between 94 and 97 per cent of the gross energy of milk was metabolized, the value being greater at the higher levels of intake. Roy *et al.* (1964) estimated 97·7 per cent of the digested energy was metabolized at very high levels of intake. Blaxter and Wood (1951c), Brisson, Cunningham and Haskell (1957) and Roy *et al.* (1964) showed respectively that 307, 268 and 302 kcal digested energy were required per 100 g gain in weight. The latter value was applicable to calves up to 115 kg live weight.

With the ruminant calf, Stobo (1964) estimated that 85 per cent of the digestible energy was metabolized at the maintenance level and up to 88 per cent at the highest level of production. Energy losses were assumed to be between 14·6 and 18·9 kcal/g N in the urine, according to the level of intake and proportion of roughage in the diet, and between 6·7 and 7·4 kcal/100 kcal gross energy as methane, according to the digestibility of the diet (Blaxter, 1962).

The net availability of the metabolizable energy of whole milk has been shown to be between 79·5 per cent (Gonzalez-Jimenez and Blaxter, 1962) and 84·5 per cent (Blaxter, 1952), whilst Tomme and Taranenko (1939) found that the net energy of whole milk was 68·5 per cent of the gross energy. The energy retention by the pre-ruminant calf is thus about 250 kcal/100 g empty body weight gain.

For the ruminant calf, Stobo (1964) using a net availability for maintenance of 74·5 per cent and for production of 63·9 per cent calculated that the energy retention was 211 kcal/100 g gain in weight. After making an allowance for gut fill, the value was 240 kcal/100 g empty body weight gain, a value similar to that for the pre-ruminant calf.

PROTEIN

The protein requirements of the calf have been estimated from feeding trials, from N balances to ascertain the minimum intake that gives maximum retention, or alternatively by the factorial method originally formulated by Mitchell (1929). Blaxter and Mitchell (1948) expressed the requirement in terms of available protein (AP) by the following equation:

$$R_{AP} = \frac{1}{(BV)} 6{\cdot}25\,(E + G)$$

where BV = biological value, E = endogenous urinary N (g/day) and
G — N retention (g/day) for a particular weight gain.

This equation was derived from the requirement of apparently digested protein (ADP) which was as follows:

$$R_{ADP} = 6{\cdot}25 \frac{1}{(BV)} (E + G + M.D) - M.D$$

where M = metabolic faecal N excretion (g/100 g dry matter ingested)
and D = dry matter intake, in units of 100 g, required to supply the energy needs of the animal.

The biological value of the protein, which takes into account the metabolic faecal N was calculated as follows:

$$BV = \frac{100 \times I - (F\text{-}M) - (U\text{-}E)}{I - (F\text{-}M)} = \frac{\text{'retained' N}}{\text{'truly digested' N}}$$

where I = N intake (g/day)
F = total faecal N (g/day)
M = metabolic faecal N (g/day)
U = total urinary N excretion (g/day)
E = endogenous urinary N (g/day)

Using this method of obtaining requirements, divergent results were obtained by Blaxter and Mitchell (1948), and by Lofgreen, Loosli and Maynard (1951). However, Lofgreen, Loosli and Maynard (1951) confirmed their findings in feeding trials. The values of Blaxter and Mitchell (1948) were higher in the range 68–159 kg live weight, and the divergencies arose partly because different biological values and different maintenance requirements were used, and partly because both groups attempted to produce requirements for 'normal' growth assuming different rates of weight gain. Blaxter and Mitchell (1948) assumed a biological value of 60 throughout the growing period whereas Lofgreen, Loosli and Maynard (1951) assumed biological values of 80 for the young calf on a liquid diet, of 75 for the 3-month-old calf, and of 65 for the 6-month-old calf.

Values for endogenous urinary N, which represent the inevitable loss of N by calves given a N-free diet, have been obtained either by direct measurement or more often by extrapolation to zero intake of apparently digested N from an equation relating apparently digested N to urinary N in calves given diets in which N is the limiting factor for growth, i.e. under conditions when biological value is at its maximum (Blaxter and Wood, 1951c). Values obtained for endogenous urinary N are given in Table 1.

TABLE 1
Endogenous Urinary Nitrogen

Diet	Weight (kg)	Endogenous urinary N (mg/kg body weight)	Reference
Liquid	32	82	Blaxter and Wood (1951a)
	34–49	65	Cunningham and Brisson (1957)
	34	64	Shillam and Roy (1963a)
	53	63	Roy *et al.* (1964)
Dry	50–176	58	Stobo (1964)
	97	41	Harris and Loosli (1944)
	145	45	Steenbock, Nelson and Hart (1915)
	168–177	33	Hart, Humphrey and Morrison (1912)
	283	25	Swanson and Herman (1943)

As for endogenous urinary N, metabolic faecal N has been measured directly on N–free diets, or with a diet containing casein labelled with radioactive phosphorus (Lofgreen and Kleiber, 1953) or alternatively by extrapolation from regression equations. However, the first method inevitably involves diets of low digestibility, which in turn tend to exaggerate metabolic faecal N output. Values obtained for metabolic faecal N are given in Table 2.

TABLE 2
Metabolic Faecal Nitrogen

Diet	Weight (kg)	Metabolic faecal N (g/100 g dry matter intake)	Reference
Liquid	32	0·43	Blaxter and Wood (1951a)
	34–49	0·33	Cunningham and Brisson (1957)
		0·27	Lofgreen and Kleiber (1953)
	53–90	0·19	Roy *et al.* (1964)
Dry (5% crude fibre)	50–176	0·33	Stobo (1964)
(30% crude fibre)		0·42	Stobo (1964)
	97	0·37	Harris and Loosli (1944)
	140–211	0·49	Harris, Work and Henke (1943)

The estimation of maximum biological value of a protein is only possible if protein and not energy intake is limiting production. The maximum biological value of whole milk approaches 80, when 1·1 kg of milk solids, which supplies sufficient energy for 1 kg weight gain, is ingested by a calf of 56 kg body weight (Roy *et al.*, 1964). Although Blaxter and Wood (1952b) obtained a biological value of 92 for the protein of skim milk, a value much greater than that found for man and rat, it would appear that this was the result of using a metabolic faecal N value of 0·43 g/100 g dry matter intake; this value appears from subsequent work to be too high for the monogastric calf. However, Brisson, Cunningham and Haskell (1957) also obtained a biological value of 92 for whole milk, even when assuming 100 per cent true digestibility of the milk protein. In assessing protein requirements for the ruminant calf, Blaxter and Mitchell (1948) originally used a value of 60·5 for biological value, based on the findings of Johnson *et al.* (1942) that biological value of protein for ruminants only varied slightly from this value. More recently, Blaxter (1959) reassessed the biological value of protein for lactating cows as 75 and considered that irrespective of protein source the biological value would be within 10 per cent of this mean figure. Preston (1963) calculated the biological values of groundnut meal and white fishmeal for calves weaned at 3 weeks of age as 70 and 83 respectively. More recently, with protein-limiting diets Stobo (1964) obtained a mean biological value of 82 for diets varying from all hay to *ad lib.* concentrates containing 12 per cent crude protein, over a range of weight gains from maintenance to 1·5 kg/day.

Values for the amount of N retained/100 g of live weight gain are given in Table 3. Preston (1963) however reported higher values (up to 4·2 g/100 g gain) for calves gaining weight slowly and attributed this finding to lower fat deposition. On the other hand, Stobo (1964) obtained higher values (up to 4·4 g/100 g gain) for calves given diets containing increasing amounts of protein, even though the calves had similar weight gains.

TABLE 3

Nitrogen Content of Live Weight Gain

Diet	Weight (kg)	N-content of live weight gain (g/100 g gain)	Reference
Liquid	32	2·62	Blaxter and Wood (1951c)
	45–69	3·22	Brisson, Cunningham and Haskell (1957)
	53	3·41	
	73	2·87	Roy *et al.* (1964)
	90	3·31	
Dry	49–53	2·88	Whitelaw, Preston and Ndumbe (1961)
	50–176	2·83	Stobo (1964)

Table 4 gives some of the results obtained for digestible crude protein requirements of calves, and shows the extent of the variability when insufficient consideration is given to weight gain.

TABLE 4

Some Estimates of Digestible Crude Protein Requirements (g/day)

Live weight gain (kg/day)	Weight (kg)				Reference
	45·4	90·8	136·2	181·6	
Based on factorial analysis					
	182	272	327	363	Armsby (1922)
0·40		232	218	209	Mitchell (1929)
0·91		440		413	Mitchell (1947)
		400–418	409–427	413–436	Blaxter and Mitchell (1948)
0·64–0·80		250		363	Lofgreen, Loosli and Maynard (1951)
Based on feeding trials					
0·60		418	486	499	Steensberg (1947)
0·41		213	327		Lofgreen, Loosli and Maynard (1951)
0·57–0·69				268–291	Bailey and Broster (1957)

More recently the minimum requirements for available and digestible protein for the ruminant calf have been determined by Stobo (1964) using protein-limiting diets. The results were obtained from balance trials with 36 calves and from feeding trials with 83 similar calves. The values are given

TABLE 5

Minimum Digestible and Available Protein Requirements (g/day) of the Ruminant Calf

Live weight (kg)	Maintenance		Maintenance + live weight gain (kg/day)								
			0·5			1·0			1·5	
	Digestible	Available	Digestible		Available	Digestible		Available	Digestible	Available
	Stobo and Roy (1963)		Roy (1959a)	Stobo and Roy (1963)		Roy (1959a)	Stobo and Roy (1963)		Stobo and Roy (1963)	
50	50	36	175	135	121	275	220	205	305	290
100	100	88	250	185	173	350	270	257	355	342
150	150	140	318	235	225	418	320	309	405	394

in Table 5 and are somewhat lower than the suggested values of Roy (1959a) which were based on the results of Hunter-Smith and Gardner (1928), Blaxter and Mitchell (1948), Blaxter and Wood (1951c), Lofgreen, Loosli and Maynard (1951), Bailey and Broster (1957) and Brisson, Cunningham and Haskell (1957). In confirmation of these results Preston *et al.* (1965) found that 267 g digestible protein were required for a weight gain of 0·9 kg/day by Friesian calves of 82 kg live weight.

There appears to be little difference in the protein requirement of the ruminant and pre-ruminant calf. The requirements of the pre-ruminant calf for available protein are given in Table 6.

TABLE 6
Available Protein Requirements (g/day) of the pre-ruminant Calf (BV = 80)

Weight (kg) Live weight gain/day (kg)	55·6 0·94	74·0 0·91	96·0 0·99	Reference
	233	239	277	From (Blaxter and Wood, 1951c)
	265	266	298	Brisson, Cunningham and Haskell (1957)
	278	238	296	Roy *et al.* (1964)

MINERALS

Calcium and Phosphorus

It is usually recommended that for maximum utilization of these minerals the Ca:P ratio of a ration should be similar to that of bone, i.e. 2·2:1, or, as more P than Ca is found in the soft tissues, somewhat less. In fact, calves have grown equally well on rations with Ca:P ratios from 6:1 to 1·2:1 when adequate vitamin D was supplied.

The maintenance requirement for these two minerals is often assessed from the endogenous faecal excretion, which represents the inevitable loss by excretion into the gut when diets devoid of Ca and P are given.

(a) *Calcium.* The daily endogenous faecal calcium loss was found to be 13·3 mg/kg body weight (Blaxter and Wood, 1952a), 9·1–28·0 mg/kg (Comar *et al.*, 1953), 11·8–18·0 mg/kg (Hansard, Comar and Plumlee, 1954; Hansard, Crowder and Lyke, 1957) and 14·5–20·4 mg/kg (Visek *et al.*, 1953). No trends with age or weight were apparent. Daily urinary losses appear to be not more than 1 mg/kg body weight.

The true absorption of dietary Ca decreases with age but is markedly affected by the source of Ca, absorption from milk being much higher than from dry food. Hansard, Crowder and Lyke (1957) obtained absorption

TABLE 7

*True Absorption Coefficients of Calcium and Phosphorus and Availability of Phosphorus**

Weight (kg)	Diet	True absorption (%) Ca	True absorption (%) P	Availability (%) P	Based on reference
33	Milk		94		Lofgreen, Kleiber and Luick (1952)
35	Milk	98			Hansard, Comar and Plumlee (1954)
40	Milk	96	99	82	Blackward, Morris and Wright (1936)
40 (2–5 weeks)	Milk	108			Smith (1957)
45	Milk	96	97	90	Blaxter and Wood (1952a)
45 (7–9 weeks)	Milk	92			Smith (1957)
61	Milk†	73	82	50	Duncan and Huffman (1934)
66	Milk	98			Hansard, Comar and Plumlee (1954)
163	Dry	41			Hansard, Comar and Plumlee (1954); Hansard, Crowder and Lyke (1957)
170	Milk	83	105	85	Hughes and Cave (1931)
183	Dry		73		Tillman, Brethour and Hansard (1959)
227	Dry		76		Tillman and Brethour (1958a)
248	Dry	51			Hansard, Crowder and Lyke (1957)
255	Dry	50			Visek *et al.* (1953)

*Endogenous faecal Ca = 15 mg/kg
urinary Ca = 1 mg/kg
Endogenous faecal P = 4·3 mg/kg
urinary P = 15 mg/kg (based on Blaxter and Wood, 1952a)

†Excessive intakes of Ca and P.

values for different sources of Ca, ranging from 68 to 41 per cent for animals of 136–227 kg and from 55 to 30 per cent for animals of 363–454 kg. Values obtained for absorption coefficients in relation to age and diet are given in Table 7. The absorption coefficients have been obtained from per cent retention values by using a value for endogenous faecal Ca of 15 mg/kg. This value would seem to be too high for the very young calf. At a maintenance level of intake the percentage of Ca retained from liquid diets was 73 per cent and increased to 92 per cent when the amount of milk was sufficient to give a weight gain of 1 kg/day (Blaxter and Wood, 1952a).

A marked difference between breeds in apparent absorption of Ca from large volumes of liquid milk substitute was observed by Roy, Stobo and Gaston (1965) who found that the percentage retention at 65 and 95 kg body weight and even at the same percentage of mature weight decreased with decreasing mature weight of the breed, the Friesian breed having a higher absorption than Jerseys or Ayrshires. The apparent absorption by the Friesian calves at 100 kg live weight was 87 per cent.

Mitchell and McClure (1937) estimated the requirements for weight gain from the N content of the tissues by assuming a uniform retention of 0·27 g P to each g N and a Ca retention equal to 1·71 × P retention. Gross requirements were obtained by assuming 70 per cent retention. Roy (1959a) analysed the results for all cattle slaughtered at different ages by Ellenberger, Newlander and Jones (1950) and estimated the Ca content/kg gain in weight to be 6·4 g at 45 kg body weight and 11·7 g at 227 kg body weight. Mitchell (1962) analysed only the results of Ellenberger, Newlander and Jones (1950) for pure-bred Holsteins and obtained a value of 14 g/kg empty-body-weight gain at 68 kg body weight and 17·7 g/kg empty-body-weight gain at 227 kg body weight. Data for the Ca content of growing cattle are also available from the work of Trowbridge, Moulton and Haigh (1919), Haigh, Moulton and Trowbridge (1920) and Moulton, Trowbridge and Haigh (1922). In the range of weight up to 227 kg, the Ca deposition, except during the very early period after birth, would appear to be about 12 g/kg weight gain. At 100 kg, a Friesian calf given a liquid diet and gaining weight at 1 kg/day would thus have a gross requirement of 14 g Ca daily, assuming a net requirement for maintenance of 1·5 g, for weight gain of 12 g and a true absorption of 95 per cent.

Nearly all feeding trials in which various levels of Ca intake have been given have been made with cattle over 160 kg live weight (Converse, 1954). However, Lindsey, Archibald and Nelson (1931) and Archibald and Bennett (1935) found that a daily intake of 17 g and of 21 g Ca was adequate for calves of 175 kg and 210–215 kg body weight respectively, whereas Otto (1938) found that the very low value of 5 g Ca/day was certainly inadequate at a live weight of 261 kg.

(b) *Phosphorus*. Endogenous faecal excretion of phosphorus has generally been obtained by isotope dilution methods. Lofgreen, Kleiber and Luick (1952) found values as low as 4·3 mg/kg body weight at 27–36 kg body weight, whereas Tillman and Brethour (1958 a, b) and Tillman, Brethour and Hansard (1959) found values varying from 12·6 to 15·6 mg/kg in the weight range 182–227 kg. From the balance trial results of Blaxter and Wood (1952a), it would appear that the value is about 5·6 mg/kg body weight at a weight of 45 kg.

The true absorption of P, as for Ca, is lower from dry foods than from milk, but it would appear that endogenous urinary losses of P are much greater than those of Ca. The results of Blaxter and Wood (1952a) indicate a value of about 15 mg P/kg body weight for endogenous urinary P and this has been used in the calculation of availability of P. Values for true absorption and availability of P are given in Table 7.

Roy (1959a) estimated the requirement of P for weight gain from the slaughter data of Ellenberger, Newlander and Jones (1950) and found a retention value of 3·7 g P/kg weight gain at 45 kg body weight, increasing to 6·2 g P/kg at 227 kg body weight. However the results of Wilson (1927) indicated a retention of about 6·6 g P/kg at both these body weights. A Friesian calf of 100 kg live weight gaining at 1 kg/day would thus have a gross requirement of 10·5 g P, based on a net requirement of 1·9 g P for maintenance, and of 7 g P/kg weight gain and an availability of 85 per cent.

In general the results of feeding trials have shown intakes of below 8–9 g P/day to be inadequate for calves up to 250 kg body weight, when assessed on the basis of normal growth and feed utilization and maintenance of normal inorganic blood P values (Huffman *et al.*, 1933; Archibald and Bennett, 1935; Beeson *et al.*, 1941; Long *et al.*, 1957; Tillman, Brethour and Hansard, 1959).

A deficiency of P causes anorexia and even a depraved appetite, and the animal thus becomes emaciated. Furthermore a deficiency in P alone may cause rickets, especially if the calf obtains excessive quantities of Ca at the same time.

Magnesium

It has been known for a number of years that whole milk supplemented with Fe, Cu, Mn and vitamin D is inadequate as the sole diet of the calf. A shortage of Mg in the diet results in hypomagnesaemia or tetany, which is associated with a blood Mg concentration of less than 0·8 mg/100 ml serum. Whole milk contains about 10–13 mg Mg/100 g, and the condition occurs, usually during the first 3 months of life with calves growing at a rapid rate on very large amounts of whole milk or milk substitute. The calf shows much excitement and is hypersensitive when touched, the head is retracted, the whites of the eyes are very apparent, owing to an upward rotation of the eyeballs, and there is often continual movement of the ears. For fatal convulsions a depletion of body Mg of 30 per cent or more was found to be necessary (Blaxter and Rook, 1954).

From a study of blood Mg values for calves gaining weight at 0·6 kg/day, Huffman *et al.* (1941) suggested that 35 mg Mg/kg body weight would maintain normal values when Mg0 was used as a supplement, whereas when hay or grain was given, less than 15 mg Mg/kg were needed. Ray (1942) concluded from a study of Mg retention in 182 kg heifers given roughages of differing Mg content, that the requirement was about 31 mg Mg/kg body weight, whereas Thomas and Okamoto (1954) found that 51 mg/kg body weight would prevent hypomagnesaemia. With 'synthetic milk' diets for calves gaining weight at 0·32 kg/day, Blaxter, Rook and MacDonald (1954) considered that about 18 mg/kg body weight were required to maintain normal blood levels. With single suckled calves an intake of 1·6 g Mg/day maintained normal serum Mg but 1·1 g/day was inadequate (Blaxter and Sharman, 1955).

The daily maintenance requirement for Mg as assessed by the value for endogenous faecal Mg of liquid-fed calves was shown by Blaxter and Rook (1954) to be about 3–4 mg/kg body weight at 3–8 weeks of age. In contrast, a value of 0·5 mg/kg body weight at 2–5 weeks and of 2·2 mg/kg at 26–32 weeks for animals weighing only 128 kg was found by Smith (1959). With diets low in Mg, the urine loss is negligible.

The availability of Mg is subject to great variation according to the age and diet of the calf. A value of 59–71 per cent for young calves on 'synthetic' liquid diets was obtained by Blaxter and Rook (1954), who showed that retention fell from about 65 per cent at the maintenance level of intake to less than 30 per cent as retention reached the maximum obtainable. Smith (1957) using Blaxter and Rook's value for endogenous faecal Mg considered that 75–90 per cent of dietary Mg was absorbed during the first 5 weeks of life, after which time the ability to utilize dietary Mg dropped sharply. At 21–34 weeks, the mean availability of Mg for milk-fed calves was only 40 per cent (Smith, 1959). With 6–9 month-old steers, given hay or maize silage, Smith (1959) estimated the mean availability of dietary Mg, using the endogenous faecal Mg value obtained with liquid fed calves, as 21–27 per cent.

The net requirement of Mg for weight gain above maintenance is of the order of 350–400 mg/kg (Hogan and Nierman, 1927) or even as much as 700 mg/kg in the very young calf (Blaxter and Rook, 1954). These latter workers, who calculated the values from the contents of Ca and N by the use of the ratio of Ca:Mg in bone of 45:1 and a ratio for N:Mg in soft tissue of 140:1, suggested a requirement of 0·96 g Mg/day for Ayrshire calves gaining weight at 0·5 kg/day during the first 8 weeks of life.

Inclusion of 875 mg Mg as Mg0/kg dry matter in a milk substitute diet for veal calves, obtaining about 1·3 g Mg daily from the basic diet, caused a significant increase in serum Mg at 12 weeks of age, but had no effect on weight gain. However, this quantity of Mg significantly reduced the dry matter content of the faeces (Roy, 1964). In the scouring calf, the faecal output of Mg is increased to 24 m–equiv. daily from 11 m–equiv. for the normal calf (Blaxter and Wood, 1953).

Sodium and Chloride

The data of Hogan and Nierman (1927) showed that the Na content of calf carcasses was about 1·4 g/kg, whilst the content of Cl was 0·9 g/kg. Theiler, Green and du Toit (1927) suggested, from observations on only two animals, that 1·5 g Na and 5 g Cl daily was adequate for normal growth of heifers, although the animals did not calve normally. According to Sheehy and Senior (1936) calves of 68–92 kg live weight consumed between 11 and 16 g NaCl daily when given a hay of poor quality and retained a considerably higher percentage of Ca and P (47–55 and 42–49 per cent respectively) than those given no additional NaCl (30–36 and 29–38 per cent respectively).

In practical rations the inclusion of 0·5–1·0 per cent NaCl in the concentrate mixture is usually recommended, or alternatively a salt lick can be provided.

In the faeces of normal calves about 5 m–equiv. Na are excreted daily in the faeces, but this value increases to about 42 m–equiv. in the scouring calf (Blaxter and Wood, 1953), and is reflected in a decline in serum Na values from 135 to 126 m–equiv./l. (Roy *et al.*, 1959).

Potassium

The requirement for K is not known, but calves kept for long periods on semi-purified diets containing about 0·11 per cent K and having a daily intake of 1·5 g K showed marked changes in their electrocardiographs, whereas diets containing 0·35 per cent K corresponding to an intake of about 4·4 g K/day gave normal results (Sykes and Alfredson, 1940). Scouring calves produce about 40 m–equiv. K daily in their faeces compared to about 2 m–equiv. K for normal calves (Blaxter and Wood, 1953). A high serum K level is associated with severe scouring in calves, and may be the cause of deaths from *Escherichia coli* localized intestinal infection (Roy *et al.*, 1959; Fisher, 1965). Bergman and Sellers (1953, 1954) injected KCl intravenously into calves, and found that gradual cardiac arrest occurred after plasma K levels of 8 m–equiv./l. had been obtained.

Iron

Calves restricted to a diet of whole milk or a milk substitute, based on fat and skim milk, unsupplemented with Fe develop a microcytic normochromic anaemia, which if allowed to continue, may develop into the hypochromic form. Fast rates of growth probably reduce the age at which a condition of severe anaemia develops. With Holstein calves gaining weight at 0·68 kg/day and having initial haemoglobin values of 11·5 g/100 ml, haemoglobin values did not fall to 5 g/100 ml until the calves were at least 289 days of age (Herman, 1936). Similarly, Knoop, Krauss and Washburn (1935) found no benefit on weight gain of a supplement of 400 mg Fe and 40 mg Cu daily during the first 112 days of life, although the haemoglobin levels of the unsupplemented group varied from 3·4–7·8 g/100 ml at this age. Blaxter, Sharman and MacDonald (1957) produced anaemia in four calves kept under laboratory conditions, and showed the beneficial effect of Fe supplementation from 6 weeks of age. In a field experiment with suckling beef calves a daily supplement of 35 mg Fe increased haemoglobin levels and mean corpuscular volume, but no measurement of the effect on performance was made. However, Niedermeier *et al.* (1959) found no significant effect of a supplement of 240 mg Fe and 7 mg Cu daily on the weight gain of calves given whole milk *ad lib.* to 39 days of age, although the haemoglobin values of the supplemented and unsupplemented calves were 15·3 and 8 g/100 ml respectively. On the other hand, Roy *et al.* (1964) using Ayrshire calves given large volumes of whole milk, obtained a significant increase in weight gain during the first 12 weeks of life by giving intramuscular injections of 1·5 g Fe as Fe–dextran. The digestibility of all nutrients was lower in anaemic than in normal calves. The calves given Fe injections had haemoglobin levels of 7 g/100 ml at 12 weeks of age, compared with 4·5 g/100 ml in those not given the Fe injections.

Even with calves having access to dry food, haemoglobin content falls during the first 3–8 weeks of life and then recovers (Wise *et al.*, 1947; Thomas *et al.*, 1954; Owen *et al.*, 1955; Wing, Jacobson and Allen, 1955).

Calves may be deficient of Fe at birth; 30 per cent of those born at Ohio experimental station had haemoglobin values of less than 9 g/100 ml (Hibbs, Conrad and Gale, 1961). Sex and breed also influences the haemoglobin content, females having a higher content than males (Thomas *et al.*, 1954). In addition, environmental temperature may affect haemoglobin levels, these being greater under cold conditions (Kazakova, 1959).

Blaxter, Sharman and MacDonald (1957) estimated that to maintain normal haemoglobin levels and to achieve adequate liver storage, the net requirement of Fe was 50–60 mg/day for a weight gain of 1·0 kg/day. Matrone *et al.* (1957) calculated that for a calf of 227 kg live weight, the net requirement was 1·2 mg Fe/day to maintain a blood value of 10 g haemoglobin/100 ml and a further 16 mg/day for a weight gain of 0·9 kg/day. Based on a 30 per cent utilization of Fe (Matrone *et al.*, 1957) the gross requirement, which contained no allowance for normal storage of Fe, would be about 56 mg/day for a calf gaining weight at 0·9 kg/day compared with a value of 100 mg/day suggested by Blaxter, Sharman and MacDonald (1957). Hansard and Foote (1961) found that the endogenous faecal loss of Fe in 16-week-old calves was

12 mg/day and the availability in milk was only 26 per cent; thus, the dietary requirement for maintenance was about 46 mg/day. The availability of Fe in the diet of ruminant calves is not known.

A detailed study of the effect of Fe-dextran injections on the haematology of calves was made by Roy *et al.* (1964). Such injections reduced the decline with age of packed cell volume, haemoglobin content, mean corpuscular volume, mean corpuscular haemoglobin concentration and total leucocyte count. Fe supplementation reduced the increase with age of erythyrocyte sedimentation rate, lymphocyte, monocyte and platelet counts, and increased the percentage of band neutrophils and eosinophils present, and enhanced the increase of eosinophils with age. The erythyrocyte count declined with age but was unaffected by Fe supplementation. The utilization of the injected Fe was estimated as between 60 and 70 per cent. To maintain blood haemoglobin levels at 10 g/100 ml, the gross requirement was estimated as about 36 mg Fe/day if given by injection, or 70 mg Fe/day if given by mouth. This latter value represented a concentration of about 116 mg Fe/kg dry matter.

Excess of Ca in the diet may impair the assimilation of Fe, and an excess of Fe may reduce the assimilation of P. Niedermeier *et al.* (1959) suggested that the addition of 240 mg Fe/day to a milk diet, given *ad lib.* to calves, had a depressing effect on the vitamin A content of the liver.

Copper

Signs of copper deficiency after the first months of life are associated in Australia and New Zealand with diets containing less than 5 mg Cu/kg dry matter, and with those which contain an excess of molybdenum or insufficient phosphorus (Neal, Becker and Shealy, 1931). In Great Britain, levels of less than 5 mg Cu/kg dry matter are rarely found in pasture. However, when grass is growing fast in the spring, its Cu content tends to be at its lowest, whilst its Mo content is at its highest. Moreover, with diets containing high levels of inorganic sulphate, Mo even at a low concentration will interfere with the assimilation of Cu (Dick, 1953).

The first sign of deficiency, which usually occurs at 2–3 months of age in beef calves at pasture, is a stilted gait, particularly of the hind legs. This is followed by a progressive loss of condition during the next 4–5 months, which may result in extreme emaciation and death. Depigmentation of the hair on the face may occur particularly around the eyes, as a result of the failure of melanin formation, for which the Cu-containing enzyme tyrosinase is required.

The Cu content of newborn calves was found by Rusoff (1941) to be 14 mg/kg, and the concentration of Cu in the liver of the newborn calf is 4–8 times that in the dam (Comar, Davis and Singer, 1948). Spring-born calves from cows, given Cu supplementation during the preceding winter, gained weight at 1 kg/day compared to 0·7 kg/day for calves of untreated cows (Field, 1953). On whole milk diets, a supplement of 6 mg Cu daily maintained normal serum Cu values, but had no effect on the haemoglobin levels of anaemic and normal calves (Matrone *et al.*, 1957).

Copper poisoning has occurred in calves given milk substitutes containing 115 mg Cu/kg dry matter (Shand and Lewis, 1957).

Manganese

Diets low in manganese content have been associated with sterility in dairy cattle. Calves reared from 90 kg body weight on a ration containing less than 10 mg Mn/kg basal diet were slightly slower to show signs of oestrus and were consistently slower to conceive than those given 40 mg Mn/kg basal diet. They also developed livers of abnormal structure, especially proliferation of the bile duct. The only organs studied that contained increased Mn after supplementation were the ovaries. No detrimental effect was observed when a daily intake of 70 mg Mn/kg basal diet was given (Bentley and Phillips, 1951). Excessively high intakes of Ca and P over a long period reduced weight gains of growing calves reared mainly on a milk diet, possibly due to interference with Mn absorption, but additional Mn (50 mg Mn/kg basal diet) tended to counteract the excess (Hawkins *et al.*, 1955).

Cobalt

Cobalt is necessary for the synthesis of vitamin B_{12} by rumen bacteria. The requirement for Co of growing cattle is not known. However, as Co deficiency occurs in Australia on pastures containing about 40 μg Co/kg dry matter, whereas cattle are normal on pastures containing 130 μg/kg dry matter (Underwood and Harvey, 1938), it has been suggested that the daily requirement is about 2·2 μg Co/kg body weight (Filmer and Underwood, 1937). The marked difference between this requirement and the toxic level of 1 mg Co/kg body weight (Ely, Dunn and Huffman, 1948; Keener *et al.*, 1949) precludes the risk of poisoning under practical conditions.

Calves born to cobalt-deficient cows are weak, and do not generally survive for more than a few days (Marston, 1952). Supplementation of the dam with Co increases the Co content of the tissues of the calf, and also that of the colostrum by about 100-fold. The tissues of newborn calves from unsupplemented dams contain from 0·1–0·22 mg Co/kg dry weight (Ward *et al.*, 1949).

Iodine

The requirement for iodine is mainly for the synthesis of the hormone thyroxine. Where iodine deficiency in calves is revealed by goitre, the use of iodized salt licks for the pregnant cows is indicated. The requirement is usually fulfilled by allowing access to stabilized iodized salt licks containing 0·0076 per cent iodine, or by inclusion of iodized salt at a level of up to 1 per cent in the concentrate mixture. The need for additional iodine exists in areas of the world where the soil has a low iodine content, but sometimes occurs because of interference to the iodine metabolism in the physiology of the plants consumed or in the physiology of the animal itself. Goitrogens are found in leguminous crops, in particular white clover (Flux *et al.*, 1956), and, also in kale (Sinclair and Andrews, 1954) and may interfere with the conversion of inorganic to organic-bound iodine.

Zinc

The respiratory enzyme carbonic anhydrase contains Zn, and the highest concentrations are found in skin and hair; a deficiency results in parakeratosis. Calves given 0·16 mg Zn/kg body weight developed signs of deficiency, but 0·88 mg/kg was found to be adequate (Miller and Miller, 1960).

Lead Toxicity

Calves are particularly susceptible to lead poisoning, the commonest sources of lead being from painted wood and metal, discarded paint tins, old lead plates of batteries, painted tarpaulins and vegetation in orchards recently sprayed with lead arsenate solution. The lethal dose for young calves is about 156–250 mg/kg body weight, but older calves may survive doses of this order. Calves tolerated 6 mg lead/kg body weight for a period of 3 years, but a single dose of 200–400 mg/kg may produce symptoms within 24 hr (Allcroft, 1950; Allcroft and Blaxter, 1950). However, a time lag of several days may occur before the onset of symptoms, the animal then dying within a few hours.

Under practical conditions, a healthy calf may be found to have died suddenly, or may have a fit from which it does not recover. In less acute cases, there is anorexia, dehydration, abdominal pain and teeth grinding with salivation. Constipation is often severe and urination ceases, although constipation is sometimes followed by diarrhoea. In the next phase, the calf attempts to push against the walls of its pen, appears to be blind and staggers around with rolling eyes and frothing mouth. After collapse, muscular spasms and tetany occur before death. In acute cases, treatment is likely to be ineffective, but calcium disodium ethylenediamine tetra-acetic acid (Ca EDTA) given daily for two periods of 6 days as a 25 per cent solution injected subcutaneously, or intravenously together with saline solution to reduce dehydration, has been partially successful in the cure of calves showing nervous symptoms resulting from lead poisoning (Holm *et al.*, 1953).

Vitamins

Carotenoids and Vitamin A

A deficiency of vitamin A in the dam may result in a premature birth, a calf born blind or so weak that it is susceptible to infection. Vitamin A maintains the integrity of the epithelial cells, a deficiency results in hyperkeratinization and lesions in the intestinal mucosa, which may have a marked effect in depressing the utilisation of protein and digestion, absorption and metabolizability of energy as a result of scouring (Ritzman *et al.*, 1945). Metaplasia of the epithelium may also occur in the cervix, the ducts of the mandibular salivary glands and the bronchi (Langham and Huffman, 1954). Scouring in calves as a result of vitamin A deficiency has been reported by Guilbert and Hart (1935), Cunningham and Addington (1938), Schmidt (1941), Hastings (1942), Creech and Seibold (1943) and Madsen and Earle (1947). Although calves are born with little reserves of vitamin A, those born of normal dams did not have a greater mortality during the first 3 weeks of life when deprived of vitamin A from birth than those given 24 mg in the first feed of life, although total liver reserves at 21 days were only 149 μg for the former compared to 233 μg for the supplemented group. Inclusion of 10 g lecithin in the first feed clearly increased the absorption of vitamin A so that the liver reserves of calves given 24 mg vitamin A were 474 μg (Aschaffenburg *et al.*, 1953). Scouring calves, however, have lower liver reserves than normal calves (Sellers and Eden, 1949), and Aschaffenburg *et al.* (1953) were able to relate liver reserves to the rate of passage of the meconium at the time when vitamin A was being given. Lundquist and Phillips (1942, 1943) and Hansen, Phillips and Rupel (1946) reported that colostrum-deprived calves could be successfully reared if given 7·5 mg vitamin A daily, but Aschaffenburg *et al.* (1949) and Aschaffenburg *et al.* (1953) were unable to protect such calves either with an initial feed of 45 mg vitamin A or with 24 mg together with 10 g lecithin. However, Roy *et al.* (1964) found that supplementation with the antibiotic chlortetracycline reduced the apparent requirement of vitamin A of veal calves up to the age of 13 weeks.

Night blindness as a result of a shortage of visual purple, or permanent blindness may result from vitamin A deficiency. The earliest symptom is the failure of the iris to contract, followed by blurring of the optic disc, fading of the tapeta lucida, and then night blindness. Finally, constriction of the optic nerve where it passes through the optic foramen results in atrophy of the optic nerve and complete blindness (Moore, Huffman and Duncan, 1935; McNutt and Wall, 1938; Moore, 1939; Wetzel and Moore, 1940; Langham and Huffman, 1954). Constriction of the optic nerve is a result of retardation of growth of the membrane bone enclosing the central nervous system and is responsible for an increase in cerebrospinal fluid pressure (Moore and Sykes, 1940, 1941).

Except for the pre-ruminant calf, whose requirement of vitamin A is mainly obtained as such, requirements for vitamin A are best expressed in units of the precursor β–carotene, the normal source of vitamin A in the ruminant. The requirement of β–carotene is about 4 to 5 times that of vitamin A itself (Boyer *et al.*, 1942) but the relative efficiency depends on the method of testing the response (Rousseau *et al.*, 1954) on the carrier used (Kon, McGillivray and Thompson, 1955; Wise *et al.*, 1958) and also on the level at which the carotene is given. Grifo *et al.* (1960) found that 3·8 μg carotene were needed to obtain the same response as 1 μg vitamin A at an intake of 88 μg carotene/kg body weight and 6·8 μg at an intake of 353 μg/kg body weight.

The most delicate test of incipient vitamin A deficiency is the failure to maintain normal cerebrospinal fluid pressure; changes in pressure will detect differences of intake of as little as 4·4 μg carotene/kg body weight (Moore, Berry and Sykes, 1943). In terms of vitamin A, Lewis and Wilson (1945) considered the requirement to prevent night blindness was 8 μg/kg body weight, for maximum growth rate 16 μg/kg, for maximum plasma levels 128 μg/kg and for maximum liver reserves 256 μg/kg. In terms of β–carotene, Boyer *et al.* (1942) found that a daily intake of 100 μg/kg was adequate for Holsteins, but not for Guernseys who required 125 μg/kg due to their poorer conversion of carotene to vitamin A. However, Moore *et al.* (1948a), using spinal fluid pressure as their criterion, found the requirements of Guernsey and Jersey calves to be 75 and 70 μg carotene/kg body weight respectively which were only slightly greater than the requirement of 66 μg/kg found for Holsteins.

For safety, maintenance of slight liver storage is necessary especially as Keener *et al.* (1942) and Moore, Berry and Sykes (1943) found that the minimum vitamin A requirement in winter may be twice as high as in summer. The maximum liver reserve attainable is not known, but Eaton *et al.* (1958) obtained liver reserves at slaughter of up to a mean of 742 μg/g in nine animals, although the reserves of the remainder of their 125 calves in the weight range 69–268 kg did not exceed 79 μg/g. The liver reserves could be related to plasma vitamin A by the equation

$$y = 9{\cdot}08x - 3{\cdot}7$$

where y = plasma vitamin A content in μg/100 ml. and
x = log liver concentration of vitamin A in μg/100 g.

Liver reserves of 14·6 μg/g were obtained in veal calves at 13 weeks of age having a daily intake of 44 μg vitamin A/kg from whole milk (Roy *et al.*, 1964.)

Prepartum feeding of carotene-rich foods causes only a slight increase in the liver reserves of vitamin A in the calf at birth, but vitamin A given prepartum has a marked effect, increasing the liver content to 43–56 μg/g (Walker *et al.*, 1949). Since the weight of liver in a young calf is about 17 g/kg body weight, and the minimum requirement is about 16 μg vitamin A/kg, high liver reserves at birth might supply the vitamin A requirement for 5–6 weeks. The main source of vitamin A for the young calf is from the ingestion of colostrum; its vitamin A content can be markedly increased by feeding vitamin A prepartum (Spielman *et al.*, 1947; Walker *et al.*, 1949). Values obtained by these workers are given in Table 8.

TABLE 8

Effect of Prepartum Diet on the Vitamin A and Carotenoid Content of Colostrum

Prepartum diet	Vitamin A content (μg/g fat) Colostrum milk			Carotene content (μg/g fat) Colostrum milk			Reference
	Day 1	Day 3	Day 7	Day 1	Day 3	Day 7	
Normal	43	15	5	34	10	3	Walker *et al.* (1949)
	91	22	10	25	9	4	Spielman *et al.* (1947)
Low carotene and vitamin A	47	17	5	11	6	2	Walker *et al.* (1949)
	48	20	8	14	6	3	Spielman *et al.* (1947)
High carotene	44	19	8	68	23	5	Walker *et al.* (1949)
	67	20	10	36	11	4	Spielman *et al.* (1947)
High vitamin A	174	48	13	30	12	3	Walker *et al.* (1949)
	165	57	14	29	7	3	Spielman *et al.* (1947)

The requirement during the last 3 months of pregnancy is considered to be some three to five times that for non-pregnant animals, but Ronning *et al.* (1953) considered a high level was necessary throughout gestation. Guilbert and Hart (1935) suggested 88 μg/kg in the last month of pregnancy, but Converse and Meigs (1938) found that a level of 178–264 μg/kg during the last 3 months of pregnancy resulted in normal calves, whereas 60 μg/kg resulted in the birth of weak calves. On the other hand, workers in Russia (Anon., 1957) considered 132–397 μg/kg during pregnancy to be insufficient to prevent bronchopneumonia and gastrointestinal disorders in the calf, but 991–1497 μg/kg was adequate. These abnormally high requirements may have been influenced by climatic conditions.

Vitamin D

The calf requires vitamin D for proper growth and bone formation, and both exposure to sunlight and the feeding of good-quality hay will normally furnish the requirement (Bechdel *et al.*, 1938). The precursor ergosterol is converted to ergocalciferol (D_2) during the sun-curing of hay, whilst exposure to sunlight results in the conversion of the precursor in the skin to choleocalciferol (D_3). However, the antirachitic power of the sun is very limited when the elevation of the sun is 35° or less, which occurs in Britain during 5–6 months of the year (Tisdall and Brown, 1929). The occurrence of rickets under practical conditions have been described by Rupel, Bohstedt and Hart (1933), Gullickson, Palmer and Boyd (1935) and Hibbs *et al.* (1945).

A shortage of phosphorus in the presence of adequate calcium and vitamin D (Thieler, 1934) or a shortage of vitamin D in the presence of adequate Ca and P will give rise to rickets (Gullickson, Palmer and Boyd, 1935; Thomas and Moore, 1951; Thomas, 1952). Increasing the Ca intake on diets deficient in vitamin D but adequate in P had only a slight effect in reducing rickets (Thomas, 1952). Huffman and Duncan (1935b) have suggested that Mg has a sparing effect on vitamin D and Smith (1957) has shown that large amounts of vitamin D given to milk-fed calves before hypomagnesaemia develops will prevent the subsequent appearance of hypocalcaemia concomitant with the fall in blood Mg. According to Colovos *et al.* (1951) a deficiency of vitamin D will reduce the digestibility and retention of protein, as well as of ash, and increase the basal metabolic rate. High rates of weight-gain may increase the requirement of vitamin D per unit of live weight.

On the basis of blood Ca and P levels and of the composition of bone ash Huffman and Duncan (1935a) considered that 0·8–1·0 i.u./kg body weight was adequate when vitamin D was given in a liquid diet to Holsteins, but Bechdel *et al.* (1938) using similar criteria, together with X-ray analysis, considered that the requirement of ruminant Holsteins was 6·6 i.u./kg daily. However, Moore *et al.* (1948b), Thomas and Moore (1951) and Thomas (1952) found that 3·3 i.u./kg body weight was borderline and 3·9 i.u./kg body weight was adequate for ruminant Holstein and Jersey calves, when the P intake was higher than that used by Bechdel *et al.* (1938). A higher requirement for pre-ruminant than for ruminant calves was postulated by Smith (1958), since although 4·5–7·0 i.u./kg body weight maintained normal serum Ca in 5–10 week old calves, Ca utilization was considerably increased when vitamin D intake was raised to the high level of 900 i.u./kg. Moreover, Keyes (1944) found that an addition of 44 i.u./kg given daily to calves receiving a milk substitute already fortified with vitamin D resulted in increased growth rate. For Holstein heifers of 187 kg body weight, 1·8–2·7 i.u./kg was adequate for normal growth (Bechtel *et al.*, 1936), a finding which suggests that the requirement may decrease with age.

Avitaminosis D in pregnant cows may result in bone abnormalities of calves at birth (Wallis, 1938, 1944). Eaton *et al.* (1947) have increased the vitamin D content of colostrum threefold by giving one million i.u. vitamin D daily to the dam for eight weeks before calving, but the supplement did not raise plasma values or liver reserves of the calves.

Vitamin D toxicity has been produced by giving $1{\cdot}0 \times 10^6$ i.u. vitamin D_3 daily to young calves (Blackburn, Blaxter and Castle, 1957). As the daily dosage was increased from 0·5 to $4{\cdot}0 \times 10^6$ i.u. growth was progressively reduced, serum Ca values rose, those of Mg fell and calcification of the vascular system became more pronounced.

Vitamin E and Selenium

The trace element Se is intimately linked with the function of vitamin E in protecting calves against muscular dystrophy, which can be of two forms.

(a) *Enzootic muscular dystrophy.* The high incidence of this condition in single-suckled calves in an area of north-east Scotland was associated with dams given a diet consisting mainly of straw and turnips (Sharman, 1954). Calves born from outwintered cows or from those given silage were rarely affected. Under such conditions, resulting in inadequate amounts of vitamin E in colostrum, 150 mg α–tocopheryl acetate given daily to calves prevented the condition, whereas 10 mg daily was insufficient (Blaxter and McGill, 1955). Further experiments showed that not less than 20 mg daily would protect the majority of calves, whilst 0·25 mg Se/day was slightly more effective than 20 mg α–tocopheryl acetate/day (Sharman, Blaxter and Wilson, 1959). A single subcutaneous injection of 15 mg Se at birth or 5 mg at 3–weekly intervals from birth afforded considerable but not as complete protection as 0·25 mg Se/day (Blaxter *et al.*, 1961). 31 per cent of the untreated calves had serum glutamic oxaloacetic transaminase activity greater than 250 units/100 ml compared with 2 to 5 per cent of the Se–treated calves. It seems probable that enzootic muscular dystrophy is associated with areas in which the soil content of Se is low (Sharman, Blaxter and Wilson, 1959) but, owing to the high toxicity of selenium, prophylactic treatment should be preferably given in the form of vitamin E. The prepartum feeding of vitamin E increases the vitamin E content of colostrum (Parrish *et al.*, 1950b). By feeding about 2 mg tocopherol/kg live weight prepartum to cows in north-east Scotland, Blaxter and Sharman (1953) largely prevented enzootic muscular dystrophy in the calves.

(b) *Muscular dystrophy induced by diets containing unsaturated fats.* The inclusion of cod liver oil, lard and other fats containing a large proportion of polyunsaturated fatty acids in diets based on skim milk may cause muscular dystrophy (Blaxter and Brown, 1952; Blaxter, Watts and Wood, 1952). 15–18 ml cod liver oil given daily were sufficient to overcome the protective action of 50 mg DL—α–tocopheryl acetate. Similarly, Maplesden and Loosli (1960) found that 200 mg of water-dispersible D—α–tocopheryl acetate were sufficient to prevent muscular dystrophy in calves given a tocopherol-deficient diet supplemented with 30 ml cod liver oil. Most other antioxidants, except methylene blue, seem to be ineffective (Blaxter *et al.*, 1953). Roy (1964) found that the inclusion of 0·0005 per cent butylated hydroxianisole did not prevent muscular dystrophy in calves given skim milk diets containing 20 per cent lard, or 10 per cent lard and 10 per cent beef dripping unsupplemented with vitamin E, but no symptoms occurred with a similar diet containing 20 per cent beef dripping. The addition of 20 mg DL—α–tocopheryl acetate/kg dry matter, however, prevented the condition and maintained serum tocopherol levels at the same level in calves given lard-containing diets as in those given beef dripping. Even so, the serum tocopherol levels resulting from these various diets may not have been really adequate for maximum performance, since a positive relationship was found between serum tocopherol level and weight gain, irrespective of the source of fat, or of differences in initial weight and food consumption. Frens (1961) recommended the inclusion of 10 mg vitamin E/kg dry matter in milk substitute diets for veal calves. Tollersrud (1963) in Norway found that 30 mg α–tocopherol/kg dry milk substitute would prevent muscular dystrophy in calves given a proprietary milk substitute containing 13 per cent lard and 6·5 per cent tallow. Cod liver oil toxicity probably does not occur to the same extent in the ruminant calf, due to the hydrogenation of unsaturated fats in the rumen (Hoflund, Holmberg and Sellmann, 1956).

In the absence of a source of polyunsaturated fatty acid, as little as 3 mg total tocopherol daily is sufficient to prevent the onset of muscular dystrophy in young calves (Blaxter, Watts and Wood, 1952; Blaxter, Wood and MacDonald, 1953; Blaxter, Brown and MacDonald, 1953).

Vitamin C, Vitamin B Complex and Vitamin K

Although vitamin C is synthesized in the body tissues, Lundquist and Phillips (1943) indicated that supplementary feeding of vitamin C may prevent infection of young calves through the navel.

The synthesis of B complex vitamins and vitamin K by the microorganisms of the rumen make their inclusion in the diet unnecessary once the rumen is fully functional. It would seem possible that the substrate may not always be ideal for synthesis of B vitamins during the first two months of life, if milk products have been completely replaced by dry feed. However, Kesler and Knodt (1951, 1952) showed that there was no difference in the content of B vitamins in the digestive tract of the calf over the range from 2 to 14 weeks or even during the period from 8 to 32 days of age.

Calves reared on purified diets have been shown to have a dietary need for riboflavine, thiamine, pyridoxine and vitamin B_{12}, but it is unlikely that acute deficiency ever occurs under practical conditions. The B vitamin content of the livers of newborn calves is given in Table 9.

(a) *Thiamine*. Johnson *et al.* (1948) established that the calf had a requirement for thiamine, and this was confirmed by Kon and Porter (1952), although the symptoms after 7 to 10 days on the deficient diet—namely, anorexia, scouring and death from dehydration—did not appear to be very specific.

TABLE 9
B Vitamin Content of Livers of Newborn Calves
(Ford and Gregory, 1964)
(*wt. of liver 816 g (720–997 g)*)

	μg/g liver	Range
Biotin	0·095	(0·08–0·12)
Folic acid	0·7	(0·3–0·9)
Nicotinic acid	74	(54–86)
Pantothenic acid	18	(16–22)
Riboflavine	9	(8–10)
Thiamine	1·2	(1·0–1·3)
Vitamin B_6 (pyridoxal)	1·9	(1·8–2·0)
Vitamin B_{12}	0·25	(0·15–0·36)

(b) *Riboflavine.* A requirement for riboflavine has been more firmly established. Wiese *et al.* (1947a) and Warner and Sutton (1948) produced symptoms of riboflavine deficiency either on a 'synthetic milk' or on 'photolysed' whole milk. As with most B vitamin deficiencies, symptoms of anorexia occur, coupled in the case of riboflavine with excessive lachrymation and salivation, and scouring.

Kon and Porter (1951), who confirmed the findings of the American workers, also found soreness in the corners of the mouth and shedding of hair. In their deficient calves, urinary secretion of riboflavine was only 0·01 μg/ml compared to 1 μg/ml for a normal calf. Death occurred after about 6 weeks on the deficient diet. Brisson and Sutton (1951), using photolysed milk diets, considered the requirement was between 35 and 45 μg/kg body weight, whereas Draper and Johnson (1952), using a purified diet, found a requirement of 1 μg/g dry matter for Holstein calves, equivalent to 14 μg/kg body weight. Similarly, Porter (1953) estimated the requirement for Shorthorn and Friesian calves to be not more than 22 μg/kg body weight.

(c) *Nicotinic acid.* Calves given liquid diets containing gelatin as a source of protein appeared to suffer from a deficiency of nicotinic acid (Blaxter and Wood, 1952b). The deficiency symptoms resemble those of black tongue in the dog. Even on diets devoid of nicotinic acid, no deficiency occurs providing proteins containing appreciable quantities of tryptophan, such as those in milk, are included (Johnson *et al.*, 1947).

(d) *Vitamin B_6 (Pyridoxine).* This vitamin has also been claimed as essential for the young calf by Johnson *et al.* (1947), and also by Kon and Porter (1952), who found that some calves showed symptoms of anorexia, scouring and fits, with deaths occurring at 3–4 weeks of age, whilst others grew poorly to 4 months without showing definite symptoms. The *post-mortem* findings showed inflammation of the lymph nodes and some degree of enteritis.

(e) *Pantothenic acid.* Johnson *et al* (1947) claimed that this vitamin was essential for the calf.

(f) *Vitamin B_{12}*. Oral doses of 0·20 and 0·28 μg/kg body weight daily were considered by Draper and Johnson (1952) to be barely sufficient to prevent vitamin B_{12} deficiency, whereas Lassiter *et al.* (1953) suggested a requirement of 20–40 μg/kg dry matter in the diet equivalent to about 0·50–1·0 μg/kg body weight. Smith and Loosli (1957) considered that the requirement was about 0·4 μg/kg body weight based on the results of single injections given by Smith, Koch and Turk (1951).

As the amount of vitamin B_{12} in milk is as low as 2·1 μg/kg (Gregory, Ford and Kon, 1958) it is possible that with calves restricted to liquid diets, a deficiency might occur. However, Roy *et al.* (1964) found no obvious signs of B_{12} deficiency in calves reared to 110 kg for veal production and obtained an average gain in weight of 1 kg/day on an estimated daily intake of 0·54 μg B_{12}/kg body weight.

(g) *Biotin*. Wiese, Johnson and Nevens (1946) claimed to have produced biotin deficiency in calves, but this has not been confirmed by Kon and Porter (1951) who failed to do so even when feeding raw egg white as a source of avidin, or the biotin antagonist, desthiobiotin. Kon and Porter (1951) suggested that the higher level of potassium and of vitamin E in their diet might be the cause of their failure to confirm the work of Wiese, Johnson and Nevens (1946).

THE COMPOSITION OF DIETS TO SATISFY THE NUTRITIONAL REQUIREMENT OF THE CALF

Factors Affecting Birth Weight

The larger the average mature weight of the breed, the larger is the average weight of the calf at birth, the faster is its potential maximum growth rate and the slower it reaches maturity. The birth weights of various breeds are given in Table 10. Within a breed, the variation in birth weight is considerable; for instance, Shorthorn bull calves vary in weight from 20 to 57 kg with a mean value of about 37 kg. Twin calves are lighter in weight than single calves, possibly due to the competition between foetuses for available nutrients (Hammond, 1944). Bull calves are heavier at birth and their gestation period is about a day longer than that of heifer calves. The relationship between birth weight and gestation length has been reviewed by Andersen and Plum (1965). The faster rate of growth of bull calves may be entirely due to their greater birth weights (Nelms and Bogart, 1956). Consecutive calves from a cow tend to increase in weight up to the third or fourth calf (Eckles, 1919) or even the fifth calf (Braude and Walker, 1949) but calves from cows of very advanced age tend to be smaller. The increase in weight from first to fifth calf in Shorthorns was only 4 kg (Braude and Walker, 1949). Gestation length

TABLE 10
Birth Weight of Various Breeds of Cattle (kg)

Breed	Origin*	Location	Bull	Heifer	Both sexes	Reference
Aberdeen-Angus		U.S.A.	30·5	28·1		Burris and Blunn (1952)
		U.S.A.	28·6	26·1		Nelms and Bogart (1956)
		U.S.A.			28·0	Meade *et al.* (1959)
		U.S.A.			25·5	Foote, Hauser and Casida (1960)
Afrikander	Lateral horned Zebu type	S. Africa			29·0	Joubert and Bonsma (1957)
Aulie-Ata	Friesian x Kirgiz	U.S.S.R.			30·0	Novikov, Startsev and Arzumanian (1950)
Ayrshire		U.S.A.	33·1	29·5		Eckles (1918)
		U.S.A.	30·9	29·5		McCandlish (1922)
		U.S.A.	34·1	31·3		Fitch, McGilliard and Drumm (1924)
		U.S.A.	36·8	35·9		Davis and Hathaway (1956, 1959)
		U.K.			33·2	Donald, Russell and Taylor (1962)
		U.S.A.	38·3	34·7		Legault and Touchberry (1962)
Bestuzhev	Kholmogor, Friesian, Simmental, Wilstermarsch and Shorthorn x local	U.S.S.R.			31·5	Novikov, Startsev and Arzumanian (1950)
Black Pied Lowland	var Friesian	Germany	41·5	38·5		Dinkhauser, Trampen and Bergmann (1944)
Brazilian Polled	? From Crioulo	Brazil	22·9	22·2		Jordão (1956)
Brown Alpine	var Brown Swiss	Italy			41·0	Bonadonna and Valerani (1946)
		Yugoslavia	41·7	39·9		Mihal (1956)
		Italy			32·0	Rognoni and Pasti (1957)
Brown Swiss		U.S.A.			45·4	Eckles (1919)
		U.S.A.			43·2	Stallcup, Horton and Brown (1956)
		U.S.A.	48·5	44·3		Legault and Touchberry (1962)
		U.S.S.R.	41·6	38·8		Novikov, Startsev and Arzumanian (1950)
Caracu	Spanish	Brazil	22·8	22·1		Jordão (1956)
Caucasian		U.S.S.R.			19·0	Novikov, Startsev and Arzumanian (1950)
Charollais		France	50·2	45·9		Dumont, Auriol and Duplan (1959)
Chebsko(Bohemian Red)		Czechoslovakia			42·0	Křížnecký and Kudlička (1942)

TABLE 10
Birth Weight of Various Breeds of Cattle (kg)—Continued

Breed	Origin*	Location	Bull	Heifer	Both sexes	Reference
Chiana	? Grey Steppe x Local	Italy			45·3	Bujatti (1956)
Czechoslovakian Red Spotted	Bernese and Simmental x Chebsko and Moravian Red	Czechoslovakia			43·6	Suchanek (1961)
Dexter	Kerry	U.K.	24·0	23·2		Joubert and Hammond (1958)
East Finnish	var Finnish	U.S.S.R.	20·0	18·5		Novikov, Startsev and Arzumanian (1950)
Egyptian		Egypt	25·8	23·9		Asker and Ragab (1952)
		Egypt	27·0	24·0		Ahmed and Tantawy (1954)
Flemish		Brazil	39·9	36·9		Jordão and Assis (1947)
Friesian		Brazil	36·7	34·7		Grossman and De Oliveira (1949)
		Italy			33·4	Rognoni and Pasti (1957)
		Holland Heifers	33·1	31·8	}	Stegenga (1961)
		Cows	38·5	36·2	}	
		U.K.			39·5	Donald, Russell and Taylor (1962)
		Holland	40·8	38·0		Van der Mey (1965)
Friesland	var Friesian	S. Africa			32·2	Joubert and Bonsma (1957)
		S. Africa	35·4	34·8		Joubert (1959)
Fulani (White)	Lyre-horned Zebu type	Nigeria			22·7	Robertson (1950)
Gir		India	25·5	23·7		Lazarus and Rangaswamy (1950)
		India	23·1	21·7		Anantakrishnan and Lazarus (1953)
Gorbatov Red	Tux x Local	U.S.S.R.			26·5	Ovsjannikov (1959)
Guernsey		U.S.A.	30·0	27·7		McCandlish (1922)
		U.S.A.	30·4	27·2		Fitch, McGilliard and Drumm (1924)
		U.S.A.	35·4	32·7		Davis and Hathaway (1956, 1959)
		U.S.A.			30·6	Stallcup, Horton and Brown (1956)
		U.S.A.	34·6	30·6		Legault and Touchberry (1962)
Hariana	Shorthorned Zebu type	India	22·3	22·5		Lazarus and Rangaswamy (1950)
		India	24·5	23·6		Sharma, Vali and Suri (1951)
		India	23·5	22·6		Kohli and Suri (1957)

TABLE 10

Birth Weight of Various Breeds of Cattle (kg)—Continued

Breed	Origin*	Location	Bull	Heifer	Both sexes	Reference
Hariana (*cont.*)		India	24·7	22·6		Singh and Desai (1959)
Hereford		U.S.A.	33·6	31·3		Gregory, Blunn and Baker (1950)
		U.S.A.	31·7	29·7		Burris and Blunn (1952)
		U.S.A.	33·1	32·1		Rice, Kelley and Lasley (1954)
		U.S.A.	32·5	31·1		Nelms and Bogart (1956)
		U.S.A.			30·2	Meade *et al.* (1959)
		S. Africa			35·3	Joubert (1959)
		U.S.A.	33·2	32·1		Lasley, Day and Comfort (1961)
		U.S.A.	36·4	34·0		Brinks *et al.* (1961)
Holstein-Friesian		U.S.A.	42·2	40·0		Eckles (1918)
		U.S.A.	44·0	40·4		McCandlish (1922)
		U.S.A.	45·9	40·9		Fitch, McGilliard and Drumm (1924)
		U.S.A.	43·4	41·1		Tyler, Chapman and Dickerson (1947)
		U.S.A.		40·6		Willard (1948)
		U.S.A.			43·5	Hilder and Fohrman (1949)
		U.S.A.	46·3	43·6		Matthews and Fohrman (1954b)
		U.S.A.	43·9	41·0		Davis, Plum and Brost (1954)
		U.S.A.		40·4		Davis and Hathaway (1955)
		U.S.A.	45·4	40·0		Davis and Hathaway (1956, 1959)
		U.S.A.			42·1	Stallcup, Horton and Brown (1956)
		U.S.A.			41·6	Foote, Tyler and Casida (1959)
		U.S.A.	44·5	41·4		Legault and Touchberry (1962)
Istoben	Kholmogor and Brown Swiss x Local	U.S.S.R.	30·0	29·0		Novikov, Startsev and Arzumanian (1950)
Jersey		U.S.A.	26·3	24·1		Eckles (1918)
		U.S.A.	25·0	23·6		McCandlish (1922)
		U.S.A.	26·7	24·5		Fitch, McGilliard and Drumm (1924)
		Brazil			22·7	Grossman and De Oliveira (1949)
		U.S.A.			25·1	Hilder and Fohrman (1949)

TABLE 10

Birth Weight of Various Breeds of Cattle (kg)—Continued

Breed	Origin*	Location	Bull	Heifer	Both sexes	Reference
Jersey (*cont.*)		U.S.A.	25·1	23·6		Arnold and Becker (1953)
		U.S.A.	27·7	25·4		Matthews and Fohrman (1954a)
		S. Africa			21·6	Joubert and Bonsma (1957)
		U.S.A.	26·3	24·5		Davis and Hathaway (1956, 1959)
		U.K.			22·5	Donald, Russell and Taylor (1962)
		U.S.A.	25·8	23·4		Legault and Touchberry (1962)
Kalmyk	Turano-mongolian type	U.S.S.R.	24·0	24·0		Novikov, Startsev and Arzumanian (1950)
		U.S.S.R.			26·6	Ritchie (1960)
Kholmogor	Friesian x Local	U.S.S.R.	39·0	37·0		Novikov, Startsev and Arzumanian (1950)
		U.S.S.R.			43·0	Ovsjannikov (1959)
Kirgiz	Turano-mongolian type	U.S.S.R.		19·0		Novikov, Startsev and Arzumanian (1950)
Kostroma	Brown Swiss and Allgau x Local	U.S.S.R.	39·7	37·0		Novikov, Startsev and Arzumanian (1950)
		U.S.S.R.			39·2	Ovsjannikov (1959)
Meuse-Rhine-Yssel		Brazil	40·3	38·0		Jordăo and Assis (1951)
		Holland	44·3	39·9		Van der Mey (1965)
Moravian Red		Czechoslovakia			58·2	Křiženecký and Kudlička (1942)
Nganda	Ankole x Zebu	Uganda	20·0	18·6		Faulkner and Brown (1953)
Nguni	Sanga type	S. Africa	26·1	22·7		Brown (1959)
		S. Africa	23·4	21·8		Reyneke, Joubert and Bonsma (1963)
Ongole	Shorthorned Zebu type	India	30·0	27·2		Littlewood (1935)
Red Danish		Norway			40·0	Venge (1948)
		U.S.A.			36·7	Hilder and Fohrman (1949)
		Denmark	39·0	36·4		Andersen (1962)
		Denmark	42·9	40·2		Nielsen (1964)
Sahiwal		India	23·2	21·8		Sayer (1936)
		India	24·0	21·8		Singh and Dutt (1961)
Shorthorn (Beef)		U.S.A.	32·5	30·3		Knapp, Lambert and Black (1940)

TABLE 10

Birth Weight of Various Breeds of Cattle (kg)—Continued

Breed	Origin*	Location	Bull	Heifer	Both sexes	Reference
Shorthorn (Beef) (*cont.*)		U.S.A.	32·8	31·2		Dawson, Phillips and Black (1947)
		U.S.A.	30·3	28·1		Burris and Blunn (1952)
		S. Africa			27·6	Joubert and Bonsma (1957)
		U.S.A.			27·2	Foote, Hauser and Casida (1960)
(Dairy)		U.S.A.	33·6	33·1		Eckles (1919)
		U.S.A.	36·8	35·2		Knapp, Lambert and Black (1940)
		U.K.	38·2	36·3		Braude and Walker (1949)
Siberian	Turano-mongolian type	U.S.S.R.	19·7	18·8		Novikov, Startsev and Arzumanian (1950)
Silesian Kuhland (Kravarsko)	var Czechoslovakia Red-spotted	Czechoslovakia			41·6	Kříženecký and Kudlička (1942)
Simmental	Bernese	U.S.S.R.	42·4	37·1		Novikov, Startsev and Arzumanian (1950)
		Yugoslavia	46·3	42·7		Petrovic (1952)
		Roumania			36·8	Negrutui, *et al.* (1958)
Sindhi		India	20·2	19·3		Rathore (1950-1)
		India	22·2	20·8		Lazarus and Rangaswamy (1950)
		India			19·1	Dave (1950)
		India	21·4	19·9		Anantakrishnan and Lazarus (1953)
		Ceylon	17·7	15·4		Mahadevan (1955)
Sinhala	Similar to Indian hill type	Ceylon	14·1	13·2		Mahadevan (1953)
South Devon		U.K.	46·2	45·0		Joubert and Hammond (1958)
Spotted Mountain	Bernese and Simmental x Local	Germany	40·1	37·4		Bachner (1951)
		Germany			39·7	Abelein and Ritter (1959)
Swedish Red and White		Norway			34·9	Venge (1948)
Tagil	Friesian, Kholmogor x Local	U.S.S.R.			29·0	Novikov, Startsev and Arzumanian (1950)
Tambov Red	Simmental x (Tux x Local)	U.S.S.R.	37·9	36·8		Novikov, Startsev and Arzumanian (1950)
Tharparkar	Lyre-horned Zebu type	India	23·8	21·1		Lazarus and Rangaswamy (1950)
		India	22·1	21·6		Chaudhuri and Sinha (1951)

TABLE 10
Birth Weight of Various Breeds of Cattle (kg)—Continued

Breed	Origin*	Location	Bull	Heifer	Both sexes	Reference
Tharparkar (*cont.*)		India	24·5	23·6		Joshi and Phillips (1953)
Ukranian Grey	Grey-steppe type	U.S.S.R.	33·0			Novikov, Startsev and Arzumanian (1950)
Ukranian Red	Red East Friesian and Angeln x Ukranian Grey	U.S.S.R.	32·0	30·2		Novikov, Startsev and Arzumanian (1950)
Ukranian Whitehead		U.S.S.R.	32·1	29·4		Novikov, Startsev and Arzumanian (1950)
Yaroslavl	Kholmogor blood	U.S.S.R.			29·8	Ovsjannikov (1959)
Yurino	Brown Swiss x (Local x Garbotov and Tux)	U.S.S.R.	35·0	31·0		Novikov, Startsev and Arzumanian (1950)
		Birth weight of Buffaloes				
Egyptian		Egypt	38·5	36·4		Asker and Ragab (1952)
		Egypt	40·3	37·3		Ahmed and Tantawy (1954)
Murrah		India	36·9	31·9		Lazarus and Rangaswamy (1950)
		India	33·7	32·8		Arunchalam, Lazarus and Anantakrishnan (1952)

*From Mason (1957).

appears to increase slightly with increased parity at calving (Andersen and Plum, 1965).

In cross-breeding, the weight of the calf tends towards the weight of the dam rather than to that of the sire. South Devon ♂ × Dexter ♀ calves weighed 26 kg compared to 33 kg for Dexter ♂ × South Devon ♀ calves (Joubert and Hammond, 1958). The birth weight of a particular cross-bred calf may be estimated by the formula

$$\tfrac{1}{2}(D_0 + S_0) + \tfrac{1}{10}(D_0 - S_0)$$

where D_0 = birth weight of the breed of the dam
and S_0 = birth weight of the breed of the sire
(Donald, Russell and Taylor, 1962). Barcroft (1944) has suggested that the size of the newborn animal is determined by the size of the placenta.

Under very poor nutritional conditions in late pregnancy, the weight of the calf will certainly be reduced (Bonsma, 1949; Joubert, 1954). A short dry period may also reduce birth weight (Krizenecky and Kudlicka, 1942). Under good nutritional conditions, steaming-up before calving, particularly of heifers, increases the birth weight of the calf, but the increase has been small (Blaxter, 1944), and when adjusted for the increased weight of the dam has invariably been non-significant.

Colostrum

Composition

Colostrum appears to be a mixture of true milk as secreted in later lactation, and the non-diffusible constituents of blood plasma that have been concentrated some ten- to fifteen-fold in their passage into the udder some time before calving. In heifers, the increase in globulin secretion occurs after the fifth month of gestation (Woodman and Hammond, 1923) but in cows, prepartum milking and isotope studies indicate that most of this concentrated plasma appears in the udder about 3–9 days before calving (Rowland *et al.*, 1953; Garner and Crawley, 1958).

The composition of colostrum and milk is given in Table 11. The high concentration of protein in colostrum, which is due largely to its content of immune lactoglobulins, falls exponentially from 14–20 per cent at the first milking, to 8–12 per cent at the second, to 5–6 per cent at the third and 4·2–4·4 per cent at the fourth (Parrish *et al.*, 1948, 1950a). Only the fat-soluble vitamins of colostrum and certain minerals are increased if given at high levels prepartum. High-protein rations before calving only lead to an increase in the non-protein nitrogen fraction (Parrish *et al.*, 1948).

TABLE 11
The Comparative Composition of Colostrum (first 24 hr after calving) and of Milk

	Colostrum		Milk*	
Fat (g/100 g)	3·6		3·5	
Non-fatty solids (g/100 g)	18·5		8·6	Ling, Kon and Porter (1961)
Protein (g/100 g)	14·3	Rowland *et al.* (1953)	3·25	
Casein (g/100 g)	5·2		2·6	
Albumin (g/100 g)	1·5		0·47	
β—lactoglobulin (g/100g)	0·80		0·30	
α—lactalbumin (g/100 g)	0·27	Based on Larson and Kendall (1957)	0·13	Rolleri, Larson and Touchberry (1956)
Serum albumin (g/100 g)	0.13		0·04	
Immune globulin (g/100 g)	5·5–6·8	Rowland *et al.* (1953); Larson and Kendall (1957)	0·09	
Lactose (anhydrous) (g/100 g)	3·10	Rowland *et al.* (1953)	4·60	Ling, Kon and Porter (1961)
Ash (g/100 g)	0·97		0·75	
Calcium (g/100 g)	0·26		0·13	
Magnesium (g/100 g)	0·04		0·01	
Potassium (g/100 g)	0·14	Garrett and Overman (1940)	0·15	Garrett and Overman (1940)
Sodium (g/100 g)	0·07		0·04	
Phosphorus (g/100 g)	0·24		0·11	
Chlorine (g/100 g)	0·12		0·07	
Iron (mg/100 g)	0·20	Luecke, Duncan and Ely (1947)	0·01–0·07	Johnstone (1944); Blaxter, Sharman and MacDonald (1957)
Copper (mg/100 g)	0·06		0·01–0·03	Sylvester and Lampitt (1935); Krenn (1937); Dills and Nelson (1942); Itzerott (1942)
Cobalt (μg/100 g)	0·5	Luecke, Duncan and Ely (1947)	0·05–0·06	Ellis and Thompson (1945), Archibald (1947)
Manganese (mg/100 g)	0·016	Sato and Murata (1932)	0·003	Sato and Murata (1932); Vallee (1959)
Carotenoids (μg/g fat)	25–45	Walker *et al.* (1949)	7	
Vitamin A (μg/g fat)	42–48	Walker *et al.* (1949)	8	Ling, Kon and Porter (1961)
Vitamin D (i.u./g fat)	0·9–1·8	Henry and Kon (1937); Eaton *et al.* (1947)	0·6	
Vitamin E (μg/g fat)	100–150	Parrish, Wise and Hughes (1947)	20	Parrish, Wise and Hughes (1947)

TABLE 11
The Comparative Composition of Colostrum (first 24 hr after calving) and of Milk—Continued

	Colostrum		Milk*	
Thiamine (μg/100 g)	60–100	Houston, Kon and Thompson (1940)	40	
Riboflavine (μg/100 g)	450	Houston, Kon and Thompson (1940)	150	
Nicotinic acid (μg/100 g)	80–100	Pearson and Darnell (1946); Lawrence, Herrington and Maynard (1946)	80	
Pantothenic acid (μg/100 g)	200	Pearson and Darnell (1946)	350	Ling, Kon and Porter (1961)
Vitamin B_6 (μg/100 g)			35	
Biotin (μg/100 g)	2–8	Lawrence, Herrington and Maynard (1946)	2·0	
Vitamin B_{12} (μg/100 g)	1–5	Collins *et al.* (1951); Anthony *et al.* (1951)	0·5	
Folic acid (μg/100 g)	0·1–0·8	Collins *et al.* (1951)	0·1	
Ascorbic acid (mg/100 g)	2·5	Kon and Watson (1937)	2·0	
Choline (mg/100 g)	37–69	Waugh, Hauge and King (1947)	13	Waugh, Hauge and King (1947)

*Proximate constituents are for Friesian milk.

Transfer of Passive Immunity

During the first 24 hr of life, the immune lactoglobulins ingested by the calf are taken up by the epithelial cells of the small intestine and pass into the lymph spaces and then into the blood circulation by way of the thoracic lymph duct (Comline, Roberts and Titchen, 1951 a, b). Not only the immune lactoglobulins, but also the other whey proteins of colostrum can pass into the blood stream, although the latter appear to be soon excreted by way of the urine (Deutsch and Smith, 1957; Bangham *et al.*, 1958).

There are two main theories as to why the epithelium of the small intestine should allow the passage of these large protein molecules during the first 24 hr of life only; firstly, that absorption is permitted owing to the absence of enzymes that would normally degrade the protein and secondly, that during this early period of life the intestine is in fact more permeable. The enzyme activity of the abomasum in the newborn calf is discussed later. However, in the present context, Hill (1956) found that the parietal cells that produce HCl in the abomasum of the lamb are not fully developed during the first 24 hr of life, and therefore the whey proteins may escape degradation. Moreover, colostrum contains an anti-tryptic enzyme which might be expected to protect the whey from the action of the proteolytic enzymes of the pancreas (Laskowski and Laskowski, 1950). However, inactivation of the enzymes in older calves allows no passage of globulin. Similarly, placing of massive amounts of colostrum directly into the small intestine, with the intention of bypassing the abomasum and swamping the enzyme activity in the intestine, does not allow globulin passage (Smith and Erwin, 1959).

Attempts to increase the period of permeability have been unsuccessful. Various hormones, including stilboestrol, progesterone, a combination of the two, cortisone and ACTH (adreno-corticotrophic hormone) given in various ways have been tried; so also has the feeding of amniotic fluid and nucleic acids (Deutsch and Smith, 1957). The development of the bacterial flora soon after birth may also be of importance for it is known that thickening of the wall occurs as a barrier against bacterial invasion.

More recently, Balfour and Comline (1962) showed that γ-globulin absorption by the calf was linked with the presence in colostral whey of a protein fraction which is not heat coagulable. This protein, together with inorganic phosphate and glucose-6-phosphate had a marked influence on the rate of absorption of γ-globulin; the authors suggest that this protein may act in a manner analogous to a surface-active agent. In this connection Aschaffenburg *et al.* (1953) found that the inclusion of lecithin in the first two feeds of life greatly increased the absorption of vitamin A, but no measurement of globulin absorption was made.

Calves that survive after being deprived of passive immunity from colostral globulins, begin producing their own autogenous antibodies to *E. coli* from about 10 days of age (Howe, 1922; Hansen *et al.*, 1946; Pierce, 1962).

Hormonal Influence on the Newborn Calf

Newborn calves are probably under the influence of a high level of sex hormones at birth. Thus, teats are large in size in bull calves but decrease rapidly with age, a decline that can be prevented by the feeding of oestrogens (Hawkins *et al.*, 1959). Colostrum is also oestrogenic (Pope and Roy, 1953) and it is probable that the oestrogens pass from the blood stream of the dam concomitant with the passage of globulin. Colostrum ingestion is not a factor in causing increased teat size, as the phenomenon also occurs in calves given only very small amounts of colostrum. Quite small amounts of synthetic oestrogens will increase the size of the accessory glands even as early as the first 3 weeks of life (Hawkins *et al.*, 1959).

The calf appears to be strongly under the influence of thyroxine at birth, for the pulse rate is very high as is the level of plasma protein-bound iodine, and both decrease exponentially with time (Lewis and Ralston, 1953; Roy, Huffman and Reineke, 1957).

The size of certain endocrine glands in the calf at 21 days are affected by the incidence of scouring during this early period of life; thyroid and adrenal weights are increased and thymus and spleen weights reduced as the incidence of scouring increases (Hawkins *et al.*, 1959).

Frequency of Sucking

Walker (1950) showed that newborn calves, left with their dams, sucked about 6 times daily and ingested 7–8 kg of colostrum on the first day of life, increasing to 10–12 kg by the fourth day. With cows at pasture, calves sucked three times daily for about 15 min (Johnstone-Wallace and Kennedy, 1944).

Calves can be fed once daily from 4 weeks of age without ill effect, but when the practice was begun at 6 days of age, growth rates were lower owing to reduced milk consumption (Anonymous, 1953, 1956).

Liquid Diets

Quality of Skim Milk

The quality of milk substitute diets given directly after the colostrum-feeding period is an important factor in predisposing calves to an *E. coli* localized intestinal infection. In this condition, coli multiply in the small intestine in which they form a carpet-like layer, but they do not invade the tissues of the calf further than the mesenteric lymph nodes. The calf scours badly, and eventually succumbs to dehydration, usually when 7–14 days of age (Smith and Orcutt, 1925; Wood, 1955).

Milk substitutes are based to a large extent on skim milk powder, which usually comprises at least 50 per cent of the total dry matter. If in the processing of skim milk, denaturation of the whey proteins has occurred to a marked extent then such milk powders will have a detrimental effect on the calf. Associated with the denaturation of the whey proteins is the reduction in ionizable Ca (Lampitt and Bushill, 1934; Verma and Sommer, 1950; Bernadoni and Tuckey, 1950; Harman and Slatter, 1950; Hilgeman and Jenness, 1951; Baker, Gehrke and Affsprung, 1954; Christianson, Jenness and Coulter, 1954; Hostettler and Stein, 1958; Davis and White, 1959), the release of SH groups (Gould and Sommer, 1939; Josephson and Doan, 1939; Townley and Gould, 1943; Larson and Jenness, 1950; Hutton and Patton, 1952; Zweig and Block, 1953), poor clotting ability by rennet, and reduced digestibility, but no effect on the biological value of the protein (Shillam and Roy, 1963a). The effect of such diets may be shown in three ways:

1. In the absence of an active infective agent, weight gain may be reduced by as much as 30 per cent in the first 3 weeks of life (Shillam and Roy, 1963a).

2. The rate of build up of 'infection' (Roy, *et al.*, 1955) will be greater when large numbers of susceptible calves are passed through a calfhouse.

3. The incidence of scouring and mortality will be higher once the 'infection' has been built up (Shillam, Roy and Ingram, 1962a) as evidenced by the dominance of one or two strains of *E. coli* (Wood, 1955).

Some of the common processing methods, together with their effect on the whey proteins, are given in Table 12. Only a mild preheating treatment before spray-drying gives a product closely similar to that of fresh milk.

TABLE 12
Effect of Heat Treatment of Milk on the Calf
(Shillam and Roy, 1961; Roy, 1964)

Treatment of milk	Per cent total milk nitrogen as non-casein nitrogen and effect on calf
	No demonstrable effect on the calf
Raw	25
Holder pasteurized (63°C for 30 min)	23
Spray-dried skim (preheating temperature 77°C for 15 sec)	22
	Detrimental to the calf especially during the first 3 weeks of life
Spray-dried skim (preheating temperature 74°C for 30 min+)	15
Roller-dried skim (110°C)	13
U.H.T. sterilized (135°C for 1–3 sec)	11

Chlortetracycline supplementation does not prevent the detrimental effect of severely heat-treated skim milk (Shillam and Roy, 1963a). Although Kastelic, Bentley and Phillips (1950) were able to induce scouring in calves under 2 weeks of age by reduction of the Ca content of a synthetic milk from 1·26 to 0·73 g/l., Shillam and Roy (1963b) obtained only a very slight beneficial effect of Ca addition to a severely heat-treated milk, even though such addition improved the clotting of the milk *in vitro*.

As might be expected from the finding that there is no change in biological value, supplementation with methionine is also ineffective. Addition of undenatured whey proteins appears to be the only factor that will improve a severely-heated skim milk powder (Shillam, Roy and Ingram, 1962b), a finding in keeping with the results of Kannan and Jenness (1961) who showed that the presence of denatured whey proteins increases the rennet clotting time of casein. The presence of denatured whey proteins thus appears to have an important detrimental effect on the rate of clotting and this is additional to the fact that heat treatment sufficient to cause denaturation also results in the binding of the Ca and a reduction in the concentration of ionizable Ca.

Blaxter and Wood (1953) suggested that scouring in calves was related to the rapid passage of milk from the abomasum without sufficient gastric digestion. This is supported by the findings of Tagari (1965) that after feeding there is a much more rapid fall in pH in the abomasum and a much lower proportion of undigested protein passing into the duodenum in the calf given a diet containing a 'mildly' preheated spray-dried skim milk, than when a diet containing a 'severely' preheated spray-dried skim milk is given.

Enzyme Activity in the Abomasum and Clotting of Casein

Impairment of rennin clotting seems to be an important factor in predisposing calves to an *E. coli* localized intestinal infection. Berridge, Davis, Kon, Kon and Spratling (1942–4) concluded that the milk-fed calf secretes only rennin and that the secretion of pepsin is not marked until the calf eats solid food. Dollar (1958) considered that the calf is poorly endowed with proteolytic digestive enzymes during the first few weeks of life and that rennin could account for almost all the proteolytic activity present. Henschel, Hill and Porter (1961) have shown that although the 1–2-week-old calf secretes little or no pepsin, by 4 weeks of age secretion of both rennin and pepsin occurs. However further studies (Porter, 1963a) have shown that the young calf may secrete either rennin or pepsin, or both, and that the exact pattern of secretion is not predictable from the age of the animal or the nature of its diet.

Kesler, Ronning and Knodt (1951) observed that the pH of the abomasum 4–6 hr after feeding was about 4·1 in newborn calves and decreased to about 3·4 at 16 days of age. Pierce (1962) found that the pH of the abomasum contents in the newborn calf, after suckling, did not reach a level sufficiently low for effective proteolysis. From *in vitro* studies it is clear that for clotting by acid alone to occur, the concentration of acid needed is twice that required in the presence of pepsin (Shillam and Roy, 1963b). It seems probable, therefore, that clotting at a very young age is not caused by the pepsin-HCl system. Jancarik (1962) has shown that pasteurized milk elicits less secretion of HCl into the abomasum of calves than does fresh milk. Ash (1964), using calves over 10 days of age, observed that the overall rate of secretion and output of acid and proteolytic activity in the abomasum increased with increasing volumes of whole milk ingested, but lagged behind the increase in intake. Thus, the effect of large volumes of whole milk, when given in two feeds, in predisposing calves to scouring during the first three weeks of life may well be the result of swamping a limited enzyme production. This effect of large volumes of milk is well known in bucket-reared calves and occurs sometimes even in single-suckled calves. Such scouring may well develop into epidemic proportions if a particular strain of *E. coli* becomes dominant.

It has been suggested in Holland (Frens, Grift and Dammers, 1961; Frens, 1961) that, due to poor rennet clotting of milk substitute diets, absorption in the small intestine of only partially broken down milk proteins may occur with the resultant formation of antibodies, so that at a later stage anaphylactic shock symptoms may be observed. Boogaerdt and Koetsveld (1961) considered this hypothesis to be unacceptable, as no anaphylactic shock occurred with calves that had been previously injected with 10 ml whole milk or skim milk powder solution, but did occur in two out of eight calves given a milk substitute diet containing the same skim milk powder, a finding that suggests that some other factor than the milk protein was involved. Van Adrichem and Frens (1965) have recently reported that calves given a synthetic milk containing 10 per cent soyabean meal produced anti-soyabean protein antibodies. No allergic reactions were observed, but growth rate was poor.

Buttermilk

Calves at pasture given reconstituted buttermilk containing about 1·3 per cent fat after the colostrum-feeding period have gained weight at nearly the same rate as those given whole milk (Flux and Patchell, 1956). For high-level feeding, reconstituted buttermilk is not entirely suitable as a sole diet, as the heat treatment given is usually severe enough to predispose to an *E. coli* localized intestinal infection, and its low fat content also predisposes to scouring at an older age (Roy, 1964). With calves at pasture given buttermilk diets in one feed/day from 3 weeks of age, Davey (1962) in New Zealand found that the reconstitution rate had a marked influence on the incidence of scouring from 3 to 9 weeks of age, those given a diet of 7 per cent dry matter had a much lower incidence than those given diets containing 12 or 22 per cent dry matter. Even so the weight gains were considerably higher for the calves given the latter two diets.

Whey

The inclusion of dried whey in milk substitutes is usually limited to less than 20 per cent of the dry matter, due to the high incidence of scouring resulting from larger amounts. This finding has been associated in part with the high mineral content of the whey, which appears to accelerate the rate of passage from the abomasum (Owen *et al.*, 1958; Bush *et al.*, 1963). Both groups found some evidence that additional fat in the diet had an ameliorating effect. Bush *et al.* (1963) considered that the fat delayed the passage from the abomasum, a finding contrary to earlier work in America (Espe and Cannon, 1935). Similarly, Ash (1964) was not able to find any consistent effect of the butterfat content of milk on rate of flow from the abomasum.

Nutritive Value of Fats

Although experiments on the effect of inclusion of vegetable or animals fats in synthetic milks (Wiese *et al.*, 1947b; Lofgreen, 1950; Kastelic, Bentley and Phillips, 1950) or in milk substitutes based on dried skim milk powder (Gullickson, Fountaine and Fitch, 1942) have been made over a number of years, it is only comparatively recently that milk substitutes consisting basically of skim milk and added fat have been used commercially.

Gullickson, Fountaine and Fitch (1942) obtained very poor results in calves given homogenized-milk diets containing 3·5 per cent of either cottonseed, soyabean or maize oil, but the animal fats, lard and tallow gave moderately good results and coconut oil gave an intermediate response. These workers reported loss of hair, first from the face, ears and neck, and later from the cannons of the rear legs of calves given the maize, cottonseed or soyabean oil. Bate, Espe and Cannon (1946), however, associated similar hair losses with unhomogenized diets, as did Roy *et al.* (1961). de Man (1951) suggested that diets containing unhydrogenated soyabean oil caused hair loss and development of eczema due to secretion of oil through the skin, whereas hydrogenated soyabean oil had much less effect. Even when homogenized, unhydrogenated soyabean oil has been found to cause a poor rate of growth, rough hair coats and a high incidence of scouring and mortality (Jacobson and Cannon, 1947; Wiese *et al.*, 1947b; Jacobson, Cannon and Thomas, 1949; Murley *et al.*, 1949; Barker, Wise and Jacobson, 1952; Adams *et al.*, 1959) but diets containing hydrogenated soyabean oil have given results nearly as good as those containing butter oil (Murley *et al.*, 1949).

Raven and Robinson (1958, 1959, 1960) studied the nutritive value of palm and palm kernel oil, and found the digestibility of both hydrogenated and unhydrogenated forms to be about 90 per cent compared with 97 per cent for butterfat. The hydrogenated palm oil appeared to have a lower protein-sparing effect than the unhydrogenated oil. The findings of Gullickson,

Fountaine and Fitch (1942) that lard and tallow appeared to be well utilized by the calf has been confirmed by many workers (Wiese *et al.*, 1947b; Blaxter and Wood, 1951a; Johnson, Hopper and Gardner, 1953; Hopper, Gardner and Johnson, 1954; Ritchey *et al.*, 1956; Thomas and Okamoto, 1956; Bell, 1958; Larsen, 1958; Comberg and Göllnitz, 1959; Kleisch and Horst, 1959; Kesler, Wilson and Moore, 1960). Jacobson, Brown and Ratcliff (1959) found that the incidence of scouring was higher with a diet containing lard than with whole milk, and Roy (1964) found a slightly higher incidence of scouring during the first 14 days of life with a diet containing tallow than with those containing lard, margarine or a mixture of lard and tallow. Brüggemann and Barth (1959) and Roy (1964) found that the digestibility of lard was superior to that of beef tallow. Hydrogenated herring oil has been used in Norway (Homb, 1960) but the calves did not grow as well as those given butterfat, though both groups were much superior to those given coconut fat, which caused severe scouring.

Increase with age in the digestibility of certain fats of low digestibility has been observed by Cunningham and Loosli (1954), Lassiter, Duncan and Christie (1957), Brüggemann and Barth (1959), Raven and Robinson (1964a) and Roy (1964). Digestibility of beef tallow increased from 83 per cent at 1–2 weeks to 89 per cent at 4–5 weeks (Raven and Robinson, 1964a), and from 85 per cent at 4 weeks to 89 per cent at 10 weeks (Roy, 1964). A mixture of equal parts of lard and tallow had a digestibility of 89 per cent at 4 weeks and 92 per cent at 10 weeks, whereas lard had a digestibility of 93 per cent at both ages (Roy, 1964).

Lloyd and Crampton (1957) showed that, at least for piglets, young guinea pigs and pups, the variation in apparent digestibility of fats could be largely accounted for by the chain length of the constituent fatty acids rather than by their saturation; thus, the shorter the chain, the greater the digestibility. Raven and Robinson (1964a) found little evidence, however, that the overall fatty acid composition had any marked effect on digestibility, and suggested that digestibility of a particular fat was related to its triglyceride make-up (e.g. the proportion of glyceryl tristearate in the total triglycerides) rather than the relative amounts of each fatty acid present.

Method of Incorporation of Fat

Fat may be incorporated into skim milk either by homogenization or by blending with skim milk powder and an emulsifying agent. Kastelic, Bentley and Phillips (1950) suggested, from slender evidence, that a fat globule size of less than 2 μ should be used in milk substitutes. Huff, Waugh and Wise (1951) observed a loss of hair in calves when glyceryl monostearate was used as an emulsifying agent in a diet containing hydrogenated cottonseed oil, but when the diet containing glyceryl monostearate was homogenized, the characteristic alopecia was absent. Hopkins, Warner and Loosli (1959) have shown with 2–3-week-old calves, that emulsification of coconut fat with crude soyabean lecithin resulted in a very large increase in digestibility of the fat, and homogenization had no additional beneficial effect. However, Roy *et al.* (1961), using calves under 3 weeks of age, found that mechanical homogenization of a margarine containing 39 per cent groundnut oil, 24 per cent palm oil and 37 per cent coconut oil gave better results than blending with lecithin. The addition of an emulsifying agent to the diet before homogenization gave no improvement in results. The mean diameter of the homogenized globule was about 3–4 μ and of the blended globule 10–20 μ. From the results of blood lipid levels, it would appear that older calves can probably tolerate fat globules of a greater size than 3–4 μ without ill effect. This is given some support by the finding of Huber *et al.* (1961a) that pancreatic lipase activity is lowest at one day of age and increases during the first week of life, although there is little change thereafter.

With fats of low digestibility, such as beef tallow, the addition of lecithin as well as homogenization results in increased digestibility (Raven and Robinson, 1964b; Roy, Stobo and Gaston, 1965).

In addition to the normal emulsifying agents used in milk substitutes, such as lecithin and glyceryl monostearate, recent experiments (Jousselin *et al.*, 1964; Le Provost, 1965) have indicated some beneficial effect of inclusion of sugar glycerides together with lecithin on the utilization of beef tallow during the first few weeks of life. Sugar glycerides are produced as the result of the transesterification of a triglyceride of natural origin with sucrose.

Vegetable Carbohydrates in Liquid Diets

Starch is poorly digested by young calves. Only 20 per cent of maize starch appeared to be digested at 4–7 days of age (Shaw, Woodward and Norton, 1918) and little response in blood reducing sugar level was obtained in calves given starch by Flipse *et al.* (1950). Their calves also suffered from diarrhoea. From the blood glucose response to test meals, Dollar and Porter (1957) deduced that the pre-ruminant calf could not digest starch or the degradation products, dextrin and maltose up to 28 days of age, a finding confirmed by Okamoto, Thomas and Johnson (1959). Dollar and Porter (1957) suggested that the ability of the calf to digest starch, as reported by Shaw, Woodward and Norton (1918) was the result of bacterial degradation of starch in the large intestine. Lactase activity in the intestinal mucosa decreases with age whilst pancreatic amylase and intestinal maltase activity begin to increase by about 9 weeks of age (Dollar and Porter, 1959). Huber *et al.* (1958) and Huber *et al.* (1961a) observed no maltase activity up to 44 days of age.

No intestinal sucrase activity was observed by Dollar and Porter (1959) and no rise in blood glucose was obtained from feeding sucrose (Okamoto, Thomas and Johnson, 1959; Velu, Gardner and Kendall, 1959; Huber and Jacobson, 1960). Huber *et al.* (1961a) could find no evidence of adaptation of enzyme activity in calves given whole milk containing either 3 per cent lactose or 2·5 per cent sucrose and 0·5 per cent corn starch after the colostrum feeding period, but it is possible that the quantity of supplements was insufficient to produce a response, or the tissue concentrations did not necessarily reflect the quantity of enzyme available during digestion. Zerebcov and Seryh (1962) in Russia, using pancreatic fistulas have found twice the concentration of amylolytic activity/kg live weight and ten times the concentration of proteolytic activity in calves first given plant food at 3 weeks of age compared to those first given such food at 7 weeks, the difference being apparent at all ages up to 6 months. However the food was given in dry form. The digestibility of various carbohydrates at 12 weeks of age, when given by nipple feeder with the intention of bypassing the rumen are given in Table 13. In interpreting these values, the authors (Huber *et al.*, 1961b) consider that the disappearance of considerable amounts of sucrose and starch resulted from microbial fermentation in the large intestine.

TABLE 13

Percentage Digestibility of Various Carbohydrates when given by Nipple Feeder to 12-week-old Calves
(Huber *et al.*, 1961b)

	Digestibility (%)
Lactose	94
Maltose	97
Sucrose	57
Amylose	83
Amylopectin	89
'Flojel' (thin boiling starch)	80
Tapioca starch	80

By feeding meal mixtures as a gruel some of the advantages of monogastric digestion can be achieved. Raven and Robinson (1961) observed a much higher retention of digested N in 14-week-old calves when a concentrate mixture was given as a gruel rather than in dry form; a finding that was probably due to the more efficient use of dietary energy in monogastric digestion. These workers (Raven and Robinson, 1964c) also found that fat homogenized into a milk powder, which was fed in a gruel together with cereals and fish meal, had a higher digestibility than when blended into the other components of the diet with lecithin. However, the digestibility was much lower than for diets in which the fat had been homogenized into skim milk, a finding they attributed to the higher intakes of Ca and Mg from the meal diets.

Vegetable Protein in Liquid Diets

The substitution of expensive milk protein by vegetable protein has not been very successful. A preliminary report from Benevenga and Ronning (1963), however, showed that whereas ground cooked soyabeans in place of skim milk gave little weight gains in a 100–day period, Drackett protein isolated from soyabean gave good results. Comparable results with A.D.M. assay protein from soyabeans have been found by Porter (1963b) who showed that the apparent digestibility and retention of N from diets containing A.D.M. assay protein were similar to those of casein. The retention was appreciably better than with soya α–protein; soyabean meal and soya flour proved to be of low digestibility and gave very poor retention. Unfortunately the values for casein were considerably lower than for skim milk, this difference possibly being due to the lack of undenatured whey proteins in the casein diet.

Supplementation of milk substitute diets containing plant products with various proteases, amylases and amyloglucosidase concentrates have had little success in increasing performance, but with the ruminant calf addition of bacterial proteases has increased the apparent digestibility of N and energy (Rust, 1963).

Dry Diets

Development of Rumination

The age at which transition to the ruminant method of digestion occurs is to a large extent dependent on the diet that a calf receives. Given a limited quantity of milk, the calf will begin to eat alternative foods, providing they are palatable, at about 7 days of age.

Schalk and Amadon (1928) showed that the nervous mechanism controlling the various muscular phenomena of rumination is very active in the 2-week-old calf and slight stimulation such as the presence of recording balloons often produces strong and frequent regurgitation effects. Similarly Roy (1959b) recorded practically normal rumen contractions in a calf as early as 4 days of age. Swanson and Harris (1958) noticed that some calves were ruminating at 5 days of age and the majority by 28 days of age, when offered a restricted milk intake together with concentrates and hay from birth. The time spent ruminating increased rapidly to average 5 hr per day at 6–8 weeks of age.

Development of Rumen Flora

The abomasum of the very young calf supports a large and diverse population of lactobacilli (Mann and Oxford, 1955) and it is possible that the backflow of milk from the abomasum to the rumen after feeding, observed by Trautmann and Schmitt (1933) and confirmed by Hoflund (1946) and Benzie and Phillipson (1957), helps to inoculate the rumen. Ingram (1962) showed that if a calf had not suckled, *E. coli* were mainly restricted to the ileum and large intestine $5\frac{1}{2}$ hr after birth, but by $8\frac{1}{2}$ hr were found in the abomasum at a level of 10^3 per ml. In the duodenum the viable count of *E. coli* reached a peak at 1–4 days of age and then declined to a very low level at 10 days, possibly due to the increasing dominance of the lactobacilli. In contrast, calves whose deaths were associated with an *E. coli* localized intestinal infection had viable counts of 10^7–10^8 per ml in the duodenum at this latter age. (See also Chapter 3.)

The typical rumen amylolytic streptococci of the adult animal do not become properly established until the rumen pH is stabilized near neutrality and, under conventional rearing methods, this does not happen until the calf is several weeks old, although in the very young calf a different amylolytic streptococcus is found which is more tolerant of a lower pH value (Mann and Oxford, 1955). Similarly, the protozoan population can become established once the pH is stabilized near neutrality at about 8 weeks of age (Mann, Masson and Oxford, 1954).

Pounden and Hibbs (1948) showed that when the proportion of grain to hay in the ration exceeded 3:1, types of bacteria associated with the digestion of hay disappeared. Similarly, Eadie, Hobson and Mann (1959) found that protozoa were encouraged by high roughage diets and lactobacilli by high concentrate diets. When calves were changed from a high-concentrate diet to one of all hay, Stobo, Roy and Gaston (1966b) found a marked reduction in the viable count of streptococci, lactobacilli and *E. coli*, but no demonstrable effect when the calves were changed from a high-roughage to an all-concentrate diet. Even with diets containing a high proportion of hay, Hibbs, Pounden and Conrad (1953) considered that there was need for rumen inoculation with the cud of an adult animal to ensure a normal microbial population. In Israel (Volcani and Levi, 1953) and in Russia (Belenki, 1953) beneficial effects of inoculation have been claimed, but later experiments have not confirmed this (Pelissier *et al.*, 1954; Davis *et al.*, 1954; Grimbergen, 1955; Williams and Jensen, 1955; Wieringa, 1956; Hardison, Miller and Graf, 1957; McArthur, 1957 a, b; Ronning, Bankston and Berousek, 1957; Brüggemann, 1958; Hibbs and Conrad, 1958; Preston, 1958). However, Bryant and Small (1956) have shown that inoculation will hasten the development of protozoa, but will not encourage the development of rumen bacteria unless calves are raised under strictly isolated conditions.

Development of Digestion in the Rumen

Concentrates or good quality grass seem to be as well digested by the young calf as by the adult ruminant (Armstrong, Preston and Armstrong, 1954;

Hogue *et al.*, 1956; Preston, Archibald and Tinkler, 1957; McArthur, 1957 a, b), but Flatt, Warner and Loosli (1959) and Mäkelä (1960) found an increased digestibility of hay or high-roughage diets with age. Stobo, Roy and Gaston (1966b) found that the digestibility of both hay and concentrates did not differ whether these were fed separately or in combination during the period 13–17 weeks of age. (See also Chapter 4.)

The concentration of total volatile fatty acids (VFA) produced by rumen fermentation increases with age of the calf, the time of the maximum value occurring about a week after weaning (Lengemann and Allen, 1955; McCarthy and Kesler, 1956; Hibbs *et al.*, 1956; Flatt, Warner and Loosli, 1959; Dinda, 1960). No difference in the concentration of total VFA in the rumen between 8 and 14 weeks of age was found by Stobo, Roy and Gaston (1966b) but as there was certainly an increase in the volume of rumen contents, the total production of VFA must have increased.

Dinda (1960), with calves weaned at 3 weeks, found a low pH at weaning but thereafter the pH increased with age. Preston (1963) has suggested that this may be due either to an increased production of lactic acid at an early age, or to a low level of saliva production. Lengemann and Allen (1955) found the greatest concentration of lactic acid in the rumen of calves at one month of age, and a slightly higher proportion was found by Stobo, Roy and Gaston (1966b) in calves given all-concentrate diets compared with those given an all-hay diet at 16 weeks.

In the developing rumen, Flatt, Warner and Loosli (1959) found that the proportion of acetic acid in the rumen was lower than in adult cows, and did not differ between calves given hay or concentrates. The proportion of butyric acid increased rapidly from 10 per cent at 4 weeks to 20–30 per cent thereafter. Similarly, Hibbs, Conrad and Pounden (1954) found values of 50 per cent for acetic acid, but as the amount of concentrates in the diet increased, butyric acid increased at the expense of propionic acid. Stobo, Roy and Gaston (1966b), on the other hand, found that the proportion of propionic acid to acetic acid in the rumen liquor at 13 weeks of age was very much higher in calves given high-concentrate diets than in those given high-roughage diets; the proportion of butyric acid was much less affected by diet. However, when a diet of hay alone was given, butyric acid concentration at 16 weeks was very much reduced compared to the value obtained with a high-roughage diet at 13 weeks. By feeding salts of VFA in place of hay and starch in a synthetic diet Martin *et al.* (1959) obtained a slight increase in growth rate.

The VFA concentration in peripheral blood increases with age (McCarthy and Kesler, 1956), but tends to be higher in calves given high-roughage or hay diets than in those given high-concentrate or all-concentrate diets (Stobo, Roy and Gaston, 1966b).

In the young calf, blood glucose levels fall from a value at birth of about 100 mg/100 ml to about 60 mg/100 ml at 6 weeks of age (McCandless and Dye,

1950; Dye and Orsini, 1952). Vandersall, Conrad and Hibbs (1957) suggested that this decline was associated with the replacement of foetal erythrocytes high in glucose with erythrocytes of lower glucose content. In the pre-ruminant calf, blood glucose rises rapidly to a peak of about 130 mg/100 ml at 1–2 hr after feeding and then declines to pre-feeding values at 5 hr after feeding (Dollar and Porter, 1957). In the ruminant calf, much higher levels are maintained by calves given high-concentrate diets than by those given high-roughage diets. At 7 weeks, milk-fed and concentrate-fed calves had mean blood glucose levels of 75 mg/100 ml whereas hay-fed animals declined to 55 mg/100 ml at 8 weeks (Warner *et al.*, 1953; Warner, Flatt and Loosli, 1956). Mean values of 93 mg/100 ml on all-concentrate diets and 68 mg/100 ml on all-hay diets were found at 16 weeks of age by Stobo, Roy and Gaston (1966b). When calves were changed from a high-roughage diet to one of all-concentrates, the increase in blood glucose concentration was greater than the decrease that occurred when calves were changed from a high-concentrate to an all-hay diet. Preston and Ndumbe (1961), giving concentrates at a lower level than Stobo, Roy and Gaston (1966b), obtained blood glucose levels of 80 mg/100 ml. Both groups found a temporary depression in plasma glucose 3 hr after feeding, whilst Preston and Ndumbe (1961) found a gradual rise after this time to maximum values 8–12 hr after food was eaten.

How far these high plasma glucose levels on concentrate as opposed to hay diets are the result of the glucogenic effect of propionate, or to glucose formed by hydrolysis of starch in the intestine is uncertain, as little is known of the extent to which concentrates can escape digestion in the rumen by rapid passage into the abomasum.

Jeffrey and Lane (1960) have shown that liver slices from 2-3 week-old calves were able to metabolize VFA at rates similar to those of adult cows, and Hawkins *et al.* (1959) found that liver slices from 3-week-old calves were able to metabolize acetate about half as efficiently as glucose. Table 14 gives a comparison of certain characteristics of 16–week old calves given concentrates or hay *ad lib.*

Concentrates

(a) *Protein quality.* Due to bacterial synthesis of protein in the rumen, the quality of protein in dry foods for calves has been considered of little importance until recently. Preston and his colleagues (Preston *et al.*, 1960; Preston and Whitelaw, 1961; Whitelaw, Preston and Dawson, 1961; Whitelaw Preston and Ndumbe, 1961; Whitelaw and Preston, 1963; Whitelaw, Preston and MacLeod, 1963) have suggested, however, that for calves given high-concentrate diets at a young age both the solubility and amino-acid composition, particularly the content of lysine and methionine, of the protein is of importance. They suggested that the protein may escape degradation in the

TABLE 14
Comparison of Certain Characteristics of 16-week-old Ayrshire Calves given Concentrates or Hay ad lib.
(Based on Stobo, Roy and Gaston, 1966b)

	Diet	
	Concentrates *ad lib.*	Hay *ad lib.*
Dry matter intake (kg/day)	2·5	1·6
Digestibility of dry matter (%)	81	58
Rumen total volatile fatty acids (m-equiv./100 ml)	11	5
Partition of rumen VFA (molar %)		
Acetic	47	74
Propionic	31	19
Butyric	18	7
Valeric and higher	4	—
Rumen pH	5·9	7·4
Plasma total VFA (mg/100 ml)	0·9	1·3
Blood glucose (mg/100 ml)	93	56
N-retention (g/day)	31	4
Live-weight gain (kg/day)	0·65	0·19

rumen and pass into the abomasum, from whence onwards quality of protein becomes as important as in a monogastric animal. The increased efficiency of utilization of protein of low solubility by the ruminant had been shown earlier by Chalmers, Cuthbertson and Synge (1954), who processed casein to reduce its solubility and thus the extent to which it was degraded into ammonia, and obtained improved N retention. The solubility of two groundnut meals tested by Whitelaw and Preston (1963) differed markedly at pH 7, but at pH 4 their solubilities were similar. These workers concluded that their failure to show a difference in N retention in calves given the two meals was due to the low pH of 5–5·5 occurring in the rumen on all-concentrate diets. However, the solubility of herring meal was not significantly affected by pH and thus 'insoluble' herring meal gave higher N retention than the 'soluble' herring meal. Moreover the higher N retention from 'insoluble' herring meal compared with that from the groundnut meals was attributed to the higher lysine and methionine content of the herring meal.

Nevertheless, satisfactory growth rates at restricted levels of concentrate intake have been reported when simple concentrate mixtures containing only one major source of protein, often of vegetable origin, have been compared with more complex mixtures containing a variety of protein-rich ingredients, irrespective of the age at weaning (Jacobson, Allen and Bell, 1951; Gardner, 1952; Hibbs, Pounden and Conrad, 1953; Holter, 1954; Lambert *et al.*, 1955; Kesler and Wilson, 1958; Murley *et al.*, 1957; Harrison *et al.*, 1960).

The inclusion of milk protein in concentrate mixtures is not necessary, but Preston (1956) with Ayrshire calves weaned at 3 weeks found that the replacement of 10 per cent linseed meal by an equal quantity of dried skim milk powder resulted in slightly improved growth rates.

Because of the danger of injury from gossypol present in cottonseed cake, it is unwise to feed this cake to calves under 3–4 months of age. Hollon *et al.* (1958) showed that cottonseed meal containing only about 0·029 per cent free gossypol could form up to 60 per cent of a concentrate mixture without ill effect.

(b) *Protein content.* Good growth rates were obtained by Harris and Loosli (1944) in heifer calves given *ad lib.* a mixture containing 18·8 per cent crude protein on a dry matter basis (16·0 per cent on an air-dried basis). Levels of 15·2 and 11·7 per cent crude protein on a dry matter basis resulted in slower growth rates, and very poor weight gains were obtained with calves given a mixture containing only 8·3 per cent crude protein. When concentrate intake was restricted, no difference was observed between diets containing 18·8 and 15·2 per cent crude protein. Brown *et al.* (1958) found that a concentrate mixture containing 16·2 per cent crude protein (air-dry basis) resulted in greater weight gains than similar mixtures containing 8·5, 13·0, 20·0 and 23·7 per cent, when given to Holstein and Jersey calves weaned at 7 weeks of age. Concentrate intake was restricted to a maximum of 2·27 and 1·81 kg/day for the two breeds respectively. Digestibility of protein tended to decline with decreasing concentration of protein in the diet as had been confirmed by Whitelaw, Preston and Ndumbe (1961) and Stobo (1964), but the highest N retention was obtained from the mixture containing 20·0 per cent crude protein. However, the most efficient use of protein was made by the calves given the 16·2 per cent mixture. Stobo, Roy and Gaston (1967b) confirmed the finding that 16 per cent crude protein (air-dry basis) in a concentrate mixture, either at restricted or *ad lib.* levels will supply the requirement of protein for maximum weight gain, but considered that after 8 weeks of age for Friesians, i.e. at a weight of 75 kg, a mixture containing 12 per cent crude protein would be sufficient.

(c) *Non-protein nitrogen.* Loosli and McCay (1943) showed that, whereas a diet containing 4·4 per cent protein equivalent did not support growth in 2-month-old calves, the addition of sufficient urea to raise the protein equivalent to 16·2 per cent resulted in an average growth rate of 0·45 kg/day between 2 and 4 months of age. At 3 months, the calves given the basal diet were in negative N balance, whilst those receiving the urea supplemented diet retained 24 to 36 per cent of dietary N. On addition of urea, apparent digestibility of dry matter was increased from 59 to 77 per cent and of carbohydrates from 62 to 77 per cent. Similarly, Brown *et al.* (1956) and Brown *et al.* (1960) indicated that calves weaned at 7 weeks of age and given a urea supplement to a basal concentrate mixture containing 6·7 per cent crude protein could utilize appreciable quantities of urea to meet their requirements for growth as early as the 6th week of life. Stobo, Roy and Gaston (1967c), on the other hand, with calves weaned at 5 weeks of age, found that the performance of calves given hay and a concentrate mixture containing 18·0 per cent crude protein, of which 33 per cent was in the form of urea, was inferior to that of calves given a concentrate mixture containing 20·2 per cent crude protein and only slightly better than the performance of calves given a concentrate mixture containing 12·0 per cent crude protein. Digestibility of the various fractions were increased by urea addition, but the digestibility of crude fibre was only slightly improved by urea addition. As later work by Stobo, Roy and Gaston (1967 a, b) showed that 12 per cent crude protein in a concentrate mixture was sufficient to supply the protein needs of calves of the weight used in the digestibility trials in their experiment on urea utilization, it is clear that the amount of N being supplied in the urea-supplemented mixture was excessive.

Roughage

In the absence of roughage, calves tend to bloat and, moreover, the voluntary intake of dry matter tends to be less (Lassiter *et al.*, 1960). This reduction in voluntary intake appears to be associated with acidosis as a result of the rapid production of acid in the rumen. Preston (1963) has suggested that the beneficial effect of hay may be in its high ash content or in the greater production of saliva concomitant with the increased rumination time on diets containing hay (Balch, 1958; Bailey, 1961). Nicholson, Loosli and Warner (1960) found that an alkaline mineral mixture was as effective as lucerne ash in increasing voluntary intake and weight gain in calves given a diet containing starch and ground maize cobs. It has been suggested by Harris (1962) that the low rumen pH in animals on all-concentrate diets may damage the mucosa and allow rumen bacteria to enter the blood stream and cause liver abscesses.

The amount of roughage that should be offered to calves depends on the growth rate required. All-roughage diets give very sub-optimal growth. For fattening calves to 6 months of age, an allowance of 1 kg hay/day is recommended by Nordfeldt and Skoog (1962).

Silage

Calves do not make satisfactory growth if weaned at 8 weeks onto a sole diet of silage of a low dry matter content, as such silage is too bulky (Roy and Shillam, 1954). In the U.S.A., calves of 3 months of age given silage *ad lib.* gained considerably more weight when the silage contained 32 per cent dry matter than when it contained 21 per cent (Newlander and Riddell, 1956).

Grass

Calves can be weaned onto good quality pasture at 5 weeks of age, but concentrates must be given until calves are at least 8 weeks of age. A number of experiments have been made in which calves have been put out at pasture from birth or from 5 weeks of age, both in Great Britain (Roy, Shillam and Palmer, 1955; Preston, 1957; Preston and MacLeod, 1958; Castle and Walker,

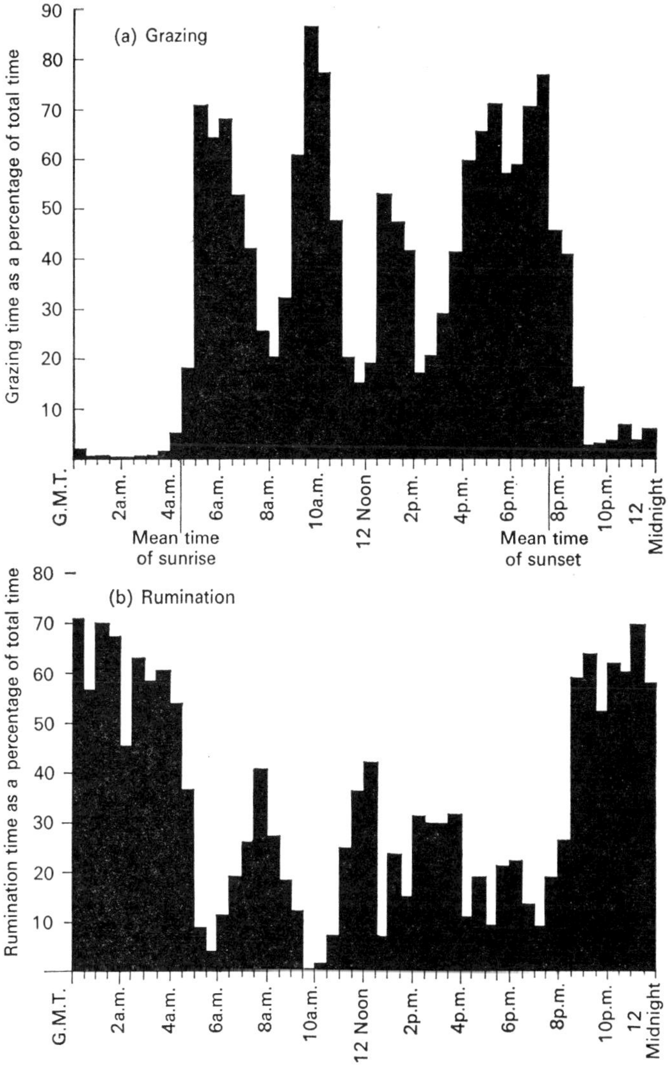

FIG. 2.—The distribution of grazing and rumination throughout the day (mean of five monthly observations).

1959; Chambers and Alder, 1960; Chambers, 1961 a, b), in New Zealand (McMeekan, 1954 a, b; McArthur, 1957 a, b) and in the U.S.A. (Noller *et al.*, 1959). In general, results are satisfactory providing maximum weight gains are not required. Most workers have found that weight gain decreased as the time of year, when calves were born and put outside, advanced from spring to midsummer. On poor quality pasture, supplementary feeding with cereals may be required after 8 weeks of age. Slightly greater weight gains have usually been found for calves put out to pasture at birth than for those put out at 5 weeks of age or later, even when given the same concentrate intake. Ayrshire calves weaned off milk at 5 weeks of age and reared on pasture from birth with an additional 2–3 kg/day barley have gained weight at the rate of 0·7 kg/day to 160 days (Roy and Shillam, 1961). Like adult ruminants, calves at pasture have a characteristic grazing and rumination pattern, as shown in Fig. 2. The time the calves spent grazing and ruminating during 24 hr at various ages is given in Table 15. Grazing seemed to start at about 5 a.m. G.M.T. irrespective of the time of sunrise and to end about 1 hr after sunset (Roy, Shillam and Palmer, 1955). One of the advantages suggested from rearing calves out of doors from birth is the development of immunity against internal parasites brought about by the ingestion of gradually increasing amounts of grass.

TABLE 15
Time Spent Grazing and Ruminating During 24 hr Periods by Calves at Pasture
(Roy, Shillam and Palmer, 1955)

Age (days)	Grazing (hr)	Ruminating (hr)
22*	1·3	1·2
51	5·6	6·8
79	8·9	7·8
115	9·4	9·0

*12 hr period only.

ANTIBIOTICS

The use of antibiotics in calf rearing is twofold; firstly, for controlling potential pathogens, particularly in the pre-ruminant calf, and secondly for their effect in increasing food intake in the young ruminant calf. With ideal management and facilities, the routine use of antibiotics in calf rearing should be unnecessary. The use of the tetracycline antibiotics at a level of 125 mg/day during the first 3 weeks of life reduced the mortality rate from *E. coli* localized intestinal infection in bought-in calves (Lang *et al.*, 1959). Moreover at a level of 250 mg/day for the first 5 days of life followed by 125 mg/day for the next

16 days, tetracycline antibiotics have been used as a substitute for colostrum if, owing to the death of the dam or owing to the removal of large volumes of prepartum secretion, colostrum is not available (Ingram *et al.*, 1958). The first oral dose should be given as soon after birth as possible, and for extra safeguard, 100 mg tetracycline should be injected intramuscularly at the same time.

Increased resistance to antibiotics of strains of *E. coli* isolated from calves that have received antibiotics has been shown to occur (Smith and Crabb, 1956 a, b; Ingram *et al.*, 1958). Moreover, these strains may pass to other calves in the community that have received no antibiotics. In the experiment of Ingram *et al.* (1958), which was made over a period of 3 months, the more resistant strains were isolated from calves, which had received no antibiotic, during the last half of the experiment only. No difference in resistance of strains of *E. coli* isolated from calves given chlortetracycline, whether they survived or died, was apparent. As no satisfactory method exists for differentiating between pathogenic and non-pathogenic strains, it was not known whether the resistant strains were greater potential pathogens than the susceptible ones. A recent survey by Glantz (1962) in the U.S.A., where tetracyclines have been used in calf foods for a number of years, showed, that, of fifty-six cultures from nine 'O' groups of *E. coli* isolated from calves, forty-seven and fifty-one cultures were insensitive (minimum inhibitory concentration >10 μg/ml) to chlortetracycline and oxytetracycline respectively, whereas only two were insensitive to furazolidone.

Procaine penicillin at a level of 250 mg/day for the first 5 days of life followed by 125 mg/day for the next 16 days of life had no protective value against *E. coli* septicaemia and in fact appeared to antagonize the beneficial effect of chlortetracycline if fed in combination with it (Ingram *et al.*, 1958).

Workers in Kansas (Radisson, Smith and Ward, 1956; Radisson *et al.*, 1956; MacFadden and Bartley, 1959) considered that the beneficial effect of antibiotics in the young calf resulted from their effect in making pathogenic bacteria more sensitive to phagocytosis.

Tetracycline incorporated in concentrate mixtures at a low level, so that the calf consumes about 15–80 mg antibiotic daily, has often been found to increase food intake and efficiency of food conversion (Lassiter, 1955). The routine feeding of antibiotics in concentrate mixtures to calves weaned onto dry food at 3 weeks of age has been considered essential for the success of the system (Preston, MacLeod and Dinda, 1959). The improvement in voluntary intake of food and in weight gain, resulting from their inclusion, is attributed by Preston (1963) to the reduction in the amount of fermentation and gas production in the rumen, with the resultant possibility that a greater proportion of the diet will be digested in the abomasum and intestine. Higher blood glucose levels in ruminant calves given chlortetracycline give some support to this theory (Murley, Allen and Jacobson, 1951; Hibbs, Conrad and

Pounden, 1952, 1954; Vandersall, Hibbs and Conrad, 1956; Everett *et al.*, 1958; Dinda, 1960).

REFERENCES

Abelein, R. and Ritter, H. C. (1959) *Mitt. bayer. Landesanst. Tierzucht* **7**, 145.

Agricultural Research Council (1965) *The Nutrient Requirements of Farm Livestock*, No. 2. *Ruminants*. Technical reviews and summaries. H.M.S.O., London.

Adams, R. S., Gullickson, T. W., Gander, J. E. and Sautter, J. H. (1959) *J. Dairy Sci.* **42**, 1552.

Ahmed, I. A. and Tantawy, A. O. (1954) *Alex. J. Agric. Res.* **2**, 1.

Allcroft, R. (1950) *J. Comp. Path. Ther.* **60**, 190.

Allcroft, R. and Blaxter, K. L. (1950) *J. Comp. Path. Ther.* **60**, 209.

Anantakrishnan, C. P. and Lazarus, A. J. (1953) *Indian J. Dairy Sci.* **6**, 23.

Andersen, H. (1962) *Årsberetn. Inst. Sterilitetsforskn. K. Vet- og Landbohøgsk. Kbh.* p. 155.

Andersen, H. and Plum, M. (1965) *J. Dairy Sci.* **48**, 1224.

Anonymous (1952) *Rep. Dep. Agric. N.Z.* p. 15.

Anonymous (1953) *Rep. Dep. Agric. N.Z.* p. 43.

Anonymous (1956) *Rep. Dep. Agric. N.Z.* p. 58.

Anonymous (1957) *Veterinariya* **34**, 54.

Anonymous (1958) *Rep. Dep. Agric. N.Z.* p. 134.

Anonymous (1965) *Dairy Fmr, Ipswich* **12**, No. 7, 78.

Anthony, W. B., Couch, J. R., Rupel, I. W., Henderson, M. B. and Brown, C. (1951) *J. Dairy Sci.* **34**, 749.

Archibald, J. G. (1947) *J. Dairy Sci.* **30**, 293.

Archibald, J. G. and Bennett, E. (1935) *J. Agric. Res.* **51**, 83.

Armsby, H. P. (1922) *The Nutrition of Farm Animals.* MacMillan, New York.

Armstrong, D. G., Preston, T. R. and Armstrong, R. H. (1954) *Nature, Lond.* **174**, 1182.

Arnold, P. T. D. and Becker, R. B. (1953) *Bull. Fla Agric. Exp. Stn* No. 529.

Arunchalam, T. V., Lazarus, A. J. and Anantakrishnan, C. P. (1952) *Indian J. Dairy Sci.* **5**, 117.

Aschaffenburg, R., Bartlett, S., Kon, S. K., Terry, P., Thompson, S. Y., Walker, D. M., Briggs, C., Cotchin, E. and Lovell, R. (1949) *Br. J. Nutr.* **3**, 187.

Aschaffenburg, R., Bartlett, S., Kon, S. K., Roy, J. H. B., Sears, H. J., Thompson, S. Y., Ingram, P. L., Lovell, R. and Wood, P. C. (1953) *Br. J. Nutr.* **7**, 275.

Ash, R. W. (1964) *J. Physiol., Lond.* **172**, 425.

Asker, A. A. and Ragab, M. T. (1952) *Indian J. Vet. Sci.* **22**, 265.

Atkeson, F. W., Warren, T. R. and Anderson, G. C. (1934) *J. Dairy Sci.* **17**, 249.

Bachner, F. (1951) *Tierzüchter* **3**, 368.

Bailey, C. B. (1961) *Br. J. Nutr.* **15** ,443.

Bailey, G. L. and Broster, W. H. (1957) *J. Agric. Sci., Camb.* **49**, 435.

Baker, J. M., Gehrke, C. W. and Affsprung, H. E. (1954) *J. Dairy Sci.* **37**, 643.

Balch, C. C. (1958) *Br. J. Nutr.* **12**, 330.

Balch, C. C., Campling, R. C., Johnson, V. W. and Roy, J. (1960) *Br. J. Nutr.* **14**, 379.

Balfour, W. E. and Comline, R. S. (1962) *J. Physiol., Lond.* **160**, 234.

Bangham, D. R., Ingram, P. L., Roy, J. H. B., Shillam, K. W. G. and Terry, P. J. (1958) *Proc. Roy. Soc. B* **148**, 184.

Barcroft, H. (1944) *Proc. Nutr. Soc.* **2**, 14.

Barker, H. B., Wise, G. H. and Jacobson, N. L. (1952) *J. Dairy Sci.* **35**, 507.

Bate, W., Espe, D. and Cannon, C. Y. (1946) *J. Dairy Sci.* **29**, 41.

Bechdel, S. I., Hilston, N. W., Guerrant, N. B. and Dutcher, R. A. (1938) *Bull. Pa Agric. Exp. Stn* No. 364.

Bechtel, H. E., Huffman, C. F., Duncan, C. W. and Hoppert, C. A. (1936) *J. Dairy Sci.* **19**, 359.

Beeson, W. M., Bolin, D. W., Hickman, C. and Johnson, R. F. (1941) *Bull. Idaho Agric. Exp. Stn* No. 240.

BELENKI, N. G. (1953) *Veterinariya* **30**, 25.
BELL, J. M. (1958) *Can. J. Anim. Sci.* **38**, 103.
BENEVENGA, N. J. and RONNING, M. (1963) *J. Anim. Sci.* **22**, 832.
BENTLEY, O. G. and PHILLIPS, P. H. (1951) *J. Dairy Sci.* **34**, 396.
BENZIE, D. and PHILLIPSON, A. T. (1957) *The Alimentary Tract of the Ruminant.* Oliver & Boyd, Edinburgh.
BERGMAN, E. N. and SELLERS, A. F. (1953) *Am. J. Vet. Res.* **14**, 520.
BERGMAN, E. N. and SELLERS, A. F. (1954) *Am. J. Vet. Res.* **15**, 25.
BERNADONI, E. A. and TUCKEY, S. L. (1950) *J. Dairy Sci.* **33**, 409.
BERRIDGE, N. J., DAVIS, J. G., KON, P. M., KON, S. K. and SPRATLING, F. R. (1942-4) *J. Dairy Res.* **13**, 145.
BLACKBURN, P. S., BLAXTER, K. L. and CASTLE, E. J. (1957) *Proc. Nutr. Soc.* **16**, xvi.
BLACKWOOD, J. H., MORRIS, S. and WRIGHT, N. C. (1936) *J. Dairy Res.* **7**, 228.
BLAXTER, K. L. (1944) *J. Agric. Sci., Camb.* **34**, 27.
BLAXTER, K. L. (1952) *Br. J. Nutr.* **6**, 12.
BLAXTER, K. L. (1959) In *Scientific Principles of Feeding Farm Live Stock.* Farmer & Stockbreeder Publications, London.
BLAXTER, K. L. (1962) *The Energy Metabolism of Ruminants.* Hutchinson, London.
BLAXTER, K. L. and BROWN, F. (1952) *Nutr. Abstr. Rev.* **22**, 1.
BLAXTER, K. L., BROWN, F. and MACDONALD, A. M. (1953) *Br. J. Nutr.* **7**, 287.
BLAXTER, K. L., BROWN, F., WOOD, W. A. and MACDONALD, A. M. (1953) *Br. J. Nutr.* **7**, 337.
BLAXTER, K. L., GRAHAM, N. MCC. and WAINMAN, F. W. (1956) *Br. J. Nutr.* **10**, 69.
BLAXTER, K. L. and HOWELLS, A. (1951) *Br. J. Nutr.* **5**, 25.
BLAXTER, K. L., HUTCHESON, M. K., ROBERTSON, J. M. and WILSON, A. L. (1952) *Br. J. Nutr.* **6**, i.
BLAXTER, K. L., MCCALLUM, E. S. R., WILSON, R. S., SHARMAN, G. A. M. and DONALD, L. G. (1961) *Proc. Nutr. Soc.* **20**, vi.
BLAXTER, K. L. and MCGILL, R. F. (1955) *Vet. Revs Annot.* **1**, 91.
BLAXTER, K. L. and MITCHELL, H. H. (1948) *J. Anim. Sci.* **7**, 351.
BLAXTER, K. L. and ROOK, J. A. F. (1954) *J. Comp. Path. Ther.* **64**, 176.
BLAXTER, K. L., ROOK, J. A. F. and MACDONALD, A. M. (1954) *J. Comp. Path. Ther.* **64**, 157.
BLAXTER, K. L. and SHARMAN, G. A. M. (1953) *Nature, Lond.* **172**, 1006.
BLAXTER, K. L. and SHARMAN, G. A. M. (1955) *Vet. Rec.* **67**, 108.
BLAXTER, K. L., SHARMAN, G. A. M. and MACDONALD, A. M. (1957) *Br. J. Nutr.* **11**, 234.
BLAXTER, K. L., WATTS, P. S. and WOOD, W. A. (1952) *Br. J. Nutr.* **6**, 125.
BLAXTER, K. L. and WOOD, W. A. (1951a) *Br. J. Nutr.* **5**, 11.
BLAXTER, K. L. and WOOD, W. A. (1951b) *Br. J. Nutr.* **5**, 29.
BLAXTER, K. L. and WOOD, W. A. (1951c) *Br. J. Nutr.* **5**, 55.
BLAXTER, K. L. and WOOD, W. A. (1952a) *Br. J. Nutr.* **6**, 1.
BLAXTER, K. L. and WOOD, W. A. (1952b) *Br. J. Nutr.* **6**, 56.
BLAXTER, K. L. and WOOD, W. A. (1953) *Vet. Rec.* **65**, 889.
BLAXTER, K. L., WOOD, W. A. and MACDONALD, A. M. (1953) *Br. J. Nutr.* **7**, 34.
BONADONNA, T. and VALERANI, L. (1946) *Zootec. Vet.* **1**, 129, 274.
BONSMA, J. C. (1949) *J. Agric. Sci., Camb.* **39**, 204.
BOOGAERDT, J. and VAN KOETSVELD, E. E. (1961) *Tijdschr. Diergeneesk.* **86**, 1287.
BOYER, P. D., PHILLIPS, P. H., LUNDQUIST, N. S., JENSEN, C. W. and RUPEL, I. W. (1942) *J. Dairy Sci.* **25**, 433.
BRAUDE, R. and WALKER, D. M. (1949) *J. Agric. Sci., Camb.* **39**, 156.
BRINKS, J. S., CLARK, R. T., RICE, F. J. and KIEFFER, N. M. (1961) *J. Anim. Sci.* **20**, 363.
BRISSON, G. J., CUNNINGHAM, H. M. and HASKELL, S. R. (1957) *Can. J. Anim. Sci.* **37**, 157.
BRISSON, G. J. and SUTTON, T. S. (1951) *J. Dairy Sci.* **34**, 28.
BRODY, S. (1945) *Bioenergetics and Growth.* Reinhold Publishing Corporation, New York.
BROWN, D. L. (1959) *Emp. J. Exp. Agric.* **27**, 276.
BROWN, L. D., JACOBSON, D. R., EVERETT, J. P., SEATH, D. M. and RUST, J. W. (1960) *J. Dairy Sci.* **43**, 1313.

BROWN, L. D., LASSITER, C. A., EVERETT, J. P. and RUST, J. W. (1956) *J. Anim. Sci.* **15**, 1125.
BROWN, L. D., LASSITER, C. A., EVERETT, J. P., SEATH, D. M. and RUST, J. W. (1958) *J. Dairy Sci.* **41**, 1425.
BROWNLEE, A. (1954) *Br. Vet. J.* **110**, 48.
BROWNLEE, A. (1956) *Br. Vet. J.* **112**, 369.
BRUGGEMANN, H. (1958) *Züchtungskunde* **30**, 179.
BRUGGEMANN, J. and BARTH, K. (1959) *Z. Tierphysiol. Tierernähr.* **14**, 284.
BRYANT, M. P. and SMALL, N. (1956) *J. Dairy Sci.* **39**, 927.
BUJATTI, P. G. (1956) *Riv. Zootec., Firenze* **29**, 264.
BURRIS, M. J. and BLUNN, C. T. (1952) *J. Anim. Sci.* **11**, 34.
BUSH, L. J., SCHUH, J. D., TENNILLE, N. B. and WALLER, G. R. (1963) *J. Dairy Sci.* **46**, 703.
CASTLE, M. E. and WALKER, R. F. R. (1959) *J. Br. Grassld Soc.* **14**, 88.
CHALMERS, M. I., CUTHBERTSON, D. P. and SYNGE, R. L. M. (1954) *J. Agric. Sci., Camb.* **44**, 254.
CHAMBERS, D. T. (1961a) *J. Agric. Sci., Camb.* **57**, 71.
CHAMBERS, D. T. (1961b) *J. Agric. Sci., Camb.* **57**, 77.
CHAMBERS, D. T. and ALDER, F .E. (1960) *J. Br. Grassld Soc.* **15**, 302.
CHAUDHURI, A. C. and SINHA, C. (1951) *Indian J. Vet. Sci.* **21**, 69.
CHRISTIANSON, G., JENNESS, R. and COULTER, S. T. (1954) *Analyt. Chem.* **26**, 1923.
COLLINS, R. A., HARPER, A. E., SCHREIBER, M. and ELVEHJEM, C. A. (1951) *J. Nutr.* **43**, 313.
COLOVOS, N. F., KEENER, H. A., TEERI, A. E. and DAVIS, H. A. (1951) *J. Dairy Sci.* **34**, 735.
COMAR, C. L., DAVIS, G. K. and SINGER, L. (1948) *J. Biol. Chem.* **174**, 905.
COMAR, C. L., MONROE, R. A., VISEK, W. J. and HANSARD, S. L. (1953) *J. Nutr.* **50**, 459.
COMBERG, G. and GOLLNITZ, L. (1959) *Tierzucht* **13**, 150.
COMLINE, R. S., ROBERTS, H. E. and TITCHEN, D. A. (1951a) *Nature, Lond.* **167**, 561.
COMLINE, R. S., ROBERTS, H. E. and TITCHEN, D. A. (1951b) *Nature, Lond.* **168**, 84.
COMLINE, R. S. and TITCHEN, D. A. (1951) *J. Physiol., Lond.* **115**, 210.
CONVERSE, H. T. (1954) *Tech. Bull. U.S. Dep. Agric.* No. 1092.
CONVERSE, H. T. and MEIGS, E. B. (1938) *J. Dairy Sci.* **21**, A114.
CREECH, G. T. and SEIBOLD, H. R. (1943) *Am. J. Vet. Res.* **4**, 353.
CRICHTON, J. A., AITKEN, J. N. and BOYNE, A. W. (1959) *Anim. Prod.* **1**, 145.
CRICHTON, J. A., AITKEN, J. N. and BOYNE, A. W. (1960a) *Anim. Prod.* **2**, 45.
CRICHTON, J. A., AITKEN, J. N. and BOYNE, A. W. (1960b) *Anim. Prod.* **2**, 159.
CUNNINGHAM, H. M. and BRISSON, G. J. (1957) *Can. J. Anim. Sci.* **37**, 152.
CUNNINGHAM, H. M. and LOOSLI, J. K. (1954) *J. Dairy Sci.* **37**, 453.
CUNNINGHAM, O. C. and ADDINGTON, L. H. (1938) *Bull. New Mex. Agric. Exp. Stn* No. 257.
DAVE, C. N. (1950) *Poona Agric. Coll. Mag.* **40**, 12.
DAVEY, A. W. F. (1962) *N.Z. J. Agric. Res.* **5**, 460.
DAVIS, D. T. and WHITE, J. C. D. (1959) *Int. Dairy Congr. XV London* **3**, 1677.
DAVIS, H. P. and HATHAWAY, I. L. (1955) *Res. Bull. Neb. Agric. Exp. Stn* No. 177.
DAVIS, H. P. and HATHAWAY, I. L. (1956) *Res. Bull. Neb. Agric. Exp. Stn* No. 179.
DAVIS, H. P. and HATHAWAY, I. L. (1959) *Res. Bull. Neb. Agric. Exp. Stn* No. 189.
DAVIS, H. P., PLUM, M. and BROST, B. (1954) *J. Dairy Sci.* **37**, 162.
DAVIS, L. R., AUTRY, K. M., HERLICH, H. and HAWKINS, G. E. Jr. (1954) *J. Dairy Sci.* **37**, 562.
DAWSON, W. M., PHILLIPS, R. W. and BLACK, W. H. (1947) *J. Anim. Sci.* **6**, 247.
DE MAN, T. J. (1951) *Tijdschr. Diergeneesk.* **76**, 175.
DEUTSCH, H. F. and SMITH, V. R. (1957) *Am. J. Physiol.* **191**, 271.
DICK, A. T. (1953) *Aust. Vet. J.* **29**, 233.
DILLS, W. L. and NELSON, J. M. (1942) *J. Am. Chem. Soc.* **64**, 1616.
DINDA, P. K. (1960) Some effects of chlortetracycline on the nutrition of the early weaned calf. Ph.D. Thesis. University of Aberdeen.
DINKHAUSER, F., TRAMPEN, K. and BERGMANN, H. (1944) *Züchtungskunde* **19**, 23.
DOLLAR, A. M. (1958) Biochemical studies of nutrition in the ruminant. Ph.D. Thesis. University of Reading.
DOLLAR, A. M. and PORTER, J. W. G. (1957) *Nature, Lond.* **179**, 1299.
DOLLAR, A. M. and PORTER, J. W. G. (1959) *Int. Dairy Congr. XV London* **1**, 185.

DONALD, H. P., RUSSELL, W. S. and TAYLOR, ST. C. S. (1962) *J. Agric. Sci., Camb.* **58,** 405.
DRAPER, H. H. and JOHNSON, B. C. (1952) *J. Nutr.* **46,** 37.
DUMONT, B. L., AURIOL, P. and DUPLAN, J. M. (1959) *C. R. hebd. Séanc Acad. Agric. Fr.* **45**, 111.
DUNCAN, C. W. and HUFFMAN, C. F. (1934) *J. Dairy Sci.* **17,** 83.
DYE, J. A. and ORSINI, D. (1952) *Fed. Proc.* **11**, 39.
EADIE, J. M., HOBSON, P. N. and MANN, S. O. (1959) *Nature, Lond.* **183**, 624.
EATON, H. D., ROUSSEAU, J. E. Jr., DICKS, M. W., TEICHMAN, R. and GRIFO, A. P. Jr. (1958) *J. Dairy Sci.* **41**, 1456.
EATON, H. D., SPEILMAN, A. A., LOOSLI, J. K., THOMAS, J. W., NORTON, C. L. and TURK, K. L. (1947) *J. Dairy Sci.* **30**, 787.
ECKLES, C. H. (1918) *Res. Bull. Mo. Agric. Exp. Stn* No. 35.
ECKLES, C. H. (1919) *J. Dairy Sci.* **2**, 159.
ELLENBERGER, H. B., NEWLANDER, J. A. and JONES, C. H. (1950) *Bull. Vt Agric. Exp. Stn* No. 558.
ELLIS, G. H. and THOMPSON, J. F. (1945) *Industr. Engng Chem. Analyt. Edn* **17**, 254.
ELY, R. E., DUNN, K. M. and HUFFMAN, C. F. (1948) *J. Anim. Sci.* **7**, 239.
EVERETT, J. P. Jr., BROWN, L. D., LASSITER, C. A., JACOBSON, D. R. and RUST, J. W. (1958) *J. Dairy Sci.* **41**, 1407.
ESPE, D. L. and CANNON, C. Y. (1935) *J. Dairy Sci.* **18**, 141.
FAULKNER, D. E. and BROWN, J. D. (1953) *Publs Colon. Advis. Coun. Agric.* No. 3.
FITCH, J. B., MCGILLIARD, P. C. and DRUMM, G. M. (1924) *J. Dairy Sci.* **7**, 222.
FIELD, H. I. (1953) *St. Vet. J.* **8**, 1.
FILMER, J. F. and UNDERWOOD, E. J. (1937) *Aust. Vet. J.* **13**, 57.
FISHER, E. W. (1965) *Br. Vet. J.* **121**, 132.
FLATT, W. P., WARNER, R. G. and LOOSLI, J. K. (1958) *J. Dairy Sci.* **41**, 1593.
FLATT, W. P., WARNER, R. G. and LOOSLI, J. K. (1959) *Mem. Cornell Univ. Agric. Exp. Stn* No. 361.
FLIPSE, R. J., HUFFMAN, C. F., DUNCAN, C. W. and WEBSTER, H. D. (1950) *J. Dairy Sci.* **33**, 557.
FLUX, D. S., BUTLER, G. W., JOHNSON, J. M., GLENDAY, A. C. and PETERSEN, G. B. (1956) *N.Z. J. Sci. Technol.* **38A**, 88.
FLUX, D. S. and PATCHELL, M. R. (1956) *N.Z. J. Sci. Technol.* **38A**, 20.
FOOTE, W. D., HAUSER, E. R. and CASIDA, L. E. (1960) *J. Anim. Sci.* **19**, 470.
FOOTE, W. D., TYLER, W. J. and CASIDA, L. E. (1959) *J. Dairy Sci.* **42**, 305.
FORD, J. E. and GREGORY, M. E. (1964) *Rep. Natn. Inst. Res. Dairy.* p. 123.
FREER, M. and CAMPLING, R. C. (1963) *Br. J. Nutr.* **17**, 79.
FRENS, A. M. (1961) *Tijdschr. Diergeneesk.* **86**, 1636.
FRENS, A. M., GRIFT, J. VAN DER and DAMMERS, J. (1961) *Tijdschr. Diergeneesk.* **86**, 255.
GARDNER, K. E. (1952) *J. Dairy Sci.* **35**, 491.
GARNER, R. J. and CRAWLEY, W. (1958) *J. Comp. Path. Ther.* **68**, 112.
GARRETT, O. F. and OVERMAN, O. R. (1940) *J. Dairy Sci.* **23**, 13.
GLANTZ, P. J. (1962) *Cornell Vet.* **52**, 552.
GONZALEZ-JIMENEZ, E. and BLAXTER, K. L. (1962) *Br. J. Nutr.* **16**, 199.
GOULD, I. A. Jr. and SOMMER, H. H. (1939) *Tech. Bull. Mich. Agric. Exp. Stn* No. 164.
GREGORY, K. E., BLUNN, C. T. and BAKER, M. L. (1950) *J. Anim. Sci.* **9**, 338.
GREGORY, M. E., FORD, J. E. and KON, S. K. (1958) *J. Dairy Res.* **25**, 447.
GRIFO, A. P. Jr., ROUSSEAU, J. E. Jr., EATON, H. D. and GOSSLEE, D. G. (1960) *J. Dairy Sci.* **43**, 1003.
GRIMBERGEN, A. H. M. (1955) *Inst. Moderne Veevoeding 'De Schothorst'* S42, 8.
GROSSMAN, J. D. (1949) *J. Am. Vet. Med. Ass.* **114**, 416.
GROSSMAN, J. and DE OLIVEIRA, W. M. (1949) *Boln. Dir. Prod. Anim.* **5**, 11.
GUILBERT, H. R. and HART, G. H. (1935) *J. Nutr.* **10**, 409.
GULLICKSON, T. W., FOUNTAINE, F. C. and FITCH, J. B. (1942) *J. Dairy Sci.* **25**, 117.
GULLICKSON, T. W., PALMER, L. S. and BOYD, W. L. (1935) *Tech. Bull. Minn. Agric. Exp. Stn* No. 105.

HAIGH, L. D., MOULTON, C. R. and TROWBRIDGE, P. F. (1920) *Res. Bull. Mo. Agric. Exp. Stn* No. 38.
HAMMOND, J. (1944) *Proc. Nutr. Soc.* **2**, 8.
HANSARD, S. L., COMAR, C. L. and PLUMLEE, M. P. (1954) *J. Anim. Sci.* **13**, 25.
HANSARD, S. L., CROWDER, H. M. and LYKE, W. A. (1957) *J. Anim. Sci.* **16**, 437
HANSARD, S. L. and FOOTE, L. E. (1961) *Fed. Proc.* **20**, 296.
HANSEN, R. G., PHILLIPS, P. H. and RUPEL, I. W. (1946) *J. Dairy Sci.* **29**, 761.
HANSEN, R. G., PHILLIPS, P. H., WILLIAMS, J. W. and SMITH, V. R. (1946) *J. Dairy Sci.* **29**, 521.
HANSSON, A. (1956) *Proc. Br. Soc. Anim. Prod.* p. 51.
HARDISON, W. A., MILLER, G. A. and GRAF, G. C. (1957) *J. Dairy Sci.* **40**, 363.
HARMAN, T. D. and SLATTER, W. L. (1950) *J. Dairy Sci.* **33**, 409.
HARRIS, A. H. (1962) *Vet. Rec.* **74**, 1434.
HARRIS, L. E. and LOOSLI, J. K. (1944) *J. Dairy Sci.* **27**, 650.
HARRIS, L. E., WORK, S. H. and HENKE, L. A. (1943) *J. Anim. Sci.* **2**, 328.
HARRISON, H. N., WARNER, R. G., SANDER, E. G., LOOSLI, J. K., SLACK, S. T. and TURK, K. L. (1960) *J. Dairy Sci.* **43**, 1084.
HART, E. B., HUMPHREY, G. C. and MORRISON, F. B. (1912) *J. Biol. Chem.* **13**, 133.
HASTINGS, C. C. (1942) *J. Am. Vet. Med. Ass.* **100**, 190.
HAWKINS, G. E., WISE, G. H., MATRONE, G., WAUGH, R. K. and LOTT, W. L. (1955) *J. Dairy Sci.* **38**, 536.
HAWKINS, G. M., ROY, J. H. B., SHILLAM, K. W. G., GREATOREX, J. C. and INGRAM, P. L. (1959) *Br. J. Nutr.* **13**, 447.
HEGLAND, R. B., LAMBERT, M. R., JACOBSON, N. L. and PAYNE, L. C. (1957) *J. Dairy Sci.* **40**, 1107.
HENRY, K. M. and KON, S. K. (1937) *Biochem. J.* **31**, 2199.
HENSCHEL, M. J., HILL, W. B. and PORTER, J. W. G. (1961) *Proc. Nutr. Soc.* **20**, xl.
HERMAN, H. A. (1936) *Res. Bull. Mo. Agric. Exp. Stn* No. 245.
HIBBS, J. W. and CONRAD, H. R. (1958) *J. Dairy Sci.* **41**, 1230.
HIBBS, J. W., CONRAD, H. R. and GALE, C. (1961) *J. Dairy Sci.* **44**, 1184.
HIBBS, J. W., CONRAD, H. R. and POUNDEN, W. D. (1952) *J. Anim. Sci.* **11**, 764.
HIBBS, J. W., CONRAD, H. R. and POUNDEN, W. D. (1954) *J. Dairy Sci.* **37**, 724.
HIBBS, J. W., CONRAD, H. R., POUNDEN, W. D. and FRANK, N. (1956) *J. Dairy Sci.* **39**, 171.
HIBBS, J. W., KRAUSS, W. E., MONROE, G. F. and POUNDEN, W. D. (1945) *J. Dairy Sci.* **28**, 525.
HIBBS, J. W., POUNDEN, W. D. and CONRAD, H. R. (1953) *J. Dairy Sci.* **36**, 717.
HILDER, R. A. and FOHRMAN, M. H. (1949) *J. Agric. Res.* **78**, 457.
HILGEMAN, M. and JENNESS, R. (1951) *J. Dairy Sci.* **34**, 483.
HILL, K. J. (1956) *Q. J. Exp. Physiol.* **41**, 421.
HOFLUND, S. (1946) *Svensk VetTidskr.* **51**, 88.
HOFLUND, S., HOLMBERG, J. and SELLMANN, G. (1956) *Cornell Vet.* **46**, 51.
HOGAN, A. G. and NIERMAN, J. L. (1927) *Res. Bull. Mo. Agric. Exp. Stn* No. 107.
HOGUE, D. E., WARNER, R. G., GRIPPEN, C. H. and LOOSLI, J. K. (1956) *J. Anim. Sci.* **15**, 788.
HOLLON, B. F., WAUGH, R. K., WISE, G. H. and SMITH, F. H. (1958) *J. Dairy Sci.* **41**, 286.
HOLM, L. W., RHODE, E. A., WHEAT, J. D. and FITCH, G. (1953) *J. Am. Vet. Med. Ass.* **123**, 528.
HOLTER, J. A. (1954) *Proc. Cornell Nutr. Conf.* p. 53.
HOMB, T. (1960) *Norg. LandbrHøgsk. Beretn. Føringsfors.* No. 99.
HOPKINS, D. T., WARNER, R. G. and LOOSLI, J. K. (1959) *J. Dairy Sci.* **42**, 1815.
HOPPER, J. H., GARDNER, K. E. and JOHNSON, B. C. (1954) *J. Dairy Sci.* **37**, 431.
HOSTETTLER, H. and STEIN, J. (1958) *Landw. Jb. Schweiz* **7**, 163.
HOUSTON, J., KON, S. K. and THOMPSON, S. Y. (1940) *J. Dairy Res.* **11**, 145.
HOWE, P. E. (1922) *J. Biol. Chem.* **53**, 479.
HUBER, J. T., HARTMAN, P. A., JACOBSON, N. L. and ALLEN, R. S. (1958) *J. Dairy Sci.* **41**, 743.
HUBER, J. T. and JACOBSON, N. L. (1960) *J. Dairy Sci.* **43**, 875.

HUBER, J. T., JACOBSON, N. L., ALLEN, R. S. and HARTMAN, P. A. (1961a) *J. Dairy Sci.* **44**, 1494.

HUBER, J. T., JACOBSON, N. L., MCGILLIARD, A. D., MORRILL, J. L. and ALLEN, R. S. (1961b) *J. Dairy Sci.* **44**, 1484.

HUFF, J. S., WAUGH, R. K. and WISE, G. H. (1951) *J. Dairy Sci.* **34**, 1056.

HUFFMAN, C. F., CONLEY, C. L., LIGHTFOOT, C. C. and DUNCAN, C. W. (1941) *J. Nutr.* **22**, 609.

HUFFMAN, C. F. and DUNCAN, C. W. (1935a) *J. Dairy Sci.* **18**, 432.

HUFFMAN, C. F. and DUNCAN, C. W. (1935b) *J. Dairy Sci.* **18**, 605.

HUFFMAN, C. F., DUNCAN, C. W., ROBINSON, C. A. and LAMB, L. W. (1933) *Tech. Bull. Mich. Agric. Exp. Stn* No. 134.

HUGHES, J. S. and CAVE, H. W. (1931) *J. Nutr.* **4**, 163.

HUNTER-SMITH, J. and GARDNER, H. W. (1928) *Agriculture, Lond.* **35**, 722.

HUTTON, J. T. and PATTON, S. (1952) *J. Dairy Sci.* **35**, 699.

INGRAM, P. L. (1962) Observations on the pathology and pathogenesis of experimental colibacillosis in calves. Ph.D. Thesis. University of London.

INGRAM, P. L., SHILLAM, K. W. G., HAWKINS, G. M. and ROY, J. H. B. (1958) *Br. J. Nutr.* **12**, 203.

ITZEROTT, A. G. (1942) *J. Aust. Inst. Agric. Sci.* **8**, 119.

JACOBSON, N. L., ALLEN, R. S. and BELL, M. R. (1951) *J. Anim. Sci.* **10**, 1050.

JACOBSON, N. L., BROWN, L. R. and RATCLIFF, L. (1959) *Proc. Distillers Feed Conf.* **14**, 10.

JACOBSON, N. L. and CANNON, C. Y. (1947) *J. Dairy Sci.* **30**, 587.

JACOBSON, N. L., CANNON, C. Y. and THOMAS, B. H. (1949) *J. Dairy Sci.* **32**, 429.

JANCARIK, A. (1962) *Sb. čsl. Akad. Zeměd Věd Živoč. Výr.* **7**, 71.

JEFFREY, J. E. and LANE, M. D. (1960) *J. Dairy Sci.* **43**, 1508.

JOHNSON, B. C., HAMILTON, T. S., MITCHELL, H. H. and ROBINSON, W. B. (1942) *J. Anim. Sci.* **1**, 236.

JOHNSON, B. C., HAMILTON, T. S., NEVENS, W. B. and BOLEY, L. E. (1948) *J. Nutr.* **35**, 137.

JOHNSON, B. C., HOPPER, J. H. and GARDNER, K. E. (1953) *J. Dairy Sci.* **36**, 599.

JOHNSON, B. C., MITCHELL, H. H., HAMILTON, T. S. and NEVENS, W. B. (1947) *Fed. Proc.* **6**, 410.

JOHNSTONE, F. A. (1944) *Fd Res.* **9**, 212.

JOHNSTONE-WALLACE, D. B. and KENNEDY, K. (1944) *J. Agric. Sci., Camb.* **34**, 190.

JORDAO, L. P. (1956) *Bolm Ind. Anim.* **15**, 23.

JORDAO, L. P. and ASSIS, F. DE P. (1947) *Bolm Ind. Anim.* **9**, 38.

JORDAO, L. P. and ASSIS, F. DE P. (1951) *Bolm Ind. Anim.* **12**, 45.

JOSEPHSON, D. V. and DOAN, F. J. (1939) *Milk Dlr* **29**, 35.

JOSHI, N. R. and PHILLIPS, R. W. (1953) *Zebu Cattle of India and Pakistan.* FAO Agricultural Studies No. 19. Rome.

JOUBERT, D. M. (1954) *J. Agric. Sci., Camb.* **44**, 5.

JOUBERT, D. M. (1959) *S. Afr. J. Sci.* **55**, 192.

JOUBERT, D. M. and BONSMA, F. N. (1957) *Scient. Bull. Dep. Agric. S. Afr.* No. 371.

JOUBERT, D. M. and HAMMOND, J. (1958) *J. Agric. Sci., Camb.* **51**, 325.

JOUSSELIN, W., LADRAT, J., CRAPLET, C. and BARRE, P. (1964) *Bull. Acad. vét. Fr.* **37**, 277.

KANNAN, A. and JENNESS, R. (1961) *J. Dairy Sci.* **44**, 808.

KASTELIC, J., BENTLEY, O. G. and PHILLIPS, P. H. (1950) *J. Dairy Sci.* **33**, 725.

KAZAKOVA, E. M. (1959) *Trudy̆ gos. Inst. éksp. Vet.* **22**, 272.

KEENER, H. A., BECHDEL, S. I., GUERRANT, N. B. and THORPE, W. T. S. (1942) *J. Dairy Sci.* **25**, 571.

KEENER, H. A., PERCIVAL, G. P., MORROW, K. S. and ELLIS, G. H. (1949) *J. Dairy Sci.* **32**, 527.

KESLER, E. M. and KNODT, C. B. (1951) *J. Dairy Sci.* **34**, 145.

KESLER, E. M. and KNODT, C. B. (1952) *Proc. Soc. Exp. Biol. Med.* **79**, 85.

KESLER, E. M., RONNING, M. and KNODT, C. B. (1951) *J. Anim. Sci.* **10**, 969.

KESLER, E. M. and WILSON, J. M. (1958) *Bull. Pa Agric. Exp. Stn* No. 634.

KESLER, E. M., WILSON, J. M. and MOORE, H. L. (1960) *Prog. Rep. Pa Agric. Exp. Stn* No. 216.

KEYES, E. A. (1944) *Bull. Pa Agric. Exp. Stn* No. 446, Suppl. No. 2, p. 4.

KLEISCH, J. and HORST, P. (1959) *Züchtungskunde* **31**, 68.

KNAPP, B. Jr., LAMBERT, W. V. and BLACK, W. H. (1940) *J. Agric. Res.* **61**, 277.

KNOOP, C. E., KRAUSS, W. E. and WASHBURN, R. G. (1935) *J. Dairy Sci.* **18**, 337.

KOHLI, M. L. and SURI, K. R. (1957) *Indian J. Vet. Sci.* **27**, 33.

KON, S. K., MCGILLIVRAY, W. A. and THOMPSON, S. Y. (1955) *Br. J. Nutr.* **9**, 244.

KON, S. K. and PORTER, J. W. G. (1951) *Rep. Natn. Inst. Res. Dairy.* p. 83.

KON, S. K. and PORTER, J. W. G. (1952) *Rep. Natn. Inst. Res. Dairy.* p. 78.

KON, S. K. and WATSON, M. B. (1937) *Biochem. J.* **31**, 223.

KRENN, J. (1937) *Mikrochemie* **23**, 149.

KRIZENECKY, J. and KUDLICKA, K. (1942) *Sb. čsl. Akad. Zeměd.* **18**, 129.

LANG, J. M., ROY, J. H. B., SHILLAM, K. W. G. and INGRAM, P. L. (1959) *Br. J. Nutr.* **13**, 463.

LANGHAM, R. F. and HUFFMAN, C. F. (1954) *Mich. St. Coll. Vet.* **14**, 122.

LAMBERT, M. R., JACOBSON, N. L., ALLEN, R. S. and BELL, M. R. (1955) *J. Dairy Sci.* **38**, 6.

LAMPITT, L. H. and BUSHILL, J. H. (1934) *Biochem. J.* **28**, 1305.

LARSEN, J. B. (1958) *Beretn. Forsøgslab.* No. 303.

LARSEN, L. H. and LARSEN, H. (1956) *Beretn. Forsøgslab.* No. 285.

LARSON, B. L. and JENNESS, R. (1950) *J. Dairy Sci.* **33**, 896.

LARSON, B. L. and KENDALL, K. A. (1957) *J. Dairy Sci.* **40**, 377.

LASKOWSKI, M. Jr. and LASKOWSKI, M. (1950) *Fed. Proc.* **9**, 194.

LASLEY, J. F., DAY, B. N. and COMFORT, J. E. (1961) *J. Anim. Sci.* **20**, 737.

LASSITER, C. A. (1955) *J. Dairy Sci.* **38**, 1102.

LASSITER, C. A., DUNCAN, C. W. and CHRISTIE, L. D. (1957) *Q. Bull. Mich. St. Univ. Agric. Exp. Stn* **40**, 282.

LASSITER, C. A., WARD, G. M., HUFFMAN, C. F., DUNCAN, C. W. and WEBSTER, H. D. (1953) *J. Dairy Sci.* **36**, 997.

LASSITER, J. W., WHITE, J. W., COOK, M. K., WARREN, E. P. and LANG, R. A. (1960) *J. Anim.Sci.* **19**, 1272.

LAWRENCE, J. M., HERRINGTON, B. L. and MAYNARD, L. A. (1946) *J. Nutr.* **32**, 73.

LAZARUS, A. J. and RANGASWAMY, M. C. (1950) *Indian J. Dairy Sci.* **3**, 46.

LEGAULT, C. R. and TOUCHBERRY, R. W. (1962) *J. Dairy Sci.* **45**, 1226.

LEITCH, I. and THOMSON, J. S. (1944-5) *Nutr. Abstr. Rev.* **14**, 197.

LENGEMANN, F. W. and ALLEN, N. N. (1955) *J. Dairy Sci.* **38**, 651.

LE PROVOST, F. (1965) *Les Saccharoglycérides.* Thesis. D.Vet.Med. Ecole Nationale Vétérinaire de Toulouse.

LEWIS, J. M. and WILSON, L. T. (1945) *J. Nutr.* **30**, 467.

LEWIS, R. C. and RALSTON, N. P. (1953) *J. Dairy Sci.* **36**, 363.

LINDSEY, J. B., ARCHIBALD, J. G. and NELSON, P. R. (1931) *J. Agric. Res.* **42**, 883.

LING, E. R., KON, S. K. and PORTER, J. W. G. (1961) In *Milk—the mammary gland and its secretion.* Editors Kon, S. K. and Cowie, A. T. Academic Press, New York and London.

LITTLEWOOD, R. W. (1935) *Agric. Live-stk India* **5**, 232.

LLOYD, L. E. and CRAMPTON, E. W. (1957) *J. Anim. Sci.* **16**, 377.

LOFGREEN, G. P. (1950) *J. Dairy Sci.* **33**, 379.

LOFGREEN, G. P. and KLEIBER, M. (1953) *J. Nutr.* **49**, 183.

LOFGREEN, G. P., KLEIBER, M. and LUICK, J. R. (1952) *J. Nutr.* **47**, 571.

LOFGREEN, G. P., LOOSLI, J. K. and MAYNARD, L. A. (1951) *J. Anim. Sci.* **10**, 171.

LONG, T. A., TILLMAN, A. D., NELSON, A. B., GALLUP, W. D. and DAVIS, B. (1957) *J. Anim. Sci.* **16**, 444.

LOOSLI, J. K. (1963) *Proc. World Conf. Anim. Prod., Rome* **2**, 355.

LOOSLI, J. K. and MCCAY, C. M. (1943) *J. Nutr.* **25**, 197.

LUECKE, R. W., DUNCAN, C. W. and ELY, R. E. (1947) *Archs Biochem.* **13**, 277.

LUNDQUIST, N. S. and PHILLIPS, P. H. (1942) *J. Dairy Sci.* **25**, 685.

LUNDQUIST, N. S. and PHILLIPS, P. H. (1943) *J. Dairy Sci.* **26**, 1023.

MACDONALD, M. A. and BELL, J. M. (1958) *Can. J. Anim. Sci.* **38**, 23.

MACFADDEN, D. L. and BARTLEY, E. E. (1959) *J. Anim. Sci.* **18**, 1378.

MADSEN, L. L. and EARLE, I. P. (1947) *J. Nutr.* **34**, 603.
MAHADEVAN, P. (1953) *Emp. J. Exp. Agric.* **21**, 55.
MAHADEVAN, P. (1955) *J. Dairy Sci.* **38**, 1231.
MAKELA, A. (1960) *Maatal. Aikakousk.* **32**, 161.
MANN, S. O. and OXFORD, A. E. (1955) *J. Gen. Microbiol.* **12**, 140.
MANN, S. O., MASSON, F. M. and OXFORD, A. E. (1954) *Br. J. Nutr.* **8**, 246.
MAPLESDEN, D. C. and LOOSLI, J. K. (1960) *J. Dairy Sci.* **43**, 645.
MARSTON, H. R. (1952) *Physiol. Rev.* **32**, 66.
MARTIN, W. G., RAMSEY, H. A., MATRONE, G. and WISE, G. H. (1959) *J. Dairy Sci.* **42**, 1377.
MASON, I. L. (1957) *A World Dictionary of Breeds, Types and Varieties of Livestock with Supplement.* Commw. Agric. Bur., Farnham Royal.
MATRONE, G., CONLEY, C., WISE, G. H. and WAUGH, R. K. (1957) *J. Dairy Sci.* **40**, 1437.
MATTHEWS, C. A. and FOHRMAN, M. H. (1954a) *Tech. Bull. U.S. Dep. Agric.* No. 1098.
MATTHEWS, C. A. and FOHRMAN, M. H. (1954b) *Tech. Bull. U.S. Dep. Agric.* No. 1099.
MCARTHUR, A. T. G. (1957a) *N.Z. J. Sci. Technol.* **38A**, 696.
MCARTHUR, A. T. G. (1957b) *N.Z. J. Sci. Technol.* **38A**, 700.
MCCANDLESS, E. L. and DYE, J. A. (1950) *Am. J. Physiol.* **162**, 434.
MCCANDLISH, A. C. (1922) *J. Dairy Sci.* **5**, 301.
MCCARTHY, R. D. and KESLER, E. M. (1956) *J. Dairy Sci.* **39**, 1280.
MCMEEKAN, C. P. (1954a) *N.Z. J. Agric.* **88**, 306.
MCMEEKAN, C. P. (1954b) *N.Z. J. Agric.* **88**, 481.
MCNUTT, S. H. and WALL, J. F. (1938) *Vet. Med.* **23**, 497.
MEADE, J. H. Jr., DOLLAHON, J. C., TAYLOR, J. C. and LINDLEY, C. E. (1959) *J. Anim. Sci.* **18**, 1149.
MIHAL, L. (1956) *Rad. poljopr-sumarsk. Fak.* **6-7**, 67.
MILLER, J. K. and MILLER, W. J. (1960) *J. Dairy Sci.* **43**, 1854.
MITCHELL, H. H. (1929) *Bull. Natn. Res. Coun. Wash.* No. 67.
MITCHELL, H. H. (1947) Quoted by L. A. Maynard (1947) *Animal Nutrition.* 2nd ed. McGraw-Hill, New York.
MITCHELL, H. H. (1962) *Comparative Nutrition of Man and Domestic Animals* Vol. 1. Academic Press, New York and London.
MITCHELL, H H. and MCCLURE, F. J. (1937) *Bull. Natn. Res. Coun. Wash.* No. 99.
MOORE, L. A. (1939) *J. Nutr.* **17**, 443.
MOORE, L. A., BERRY, M. H. and SYKES, J. F. (1943) *J. Nutr.* **26**, 649.
MOORE, L. A., HUFFMAN, C. F. and DUNCAN, C. W. (1935) *J. Nutr.* **9**, 533.
MOORE, L. A. and SYKES, J. F. (1940) *Am. J. Physiol.* **130**, 684.
MOORE, L. A. and SYKES, J. F. (1941) *Am. J. Physiol.* **134**, 436.
MOORE, L. A., SYKES, J. F., JACOBSON, W. C. and WISEMAN, H. G. (1948a) *J. Dairy Sci.* **31**, 533.
MOORE, L. A., THOMAS, J. W., JACOBSON, W. C., MELIN, C. G. and SHEPHERD, J. B. (1948b) *J. Dairy Sci.* **31**, 489.
MOULTON, C. R., TROWBRIDGE, P. F. and HAIGH, L. D. (1922) *Res. Bull. Mo. Agric. Exp. Stn* No. 55.
MURLEY, W. R., ALLEN, R. S. and JACOBSON, N. L. (1951) *J. Anim. Sci.* **10**, 1057.
MURLEY, W. R., JACOBSON, N. L., WISE, G. H. and ALLEN, R. S. (1949) *J. Dairy Sci.* **32**, 609.
MURLEY, W. R., MOCHRIE, R. D., EDWARDS, J. R. and HOLLON, B. F. (1957) *J. Dairy Sci.* **40**, 1390.
NEAL, W. M., BECKER, R. B. and SHEALY, A. L. (1931) *Science, N.Y.* **74**, 418.
NEGRUTUI, E., PAP, S., MECEA, P., POP, A., BALOGH, G., EGRY, L. and BALTAN, G. (1958) *Studii Cerc. Agron. Cluj.* **9**, 171.
NELMS, G. E. and BOGART, R. (1956) *J. Anim. Sci.* **15**, 662.
NEWLANDER, J. A. and RIDDELL, W. H. (1956) *J. Dairy Sci.* **39**, 930.
NICHOLSON, J. W. G., LOOSLI, J. K. and WARNER, R. G. (1960) *J. Anim. Sci.* **19**, 1071.
NIEDERMEIER, R. P., ALLEN, N. N., LANCE, R. D., RUPNOW, E. H. and BRAY, R. W. (1959) *J. Anim. Sci.* **18**, 726.

NIELSEN, E. (1964) *Beretning Forsøgslab.* No. 345.
NOLLER, C. H., STILLIONS, M. C., CROWL, B. W., LUNDQUIST, N. S. and DELEZ, A. L. (1959) *J. Dairy Sci.* **42**, 1592.
NORDFELDT, S. and SKOOG, H. (1962) *K. Lantbrukshögsk. Statens Husdjursförsök Särtryck Forlandsmedd.* No. 153.
NOVIKOV, E. A., STARTSEV, D. I. and ARZUMANIAN, E. A. (1950) *Breed Improvement in Cattle Breeding.* State Publishing House for Agricultural Literature, Moscow.
OKAMOTO, M., THOMAS, J. W. and JOHNSON, T. L. (1959) *J. Dairy Sci.* **42**, 920.
OTTO, J. S. (1938) *Onderstepoort J. Vet. Sci.* **10**, 281.
OVSJANNIKOV, G. F. (1959) *Zhivotnovodstvo, Mosk.* **21**, 59.
OWEN, F. G., JACOBSON, N. L., ALLEN, R. S. and HOMEYER, P. G. (1958) *J. Dairy Sci.* **41**, 662.
OWEN, F. G., VOELKER, H. H., JACOBSON, N. L. and ALLEN, R. S. (1955) *J. Dairy Sci.* **38**, 891.
PALOHEIMO, L. and MAKELA, A. (1959) *Suom. Maatal. Seur. Julk.* **94**, 15.
PARRISH, D. B., WISE, G. H. and HUGHES, J. S. (1947) *J. Dairy Sci.* **30**, 849.
PARRISH, D. B., WISE, G. H., HUGHES, J. S. and ATKESON, F. W. (1948) *J. Dairy Sci.* **31**, 889.
PARRISH, D. B., WISE, G. H., HUGHES, J. S. and ATKESON, F. W. (1950a) *J. Dairy Sci.* **33**. 457.
PARRISH, D. B., WISE, G. H., LATSCHAR, C. E. and HUGHES, J. S. (1950b) *J. Nutr.* **40**, 193.
PEARSON, P. B. and DARNELL, A. L. (1946) *J. Nutr.* **31**, 51.
PELISSIER, C. L., SLACK, S. T., TRIMBERGER, G. W., TURK, K. L. and LOOSLI, J. K. (1954) *Fm Res.* **20**, No. 2.
PETROVIC, V. (1952) *God. poljopr. Fak. Zemun* **4**, 363.
PETTYJOHN, J. D., EVERETT, J. P. Jr. and MOCHRIE, R. D. (1963) *J. Dairy Sci.* **46**, 710.
PIERCE, A. E. (1962) In *Animal Health and Production*, p. 189. Butterworth, London.
POPE, G. S. and ROY, J. H. B. (1953) *Biochem. J.* **53**, 427.
PORTER, J. W. G. (1953) *Rep. Natn. Inst. Res. Dairy.* p. 88.
PORTER, J. W. G. (1963a) Personal communication.
PORTER, J. W. G. (1963b) *Rep. Natn. Inst. Res. Dairy.* p. 126.
POUNDEN, W. D. and HIBBS, J. W. (1948) *J. Dairy Sci.* **31**, 1051.
PRESTON, T. R. (1956) *Proc. Br. Soc. Anim. Prod.* p. 67.
PRESTON, T. R. (1957) *J. Br. Grassld Soc.* **12**, 178.
PRESTON, T. R. (1958) *Proc. Br. Soc. Anim. Prod.* p. 33.
PRESTON, T. R. (1963) *Wld Rev. Nutr. Diet.* **4**, 117.
PRESTON, T. R., ARCHIBALD, J. D. H. and TINKLER, W. (1957) *J. Agric. Sci., Camb.* **48**, 259.
PRESTON, T. R. and MACLEOD, N. A. (1958) *Agric. Rev.* **4**, 32.
PRESTON, T. R., MACLEOD, N. A. and DINDA, P. K. (1959) *Anim. Prod.* **1**, 13.
PRESTON, T. R. and NDUMBE, R. D. (1961) *Br. J. Nutr.* **15**, 281.
PRESTON, T. R. and WHITELAW, F. G. (1961) *Proc. VIIIth Int. Congr. Anim. Prod., Hamburg* **1**, 203.
PRESTON, T. R., WHITELAW, F. G., MACLEOD, N. A. and PHILLIP, E. B. (1965) *Anim. Prod.* **7**, 53.
PRESTON, T. R., WHITELAW, F. G., NDUMBE, R. D. and CHARLESON, E. B. (1960) *Proc. Nutr. Soc.* **19**, v.
RADISSON, J. J., BARTLEY, E. E., LORD, T. H. and SWENSON, M. J. (1956) *J. Dairy Sci.* **39** 1386.
RADISSON, J. J., SMITH, C. K. and WARD, G. M. (1956) *J. Dairy Sci.* **39**, 1260.
RATHORE, A. (1950-1) *Iowa St. Coll. J. Sci.* **25**, 345.
RAVEN, A. M. and ROBINSON, K. L. (1958) *Br. J. Nutr.* **12**, 469.
RAVEN, A. M. and ROBINSON, K. L. (1959) *Br. J. Nutr.* **13**, 178.
RAVEN, A. M. and ROBINSON, K. L. (1960) *Br. J. Nutr.* **14**, 135.
RAVEN, A. M. and ROBINSON, K. L. (1961) *Nature, Lond.* **192**, 1256.
RAVEN, A. M. and ROBINSON, K. L. (1964a) *J. Soc. Dairy Technol.* **17**, 5.
RAVEN, A. M. and ROBINSON, K. L. (1964b) *J. Sci. Fd Agric.* **4**, 214.

RAVEN, A. M. and ROBINSON, K. L. (1964c) *J. Sci. Fd Agric.* **4**, 219.
RAY, S. N. (1942) *Indian J. Vet. Sci.* **12**, 204.
REID, J. T., LOOSLI, J. K., TRIMBERGER, G. W., TURK, K. L., ASDELL, S. A. and SMITH, S. E. (1964) *Bull. Cornell Univ. Agric. Exp. Stn* No. 987.
REYNEKE, J., JOUBERT, D. M. and BONSMA, J. C. (1963) *Proc. S. Afr. Soc. Anim. Prod.* **2**, 107.
RICE, F. J., KELLEY, A. M. and LASLEY, J. F. (1954) *J. Anim. Sci.* **13**, 961.
RIEK, R. F. (1954) *Aust. Vet. J.* **30**, 29.
RIGGS, J. K., COLBY, R. W. and SELLS, L. V. (1953) *J. Anim. Sci.* **12**, 379.
RITCHEY, S. J., HOPPER, J., GARDNER, K. E. and JOHNSON, B. C. (1956) *J. Dairy Sci.* **39**, 1070.
RITCHIE, G. R. (1960) *Anim. Breed. Abstr.* **28**, 351.
RITZMAN, E. G. and BENEDICT, F. G. (1924) *Tech. Bull. New Hamps. Agric. Exp. Stn* No. 26.
RITZMAN, E. G. and COLOVOS, N. F. (1943) *Tech. Bull. New Hamps. Agric. Exp. Stn* No. 80.
RITZMAN, E. G., COLOVOS, N. F., KEENER, H. A. and TEERI, A. E. (1945) *Tech. Bull. New Hamps. Agric. Exp. Stn* No. 87.
ROBERTSON, A. (1950) Conference on the improvement of livestock under tropical conditions. Edinburgh.
ROGNONI, G. and PASTI, C. (1957) *Atti Soc. ital. Sci. vet.* **11**, 338.
ROLLERI, G. D., LARSON, B. L. and TOUCHBERRY, R. W. (1956) *J. Dairy Sci.* **39**, 1683.
RONNING, M., BANKSTON, B. G. and BEROUSEK, E. R. (1957) *Tech. Bull Okla. Agric. Exp. Stn* No. T68.
RONNING, M., BEROUSEK, E. R., KUHLMAN, A. H. and GALLUP, W. D. (1953) *J. Dairy Sci.* **36**, 52.
ROUSSEAU, J. E. Jr., EATON, H. D., HELMBOLDT, C. F., JUNGHERR, E. L., ROBRISH, S. A., BEALL, G. and MOORE, L. A. (1954) *J. Dairy Sci.* **37**, 889.
ROWLAND, S. J., ROY, J. H. B., SEARS, H. J. and THOMPSON, S. Y. (1953) *J. Dairy Res.* **20**, 16.
ROY, J. H. B. (1959a) In *Scientific Principles of Feeding Farm Live Stock.* Farmer & Stockbreeder Publications, London.
ROY, J. H. B. (1959b) *Outl. Agric.* **2**, 219.
ROY, J. H. B. (1964) *Vet. Rec.* **76**, 511.
ROY, J. H. B., GASTON, H. J., SHILLAM, K. W. G., THOMPSON, S. Y., STOBO, I. J. F. and GREATOREX, J. C. (1964) *Br. J. Nutr.* **18**, 467.
ROY, J. H. B., HUFFMAN, C. F. and REINEKE, E. P. (1957) *Br. J. Nutr.* **11**, 373.
ROY, J. H. B., PALMER, J., SHILLAM, K. W. G., INGRAM, P. L. and WOOD, P. C. (1955) *Br. J. Nutr.* **9**, 11.
ROY, J. H. B. and SHILLAM, K. W. G. (1954) *Rep. Natn. Inst. Res. Dairy.* p. 47.
ROY, J. H. B. and SHILLAM, K. W. G. (1961) *Rep. Natn. Inst. Res. Dairy.* p. 49.
ROY, J. H. B., SHILLAM, K. W. G., HAWKINS, G. M. and LANG, J. M. (1958) *Br. J. Nutr.* **12**, 123.
ROY, J. H. B., SHILLAM, K. W. G., HAWKINS, G. M., LANG, J. M. and INGRAM, P. L. (1959) *Br. J. Nutr.* **13**, 219.
ROY, J. H. B., SHILLAM, K. W. G. and PALMER, J. (1955) *J. Dairy Res.* **22**, 252.
ROY, J. H. B., SHILLAM, K. W. G., THOMPSON, S. Y. and DAWSON, D. A. (1961) *Br. J. Nutr.* **15**, 541.
ROY, J. H. B. and STOBO, I. J. F. (1965) Unpublished.
ROY, J. H. B., STOBO, I. J. F. and GASTON, H. J. (1965) *Rep. Natn. Inst. Res. Dairy.* p. 43.
ROY, J. H. B., STOBO, I. J. F. and GASTON, H. J. (1966) To be published.
RUPEL, I. W., BOHSTEDT, G. and HART, E. B. (1933) *Res. Bull. Agric. Exp. Stn Univ. Wis.* No. 115.
RUSOFF, L. L. (1941) *Bull. Fla Agric. Exp. Stn* No. 356.
RUST, J. W. (1963) *Diss. Abstr.* **24**, 8.
SANDER, E. G., WARNER, R. G., HARRISON, H. N. and LOOSLI, J. K. (1959) *J. Dairy Sci.* **42**, 1600.
SATO, M. and MURATA, K. (1932) *J. Dairy Sci.* **15**, 461.

SAYER, W. (1936) *Agriculture Live-Stk India* **6**, 795.
SCHALK, A. F. and AMADON, R. S. (1928) *Bull. N. Dak. Agric. Exp. Stn* No. 216.
SCHMIDT, H. (1941) *Am. J. Vet. Res.* **2**, 363.
SELLERS, K. C. and EDEN, E. (1949) *J. Comp. Path. Ther.* **59**, 205.
SETTLEMIRE, C. J., HIBBS, J. W. and CONRAD, H. R. (1964) *J. Dairy Sci.* **47**, 875.
SHAND, A. and LEWIS, G. (1957) *Vet. Rec.* **69**, 618.
SHARMA, G. P., VALI, K. N. and SURI, K. R. (1951) *Res. Bull. E. Panjab Univ.* No. 18, p. 57.
SHARMAN, G. A. M. (1954) *Vet. Rec.* **66**, 275.
SHARMAN, G. A. M., BLAXTER, K. L. and WILSON, R. S. (1959) *Vet. Rec.* **71**, 536.
SHAW, R. H., WOODWARD, T. E. and NORTON, R. P. (1918) *J. Agric. Res.* **12**, 575.
SHEEHY, E. J. and SENIOR, B. J. (1936) *J. Dep. Agric. Repub. Ire.* **34**, 1.
SHILLAM, K. W. G. and ROY, J. H. B. (1961) *Proc. VIII Int. Congr. Anim. Prod., Hamburg* **3**, 276.
SHILLAM, K. W. G. and ROY, J. H. B. (1963a) *Br. J. Nutr.* **17**, 171.
SHILLAM, K. W. G. and ROY, J. H. B. (1963b) *Br. J. Nutr.* **17**, 183.
SHILLAM, K. W. G., ROY, J. H. B. and INGRAM, P. L. (1962a) *Br. J. Nutr.* **16**, 585.
SHILLAM, K. W. G., ROY, J. H. B. and INGRAM, P. L. (1962b) *Br. J. Nutr.* **16**, 593.
SINCLAIR, D. P. and ANDREWS, E. D. (1954) *N.Z. Vet. J.* **2**, 72.
SINGH, S. B. and DESAI, R. N. (1959) *Indian J. Dairy Sci.* **12**, 51.
SINGH, S. B. and DUTT, M. (1961) *Indian J. Dairy Sci.* **14**, 37.
SMITH, H. W. and CRABB, W. E. (1956a) *Vet. Rec.* **68**, 274.
SMITH, H. W. and CRABB, W. E. (1956b) *J. Gen. Microbiol.* **15**, 556.
SMITH, R. H. (1957) *Biochem. J.* **67**, 472.
SMITH, R. H. (1958) *Biochem. J.* **70**, 201.
SMITH, R. H. (1959) *Biochem. J.* **71**, 306.
SMITH, S. E., KOCH, B. A. and TURK, K. L. (1951) *J. Nutr.* **44**, 455.
SMITH, S. E. and LOOSLI, J. K. (1957) *J. Dairy Sci.* **40**, 1215.
SMITH, T. and ORCUTT, M. L. (1925) *J. Exp. Med.* **41**, 89.
SMITH, V. R. and ERWIN, E. S. (1959) *J. Dairy Sci.* **42**, 364.
SPIELMAN, A. A., THOMAS, J. W., LOOSLI, J. K., WHITING, F., NORTON, C. L. and TURK, K. L. (1947) *J. Dairy Sci.* **30**, 343.
STALLCUP, O. T., HORTON, O. H. and BROWN, C. J. (1956) *Bull. Ark. Agric. Exp. Stn* No. 576.
STEENBOCK, H., NELSON, V. E. and HART, E. B. (1915) *Res. Bull. Agric. Exp. Stn Univ. Wis.* No. 36.
STEENSBERG, V. (1947) *Br. J. Nutr.* **1**, 139.
STEGENGA, T. (1961) *Veeteelt-en Zuivelberichten* **4**, 319.
STOBO, I. J. F. (1964) Studies in the nutrition of young cattle with special reference to rumen development and protein requirements of the early weaned calf. Ph.D. Thesis. University of Reading.
STOBO, I. J. F. and ROY, J. H. B. (1963) *Rep. Natn. Inst. Res. Dairy.* p. 47.
STOBO, I. J. F., ROY, J. H. B. and GASTON, H. J. (1966a) *Br. J. Nutr.* **20**, 171.
STOBO, I. J. F., ROY, J. H. B. and GASTON, H. J. (1966b) *Br. J. Nutr.* **20**, 189.
STOBO, I. J. F., ROY, J. H. B. and GASTON, H. J. (1967a) *Anim. Prod.* **9**, 7.
STOBO, I. J. F., ROY, J. H. B. and GASTON, H. J. (1967b), *Anim. Prod.* **9**, 23.
STOBO, I. J. F., ROY, J. H. B. and GASTON, H. J. (1967c) *Anim. Prod.* **9**, 155.
SUCHANEK, B. (1961) *Věd. Pr. Výzk. Ust. Chov. Skotu ČSAZV Rapotine* **1**, 27.
SWANSON, E. W. and HARRIS, J. D. (1958) *J. Dairy Sci.* **41**, 1768.
SWANSON, E. W. and HERMAN, H. A. (1943) *Res. Bull. Mo. Agric. Exp. Stn* No. 372.
SYKES, J. F. (1955) *The Yearbook of Agriculture, Water* U.S.D.A. p. 14.
SYKES, J. F. and ALFREDSON, B. V. (1940) *Proc. Soc. Exp. Biol. Med.* **43**, 575.
SYLVESTER, N. D. and LAMPITT, L. H. (1935) *Analyst, Lond.* **60**, 376.
TAGARI, J. H. (1965) Unpublished.
TAMATE, H., MCGILLIARD, A. D., JACOBSON, N. L. and GETTY, R. (1962) *J. Dairy Sci.* **45**, 408.
THEILER, A. (1934) *Vet. J.* **90**, 143.

THEILER, A., GREEN, H. H. and DU TOIT, P. J. (1927) *J. Agric. Sci., Camb.* **17**, 291.
THOMAS, J. W. (1952) *J. Dairy Sci.* **35**, 1107.
THOMAS, J. W. and MOORE, L. A. (1951) *J. Dairy Sci.* **34**, 916.
THOMAS, J. W. and OKAMOTO, M. (1954) *J. Dairy Sci.* **37**, 656.
THOMAS, J. W. and OKAMOTO, M. (1956) *J. Dairy Sci.* **39**, 928.
THOMAS, J. W., OKAMOTO, M., JACOBSON, W. C. and MOORE, L. A. (1954) *J. Dairy Sci.* **37**, 805.
TILLMAN, A. D. and BRETHOUR, J. R. (1958a) *J. Anim. Sci.* **17**, 100.
TILLMAN, A. D. and BRETHOUR, J. R. (1958b) *J. Anim. Sci.* **17**, 104.
TILLMAN, A. D., BRETHOUR, J. R. and HANSARD, S. L. (1959) *J. Anim. Sci.* **18**, 249.
TISDALL, F. F. and BROWN, A. (1929) *J. Am. Med. Ass.* **92**, 860.
TKACEV, I. and TARANENKO, G. A. (1963) *Vest. sel'.-khoz. Nauki, Mosk.* **5**, 65.
TOLLERSRUD, S. (1963) *Nord. VetMed.* **15**, 543.
TOMME, M. F. and TARANENKO, G. A. (1939) *Dokl. vses. Akad. sel'.-khoz. Nauk* No. 18, p. 36.
TOWNLEY, R. C. and GOULD, I. A. (1943) *J. Dairy Sci.* **26**, 843.
TRAUTMANN, A. and SCHMITT, J. (1933) *Arch. Tierernähr. Tierz.* **9**, 11.
TROWBRIDGE, P. F., MOULTON, C. R. and HAIGH, L. D. (1919) *Res. Bull. Mo. Agric. Exp. Stn* No. 30.
TYLER, W. J., CHAPMAN, A. B. and DICKERSON, G. E. (1947) *J. Dairy Sci.* **30**, 483.
UNDERWOOD, E. J. and HARVEY, R. J. (1938) *Aust. Vet. J.* **14**, 183.
VALLEE, B. L. (1959) *Physiol. Rev.* **39**, 443.
VAN ADRICHEM, P. W. M. and FRENS, A. M. (1965) *Tijdschr. Diergeneesk.* **90**, 525.
VAN DER MEY, G. J. W. (1965) *Eur. Ass. Anim. Prod. 8th Study Meeting*, Noordwijk.
VANDERSALL, J. H., CONRAD, H. R. and HIBBS, J. W. (1957) *J. Anim. Sci.* **16**, 1111.
VANDERSALL, J. H., HIBBS, J. W. and CONRAD, H. R. (1956) *J. Dairy Sci.* **39**, 929.
VELU, J. G., GARDNER, K. E. and KENDALL, K. A. (1959) *J. Dairy Sci.* **42**, 920.
VENGE, O. (1948) *Nord. JordbrForsk.* No. 208.
VERMA, I. S. and SOMMER, H. H. (1950) *J. Dairy Sci.* **33**, 397.
VISEK, W. J., MONROE, R. A., SWANSON, E. W. and COMAR, C. L. (1953) *J. Nutr.* **50**, 23.
VOLCANI, R. (1963) *Proc. Wld Conf. Anim. Prod., Rome* **2**, 355.
VOLCANI, R. and LEVI, D. (1953) *Refuah vet.* **10**, 54.
WALKER, D. M. (1950) *Bull. Anim. Behav.* **8**, 5.
WALKER, D. M., THOMPSON, S. Y., BARTLETT, S. and KON, S. K. (1949) *Int. Dairy Congr. XII Stockholm* **1**, 83.
WALLIS, G. C. (1938) *J. Dairy Sci.* **21**, 315.
WALLIS, G. C. (1944) *Bull. S. Dak. Agric. Exp. Stn* No. 372.
WARD, G. M., BENNE, E. J., WEBSTER, H. D., DUNCAN, C. W. and HUFFMAN, C. F. (1949) *J. Anim. Sci.* **8**, 632.
WARNER, R. G., BERNHOLDT, H. F., GRIPPIN, C. H. and LOOSLI, J. K. (1953) *J. Dairy Sci.* **36**, 599.
WARNER, R. G., FLATT, W. P. and LOOSLI, J. K. (1956) *J. Agric. Fd Chem.* **4**, 788.
WARNER, R. G., GRIPPIN, C. H., FLATT, W. P. and LOOSLI, J. K. (1955) *J. Dairy Sci.* **38**, 605.
WARNER, R. G. and SUTTON, T. S. (1948) *J. Dairy Sci.* **31**, 976.
WATSON, R. H. (1944) *Res. Bull. Commonw. Aust. Coun. Sci. Industr. Res.* **180**, 1.
WAUGH, R. K., HAUGE, S. M. and KING, W. A. (1947) *J. Dairy Sci.* **30**, 457.
WESTER, J. (1930) *Vet. J.* **86**, 401.
WETZEL, J. O. and MOORE, L. A. (1940) *Am. J. Ophthal.* **23**, 499.
WHITELAW, F. G. and PRESTON, T. R. (1963) *Anim. Prod.* **5**, 131.
WHITELAW, F. G., PRESTON, T. R. and DAWSON, G. S. (1961) *Anim. Prod.* **3**, 127.
WHITELAW, F. G., PRESTON, T. R. and NDUMBE, R. D. (1961) *Anim. Prod.* **3**, 121.
WHITELAW, F. G., PRESTON, T. R. and MACLEOD, N. A. (1963) *Anim. Prod.* **5**, 227.
WIERINGA, G. W. (1956) *Tijdschr. Diergeneesk.* **81**, 242.
WIESE, A. C., JOHNSON, B. C., MITCHELL, H. H. and NEVENS, W. B. (1947a) *J. Nutr.* **33**, 267.

WIESE, A. C., JOHNSON, B. C., MITCHELL, H. H. and NEVENS, W. B. (1947b) *J. Dairy Sci.* **30**, 87.
WIESE, A. C., JOHNSON, B. C. and NEVENS, W. B. (1946) *Proc. Soc. Exp. Biol. Med.* **63**, 521.
WILLIAMS, J. B. and JENSEN, C. (1955) *Bi-m. Bull. N. Dak. Agric. Exp. Stn* **17**, 91.
WILLARD, H. S. (1948) *Bull. Wyo. Agric. Exp. Stn* No. 286.
WILSON, J. (1927) *The Principles of Stock Feeding*, p. 170, Vinton & Co., London.
WINCHESTER, C. F. and MORRIS, M. J. (1956) *J. Anim. Sci.* **15**, 722.
WING, J. M., JACOBSON, N. L. and ALLEN, R. S. (1955) *J. Dairy Sci.* **38**, 1006.
WISE, G. H., CALDWELL, M. J., PARRISH, D. B., FLIPSE, R. J. and HUGHES, J. S. (1947) *J. Dairy Sci.* **30**, 983.
WISE, G. W., JACOBSON, N. L., ALLEN, R. S. and YANG, S. P. (1958) *J. Dairy Sci.* **41**, 143.
WOOD, P. C. (1955) *J. Path. Bact.* **70**, 179.
WOODMAN, H. E. and HAMMOND, J. (1923) *J. Agric. Sci., Camb.* **13**, 180.
ZEREBCOV, P. I. and SERYH, M. M. (1962) *Izv. timiryazev. sel'.-khoz. Akad.* **4**, 214.
ZWEIG, G. and BLOCK, R. J. (1953) *J. Dairy Sci.* **36**, 427.

CHAPTER 19

Nutrition of the Dairy Cow

J. F. D. GREENHALGH
The Rowett Research Institute, Bucksburn, Aberdeen

INTRODUCTION

In 1963 there were nearly 1,000 million cattle in the world and a further 100 million buffalo (Table 1). The proportion of the cattle which can be classified as dairy cows is rather difficult to estimate, because in many parts of the world milk is regarded as the by-product of an animal kept mainly as a source of power or even for reasons of religion or social status. In countries where agriculture is better developed there is a clearer division between cattle kept for milk production and those kept for beef production. In the U.S.A., for example, out of a total of 104 million cattle, 49 million are classified as cows. Of these, 19 million are classified as dairy cows, the remainder being cows used for providing and rearing the calves required for beef production. In Britain about 30 per cent of the cattle population consists of dairy cows and in the Netherlands the proportion is nearly 50 per cent.

TABLE 1
World Cattle Numbers and Milk Production

Region	No. of cattle (millions)	Annual milk production (million metric tons)
Europe	117·9	130·5
U.S.S.R.	86·9	59·9
North America	115·0	65·0
Latin America	212·0	21·9
Asia*	257·2	19·7
Africa	120·9	11·2
Oceania	25·6	12·4
World*	935·5	320·6

*Excluding China.
Source: *FAO Production Yearbook*, Vol. 18, 1964

Table 1 shows that it is in those parts of the World where a large proportion of the cattle population is kept specially for milk production that most of the

World's milk is produced. Europe, the U.S.S.R. and North America together possess one-third of the World's cattle but account for 80 per cent of total milk production. This lack of concordance between cattle numbers and milk production is due not only to variations from region to region in the proportion of cattle reserved for milk production, but also to great differences in milk yield per cow (Table 2).

The lower milk yields of cattle kept in the tropics are in part a reflection of the lower genetic potential for milk production of zebu cattle (*Bos indicus*) as compared with cattle of European origin (*Bos taurus*). They also reflect differences in climate, in the incidence of disease and, particularly, in nutrition. Dairy cattle in the tropics are often expected to live on nothing but small quantities of poor-quality pasture herbage or preserved roughages. In contrast, dairy cattle in temperate regions either graze on pastures of high quality or else are fed indoors on preserved roughages supplemented with large quantities of cereal grains and oilseed residues.

The 'dairy cow' whose nutrition is to be considered in this chapter may therefore be a 300-kg zebu animal, consuming odd patches of vegetation plus rice straw and producing 1–2 kg milk per day, or it may be a 600-kg Friesian, given 6 kg hay and 12 kg concentrated foods, and yielding 30 kg milk per day. Because of such enormous variations in feeding practice, this chapter must for the most part be restricted to the principles on which the feeding of the dairy cow is based. Where examples of feeding practices are given they relate to high-yielding European-type cattle, mainly because 80 per cent of the World's milk is produced by such animals.

TABLE 2

The Annual Milk Production of Milking Cows in Selected Countries

Country	Milk yield (kg)
Netherlands	4100
United Kingdom	3660
United States	3420
New Zealand	2700
Chile	2500
Italy	2050
U.S.S.R.	1690
Kenya	1200
Pakistan	420
Uganda	280

Source: *FAO Production Yearbook*, Vol. 18, 1964

The primary objectives in feeding the specialized dairy cow are to allow her to maintain herself (and to grow if she has not already achieved her mature weight), to provide for the production of one calf *per annum*, and to promote the secretion of milk satisfactory in both quantity and quality. The first part of the chapter is concerned with nutrient requirements for maintenance and milk production in the cow (reproduction is discussed in Chapter 12). In later sections, the relations between level of feeding and milk yield and composition are described.

The space devoted to each topic has been determined in part by whether or not that topic is discussed elsewhere in this volume, and also by the availability of suitable review articles. References to reviews are given whenever possible.

ENERGY NUTRITION

Categories of Energy

The energy values of foods for ruminant animals and the energy requirements of these animals may be expressed in terms of digestible, metabolizable or net energy. The relation of each of these categories to the others is shown in Fig. 1. A further, indirect, measure of the energy value of foods is their Total Digestible Nutrients (TDN) content, defined as the sum of the weights

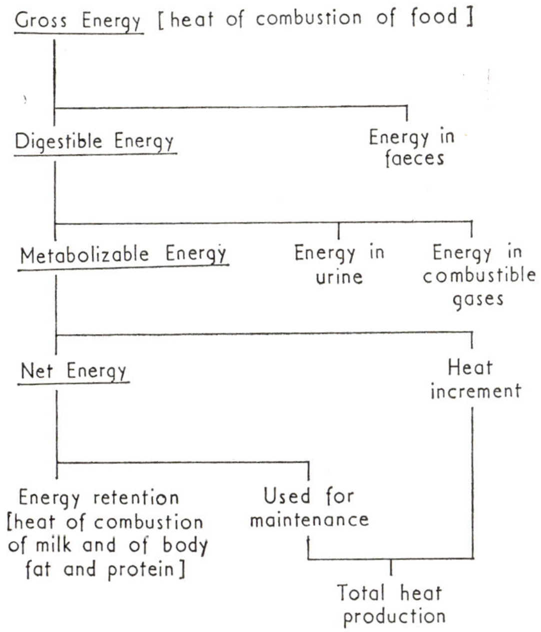

Fig. 1.—The partition of food energy in the dairy cow.

in 100 units of food of digestible crude protein, crude fibre and nitrogen-free extractives, plus 2·25 times the weight of digestible ether extract. In foods for ruminants 1 kg TDN is equivalent to 4400 kcal digestible energy (Swift, 1957) or 3575 kcal metabolizable energy (Reid, 1961). Net energy is usually expressed as a weight of Starch Equivalent (SE). The SE of a food may be defined as the weight of starch which promotes the same retention of energy in a mature fattening ruminant as 100 weight units of the food. The net energy value of starch when used to promote fat storage in ruminants is 2360 kcal per kg.

As the proportions of food energy lost in faeces and urine are easily measured and as the proportion lost as combustible gas is reasonably constant (about 8 per cent of gross energy), the digestible and metabolizable energy values of foods are readily determinable. The metabolizable energy (ME) values of the foods most commonly given to dairy cows, such as pasture herbage, good quality conserved roughages, cereal grains and oilseed residues, are in the range 2000–3500 kcal per kg, although poor hays and cereal straws contain as little as 1300 kcal ME per kg. The metabolizable energy value of a particular food usually falls as the quantity consumed increases; for example, Armstrong (1964) found that of mature ryegrass to fall from 2370 kcal per kg when the grass was eaten in quantities sufficient for maintenance to 2140 kcal per kg when eaten in quantities three times as great.

The proportion of the metabolizable energy which is lost as heat is obviously quite difficult to measure (see Chapters 14 and 15), and for a particular food it varies according to the physiological function for which the metabolizable energy is used by the animal. For foods in general, the efficiency with which metabolizable energy is converted into net energy is 70–80 per cent if the metabolizable energy is used for the function of maintenance (i.e. the proportion lost as heat is 30–20 per cent), but the efficiency is only 30–65 per cent if the metabolizable energy is used for the synthesis of body fat. The efficiency with which metabolizable energy is used for milk synthesis is particularly difficult to measure, for reasons which are discussed later, but is believed to lie in the range 60–70 per cent. In general, the efficiency of utilization of metabolizable energy is related to the concentration of metabolizable energy in the food, as is illustrated in Fig. 2. It should be noted, however, that the concentration of metabolizable energy has a greater effect on the efficiency with which it is utilized for body fat synthesis than that with which it is utilized for maintenance. This means that for foods low in metabolizable energy the ratio of net energy for maintenance to net energy for fattening is wider than it is for foods high in metabolizable energy.

It is highly desirable that the energy values of foods and the energy requirements of animals should be expressed in terms of the same category of energy. The energy values of foods are better expressed as metabolizable (or digestible) energy than net energy, because the latter is more difficult to determine and depends upon the use to which the food is put. For just the same reasons, however, the energy requirements of animals are better expressed in terms of net energy. Requirements expressed as net energy are direct measures of the energy expended or retained by the animal. For example, if a cow produces 10 kg milk having a heat of combustion of 750 kcal per kg, her requirement of net energy for milk synthesis is 7500 kcal. To express this

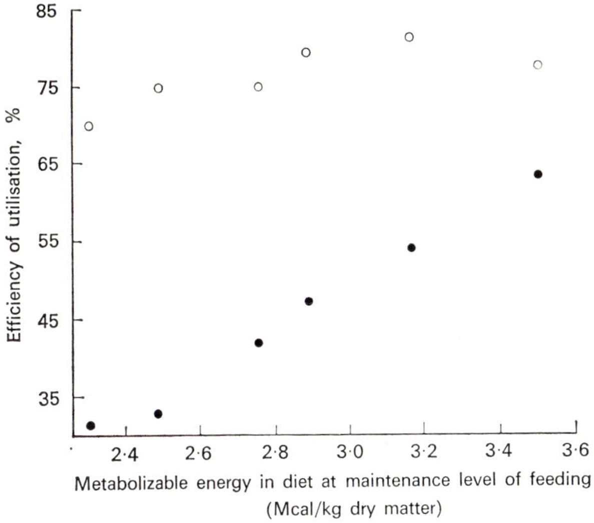

FIG. 2.—The effect of the concentration of metabolizable energy in the diet on the efficiency of utilization of metabolizable energy by steers for maintenance (open circles) or fattening (closed circles). (From data given by Blaxter and Wainman, 1964.)

requirement in terms of metabolizable or digestible energy one has to assume average values for the efficiency with which metabolizable or digestible energy can be utilized in milk synthesis. The ranges in efficiency are quite large, however, and so the assumption of average values may lead to large errors.

The two main systems for expressing the energy requirements of ruminant animals and the energy values of their foods are the Total Digestible Nutrients system which is used in North and South America and the Starch Equivalent system which is used in Europe. (The Scandinavian Feed Unit system is now closely related to the Starch Equivalent system.)

In the TDN system it is assumed that for any one function of the animal, digestible or metabolizable energy is utilized with constant efficiency regardless of the nature of the food. In the Starch Equivalent system, where the category of energy used is net energy for fattening, the assumption made is that the ratios between net energy values for maintenance, for milk synthesis and for body fat synthesis are the same for all foods. For these three functions 1 kg SE is assumed to provide 3100, 2950 and 2360 kcal net energy, respectively. The requirement of starch equivalent for milk synthesis, for example, is then calculated from the heat of combustion of milk (750 kcal per kg) to be

$$\frac{750}{2950} = 0{\cdot}25 \text{ kg SE per kg milk.}$$

From earlier discussion it should now be clear that the assumptions involved in both systems are unsound. Detailed criticisms of the systems have been given by McDonald, Edwards and Greenhalgh (1966) and Blaxter (1962a). In Britain a new system has now been proposed (Agricultural Research Council, 1965) in which (a) the energy values of feeds are expressed in terms of metabolizable energy and (b) the energy requirements of animals are expressed initially as net energy but are then translated into corresponding allowances of metabolizable energy for diets having various levels of metabolizable energy concentration. The translation is made by assuming that the efficiency with which metabolizable energy is utilized for each function of the animal can be predicted from the concentration of metabolizable energy in the diet. Allowance is also made for the fact mentioned earlier (page 720) that as food intake increases the metabolizable energy value of the food decreases. In this system the requirement of metabolizable energy by a 500-kg cow producing 15 kg milk per day, for example, is calculated to be 32,600 kcal (32·6 Mcal) when the diet contains 2·2 Mcal metabolizable energy per kg dry matter, but 28·2 Mcal when the metabolizable energy concentration is 3·4 Mcal per kg.

In the paragraphs which follow the energy requirements of dairy cows are considered as net energy, and estimates are given of the efficiency with which metabolizable energy may be utilized in meeting these requirements.

Energy Requirements for Maintenance

When a cow has been fasted until it is no longer metabolizing nutrients absorbed from its digestive tract, is kept in a thermoneutral environment, and produces no milk, its heat production is known as its fasting metabolism. The fasting metabolism is an estimate of the minimal quantity of net energy which must be supplied to the animal to promote energy equilibrium. It is a minimal estimate because when cows are kept under normal feeding conditions they are likely to expend more energy for voluntary muscular activity than when they are kept in a respiration chamber or calorimeter and some of the energy required for this activity is considered to be a part of the maintenance requirement. The term ‘activity increment’ is sometimes used to describe the addition made to estimates of fasting metabolism when calculating requirements of energy for maintenance.

The term ‘fasting metabolism’ is preferred to ‘basal metabolism’ as used in human nutrition, because farm animals do not achieve the state of muscular relaxation required for a true estimate of basal metabolism. The steers of Forbes, Kriss and Braman (1927), for example, stood on their feet for about 3 hours per day, even after 3–7 days of fast. In ruminant animals, digestion and absorption continue for a long time after fasting begins, and a period of 2–4 days is needed for the post-absorptive state to be reached.

In cattle more determinations of fasting metabolism have been made with castrated males than with cows, but the results for these two sexes are similar if the cows are not more than 7 months pregnant. The summary prepared by the Agricultural Research Council (1965) shows that fasting metabolism falls from 95 kcal per kg metabolic liveweight ($W^{0\cdot73}$) per day at two years of age to 80 kcal at four years and above (equivalent to 7·5 Mcal per day for a cow of 500 kg). The value of 80 kcal per kg $W^{0\cdot73}$ per day is higher than the 70 kcal considered by Brody (1945) to be the mean value for species ranging in size from mice to elephants. There is evidence that in the later stages of pregnancy fasting metabolism increases by 20 per cent, and also that there are differences between breeds in fasting metabolism. Lactating cows have not often been subjected to a fast, but in two animals so treated, Ritzman and Benedict (1938) found that after 4–5 days of fasting, when milk synthesis appeared to have ceased, fasting metabolism was about 30 per cent higher than in the same cows when non-lactating.

The extent to which muscular activity is greater in fed animals than in those fasting is obviously quite variable. At one extreme, a cow tied up in a stall has no greater opportunity, and perhaps no greater inclination, to expend energy in muscular activity than one kept in a calorimeter. At the other extreme, a cow walking to and from a pasture twice a day, and having to seek out and harvest its food, is likely to expend a considerable amount of energy in these activities. Graham (1965) estimated that sheep at pasture might expend five times as much energy in voluntary muscular activity as sedentary sheep indoors, and, as a result, have a maintenance requirement 40 per cent greater. Much of this difference was due to the estimated cost of harvesting pasture herbage, and it is perhaps more reasonable to regard energy expended in this way as a deduction from the energy value of the food than as an addition to the energy requirement of the animal. The energy cost of walking on the level is relatively small. In cattle it was found by Hall and Brody (1934) to be 0·5 cal per kg per metre. A 500-kg cow walking 5 km per day would therefore require 1250 kcal per day—about 17 per cent of its fasting metabolism.

From the results of trials made with both cattle and sheep, Blaxter (see Agricultural Research Council, 1965) calculated that the efficiency of utilization of metabolizable energy for maintenance (k_m) was related to the percentage of the gross energy of the food which was metabolizable (Q_m) by the equation:

$$k_m = 54\cdot6 + 0\cdot30\, Q_m \tag{1}$$

If all foods are assumed to contain 4·4 Mcal gross energy per kg dry matter, the equation becomes:

$$k_m = 54\cdot6 + 6\cdot82\ \mathrm{ME} \tag{2}$$

(where ME = the metabolizable energy value of the food in Mcal per kg dry matter).

A diet containing 2·2 Mcal ME per kg would therefore be utilized with 69·6 per cent efficiency and one containing 3·2 Mcal, with 76·4 per cent efficiency.

The efficiency with which metabolizable energy is utilized for muscular work is believed to be the same as that with which it is used for maintenance.

Energy Requirements for Milk Synthesis

About half the energy of cow's milk is contained in the fat fraction and a quarter each in the protein and lactose fractions. As the fat and protein contents of milk are positively correlated, total energy content may be predicted from fat content alone. An equation for this purpose is that of Gaines and Davidson (1923):

$$\text{kcal per kg milk} = 304{\cdot}8 + 114{\cdot}1 \text{ (per cent fat)}, \qquad (3)$$

which predicts milk containing 4 per cent fat to have an energy value of 761 kcal per kg.

Milk yields are frequently converted to what is called '4 per cent fat-corrected milk' (FCM), which is the weight of milk containing 4 per cent fat that contains the same quantity of energy as the yield observed. For example, milk with 3 per cent fat is predicted from equation (3) to contain 647 kcal per kg. Ten kg of such milk would therefore contain the same quantity of energy as $10 \times \frac{647}{761} = 8{\cdot}5$ kg milk with 4 per cent fat. A simple formula for calculating FCM is:

$$\text{FCM} = 0{\cdot}4 \text{ (weight of milk)} + 15 \text{ (weight of fat)} \qquad (4)$$

Tyrrell and Reid (1965) have shown that while equation (3) is accurate for milks of normal fat content, it underestimates the energy content of milks containing less than about 3 per cent fat and overestimates that of milks with more than 6 per cent fat. They have proposed equations for predicting energy content from contents of both fat and solids-not-fat.

Lactating cows normally use energy for the three functions of maintenance, milk production and liveweight gain simultaneously (or alternatively, liveweight losses may contribute to the energy available for maintenance and milk synthesis). In either case it is difficult to determine how much of the metabolizable energy intake is used specifically for the synthesis of milk. The usual approach has been to calculate the metabolizable energy needed for maintenance and liveweight change from the results of calorimetric trials made with non-lactating animals. After considering calorimetric experiments of this kind, and also some trials in which complete energy balances were not measured, the Agricultural Research Council (1965) concluded that the efficiency of utilization of metabolizable energy for milk synthesis was 65–70 per cent for diets within the normal range of metabolizable energy concentration (2·2–3·0 Mcal ME per kg dry matter). For diets beyond both limits of this range, efficiency of utilization declined to about 60 per cent. These conclusions have been confirmed in goats by Armstrong and Blaxter (1965), although Coppock, Flatt and Moore (1964) have found lower efficiency values, particularly when the diet was composed entirely of roughage.

Energy Requirements for Liveweight Gain

Cows normally lose weight in early lactation and gain during the later part of the lactation and during the 2–3 months dry period between lactations. Gains made during the last two months of pregnancy are caused largely by the growth of the foetus and the structures associated with it. These are discussed later in this chapter and also in Chapter 12.

Little is known about the energy content of the liveweight gains or losses of dairy cows. In part, weight changes during lactation may be due merely to changes in the weight of intestinal contents, particularly those associated with secular changes in food intake (see below). Recently, Bath, Ronning, Meyer and Lofgreen (1965) have used a comparative slaughter technique, coupled with carbon and nitrogen balances, to determine the energy content of liveweight losses in heifers during the first four months of lactation. The mean value obtained was 4985 kcal per kg but there was considerable variation between individual animals (from 3257 to 18,755 kcal per kg in the case of estimates made by carbon and nitrogen balance—the latter is obviously an impossible value).

The data of Ellenburger, Newlander and Jones (1950) for the energy content of the bodies of dairy cattle slaughtered at different weights suggest that each 1 kg of gain contains about 3500 kcal. In the study of Brody and Procter (1935), which is referred to in more detail later, a change in liveweight of 1 kg was associated with a change in ME intake of 7500 kcal. If ME is utilized for growth or fattening with an efficiency of 50–60 per cent (see below) this figure corresponds to 3750–4500 kcal per kg.

The efficiency of utilization of metabolizable energy for growth and fattening (k_f) has been studied quite extensively in non-lactating, growing cattle, but not in dairy cows. In the former, it is related to the percentage of the gross energy in the diet which is metabolizable (Q_m). The Agricultural Research Council (1965) give the following equation:

$$k_f = 0{\cdot}81\, Q_m + 3{\cdot}0 \qquad (5)$$

If foods are assumed to contain 4·4 Mcal gross energy per kg dry matter, this becomes:

$$k_f = 18{\cdot}4\,\text{ME} + 3{\cdot}0 \qquad (6)$$

A diet containing 2·2 Mcal ME per kg would therefore be utilized with 43 per cent efficiency and one with 3·2 Mcal per kg, with 62 per cent efficiency. (Note that these efficiency values are lower, and also more widely spaced, than those predicted by eq. (2) for the function of maintenance.)

When cattle are losing weight, and are using the energy of body fat and protein for maintenance purposes, the efficiency with which that energy is utilized is about 70 per cent (Van Es, 1961).

A COMPARISON OF REQUIREMENTS CALCULATED BY DIFFERENT SYSTEMS

The Starch Equivalent, Total Digestible Nutrients and several other systems used for dairy cows were comprehensively reviewed and compared with one another by Blaxter (1950). Fig. 3 shows a comparison of the new system proposed by Blaxter (1962a) (see also Agricultural Research Council, 1965) with two major existing systems. Two diets containing 2·2 and 3·0 Mcal metabolizable energy per kg dry matter have been used in the comparison. Their TDN contents were calculated on the assumption that 1 kg TDN contains 3·58 Mcal ME. Their starch equivalents were estimated by predicting with eq. (6) the efficiency with which metabolizable energy would be utilized for body fat synthesis, calculating their net energy value, and expressing this as a percentage of the net energy value of starch (2·36 Mcal per kg). The

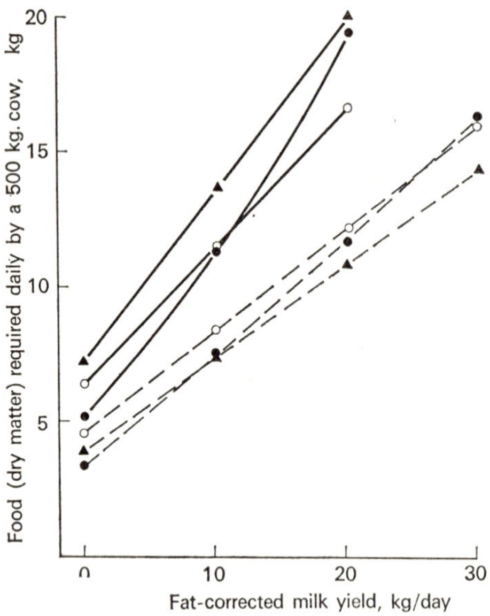

FIG. 3.—The quantities of food needed by dairy cows to supply their energy requirements for maintenance and milk production, as calculated for two foods by three systems of energy evaluation.

[Continuous lines=food with 2·2 Mcal metabolizable energy per kg; broken lines=food with 3·0 Mcal per kg. Triangles=Starch Equivalent system (Evans, 1960); open circles=Total Digestible Nutrients system (NRC, 1958); closed circles=new system (ARC, 1965)]

amounts of food dry matter needed to meet the requirements of cows having differing yields were then calculated according to the three systems. The tendency for the TDN system to overvalue poorer quality foods is illustrated by the relatively small difference in elevation of the lines for the two foods. In Blaxter's system the relation between food intake and milk yield is non-linear because an allowance is made for the decline in metabolizable energy value which accompanies increasing intake.

Energy Requirements—the Whole as the Sum of the Parts

Up to now we have considered energy requirements in a factorial manner; that is, we have estimated the requirement for each function separately, and we assume that the cow's total requirement is the sum of the requirements for the separate functions. The energy requirements of cattle, however, have often been examined in the reverse manner, by measuring total energy intake and partitioning this between maintenance and the productive functions.

Van Es (1961) analysed the previously-published results of 236 calorimetric trials made with non-lactating cattle fed at about the maintenance level. He calculated equations of the kind:

$$y = aW^{0 \cdot 8} + bG \qquad (7)$$

(where y = metabolizable energy intake, W = liveweight, G = the energy gained or lost by the animal, and a and b are coefficients corresponding to metabolizable energy requirements for maintenance and energy gain or loss). The value of a for cows losing energy was found to be slightly different from those gaining, but the average was equivalent to a requirement for a 500-kg cow of 11·6 Mcal metabolizable energy per day. In experiments of his own, which were made exclusively with dairy cows, Van Es (1961) obtained a value of 12·6 Mcal. The value calculated from fasting metabolism (80 kcal per kg $W^{0 \cdot 73}$ per day), assuming metabolizable energy to be used with an efficiency of 74 per cent, is $80 \times 93 \cdot 4 \times {}^{100}/_{74} = 10{,}100$ kcal or 10·1 Mcal metabolizable energy per day. The difference between this last figure and those of Van Es is unlikely to reflect a difference in voluntary muscular activity, because the animals considered by Van Es were kept in calorimeters.

The approach used by Van Es had been employed earlier by Brody and Procter (1935) who partitioned by multiple regression analysis the annual TDN intake of 243 lactating cows. If 1 kg TDN supplies 3575 kcal metabolizable energy, the estimates of requirement obtained were 154 kcal per kg $W^{0 \cdot 73}$ per day for maintenance, 1090 kcal per kg FCM produced and 7500 kcal per kg liveweight change. The first of these values is equivalent to 14·4 Mcal metabolizable energy per day for the maintenance of a 500-kg cow; the second implies an efficiency of conversion of metabolizable energy to milk energy of 70 per cent.

Regression analysis has been used particularly to estimate the requirements of grass-fed cows, both when grazing in the field and when fed on cut herbage indoors. In experiments of this kind, metabolizable energy intake can be estimated only with appreciable error, especially in animals grazing in the field, and changes in the body weight of cows may be confused by large changes in the weights of intestinal contents. The results obtained, which have been discussed by Holmes and Jones (1964) and Blaxter (1964), must therefore be treated with caution. In general, they suggest that the maintenance requirements of lactating cows are higher than would be estimated from fasting metabolism data alone. For housed animals, the regression estimates, like those of Van Es (1961) and Brody and Procter (1935), are up to 40 per cent greater, but for grazing animals they are up to 100 per cent greater. The apparently high maintenance requirements of grazing cows have been attributed to the energy costs of harvesting grass (see page 507), but seem too large to be completely accounted for in this way. Also, if the energy cost of harvesting herbage is high, one would expect regression estimates of energy requirements for production to be similarly inflated, but these are equal to, and in some cases less than, those estimated from the energy of the products and the appropriate efficiency factors for the utilization of metabolizable energy.

For dairy cows, and indeed for all ruminants, energy requirements estimated synthetically, by the factorial approach, appear not to agree entirely with those calculated by the analytical method just described. The disagreement is no doubt due in part to the difficulties of measuring the energy transactions of cows kept under practical conditions, but it may also reflect inaccuracies in the values employed in factorial calculations. In particular, the maintenance requirement of a cow may be greater when it is lactating and fully fed than when it is fasting or fed only at a non-productive level. It seems important that factorial estimates should be thoroughly tested under practical conditions.

PROTEIN NUTRITION

Categories of Protein

The protein values of foods for ruminants and the protein requirements of ruminants are usually expressed in terms of digestible crude protein (DCP). The DCP content of a food is calculated as 6·25 times the weight of nitrogen per unit weight of food less 6·25 times the weight of faecal nitrogen per unit weight of food consumed. In dairy cows the maximum proportion of the nitrogen apparently absorbed which may be 'retained' as milk or flesh (i.e. the biological value of the dietary protein) is in the range 60–80 per cent. The range is narrow because the activities of rumen microorganisms ensure that the proportions of the amino-acid mixtures absorbed remain reasonably constant regardless of the composition of the food proteins (see Chapter 4).

In the faeces of ruminants a large proportion of the nitrogen is considered to be of endogenous origin (metabolic faecal nitrogen). If the diet contains 10 per cent protein about two-thirds of the faecal nitrogen will be metabolic faecal nitrogen, and if the diet contains 20 per cent protein this proportion falls to about one-half. In cattle, approximately 0·5 g metabolic faecal nitrogen is excreted for each 100 g of dry matter consumed (Agricultural Research Council, 1965). If the proteins giving rise to this nitrogen are truly endogenous* they must be synthesized by the cow from absorbed amino acids, and if the latter are used for this purpose with an efficiency (biological value) averaging 70 per cent, then the quantity of nitrogen which must be absorbed to replace that lost as metabolic faecal nitrogen is $0{\cdot}5 \times \frac{100}{70} = 0{\cdot}71$ g per 100 g dry matter consumed. This means that of the protein apparently absorbed by the animal (DCP) from each 100 g of food dry matter, $6{\cdot}25 \times (0{\cdot}71–0{\cdot}5) = 1{\cdot}3$ g must be set aside as protein 'wasted' in the replacement of losses of metabolic faecal nitrogen. The DCP content of a food therefore overestimates the protein in it which will be available to the animal for maintenance and production; in addition, protein requirements expressed in terms of DCP vary according to the food intake of the animal (see below).

To avoid these disadvantages of DCP, Blaxter and Mitchell (1948) proposed that protein for ruminants should be expressed as available protein, this being DCP less the protein wasted in replacing losses of metabolic faecal nitrogen. In general terms, the available protein (AP) content of a food is calculated as:

$$AP = 6{\cdot}25\,(F - U - \frac{100M}{B}) \qquad (8)$$

where: F = N in food
U = undigested food N } g/100 g food dry matter
M = metabolic faecal N }
B = biological value of N absorbed (%)

If M and B are assumed to be 0·5 g and 70 per cent, respectively, % AP = % DCP − 1·3. A requirement expressed as DCP may be converted into AP by deducting 1·3 g per 100 g food dry matter.

*This is debatable; much metabolic faecal nitrogen may be derived from intestinal bacteria.

Requirements for Maintenance

The protein required for the maintenance of cattle consists of that needed to balance the endogenous urinary losses of nitrogen, together with a much smaller quantity needed to replace dermal losses in hair and scurf. The endogenous urinary loss of nitrogen is measured in a manner analogous to that used to measure fasting metabolism. When an animal is placed on a nitrogen-free diet all urinary nitrogen is assumed to be endogenous in origin. The data reviewed by the Agricultural Research Council (1965) showed the endogenous urinary nitrogen excretion of adult cattle to vary from 0·10 to 0·21 g per kg $W^{0\cdot73}$. The mean value for cattle over 200 kg liveweight was considered to be 0·12 g per kg $W^{0\cdot73}$. This is considerably lower than the value of 0·16 g estimated from the relationship proposed by Smuts (1935) that 2 mg of endogenous urinary nitrogen is excreted per kcal basal metabolism. The discrepancy may be due either to the relatively high fasting metabolism of cattle (see page 723) or to the ability of ruminants to conserve urea nitrogen by re-cycling it through the rumen (see Chapter 4). The second suggestion is supported by the fact that in ruminants the proportion of endogenous urinary nitrogen contributed by urea and ammonia may be less than 20 per cent (Kehar, Mukherjee and Sen, 1943). In non-ruminants, including the pre-ruminant calf (Blaxter and Wood, 1951) and lamb (Walker and Faichney, 1964) this proportion is 50–80 per cent.

Dermal losses of nitrogen amount to 0·02 g per kg $W^{0\cdot73}$ per day (Blaxter—see Agricultural Research Council, 1965).

The daily requirement for available protein for the maintenance of a 500-kg cow (93·4 kg $W^{0\cdot73}$) is calculated as:

$$6\cdot25\,(0\cdot14 \times 93\cdot4 \times \frac{100}{70}) \simeq 115 \text{ g/day}$$

Requirements for Milk Synthesis

The protein content (% N × 6·38) of cow's milk is normally in the range 3 to 4 per cent, and is directly proportional to fat content. Overman, Sanmann and Wright (1929) described the relationship between protein and fat by the equation:

$$\%\,\text{protein} = 1\cdot965 + 0\cdot406\,(\%\,\text{fat}) \qquad (9)$$

although Vanschoubroek (1963) concluded that the regression coefficient should be 0·2 rather than 0·4. The protein : energy ratio of milk varies from 43 to 49 mg per kcal, being lower in high-fat milks. Four per cent FCM (760 kcal per kg) would therefore contain 3·3 to 3·7 per cent protein.

In the calculations which follow a value of 3·5 per cent protein will be employed, but it should be recognized that the protein content of the high-fat milk yielded by Channel Island breeds of cow will be higher than this.

The essential amino acid proportions of milk proteins are similar to those of bovine muscle proteins (Table 3). Milk proteins are also similar in this regard to the microbial proteins from whose amino acids they are fabricated. It is therefore likely that the biological value of microbial protein for milk protein synthesis will be high, and that it will be similar to that for growth.

The quantity of available protein required to produce 1 kg milk containing 3·5 per cent protein is calculated to be:

$$\frac{35 \times 100}{70} = 50 \text{ g}$$

TABLE 3

The Amino Acid Composition of Milk, Bovine Muscle and Microbial Proteins (g/16 g N)

Amino acid	Protein source			
	Whole milk*	Bovine muscle†	Rumen bacteria‡	Rumen protozoa‡
Arginine	3·5	7·7	4·6	4·6
Histidine	2·7	3·3	1·7	1·8
Isoleucine	6·5	6·0	5·7	6·7
Leucine	9·9	8·0	6·9	7·8
Lysine	8·0	10·0	6·8	10·5
Methionine	2·4	3·2	2·6	2·4
Cystine	0·9	1·2	1·1	1·6
Phenylalanine	5·1	5·0	4·5	6·0
Tyrosine	4·9	4·0	4·3	5·0
Threonine	4·7	5·0	5·2	5·0
Tryptophan	1·3	1·4	—	—
Valine	6·7	5·5	6·0	5·3

* Block and Weiss (1956).

† Harvey (1956).

‡ Calculated from the values given by Weller (1957) for grazing sheep.

REQUIREMENTS FOR MAINTENANCE AND MILK PRODUCTION COMBINED

A cow of 500 kg liveweight producing 15 kg milk per day would require 115 + 750 = 865 g available protein per day. If the cow consumed 12 kg dry matter per day this would be equivalent to 865 + (1·3 × 120) = 1020 g DCP per day. If, however, the cow consumed 15 kg dry matter per day, her requirement for DCP would be 865 + (1·3 × 150) = 1060 g per day.

In Table 4 these figures are compared with other estimates of the protein requirements of dairy cows.

FOOD INTAKE IN DAIRY COWS

In high producing cows, requirements for nutrients may be very large in relation to the size of the animal. For example, a cow producing 35 kg milk per day requires about 55 Mcal ME per day, or 4–5 times its requirement for

TABLE 4

The Protein Requirements of Dairy Cows

Reference	Requirement of DCP (g)		
	For daily maintenance of 500-kg cow	For production of 1 kg milk containing 4% fat	For 500-kg cow producing 15 kg milk/day
Evans (1960)*	365	60	1265
N.R.C. (1958)	295	70	1345
A.R.C. (1965)	—	—	1020–1060†

* These standards were originally stated in terms of protein equivalent. The protein equivalent of a food is the mean of its percentages of digestible true protein and digestible crude protein. Here it is assumed that 1 kg protein equivalent = 1·15 kg DCP.

† Assuming dry matter intakes of 12–15 kg/day.

maintenance. In contrast, a fattening steer gaining at the high rate of 1·5 kg liveweight per day would require only 2–2½ times its requirement for maintenance. To satisfy her requirement for energy, a cow producing 35 kg milk per day would need to consume 16–30 kg dry matter per day, the smaller quantity applying with a diet at the top end of the normal range in metabolizable energy concentration (3·4 Mcal per kg). As a general rule the maximum daily intake of dairy cows does not exceed 3·5 per cent of liveweight, or 17·5 kg dry matter for a 500-kg cow. High-yielding cows are, therefore, often unable to eat enough to satisfy their requirements for energy, particularly if the energy concentration of their diet is low. For nutrients other than those supplying energy the problem is less serious, because their concentrations in the diet may be increased more readily.

Food intake in cattle is obviously related to body size. Baker (1964), analysing collected data on intake in grazing cattle, found the relation to be approximately linear over the liveweight range 200–600 kg, although others have suggested that intake in ruminants varies with liveweight raised to a power <1 and perhaps close to 0·73 (Holmes, Jones and Drake-Brockman, 1961; Blaxter, Wainman and Wilson, 1961; Corbett, Langlands and Reid, 1963). This point is of some importance, for if intake varies with $W^{1 \cdot 0}$ and fasting metabolism with $W^{0 \cdot 73}$, large cows will use a smaller proportion of their energy intake for maintenance purposes than will small cows.

Food intake is also related to milk yield. In general, lactating cows eat 40–50 per cent more than dry cows (Hutton, 1963; Holmes, Jones, Drake-Brockman and White, 1965). It is often difficult, however, to show a clear relation among cows between level of yield and intake, even when differences due to animal liveweight are reduced by expressing both intake and yield in relation to $W^{0 \cdot 73}$ (Fig. 4). Any relation between intake and milk yield raises the question as to which is cause and which, effect. Do cows with large appetites become high yielders, or is it the other way round? Within cows (i.e. at different stages of lactation) there is a poor relationship between yield and intake, for the latter reaches its peak several weeks after the former and then declines more slowly. This is shown in Fig. 5, which relates to 24 lactation records obtained from 15 cows fed solely on alfalfa hay. The conditions of the experiment were unusual in that the cows lactated for 365 days and had an average calving interval of about 16 months. But similar curves have been obtained with cows calving at shorter intervals and fed on pasture herbage (Hutton, 1963) or on a completely-milled diet containing a high proportion of concentrates (Aitken and Preston, 1964). It is not known why cows should fail to achieve maximal intakes when milk yields (and hence nutrient requirements) are at a maximum; clearly, the failure of the cow to match

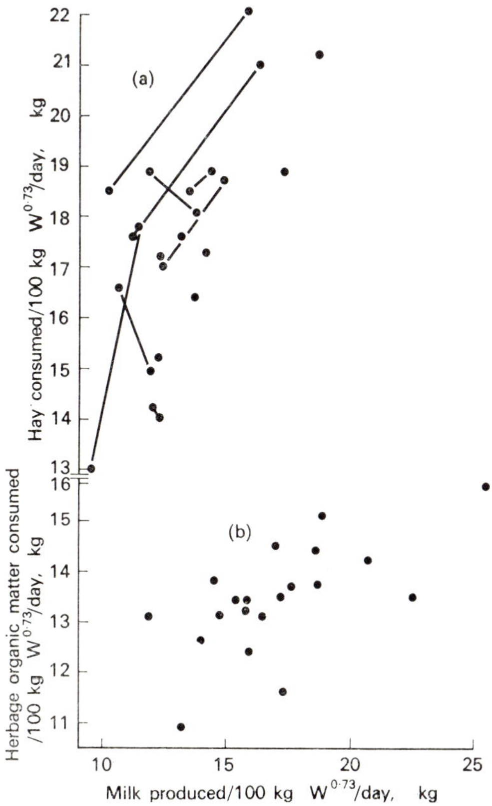

FIG. 4.—Food intake and milk yield in dairy cows.

a) Friesian cows fed solely on alfalfa hay for complete lactations (the lines join points for the same cow in different lactations.) (Graves *et al.*, 1938).

b) Grazing Ayrshire cows studied for four 7-day periods spread over the first 6-8 months of lactation (Greenhalgh *et al.*, 1966.)

intake with yield increases its chances of being underfed in early lactation. In their experiment with alfalfa hay as the sole feed, Graves *et al.* (1938) calculated that TDN intake was only 74 per cent of TDN requirement in the first month of lactation, and that intake did not equal the requirement until the fourth month. Cows fed *ad libitum*, as when grazing, are able to replenish in late lactation any reserves depleted earlier, but this is not the case with cows rationed according to milk yield. The latter have to wait for the replenishment of their reserves until the dry period, when they may be fed lavishly or 'steamed up', as discussed later.

Food intake in dairy cows depends, of course, on characteristics of the feed as well as characteristics of the animal (see Chapter 17 and the review of Balch and Campling, 1962). It is now well established that ruminant animals eat less of foods low in digestibility than of more digestible materials (Blaxter *et al.*, 1961; Blaxter and Wilson, 1962). The relationship is considered to be due to the fact that highly digestible foods are digested rapidly, and the gut space made available from one meal to the next is therefore greater than with foods which are digested more slowly. Dairy cows are normally fed on diets of high digestibility and for these the effect of digestibility on intake is less clear cut than for less digestible diets. Thus Corbett *et al.* (1963) found that as the digestibility of the herbage grazed by dairy cows declined from 79 to 67 per cent, their dry matter intakes fell only slightly, from 11·7 to 11·0 kg per day. Conrad, Pratt and Hibbs (1964) concluded that with diets less than about 67 per cent digestible, digestibility was an important determinant of the

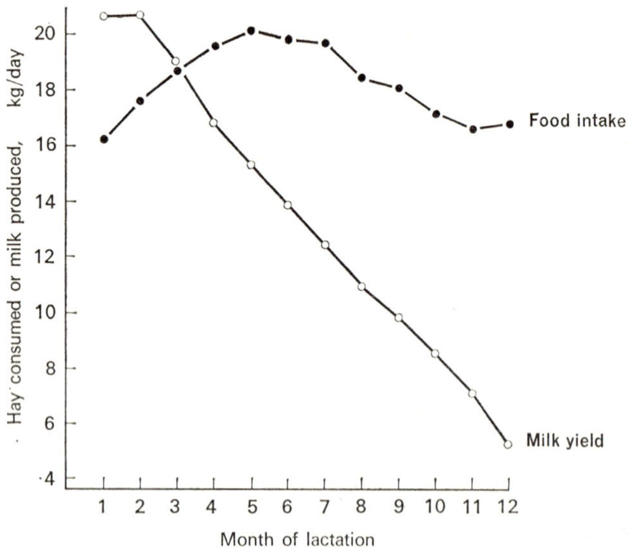

FIG. 5.—Food intake and milk yield at different stages of lactation in cows fed on alfalfa hay alone (Graves *et al.*, 1938).

intake of dairy cows, but that with diets more than 67 per cent digestible intake was determined more by liveweight and milk yield.

Cattle generally consume less dry matter from silage than from hay of the same digestibility (Moore, Thomas and Sykes, 1960).

Relationships between intake and digestibility have been studied for the most part with diets composed solely of roughages, but dairy cows are usually given a diet which contains concentrated foods also. In grazing cattle, Holmes and Jones (1964) calculated that the effect on 'total feed intake' (I) of a supplement of 1 lb (0·45 kg) concentrates depended on the organic matter digestibility (D) of the herbage, the two being related by the equation:

$$I = 2{\cdot}8 - 0{\cdot}034D \qquad (10)$$

Thus if the grass were 80 per cent digestible a supplement of 1 lb would increase total intake by only 0·1 lb (there being a reduction of 0·9 lb in grass consumption). If the grass were only 60 per cent digestible, however, the addition of a supplement would cause a reduction of only 0·2 lb in herbage intake. A similar relationship has been found when concentrates are added to diets of hay and silage (Holmes, Jones, Drake-Brockman and White, 1965). Since concentrates are about 80 per cent digestible themselves, their addition to roughages of low digestibility has a much greater effect on the digestibility of the complete diet than their addition to better-quality roughages, and their effects on total intake are partly explainable in this way. The results of Blaxter and Wilson (1963), however, suggest that the addition of fixed amounts of concentrates to hay eaten *ad libitum* (by sheep) increases total intake much more than would be expected from the improvement in digestibility alone, particularly with hay of low digestibility. In one instance sheep given a diet of poor hay and concentrates ate 30 per cent more than when given a diet of equal digestibility but composed of medium quality hay without concentrates. A similar effect in cows was found by Holmes and Jones (1965).

MINERAL NUTRITION

The energy and protein requirements of dairy cattle have been discussed in considerable detail because they are not covered elsewhere in this volume. Mineral nutrition, however, receives a chapter to itself (Chapter 11), and only a brief account of the mineral requirements of dairy cattle will be given here. The emphasis will be on the special demands for minerals which are imposed by lactation.

Minerals in Milk

The mineral composition of cow's milk is given in Table 5. Mean values are shown for major elements because the concentrations of these are stable or vary in a predictable way (see p. 736). Trace element concentrations are more variable, partly, no doubt, because milk samples are readily contaminated by trace elements after leaving the cow.

The calcium content of milk is positively related to its fat and protein contents. For each 1 per cent change in fat content, calcium content changes by about 0·008 per cent. The calcium content of milk declines slightly in early lactation and then rises towards the end of lactation. The phosphorus content of milk does not vary with fat and protein contents but tends to decline slightly as lactation advances. The magnesium content of milk is also fairly stable (Rook and Storry, 1962) and is not depressed by hypomagnesaemia in the cow. Chloride in milk increases as lactose content falls; it therefore rises during lactation.

The contents in milk of most trace elements vary with stage of lactation (see Underwood, 1962). Those of cobalt, zinc, manganese, molybdenum and iodine can be increased by giving supplements to the cow, iodine being particularly sensitive in this regard. In contrast, copper and iron contents may not be increased in this way, although the copper content of milk is reduced in copper-deficient animals.

TABLE 5

*Representative Values for the Minerals in Cow's Milk**

Major minerals (g/kg)		Trace elements (mg/kg)	
Calcium	1·2†	Iron	0·4–0·5
Phosphorus	1·0	Copper	0·05–0·12
Magnesium	0·1	Cobalt	0·0005–0·0006
Sodium	0·5	Zinc	3–5
Potassium	1·5	Manganese	0·04–0·37
Chlorine	1·1	Molybdenum	0·08–0·15
		Iodine	0·02–0·08‡

* Ling, Kon and Porter (1961) and Underwood (1962).
† For milk containing 4 per cent fat.
‡ In non-goitrous areas.

MINERAL REQUIREMENTS

The requirements calculated by the U.K. Agricultural Research Council (1965) are shown in Table 6. Requirements for major minerals were calculated factorially, by summation of endogenous losses of each element and the amounts secreted in milk, and division of the sum (the net requirement) by the estimated availability of the element in foods. For trace elements, insufficient information was available for factorial calculations, and the standards of Table 6 are based mainly on the results of feeding trials in which the minimum intakes needed to prevent signs of deficiency were determined.

Less detailed standards have been devised by the U.S. National Research Council (1958); for the major minerals these are based mainly on the results

TABLE 6

The Dietary Requirements of Dairy Cows for Minerals

(*a*) *Major elements*

Element	Maintenance (g/100 kg liveweight per day)	Milk production (g/kg milk)
Calcium	3·6	2·8*
Phosphorus	5·1	1·7
Magnesium	1·5	0·63
Sodium	1·7	0·63
Chlorine	2·6	1·15

*For milk containing 4 per cent fat; the requirement increases to 3·2 mg Ca/kg milk when milk contains 6 per cent fat.

(*b*) *Trace elements*

Element	Dietary concentration (mg/kg dry matter)
Iron	30
Cobalt	0·1
Copper	10
Zinc	50
Manganese	40
Iodine	0·8†

*2·0 mg I/kg dry matter if the feed contains goitrogens.

(Agricultural Research Council, 1965.)

of feeding and balance trials and for some elements they provide an interesting contrast for the Agricultural Research Council's factorial estimates.

The requirements for lactation of the major minerals are relatively large in relation to those for maintenance, since animals possess mechanisms for minimizing endogenous losses of minerals. In the course of a lactation a cow yielding 4000 kg milk will secrete in its milk quantities of calcium and phosphorus which are as great as the total amounts normally present in its body. In the case of magnesium the quantity in milk is twice that in the body.

Calcium and phosphorus. Many balance trials made with cows have shown that in early lactation cows are more often than not in negative balance with respect to calcium and phosphorus, and that in late lactation they are usually in positive balance. When reviewing these trials Duncan (1958) showed that in four successive three-monthly periods of lactation the proportions of calcium balances which were positive were 29, 66, 86 and 85 per cent, respectively. The cow therefore depletes her reserves of these elements in early lactation and restores them in late lactation (and during the dry period). Duncan calculated that a cow might lose 20 per cent of its body calcium during the phase of depletion. Over the complete lactation cycle there is usually a net positive balance, although Duncan considered that this might, for technical reasons, be over-estimated in balance trials.

The pattern of depletion and repletion is, no doubt, broadly related to the changes which occur during lactation in the quantities of calcium and phosphorus secreted in milk, but there is no very close relation between the dietary intakes of these elements and the balances observed. Thus Duncan (1958) found that in 134 balance trials which were positive with respect to calcium and in 112 which were negative, the mean intakes (and ranges) of calcium were 85 g (36–186 g) and 82 g (18–182 g), respectively. The position was similar for phosphorus. It has been suggested that the availability of dietary calcium is influenced by the animal's calcium reserves. Initially, the latter are high and availability is low. It is generally considered that the pattern of depletion and repletion is a natural and often inevitable course of events in the lactating cow. This means that requirement data for calcium and phosphorus are significant only for the lactation as a whole.

The requirements for calcium and phosphorus given in Table 6 are calculated on the assumption that the availability of these elements is 45 and 55 per cent, respectively, these values being lower than those taken by others (see Mitchell and McClure, 1937). In addition, the endogenous losses of the elements are considered to be greater than hitherto assumed. The requirements of Table 6 are therefore appreciably greater than those suggested, for example, by the U.S. National Research Council (1958). The latter calculate the maintenance requirements of a 500-kg cow to be 9 g Ca and 9 g P per day and those for milk production to be 2·2 g Ca and 1·5 g P per kg milk. According to these standards, a cow giving 15 kg milk per day would need 42 g Ca and 32 g P per day, or considerably less than the quantities 60 g Ca and 50 g P given in Table 6. There seems to be a great need for long-continued and carefully-conducted balance trials with cows to determine, first, what quantities of calcium and phosphorus are needed over the lactation cycle as a whole, and second, how these quantities should be distributed within the cycle.

From the Agricultural Research Council's estimates of calcium and phosphorus requirements one may calculate that an average cow, weighing 500 kg, yielding 15 kg milk and eating 12 kg dry matter per day, would require a diet containing 0·5 per cent calcium and 0·4 per cent phosphorus in its dry matter. Only very cautious generalizations may be made about the mineral composition of foods, because these are so dependent on the mineral status of the soils on which they are grown, but it seems unlikely that many of the diets normally given to dairy cows will contain *both* 0·5 per cent calcium and 0·4 per cent phosphorus. Animals fed mainly on pasture herbage, fresh or preserved, are likely to receive sufficient calcium but not enough phosphorus. Conversely, animals given a diet composed largely of concentrates are likely to be calcium, but not phosphorus, deficient. Within both classes of diet, however, there are many rations which are deficient in both elements. Pasture herbage containing little or no legumes may well be deficient in calcium as well as phosphorus (see Thompson, 1953), and concentrate diets based on maize, in addition to being very low in calcium, may also be deficient in phosphorus. It seems that if the requirements specified by the Agricultural Research Council (1965) are to be met, natural diets for dairy cows will nearly always need to be supplemented with either or both calcium and phosphorus.

The commonest clinical sign of calcium or phosphorus deficiency in dairy cattle is osteomalacia, in which severe depletion of bone minerals increases liability to displacement or fracture. Phosphorus deficiency also causes a reduced, and perhaps depraved, appetite, and a reduction in milk yield. The evidence for calcium deficiency depressing milk yield is less convincing. The condition known as milk fever, which is characterized by a depression in blood calcium concentration, is described in Chapter 33.

Magnesium. The most distinctive feature of the magnesium nutrition of ruminants is the poor efficiency with which they absorb magnesium from foods (see Rook and Storry, 1962; Care, 1965). The average 'true' availability of magnesium has been taken as 20 per cent by the Agricultural Research Council (1965). A cow giving 15 kg milk per day is calculated to require 17 g magnesium, or 0·14 per cent in a dry matter intake of 12 kg per day. The diets of dairy cows usually contain more than this, the main exception being pasture herbage which in early spring may contain no more than 0·1 per cent (Rook and Balch, 1958). Even if herbage contains an apparently adequate concentration of magnesium, any depression in food intake, such as may occur in grazing cattle in early spring (see Corbett *et al.*, 1963), may cause magnesium intake to fall below the requirement. It is at this time of year that the hypomagnesaemia associated with 'grass tetany' or 'grass staggers' is most prevalent in grazing dairy cows. While grass tetany is largely prevented by supplements of magnesium (e.g. by 50 g MgO per cow per day), it is regarded more as a 'metabolic' disorder than a nutritional one, and is therefore discussed in Chapter 33.

Potassium, sodium and chlorine. Foods contain so much potassium in relation to the animal's requirement for this element that it need not be considered here. According to the Agricultural Research Council's standards (which are derived from rather limited data) a cow yielding 15 kg milk per day would require 18 g sodium and 29 g chlorine per day, over 80 per cent of these amounts being used to replace those secreted in milk. The proportions in which sodium and chlorine are required approximate to those of sodium chloride, and requirements are often stated as sodium chloride. In a dry matter intake of 12 kg per day 18 g sodium and 29 g chlorine represent concentrations of 0·15 and 0·25 per cent, respectively. Many feeds contain lower concentrations than these, with roughages tending to be deficient in sodium, oilseed residues in chlorine, and cereals in both. On the whole, sodium deficiency seems more likely than chlorine deficiency, although the supplement usually employed is sodium chloride.

The signs of sodium deficiency in dairy cattle are a depraved and reduced appetite, loss of bodyweight and reduced milk yield (Smith and Aines, 1959).

Trace elements. Factorial calculations of the kind used for major minerals cannot be used to estimate the trace element requirements of dairy cows, because the trace element contents of milk are less well characterized—and probably more variable—than the contents of major minerals, and because information on endogenous losses and the availability of trace elements is either very limited or non-existent.

Lactation, however, appears to increase requirements for trace elements much less than it does the requirements for major elements. In general, the amounts of the trace elements secreted in milk are small in relation to the minimum dietary intakes considered necessary. For example, a cow yielding 15 kg milk per day would secrete in its milk about 6·0 mg iron and 1·5 mg copper, but the dietary concentrations of these elements which are considered necessary for health would provide minimum intakes of about 360 mg iron and 60 mg copper per day. In contrast, the cow would secrete 18 g calcium per day in milk, but its minimum daily intake would be 60 g.

As the special demands for trace elements imposed by lactation cannot yet be assessed, the trace elements will not be considered individually here. General features of trace element nutrition are described in Chapter 11 of this volume, and also by Underwood (1962). It should be noted in passing, however, that nutritional deficiencies of the trace elements are probably less common in dairy cows than in other cattle or in sheep, because it is less usual for the diet of dairy cows to be restricted to the produce of a particular soil type. The deficiency disease most likely to be encountered in dairy cows is that associated with an imbalance of copper and molybdenum (and sulphate) levels in pasture herbage (Allcroft, 1963).

VITAMIN NUTRITION

Vitamin A

In common with other ruminants, dairy cows obtain practically all their vitamin A from the β-carotene and, to a much lesser extent, the other precursors of the vitamin which are present in their feeds. The carotene is absorbed and converted to vitamin A in the wall of the intestine, and possibly elsewhere (see Moore, 1957). In cattle much carotene escapes this conversion and, together with vitamin A, circulates in the blood and finds its way into the milk.

The structures of carotene and vitamin A and the chemistry of the conversion process suggest that, weight for weight, carotene and pre-formed vitamin A in the diet might be equally effective as a source of vitamin A to the animal. In practice, however, carotene is always less effective. In rats dietary carotene may be as potent as half its weight of the pre-formed vitamin, but in ruminants its maximum potency is about one-fifth. At high levels of intake the relative potency of carotene is much reduced (see Moore, 1957). One reason for the lower potency of carotene is that it is less readily absorbed than the pre-formed vitamin.

In the discussion that follows, unit weight of β-carotene (henceforth referred to simply as carotene) will be considered equivalent to one-fifth that weight of vitamin A alcohol (henceforth called vitamin A). The evaluation of the vitamin A potency of milk on this basis will be satisfactory when the milk is to be consumed by the calf, but as carotene has a higher potency in the rat and perhaps also in man (Hume and Krebs, 1949), the vitamin A potency of milk for man may well be undervalued. One international unit of vitamin A is taken to be equal to 0·300 μg vitamin A alcohol or 0·344 μg vitamin A acetate. The vitamins in milk have recently been reviewed by Hartman and Dryden (1965).

The relative proportions of carotene and vitamin A in milk vary according to the breed of cow and to its carotene intake. When carotene intake is high, as it is in grazing cows, the contribution of carotene to the total vitamin A activity of milk is, at one extreme, 50 per cent or more for Guernsey cows and at the other, 20 per cent or less for Friesians. The vitamin A content of milk is less susceptible than the carotene content to changes in the carotene intake of the cow, and as the latter declines the contribution of carotene to the total vitamin A activity of milk becomes relatively smaller.

The vitamin A activity of milk is dependent on the fat content, or more particularly, on the content of long chain fatty acids in milk (National Institute for Research in Dairying, 1964). The vitamin A activity of milk is therefore commonly expressed per unit of fat or of butter (butter contains about 82 per cent fat); this mode of expression makes differences between breeds of cow quite small.

Some typical values for the vitamin A activity of cow's milk fat are shown in Table 7. The table shows that in most countries vitamin A activity is much lower in winter than in summer, the difference reflecting the relative carotene

TABLE 7
The Vitamin A Activity of Butterfat (μg/g)

Country	Winter	Summer
New Zealand	15·5	12·2
Great Britain	8·5	12·2
Sweden	5·5	9·0
Denmark	6·1	12·2
United States	7·7	12·7
Holland	8·7	13·7

(Calculated from the data of Thompson *et al.* (1964) by assuming that butter contains 82 per cent fat)

contents of the fresh pasture herbage grazed in summer and of the preserved roughages eaten in winter. In New Zealand, however, where cattle are at pasture throughout the year, there is no winter decline. In fact, New Zealand milk fat contains more vitamin A in winter—a phenomenon which has not been adequately explained (see McGillivray, 1960). There may also be a mid-summer decline in milk vitamin A activity, which again is not fully understood.

The relation between the vitamin A activity of milk and the carotene intake of the cow has been described by the Agricultural Research Council (1965) in the form of the equation

$$y = 4{\cdot}40x - 5{\cdot}24 \qquad (11)$$

(where y = vitamin A activity of milk fat, in μg/g, and x = log carotene intake, in μg/kg liveweight per day).

This equation shows that even at low intakes of carotene the proportion finding its way into the milk as either vitamin A or carotene is small. For example, if produced by a 500-kg cow receiving 100 μg carotene per kg liveweight per day (a total of 50 mg carotene), 15 kg milk containing 0·6 kg fat would be predicted to have a total vitamin A activity of 2·1 mg; this would be equivalent to perhaps 4–10 mg carotene. At higher intakes the proportion becomes very much smaller; a cow at pasture, taking in 5 g carotene per day would secrete the equivalent of no more than 35 mg carotene in 15 kg milk. McGillivray (1960) considered that a 'ceiling' level for the vitamin A activity of milk was reached at a carotene intake of about 2·5 g per day. But the values used to derive eq. (11) suggest that the level continues to increase up to intakes of at least 5 g carotene per day. However, as the logarithmic form of the equation implies, increments in carotene intake at these high levels have little effect on the milk; as intake doubles, from 2·5 to 5·0 g per day, the vitamin A activity of milk is predicted to increase by only 12 per cent.

Feeding cows with massive doses of pre-formed vitamin A can increase the vitamin A activity of milk fat to as much as 100 μg per g (Blaxter, Kon and Thompson, 1944) but no more than 5 per cent of the dose is accounted for in the milk.

Of the carotene consumed but not accounted for in milk, the greater part is destroyed in the digestive tract or excreted in faeces. Only a small amount of vitamin A is stored in the liver of the dairy cow. Moore (1957) gives a representative value of 150 i.u. per g liver or about 750,000 i.u. per cow, but greater amounts (*c.* 600 i.u. per g) are found in the livers of cows grazing or fed on carotene-rich roughages (Byers, Weswig, Bone and Jones, 1955).

The metabolism of vitamin A and carotene in lactating animals has been reviewed by McGillivray (1960) and Owen (1965).

Requirements for vitamin A. The vitamin A requirements of animals are difficult to assess because they vary according to the criteria of adequacy employed. For example, Lewis and Wilson (1945) found that while calves needed as little as 8 μg vitamin A per kg liveweight to prevent obvious signs of deficiency, they required 16 μg per kg for maximum growth rate and as much as 64 μg per kg for the consistent promotion of moderate storage in the liver. Few assessments have been made with lactating cows, but the Agricultural Research Council (1965) suggest a basic allowance of 80 μg carotene per kg liveweight per day, *plus* 50 μg per kg liveweight for lactation. The total daily allowance is therefore 130 μg per kg or 65 mg for a 500-kg cow. The National Research Council (1958) recommends a minimum allowance of 44 mg carotene per day for a 500-kg cow, or 200 mg if the vitamin A potency of milk is to be 'optimal'. The Agricultural Research Council's recommendation of 130 mg carotene per day would promote a vitamin A activity in milk fat of only 4 μg/g (eq. (11)), which is much lower than the activity usually found in milk (Table 7). Thus what is adequate for the cow may be less adequate for the consumer of its milk, whether this be the calf, which may have no other dietary source of vitamin A in early life, or man, who obtains as much as 50 per cent of his vitamin A from milk and milk products (McGillivray, 1960).

A requirement of 65 mg carotene per day for a 500-kg cow implies a dietary concentration of about 5 mg per kg dry matter. Of the concentrate foods given to dairy cows only yellow maize contains an appreciable content of carotene (4 mg per kg) or other precursors of vitamin A, and practically all the carotene supplied to the cow is contained in fresh or preserved pasture herbage or other green crops. Fresh pasture herbage contains as much as 500 mg carotene per kg dry matter, but preserved roughages can contain anything between this and no carotene at all (Chapter 17). As carotene is so susceptible to oxidation, particularly photochemical oxidation, sun-dried roughages may contain less than 5 mg per kg. Kale, although a rich source of carotene, also contains a goitrogen which by suppressing thyroxine production may reduce the absorption of carotene (see Owen, 1965; Thompson *et al.*, 1964). Modern developments in the winter-feeding of dairy cattle are tending to give diets either much higher or much lower in carotene than was previously the case. On the one hand, the tendency on many farms for silage and barn-dried hay to replace sun-dried hay should increase the carotene intake of the cow; on the other, increasing reliance on concentrates, perhaps to the complete exclusion of roughages other than straw, has made it necessary for cows to be given dietary supplements of carotene or pre-formed vitamin A.

A deficiency of carotene in the diet of dairy cows is likely to manifest itself by affecting reproductive performance before inducing the well-known eye lesions. There may be difficulty in getting cows in calf, and calves may be dead, weak or blind at birth (Ronning, Berousek, Kuhlman and Gallup, 1953). The Agricultural Research Council (1965) recommends a higher allowance of carotene (150 μg per kg liveweight or 75 mg for a 500–kg cow) for pregnant than for lactating cows, and the recommendation of the National Research Council (1958) is the same.

Vitamin D

Cattle receive vitamin D_2 (ergocalciferol) in their food, since this is produced in dead plant cells by the ultraviolet irradiation of ergosterol, and they also obtain vitamin D_3 (cholecalciferol) by the irradiation of the 7-dehydrocholesterol present in their skin. In theory, the former should be their major source in winter, when they eat sun-dried hay but have little access to sunlight themselves, and D_3 should predominate in summer. Milk can contain both D_2 and D_3, but according to Günther *et al.* (see Günther, 1964) D_3 is the usual form, even in winter milk. Vitamin D_2 is found in milk when the cow is given massive doses of this form of the vitamin (see below). The total vitamin D content of milk is in the range 0·1 to 1·0 i.u. per g milk fat, being at the bottom end of this range in winter and 3–4 times as great in summer as in winter (Thompson *et al.*, 1964).

The factor apparently having the greatest influence on the vitamin D content of milk is the insolation of the cow, although the effects of insolation are not readily predictable. Wallis (1943) found that cows previously depleted of vitamin D and then kept out-of-doors but on winter rations had low levels of vitamin D in milk. Campion, Henry, Kon and Mackintosh (1937) showed that grazing cows gave milk richer in vitamin D than cows kept out-of-doors on winter rations. Also, the vitamin D content of milk is not always related to the amount of ultraviolet light which the cow should have received (Thompson *et al.*, 1964).

Milk is a relatively poor source of vitamin D for Man. It contributes only a small proportion of his total intake of the vitamin, and infants reared largely on milk require an additional source. For this reason there have been attempts to increase the potency of cow's milk by giving the animals rich sources of vitamin D such as irradiated yeast. The vitamin D content of milk can be increased to over 10 i.u. per g fat in this way but, as with vitamin A, only a small proportion of the dose given (1–2 per cent) is recovered in milk. An increase in potency is more readily made by either irradiating the milk or adding the vitamin to it.

Vitamin D deficiency is practically unknown in dairy cows although it was produced experimentally by Wallis (1938), who kept cows away from sunlight and on a diet devoid of the vitamin. This treatment rapidly reduced the vitamin D content of milk to zero and brought about steady declines in blood calcium and phosphorus levels. After 4–8 months, signs of stiffness and swollen joints appeared, and bone fractures occurred. A single dose of about 42,000 i.u. vitamin D cured the stiffness and prevented its recurrence for a further 4 months.

It is known that the vitamin D requirements of adult animals are small provided that they receive calcium and phosphorus in adequate quantities and the correct proportions. The Agricultural Research Council (1965) suggest that pregnant or lactating cows should receive daily 10 i.u. per kg liveweight.

In feeds vitamin D was formerly considered to occur only in those of animal origin, such as fish meal, and in sun-dried roughages. More recently it has been found in green pasture herbage, sometimes at levels sufficient to meet the requirement mentioned above and, exceptionally, at a level of 1800 i.u. per kg dry matter (Keener, 1954). Autumn herbage, which contains much dead plant material, was found to be considerably richer in vitamin D than spring growths. Hays contain 150 to 3000 i.u. per kg (Wallis, Kennedy and Fishman, 1958); while those cured out-of-doors tend to be richer than those dried indoors there is considerable unexplained variation in vitamin D content. The vitamin D content of silage may increase during storage (Keener, 1954). It is possible that unexplained variations in the vitamin D content of grass products may be due to the presence of an anti-vitamin D factor. A rachitogenic factor in pasture herbage was thought to be carotene (Grant and O'Hara, 1957), but there is some evidence against this (see Henry, Thompson, McCallum and Stewart, 1959).

VITAMIN E

The vitamin E activity of cow's milk is due mainly to α-tocopherol, which is normally present to the extent of 10–50 μg per g fat. The tocopherol content of milk is strongly influenced by the tocopherol intake of the cow, even though less than 1 per cent of that present in the diet may appear in milk (Krukovsky, Trimberger, Turk, Loosli and Henderson, 1954). As pasture herbage is richer in tocopherol than winter diets, there are seasonal fluctuations in the tocopherol content of milk, which may be twice as high in summer as in winter (Anglin, Mahon and Chapman, 1955). The lowest values are reached in late winter, and in beef cows fed on diets exceptionally deficient in the vitamin, milk tocopherol may fall to less than 5 μg per g fat (Blaxter, 1957). Gullickson *et al.* (1949) found that if dairy cattle were fed for several years on diets very low in tocopherol they showed none of the signs of reproductive failure which such diets promote in rats, but many animals died suddenly from heart failure. These deaths occurred shortly before or after parturition. The vitamin E activity of feeds varies from zero in some concentrates and in roots to the equivalent of 400 mg per kg α-tocopherol in young pasture herbage. There may be extensive destruction of the vitamin in hay making, and hays can contain as little as 10 mg per kg. However, the requirements of adult cattle seem to be quite low and vitamin E deficiency does not occur in dairy cows in practice.

Requirements for tocopherol depend on the presence or absence of other substances in the diet. Thus, unsaturated fatty acids increase requirements and selenium may reduce them (Blaxter, 1962b). At present it does not seem possible to state the requirements of either adult or young cattle.

Vitamins B

Adult ruminants have no dietary requirements for B vitamins as these are synthesized by alimentary microorganisms. The levels of B vitamins in cow's milk are not affected by the diet (Ling *et al.*, 1961). A possible exception to this rule is that a dietary deficiency of cobalt may reduce the vitamin B_{12} content of cow's milk, since it has this effect in ewes (see Underwood, 1962).

THE EFFECTS OF LEVEL OF FEEDING OF DAIRY COWS ON THEIR MILK PRODUCTION

Up till now we have been concerned largely with defining the nutrient requirements of cows for milk production. The situation has been a static one in the sense that our interest has been in meeting the needs of animals of stated liveweight producing at stated levels of production. The farmer, however, must often use a more dynamic approach. While he may be interested to know the quantity of food needed by a 500-kg cow giving 15 kg milk per day, he may be equally interested in knowing what is likely to happen when the cow is given more or less than this quantity. His interest will extend from the short-term to the long-term effects of under and over feeding and, since foods may be less plentiful at one time than at another, he may also wish to know whether the effect of underfeeding at one time may be counterbalanced by overfeeding at another.

Many experiments have been made to answer these questions. In a typical level of feeding experiment the levels compared may involve different quantities of a single food or of a diet of fixed composition, but more commonly they involve differences in the composition of rations as well as in their size. Thus, varying quantities of concentrates may be added to a fixed quantity of roughage, or there may be a stepwise replacement of roughage by concentrates. As a result, the differences between one level and another usually involve all nutrients, but some more than others. For example, the addition of concentrates to a basal level of roughage is likely to increase phosphorus intake by a greater proportion than calcium intake. And so, while level or feeding experiments have the advantage of simulating the choices open to the farmer, they have the potential disadvantage of being difficult to interpret, in that the responses observed may not be attributed to specific nutrients. This disadvantage may be more potential than real, however, for in most levels of feeding experiments, protein, minerals and vitamins have been given in amounts greater than requirements, even at the lower levels of feeding, and the factor responsible for any differences between levels in their effects on the cows has been energy intake. Even where low levels of feeding have been inadequate with respect to protein, minerals and vitamins they have usually been still more inadequate with respect to energy.

The milk yield of the cow may be affected by level of feeding in early life (from birth to first calving), in the dry period which precedes each lactation, and during each lactation.

The Effects of Level of Feeding in Early Life

The general plan of the experiments to be referred to in this section has been to impose two or more levels of feeding from birth (or a little later) to first calving (or a little earlier) and then to feed all animals according to milk yield during lactation. The levels have usually been chosen so that one provides a growth rate considered normal by Ragsdale's (1934) or other standards. The experiments of Crichton, Aitken and Boyne (1960) and Evans (1964) were unusual in that half the animals were subjected to a reversal of the treatments half-way through the rearing period. In all experiments except those of Hansson (1956) the levels could be imposed accurately only during the winter, when the animals were indoors; during the grazing season the treatment differentials were either discontinued (Evans, 1964) or were maintained in an approximate fashion by the use of pastures of different quality (Crichton *et al.*, 1960) or by supplementary feeding (Reid, Loosli, Trimberger, Turk, Asdell and Smith, 1964). The treatments have been evaluated mainly in terms of milk yield and composition in first and subsequent lactations although observations have also been made of reproductive performance (e.g. by Reid *et al.*, 1964) and of the longevity of the animals.

The seven main experiments that have been made are summarized in Table 8. There are others, but they are either small (e.g. Swanson, 1960; Swanson and Hinton, 1964; Belic, Pavlovic and Bokorov, 1963) or incompletely described (New Zealand Department of Agriculture, 1958). There are also the reviews of Burt (1956) and of Breirem, Ekern and Homb (1961). The collation of the main experiments is difficult for a variety of reasons, the main one being that there is no common definition of level of feeding. Where food or energy intakes are given, the values often include estimates of intake during the grazing season, and they are expressed in a variety of units—starch equivalent, total digestible nutrients and Scandinavian feed units. The cattle used differed in breed and hence in size and milk production. They also differed in age at calving; in some experiments all animals calved at the same age but in others the poorly-fed animals calved later.

In the absence of adequate data for energy intake, the liveweight gains probably provide the most reliable estimates of the relative planes of nutrition. These range from 0·70 kg per day for the high-plane animals of Reid *et al.* (1964) down to 0·30 kg per day for those of Hansson (1956) which were fed at 40 per cent of the normal level. (The latter were found difficult to get in calf.) The differences between treatments in liveweight gain are much smaller than might be predicted from the differences in food intake. In part this is to be expected, because the faster growing animals probably retained more

TABLE 8

The Influence of Level of Feeding during the Rearing Period on Milk Production in Heifers

Reference	Animals (no. with first lactations)	Treatments	Daily live-weight gain to 2 years of age (kg)	Liveweight (kg)		First lactation[a]	
				At 24±1 months	After first parturition	FCM yield (kg)	Fat (per cent)
Bonnier *et al.* (1948)	18 Swedish Red and White twins	Low: 1928 SFU[b] (1–27 months; includes pasture)	0·38	317	—	3240	3·57
		High: 3705 SFU	0·52	415	—	3530	3·74
Hansson (1956)	64 Swedish Red and White twins	40:1330 SFU, 1–25 months	0·30[c]	—	—	—	—
		60:1620 SFU, 1–25 months	0·44	330	—	3460	3·86
		80:2100 SFU, 1–25 months	0.54	380	—	3490	3·89
		100:2600 SFU, 1–25 months	0·62	430	—	3350	3·97
		120:3050 SFU, 1–25 months	0·69	450	—	3180	4·09
		140:3350 SFU, 1–25 months	0·69	460	—	2860	4·03
Steensburg (1947); Hansen and Steensburg (1950)	86 Red Danish	Low: 2·17 kg SE per day	0·46	371	372	2839	3·99
		Normal: 2·65 (Birth to calving)	0·54	425	405	2631	4·05
		High: 3·06	0·58	453	439	3050	4·05
Eskedal and Klausen (1958)	122 Jersey	Low: 2094 SFU (2 months to calving; includes pasture)	0·38	298	328	3739	6·06
		Normal: 2387 SFU	0·42	322	336	3580	5·96
		High: 2461 SFU	0·46	348	337	3406	6·25

[a] Of about 300 days. The yields given by Bonnier *et al.* (1948) and Hansson (1956) are for 252 days and have been increased by 11 per cent.

[b] Scandinavian feed units.

[c] Gains from 1 to 19 months.

TABLE 8—*continued*

The Influence of Level of Feeding during the Rearing Period on Milk Production in Heifers

Reference	Animals (no. with first lactations)	Treatments		Daily live-weight gain to 2 years of age (kg)	Liveweight (kg)		First lactation	
					At 24±1 months	After first parturition	FCM yield (kg)	Fat (per cent)
Crichton *et al.* (1960)	48 Ayrshire and Friesian twins		*0–44, 44–104 weeks*					
		LL[d]	453+1071 kg TDN	0·42	336	370	2748	3·74
		LH	453+1880 kg TDN	0·52	414	398	2797	3·85
		HL	710+1054 kg TDN	0·48	381	395	2357	3·81
		HH	710+1871 kg TDN	0·57	447	429	2798	3·92
Evans (1964)	63 Shorthorn	LL[e]	(L to gain 0·44 kg/day	0·41	321	—	2020[f]	3·50[f]
		LH	in first winter or	0·40	319	—	2610	3·55
		HL	0·07 in second; H to	0·43	349	—	2350	3·45
		HH	gain 0·65 or 0·36 kg/day)	0·44	360	—	2480	3·60
Reid *et al.* (1964)	98 Holstein	Low:	2·44 kg TDN per day	0·39	311	384	3843	3·62
		Medium:	3·95 (Birth to first calving;	0·57	434	482	3986	3·77
		High:	5·92 *plus* pasture)	0·70	526	548	4086	3·81

[d] At 44 weeks, half the animals on treatment L (low) were changed to H (high), and half those on H were changed to L.

[e] For the second winter, half the animals which had received treatment L in the first winter were changed to H, and half those previously on H were changed to L. During the summers all animals were treated alike.

[f] Approximate values (exact fat percentages not stated).

energy per unit of gain than the slower growing (and thus became *fatter* rather than skeletally *larger*), and because they required more feed for maintenance. However, low plane animals often seem to have overcome the handicap put upon them during the winter by growing faster at pasture. This compensatory growth is most obvious in the experiment of Evans (1964), in which all animals grazed together, but it occurred also when an attempt was made to maintain the treatment differentials at pasture (e.g. by Crichton *et al.*, 1960). Further compensatory growth occurred in late pregnancy, whether or not all animals were fed alike then, and after calving, differences in liveweight were always smaller than at 2 years of age.

There are large differences in milk yield from one experiment to another, but within experiments the differences are quite small. Of those experiments having the widest ranges in plane of nutrition, the Swedish ones (Bonnier *et al.*, 1948; Hansson, 1956) suggest that both very low and high planes during rearing have deleterious effects on milk production, but that of Reid *et al.* (1964) showed no significant differences in milk production. The Jersey cows of Eskedal and Klausen (1958) seemed to be affected by quite small differences in plane of nutrition. Swanson (1960) and Swanson and Hinton (1964) found Jerseys to be adversely affected by both low and high levels of feeding.

The reversal experiments of Crichton *et al.* (1960) and Evans (1964) suggest that level of feeding in the later stages of rearing is more important than in the earlier stages. The first-named found that heifers changed from a high to a low plane for the last 14 months of rearing, gained less during that period than those kept on a low plane throughout, and gave the least milk of all four treatment groups. Evans's LH group made exceptionally rapid gains during their second winter and gave most milk.

The performance of cows in later lactations was recorded by most investigators; Reid *et al.* (1964) give yields up to the fourth lactation for at least 28 cows per treatment. In general, differences between treatments in liveweight tended to disappear almost completely as the cows aged, and differences in milk yield also became smaller. However, the larger differences in yield found by Crichton *et al.* (1960) and Evans (1964) tended to persist until the third and later lactations.

Nearly all the experiments agree in showing that a high plane of nutrition during rearing promotes an increased percentage of fat in milk. In several cases this effect was still evident in lactations after the first.

Several of the Scandinavian experiments showed that animals reared on a high plane tended to have shorter useful lives, because they were the more frequently culled for low yield, susceptibility to disease and reproductive difficulties. Thus Hansen and Steensberg (1950) found the average life span of their low, normal and high plane cows to be 86·7, 80·1 and 67·2 months, respectively. In the Cornell experiment (Reid *et al.*, 1964), 80–85 per cent of the animals reached their fifth calving, regardless of treatment.

Perhaps the firmest conclusion one may draw from these experiments is that cattle are remarkably tolerant to departures from the normal in their level of feeding during the rearing period. When offered the opportunity they compensate for periods of slow growth, and even if they are much below the normal weight at their first calving, they still produce satisfactory quantities of milk. In other words, the level of feeding during rearing can vary over quite a wide range without affecting subsequent milk production, although very low or very high levels may have adverse effects. This conclusion is supported by studies made with large numbers of cows of the relation between milk production and age or weight at first calving. For example, Erb and Ashworth (1961) found that at constant age an increase in weight of 45 kg increased lactation yield by only 107 kg—a non-significant improvement; Clark and Touchberry (1962) found yield to increase by less than 2 per cent for a 45-kg increase in weight; and Miller and McGilliard (1959) concluded that only 2 per cent of intra-herd variations in yield were explained by the liveweight of the cows.

The optimum level of feeding may therefore be determined by factors other than milk yield. A relatively low level may favour longevity, but it may also delay sexual maturity and hence increase age at first calving (Reid *et al.*, 1964). The effects of nutrition on longevity in cows are certainly worth further investigation—and the final outcome of the Cornell experiment is awaited with great interest—but studies lasting 20 years or so are needed. Studies of the effects on milk production of nutrition in the first pregnancy would also be valuable.

The Effects of Level of Feeding during the Dry Period

Ideally, a dairy cow should calve at yearly intervals and should lactate for about 300 days, but in practice calving intervals are often longer or lactations shorter. The cow therefore has a dry period between lactations which in the ideal situation is about two months but in practice may be longer. During this dry period the cow should complete the replenishment of body reserves lost in early lactation, and will require nutrients also to provide for the rapid growth of the product of conception which occurs during the later stages of pregnancy, and for the regeneration of mammary tissue. Requirements for these processes are small in relation to those for lactation, however, and when the cow calves she has to adapt herself to a sudden increase in food intake.

The levels of feeding normally imposed during the dry period are quite variable. Under the semi-natural conditions of New Zealand dairy farming, cows generally calve in the early spring, and the dry period comes at a time when their food—pasture herbage—is most scarce. Elsewhere, dry cows may be kept apart from the main herd until shortly before calving and fed on poorer quality feeds—inferior pasture for those calving in autumn and lower quality roughages for spring-calving cows. In the practice known as 'steaming up', however, dry cows are offered quantities of concentrates which increase gradually during the last six weeks of pregnancy. By the time of calving, the amounts of concentrates given are about 75 per cent of those the cow is anticipated to require in early lactation. Steaming up is claimed to increase milk production, in part by preparing the cow for the large rations of concentrates she may receive in early lactation.

The questions raised about feeding during the dry period are, therefore, whether a low plane of nutrition will have adverse effects on subsequent milk production, and whether a high plane, promoting considerable storage of nutrients, is beneficial. These questions are obviously complicated by the condition of the cow at the end of the lactation preceding the dry period; if she has experienced no net loss of weight during lactation then she is less likely to be affected by a low level of feeding during the dry period than if lactation has made appreciable inroads on her reserves. In heifers, the response to feeding during the dry period is likely to depend on the level imposed previously, during the rearing period.

The general plan of the experiments to be referred to in this part of the chapter has been to impose two levels of feeding for a period varying from 2 to 10 (but usually 6) weeks before calving, and during lactation to feed all cows according to the same standards for maintenance and production. In experiments specifically concerned with underfeeding the feeds used have usually been pasture herbage and conserved roughages. In those concerned with steaming up, the lower level of feeding has been provided by roughages and the higher level(s) by adding concentrates to roughages. In some experiments the concentrates have been given at a constant rate; in others the daily rate has been gradually increased over the dry period.

As in the studies of level of feeding during the rearing period, records of feed intake are often incomplete and difficult to compare with one another in terms of energy intake. Once again, liveweight changes provide an indication of the planes of nutrition, although one should bear in mind that in the experiments to be considered weight changes were measured over periods of only a few weeks, and that they may have been biassed by large changes in the weights of intestinal contents. The data of Jakobsen (1957) suggest that in the later stages of pregnancy the weight of the uterus and its contents increases as shown in Table 9. The values given by Swett, Matthews and Fohrman (1948), however, are rather lower than this; they range from about 24 to 34 kg for the last 60 days of pregnancy, the lower amount being for small cows (Jerseys) and the higher for large animals (Holsteins). Patchell (1957), working from the unpublished data of Dr. L. R. Wallace, calculated that in Jersey cows the uterus and its contents increases by about 30 kg during the last 10 weeks of pregnancy. It may be presumed that if pregnant cows gain less than the amounts shown in Table 9 (or somewhat lesser amounts in the case of small breeds) then their own body weight (hereafter called maternal liveweight) is declining. Provisionally, we may assume that any level of feeding during the dry period which leads to apparent losses in maternal liveweight is underfeeding.

TABLE 9

The Weight of the Uterus and its Contents in Cows during the Later Stages of Pregnancy

Weeks before calving	Weight (kg)	Average rate of gain to calving (kg/day)
10	24·5	0·65
9	27·2	0·68
8	30·2	0·71
7	33·6	0·74
6	37·3	0·78
5	41·4	0·82
4	46·0	0·86
3	51·0	0·90
2	56·7	0·95
1	63·0	1·00
0	70·0	—

Calculated from the data of Jakobsen, 1957

The main experiments concerned with nutrition in the dry period are summarized in Table 10. Some were reviewed by Burt (1956) and Broster (1958). The liveweight gains observed (–0·3 to 1·1 kg per day) suggest that changes in maternal liveweight ranged from a loss of as much as 1 kg per day (Foot, Line and Rowland, 1963) to a gain of about 0·3 kg. In some experiments undernutrition, as defined by maternal losses in liveweight, has been compared with adequate nutrition, or maternal gains of zero (Flux, 1950; Patchell, 1957). In others, adequate feeding has been compared with overfeeding, or steaming up (Greenhalgh and Gardner, 1958; Nordfeldt, Hoflund, Ånstrand and Hallgren, 1959). The experiments of Broster, Ridler and Foot (1958) and Castle and Watson (1961) probably come into this category, although liveweight changes are not given. In between these two categories are studies in which the liveweight gains observed lie on both sides of the figure considered to represent the gain in products of conception, and therefore in which underfeeding was compared with overfeeding. The extreme example of this is the trial of Foot *et al.* (1963), but that of Swanson and Hinton (1962), which was really concerned with steaming up, probably comes into the same category. In the three-treatment experiments of Blaxter (1944) and of Schmidt and Schultz (1959) some of the comparisons are also of under- with overfeeding. In some instances, the aim of achieving widely-contrasting levels by adding concentrates to roughage diets seems to have been missed because large rations of concentrates were not fully consumed and also depressed roughage intake. Thus Schmidt and Schultz (1959) found their highest-level cows to eat only 5·8 of the 6·8 kg concentrates offered, and they ate less hay than the

TABLE 10

The Influence of Level of Feeding during the Dry Period on Milk Production in the Following Lactation

Reference	Animals	Length of period (days)	Treatments (Feed quantities in kg/day)	Liveweight gain (kg/day)	FCM yield (kg) (and days)	Milk composition (per cent) Fat	SNF
(1) Flux (1950)	10 Jersey-type twin cows	70	(1) Poor grass+<2·7 hay	—0·1	1823 (270)	4·71	—
			(2) Grass, hay+silage *ad lib.*	0·5	2226	4·88	—
(2) Patchell (1957)	36 Jersey twin cows	70	(1) Bare pasture+2·7 hay, 1·8 silage	—0·6[a]	2172[b] (252)	—	9·43
			(2) Better pasture+roughages *ad lib.*	0	2505[b]	—	9·54
(3) Foot *et al.* (1963)	46 Friesian-type heifers	60[c]	(1) Restricted, sparse pasture	—0·3	2103 (189)	3·59	8·51
			(2) Good pasture+3·6 concentrates	1·0	2587	3·75	8·63
(4) Blaxter (1944)	51 cows and heifers, mixed breeding	42	(1) Limited roughage	0·2	1600 (133)[d]	3·80	—
			(2) Liberal roughage	0·9	1740	3·99	—
			(3) Roughage+0·9–2·7 concentrates ((2) and (3) equal in energy intake)	1·1	1980	3·99	—
(5) Swanson and Hinton (1962)	150 Holstein and Jersey cows	42	(1) Roughage *ad lib.*	0·4	3417 (210)	4·02	—
			(2) Roughage+3·6 concentrates	0·7	3618	4·00	—
(6) Greenhalgh and Gardner (1958)	36 cows and heifers, mainly Holsteins	42	(1) Silage 0·9, hay 0·7/45 kg liveweight	0·6	1672 (84)	3·83	—
			(2) As (1)+2·0–4·8 concentrates	1·0	1657	3·90	—
(7) Schmidt and Schultz (1959)	63 cows, mainly Holsteins	56	*Silage* *Hay* *Concentrates*				
			(1) 12·8 8·8 0	0·4	5512 (300)	4·06	—
			(2) 11·8 8·4 2·7	0·6	5510	4·07	—
			(3) 12·3 6·4 5·8	0·9	5518	3·99	—
(8) Nordfeldt *et al.* (1959)	40 cows of about 600 kg liveweight	*ca* 40	(1) 'Normal feeding'	0·7	3010 (130)	4·02	—
			(2) 32–45 per cent more SFU than (1)	1·2	2909	4·07	—

[a] Corrected for estimated gain of uterus and contents; actual gains about 0·4 kg/day greater.

[b] Milk, not FCM.

[c] Concentrates given for last 42 days.

[d] Yields estimated from graphs.

TABLE 10—*continued*

The Influence of Level of Feeding during the Dry Period on Milk Production in the Following Lactation

Reference	Animals	Length of period (days)	Treatments (Feed quantities in kg/day)	Liveweight gain (kg/day)	FCM yield (kg) (and days)	Milk composition (per cent) Fat	SNF
(9) Broster *et al.* (1964)	40 Friesian heifers	35	*SE* *DCP*				
			(1) 3·8 0·30 } supplied by hay	0·2	1149 (66)	3·9	8·8
			(2) 4·1 0·53 } *ad lib.* and vary-	0·7	1164	4·1	8·9
			(3) 4·5 0·44 } ing amounts of	1·1	1179	4·2	9·0
			(4) 4·7 0·97 } concentrates	0·9	1170	4·0	8·8
(10) Hancock (1953)	14 Jersey twins	28	(1) Pasture + silage	0·8	1052 (84)	4·61	—
			(2) As (1) + 2·5 concentrates	0·9	1179	4·66	—
(11) Wallace (1958)	36 twin cows	42	(1) Pasture, intake 4·1 digestible OM	0·3	807 (56)	5·01	—
			(2) Pasture, intake 4·9 digestible OM	0·5	862	4·93	—
			(3) Pasture, intake 5·6 digestible OM	0·6	837	5·06	—
			(4) Pasture, intake 6·3 digestible OM	0·9	868	5·78	—
(12) Broster *et al.* (1958)	88 Shorthorn and Friesian heifers		*Before calving (concentrates/day)* *After calving (concentrates/10 kg milk)*				
		14	(1) 1·8 3	—	2924 (305)	3·69	8·98
		14	(2) 1·6 5	—	3390	3·48	8·97
		21	(3) 5·2 3	—	3539	3·69	9·04
		21	(4) 5·4 5	—	3385	3·38	9·02
			(all + grazing, hay and silage)				
(13) Castle and Watson (1961)	32 Ayrshire cows	35	*Before calving (concentrates/day)* *After calving (concentrates/10 kg milk)*				
			(1) 2·3 2	—	3694 (283)	4·17	8·63
			(2) 2·3 4	—	3978	4·33	8·78
			(3) 4·8 2	—	3574	4·36	8·63
			(4) 4·8 4	—	4121	4·19	8·67
			(all + grazing, or 4·5 hay and silage *ad lib*).				

other cows. Holmes *et al.* (1965) and Holmes and Jones (1965), who made a special study of food intake in steamed up cows, found that 1 kg concentrate dry matter depressed roughage dry matter intake by up to 0·4 kg.

Several investigators have measured liveweight changes after calving. In general, the treatment groups which gained most in the dry period lost most in the early part of lactation. For example, treatment groups (1) to (3) of Schmidt and Schultz (1959) lost 0·3, 0·4 and 0·8 kg per day, respectively, after calving, and after 84 days of lactation the weight advantages of the higher level groups had disappeared.

The milk yields given are for the longest period over which measurements were made. In some of the original papers, records are given separately for the first 6–12 weeks of lactation and for the complete lactation. In these cases, differences between treatments were often greater in early lactation than overall. It seems quite clear that underfeeding during the dry period depresses milk production. The New Zealand experiments (Flux, 1950; Patchell, 1957; Wallace, 1958) show this effect most clearly but there is also some evidence for it in Blaxter's experiment (treatments 1 v. 2) and that of Swanson and Hinton (1962). However, Schmidt and Schultz's groups (1) and (2) gave similar yields, and so did treatment groups (1) and (2) of Broster *et al.* (1964). How underfeeding depresses milk production is not known, although a poorer development of mammary tissue may be involved.

The effects of steaming up are less clearly established. In two of the four experiments in which it had a positive effect on yield (Blaxter, 1944; Swanson and Hinton, 1962), the practice may have been compared with underfeeding, and no weight changes are given by Broster *et al.* (1958). The steaming up treatment of Hancock (1953) had little effect on liveweight gain but nevertheless increased milk production. It is therefore possible that Hancock's control cows ate more roughage than those which were steamed up, and his two treatments may correspond to treatments (2) and (3) of Blaxter (1944). If so, the two experiments support the suggestion that concentrates given during the dry period have some special merit not possessed by additional roughage in preparing the cow for concentrate feeding after calving. In this regard, it is significant that Wallace (1958) was unable to stimulate milk production by additional pasture herbage, even though liveweight gain was increased from 0·5 to 0·9 kg per day (treatments 2–4). In most of the experiments specifically concerned with steaming up, and in which the addition of concentrates to a basal roughage ration caused appreciable gains in maternal liveweight, the practice failed to improve milk yields. Broster *et al.* (1958) found steaming up to have a positive effect only when it was followed by a lower than normal level of feeding after calving. They suggest that steaming up may permit lower levels of concentrate feeding during lactation, with a consequent saving of feed costs (cf. their treatments (2) and (3)). Castle and Watson (1961), however, found no interaction between levels of feeding before and after calving; in their experiment a low level after calving depressed milk yield regardless of the level imposed during the dry period.

Underfeeding during the dry period depresses the fat and solids-not-fat content of milk, but steaming up seems to have little or no beneficial effect.

Two points need to be borne in mind when one considers the effects of steaming up. The first is that in many of the experiments listed in Table 10 the variations in yield between similarly-treated cows were quite high, and coefficients of variation were commonly 15–20 per cent. The second point is that made earlier, that the response of the cow to steaming up probably depends on her body condition at the beginning of the dry period. Restoring the reserves of a thin cow may well have a greater effect on subsequent milk production than increasing the reserves of an already fat cow.

While the case for avoiding underfeeding during the dry period seems to be proved, that for encouraging liberal feeding remains not proven. Possibly, the decision whether or not to steam up should be based on the individual cow's condition at the beginning of the dry period.

THE EFFECTS OF LEVEL OF FEEDING DURING THE LACTATION

The rations of dairy cows are commonly calculated in the following manner. From time to time during the lactation—perhaps at the end of each week—the milk yield of each cow is measured and her liveweight assessed. The quantities of nutrients required to produce that yield and to maintain that liveweight are then calculated and used to determine the ration for the next period of lactation. An obvious disadvantage of this procedure is that rationing is according to previous rather than current yield, which means that cows are likely to be underfed in early lactation, when yield is rising, and overfed when yield is declining. To overcome this, good cowmen try to feed for anticipated future yields, especially in early lactation. There is a close association between the cow's peak yield, usually achieved in the second month, and her total lactation yield, and so it is considered important that the cow should not be underfed in early lactation. Nevertheless, cows are often quite resistant to underfeeding in early lactation. Kajanoja (1944) found that the reduction in milk yield resulting from a decrease in energy intake was less than half that predicted from feeding standards, and also that cows underfed for 2 or 3 weeks quickly recovered to give normal yields. When Broster *et al.* (1958) gave cows either 2·0 or 3·2 kg SE per 10 kg milk (as compared with a normal allowance of 2·6 kg) during the first 12 weeks of lactation, those given the higher allowance gave only 2 per cent more FCM in that period (but about 7 per cent more milk). Subsequently the cows were all fed alike and over the whole lactation those fed at the higher level initially gave 5 per cent more FCM.

In grazing trials, Flux and Patchell (1954) and Patchell (1957) divided pairs of twins between high and low levels of feeding for a 6-week period at or near the beginning of lactation. The yields they obtained in these periods and in the remainder of the lactation are shown in Table 11. In Patchell's experi-

TABLE 11

The Effects of Varying the Level of Feeding for Six Weeks Early in the Lactation on Current, Subsequent and Total Yield

Reference	Character	Level of feeding	Yield (kg) in:		
			6-week period	Rest of lactation	Complete lactation
Patchell (1957)	Milk (1952–3)	Low	492	2037	2529
		High	551	2003	2554
	Milk (1953–4)	Low	449	2036	2485
		High	532	2179	2711
Flux and Patchell (1954)	Milk	Low	262	1302	1564
		High	355	1497	1852
	FCM	Low	324	1535	1859
		High	411	1774	2185

ments, depressing milk yield by 12 or 16 per cent in early lactation had little effect on subsequent production. In Flux and Patchell's trial the initial depression in milk yield was about 40 per cent and the subsequent effect was considerable, although statistically non-significant. The differences in FCM production, however, were smaller. (It is quite usual for short-term depressions in milk yield to be accompanied by increases in fat percentage.)

It should be noted that in early lactation there may be difficulty in getting cows to eat the quantities of food they require, because their requirements are then high and their appetite less than maximal. As a result, cows usually lose weight in early lactation.

A second disadvantage of the usual procedure for rationing cows is that the emphasis is on yield determining level of feeding rather than the reverse. The basic assumption made is that the yield of the cow is controlled by factors other than the current level of feeding, such as genetic potential and previous nutrition, and that feeding during lactation should conform to yield rather than determine it. The fact that inherited differences in yield between cows are much smaller than those due to environment (including feeding) has been shown in experiments in which pairs of twin cows were divided between herds with contrasting environments (Brumby, 1959; Wiener, 1960). Nevertheless, there may be large differences in milk yield among cows managed together and having the same plane of nutrition (see Broster, 1963).

The question often arises whether the yield of a particular cow could be improved by increasing its food intake above the recommended standards (and also how the value of any extra milk produced would compare with the cost of extra food). Numerous experiments on this subject have been conducted; a few were reviewed by Yates, Boyd and Pettit (1942) and a greater number by Burt (1957a). In these experiments, groups of cows are generally given roughage either *ad libitum* or at a limited level sufficient to provide energy for maintenance, and varying amounts of concentrates. The quantities of concentrates given are usually related to the current yields of individual cows, typical treatments being 3, 4 or 5 kg concentrates per 10 kg milk produced. Alternatively, all the cows on each treatment may be given the same amount of concentrates per day, with the daily amounts being related to some hypothetical lactation curve rather than to actual yield. The justification for varying only the concentrate part of the ration is that this represents the usual choice open to the farmer. The addition of varying amounts of concentrates to a fixed level of roughage, however, produces rations varying in quality as well as quantity, and variations in quality may be widened further if the concentrate additions reduce roughage intake. Nevertheless, in very few of these experiments has a diet of constant composition been compared at different levels of intake (for examples, see Bloom, Jacobson, McGilliard, Homeyer and Heady, 1957; Nordfeldt and Claesson, 1964). Burt (1957a) used the food intakes quoted for the experiments he reviewed to calculate the starch equivalent intakes of the cows. He also sub-divided the experiments into long-term trials covering a complete lactation or longer, and short-term trials, in which it was usual for each animal to be exposed to each level of feeding within one lactation.

Long-term trials. The main long-term trials considered by Burt (1957a) were those of Jensen, Klein, Rauchenstein, Woodward and Smith (1942). In these the quantities of production starch equivalent (i.e. quantities in excess of those required for maintenance) varied from 2·5 to 4·1 kg per 10 kg 4 per cent FCM, there being a total of six levels of feeding. The normal standards are 2·6 kg SE in Britain (Evans, 1960) or 4 SFU (2·8 kg SE) in Scandinavia; the latter is close to the Total Digestible Nutrients standards used in the United States (see Blaxter, 1950). The experiments of Jensen *et al.* were therefore largely concerned with overfeeding by current standards. The main conclusion drawn was that the response in terms of extra milk was generally less than half that predicted by the standards, of about 3·7 kg FCM per kg additional starch equivalent. The best response, of 1·8 kg FCM per kg SE, was given by the first increment, and the worst (0·6 kg FCM per kg SE) by the last increment. In other, Scandinavian, experiments the allowance varied from 2·0 to 3·4 kg SE per 10 kg FCM; in these, some of the later increments failed to increase milk production at all.

Of the long-term experiments published since Burt's (1957a) review the most interesting are those of Bloom *et al.* (1957), Nordfeldt and Ruudvere (1963) and Nordfeldt and Claesson (1964). Each of these experiments was of a

3 × 3 × 4 factorial design. Cows drawn from three strata of productive ability, as determined by yield in early lactation, were fed at three levels of intake on rations with four ratios of roughage to concentrates. Bloom *et al.* (1957) fed their cows according to fixed scales in which the quantities of net energy given followed the estimated lactation curve. The average intakes of net energy above maintenance were 4·7, 5·8 and 6·8 kg SE per day. Nordfeldt fed his cows according to yield (0·3, 0·4 or 0·5 SFU per kg FCM). With one exception, there were no significant interactions between the factors studied. The three levels of feeding employed by Bloom *et al.* (1957) gave yields of 15·0, 16·1 and 16·1 kg FCM per day, respectively; thus the first increment of energy intake produced a response of 1·0 kg FCM per kg SE and the second, no response. In a more detailed study of these results (Heady, Schnittker, Jacobson and Bloom, 1956) it was shown that the marginal response to increments of net energy declined from about 1·8 kg FCM per kg SE at low levels of intake to 0·7 kg at high levels, a conclusion quite similar to that drawn from the results of Jensen *et al.* (1942). In Nordfeldt's experiments there tended to be a negative response to feeding less than the standard of 0·4 SFU per kg FCM, but little or no response to feeding above the standard.

Short-term experiments. In the trials summarized by Burt (1957a) there were great differences from one to another in the ratio of production starch equivalent to milk produced. For example, in one trial cows receiving 4·4 kg production SE per day yielded 19·0 kg FCM per day, whereas in another, cows given 4·3 kg SE yielded only 13·3 kg FCM. There are several probable reasons for this. The methods used to calculate starch equivalent could not be very accurate; the trials were carried out at different stages of lactation; the short duration of some of the trials may have prevented the treatments from exercising their full effects. These difficulties must be borne in mind also when treatments within experiments are compared.

These trials were for the most part concerned with overfeeding by comparison with present standards. The responses tended to be smaller than those observed in the long-term trials, that is, they were generally less than 1 kg FCM per kg SE. There was also a tendency for the response to decline as the intake of starch equivalent per kg milk increased. Many short-term trials have been completed since Burt's review. There is not space enough here to refer to all of them, but generally they confirm the results of those done earlier. Holmes, Arnold and Provan (1960) found a linear increase in yield of 0·8 kg FCM per kg SE in a trial where 1·7–3·7 kg SE were given for each kg milk produced. In a second experiment the response was 0·6 kg FCM per kg SE. Castle, Drysdale and Waite (1961) found milk yield to increase by 0·7 kg FCM per kg SE when the allowance per 10 kg milk was increased by a supplement of fodder beet from 3·0 to 3·4 kg SE. A further increase, to 4·0 kg SE per 10 kg milk, failed to improve the yield. Dijkstra and Frens (1963) obtained responses of 1·0 to 1·4 kg milk per kg SE in three trials lasting 77 or 92 days.

Grazing trials. In a short-term experiment made with cows grazing pasture herbage alone, the response to changes in energy intake was about 0·65 kg FCM per kg SE (Greenhalgh *et al.*, 1966). Grazing cows are often given supplements of concentrates, with the aims of increasing energy intake and of balancing the high protein content of young pasture herbage. In most investigations of this practice the response to extra starch equivalent has been shown to be very low. Some typical responses (per kg supplementary SE) have been as follows: 0·1 or 0·4 kg FCM (Corbett and Boyne, 1958); about 0·9 kg FCM (Nordfeldt and Hansson, 1957); 0·7 or 1·5 kg milk (Laird and Walker-Love, 1962); 0·5 kg milk (Castle, Drysdale and Watson, 1964). As discussed earlier, concentrate supplements depress herbage intake, and so the responses given above underestimate the change in yield per unit change in *total* SE intake. From the data of Corbett and Boyne (1958), who estimated the herbage consumption of their cows, one may calculate the response to be 0·3 or 0·6 kg FCM per kg change in total SE. The responses obtained in grazing trials are therefore of the same magnitude as those found in experiments on winter feeding.

Level of yield and the response to extra energy. Blaxter (1956) recalculated some of the data of Yates *et al.* (1942) and Jensen *et al.* (1942) to show that the response of cows to extra energy is directly proportional to their general level of yield. If the allowance of starch equivalent per unit of milk is increased, high yielding cows show a greater response, in terms of units of milk per unit of extra starch equivalent, than do low yielders. Later, Blaxter (1962a) produced some evidence for this from his own work. In his earlier paper he also suggested that the relationship applied within as well as between cows, meaning that a cow at the peak of her lactation would show a greater response to additional dietary energy than she would later, when her yield was lower. Burt (1957b) also found relationships between the response to extra SE and level of yield, but not between response and stage of lactation.

Two criticisms must be made of apparent relationships between level of yield and response to extra SE. The first is that while responses in low yielding cows have usually been measured when they were overfed in relation to feeding standards, those in high yielders have more commonly been measured when they were fed at levels less than, or approximating to, the standards. The latter would therefore be expected to show a bigger response. This criticism probably applies to Blaxter's (1962) data, and it certainly applies to those of Burt (1957b). The second criticism is that the relationships depend very much on the starch equivalent system providing an accurate measure of the relative energy values of different types of food, because high yielding cows are usually given a higher proportion of their energy intake in the form of concentrates. It is noteworthy that Nordfeldt and Ruudvere (1963), Nordfeldt and Claesson (1964) and Bloom *et al.* (1957), who kept their roughage: concentrate ratios constant. found no significant tendency for cows of higher 'production ability' (as defined by initial yield) to respond more to higher levels of feeding.

Conclusion. Varying the energy intake of cows appears to have less effect on their milk yields than would be predicted from feeding standards. Why should this be so? In the first place, calculations based on starch equivalent may fail to describe accurately the energy intakes of cows. It was explained earlier that, as food intake increases, metabolizable energy per unit weight of food declines. The main cause of this decline is an increase in faecal losses of energy or in other words, a decrease in digestibility. Moe, Reid and Tyrrell (1965) found that as the plane of energy nutrition of dairy cows increased from maintenance to five times maintenance the total digestible nutrients in the diet declined by 4 per cent per unit increase in the plane. This means that when calculations are based on the total digestible nutrients measured at maintenance, a nominal increase in intake of 4 kg TDN, from 12 to 16 kg, would cause an actual increase of only about 3 kg. (It should be noted that the new scheme discussed earlier for calculating energy intake in terms of metabolizable energy attempts to take into account the depression in energy value with increasing intake.)

The effects on milk production of either increases or decreases in energy intake tend to be buffered by changes in the cow's body reserves. In the experiments considered earlier in this section the cows fed at higher levels tended to gain more weight (or lose less) than those fed at lower levels, although the trends were often obscured by the usual variability in animal liveweight and may have been biassed by changes in intestinal contents. It seems likely that as energy intake increases, a successively smaller proportion of each increment is used for the synthesis of milk and a successively greater proportion is retained by the cow. This partition of energy increments between milk production and liveweight differs from one cow to another, those destined to give high yields using relatively more of each increment for milk synthesis. The partition may also depend on the stage of lactation.

The liveweight response to changes in energy intake must nevertheless be viewed as a temporary one. Cows which consistently gain in weight will increase their maintenance requirements and, if maintenance allowances are not adjusted accordingly, the energy available for production will decrease. This seems to have happened in the trials of Jensen *et al.* (1942), where the differences in liveweight change among the treatment groups were much smaller in the second year than the first.

Reid (1956) recalculated the relation between TDN intake and milk yield found by Jensen *et al.* (1942) after making allowance for (a) the decline in the total digestible nutrients content of the diet with increasing intake, (b) the increased maintenance requirements of the better-fed cows and (c) the energy retained as liveweight gain; there was still a discrepancy between the milk produced per increment of total digestible nutrients available for milk production and that expected from feeding standards. Such calculations cannot be made accurately, however, unless the energy content of liveweight gain is known. It must be borne in mind also that the efficiency with which metabolizable energy is used for body fat synthesis is less than that with which it is used for milk synthesis. The partition of energy increments between milk and body tissues needs to be studied in calorimetric trials.

There are two possible reasons why milk yield responses to additional energy tend to be greater in long- than in short-term trials. It is possible that in short-term trials insufficient time is allowed for the changes in intake to have their full effects. The second reason is that the long-term trials began early in lactation, when the response may be greatest, whereas most short-term experiments were made during the declining phase. In those longer-term experiments which began in the declining phase (after an initial standardization period) the responses were small (Bloom *et al.*, 1957; Nordfeldt and Ruudvere, 1963; Nordfeldt and Claesson, 1964).

THE EFFECTS OF THE NUTRITION OF THE COW ON THE COMPOSITION OF MILK

In many countries milk is produced mainly for the manufacture of butter and cheese and its value for this purpose is obviously determined by its composition, particularly its fat content. In New Zealand milk yields are commonly expressed as weights of fat rather than of liquid milk. Milk produced for consumption as a liquid was originally sold simply by volume; the adulteration with water which this encouraged and a growing awareness of the high nutritive value of milk constituents have led in most countries to some control of milk composition. There may be legal minima for contents of fat and solids-not-fat, and the price paid to the producer for milk may vary according to its composition. In some countries milk is brought to a standard fat content before sale to the consumer. The introduction of legal minima, although obviously valuable from a nutritional point of view, was fostered by the belief in milk as a 'natural' food which must not be adulterated. As a result, milk quality was at first assessed mainly in terms of that constituent, fat, whose concentration could be gauged by the consumer; interest in solids-not-fat is of more recent origin. This section of the chapter is restricted to fat, protein and lactose, since the mineral and vitamin contents of milk were discussed earlier. It is restricted also to the gross proportions of these three constituents of milk; the chemical constitution of milk fats is considered in Chapter 10. Protein and lactose (and ash) are often measured together as solids-not-fat. Changes in the solids-not-fat content of milk which are due to the nutrition of the cow are brought about mainly by changes in protein content, lactose being less susceptible to nutritional influences.

It is important to realize that milk composition is determined by many factors other than nutritional ones. For example, the breed, breeding and age of the cow affect milk composition, and composition varies also according to the stage of lactation. All these, and the nutritional influences, have been reviewed by Ling, Kon and Porter (1961) and Rook (1961).

Milk Fat

The milk fat of cattle is characterized by its high proportion of the residues of short-chain fatty acids. As shown in Chapter 10, butyric acid may contribute 10 per cent of the total fatty acid molecules and the C_6–C_{14} acids, a further 20 per cent. These short-chain acids are believed to be synthesized in the mammary gland from acetate and also from β-hydroxybutyrate. The longer-chain acids arise from the plasma triglycerides, and glycerol arises from the same source and possibly by synthesis from glucose (Rook and Storry, 1964).

Changes in food or energy intake which have an almost immediate effect on the quantity of milk secreted have a much slower effect on the quantity of fat secreted, and so a short-term increase or decrease in milk yield is frequently accompanied by a fall or rise, respectively, in the fat content of milk. This effect of the ration on milk fat content must always be borne in mind when one considers the effect of its composition. For example, the decrease in milk fat content encountered when cows are changed from winter rations to pasture grass is probably due both to an increase in energy intake (and hence in milk yield), and to the change to a less fibrous diet (see below).

Prolonged underfeeding eventually reduces the fat content of milk (Isaachsen, Ulvesli and Breirem, 1956). Cows with larger body reserves of fat at the beginning of lactation often give milk of higher fat content (see Table 10). The effects of giving additional dietary fat depend particularly on whether or not its component fatty acids are saturated or unsaturated. Saturated fats frequently increase milk fat content (for references see Van Soest, 1963) but unsaturated ones often depress fat content, for a reason to be considered later.

The most profound effect of feeding on milk fat is the depression observed when diets low in fibre are given. Such diets include very young pasture herbage and those low in roughage and high in concentrates (e.g. 3 kg or less of roughage together with 10 kg or more of concentrates). For example, Balch *et al.* (1954) found that replacing 3·6 or 1·8 kg of an allowance of 5·4 kg hay with iso-caloric amounts of concentrates caused the milk fat percentage to fall from about 3·8 to 2·5 or less. A similar effect is sometimes, but not always, obtained when hay is ground instead of being given long. For example, in the experiment of Balch *et al.* (1954) referred to above, a diet containing 3·6 kg per day of ground hay reduced the fat percentage of milk to less than 2·0. The heifers of Ronning (1960) gave milk containing only 2·6 per cent fat when fed solely on 16 kg finely-ground and pelleted hay per day.

More recently the fat content of milk has been shown to depend on the type of concentrated food given, both when the diet contains only 1·8 kg hay (Balch *et al.*, 1955a) and when larger amounts (at least 4·5 kg) are given (Jorgenson, Schultz and Barr, 1965). Maize, especially when ground or cooked, causes a particularly marked fall in milk fat content (Balch, Broster, Rook and Tuck, 1965).

Only a brief outline can be given here of the possible mechanisms linking the diets described above with a depression in milk fat content; detailed accounts are given in the reviews of Van Soest (1963) and Rook and Storry (1964). Acetate, with its role as a precursor of the shorter-chain fatty acids of milk fat, seemed to be the obvious intermediary, although more recent experiments have shown that the output in milk of both long- and short-chain acids is affected by low-fibre diets (Storry and Rook, 1965). The fat-depressing diets referred to above, and also those containing unsaturated fats which were mentioned earlier, were often found to reduce the proportion of acetic acid in rumen volatile fatty acid mixtures and to increase the proportion of propionic acid. For example, Balch *et al.* (1955b) showed that changing from a daily ration of 7·3 kg hay and 9·1 kg concentrates to one of 0·9 kg hay and 10·9 kg concentrates reduced the proportion of acetic acid in rumen liquor from 50 to 33 per cent, and increased that of propionic acid from 25 to 38 per cent. It is not certain, however, what these changes in proportions mean in terms of the quantities of the acids absorbed and metabolized. Sometimes, but not always, a reduction in the proportion of acetic acid in the rumen liquor is accompanied by a reduction in the concentration of acetate in peripheral blood.

The significance of acetate has been investigated further by adding salts of the fatty acids to both normal and fat-depressing diets. In general, supplements of sodium acetate equivalent to no more than 1 kg acid per day had no effect with normal diets, but caused a partial recovery in milk fat content when added to fat-depressing diets (Balch and Rowland, 1959). However, intravenous infusions of acetate had no effect on milk fat (Rook and Storry, 1964). Feeding with propionate tends to reduce the fat content of milk to a limited extent. Relatively large amounts of acetic, propionic and butyric acids (up to 2 kg per day) have recently been given to cows by continuous intraruminal infusion (Rook, Balch and Johnson, 1965). Acetic and butyric acids increased milk fat content and propionic decreased it slightly.

The effects of low-fibre diets cannot, therefore, be wholly explained in terms of their causing an absolute deficiency of acetic acid. Either the ratio of acetic acid to propionic is an important factor or else propionic acid itself may have a specific effect in depressing fat content. It has been suggested that any increase in the amounts of propionate or other glucogenic substances metabolized by the cow will reduce the quantity of β-hydroxybutyrate reaching the mammary gland, although intravenous infusions of β-hydroxybutyrate do not increase milk fat content (Rook and Storry, 1964). Also, propionate, like glucose, may depress plasma triglyceride levels and hence affect the secretion of the longer-chain fatty acids in milk (McClymont and Vallance, 1962).

Solids-not-fat (SNF)

The caseins, β–lactoglobulins and α–lactalbumin, which together comprise over 90 per cent of the proteins of normal cow's milk, are now believed to be synthesized in the mammary gland largely from the amino acids of blood serum. The globulins and serum albumin, on the other hand, are believed to be transferred directly from blood. Lactose is synthesized in the mammary gland from blood glucose (see Kon and Cowie, 1961).

Varying the protein intake of the dairy cow has little effect on the protein content of milk, although severe reductions below the standard of 0·5 kg digestible crude protein per 10 kg milk cause milk yield to fall. Milk protein content is much more sensitive to variations in energy intake, however, and particularly to reductions below the accepted standards. Rook and Line (1961) assembled the results of several experiments to show that a 40 per cent reduction in energy intake, from 2·5 to 1·5 kg SE per 10 kg milk, caused milk solids-not-fat to decline by 0·35 percentage units. A 40 per cent increase, to 3·5 kg SE, caused solids-not-fat to rise by about 0·2 percentage units. Probably 75 per cent of the fall in solids-not-fat would have been due to protein and the remainder to lactose; all the increase would probably have been due to protein. It is usual for the solids-not-fat in milk to increase when cows are first put out to graze in the spring; this has been shown by Rook, Line and Rowland (1960) to be due to the improvement in energy intake resulting from the change from winter rations to pasture, rather than to a specific effect of pasture herbage.

The solids-not-fat content of milk is also affected by plane of nutrition before calving. For example, Foot *et al.* (1963), in the experiment referred to in Table 10, found that heifers very well fed before calving gave milk in the first 8 weeks of lactation which was 0·24 percentage units higher in solids-not-fat than that of heifers very poorly fed. There continued to be a difference throughout the lactation.

Dietary supplements which promote a propionic-acid type of fermentation in the rumen, such as flaked maize, appear to be more effective than others in improving the protein content of milk. Rook *et al.* (1965) have shown that ruminal infusions of propionic acid, and also of lactic acid, increased the protein content of milk. Acetic and butyric acids did not have this effect.

The mechanisms connecting energy intake and milk protein concentration have not been determined. Rook and Storry (1964) suggest that the degradation and synthesis of amino acids in the liver, and hence plasma amino acid levels, are affected by the quantity of propionate metabolized in the liver. It is perhaps rather surprising that propionate intake should not, by affecting blood glucose levels, have a more pronounced influence on milk lactose. It appears, however, that the lactose content of milk is insensitive to variations in blood glucose within the normal range of 40–80 mg per 100 ml.

POSTSCRIPT

Since this chapter was completed the National Research Council of the United States have revised their 1958 version of *Nutrient Requirements of Dairy Cattle*, and so the comparisons I have made between these and other standards—as in Table 4 and Fig. 3—are not quite correct. A notable feature of the 1966 version is that nutrient requirements per kg milk are considered to increase with increasing daily yield.

REFERENCES

AGRICULTURAL RESEARCH COUNCIL (1965) *The Nutrient Requirements of Farm Livestock. No. 2 Ruminants*, Agricultural Research Council, London.
AITKEN, J. N. and PRESTON, T. R. (1964) *Anim. Prod.* **6**, 260.
ALLCROFT, R. (1963) In *Animal Health, Production and Pasture* (Editors, A. N. Worden, K. C. Sellers and D. E. Tribe), Longmans, London.
ANGLIN, C., MAHON, J. H. and CHAPMAN, R. A. (1955) *J. Dairy Sci.* **38**, 333.
ARMSTRONG, D. G. (1964) *J. Agric. Sci.* **62**, 399.
ARMSTRONG, D. G. and BLAXTER, K. L. (1965) In *Energy Metabolism* (Editor, K. L. Blaxter), Academic Press, London.
BAKER, R. D. (1964) *J. Br. Grassld. Soc.* **19**, 149.
BALCH, C. C., BALCH, D. A., BARTLETT, S., HOSKING, Z. D., JOHNSON, V. W., ROWLAND, S. J. and TURNER, J. (1954) *J. Dairy Res.* **21**, 172.
BALCH, C. C., BALCH, D. A., BARTLETT, S., HOSKING, Z. D., JOHNSON, V. W., ROWLAND, S. J. and TURNER, J. (1955a) *J. Dairy Res.* **22**, 10.
BALCH, C. C., BALCH, D. A., BARTLETT, S., BARTRUM, M. P., JOHNSON, V. W. and TURNER, J. (1955b) *J. Dairy Res.* **22**, 270.
BALCH, C. C., BROSTER, W. H., ROOK, J. A. F. and TUCK, V. J. (1965) *J. Dairy Res.* **32**, 1.
BALCH, C. C. and CAMPLING, R. C. (1962) *Nutr. Abstr. Rev.* **32**, 669.
BALCH, C. C. and ROWLAND, S. J. (1959) *J. Dairy Res.* **26**, 162.
BATH, D. L., RONNING, M., MEYER, J. H. and LOFGREEN, G. P. (1965) *J. Dairy Sci.* **48**, 374.
BELIC, M., PAVLOVIC, M. and BOKOROV, T. (1963) *Stočarstvo* **17**, 548.
BLAXTER, K. L. (1944) *J. Agric. Sci.* **34**, 27.
BLAXTER, K. L. (1950) *Nutr. Abstr. Rev.* **20**, 1.
BLAXTER, K. L. (1956) *Proc. Brit. Soc. Anim. Prod.* p. 3.
BLAXTER, K. L. (1957) *Vet. Rec.* **69**, 1150.
BLAXTER, K. L. (1962a) *The Energy Metabolism of Ruminants*, Hutchinson, London.
BLAXTER, K. L. (1962b) *Vitam. and Horm.* **20**, 633.
BLAXTER, K. L. (1964) *Proc. Nutr. Soc.* **23**, 62.
BLAXTER, K. L., KON, S. K. and THOMPSON, S. Y. (1944) *J. Dairy Res.* **14**, 225.
BLAXTER, K. L. and MITCHELL, H. H. (1948) *J. Anim. Sci.* **7**, 351.
BLAXTER, K. L. and WAINMAN, F. W. (1964) *J. Agric. Sci.* **63**, 113.
BLAXTER, K. L., WAINMAN, F. W. and WILSON, R. S. (1961) *Anim. Prod.* **3**, 51.
BLAXTER, K. L. and WILSON, R. S. (1962) *Anim. Prod.* **4**, 351.
BLAXTER, K. L. and WILSON, R. S. (1963) *Anim. Prod.* **5**, 27.

BLAXTER, K. L. and WOOD, W. A. (1951) *Br. J. Nutr.* **5**, 29.
BLOCK, R. J. and WEISS, A. B. (1956) *Amino Acid Handbook: Methods and Results of Protein Analysis*, Thomas, Illinois.
BLOOM, S., JACOBSON, N. L., MCGILLIARD, L. D., HOMEYER, P. G. and HEADY, E. O. (1957) *J. Dairy Sci.* **40**, 81.
BONNIER, G., HANSSON, A. and SKJERVOLD, H. (1948) *Acta Agric. Suec.* **3**, 1.
BREIREM, K., EKERN, A. and HOMB, T. (1961) *Fed. Proc.* **20** (Suppl. 1), 275.
BRODY, S. (1945) *Bioenergetics and Growth*, Reinhold, New York.
BRODY, S. and PROCTER, R. C. (1935) *Res. Bull. Mo. Agric. Exp. Sta.* No. 222.
BROSTER, W. H. (1958) *N.A.A.S. q. Rev.* No. 42, p. 47.
BROSTER, W. H. (1963) *N.A.A.S. q. Rev.* No. 62, p. 76.
BROSTER, W. H., RIDLER, B. and FOOT, A. S. (1958) *J. Dairy Res.* **25**, 373.
BROSTER, W. H., TUCK, V. J. and BALCH, C. C. (1964) *J. Agric. Sci.* **63**, 51.
BRUMBY, P. J. (1959) *N.Z. J. Agric.* **99**, 194.
BURT, A. W. A. (1956) *Dairy Sci. Abstr.* **18**, 883.
BURT, A. W. A. (1957a) *Dairy Sci. Abstr.* **19**, 435.
BURT, A. W. A. (1957b) *J. Dairy Res.* **24**, 285.
BYERS, J. H., WESWIG, P. H., BONE, J. F. and JONES, I. R. (1955) *J. Dairy Sci.* **38**, 657.
CAMPION, J. E., HENRY, K. M., KON, S. K. and MACKINTOSH, J. (1937) *Biochem. J.* **31**, 81.
CARE, A. D. (1965) *Proc. Nutr. Soc.* **24**, 99.
CASTLE, M. E., DRYSDALE, A. D. and WAITE, R. (1961) *J. Dairy Res.* **28**, 67.
CASTLE, M. E., DRYSDALE, A. D. and WATSON, J. N. (1964) *J. Br. Grassld. Soc.* **19**, 381.
CASTLE, M. E. and WATSON, J. N. (1961) *J. Dairy Res.* **28**, 231.
CLARK, R. D. and TOUCHBERRY, R. W. (1962) *J. Dairy Sci.* **45**, 1500.
CONRAD, H. R., PRATT, A. D. and HIBBS, J. W. (1964) *J. Dairy Sci.* **47**, 54.
COPPOCK, C. E., FLATT, W. P. and MOORE, L. A. (1964) *J. Dairy Sci.* **47**, 1330.
CORBETT, J. L. and BOYNE, A. W. (1958) *J. Agric. Sci.* **51**, 95.
CORBETT, J. L., LANGLANDS, J. P. and REID, G. W. (1963) *Anim. Prod.* **5**, 119.
CRICHTON, J. A., AITKEN, J. N. and BOYNE, A. W. (1960) *Anim. Prod.* **2**, 159.
DIJKSTRA, N. D. and FRENS, A. M. (1963) *Versl. landbouw. Onderz.* No. 69, 18.
DUNCAN, D. L. (1958) *Nutr. Abstr. Rev.* **28**, 695.
ELLENBURGER, H. B., NEWLANDER, J. A. and JONES, C. H. (1950) *Bull. Vermont. Agric. Exp. Sta.* No. 558.
ERB, R. E. and ASHWORTH, U. S. (1961) *J. Dairy Sci.* **44**, 515.
ESKEDAL, H. W. and KLAUSEN, S. (1958) *Beretn. Forsøgslab. Kobnhavn* No. 305.
EVANS, R. E. (1960) *Bull. Minist. Agric. London*, No. 41.
EVANS, W. M. R. (1964) *Expl. Husb.* No. 10, p. 1.
FLUX, D. S. (1950) *J. Agric. Sci.* **40**, 177.
FLUX, D. S. and PATCHELL, M. R. (1954) *J. Agric. Sci.* **45**, 246.
FOOT, A. S., LINE, C. and ROWLAND, S. J. (1963) *J. Dairy Res.* **30**, 403.
FORBES, E. B., KRISS, M. and BRAMAN, W. W. (1927) *J. Agric. Res.* **35**, 946.
GAINES, W. L. and DAVIDSON, F. A. (1923) *Bull. Ill. Agric. Exp. Sta.* No. 245.
GRAHAM, N. MCC. (1965) In *Energy Metabolism* (Editor, K. L. Blaxter), Academic Press, London
GRANT, A. B. and O'HARA, P. B. (1957) *N.Z. J. Sci. Tech. A.* **38**, 548.
GRAVES, R. R., DAWSON, J. R., KOPLAND, D. V., WATT, A. L. and VAN HORN, A. G. (1938) *Tech. Bull. U.S. Dept. Agric.* No. 610.
GREENHALGH, J. F. D. and GARDNER, K. E. (1958) *J. Dairy Sci.* **41**, 822.
GREENHALGH, J. F. D., REID, G. W., AITKEN, J. N. and FLORENCE, E. (1966) *J. Agric. Sci.* **67**, 13.
GULLICKSON, T. W., PALMER, L. S., BOYD, W. L., NELSON, J. W., OLSON, F. C., CALVERLEY, C. E. and BOYER, P. D. (1949) *J. Dairy Sci.* **32**, 495.
GUNTHER, K. (1964) *Z. Tierphysiol. Tierernähr. Futtermittelk.* **19**, 156.
HALL, W. C. and BRODY, S. (1934) *Res. Bull. Mo. Agric. Exp. Sta.* No. 208.
HARTMAN, A. M. and DRYDEN, L. P. (1965) *Vitamins in Milk and Milk Products*, American Dairy Science Association.
HARVEY, D. (1956) *Tech. Comm. Commonw. Bur. Anim. Nutr.* No. 19.

HANCOCK, J. (1953) *N.Z. J. Sci. Tech. A.* **35**, 97.
HANSEN, K. and STEENSBURG, V. (1950) *Beretn. Forsøgslab. Kobnhavn* No. 246.
HANSSON, A. (1956) *Proc. Br. Soc. Anim. Prod.* p. 51.
HEADY, E. O., SCHNITTKER, J. A., JACOBSON, N. L. and BLOOM, S. (1956) *Res. Bull. Iowa Agric. Exp. Sta.* No. 444.
HENRY, K. M., THOMPSON, S. Y., MCCALLUM, J. W. and STEWART, J. (1959) *Proc. XVth Int. Dairy Congr., London*, **1**, 172.
HOLMES, W., ARNOLD, G. W. and PROVAN, A. L. (1960) *J. Dairy Res.* **27**, 191.
HOLMES, W. and JONES, J. G. W. (1964) *Proc. Nutr. Soc.* **23**, 88.
HOLMES, W. and JONES, J. G. W. (1965) *Anim. Prod.* **7**, 39.
HOLMES, W., JONES, J. G. W. and DRAKE-BROCKMAN, R. M. (1961) *Anim. Prod.* **3**, 251.
HOLMES, W., JONES, J. G. W., DRAKE-BROCKMAN, R. M. and WHITE, N. (1965) *Anim. Prod.* **7**, 27.
HUME, E. M. and KREBS, H. A. (1949) *Spec. Rep. Med. Res. Counc., London*, No. 264.
HUTTON, J. B. (1963) *Proc. N.Z. Soc. Anim. Prod.* **23**, 39.
ISAACHSEN, H., ULVESLI, O. and BREIREM, K. (1956) *Acta Agric. Scand.* **6**, 3.
JAKOBSEN, P. E. (1957) *Beretn. Forsøgslab. Kobnhavn* No. 299.
JENSEN, E., KLEIN, J. W., RAUCHENSTEIN, E., WOODWARD, T. E. and SMITH, R. H. (1942) *Tech. Bull. U.S. Dep. Agric.* No. 815.
JORGENSON, N. A., SCHULTZ, L. H. and BARR, G. R. (1965) *J. Dairy Sci.* **48**, 1031.
KAJANOJA, P. (1944) *Suom. maatal. Seur. Julk.* No. 56.3.
KEENER, H. A. (1954) *J. Dairy Sci.* **37**, 1337.
KEHAR, N. D., MUCKERJEE, R. and SEN, K. C. (1943) *Ind. J. Vet. Sci.* **13**, 257.
KON, S. K. and COWIE, A. T. (1961) *Milk: the Mammary Gland and its Secretion* (2 vols), Academic Press, London.
KRUKOVSKY, V. N., TRIMBERGER, G. W., TURK, K. L., LOOSLI, J. K. and HENDERSON, C. R (1954) *J. Dairy Sci.* **37**, 1.
LAIRD, R. and WALKER-LOVE, J. (1962) *J. Agric. Sci.* **59**, 233.
LEWIS, J. M. and WILSON, L. T. (1945) *J. Nutr.* **30**, 467.
LING, E. R., KON, S. K. and PORTER, J. W. G. (1961) In *Milk: the Mammary Gland and its Secretion* (Editors, S. K. Kon and A. T. Cowie), Vol. II, p. 195, Academic Press, London.
MCCLYMONT, G. L. and VALLANCE, S. (1962) *Proc. Nutr. Soc.* **21**, xli.
MCDONALD, P., EDWARDS, R. A. and GREENHALGH, J. F. D. (1966) *Animal Nutrition*, Oliver and Boyd, Edinburgh.
MCGILLIVRAY, W. A. (1960) *Wld. Rev. Nutr. Diet.* **2**, 131.
MILLER, R. H. and MCGILLIARD, L. D. (1959) *J. Dairy Sci.* **42**, 1932.
MITCHELL, H. H. and MCCLURE, F. J. (1937) *Bull. Nat. Res. Counc., Washington*, No. 99.
MOE, P. W., REID, J. T. and TYRRELL, H. F. (1965) *J. Dairy Sci.* **48**, 1053.
MOORE, L. A., THOMAS, J. W. and SYKES, J. F. (1960) *Proc. VIIIth Int. Grassld. Congr.* p. 701.
MOORE, T. (1957) *Vitamin A*, Elsevier, Amsterdam.
NATIONAL INSTITUTE FOR RESEARCH IN DAIRYING (1964) *Ann. Rep.* p. 51.
NATIONAL RESEARCH COUNCIL (1958) *Publ. Nat. Res. Counc., Washington*, No. 464.
NEW ZEALAND DEPARTMENT OF AGRICULTURE (1958) *Ann. Rep. Anim. Res. Div.*, 1957–58, p. 21.
NORDFELDT, S. and CLAESSEN, O. (1964) *LantbrHögsk. Annlr.* **30**, 517.
NORDFELDT, S. and HANSSON, L. (1957) *K. LandbrHögsk. och St. LantbrFörs Medd.* No. 63.
NORDFELDT, S., HOFLUND, S., ÄNSTRAND, G. and HALLGREN, W. (1959) *K. LantbrHögsk. och St. LantbrFörs Medd.* No. 71.
NORDFELDT, S. and RUUDVERE, A. (1963) *LantbrHögsk. Annlr.* **29**, 345.
OVERMAN, O. R., SANMANN, F. P. and WRIGHT, K. E. (1929) *Bull. Ill. Agric. Exp. Sta.* No. 325.
OWEN, E. C. (1965) *Wld. Rev. Nutr. Diet.* **5**, 133.
PATCHELL, M. R. (1957) *N.Z. J. Sci. Tech. A.* **38**, 682.
RAGSDALE, A. C. (1934) *Bull. Mo. Agric. Exp. Sta.* No. 336.
REID, J. T. (1956) *Mem. Cornell Agric. Exp. Sta.* No. 344.

REID, J. T. (1961) In *Milk: The Mammary Gland and its Secretion* (Editors, S. K. Kon and A. T. Cowie), Vol. II, p. 47, Academic Press, London.
REID, J. T., LOOSLI, J. K., TRIMBERGER, G. W., TURK, K. L., ASDELL, S. A. and SMITH, S. E. (1964) *Bull. Cornell Agric. Exp. Sta.* No. 987.
RITZMAN, E. B. and BENEDICT, F. G. (1938) *Publ. Carnegie Instn.* No. 494.
RONNING, M. (1960) *J. Dairy Sci.* **43**, 811.
RONNING, M., BEROUSEK, E. R., KUHLMAN, A. H. and GALLUP, W. D. (1953) *J. Dairy Sci.* **36**, 52.
ROOK, J. A. F. (1961) *Dairy Sci. Abstr.* **23**, 251.
ROOK, J. A. F. and BALCH, C. C. (1958) *J. Agric. Sci.* **51**, 199.
ROOK, J. A. F., BALCH, C. C. and JOHNSON, V. W. (1965) *Br. J. Nutr.* **19**, 93.
ROOK, J. A. F., LINE, C. and ROWLAND, S. J. (1960) *J. Dairy Res.* **27**, 427.
ROOK, J. A. F. and LINE, C. (1961) *Br. J. Nutr.* **15**, 109.
ROOK, J. A. F. and STORRY, J. E. (1962) *Nutr. Abstr. Rev.* **32**, 1055.
ROOK, J. A. F. and STORRY, J. E. (1964) *Chemy Ind.* No. 43, p. 1778.
SCHMIDT, G. H. and SCHULTZ, L. H. (1959) *J. Dairy Sci.* **42**, 170.
SMITH, S. E. and AINES, P. D. (1959) *Bull. Cornell Agric. Exp. Sta.* No. 938.
SMUTS, D. B. (1935) *J. Nutr.* **9**, 403.
STEENSBURG, V. (1947) *Br. J. Nutr.* **1**, 139.
STORRY, J. E. and ROOK, J. A. F. (1965) *Br. J. Nutr.* **19**, 101.
SWANSON, E. W. (1960) *J. Dairy Sci.* **43**, 377.
SWANSON, E. W. and HINTON, S. A. (1962) *J. Dairy Sci.* **45**, 48.
SWANSON, E. W. and HINTON, S. A. (1964) *J. Dairy Sci.* **47**, 267.
SWETT, W. W., MATTHEWS, C. A. and FOHRMAN, M. H. (1948) *Tech. Bull. U.S. Dep. Agric.* No. 964.
SWIFT, R. W. (1957) *J. Anim. Sci.* **16**, 753.
THOMPSON, A. (1953) *Agric. Progr.* **28**, 52.
THOMPSON, S. Y., HENRY, K. M. and KON, S. K. (1964) *J. Dairy Res.* **31**, 1.
TYRRELL, H. F. and REID, J. T. (1965) *J. Dairy Sci.* **48**, 1215.
UNDERWOOD, E. J. (1962) *Trace Elements in Human and Animal Nutrition*, 2nd ed., Academic Press, New York.
VAN ES, A. J. H. (1961) *Versl. Landbouw. Onderz.* No. 67.5.
VANSCHOUBROEK, F. X. E. J. (1963) *Neth. Milk and Dairy J.* **17**, 12.
VAN SOEST, P. J. (1963) *J. Dairy Sci.* **46**, 204.
WALKER, D. M. and FAICHNEY, G. J. (1964) *Br. J. Nutr.* **18**, 187.
WALLACE, L. R. (1958) *Proc. Ruakura Fmrs' Conf. Week* p. 216.
WALLIS, G. C. (1938) *J. Dairy Sci.* **21**, 315.
WALLIS, G. C. (1943) *J. Dairy Sci.* **26**, 743.
WALLIS, G. C., KENNEDY, G. H. and FISHMAN, R. H. (1958) *J. Anim. Sci.* **17**, 410.
WELLER, R. A. (1957) *Aust. J. Biol. Sci.* **10**, 384.
WIENER, G. (1960) *Anim. Prod.* **2**, 117.
YATES, F., BOYD, D. A. and PETTIT, G. H. N. (1942) *J. Agric. Sci.* **32**, 428.

CHAPTER 20

The Use of Traditional Foods in the Feeding of Beef Cattle

T. L. DODSWORTH

The North of Scotland College of Agriculture, Aberdeen

THIS chapter is based on practices in the United Kingdom and in particular on the experiences of the author. This form of approach is unavoidable when the use of traditional foods and feeding systems are discussed if they are to be dealt with authoritatively. However, the principles elucidated are universally applicable and it is for the reader to apply these principles in his own conditions of crop and animal production.

HISTORICAL INTRODUCTION

Robert Wallace, Professor of Agriculture and Rural Economy in the University of Edinburgh from 1885 to 1922, stated in *Farm Live Stock of Great Britain* that 'The great function of livestock on a farm is not the making of so much profit on so many stones of beef—for this, unless side-issues and secondary advantages are estimated, is frequently absent—but (a) the formation of a home demand for crude, bulky, and unmarketable farm produce, and (b) the making of farmyard manure to maintain the fertility of the land' (Wallace, 1907).

This was the opinion of one authority at the beginning of this century when for nearly 100 years beef production had followed a similar pattern in which cattle utilized grassland in summer and in winter consumed diets based on roots, straw and hay. In this way 'the crude, bulky and *unmarketable farm produce*' referred to by Wallace, achieved a sale value and contributed to the overall profitability of the farm.

This method of producing beef only became possible after the introduction and adoption of turnips and clover as farm crops during the latter part of the eighteenth century. Previously any deliberate fattening of cattle must have occurred in summer on such natural pasture as was available. Animals surplus to draught and breeding requirements were slaughtered in the autumn since the only animal food available during the winter would be hay and such supplies as were available would largely be required for the draught and breeding stock. This situation changed as the turnip and clover crops were more widely adopted in rotations throughout the country together with a general increase in the output of cereals. In Orwin's words: 'The clover crop meant more hay and more grazing; the turnip crop meant winter keep; the combination of all meant more livestock; more livestock meant more meat and more manure—and more manure meant more corn' (Orwin, 1949). These changes in the quantity and quality of foods available enabled farmers to keep more cattle and to fatten them throughout the year permitting them to meet the growing demand for beef made by the increasing urban population.

The eighteenth-century butcher seems to have been as suspicious of change as is his counterpart today. In 1764 the members of the Gordon's Mill Farming Club in Aberdeen were told that London butchers said that cattle and sheep fed on turnips yielded bad meat (Smith, 1962). Such prejudices must have been overcome for when Ingle reviewed 200 feeding trials performed during the years 1833–1908 the rations were nearly all based upon roots, straw and hay usually supplemented by grain and protein cake (Ingle, 1909, 1910). An example of a ration fed in a trial reviewed by Ingle is

Linseed cake	4·3 lb
Cotton cake	4·3 lb
Hay and straw chaff	8 lb
Roots	112 lb

(1 lb = 0·455 kg)

This ration was fed by the late T. B. Wood to animals of 1138 lb liveweight. The nutritive ratio of the 200 rations reviewed ranged from 1:2·7 to 1:69·5 the average being 1:9·8. The ration having a nutritive ratio of 1:2·7 was

Decorticated cotton cake	10·85 lb
Wheat straw chaff	13·60 lb
Mangolds	10·85 lb

and that with a ratio of 1:69·5

Wheat straw chaff	10·4 lb
Mangolds	41·6 lb

Ingle concluded that the following ration having an albuminoid ratio of 1:5·45 would be suitable for fattening a bullock of 1000 lb liveweight.

Linseed cake	3 lb
Decorticated cotton cake	3 lb
Clover hay chaff	10 lb
Maize or barley meal	2 lb
Oat straw chaffed	5 lb
Roots	40 lb

Doubts regarding the profitability of this system of winter fattening were expressed during the latter years of the nineteenth century. Bear claimed that during the period 1877–83 meat production was not directly profitable but that winter fattening was continued as the farmer expected an indirect profit through the crops receiving dung from cattle fattening yards (Bear, 1888).

This somewhat pessimistic view of beef production was maintained throughout the present century. When the 'enterprize cost' system is applied to fattening beef cattle in winter the results almost invariably show a direct financial loss. There has been a tendency to adopt this as the whole truth. Wallace's teaching that one function of beef cattle is to convert unsaleable farm products into cash has been forgotten or overlooked. However, an 'enterprize costing' clearly does not tell the whole story of beef production since this has continued on many farms and in certain areas, e.g. in north-east Scotland, beef cattle contribute substantially either directly or indirectly to the farm income. On some farms such as upland and hill farms, the only way of exploiting the 'crops' which can be grown, viz., poor grass and heather, is by means of livestock.

These contentions do not imply that there is no need to attempt to develop beef production systems which are directly profitable. Indeed steps have been taken in this direction.

REDUCTION IN AGE AT SLAUGHTER

One factor affecting profitability relates to speed of turnover of capital over the lifetime of the animal. This was very slow when animals were slaughtered at 3 years of age and over thus adding to the cost of meat production. The move toward killing at younger ages was becoming marked by 1939 by which time 500,000 animals were slaughtered annually as 'baby beeves', i.e. at 18 months of age or less. Obviously, providing the same profit per animal can be obtained, reduction in age at slaughter must lead to an increase in overall profitability.

This trend had been stimulated by the work reported by Duckham (1926), Hunter-Smith (1925), Hunter-Smith and Gardner (1928), Gardner and Hunter-Smith (1932) and Common (1931). Duckham was already suggesting in 1926 that dairy heifers should be mated to bulls of beef breeds with *all* the offspring to be finished as baby beef in order to facilitate selection within the dairy herd and to lead to an increase in the supply of home produced beef. Duckham also noted that there was less difference in conformation between animals of the dual purpose and beef breeds when fed for baby beef than when they were 'stored' and slaughtered at greater ages. These workers' results indicated that it was relatively more profitable to produce baby beef than mature beef.

The slaughter of a proportion of animals at younger ages was halted in 1939 but recommenced in 1954 when meat marketing was decontrolled. Whereas some 66 per cent of animals slaughtered in the Aberdeen area in 1954 were 3 years of age or more, in 1965 some 66 per cent were killed at

18 months or less. This can be partly explained by improvements in the quantity and quality of foods available, e.g. hay is no longer sold for feeding to town horses and conservation techniques have improved. Furthermore, current retail butchering practice leads to a general demand for an animal at about 1000 lb liveweight preferably under 2 years of age carrying little or no surplus fat. It must therefore grow from birth to slaughter gaining on average 1·5 lb daily or more. If the farmer is to meet these requirements successfully whilst using traditional foods, then it is necessary that he should understand the principles of growth and development in the animal body and appreciate that the quality of rations offered to cattle affects the level of food intake, efficiency of food conversion and rate of liveweight gain. These factors must be applied so that economy of production is achieved and they must be considered together with some principles of management since it is not possible to divorce these from the principles of feeding.

FACTORS AFFECTING SUCCESSFUL REDUCTION IN AGE AT SLAUGHTER

Growth and Development in the Animal Body

A high proportion of the weight of the carcass in the prime cuts is only efficiently achieved when the animal has grown from birth to slaughter without suffering any check to growth. Early in life the rate of development of the extremities is relatively faster than that of the rest of the body. Subsequently as the animal grows development proceeds in what has been described as a growth wave which begins at the distal part of the limbs moving through the body towards the loin region as age increases, provided that nutrition is adequate (Hammond, 1940). Therefore the proportion of carcass weight in the sirloin and rump is lowest in calves and in animals whose growth has been restricted. This difference in the proportion of carcass weight found in the different parts of the body seems to be due to variations in the rate at which different tissues develop (Elsley, McDonald and Fowler, 1964).

The tissues, bone, lean and fat constitute the major part of the weight of the carcass and these tissues have priority for nutrients in that order. Thus when the plane of nutrition is low only bone can develop at its fullest rate. The supply of nutrients must exceed that required by bone and other essential tissues before significant development of muscle can occur. Fat deposition will be minimal until the supply of nutrients is adequate for these other tissues. By offering a high plane of nutrition it is possible to encourage all three tissues to develop together.

In practice all feeding systems permit some fat deposition except when maintenance requirements are not exceeded by the nutrients supplied. Since lower planes of nutrition do not allow equal rates of fat deposition in all parts of the body differences in the proportion of fat in the joints occur. Because some minimum proportion of fat is required in all joints for culinary purposes the plane of nutrition should permit the achievement of this level of fatness in the high value cuts by the time the animal is slaughtered.

If animals are to be slaughtered at an early age, calf-rearing systems must be designed to avoid imposing checks and the animal must be so managed subsequent to the calf-rearing period to ensure that it grows and gains weight continuously. Sudden changes in the level and type of feeding and management must be avoided since both impose checks on growth. Beef cattle are most likely to suffer a check other than by changes in the level of nutrition when they are weaned from liquid feeds, when they are turned out to graze in spring and again when they are housed in the autumn. Sudden dietary and management changes of this nature result in a loss in liveweight and about 4 weeks usually pass before this is regained. These checks not only arrest growth, but also reduce profits since the cost of feeding the animal continues with no return until the lost weight is recovered. An animal turned out and housed on three occasions before it is slaughtered at 2 years old may have spent 3 months or one-eighth of its life during which the cost of producing the fat beast is increased without bringing any additional financial gain.

For these reasons turning out to grass should be a gradual process. The animals should be allowed out for an hour per day initially and this time increased over a period of about 2 weeks before they are left out at night. The winter ration should continue to be offered until the cattle are fully turned out and until a gradual change in diet has been affected. Similarly at least part of the winter diet should be fed in the field for 1 or 2 weeks before they are housed in the autumn. If possible, the animals should be housed only at night for about a week before they are kept housed both night and day.

When cattle are housed in courts or similar accommodation the size of groups should be kept as small as possible. The bigger the number per pen or court, the more important it is that the animals are evenly matched. A greater return in terms of liveweight increase will be obtained from a given weight of food fed to, say, fifty animals divided into five pens of ten each, than from fifty animals running together.

The Effects of Food Quality on Intake and Performance

The achievement of slaughter condition at an early age is dependent on the animal making rapid liveweight gain from birth to slaughter. This in turn depends on a uniformly high intake of digestible nutrients which is directly related to the concentration of nutrients in the diet. The change in quantity and quality of feeds available has made possible some changes in traditional feeding systems to permit this. Thus a high level intake could not be achieved when the diet consisted solely or mainly of low-quality foods, e.g. straw, low-quality grass silage or poor-quality grass.

This was fully demonstrated by the results of work done on the fattening value of grass silages in the early 1950's. When cattle were fattened on rations including silage having a starch equivalent of about 50 in the dry matter, only moderate liveweight gains were obtained and when these silages were fed to appetite the daily intake was around 70 lb. When high-quality silages were used in other trials the daily intake exceeded 100 lb and good liveweight gains were recorded. The starch equivalent in the dry matter of these silages ranged between 60 and 70 (Dodsworth, 1952, 1954; Dodsworth, Allan and Campbell, 1952; Dodsworth and Campbell, 1952; Dodsworth and Campbell, 1953 a and b; Morrison and Heaney, 1949; Morrison and Stephenson, 1950).

The effect of food quality on intake was also noted by Ball and Graham (1961) when they recorded the performance of beef cattle during two winters on a number of farms where self feeding was practised. Ball and Graham found that silage intake averaged 10 lb per live hundredweight increasing to 11 or 12 lb per live hundredweight with good quality silage and falling to 8 lb or less when silage quality was poor. The same type of relationship was found by Blaxter when working with both sheep and cattle (Blaxter, Wainman and Wilson, 1961; Blaxter and Wilson, 1962). Thus high intakes of dry matter are only achieved when the dry matter is of high quality.

In traditional feeding systems many of the foods used may be classed as succulents. Such foods are sometimes regarded as being of low quality because of their low dry-matter content. It should, however, be noted that the starch equivalent of the dry matter is often high, e.g. 65·4 in fodder beet and it may exceed 70 in swedes and grass silage. The animal's ability to ingest large quantities of this dry matter is, however, partly limited by the volume of associated water in these foods. It is therefore important that the dry-matter percentage in succulents should be high. The significance of this was illustrated when one group of fattening cattle received a ration including fodder beet (of the order of 20 per cent D.M.) in lieu of swedes (10 per cent D.M.) in the ration fed to a control group of animals. Both groups consumed about the same weight of fresh roots. The animals receiving fodder beet consequently achieved a higher dry-matter intake and made bigger liveweight gains than the animals receiving swedes (Dodsworth, 1956).

The dry-matter percentage of succulents should not only be high, it should also be uniform or reasonably constant since abrupt changes have been shown to depress the digestibility of the dry matter ingested leading to depression of the rate of liveweight gain.

Consideration of these points has led to an appreciation of the importance of quality of foods traditionally used in beef production. Straw and poor-quality hay are not now used as freely and great emphasis is placed on the quality of silage and roots fed in winter and grass used for rearing and fattening in summer.

The Importance of Feeding to Appetite

It was common practice, in areas where roots traditionally provided the major part of winter fattening rations, to feed that weight of roots which could be eaten in a certain time, e.g. 20 minutes was a frequent allocation. The animals were then left to browse on straw or possibly hay until the next feed time. Under these circumstances rapid liveweight gains cannot be obtained without the liberal use of concentrates; but this is wasteful since cattle, unlike pigs, are not particularly efficient converters of concentrates into meat, and hence money. It seems desirable, therefore, to restrict concentrate feeding to beef cattle to a minimum and to utilize fully their ability to convert farm-grown roughage into meat. The maximum intake of roughage is encouraged by self feeding and when trough feeding must be practised the animals should be offered that weight of roots or silage which they will clean up by the next feed.

Beef cattle should also be fed to appetite if the object is to obtain the maximum possible liveweight gain per unit of food. If the amount eaten is restricted, a bigger proportion of the food is utilized for maintenance and a smaller liveweight gain return is obtained. Consider the feeding of 150 tons swedes, 15 tons hay and $11\frac{1}{4}$ tons oats to fatten cattle of 9 cwt liveweight (1 ton=2240 lb). Either of the following two rations could be used:

	A	B
Swedes	80 lb	60 lb
Hay	8 lb	6 lb
Oats	6 lb	$4\frac{1}{2}$lb

Ration A would provide for maintenance plus 2 lb liveweight increase daily, and ration B for maintenance plus a gain of 1 lb liveweight daily. The level of feeding in ration B is three-quarters of that in ration A but it produces only half the daily gain. The tonnages available would last 20 weeks if ration A was fed to thirty animals and would produce a total liveweight increase of 8400 lb liveweight gain. Ration B should feed forty animals for 20 weeks but would produce only 5600 lb liveweight gain. An increase of one-third in the weight of food given in ration B leads to a 50 per cent increase in the total liveweight gain made from a smaller number of animals. This is because a smaller proportion of the food is expended on maintenance.

The relationship between rate of liveweight gain and the cost per pound of liveweight gained is simply illustrated in the following theoretical example. Consider feeding animals of 1000 lb liveweight solely on barley grain at three levels on which they could maintain their liveweight (M), gain at the rate of 1 lb daily or at 2 lb daily.

Cost per pound Liveweight Gain

	Animals 1000 lb liveweight fed for		
	M only	M+1 lb gain	M+2 lb gain
Starch Equivalent required lb	6	9	12
Barley required to provide this Starch Equivalent lb	8·6	12·9	17·1
Cost if barley is £20 per ton	18*d.*	27·5*d.*	36·5*d.*
Cost per pound liveweight gain	—	27·5*d.*	18·25*d.*

The cost per pound of liveweight gained is reduced by 33 per cent when the rate of liveweight gain increases from 1 to 2 lb daily.

These examples demonstrate that a better return per unit of food used and better financial returns are usually obtained when animals are fed to appetite than when food intake is deliberately restricted.

In practice feeding bulky farm produced foods to appetite in winter is often prohibited by the amounts available due to year to year fluctuations in crop yields. The farmer should make an estimate each autumn of the amounts of the various foods available for winter feeding. The amount of each food available for feeding daily to each class of stock on the farm may then be calculated together with the degree of supplementation required to ensure performance at the desired level. This supplementation may involve purchasing additional food and a basis is required for determining which food is to be purchased. The choice may vary according to the class of stock to be fed and the nature of foods already on the farm. However, when foods have to be bought the 'unit value' system should always be used. This indicates the relative monetary value of the various foods available at the time.

The 'Unit Value' System

If mineral and vitamin supplements are excluded foods are usually bought either as a source of energy or of protein and may be divided into two groups accordingly. Foods purchased as a source of energy should be valued according to the cost of the unit of energy provided. This is calculated by dividing the cost of a unit weight of the food by the appropriate energy value of that food, e.g. in Britain the cost per ton is divided by the Starch Equivalent of the food. It is usually found that the cost of one unit of energy varies, e.g.:

Calculation of cost of one unit of energy

Food	Starch Equivalent	Price/ton	Cost of one unit of Starch Equivalent
Barley	71·4	£25. 5*s.* 0*d.*	7*s.* 1*d.*
Oats	59·5	£22. 0*s.* 0*d.*	7*s.* 5*d.*
Wheat	72·0	£26.10*s.* 0*d.*	7*s.* 4½*d.*
Maize	77·6	£28. 0*s.* 0*d.*	7*s.* 2½*d.*
Flaked Maize	84·0	£33. 7*s.* 6*d.*	7*s.*11½*d.*

The same unit of weight and fully comparable prices must be used throughout the calculation, i.e. either all foods *ex* mill or preferably all delivered home to the farm. In this example barley is cheapest while oats and wheat are relatively expensive as sources of energy. Barley should therefore be used to the maximum extent possible within the limits of any husbandry factors which restrict its use. Flaked maize is the most expensive and should not be used in this cost situation except when there are good reasons for so doing. For example, its high palatability would probably justify its use in a calf concentrate mixture but the level used must be the minimum consistent with obtaining the necessary degree of palatability because of the high cost per unit of energy.

Having calculated the Starch Equivalent unit value for each food on offer it is then possible to calculate a comparative value for the foods. This is done by multiplying the cost of the cheapest unit of S.E. available (i.e. 7*s.* 1*d.* as in barley in the table above) by the S.E. of each food in turn as shown below.

Comparative values of foods when one unit of S.E. costs 7s. 1d.

Food	Starch Equivalent	Comparative value per ton
Barley	71·4	£25. 5*s.* 0*d.*
Oats	59·5	£21. 1*s.* 6*d.*
Wheat	72·0	£25.10*s.* 0*d.*
Maize	77·6	£27. 9*s.* 8*d.*
Flaked maize	84·0	£29.15*s.* 0*d.*
Hay	25·0	£8.17*s.* 1*d.*
	30·0	£10.12*s.* 6*d.*
	35·0	£12. 8*s.* 0*d.*

This table shows that when barley can be bought at £25.5*s.*0*d.* per ton the other foods should be procurable at the prices shown if they are all to provide energy at the same unit cost. The use of this system is to be recommended strongly when buying feeding stuffs.

Hay is a very interesting example to consider. If hay is being bought as a source of energy its value is calculated as above, and in the example figures for poor, average and good hays are given. They show that hay is seldom worth the price asked on the market in Britain. If hay must be bought, e.g. for feeding to calves whilst being bucket reared, then only the minimum necessary for the specific purpose in mind should be purchased.

Such calculations also indicate the opportunity to profit from market situations. Taking an example from the table above, if hay is making £15 per ton then the farmer should sell the maximum amount possible from his farm, barley being bought back to replace the energy sold. Similarly it may pay to sell oats and to buy back barley.

The same technique should be used when protein-rich foods are purchased but care must be taken to use the same 'protein value' when calculating the cost of the unit. The protein content of a food may be given in food tables as

'Crude Protein', 'Digestible Crude Protein' or 'Protein Equivalent'. Of these the Digestible Crude Protein (D.C.P.) value should be used for preference, provided that it is given for all the foods. The cost of the unit of protein is then calculated by dividing the cost per ton by the Digestible Crude Protein (D.C.P.) of the food.

Calculation of cost of unit of D.C.P.

Food	Digestible Crude Protein	Price/ton	Cost of one unit of D.C.P.
Decorticated earth nut flakes	42	£48.10*s*. 0*d*.	£1. 3*s*. 1½*d*.
Soya bean meal	40	£44.10*s*. 0*d*.	£1. 2*s*. 3*d*.
White fish meal	60	£72. 0*s*. 0*d*.	£1. 4*s*. 0*d*.

In this situation soya bean meal should be bought. Since, after the pre-ruminant stage, beef cattle utilize animal and vegetable protein with equal efficiency the cheapest source of protein can be fully exploited. More expensive units of protein should only be purchased for some valid reason as when fish meal is incorporated in creep feed offered to suckler calves before sale in order to impart bloom to the coat.

Breeding Important rather than Breed

It is important to remember that complete success will not follow upon the practical application of these principles of animal feeding unless due regard is also given to the importance of breeding.

If the object of feeding beef cattle between birth and slaughter is to obtain the maximum return in terms of liveweight gain per unit of food used, it follows that the animals employed should be capable of gaining weight at the maximum rate which the available feeds permit. As the quality of feed available for beef cattle has improved, so the traditional beef breeds have been exposed to the criticism that they are not capable of sufficiently rapid growth to achieve the reduction in food conversion required if high-quality and expensive diets are to be used economically. Low potential growth rates are associated with excessively early fat deposition, which makes it necessary to slaughter animals at extremely light weights. This may be acceptable when the cost of the calf is relatively low but a serious disadvantage when the initial cost of the calf is high due to shortage of supply when it is desirable to spread this cost over as large a carcass as possible.

Search for improvement in growth rate has taken two forms, the examination of alternative breeds such as the Friesian and the Charollais and improvement in the performance of existing breeds. Study of inheritance of liveweight gain in cattle indicates a high heritability of this character, i.e. a bull which gains weight rapidly will sire calves which will gain weight rapidly. The converse is of course equally true and may be of greater importance since a slow-growing bull sires slow-growing calves. This must be of importance to the feeder of cattle whose chance of profit increases if he can select one lot of animals which, given the same food, will gain 0·25 to 0·50 lb more liveweight daily than another lot without depression in carcass and other qualities. The practical significance of this has been widely appreciated in the U.S.A. and Canada and has been demonstrated in the U.K. by Dodsworth, Ball and Ravell (1963), Ball (1965) and Jones and Francis (1963). Dodsworth, Ball and Ravell found that the offspring of one bull realized £3.4*s*.0*d*. more per head at slaughter than the offspring of a slower-growing bull. In this case the faster-growing bull gained only 0·25 lb daily more than the slower-growing bull. Bigger differences have been recorded in many other groups of bulls performance-tested for rate of liveweight gain. It is, therefore, clear that the financial return on the carcass of the progeny of one bull may well exceed that on the progeny of another bull by £5 or more. It is therefore imperative that effort be made to ensure that animals having genetic potential for rapid liveweight gain are used when the full benefits of good feeding and management are desired.

However, it should never be forgotten that the traditional early maturing beef breeds were developed against a background of low-quality diets and as long as this kind of diet is available there will be a place for them, indeed it would be a great pity if through success in breeding for high liveweight gains it became no longer possible to convert poor-quality diets into meat without carrying animals to excessive ages and weights.

CURRENT SYSTEMS OF FEEDING BEEF CATTLE

The principles of feeding which have been discussed above are successfully incorporated in various systems of feeding beef cattle now practised and which are described below.

The Feeding of Single-suckled Calves

Traditionally, single-suckled calves stayed with their dams on upland and hill farms until they were weaned at the time of calf sales in September and October. The calves were sold to lowland farms where they would spend their first winter and receive a ration generally consisting of about 30 lb of swedes or turnips plus oat straw to appetite, together with perhaps 1–2 lb oat grain per head daily. The object was to take the calves through the winter gaining about $\frac{3}{4}$ lb liveweight daily. The calves would be turned out at a year of age and be housed again in the autumn weighing 800-900 lb. During the second winter these animals might again be maintained on a ration of turnips and straw and be turned out to graze and fatten during their third summer. These animals would go fat at 2½ years of age and weigh 1100–1200 lb liveweight. Some may of course be fattened during the second winter going fat at 2 years of age.

The situation is now different in that a big proportion of single, suckled calves are fattened during their first winter reaching the butcher at 14 to 16 months of age. All steers weighing 500 lb or more and heifers weighing over 450 lb may be so treated and to ensure success they should receive a concentrate for about 6 weeks before weaning. The mixture need only consist of cereals. Suckled calves whose diet has been so supplemented should settle more quickly when housed because they have already been introduced to part of their winter ration. In place of the traditional store ration which was fed to weaned calves consisting of 30 lb swedes with 1 lb grain the calf now gets 30 lb swedes plus 30 lb grass silage plus grain rising to 8 or 10 lb daily. As a result of this change in feeding practice there is now a much more rapid turnover in animals and it is possible to intensify beef production. This is well illustrated by referring to a specific example. The stock carried on a farm consisting of 283 acres of arable plus 346 acres (2·5 acres=1 hectare) of rough-hill grazing was 250–300 ewes plus about forty-five weaned calves which were bought in the autumn. During the 1950's the weaned calves were fed a store ration through the first winter. They were turned out to graze in spring at a year of age some reaching slaughter condition on pasture at about 18 months while the balance were again housed and slaughtered at the end of the second winter at 2 years of age. The number of fat cattle sold annually averaged about forty. The output of fat cattle was first doubled by introducing a herd of breeding cows to the farm which was built up to forty by 1960. Seventy to eighty cattle were then sold annually forty having been bred on the farm and forty bought in as weaned calves. The rate of turnover was increased in 1961 when all the young stock were sold fat off pasture at 18 months of age. The intensity of feeding was then stepped up further in 1962 when twenty-eight forward calves were sold at 15 months of age having been fattened during their first winter: the heifers weighed 800–850 lb, the steers 900–950 lb. The winter feeding system practised with these animals was:

From housing in October to January	Kale	10 rising to 30 lb
	Oats	2 rising to 8 lb
	Hay	4 rising to 5 lb
From January	Swedes	15 rising to 25 lb
	Silage	12 rising to 20 lb
	Oats	8 lb

Another seventy-eight animals were sold fat off pasture at 15 to 18 months of age leaving only seven animals to carry forward to the next winter. Thus the outpnt of fat cattle was more than doubled while the output from sheep was maintained at its original level. This was possible because of better land utilization and grassland management and the production of better-quality roughage for feeding in winter.

Feeding Autumn-born Beef Cross Dairy Calves for Beef

During the 1950's there was a fall in the amount of beef imported into the U.K. leading to an increase in the demand for home-produced beef. This increase was not met by the production of beef from more beef-bred calves because beef cow numbers were inadequate to meet the demand. Fortunately there was available an alternative source of supply of calves which could be reared, the 800,000 dairy-bred bull calves which were slaughtered at birth. These calves had been considered unsuitable for rearing for beef, the suggestion made by Duckham in the 1920's having been forgotten or largely ignored. Work then began in earnest to devise feeding systems which could successfully utilize these dairy-bred calves for beef production. Various semi-intensive systems of feeding were developed in different parts of the country and the following system was developed in Aberdeenshire.

Most dairy farmers arrange to calve the majority of their cows in the autumn in order to obtain more than half of the total milk during the winter period when high milk prices prevail. The following system was therefore designed for the rearing of autumn-born calves bearing in mind the basic principle of calf rearing, i.e. that the calf should suffer little or no check to growth during the early months of life and that growth should be continuous. The following points also were considered:

Glucose and lactose are the only carbohydrates which the pre-ruminant calf can utilize. It is wasteful, therefore, to feed other carbohydrates such as maltose and sucrose, dextrin or starch, which may lead to bacterial fermentation in the gut and thus to scouring. Furthermore, fat is essential as a source of energy in addition to glucose or lactose for there is a limit to the amount of sugar which can be fed without causing scouring and unthriftiness. No suitable replacement has been found for milk protein for the pre-ruminant calf.

In view of these facts it is essential to feed whole milk to the calf for about 3 weeks. The milk is then gradually replaced by a good milk substitute which should be based on dried skim milk. Concentrates and hay should be introduced during the second week, clean water being available at all times.

Calves may be weaned from liquid food according to a variety of criteria, e.g. on liveweight, age or level of food intake. Weaning on the basis of liveweight and especially on age are not uniformly satisfactory because the stage of physiological development reached varies between individual animals at a given liveweight or chronological age, e.g. at 8 weeks of age one calf may be strongly developed and full of vigour whilst another is still weakly and is unable to grow at a satisfactory rate without receiving milk or high-quality liquid food. Experience indicates for such reasons that it is better to wean when concentrate intake reaches some given level—a safe level being 2·5 lb daily. Concentrate intake then rises almost at once to 3 or 4 lb daily. The additional concentrates ingested replace the 1–1·5 lb milk substitute powder withdrawn from the ration and the calf should suffer no setback since it has not sustained any diminution in food intake.

The inclusion of finely-ground meals in the concentrate mixture should be avoided as they may aggravate indigestion and bloat. These are reduced to a minimum when the concentrate mixture consists of flaked ingredients which are less prone than meals to rapid bacterial digestion and consequently excessive gas production. Furthermore, the use of flakey ingredients helps to ensure that the mixture is highly palatable and consumed freely. The calf-rearing system can then be summarized as follows:

25 gal whole milk   1 gal.=4·55 l
50 gal milk substitute
Concentrates and hay from 1 week
Calf weaned when eating 2½ lb concentrates daily

A suitable concentrate mixture is:

A concentrate mixture for bucket-reared calves
5 cwt decorticated earthnut expeller flakes   1 cwt=50·8 kg
5 cwt Flaked maize
6 cwt Bruised oats
3 cwt Flaked or bruised barley
1 cwt Broad bran
Vitamin A to supply 100 i.u. per pound
Vitamin D to supply 75 i.u. per pound
Minerals including a magnesium supplement to supply 0·5 g Mg per pound

Concentrate intake is allowed to increase with appetite to a maximum of 6 lb daily and is then held at this level for the rest of the winter. Silage and swedes are offered from 12 to 16 weeks of age and are fed to appetite the maximum intake varying between 15 and 20 lb of each daily by the time the calves are turned out in spring. Autumn-born calves reared in this manner weigh 450 lb when turned out in spring. The attainment of this liveweight is desirable since it appears that calves weighing less than 400 lb cannot ingest sufficient dry matter in the form of low dry-matter succulents to permit satisfactory liveweight gain. They may make satisfactory liveweight gain when grazing if the dry-matter content of the sward exceeds 20 per cent, when the digestibility of the dry matter remains high (over 75 per cent) and when reasonable weather prevails, but these conditions do not usually occur in many parts of Britain.

Every effort is made to keep the calves on short-leafy grass throughout the grazing season. The calves should receive 2 lb daily of the cheapest cereal available throughout the grazing season. This is done partly as an insurance against the onset of a long spell of wet weather which would result in low dry matter in the grass. It also insures against fluctuations in the quantity of grass available to the calves. Such a fluctuation can be due, for example, to calves having to give way to the more immediately productive fattening cattle or dairy cows on a farm. In addition some evidence indicates that animals of this age settle more rapidly when housed again for finishing, after receiving some grain during the grazing season.

The animals are housed again on or about 1st October, i.e. before the onset of cold nights, having received wet distillers' grains and swedes in the field for 1 or 2 weeks. The fattening ration consists of:

Swedes rising to 40 lb daily
Silage rising to 40 lb daily
Wet distillers' grains 10 lb daily
Oats and/or barley 4 lb daily

The cattle are ready for slaughter at 18 months of age or less at average liveweights ranging from 900 to 1120 lb according to the breeding of the beef sire used to get the calves.

The lifetime consumption of foods produced on the farm has averaged over a number of years silage 2·5 tons, swedes 3·0 tons, cereal 0·7 tons plus 0·2 acres of grazing. Under Scottish conditions this represents the produce of about 1·3 acres, straw from the cereal acreage being available for use as bedding or for sale. The margin above food costs is such as to leave a net profit per acre comparable with many other farm enterprises.

This system has been widely adopted as a means of utilizing dairy-bred calves for beef production. It can be amended to utilize foods produced in various farming systems and should be successful provided that a sound-calf-rearing method is practised and attention is paid to the production of high-quality roughages.

Winter Fattening of Animals bought in the Autumn

A potential criticism of the two semi-intensive systems described above is that both lead to the production of beef in quantity during one period of the year. The importance of this criticism is reduced by the fact that all calves do not conveniently fit into such systems, e.g. because of birth date. Neither is it convenient to fit these systems into all farms. It is still convenient to fatten some animals at a greater age on some farms, particularly where large quantities of low-quality by-products are available, the monetary value of which can only be realized by conversion into beef. Winter fattening of older animals is therefore still practised and these animals are usually bought in the late summer or autumn. It is commonly said that the chance of making a profit on winter fattening depends on appreciation in the value of the unit of liveweight over the fattening period since the value of the increase in liveweight only equals the cost of food consumed, labour and overheads. Thus an animal bought at 7 cwt fed for 140 days, gaining 2 lb. daily and sold at 9½ cwt costs £63 and realises £85.10*s*.0*d*.—if bought and sold at £9 per cwt. The margin between buying and selling prices is £22.10*s*.0*d*.—which equals the value of food, labour and overheads for the 20-week feeding period when these cost 22*s*. 6*d*. per week. Profit must therefore depend either on an increase in the value of the unit of liveweight or on reducing the costs of food, labour and overheads. The former depends on the skill of the buyer and the opportunity arising permitting the exercise of that skill. Success in reducing food costs depends on a number of factors including the application of the principles discussed in the early part of this chapter, viz. the spreading of maintenance requirements over large live-weight gains, avoiding checks in rate of liveweight gain, and the skilful use of the 'unit value' system when buying foods. To these must be added the production of heavy yields of food crops grown on the farm as the cost of growing 1 acre of a crop such as swedes is fairly constant irrespective of yield and so the cost per ton is approximately halved if the yield is doubled.

These exercises are only of interest when the merits of the enterprise cost system are debated, they may be of little value to the farmer whose interest lies in the evaluation of a farming system which includes winter fattening. An enterprise costing may show a direct loss but costing the farm as a complete unit may indicate a reduction of total farm profit if the winter fattening enterprise is removed and is not replaced by some other means of converting crops and crop by-products having no sale value into cash.

It is difficult to avoid an abrupt check in rate of liveweight gain when cattle are bought in. Regular weighing usually demonstrates that such cattle lose liveweight for 1 to 2 weeks before they begin to gain liveweight. About 4 weeks pass before the animals regain the lost weight; food costs mount during this period with no compensating gain in weight. This loss in liveweight may be due to changes either in the diet or in environment or both and it is difficult if not impossible to obviate these completely when dealing with animals bought in markets. It may be possible to reduce the effects of change in diet by buying from a known source, preferably directly from the farm of origin if the previous owner can be persuaded to feed part of the winter diet in the field before the animals are sold. The effect of a sudden change in environment is most severe when animals are tied up by the neck in a byre. This may be reduced by housing animals in courts or pens where they have more freedom of movement but this does not usually eliminate entirely loss in liveweight on housing. There may be a case for the use of a 'tranquillizer' during this period if a satisfactory and economical material can be identified. In one case where a tranquillizing material was tested the results were not satisfactory (Dodsworth and Ball, 1961).

If, as seems to be the case, this check cannot be fully avoided, it appears essential to keep food costs to a minimum while the animals are losing weight. There may indeed be a case for feeding a maintenance ration only during the first 2 weeks after 'bought in' cattle are housed. When this is done it is important that the diet should still include all the foods which are to be used in the fattening ration so that it is not necessary to add other foods when the animals begin to gain in weight. Changes in diet which may cause temporary upsets to the digestive system and hence to rate of liveweight gain, are thus avoided.

If the animals are housed in courts or pens the numbers in individual groups should not exceed ten and the animals in groups must be evenly matched to reduce competition and bullying to a minimum.

The importance of regular weighing. Regular weighing is generally accepted as an integral part of some systems of animal production, e.g. most farmers agree that rationing and selection of breeding stock cannot be performed efficiently either in dairy or in pig herds without routine weighing. This aid to management has not yet been accepted as a necessity at all stages of beef production. It seems, however, to be essential before the food costs of winter fattening can be reduced to a minimum.

When cattle are bought in for fattening in winter they will almost certainly include a small proportion which do not increase in liveweight. Experience indicates that this occurs however skilful the buyer may be. Such animals often appear to be thrifty, they consume food throughout the fattening period which may amount to 20 weeks but show little or no return in terms of liveweight gain. The food cost of such an animal would amount to £20 or more and this cost must be borne by the rest of the group. The fact that the animal is not gaining liveweight is not apparent to the eye even of the skilled cattleman and is usually only revealed by routine weighing. When this is done such an animal can be sold for slaughter before all these costs are incurred.

Routine weighing should also be used to reduce food costs as the animals reach slaughter condition when some animals will begin to gain weight at a lower rate while still consuming a full ration. Efficiency of food conversion then falls and food costs per pound of liveweight gain rise. These animals should be slaughtered as soon as there is a noticeable fall in the rate of gain and the animal is in a marketable condition. This can only be done satisfactorily when a weighing machine is used since the change in rate of gain can not be detected with certainty by eye.

It is of course important to keep labour costs involved in weighing to a minimum. This can only be done when pen design and layout permit efficient stock movement and weighing.

Efficient use of labour. Labour charges account for only 15 per cent of the total costs of winter-fattening cattle. However, every effort must be made to reduce this charge. An early criticism of fattening cattle tied up in byres was that the labour charge was higher than when cattle were fattened in courts. This criticism led to the substitution of court for byre fattening in areas where the traditional type of farm building was not very substantial and could be modified economically. Other steps have been taken in areas where the traditional buildings are very substantial and cannot be modified to permit court feeding without heavy capital expenditure.

One cattleman could look after thirty-six to forty fattening cattle in a traditional byre in which the animals were tied up at 3 ft 9 in. centres. Food was carried in to each animal by hand and dung was removed manually. This layout has now been amended in many cases to include a feeding passage in front of the animals giving access to a feed cart which may be motorized and self-emptying. The dung passage is either cleaned out by a powered dozer blade or is slatted. In the latter case dung is removed from the cellar below as is necessary by a variety of methods. The animals are tied up at 2 ft 6 in. centres hence the system is named the 'close tied byre'. Under these conditions the cattleman can attend to as many as 160 fattening cattle.

The labour cost for fattening cattle in courts can be reduced when self feeding is practised. Self feeding is possible when grass silage constitutes all or the major part of the diet. Animals of 1000 lb liveweight require 6 in. run of feed face per head given 24-hour access to the silage. The labour requirement is reduced to a minimum per head per week, time is only required for daily inspection of animals, trimming the silo face, feeding any supplement given and for spreading litter. The latter requirement is removed if the court is slatted or where individual cubicles are provided. It is imperative to remember that the rate of liveweight gain attained by animals self feeding on grass silage is a direct reflection of the quality of that silage where this provides the major or sole source of nutrients.

Such changes in the management of cattle have helped to reduce the effects of increases in the cost of labour.

Fattening Cattle on Pasture

In the traditional grassland fattening system cattle were sold fat from grass for slaughter at 2½ to 3 years of age. Raymond (1964) showed that cattle slaughtered at this age had required 5000 lb of Starch Equivalent to take them from 200 to 1000 lb liveweight. He also showed that 3800 lb of Starch Equivalent was required for this liveweight gain when cattle were slaughtered off grass at 20 months of age and only 2700 lb Starch Equivalent when cattle were slaughtered at 1 year of age having been fed on an all-concentrate diet, i.e. as barley beef. Thus there was need as with other traditional feeding systems to develop more intensive grazing methods which would permit slaughter at a younger age (see Chapter 21).

The Development of Intensive Grazing Systems

A common traditional grazing system used with beef cattle was to stock a fattening pasture with that number of beef cattle which it was anticipated could be sold fat off that field without having to move the cattle to another. Some sheep or other cattle were also grazed in the field; the stock additional to fattening cattle being increased or decreased according to the growth rate of the sward. A common rate of stocking at the beginning of the season was one cattle beast plus one sheep per acre. Under these conditions the live-weight gain obtained per acre was not very high. It was maintained by many that the old permanent pastures in the Midlands of England and other areas gave the greatest output of meat per acre. Davies (1942) stated that an output of 550 lb liveweight increase per acre was claimed when two animals were fattened per acre but that a more usual figure was about 350 lb per acre. In a survey by Davies and Williams (1948) the maximum output recorded from permanent grass was 388 lb per acre and this result was obtained on a pasture classified as being of 'super' quality. Results obtained by Roberts (1931), Bywater and Robertson (1946) and Findlay (1945) all indicated that the average output obtained from a temporary ley was 350–450 lb liveweight per acre. These workers all practised set stocking, a stocking rate of one animal per acre was employed and the level of manuring would not be comparable with that practised today.

Rotational grazing was advocated in the 1920's and was practised by some dairy farmers as a means towards increasing output. At least one farmer tried rotational grazing as a means of increasing output from grass grazed by beef cattle. Low (1930) showed that he had more than doubled the output from grass grazed by beef cattle when he divided a grass field into seven plots which were grazed in rotation. Low concluded that while the results did not then yield a commercial profit it might in the future pay the farmer to adopt such a grazing system as a means towards increasing the food supply of the nation. Interest in intensifying beef cattle-grazing systems was revived during the 1950's, partly because of the successful use of the electric fence in intensifying grazing systems with dairy cows.

When beef producers attempted to practice the strip-grazing technique with beef cattle many found that whilst the output per acre was increased the performance of individual animals was poorer than had been obtained with traditional set stocking. This could often be attributed to poor grazing management. When the system was introduced on a farm the electric fence would be moved daily at a set time, say at 7 a.m. Later, however, when staff became heavily involved in other field work, movement of the electric fence became more haphazard, varying throughout the morning or afternoon. Animals become very restive under these conditions expecting anyone who passes the field to move the fence. Thus the same meticulous attention may not be paid to feeding time in the field as is given to the feeding of housed cattle. Because of the difficulty of achieving the necessary level of management some farmers are reverting to a rotational grazing system which is not so demanding of labour.

When the rotational grazing system was first introduced the area grazed was divided into four or six paddocks, each being grazed in turn to give a grazing cycle of 20 to 30 days according to the growth rate of the sward. The object was to enable the animals to graze each paddock as the sward reached the full leafy stage of growth. This system suffered from the fault that given four or six paddocks, each had to be grazed for 4 to 6 days to give the necessary 20- to 30-day cycle. This meant that the animals spent one day or more cleaning up a paddock before they could be allowed to move forward. The animals therefore suffered constant changes in plane of nutrition leading to a 'saw tooth' effect in level of performance. This effect has now been removed since the rotational grazing system has been changed to include as many as ten paddocks each being grazed for 2 or 3 days only. The same 20- to 30-day cycle is achieved but the animals spend half a day or less cleaning up a paddock provided the stocking rate is high enough. Given appropriate manuring and a stocking rate of two and a half cattle per acre over the whole field, a liveweight gain of 800 lb per acre has been achieved.

The Feeding of Supplements on Grass

Rapid liveweight gains are dependent on a high intake of digestible nutrients which in turn depends on the quality of the ration offered. It may therefore seem logical to advocate the feeding of supplements to grazing cattle in order to ensure the availability of a high-quality ration. The feeding of supplements in spring and autumn has been recommended to ensure a gradual change in diet when cattle are turned out to graze and when they are housed. Supplementation of pasture was also recommended if cattle scoured badly when turned out in spring and in this case the supplement recommended was undecorticated cotton cake fed at the rate of 2–4 lb per head daily which was said to alleviate scouring. The use of supplements was further suggested in order to expedite fattening.

The use of supplements in spring and in autumn to ensure a gradual change in diet and during times of drought when the amount of grass does not meet the animals' requirements can still be recommended. However, the use of cotton cake to alleviate scouring and the feeding of supplements to enhance growth rate must be questioned. Dodsworth and Ball (1962a) found no advantage from feeding 3 lb undecorticated cotton cake daily to fattening cattle. Furthermore, they showed that the feeding of cotton cake to sheep had no effect on scouring in that the dry-matter percentage of the faeces of sheep receiving cotton cake was not affected. In nine feeding trials they found no benefit in terms of liveweight gain, carcass finish or financial return from the feeding of supplements to fattening cattle. The swards used in these trials were all held at the long leafy stage of growth and the Starch Equivalent of the dry matter of the swards probably never fell much below 60. Under such circumstances the feeding of oat grain with Starch Equivalent of 60 or of undecorticated cotton cake with a Starch Equivalent of 41 cannot be expected to lead to an improvement in performance unless the supplement is eaten and digested in addition to the normal grass intake. Since the animals receiving supplements did not show any improvement in performance, either the supplement eaten replaced grass or it depressed the digestibility of the total diet. These results and the results of other workers such as Alder, Head and Berting (1956) all indicate that when a concentrate is offered in addition to a succulent then the concentrate may be eaten in addition to a 'full' intake of the succulent when the succulent has a low dry matter. As the dry-matter percentage of the succulent rises the concentrate eaten replaces part of the succulent and when the nutritive value of the concentrate is lower than that of the grass replaced, animal performance must be expected to suffer. It follows that the feeding of supplements to fattening cattle cannot be recommended when maximum liveweight gain is required so long as the animals are grazing good swards.

CONCLUSIONS

The function of a farmer is usually to utilize his land in order to obtain the best financial return. If circumstances permit this is most readily achieved by growing cash crops over the whole farm. When this is done crop by-products remain which have no sale value, e.g. cereal straw and sugar beet tops. Such complete cash cropping is often considered undesirable, other crops being introduced for a variety of reasons, e.g. disease gapping; in high rainfall areas little or no cash cropping is possible and the whole farm is occupied by grass. Crops introduced such as fodder roots and grass have no direct sale value and must be converted into cash through animals and these crops have come to be regarded as the traditional feeds for beef cattle. The use of beef cattle for converting such crops and cash crop by-products into cash has for long been regarded as being directly unprofitable but the suggestion made by Wallace 60 years ago that beef cattle could in this way contribute to the overall farm income is still valid. Furthermore, if the principles and practices concerning rearing, feeding and management described are observed the use of these traditional foods in the feeding of beef cattle can now be directly profitable.

ACKNOWLEDGEMENTS

The author wishes to acknowledge the assistance given by his colleagues Mr. C. Ball and Dr. P. J. Broadbent in the preparation of this chapter.

REFERENCES

ALDER, F. E., HEAD, M. J. and BERTING, J. F. R. (1956) *Proc. Brit. Soc. Anim. Prod.* 79.
BALL, C. (1965) *Scot. Agric.* **44**, 190.
BALL, C. and GRAHAM, T. C. (1961) *Scot. Agric.* **40**, 122.
BEAR, W. E. (1888) *The British Farmer and His Competitors*, Cassell, London.
BLAXTER, K. L., WAINMAN, F. W. and WILSON, R. S. (1961) *Anim. Prod.* **3**, 51.
BLAXTER, K. L. and WILSON, R. S. (1962) *Anim. Prod.* **4**, 351.
BYWATER, T. L. and ROBERTSON, G. C. A. (1946) Univ. of Leeds and Yorks. Council for Agric. Education Booklet 209.
COMMON, R. H. (1931) *J. Minist. Agric. N. Ireland*, **3**, 34.
DAVIES, W. (1942) *J. Roy. Agric. Soc. Engl.* **103**, 125.
DAVIES, W. and WILLIAMS, T. E. (1948) *J. Roy. Agric. Soc. Engl.* **109**, 148.
DODSWORTH, T. L. (1952) *Scot. Agric.* **31**, 124.
DODSWORTH, T. L. (1952b) *J. Agric. Sci.* **42**, 395; **43**, 166.
DODSWORTH, T. L. (1954) *J. Agric. Sci.* **44**, 383.
DODSWORTH, T. L. (1956) *Scot. Agric.* **35**, 164.
DODSWORTH, T. L., ALLAN, R. T. P. and CAMPBELL, W. H. McK. (1952) *Scot. Agric.* **31**, 219.
DODSWORTH, T. L. and BALL, C. (1961) *Anim. Prod.* **3**, 315.
DODSWORTH, T. L. and BALL, C. (1962a) *Anim. Prod.* **4**, 221.
DODSWORTH, T. L. and BALL, C. (1962b) *Anim. Prod.* **4**, 227.
DODSWORTH, T. L., BALL, C. and RAVELL, J. (1963) *Scot. Agric.* **42**, 194.
DODSWORTH, T. L. and CAMPBELL, W. H. McK. (1952) *J. Agric. Sci.* **42**, 395.
DODSWORTH, T. L. and CAMPBELL, W. H. McK. (1953a) *Scot. Agric.* **32**, 137.
DODSWORTH, T. L. and CAMPBELL, W. H. McK. (1953b) *Scot. Agric.* **43**, 166.
DUCKHAM, A. N. (1926) *J. Roy. Agric. Soc. Engl.* **87**, 123.
ELSLEY, F. W. H., McDONALD, I. and FOWLER, V. R. (1964) *Anim. Prod.* **6**, 141.
FINDLAY, W. M. (1945) *Scot. Agric.* **25**, 171.
GARDNER, H. W. and HUNTER-SMITH, J. (1932) *J. Minist. Agric. Engl.* **38**, 993.
HAMMOND, J. (1940) *Farm Animals*, Edward Arnold, London.
HUNTER-SMITH, J. (1925) *J. Minist. Agric. Engl.* **32**, 779.
HUNTER-SMITH, J. and GARDNER, H. W. (1928) *J. Minist. Agric. Engl.* **35**, 722.
INGLE, H. (1909) *Trans. R. Highl. Agric. Soc. Scot.* **21**, 196.
INGLE, H. (1910) *Trans. R. Highl. Agric. Soc. Scot.* **22**, 168.
JONES, E. L. and FRANCIS, A. L. (1963) *Experimental Husbandry*, **9**, 52.
LOW, W. (1930) *Trans. R. Highl. Agric. Soc. Scot.* **42**, 113.
MORRISON, J. and HEANEY, I. H. (1949) *Agriculture, London*, **56**, 71.
MORRISON, J. and STEPHENSON, W. A. (1950) *Agriculture, London*, **57**, 251.
ORWIN, C. S. (1949) *A History of English Farming*, Thomas Nelson, London.
RAYMOND, W. F. (1964) edited by Dempster, D. G. (1964) *Advanc. Sci.* **21**, 93.
ROBERTS, E. J. (1931) *J. Agric. Wales*, **7**, 187.
SMITH, J. H. (1962) *The Gordon's Mill Farming Club* 1758–1764, Oliver and Boyd, Edinburgh.
WALLACE, R. (1907) *Farm Live Stock of Great Britain*, Oliver and Boyd, London.

CHAPTER 21

Intensive Beef Production

T. R. PRESTON

Instituto de Ciencia Animal, Catalina de Güines, Habana, Cuba

In the United Kingdom intensive beef production has come to mean almost any system which is new and not traditional. It is, however, necessary to have a more exact definition. In the sense in which it is most used it has come to mean a method of production in which mainly dairy-type cattle grow at a very fast rate by virtue of being given high-energy foods; it is also implied that the animals are housed in a fairly confined space and that this system operates with a very low labour requirement. Systems of beef production which are not intensive are characterized by only moderate rates of growth, the use of grass and forages, more complex methods of feeding—usually with rationing of concentrates, reduced density of stocking and more emphasis on the merits of the specialized beef breeds as exemplified by Aberdeen-Angus, Hereford and Beef Shorthorn.

As practised in the United Kingdom intensive beef production usually involves four distinct enterprises: (1) calf procurement; (2) rearing; (3) finishing; and (4) marketing. Procurement of the calf, usually from a dairy herd, is when it is between three and seven days of age; it is then dispatched to the rearer. The rearer keeps the calf until it is weaned off liquid food and established on *ad libitum* concentrates. This phase usually takes about 12 weeks, by which time the calf weighs between 80 and 100 kg live weight. The finishing stage is the most important part of the programme and it is characterized by *ad libitum* feeding of grain, usually barley, and a protein-mineral-vitamin-supplement, until the animal reaches slaughter condition at around 380 kg live weight and at slightly less than one year of age. Marketing involves supervision of slaughtering and dispatch to a wholesale meat market where carcasses are sold usually on a commission basis.

These activities may all be located in the same general area but due to the different circumstances applying to each phase of production, it is more usual to find them in different parts of the country. For instance, calf procurement must be based in an area where there is fairly intensive dairying because in these regions calf prices are usually lowest. The third, finishing stage is, however, most frequently located adjacent to, or in, the grain-growing areas, since work study considerations quickly reveal that the heaviest item in the programme is grain and therefore this should be moved as short a distance as possible. Factors which govern rearing operations are first that large-scale rearing of calves can be quite hazardous because of disease and, therefore, this stage tends to be based in an area which is characterized by small farms. Second, calves tend to suffer more from transit over long distances as they become older, therefore there is a tendency for the rearing scheme to be located fairly close to where the cattle are going to be finished. With regard to marketing, the fact to be most borne in mind is that the product of an intensive beef enterprize is characterized by leanness, tenderness and uniformity. It thus sells much more readily in industrial cities which specialize in modern methods of selling such as supermarkets and self-service stores; and it is less suited to the requirement of the luxury trade which puts more emphasis on flavour and taste.

A typical intensive beef finishing unit having slatted floors and mechanical ventilation is shown in Plate 1.

INPUT-OUTPUT RATIOS

The function of management in any intensive beef operation is to maximize profit. To do this it is necessary to know the relationship which exists between the value of the product and the costs of the various inputs. By expressing this relationship in the form of an equation it is possible to assess with some accuracy which combination of factors, i.e. inputs, will give maximum profit. The output is made up of the weight of the carcass meat produced and its sale value per unit of weight. The inputs are many and can be grouped together as follows: (a) feeds; (b) calves; (c) interest on invested capital and depreciation of equipment and housing; (d) labour; (e) miscellaneous costs (veterinarian, telephone, etc.). The most important input is feed for not only does it account for the greater part of the costs (it is approximately equal to 50 per cent of the value of the carcass) but also it controls the amount of the output to a very much greater extent than any of the other inputs. For all practical purposes the inputs other than feeds can be considered as a constant fraction dependent on the conditions operating on any one particular unit. We are then left with only two variables, namely, the amount of feed consumed by the animal and the amount of live weight or carcass weight increase that this feed produces. This relationship cannot be linear due to the biological changes which take place in the animal during growth and fattening. The amount required for each consecutive increase of live weight or carcass weight becomes steadily larger until the point is reached when the sale value of the extra carcass weight is exactly equal to the cost of feed required to produce it. From the point of view of efficient management this must be the point in time when the animal should be slaughtered.

The relationship between live weight increase and successive increments of food has been examined by Godsell and Preston (1963). They examined data from approximately 60 animals fed similar diets (which produced no significant difference in performance) under conditions of individual feeding.

A second degree function provided the best fit for the six treatments and for the aggregate results. The combined equation was:

$$y = -0{\cdot}013 + 0{\cdot}251x - 0{\cdot}004x^2$$
$$(r = 0{\cdot}991)$$

where y is live weight gain and x is quantity of diet, both in units of 100 kg. The fitted curve is given in Fig. 1.

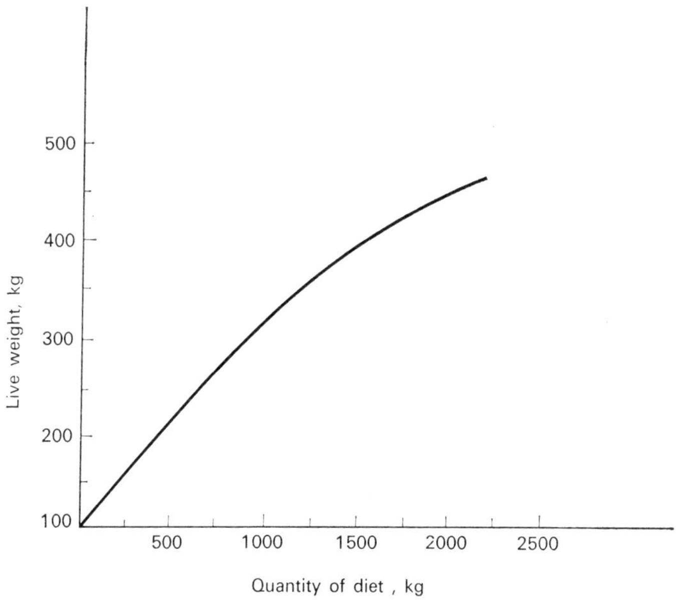

FIG. 1

Similar relationships but with data taken from commercial farms and representing over 200 animals on 12 separate units have been determined by Weiser (1966). The nature of the relationship was similar to that determined by Godsell and Preston (1963) although the animals in the second trial performed better because of being kept in groups instead of being tied individually. With the aid of these equations and knowing at any one time the value of the live weight gain and the cost of the feed, it is a relatively simple matter to decide at what stage of growth it is best to slaughter the animals for sale

SPEED OF GROWTH

In the previous section I described the relation between live weight gain and feed consumption and how this relationship changed as the animal became heavier. This relationship, in effect, represents the efficiency of utilization of feed by the animal. Efficient and profitable beef production requires that this efficiency factor or rather its reciprocal, feed conversion ratio, is as low as possible. We have seen that this can be reduced for individual animals by reducing the weight at slaughter. However, efficiency is not determined solely by the stage of development of the animal. It is affected considerably by the speed at which the animal grows and to a lesser extent according to the chemical and physical composition of the food that is given.

An increase in the rate of growth obviously has a favourable influence on the proportion of the total food which is used by the animal for production in comparison to that used for maintenance. It is thus clear that when all other things are equal then animals which grow quickly are more efficient than those which grow slowly.

The rate of growth of an animal is limited by its genotype. An animal inherits from its parents genes which determine its sex, and also the rate of secretion of somatropin (growth hormone) and other hormones which also have a modifying influence on growth. The genotype of the animal thus sets the limits, i.e. it determines the potential that an animal possesses in terms of growth rate. It is, however, extremely important to realize that the extent to which the animal expresses this potential is entirely determined by its ability to consume sufficient nutrients, by far the most important of these is energy.

Knowledge of the mechanism by which hormones control growth is not important in the context of this particular chapter. It is sufficient to emphasize the need to select, for intensive beef production, those animals which have a high rate of secretion of these hormones. Genetic selection for these characteristics can have fairly rapid results since the manifested trait, i.e. growth rate, can be measured in both males and females. The method used is commonly referred to as 'performance testing' whereby animals are fed *ad libitum* on uniform, high-energy diets and those which exhibit the fastest growth rate are selected to be the parents of future generations. It has been found that animals which possess high rates of secretion of growth hormone are usually characterized by having a high birth weight and also a large size at maturity; in other words big animals tend to grow faster than small ones. It is a sad reflection on our times to realize that countries generally considered to be highly developed in beef production have evolved breeds of beef cattle, many of which have low potential rates of growth. This is because selection has been by eye on the basis of conformation and early fat deposition; characteristics which in the main are antagonistic to rapid growth rate. In less developed countries on the other hand, particularly where animals are considered to be 'multi-purpose', the genetic potential for rate of growth in the indigenous breeds is frequently high. In many cases this stems from the use of cattle for draught purposes since natural selection for such animals has tended to favour the biggest and strongest; and, at the same time, quite accidentally, this has resulted in the selection of animals with high potential growth rates.

SEX HORMONES

The influence of sex on rate of growth is well established and in the general context of intensive beef production there are many experiments indicating that entire males grow faster than castrated males, which in turn, grow faster than females (Turton, 1962; Aitken *et al.*, 1963). The androgenic hormones secreted by the testes and to a lesser extent by the adrenal glands are known to increase rates of protein synthesis and hence growth. Although castration of the male deprives it of most of its androgenic hormones the balance of androgenic to oestrogenic hormones still give it an advantage, from the point of view of growth rate, over the female where the predominating secretion of oestrogen from the female gonads tends to have a depressing effect on growth (Albert, 1942; Mathews *et al.*, 1942).

Besides the natural sex hormones, however, we have to consider the administration of synthetic sex hormones and the effects that these, too, can have on rate of growth. The synthesis of the steriod hormones diethylstilboestrol and hexoestrol, and the demonstration that their properties resemble those of natural oestrogens, has been well documented (Burrows, 1949; Pincus and Thimann, 1955). The knowledge that their administration, either orally or subcutaneously, to castrated male cattle has a marked accelerating effect on rate of growth of up to 20 per cent is also well known (Sykes *et al.*, 1953). Administration of these hormones obviously affects the balance of natural hormones in the animal but whether the increase in rate of growth results from an increase in production of androgen from the adrenal glands in an attempt to compensate for the added oestrogen (Clegg and Cole, 1954; Preston *et al.*, 1960), or whether the administered oestrogen affects directly the anterior pituitary in some way so as to release more somatropin from this gland (Martin and Lamming, 1958) is still open to question. The important thing to note is that for intensive beef production in particular there is a consistent economic advantage in favour of administration of synthetic oestrogen to castrated males (Rowett Research Institute, 1964). Results from hormone treatment of bulls for conventional systems of feeding are equivocal (Klosterman *et al.*, 1955; Cahill *et al.*, 1956; Voelker and Dracy, 1956).

Folman (1965, personal communication) claimed better growth from hormone implantation of intensive Holstein bulls but in the absence of confirmatory evidence such treatment cannot be recommended for commercial practice.

VOLUNTARY INTAKE OF FEED

The importance in intensive beef production of securing a high voluntary intake of nutrients has already been emphasized. Without this assurance animals with high genetic potential for growth perform no better than those which are less developed in this respect. Of the nutrients needed by cattle, energy is by far the most important. This is not to deny the need for other nutrients in proportions related to the energy consumed by the animal but this aspect is more properly the subject of the subsequent section on nutrient requirements.

The factors governing voluntary intake of feed by ruminants are complex and have been reviewed by Balch and Campling (1962). Their conclusions, widely accepted in respect to low energy diets where digestibility is the chief limiting factor, offer little guidance to the students of intensive beef production. For in the context of high energy diets they suggest only that appetite control is determined probably by chemostatic or thermostatic mechanisms.

With regard to thermostatic control it appears that irrespective of the type of diet growing cattle regulate their intake so that their daily heat production remains about the same (Fontenot and Theurer, 1965). The heat increment of the feed is more or less a function of the amount of acetic acid produced relative to propionic and butyric acids. Fermentation of the structural carbohydrates, i.e. cellulose and hemicelluloses, favours acetic acid production while the more soluble carbohydrates produce greater proportions of propionic and butyric acids; thus cattle tend to consume more metabolizable energy when their diet is high in concentrates than when it is predominantly forage. The pH of the rumen liquor determines to some extent the species of bacteria predominating in the rumen and these in turn influence the types of fatty acids produced. It is known that rumen pH is affected by inclusion in the diet of buffers (Preston *et al.*, 1962) and of anti-bacterial agents such as antibiotics (Preston *et al.*, 1959); this could prove to be a fertile field of research as a means to securing greater voluntary intakes.

Heat production by the animal is also affected by the air temperature since the possibility of dissipating heat from the body by radiation, convection and conduction becomes progressively less as the environmental temperature rises. In order to secure high intakes in hot climates one must try, on the one hand, to use feeds low in structural carbohydrates (which have high heat increments) and, on the other, provide a microenvironment which is conducive to removal of heat. Keeping animals in the shade, encouraging air movement by good building design, spraying with cold water, all these are means of elevating voluntary intake and hence productive rate.

COMPOSITION OF FEED AND EFFICIENCY OF UTILIZATION

Variation in heat increment resulting from different proportions of structural and soluble carbohydrate in the feed does not solely affect voluntary feed intake. This heat increment resulting from the process of digestion and metabolism represents energy which is lost for productive purposes: thus there is a direct relation between composition of feed and efficiency of utilization. These relationships have been set out in some detail by Armstrong and Blaxter in a series of papers produced from the Hannah Dairy Research Institute (see Blaxter, 1962). The consensus of their findings was that for lipogenesis, i.e. fattening, the mixture of fatty acids resulting from fermentation in the rumen is utilized with increasing efficiency as the proportion of acetic acid, relative to the other acids, decreases. Whether the efficiency with which protein is synthesized is affected in the same way as is the process of lipogenesis is still not known although the fact that both processes occur simultaneously in growth probably means that the overall efficiency of the growth and fattening process is also influenced favourably by types of fermentation which result in high proportions of propionic and butyric acids.

NUTRIENT REQUIREMENTS

It is clear that it is not only the energy components of the feed which influence the efficiency of utilization. It is essential that all the other required nutrients are present in the diet in amounts needed by the animal. Moreover, since it is the intake of energy which determines the animal's productive rate it would appear that the most logical method of expressing the requirements for other nutrients is as a proportion of the energy value of the diet. This may not be true for all feeding regimes but it is certainly appropriate in the case of intensive beef production where rationing is not practised and energy is always available *ad libitum*.

PROTEIN

Although it is still conventional to talk about the ruminant's needs for protein this is not strictly correct since the protein needs of the animal are governed to a very large extent by the transforming action of the rumen microorganisms. We know for example that ruminants can produce satisfactorily on synthetic diets which supply no pre-formed protein and where all the nitrogen is in the form of urea (Virtanen, 1963).

In theory then, one could dispense entirely with the need for dietary protein and instead define requirements solely in terms of compounds which in the rumen hydrolyse to form ammonium ions that can be metabolized directly by bacteria for synthesis of bacterial protein. In fact for certain strains of bacteria ammonium ions are essential for growth (Bryant and Robinson, 1961). It is, however, equally true that up to the present time the performance of animals given diets which do not contain protein, tends to be less than optimum and the efficiency of feed utilization is less than when a more varied diet is given. The reason for this is not fully known at the present. The most likely explanation is that the microorganisms themselves have a requirement for certain amino acids and that if these are not supplied in the feed then the rate of multiplication of the microorganisms, and hence the synthesis of protein, is retarded and synthesizing bacteria do not operate at full capacity. The other possibility is that the bacterial protein itself, which, in the absence of feed protein has to supply all the animal's needs, may be deficient in certain amino acids and because of this the animal's productive capacity is not fully attained.

In practical diets for intensive beef production the question of supplying all the animal's needs for protein through the medium of non-protein nitrogen is purely hypothetical because in any such diet the major component of the diet is cereal, the protein content of which rarely is less than 9 per cent. As the total requirement for protein tends to vary between 13 per cent and 15 per cent of the diet it is clear that in most circumstances it is not necessary to supply more than 40 per cent of the total nitrogen from sources other than protein and the more usual requirement will be of the order of 30 per cent of the total. In large-scale trials with 250 animals (Rowett Research Institute, 1965) it was shown that at least half the supplemental nitrogen needed to balance the cereal in an all-concentrate diet for beef cattle could be in the form of urea. There were no cases of toxicity and performance was the same on both control and experimental diets containing the non-protein nitrogen. In more recent experiments, intensive beef cattle were given cereal (barley), and all the supplemental nitrogen in the form of mixed organic salts of ammonia dissolved in the drinking water (minerals were also given as organic salts in the drinking water and vitamins were added to the feed); their performance was the same as control cattle given a conventional protein supplement of soya bean meal (Preston *et al.*, 1965a). Similar findings on the exclusive use of urea nitrogen for supplementation of cereals in high energy diets for beef cattle have also been reported from the U.S.A. (Wise *et al.*, 1961; Beeson *et al.*, 1954) and from Australia (Ryley, 1964, personal communication) and from Cuba (Instituto Ciencia Animal, 1966). The animals were at least 12 weeks old in all these experiments and since no large-scale experiments have been made on the use of non-protein nitrogen with young animals it seems reasonable to consider this as the minimum age at which to begin to use non-protein nitrogen instead of protein.

One cannot generalize about the efficiency with which non-protein nitrogen is used as a substitute for protein in ruminant diets but it is certain that this will vary according to the other constituents. For example, in situations where urea-containing diets have been inferior to diets containing protein, it has been because urea hydrolysis has proceeded at a faster rate than microbial synthesis (Chalupa, Evans and Stillions, 1964). Thus the main limiting factor to the efficiency of utilization of non-protein nitrogen is the extent to which the rest of the diet can support fast growth of microorganisms. Since starch is one of the most favoured substrates for bacteria and protozoa it is axiomatic that all-cereal diets, supplemented with minerals and vitamins, form the best possible medium for the utilization of non-protein nitrogen.

The rate of protein metabolism is another important factor influencing use of non-protein-nitrogen and of course this is very much influenced by the rate of microbial synthesis. In ruminating calves between 50 kg and 90 kg live weight the rate of weight gain and of nitrogen retention decreased progressively with increasing substitution of protein by urea (Rowett Research Institute, 1965). Moreover in this type of animal even when only pre-formed protein is given, its amino acid composition has been shown to influence nitrogen retention and weight gain (Whitelaw *et al.*, 1961; Whitelaw *et al.*, 1962; Whitelaw *et al.*, 1963; Whitelaw and Preston, 1963). Reduced rates of production in very high yielding dairy cows (average of 25 kg milk daily) when urea replaced protein have also been reported by Drori (personal communication, 1966). Young growing calves and high producing dairy cows both have higher rates of protein metabolism than fattening steers. This difference in rate of protein metabolism might account for urea substitution being less effective with the former than the latter.

Quantitative aspects of protein requirements have been reviewed periodically by the National Research Council of America in their published Nutrient Requirements of Beef Cattle (see National Research Council, 1958, 1963). The methods so far adopted have been to base requirements on those amounts found to give satisfactory performance under field trial conditions. A more comprehensive investigation of the protein requirements of cattle and sheep has recently been made by the Agricultural Research Council of Great Britain (Agricultural Research Council, 1965). In this case the minimum protein requirements were estimated by a factorial method in which the requirement was computed from estimates of the separate factors describing the inevitable losses of protein from the body which have to be replaced, the rates of storage of protein during growth and pregnancy, the output of protein in milk during lactation, and the biological value of dietary protein for providing these net requirements. Unquestionably this approach is comprehensive and gives a common basis for expressing protein requirements for widely different functions. On the other hand, its very comprehensiveness implies a loss of precision in terms of requirement for any specific function or purpose. For example, in the case of intensive beef production the published minimum requirements are little more than half of what has been found to be necessary in commercial practice (Agricultural Research Council, 1965).

Factorially derived estimates are hypothetical and their usefulness as standards of adequacy depends on the accuracy of the estimates of the various factors used. These in turn depend on whether these factors are true biological constants, in which case their reliability is a function of the accuracy with which they can be measured, or if on the other hand they are dynamic relationships which vary according to any number of factors such as diet, species, body function, environment, etc. In the latter case the value of the factorial approach is open to question; for to incorporate the necessary correction factors would make the method completely unwieldly. For example, many of the factors are computed on the basis of body weight and while this is acceptable in animals at, or near, maintenance levels of nutrition, it is highly probable that in high-producing animals the endogenous losses of nitrogen are more properly a function of energy intake relative to body weight than body weight alone. And it is certain, too, that biological value decreases as the protein supply increases, and before it ceases to be limiting (Roy and Stobo, 1963; Preston and McDonald, 1963, unpublished data) and also appears to decrease with increasing age and weight of animal (Preston and McDonald, 1963, unpublished data). Nevertheless, because of its rational basis the factorial method of estimating requirements under given sets of conditions at least provides a base line, a minimum below which, in the initial approach, the protein allowance should not fall. But it should also be looked on as a point of departure from which a more precise estimate of protein needs for particular circumstances can be derived. Ultimately, too, for commercial intensive beef production, the needs of protein will be determined not solely in terms of the nutritional adequacy of the diet, or in terms of whether performance is optimum, but rather in terms of the increase in output (carcass weight), or improvement in feed utilization efficiency (both expressed in monetary terms) as a function of the cost of supplying the additional protein needed to obtain the extra output.

At the moment input/output data for protein, that can be interpretated in terms of economics, are not available. For the time being we must therefore consider the existing information relating to the effect of different levels of dietary protein on production characteristics, in particular the daily live weight gain, feed utilization efficiency and carcass composition.

In the early trials on intensive beef production (Preston *et al.*, 1963a) the level of protein used in an all-barley ration was approximately 16 per cent in dry matter and this level was kept constant over the live weight range 100 kg to 400 kg. These levels were simply estimates of the likely requirements in the light of what had been found necessary for cattle fed by traditional methods. In 1962, Bowers and Preston began a series of experiments to obtain more precise information on protein requirements for intensive systems of feeding. Peruvian fish meal was used in the first experiment since earlier experience with early-weaned calves given all-concentrate diets had indicated that this was the most efficient type of protein to use (Whitelaw *et al.*, 1961). In the first trial with Friesian steers at 150 kg live weight the retention of nitrogen was highest when the diet had 22 per cent crude protein in dry matter. In the second trial at 240 kg live weight maximum values both for daily and percentage nitrogen retention were obtained with a diet having 20 per cent crude protein in dry matter. Urinary nitrogen increased significantly with increasing content of protein in the diet and was also higher at the heavier live weight. The suggestion from these results that protein requirements for cattle fed high-barley diets were between 20 per cent and 22 per cent protein in dry matter was, however, considerably at variance with the level of 16 per cent in dry matter which was giving excellent results on commercial farms. It seemed that there might be some factors associated with the fish meal which was leading to the high urinary losses of nitrogen and consequent inefficient use of the protein (Bowers *et al.*, 1965a).

To test this hypothesis four different protein sources were compared, namely, soya bean meal, Peruvian fish meal, ground-nut meal and cotton seed meal (Bowers *et al.*, 1965b). The eight steers used weighed 280 kg live weight at the start. The most interesting feature of the results was that the average daily retention of nitrogen for all the diets, with the exception of fish meal, was higher (43 g N daily) than the maximum values recorded in the previous experiments at 240 kg live weight (37 g N daily) and 150 kg live weight (40 g N daily). In particular retention on the fish meal was only 34 g daily. Closer examination of the results indicated that the reason for this difference was because the daily consumption of the fish meal diet was significantly less (5·7 kg) than for the other diets containing the vegetable proteins (6·5 – 6·9 kg). The experiment thus showed that the poor results with the fish meal diet were due, not to inefficient utilization of the protein, but to the depressing effect that the fish meal had on voluntary feed intake.

In the subsequent experiments soya bean meal was used as the protein source (Rowett Research Institute, 1965). In one trial the levels of protein were 13 per cent, 16 per cent and 19 per cent of dry matter and the starting weight was 120 kg. There were no differences either in rate of live weight gain or in feed utilization efficiency between animals given the different levels of protein. There was, however, a significant increase in fat content of the carcass with increase in protein level. A further trial was made with protein levels ranging from 11 per cent to 20 per cent and with animals starting at

90 kg live weight (Rowett Research Institute, 1965). In this experiment nitrogen retention, weight gain and feed efficiency were all best when there was 17 per cent crude protein in the dry matter of the diet over the live weight range of 92 kg to 220 kg, but beyond this range equally good results were obtained with only 14 per cent crude protein in dry matter; data on carcass composition are not yet available. The best interpretation of all these trials would seem to be that after the calf stage (starting at 90 kg live weight) and until 250 kg live weight the diet should have approximately 17 per cent crude protein in dry matter and that subsequently, to slaughter, the level could safely be reduced to 13 per cent; and that the protein source should not be fish meal.

Detailed experiments have not been made to determine the exact requirements of nitrogen when the cereal component of the diet is supplemented only with non-protein nitrogen; however, in the experiment reported earlier (Preston *et al.*, 1965), when both ammonium salts of short chain fatty acids and urea were compared against the standard protein supplement based on soya bean, the level of 'protein' (N × 6·25) for the diets supplemented with ammonium salts or urea was equivalent to 14 per cent in dry matter. The live weight range was 130 kg – 380 kg. Since performances on these diets were comparable with that on the standard protein supplement it seems a reasonable assumption at this stage that requirements for 'nitrogen × 6·25' will be approximately the same irrespective of whether that nitrogen originates all, or in part, from protein or non-protein sources.

Vitamins

In the first experiments on intensive beef production (Preston *et al.*, 1963b) and in subsequent trials on farms the level of vitamin A in the diet was of the order of 4000 i.u. per kg. This level was twice the minimum requirement proposed by the Agricultural Research Council of Great Britain (Agricultural Research Council, 1965) and four times the amount recommended by the National Research Council of the United States (National Research Council, 1963). Despite this, frequent opinions were expressed concerning the possibility of sub-clinical vitamin A deficiency in certain intensive beef units, the reported deficiency symptoms varying from loss of skin condition to impaired vision (see Foll, 1964). Beaumont and Parr (1963) first drew attention to the low levels of vitamin A in livers from a group of steers fed all-concentrate diets. The results were later confirmed by Quarterman and Mills (1964) who reported that out of 15 livers from a group of steers fed all-concentrate diets, 7 had vitamin A concentrations in the liver below the lower limit of the normal range suggested by Moore and Payne (1942), while 5 had vitamin A levels normally associated with clinical signs of vitamin A deficiency. Doubts about the stability of vitamin A in all-concentrate diets were refuted by Foll (1964), who pointed out that losses in mixing and storage were not more than 5 per cent. He further showed that if the level of vitamin A in the diet was increased to 10,000 i.u. per kg this resulted in vitamin A levels in the liver averaging 374 i.u. per gramme of fresh tissue compared with the values of 31 and 14 reported by Beaumont and Parr (1963) and Quarterman and Mills (1964), respectively. It is not known to what extent these high levels of vitamin A, which are apparently needed for intensive beef cattle, are due to a high requirement *per se* as a consequence of the high productive rate of these animals, or to destruction of vitamin A in the rumen or in the intestines, or to both causes. Although 4000 i.u. vitamin A per kg of feed is known to be limiting in many cases it is probable that 10,000 i.u. is unnecessarily high for in a trial involving 220 intensive beef cattle on 11 commercial farms in Scotland there were no improvements in live weight gain from monthly injections of vitamin A where the diet already contained 6000 i.u. per kg (Rowett Research Institute, 1965).

There is no information about the requirements for vitamin D in diets for intensively fed cattle. However, in view of the known fact that excess vitamin A can produce a rachitogenic effect (Grant, 1955), it seems wise to maintain the level of vitamin D in the generally accepted proportion of one quarter of the amount of vitamin A.

The situation with regard to vitamin E is much more confused. Burroughs *et al.* (1963) reported significantly higher live weight gains when all-concentrate diets were supplemented with vitamin E. Similar results were obtained by the addition of selenium at a level of 0.1 mg per kg of diet. Despite this, Preston and MacDearmid (Rowett Research Institute, 1965) found no beneficial results from monthly injections of vitamin E combined with vitamin A or with selenium. The nature of the diets used may account for these differences, since Preston and MacDearmid used rolled barley and a mixed supplement of soya bean meal, fish meal and ground-nut meal, whereas Burroughs *et al.* (1963) used maize and soya bean meal. It is also possible that the vitamin A level in the diet used by Burroughs *et al.* (1963) was low and that the added vitamin E was having a sparing effect on vitamin A. Nevertheless, at the present time there appears to be no justification for giving supplemental vitamin E in all-concentrate diets for beef cattle although the situation may well change if the use of non-protein nitrogen supplements becomes more prevalent since the protein meals which they replace are normally quite good sources of vitamin E.

With regard to vitamins of the B complex, it can be assumed that supplementary sources are not necessary. There is some evidence that rate of synthesis of these vitamins by rumen microorganisms is less on high-concentrate diets compared with those containing more roughage (Tillman and McAleese, 1963), however, the amounts produced appear to be adequate, for there have been no reports of deficiencies, or of beneficial results from vitamin B supplementation (Rowett Research Institute, 1965).

Trace Elements

On the basis of analyses of cereals and information on requirements (admittedly obtained with roughage-containing diets), the only two trace elements that are likely to be limiting are cobalt and iodine. Suggested rates of addition for cobalt are 2 mg per kg and the same for iodine (Agricultural Research Council, 1965).

Major Elements

On the basis of requirements established for traditional methods of feeding, most cereals supply the animal's needs for phosphorus and magnesium; this is borne out by the absence of any suggestion of deficiency arising from these causes in commercial feeding practice. Cereals, however, are poor sources of calcium, sodium and chlorine. The suggested rates of addition are 2·5 g calcium per kg of diet and 5 g of sodium chloride per kg of diet (Agricultural Research Council, 1965).

IMPORTANCE OF ROUGHAGE

Roughage has two functions in the diet of ruminants. One is nutritional but the other concerns the health of the animal; for under certain circumstances, which have particular significance when high energy diets are given, the peculiar nature of rumen microbial digestion imposes its own restraints on voluntary feed intake and on the incidence of certain metabolic deseases, particularly bloat and acidosis. Bloat is the rapid accumulation of gases in the rumen as a result of an excessive rate of fermentation and/or failure to get rid of all the gas by belching. An excessive rate of fermentation may occur because the animal consumes too much food at one meal, or because the surface area of the food particles is increased by fine grinding, or the starch in the food is partially degraded or hydrolysed before the animal eats it. Difficulty in belching is often the result of inadequate rumen motility which may be caused by the absence of sufficient coarse particles in the food or by physical restraint on the animal, as for example when it is tethered. It used to be thought that the physical characteristics of roughage could only be supplied by forages such as grass, hay, silage, straw, etc. It is now known however that the husks of certain cereals can, if the correct method of processing is used, also serve as roughage (Guerin *et al.*, 1959; Preston *et al.*, 1963b). Some observations on this particular problem were made by Preston and co-workers (Rowett Research Institute, 1964). They compared different methods of processing barley, namely: rolling, grinding (2 mm screen of a hammer mill), pelleting (9 mm die) of whole barley grain and pelleting of barley which had previously been ground. Observations were also made of animals tied individually as compared with others kept loose in small groups. Some animals were also kept on litter, made from sawdust or straw, while others were kept on a concrete slatted floor. (For description of slatted floors see Baxter and Soutar, 1964.) The essential features of the results were that: (1) rates of growth were highest and there was complete freedom from bloat when the cattle were given rolled barley and were kept loose in small groups; (2) pelleting of whole barley was a satisfactory processing method and gave results comparable with when the barley was rolled but that ground barley and pelleted ground barley gave results which were inferior; (3) performance of cattle on slatted floors was comparable with those kept on sawdust or straw litter provided the grain in the diet was either rolled or pelleted whole; (4) the incidence of bloat was much greater (60 per cent) when pelleted ground barley was fed than when the barley was ground and not pelleted (32 per cent); with rolled or pelleted whole barley the incidence was between 5 per cent and 8 per cent. This incidence of bloat in cattle given rolled barley is because there was a 10 per cent incidence when the animals were tied individually in stalls (as indicated previously the incidence was nil when the animals were loose in small groups); (5) with ground barley the incidence of bloat was high generally and not affected by the animals being tied or loose; (6) giving supplementary forage, in the form of hay, to animals already receiving rolled barley had no effect on the incidence of bloat. Unfortunately, such supplementation was not studied with animals receiving ground grain.

Intensive beef production, as defined in this article, has in the main been restricted to Europe. For this reason barley has been the cereal of choice and there has been little reported work with other grains.

Ground maize with no added roughage would seem to be particularly unsuitable for intensive beef production in view of the report by Preston *et al.* (1963b) that out of a group of 15 animals given only ground maize and protein supplement, 5 (33 per cent) died from bloat. In subsequent trials the incidence of bloat was almost nil, however, when the ground maize was mixed with rolled oats in the ratio of 2 maize : 1 oats. Dried sugar beet pulp at a level of 25 per cent of the diet was very bad and the incidence of bloat was 75 per cent. It is possible that the beet pulp expanded rapidly in the rumen, due to absorption of water, and by so reducing the effective rumen volume may well have exacerbated the bloat problem (Rowett Research Institute, 1964).

When maize is grown on the farm where it is to be fed the simplest method for supplying the additional roughage is to use the complete ear including both grain and cob. Live weight gains of up to 1·5 kg per day and complete freedom from bloat have recently been reported by Preston (Instituto Ciencia Animal, 1966) when such diets were fed to bulls starting at 100 kg live weight and finishing at 400–450 kg live weight.

Low-roughage diets for finishing grass-reared cattle are widely used in the United States of America. The system is not strictly comparable to the European method of high-energy feeding from birth, nevertheless the American experience is of considerable interest. For example Ellis (1965) has analysed many of the experiments in this field and found equivocal findings relating to the need for roughages. Ellis (1965) quoted results from 3300 cattle in 27 trials where overall mean daily gains were 2·64 lb per day (2·2 lb = 1 kg) and 2·54 lb per day and conversion ratios were 7·2 and 8·8 for cattle receiving all-concentrate diets or diets containing some roughage. Unfortunately he does not give details of the kinds of grain nor the method of processing. It may well be that cattle reared on roughage for their first 6 to 9 months are better able later to withstand all-concentrate diets which in any case are usually given over only a short finishing phase rarely exceeding 120 days. By comparison the European system, with which we are most concerned in this chapter, relates to cattle given all-concentrate diets for almost 300 days continuously beginning at an age of two months. The importance of preserving the roughage characteristics of barley husks, by rolling, was first stressed by Guerin *et al.* (1959). However, Wise *et al.* (1961) and others obtained good results using only ground maize while ground sorghum was used successfully by Lofgreen (1961), Pope *et al.* (1963), Durham *et al.* (1963) and Kercher and Bishop (1963).

But for intensive beef production on the European system, with high-energy feeding from birth to slaughter at less than one year, the evidence would seem to indicate that there is a distinct need for some form of 'roughage', either 'built-in' in the form of husk or as some form of supplementary forage (Rowett Research Institute, 1964). But whether the beneficial effect of such roughage can be entirely ascribed to the stimulatory effect of the coarse fibres on rumen motility is less clear. For example the bulkiness or density of the diet may also be important. Hughes *et al.* (1964) reported increased feed consumption and weight gains when ground polythene was included in an all-concentrate diet and Oltgen *et al.* (1960) noted similar findings with a purified diet for lambs. Thus it may be that an important role of roughage is in making the diet more 'open' and in creating 'bulk' in the rumen. The importance of density must be stressed because, for example, ground polythene has a much lower density than most ground roughages and this may be one reason for its greater effectiveness. Some workers have hypothesied that an important quality of roughage may be its abrasive nature and to this end have included sand in the diet. Cooley and Burroughs (1962) found that a supplement of sand at a level of 2 per cent gave some improvement in performance when added to a predominantly ground maize diet. Similar results were obtained by Preston and White (1964) when sand was added to an early-weaning diet, given to calves between 40 kg and 90 kg live weight, comprising only ground maize and soya bean meal. On the other hand there were no advantages from including sand in a rolled-barley/soya bean meal diet given to Holstein steers between 100 kg and 400 kg live weight (Rowett Research Institute, 1964).

The nutritional role of roughage stems primarily from its high content of cellulose since on fermentation this produces mainly acetic acid, whereas concentrates with their high content of starch and protein ferment to mixtures of fatty acids containing more propionic acid and smaller quantities of acetic acids (Balch and Rowland, 1957). Since mixtures high in propionic acid are used more efficiently for lipogenesis than mixtures high in acetic acid (Armstrong *et al.*, 1958), it can be argued that the inclusion of roughage in the diet merely decreases the efficiency of feed utilization in fattening animals. Lipogenesis, however, is not the sole criterion of growth; in fact protein synthesis is much more important. Unfortunately the relative energetic efficiencies with which acetate and propionate contribute to protein synthesis are not well understood. Rook *et al.* (1960) reported that nitrogen retention and gain in empty live weight were higher when acetic acid was infused into the rumen of heifers than when propionic acid was given. Preston and Bowers (see Rowett Research Institute, 1964) noted that feeding intensive Holstein steers on diets high in protein gave fatter carcasses than when lower levels of protein were used. Preston *et al.* (1965a, and unpublished data) also found significantly less fat in intensive Holstein steers given ammonium acetate as a supplement to barley as compared with control animals given barley and soya bean protein. And the fermentation of protein is known to increase the proportion of propionic acid relative to acetic acid (Blaxter and Martin, 1962). There is thus some evidence that acetic acid may be more favourable for protein synthesis than propionic acid. Nevertheless, since in growing animals protein synthesis and lipogenesis are inseparable, one cannot be sure of the exact effect of varying acetate/propionate ratios (which result from varying amounts of roughage in the diet) on the energetic efficiency of the overall growth process.

Most roughages are also rich in trace minerals and in some experiments their reported benefits in all-concentrate diets were traced to these particular factors (Pope *et al.*, 1962; Renbarger *et al.*, 1964).

Another important aspect of the composition of roughages is their content of alkali elements—particularly calcium, potassium and sodium. In the early work on the development of purified diets for ruminants, Matrone *et al.* (1957, 1959) found that sodium and potassium ions given in the form of salts of fatty acids or bicarbonates led to highly significant improvements in voluntary feed intake and in performance when there was no roughage in the diet. These workers suggested that one important contribution of roughage was a buffering action resulting from its content of alkali elements present as organic complexes which latter being metabolizable provided a net increase in alkalinity. Nicholson *et al.* (1960) found similar advantages from bicarbonate salts of potassium, sodium, magnesium and calcium when formulating semi-purified diets for calves.

Yet despite these good results with synthetic diets the use of buffers in more practical rations has not been successful. Preston *et al.* (1963b) reported a 20 per cent mortality from bloat in cattle given ground maize, protein supplement and 2·5 per cent sodium bicarbonate. In a subsequent experiment with zero, 1 per cent, 2 per cent or 3 per cent sodium bicarbonate included in a rolled-barley diet results were best for the control ration; moreover at slaughter the kidneys of the supplemented cattle showed considerable hemorragic areas and there were some cases of urinary blockage due to calculi (Rowett Research Institute, 1964). Similar findings were observed in lambs by Ellis *et al.* (1963) who had to abandon feeding of 1·25 per cent sodium bicarbonate in an 'all-concentrate' sorghum ration due to an extremely high incidence of urinary calculi. Wise *et al.* (1961) found no benefit from sodium and potassium acetate in a ground maize diet containing urea and suggested that the release of NH_4 ions from the urea might have provided the needed buffering action. Nicholson *et al.* (1961, 1962, 1963) reported small improvements in some experiments from addition of buffers but on average there were no significant advantages.

The age of the animal might well be a factor influencing response to buffers for it is known that salivary secretions are poorly developed in the young ruminant (Kay, 1963). Preston *et al.* (1962) observed significant increases in dry matter intake and in nitrogen retention when 7·5 per cent sodium bicarbonate was added to an all-concentrate diet given to calves over the age range of 4 to 16 weeks. The good results with buffers reported by Matrone *et al.* (1957, 1959) were also obtained with young lambs.

FEED PROCESSING

The objectives when processing feed for intensive beef production are on the one hand to preserve the roughage characteristics of the outer portion —the 'husk'—of the grain, and on the other to ensure maximum digestibility and efficiency of utilization of the grain carbohydrate. Some aspects of feed processing in connection with the need to retain the 'built-in' roughage of the grain husk have already been discussed in an earlier section. The conclusion from this was that rolling is the best way of processing. However, the degree of processing achieved by rolling depends on many factors such as the design of the mill, size of the grain and particularly its moisture content. For barley this is quite critical because destruction of the husk is considerable when the moisture content is less than 16 per cent. Unfortunately grain cannot be stored aerobically (the normal method) at 16 per cent moisture content and it is more usual to store at 14 per cent (which is satisfactory) and then moisten the grain with water immediately prior to rolling. Either water can be added and the moistened grain allowed to stand for 24 hr, so that the added water is assimilated into the grain; or steam can be injected into the grain immediately before it enters the rolling mill. The latter system is generally preferred on account of its flexibility and the fact that there is no need to hold the grain for any length of time before rolling. An excellent review of rolling and steaming grain with particular emphasis on the mechanical and engineering aspects has been made by Roskamp (1965).

Another method for securing adequate moisture in the grain before rolling is to harvest when the grain has between 16 per cent and 22 per cent moisture and then store in an hermetically-sealed silo. Slight fermentation takes place while the existing oxygen amongst the grain is converted to carbon dioxide, but this is often advantageous since it imparts to the grain a slightly 'beery' smell which tends to enhance palatability. When the grain is taken out of the silo and is exposed to the air it begins to ferment again and the higher the air temperature, and the more moisture there is in the grain, then the more rapid is the fermentation. It is therefore extremely important that having been removed from the silo, the grain is processed and given to the cattle as rapidly as possible. In extracting high-moisture grain from the silo it is also important to keep to the minimum any further ingress of air.

It is often claimed that high moisture barley has a greater feeding value than barley stored dry but there is no justification for this belief. For example Preston *et al.* (1965b) found no difference in digestibility or nitrogen retention between rolled, dry barley or rolled, high-moisture barley. In a second experiment (Rowett Research Institute, 1964) they reported similar findings and the treatments included barley which had been steamed prior to rolling as well as the previously mentioned treatments. An interesting observation was that the digestibility of high-moisture barley which had received no processing at all was almost as high (73 per cent) as the same barley which had been rolled (80 per cent).

That the pelleting of whole barley can be as satisfactory a method of processing as rolling was shown in the earlier section. The effect of this treatment on nutritive value, and of grinding compared with rolling, was also studied by Preston and co-workers (Rowett Research Institute, 1964). The concensus of their findings was that there were no significant differences either in dry matter intake, dry matter digestibility or nitrogen retention between the different methods of processing.

In general, then, one might say that the importance of feed processing lies more in the effect that this has on the roughage properties of the diet and has less consequence in terms of nutritive value *per se*. However, recent research at the University of Arizona suggests that improvements in nutritive value of grain can be brought about but that the necessary conditions for achieving this are quite specific.

Sorghum has been studied most since it has been recognised for some time that both the seed coat and the starch are more resistant to digestion than is the case with other grains (Saba *et al.*, 1964). Two methods have been used to treat the grain. The first is to apply moist steam at low pressure in a large tempering chamber for approximately 20 min, by which time the temperature of the grain reaches about 100°C. It is then rolled. To make a good flake from the sorghum it is necessary that it contains at least 16 per cent moisture as it is rolled. A comparison of sorghum processed by this method or by the more conventional dry-rolling method without steam showed that daily gain and food conversion ratio were improved significantly for the steam-processed sorghum relative to the dry-rolled sorghum. A similar trial with barley also showed significantly better live weight gain (7 per cent) for the steam-processed compared with dry-rolled grain but no difference in feed utilization efficiency. Digestibility trials with the sorghum rations showed significant improvements in digestibility of dry matter and nitrogen-free extract as a result of steam processing with the consequence that the calculated Total Digestible Nutrients of the sorghum was significantly increased; protein digestibility was not affected (Hale *et al.*, 1965; Hale, 1965). It must be stressed that the efficiency of rolling appears to be extremely important in bringing about this increased nutritive value of the sorghum. According to Hale (1965) a very flat flake has to be produced as demonstrated by the fact that the percentage disappearance of grain dry matter from nylon bags inserted in the rumen of a steer was 55 per cent in 2 hr for steam-processed sorghum rolled into a flat flake, as compared with 21 per cent when the grains were only crushed. Grinding the very flat flake gave no further improvement in rate of disappearance of dry matter, thus it would seem that the increased digestibility of the processed grain was due to the heat and pressure applied during the steaming and rolling processes rather than the increased surface area of the flat flake compared with the crushed grain. The importance of flaking has been further demonstrated by Husted *et al.* (1966) who found that steam-processed sorghum cut into small pieces by a decorticator was only slightly better than dry-rolled millo but that steam-processed sorghum which had been flaked was significantly more digestible than the steam-processed sorghum which had only been cut. It was found that the adequacy of the processing was best described by the final density of the product. For example, steam processed sorghum which has been properly flaked has a density approximately 50 per cent of that of whole grain (Husted *et al.*, 1966). Gas production studies made in an *in vitro* system have further confirmed the importance of satisfactory flaking after heat treatment if maximum utilization of the grain is to be obtained (Trei *et al.*, 1966).

The other method for applying the necessary heat treatment to the grain is pressure-cooking. At the University of California (reported by Hale, 1966) maize, sorghum and barley were either rolled dry, or first subjected to steam at 20 lb pressure for 1·5 min. The degree of improvement was greater for sorghum than for the other grains and for the sorghum the 11 per cent greater live weight gain by pressure cooking was of a similar order to that noted for sorghum processed by low pressure steam for the longer period of 20 min. Too much heat can be detrimental since it was found that cooking for 1·5 min at 60 lb pressure (2·2 lb = 1 kg) gave no better results than when the grain was rolled dry; similarly, extending the heat treatment until the sorghum was completely gelatinized gave weight gains which were less than when dry sorghum was only ground (see Hale, 1966).

Thus it appears that even at this stage there is sufficient evidence to justify heat treatment and flaking of sorghum and that provided the specified conditions are adhered to, then economic improvements in performance of cattle can be expected. The same may well hold true for other grains, particularly maize and barley, but more research is needed before it can be predicted with certainty that heat treatment and flaking will be an economic proposition for these as well as for sorghum.

CARCASS COMPOSITION

Consumer requirements for meat have been increasingly orientated towards a need for cuts and joints which can be prepared easily (i.e. can be grilled or roasted rather than stewed), for minimum waste fat, guaranteed tenderness, and simplified shopping through pre-packaging and self-service supermarkets. Flavour, although still important in the specialist luxury trade, tends to have a lower priority than the previous items.

As a result of fast growth and thus the possibility of slaughter at, or at less than twelve months of age, the intensive beef animal almost always has meat which is tender and with this the possibility that the maximum proportion of the carcass can be used for grilling and roasting. Little can be done about developing a full flavour since this characteristic is dependent mainly upon age, and to extend the life of the intensive beef animal inevitably depresses feed efficiency, and hence profitability, to a point which is unlikely to be compensated for by a higher price for the product.

The amount of lean meat in the more valuable parts of the carcass (from loin, rump and thigh) is affected to only a slight degree by the conformation of the animal, unless there is muscular hypertrophy (Mason, 1956; Pomeroy and Williams, 1962; Lawrie *et al.*, 1963).

The proportion of fat to lean in the carcass of the intensive beef animal is, however, a more labile characteristic. For instance an increase in age and weight at slaughter will increase the fat percentage (Callow, 1944). At the same age and weight heifers have more fat than castrated males which in turn are fatter than bulls (Aitken *et al.*, 1963; Preston *et al.*, 1963a). Administration of the synthetic oestrogens, hexoestrol and stilboestrol to castrated males also depresses fat percentage (see Aitken and Crichton, 1956; Preston *et al.*, 1957). Breed is important, too; and the traditional beef races, e.g. the Aberdeen Angus and their crossbreds, produce fatter carcasses than dairy animals such as Holsteins (Preston *et al.*, 1963a).

With respect to nutrition the determining factors are the concentration of metabolizable energy in the diets (kcal M.E./kg of dry matter), the ratio of protein to energy and the extent to which protein is replaced by non-protein nitrogen. For example, intensive beef steers given a maize diet (M.E. in dry matter (D) equals 3326 kcal/kg) were fatter than animals of the same weight which had had barley (M.E./D equals 3106 kcal/kg) (Preston *et al.*, 1963b); similar results were reported by Martin *et al.* (1966).

In an investigation of protein requirements of intensive beef steers Bowers and Preston (Rowett Research Institute, 1964) compared protein levels of 13 per cent, 16 per cent and 19 per cent in dry matter by varying the amount of soya bean meal in a barley diet. When slaughtered at 400 kg live weight the fat content in the tenth rib cut increased significantly with increasing protein level in the diet. In two subsequent trials with similar animals and the same basal barley diet, Preston *et al.* (Rowett Research Institute, 1965) compared the standard protein supplement of soya bean meal with a synthetic supplement comprising mainly ammonium acetate administered as a solution in the drinking water. Performance was similar for both treatments but the fat content of the tenth rib cut was significantly less on the diet with ammonium acetate than when soya bean meal protein was used.

HOUSING AND ENVIRONMENT

By definition intensive beef production implies confined feeding and whatever the climate some form of building construction is necessary. Housing costs are therefore an important input in the production function (as discussed at the beginning of this chapter) and building design must be studied carefully not only in relation to the direct depreciation and interest charges that are incurred, but because housing indirectly affects other inputs. For example, the environment created within a building can influence the performance and hence the feed requirement of the animal; environment also influences incidence of disease and the need for veterinary treatment—another important input. Labour costs in an intensive unit are very closely related to building design, particularly with respect to cleaning operations. Finally, consideration must be given, particularly in large units, to providing facilities for research and development; this is a particularly important input in the production function for increased efficiency necessitates a continuous programme of evaluation of new research findings with a view to incorporating in the system those found to be beneficial.

ENVIRONMENT

Cattle maintain their body temperature within narrow limits by means of various mechanisms for controlling production or loss of heat. For example, at air temperatures below the critical temperature the animal first reduces heat loss by vasoconstriction in the skin and by inducing greater insulation through erection of the hair coat. Beyond this, thermo-neutrality is maintained by catabolism of energy from body stores or from absorbed nutrients; this use of energy solely for heat production inevitably brings about a concomitant reduction in feed efficiency for productive purposes. Above the upper critical temperature the animal reduces heat production primarily through reducing feed intake. This also causes a decline in feed utilization efficiency since a reduced feed intake automatically means less production relative to maintenance.

FACTORS GOVERNING HEAT PRODUCTION

At 10°C (an average air temperature in the United Kingdom) approximately 75 per cent of heat lost is dissipated by radiation, convection and conduction. In these circumstances close confinement, as in intensive beef production, will reduce the efficiency of heat loss as compared with low density housing used in more extensive systems and, according to Morgan and Parker (1964), this could be equivalent to a rise in environmental temperature of about 5°C.

Above 20°C Findlay (1959) concluded that heat loss must be primarily by moisture vaporization. There is, in fact, a direct relation between environmental temperature and the proportion of total heat lost as latent heat, i.e. evaporation of moisture from the skin and lungs and by water exchange.

Other factors influencing heat loss are air movement and coat covering. Blaxter and Wainman (1961) showed that increasing air speed from 31 to 144 ft^3 per minute increased heat loss by about 6 per cent when the environmental temperature was 0°C but had no effect at 20°C.

Heat production is also affected by the heat increment resulting from digestion of food and absorption and metabolism of nutrients. For example, increases in the level of feeding (Forbes *et al.*, 1928, 1930; Kibler, 1957), or in the proportion of forage relative to concentrate in the diet, bring about increases in the heat increment (Blaxter, 1962).

Environment and Building Design

The upper limit of air temperature above which production rate might be reduced is thought by Yeck and Stewart (1959) to be about 25°C; for the lower limit Blaxter and Wainman (1961) recorded a critical temperature for steers at maintenance in a respiration chamber to be about 6°C. For intensive beef cattle exposed to air movement (which raises the critical temperature) but fed on a high plane (lowers critical temperature) the desired minimum air temperature for efficient production is likely to be about the same, that is 6°C.

Relative humidity is generally thought to have little effect on production except at air temperatures in excess of 25°C (Yeck and Stewart, 1959). At this point these authors recorded a loss of 31 per cent production when relative humidity was 90 per cent with an air temperature of 44°C. Relative humidity plays an important role in transmission of disease, which affects production albeit indirectly, but this will be discussed later.

Ventilation is necessary to remove moisture, carbon dioxide and ammonia. It is of little significance in extensive feeding systems when there is a low density of stocking. However, for intensive beef production adequate ventilation is of extreme importance. With such systems ventilation is also a means of controlling temperature and this generally determines the limit for the maximum rate of ventilation likely to be required. Methods for the calculation of both minimum and maximum rates of ventilation are given by Morgan and Parker (1964) who also emphasize the importance of insulation, particularly of the roof, as a means for retaining heat and hence maintaining air temperature; and also for allowing precision in ventilation systems.

The recommendations of Morgan and Parker (1964), taking into account both theoretical principles and their own practical experience, are as follows:

Live weight (lb) (2·2 lb=1 kg)	Ventilation rate (cfm) 'Minimum'	'Maximum'	Optimum Temperature (°F)
85	7	40	60°F falling to 55°F
250	15	87	50°F—55°F
300	16	97	
350	18	107	
400	20	117	
450	22	126	
500	23	135	
550	25	145	
600	26	153	
650	28	162	
700	29	170	
750	31	178	
800	32	186	
850	33	193	
900	35	200	
1000	37	215	

Building Design

Adequate ventilation is one of the most important factors in any climate and for this reason cattle buildings should preferably be long and narrow, with a maximum width of 9–10 m. In countries with a long, cold or wet winter, as in the United Kingdom and much of Northern Europe, the desired environmental conditions can only be maintained if the walls and roofs are totally enclosed and of insulated construction. Natural ventilation systems using the 'stack effect' can be used and in this case extraction would be by a continuous unobstructed ridge or a series of chimneys having a total area of 50–60 in.^{2} per animal (1 in. = 2·54 cm). Mechanical ventilation is, however, preferred, according to the standards proposed by Morgan and Parker (1964).

According to Morgan and Parker (1964) air inlets should be located 18 in. to 2 ft (1 ft = 30·48 cm) below the level of the eaves and should be both baffled and insulated. The inlet area should be equivalent to 5 ft^2 per 1000 ft^3 per minute of fan extraction capacity, this area should be distributed around at least half and preferably two-thirds of the wall length. However, many variations are possible, for example, the one used by Ffiske (see Baxter and Soutar, 1964) whereby air is drawn into the building and extracted mechanically at floor level. Insulation is important because without it it is often extremely difficult to maintain the minimum ventilation rate and yet not lower the temperature within the building below what is desired. What frequently happens is that this desired temperature is maintained by reducing ventilation rate below the minimum with the result that respiratory troubles begin to develop.

In Mediterranean, and particularly tropical climates, however, the main problem is to dissipate heat and keep the cattle cool. The building should then be completely open at the sides and the roof should overhang at the eaves for at least 1 or 1·5 m in order to give shade and adequate protection against rain. A narrow building is particularly important under these conditions for it considerably facilitates air movement.

Flooring

In the author's experience, slatted floors have proved the most desirable type of construction for intensive beef production for almost all climates. The cost of providing, spreading and subsequently removing conventional straw litter in itself almost offsets the greater capital cost of such a floor. And the greater cattle density which can be achieved on slats (2 m^2 per animal, compared with at least 4 when litter is used) means that building costs are divided among more animals. Environmental control and management is easier in that the liquid excreta which collects under the slats can be pumped out without disturbing the animals. Studies in Scotland (Rowett Research Institute, 1964) and in the U.S.A. (Henderson *et al.*, 1965; Klosterman, 1966) have shown that beef cattle have the same performance on slatted floors as on solid floors with litter.

In tropical regions raised slatted floors seem to facilitate greater air circulation and thus help to keep the animals cooler. They are particularly suitable in high rainfall areas where little litter is available. They take less space and there are fewer problems with odours and insects (Preston, 1966, unpublished observations). There are many ways of constructing slatted floors and different materials can be used. But in general concrete is preferred and the floor should be arranged so that individual slats have a width of about 10 cm and a distance between them of about 3·5 cm. An excellent review on the subject of slatted floors for intensive beef cattle has been made by Baxter and Soutar (1964).

SIZE OF PENS

Experience of intensive beef production in widely different countries, for example Israel, Yugoslavia, United States, United Kingdom and Cuba (Preston, 1966, unpublished observations) favours having only small numbers, from 10 to 15 animals, in each pen. Baxter and Soutar (1964) also reported that the average number of animals per pen on 12 intensive beef farms in Scotland was between 10 and 12. On the basis of this same survey they recommended a floor area per animal of 1 m^2 for animals with live weights of 100 kg rising to 2 m^2 at live weights of 400 kg. Thus a building approximately 8 m wide and having cross divisions every 3 m offers a convenient configuration. In intensive beef units feeding passages are not really necessary, since the filling of self-feeders with concentrates is easily accomplished mechanically. In enclosed buildings raised cat walks can be used for ease of observation. Entry to the pens can then be directly from the outside, for example, with one outside door to serve two adjacent pens. This exit should communicate directly with a race leading to a weighing machine, for regular weighing is an indispensable feature of intensive beef management.

FEEDING AND WATERING

Self-feeders for intensive beef cattle should have 7·5 cm lateral feeding space and a minimum capacity of about 10 kg food per animal. They should be sited away from watering points and designed with adjustable openings to avoid wastage. Details of suitable self-feeders are given by Baxter and Soutar (1964). An increase in the depth of the trough to about 25 cm overall is suggested as a further safeguard against feed wastage.

Water bowls should be sufficiently deep to avoid splashing and wastage. Such a bowl, developed at the Rowett Research Institute, is described in Farm Building Progress (1965). In their survey Baxter and Soutar (1964) found that the average number of animals to each water bowl was about twelve; however, it is an advantage to have two for each pen as a precaution in the event of one failing. For lack of water, or feed, even for short periods of time can often predicate metabolic upsets such as bloat and acidosis. In buildings with open sides the bowls should be sited outside the pens so that cattle have to reach through the fence in order to drink. This avoids fouling with excreta. When water bowls have to be sited inside the pen it is often an advantage to surround them with a guard ring sited about 30 cm from the bowl. In hot climates, water bowls must be shaded and have the connecting water pipe below ground for the drinking of cold water is an important mechanism for dissipating excess heat.

DISEASE

This section is not intended as a general treatise on cattle diseases. Rather the aim is to draw attention to those conditions most likely to be manifested in an intensive beef enterprise.

Respiratory Disease

A number of diseases can affect the respiratory organs of cattle and the causal agents may be bacteria or viruses. In intensive beef units respiratory diseases tend to follow a certain pattern, the pertinent points of which have been described by Loosmore (1964) as follows. Most respiratory disease in intensive beef cattle is caused by viruses with bacteria acting as associated or secondary invaders. Viruses do not survive long in the environment and the degree of infection is generally a function of the number of organisms held in the air in droplets of water, and the proximity of animals. Air temperature has little effect on the spread of such disease but ventilation is paramount since it controls the rate of removal of such infectious droplets from the immediate environment around the animal; it also removes irritants, such as ammonia and dust, which cause the animal to cough and so to spread the droplet infection.

In the United Kingdom viruses causing respiratory infection are widely spread and antibodies to invading organisms can be found in 30–50 per cent of all cattle. Calves usually encounter this disease early in life and develop active immunity to the disease whilst they are still protected by the passive immunity obtained from the dam's colostrum. Not all calves encounter the disease at this time, however, and it is these animals which, when mixed later with calves from other sources (and by which time their own passive immunity has been lost), become actively infected. This explains the severe outbreaks of pneumonia which occur in 3–4-month-old calves brought together for the first time from different sources, or suddenly introduced into an intensive beef unit already holding older cattle. The presence of a few susceptible calves in a group quickly multiplies the infection to a point where immune calves also succumb.

In the long-term, control is most likely to be effected by the development of vaccines for use before such calves are mixed with others from another source. Whenever possible calves being introduced to an intensive beef unit should be put in a part which has been rested from older cattle. Separation of buildings in large units helps to facilitate this. Mixing with older cattle should be avoided until they are beyond 6–8 months of age. With smaller units this degree of organization is difficult and probably the best procedure here is to rear the calf, from the earliest possible age, in close proximity to the older animals in the rest of the unit.

In the short-term prevention of this condition probably lies with bringing together calves from different sources at the earliest possible age and thus introducing them to the infection when they still have passive maternal immunity (Loosmore, 1964).

In general terms, when cattle are mixed together and thus exposed to disease challenges, it is important to avoid stress. Transit fever is a name given to respiratory disease which is triggered off by travelling. It is usually encountered immediately after weaned calves arrive at their new destination—after having been moved considerable distances, for example, 500 km or more. It is quite prevalent in intensive beef production operations; if recognized at an early stage it usually responds quite rapidly to treatment. It rarely affects very young calves, for example those less than one week of age, and is most evident in calves between 3 and 4 months. Overcrowding while travelling should be avoided and if possible an independent ventilation system should be installed for use, particularly when the transport is stationary.

Liver Abscesses

The average incidence of liver abscesses in intensive beef cattle in the United Kingdom is about 28 per cent (Weiser *et al.*, 1965). An even higher incidence in older cattle fed all-concentrate diets has been reported recently from the United States of America (Ellis *et al.*, 1963; Harvey *et al.*, 1965; Haskins *et al.*, 1965). It has been suggested by Harris (1962) that the condition is probably linked to a low rumen pH (Rowett Research Institute, 1964) causing damage to the mucosa and thus allowing entry of ruminal bacteria, mainly *Corynebacterium pyogenes* and *Fusiformis* spp., to the blood and then to the liver. A similar etiology to account for liver abscesses in 'feed lot' cattle in the U.S.A. was propounded by Jensen *et al.* (1954). The only preventative method is to feed an antibiotic at a low level in the diet; for example, Weiser *et al.* (1965) noted that administering chlortetracycline in the diet at a level of 20 mg/kg reduced from 28 per cent to 11 per cent the incidence of liver abscesses in Holstein steers fed rolled barley. Ellis *et al.* (1963) reported an even more marked reduction when the antibiotic was added to an all-concentrate diet based on ground sorghum; while Harvey *et al.* (1965) noted a similar improvement when feeding chlortetracycline in an all-concentrate diet based on maize.

An often-voiced opinion is that liver abscesses can be prevented by giving small quantities of roughage but Harvey *et al.* (1965) noted that the addition to the diet of either hay or rice husks had no effect on the incidence of liver lesions. Preston and co-workers (Rowett Research Institute, unpublished data) also could find no relationship between hay feeding and the incidence of abscesses.

Lesions and necrosis in the kidneys of intensive beef cattle have also been reported but in both cases these were associated with the feeding of alkaline salts (mainly sodium bicarbonate) in the all-concentrate diet (Preston and Fell, Rowett Research Institute, unpublished data; Nicholson and Cunningham, 1961).

Acidosis

'Acidosis' is a condition of imbalance whereby too much lower fatty acid is produced by fermentation in the rumen for the animal to utilize in its blood and tissues. The condition was first described in intensive beef cattle by McDonald *et al.* (1962), and subsequently by others (e.g. Preston, 1963; Uhart and Carroll, 1966), and has been associated with both young and older cattle which previously had been on pasture, or forage, and were then changed rapidly to *ad libitum* feeding of high-grain diet such as rolled barley. Symptoms of the disease are noted frequently within one or two days of the change and consist of lack of appetite, dullness, posterior incoordination and finally, recumbency and death. These symptoms are typical of 'acidosis' caused by acute over-eating of cereals as described by Broberg (1960). He reported that such an acidosis was characterized by a deficiency of thiamine which plays an essential role in the metabolism of fatty acids for energy. Broberg obtained rapid recoveries by giving injections of this vitamin to animals severely affected by acidosis and we have also had good results from following the same procedure (Preston, 1965; Rowett Research Institute, unpublished data). Prevention is much better than treatment, however, and to this end a special 'high fibre' diet was devised to serve as an intermediate stage in the change from forage to cereal (Preston, 1963). One such ration which proved satisfactory in many trials and in commercial practice contained 25 per cent undecorticated cotton seed meal, 25 per cent oat husks, 20 per cent rolled oats, 28 per cent rolled barley and minerals and vitamins. Other diets, for example, 50 per cent ground maize cobs and 50 per cent ground ear maize; or 25 per cent bagacillo and 75 per cent ground maize have also been used with success (Instituto Ciencia Animal, 1966). It is important that the changeover diet contains about 20 per cent crude fibre. The feeding plan recommended by Preston (1963) is to give the changeover diet *ad libitum* from a self-feeder for the first seven days after the cattle come from grass or forage. The all-concentrate diet is then substituted gradually over the next seven to ten days so that at least fourteen days elapse before cattle are maintained solely on the all-concentrate diet. In special cases (for example, with range cattle which have never previously had concentrates) it is an advantage to give small amounts of cut forage or hay for the first three to five days in addition to the changeover diet (Instituto Ciencia Animal, 1966).

Clostridial infections

Often in intensive beef units there are cases of sudden death for which there is no obvious cause. In recent investigations, however, by the Burroughs Wellcome Laboratories, Beckenham (personal communication) in connection with the Veterinary Investigation Services in the United Kingdom, close examinations by new staining techniques showed the presence of *Clostridium oedomatiens* in 20 per cent of intensive beef animals which had died suddenly; these organisms were also found in the kidneys of some animals at the Rowett Research Institute which also had died suddenly (Fell, 1965, personal communication). More information is needed concerning the extent of clostridial disease in intensive beef units but routine vaccination with combined attenuated strains of Cl. *oedematiens* and Cl. *chauvei* may well prove to be an economic proposition.

Bloat

Bloat is always a problem in intensive beef production. Ways of reducing this were suggested by Preston and co-workers (Rowett Research Institute, 1964) and include: 1. Providing some roughage-like material in the diet either by preserving the husks of the grain, as for example by carefully rolling of barley, or by adding about 5–10 per cent of a high-fibre product such as oat husks, cotton seed husks, bagacillo or ground roughage; 2. Maintaining a fresh feeding face in the self-feeder and checking feeders and waterbowls regularly to remove dirt and excreta; 3. Having cattle loose and not secured in stanchions. To relieve acute bloat it is much preferred to use stomach tubes rather than a trochar and cannula. Chronic bloaters should always be slaughtered or changed to less intensive feeding.

REFERENCES

AGRICULTURAL RESEARCH COUNCIL (1965) *Nutrient Requirements of Farm Livestock*. No. 2, Ruminants, p. 153. Agricultural Research Council, London.

AITKEN, J. N. and CRICHTON, J. A. (1956) *Brit. J. Nutr.* **10**, 220.

AITKEN, J. N., PRESTON, T. R., MACDEARMID, A. and PHILIP, E. B. (1963) *Anim. Prod.* **5**, 215.

ALBERT, S. (1942) *Endocrinology*, **30**, 454.

ARMSTRONG, D. G., BLAXTER, K. L., GRAHAM, N.McC. and WAINMAN, F. W. (1958) *Brit. J. Nutr.* **12**, 177.

BALCH, D. A. and ROWLAND, S. J. (1957) *Brit. J. Nutr.* **11**, 288.

BALCH, C. C. and CAMPLING, R. C. (1962) *Nutr. Abstr. & Rev.* **32**, 669.

BAXTER, S. H. and SOUTAR, D. S. (1964) *Farm Buildings*, Autumn, 1964.

BEAUMONT, W. H. and PARR, W. H. (1963) *Vet. Rec.* **75**, 1,344.

BEESON, W. H. (1954) *J. Anim. Sci.* **23**, 867 (Abs.)

BLAXTER, K. L. and WAINMAN, F. W. (1961) *J. Agric. Sci.* **56**, 81.

BLAXTER, K. L. (1962) *The Energy Metabolism of Ruminants*. Hutchinson, London.

BLAXTER, K. L. and MARTIN, A. K. (1962) *Brit. J. Nutr.* **16**, 397.

BOWERS, H. B., PRESTON, T. R., McDONALD, I., MACLEOD, N. A. and PHILIP, E. B. (1965a) *Anim. Prod.* **7**, 19.

BOWERS, H. B., PRESTON, T. R., MACLEOD, N. A., McDONALD, I. and PHILIP, E. B. (1965b) *Anim. Prod.* **7**, 303.

BROBERG, G. (1960) Acute Overheating in Cereals. Disertation. A. B. Louisa Nya Tryckeri-Louisa.

BRYANT, M. P. and SMALL, N. (1956) *J. Dairy Sci.* **39**, 927.

BURROUGHS, W., KOHLMEIER, R., BARRINGER, R., KAWASHIMA, R. and TRENKLE, A. (1963) *J. Anim. Sci.* **22**, 929.

BURROWS, H. (1949) *Biological Actions of Sex Hormones*, 2nd ed. Cambridge: Cambridge University Press.

CAHILL, V. R., KUNKLE, L. E., KLOSTERMAN, E. W., DEATHERAGE, F. E. and WIERBICKI, E. (1956) *J. Anim. Sci.* **15**, 701.

CALLOW, E. H. (1944) *J. Agric. Sci.* **34**, 177.

CHALUPA, W., EVANS, J. L. and STILLIONS, M. C. (1964) *J. Nutr.* **84**, 77.

CLEGG, M. T. and COLE, H. H. (1954) *J. Anim. Sci.* **13**, 108.

COOLEY, J. R. and BURROUGHS, W. (1962) *J. Anim. Sci.* **21**, 991 (Abs.).

DURHAM, R. M., HARBOUGH, F. G., STOVELL, R., ELLIS, G. F. Jr. and McGIRTY, D. (1963) Texas Tech. Coll. Livestock and Feeders' Day Rep.

ELLIS, G. F. Jr., DURHAM, R. M. and STOVELL, R. (1963) Texas Tech. Coll. Livestock & Feeders' Day Rep.

ELLIS, G. F. Jr. (1965) *Feedstuffs*, June 12, 50.

FARM BUILDING PROGRESS (1965) North of Scotland Coll. Agric. Aberdeen.

FINDLAY, J. D. (1959) *Agric. Prog.* **34**, 74.

FOLL, N. J. (1964) *Vet. Rec.* **76**, 460.

FONTENOT, J. P. and THEURER, C. B. (1965) *Feedstuffs*, May 8, 80.

FORBES, E. B., BRAMAN, W. W. and KRISS, M. (1928) *J. Agric. Res.* **37**, 253.

FORBES, E. B., BRAMAN, W. W. and KRISS, M. (1930) *J. Agric. Res.* **40**, 37.

GODSELL, T. E. and PRESTON, T. R. (1963) Input–output relationships in beef production, Proc. Seminar Cooperation between Research in Agricultural Natural Sciences and Agricultural Economics in production of input/output data. O.E.C.D. Paris (1963).

GRANT, A. B. (1955) *Vet. Rev. & Annot.* **1**, 115.

GUERIN, H. B., WILLIAMSON, J. L., THOMPSON, J. C., WILCKE, H. L. and BETHKE, R. M. (1959) *J. Anim. Sci.* **18**, 1,489 (Abs.).

HALE, W. H. (1965) *Feedstuffs* July 3, 40.

HALE, W. H. (1966) Conference at Harrogate, 24–25 March 1966. Sponsored by U.S. Feed Grains Council, London.

HALE, W. H., TAYLOR, B., SABA, W. J., CUITON, L. and THEURER, B. (1965) *J. Anim. Sci.* **24**, 883 (Abs.).

HARRIS, A. H. (1962) *Vet. Rec.* **74**, 1,434.
HARVEY, R. W., WISE, M. B. and BARRICK, E. R. (1965) *J. Anim. Sci.* **24**, 885 (Abs.).
HASKINS, R. R., WISE, M. B. and BARRICK, E. R. (1965) *J. Anim. Sci.* **24**, 985 (Abs.).
HENDERSON, H. E., NEWLAND, H. W. and ESMAY, M. L. (1965) *Mich. Agric. Expt. Sta. Rep.* AH–BC 643.
HUGHES, J., EWING, S. A., STOBOE, L. and NELSON, E. (1964) *Okla. Agric. Exp. Sta. Misc. Publ.* MP 74.
HUSTED, W. T., HALE, W. H. and THEURER, B. (1966) *J. Anim. Sci.* **25**, 903 (Abs.).
INSTITUTO CIENCIA ANIMAL (1966) *Ann. Rep.* No. 1 La Habana.
JENSEN, R., DEANE, H. M., COOPER, L. J., MILLER, V. A. and GRAHAM, R. W. (1954) *Amer. J. Vet. Res.* **15**, 202.
Kay, R. N. B. (1963) *J. Dairy Res.* **30**, 261.
KERCHER, C. J. and BISHOP, D. V. (1963) *J. Anim. Sci.* **22**, 839 (Abs.).
KIBLER, H. H. (1957) *Miss. Agric. Expt. Sta. Res. Bull.* No. 643.
KLOSTERMAN, E. W., CAHILL, V. R., KUNKLE, L. E. and MOXAN, A. N. (1955) *J. Anim. Sci.* **14**, 1,050.
KLOSTERMAN, E. W. (1966) *Feedstuffs*, Jan. 15, 38.
LAWRIE, R. A., POMEROY, R. W. and WILLIAMS, D. P. (1963) *Anim. Prod.* **5**, 218 (Abs.).
LOFGREEN, G. P. (1961) *Feedstuffs*, Aug. 12, p. 50.
LOOSMORE, R. M. (1964) Proc. Conf. on Intensive Housing of Beef Cattle, Farm Buildings Centre, Warwickshire, England, 11 Nov. (1964).
MACDONALD, D., BRADLEY, R. and MCCREA, C. T. (1962) *Vet. Rec.* **74**, 1,280.
MARTIN, E. M. and LAMMING, G. E. (1958) *Proc. Nutr. Soc.* **17**, xlviii.
MARTIN, T. G., HOWARD, R. D., LANE, G. T., JUDGE, M. D. and ALBRIGHT, J. L. (1966) *J. Anim. Sci.* **25**, 885 (Abs.).
MASON, I. L. (1956) *Proc. Brit. Soc. Anim. Prod.*
MATRONE, G., RAMSAY, H. A. and WISE, G. H. (1957) *Proc. Soc. Exp. Biol. and Med.* **95**, 731.
MATRONE, G., RAMSAY, H. A. and WISE, G. H. (1959) *Proc. Soc. Exp. Biol. and Med.* **100**, 8.
MATHEWS, C. S., SCHWABE, E. L. and EMERY, F. E. (1942) *Growth* **6**, 7.
MOORE, T. and PAYNE, J. E. (1942) *Biochem. J.* **36**, 34.
MORGAN, J. T. and PARKER, J. E. (1964) Proc. Conf. on Intensive Housing of Beef Cattle, Farm Buildings Centre, Warwickshire, England, 11th November 1964.
NATIONAL RESEARCH COUNCIL (1958) *Publ. Nat. Res. Coun., Wash.* N.o 579.
NATIONAL RESEARCH COUNCIL (1963) *Publ. Nat. Res. Coun., Wash.*, No. 1137.
NICHOLSON, J. W. G., LOOSLI, J. K. and WARNER, R. G. (1960) *J. Anim. Sci.* **19**, 1,071.
NICHOLSON, J. W. G. and Cunningham, H. M. (1961) *Canad. J. Anim. Sci.* **41**, 134.
NICHOLSON, J. W. G., Cunningham, H. M. and FRIEND, D. W. (1962) *Canad. J. Anim. Sci.* **42**, 75.
NICHOLSON, J. W. G., Cunningham, H. M. and FRIEND, D. W. (1963) *J. Anim. Sci.* **22**, 368.
OLTJEN, R., SIRNY, R. and TILLMAN, A. D. (1960) *J. Anim. Sci.* **19**, 1,282 (Abs.).
PINCUS, G. and THIMANN, K. V. (1955) *The Hormones, Physiology, Chemistry and Applications*, Volume 3, Academic Press: New York.
POMEROY, R. W. and WILLIAMS, D. R. (1962) *Anim. Prod.* **4**, 302 (Abs.).
POPE, L. S., HARPER, O. F., STEPHENS, D. and WALTER, G. (1962) *Okla. Agric. Exp. Sta. Misc. Publ.* MP 67.
POPE, L. S., WALTERS, L. E., WALTER, G. R. and CAMPBELL, W. D. (1963) *Okla. Agric. Exp. Sta. Misc. Publ.* MP 70.
PRESTON, T. R., GEE, I. and CRICHTON, J. A. (1957) *Agric. Rev.* November, **39**.
PRESTON, T. R., DINDA, P. K. and MACLEOD, N. A. (1959) *Proc. Nutr. Soc.* **18**, xx.
PRESTON, T. R., GREENHALGH, I. and MACLEOD, N. A. (1960) *Anim. Prod.* **2**, 11.
PRESTON, T. R., WHITELAW, F. G., MACLEOD, N. A. and CHARLESON, E. B. (1962) *Anim. Prod.* **4**, 299 (Abs.).
PRESTON, T. R. (1963) *Vet. Rec.* **75**, 125.
PRESTON, T. R., WHITELAW, F. G., AITKEN, J. N., MACDEARMID, A. and CHARLESON, E. B. (1963a) *Anim. Prod.* **5**, 47.

PRESTON, T. R., AITKEN, J. N., WHITELAW, F. G., MACDEARMID, A., PHILIP, E. B. and MACLEOD, N. A. (1963b) *Anim. Prod.* **5**, 245.
PRESTON, T. R. and WHITE, F. (1964) Rowett Research Institute. Unpublished data.
PRESTON, T. R., KAY, M., WALKER, T., BOWERS, H. B., MACLEOD, N. A., MACDEARMID, A., PHILIP, E. B. and HARGRAVE, K. (1965a) *Anim. Prod.* **7**, 288 (Abs.).
PRESTON, T. R., BOWERS, H. B., MACLEOD, N. A. and PHILIP, E. B. (1965b) *Anim. Prod.* **7**, 385.
QUARTERMAN, J. and MILLS, C. F. (1964) *Proc. Nutr. Soc.* **23**, x.
RENBARGER, R., POPE, L. S. and WALLER, G. (1964) *Okla. Agric. Exp. Sta. Misc. Publ.* MP 74.
ROOK, J. A. F., BALCH, C. C. and CAMPLING, R. C. (1960) *Proc. Nutr. Soc.* **19**, i.
ROSKAMP, H. (1965) *Feedstuffs*, Oct. 9, p. 55.
ROY, J. A. B. and STOBO, I. J. F. (1963) Unpublished data from National Institute Research Dairying, Shinfield.
ROWETT RESEARCH INSTITUTE (1964) Annual Open Day Cattle Section, Rowett Research Institute, Aberdeen.
ROWETT RESEARCH INSTITUTE (1965) Annual Open Day Cattle Section, Rowett Research Institute, Aberdeen.
SABA, W. J., HALE, W. H., HUBBERT, F. Jr., KIERNAT, J. and TAYLOR, B. (1964) *J. Anim. Sci.* **23**, 533.
SYKES, J. F., ANDREWS, F. N., HILL, F. W., LORENZ, F. W., THOMAS, J. W. and WINCHESTER, C. F. (1953) *Report of Committee on Animal Nutrition*, N.R.C. Publication No. 266, 54. Washington.
TILLMAN, A. D. and MCALEESE, D. M. (1963) *Sci. Proc. Roy. Dublin Soc.* **1**, 93.
TREI, J. E., HALE, W. H. and THEURER, B. (1966) *J. Anim. Sci.* **25**, 910 (Abs.).
TURTON, J. D. (1962) *Anim. Breed. Abs.* **30**, 447.
UHART, B. A. and CARROLL, F. D. (1966) *J. Anim. Sci.* **25**, 911 (Abs.).
VIRTANEN, A. I. (1963) *Proc. XII Congr. Assn. Scand. Agric. Scientists*, Helsinki.
VOELKER, H. H. and DRACY, A. E. (1956) *J. Dairy Sci.* **39**, 929.
WEISER, M. F., PRESTON, T. R., MACDEARMID, A. and ROWLAND, A. (1965) *Anim. Prod.* **7**, 282 (Abs.).
WEISER, M. F. (1966) M.Sc. Thesis, University of Edinburgh.
WHITELAW, F. G., PRESTON, T. R. and DAWSON, G. S. (1961) *Anim. Prod.* **3**, 127.
WHITELAW, F. G., PRESTON, T. R. and MACLEOD, N. A. (1962) *Anim. Prod.* **4**, 300 (Abs.)
WHITELAW, F. G. and PRESTON, T. R. (1963) *Anim. Prod.* **5**, 131.
WHITELAW, F. G., PRESTON, T. R. and MACLEOD, N. A. (1963) *Anim. Prod.* **6**, 227.
WISE, M. B., BLUMER, T. N., MATRONE, G. and BARRICK, E. R. (1961) *J. Anim. Sci.* **20**, 561.
YECK, R. G. and STEWART, R. E. (1959) *Trans. Aus. Soc. Agric. Eng.* **2**, 71.

CHAPTER 22

Nutrition in Relation to Wool Production and Body Growth in Sheep

S. B. SLEN

Canada Department of Agriculture, Research Station, Lethbridge, Alberta

NUTRITION AND WOOL PRODUCTION

Protein and Wool Production

Wool is a fibre or modified hair that grows from the skin of sheep. It is formed as a living substance and, as such, its growth is regulated by the inherited characteristics of the sheep and by the general condition of the sheep producing it. It is made up almost entirely of the protein keratin, there being a small amount of carbohydrate. Consequently, the relationship of the level of protein intake to wool production is critical.

During the latter half of the nineteenth century, interest concerning the relationship of protein to wool production began to develop and as a result many experimental studies were initiated. One of the earliest experiments was conducted by Henneberg (1864) in which he compared the effects of three relatively high levels of protein intake (25–30 per cent of the ration) with a 10–11 per cent intake. He concluded that fattening feeds, characterized by an increase in protein, had only a slight effect on the amount of wool grown. It is now well established that the low level used was adequate for maximum wool growth and consequently little response would be expected from the higher levels. Subsequently, this was confirmed by a number of other workers.

It was observed by a number of workers that increasing the protein content of the ration increased wool production, but decreasing this constituent did not decrease wool production. It was not until Fraser reported his work (1931) at the Rowett Research Institute, that it was realized that a protein deficient ration produced a reduction in wool production.

It should be pointed out that the relationship between protein intake and wool production is complicated by the state of energy balance. When sheep are in a serious negative energy balance, the amino acids from the breakdown of its tissues provide the substrate for some wool growth. Thus, a fasting sheep living at the expense of its tissue would grow wool at a faster rate than it would if the demands for energy were met by a low-protein ration. When a sheep is in a positive energy balance, rate of wool growth is proportional over a fairly wide range to the protein intake.

Although a considerable amount of work was done on the relation of protein intake to wool production, virtually none was conducted on the effect of feeding rations of equal caloric value and varying only the protein content. Most of the work confounded energy with protein in the ration and it was not possible to determine which caused the actual response in wool growth.

Using pelleted isocaloric rations Slen and Whiting (1952) at the Research Station, Lethbridge, Alberta, conducted a number of experiments to determine the relation of protein levels in the ration to wool production. Feeding levels of 7, 10, and 13 per cent protein it was found that the higher protein intake resulted in considerable increase in total wool production.

Grease fleece weights were increased significantly by each addition of 3 per cent protein. These increases amounted to approximately 16 and 25 per cent, respectively. The latter increase resulted primarily from greater impurities in the fleece. Clean or scoured fleece weights responded significantly to the first increase in protein but no further increase was evident from the 13 per cent protein ration. These results were confirmed from samples of clean wool taken from measured shoulder areas during the test period.

Fibre thickness and fibre length are affected markedly by level of protein. Recent results have shown that increasing the protein intake from 7 to 13 per cent of the ration increased the fibre thickness significantly by about 8 per cent. Work at the University of Adelaide (Anonymous, 1936) has shown that, using similar levels it was possible to change the grade of the wool based on fibre thickness from 70'S to 56'S. In general, maximum expression for fibre thickness is reached at the 13 per cent level or about 0·29 lb of protein per day. Fibre length also is increased significantly by increasing the level of protein intake. This is most logical since this provides additional substrate, which is available for growth. It appears that longitudinal fibre growth responds to a greater extent than lateral growth, i.e. fibre thickness. Both fibre thickness and length are reduced during periods of stress caused by changes in temperature, pregnancy, and lactation. One other factor related to wool growth is that on low protein or low plane rations a proportion of the fibres cease to function, which in turn reduces production.

Since protein is the most expensive portion of the animal's diet, critical consideration must be given to its use. In view of this, a number of experiments in Australia and North America have been conducted to ascertain the most economical levels of feeding. During periods of cold weather, which, in the northern hemisphere coincide with pregnancy, protein supplementation is practised only during the advanced stages of pregnancy and early lactation. It has been shown by Franklin (1952) and Smoliak and Slen (1958) that supplemental feeding once per week has been satisfactory for such conditions. In general, this does not improve wool production but it has a significant beneficial effect on lamb production.

CARBOHYDRATE AND WOOL PRODUCTION

In a host of experiments it has been shown that plane of nutrition has a marked effect on wool production. Hill (1921) found that a group of wethers sheared 4·1 lb of clean wool under Wyoming range conditions, but when fed under feedlot conditions for the following 3 years they averaged 5·2 lb. He concluded that range sheep did not ordinarily get sufficient feed for maximum wool production.

Dowell and Bowstead (1923) conducted a series of experiments on pregnant ewes and concluded that any roughage or combination of feeds that maintained the ewes in a good, vigorous condition tended to produce a heavier fleece than did less satisfactory feeds.

Joseph (1926) did not believe that wool growth could be affected easily by changes in levels of feeding. As a result of his experiments he concluded that in sheep of fine-wool breeding, the organs concerned with the secretion of the wool fibre were not easily subject to such influences as changes in level of feeding, especially for periods of 5 or 6 hr, if the sheep remained in normal health. He advocated that the level of winter feeding should be based on the needs of the sheep for health and conditions for withstanding vigorous winter climatic conditions rather than for wool.

Wilson (1931) studied the influence of the plane of nutrition upon various factors related to wool production. Three Romney wethers (6 months of age) were closely shorn and placed on a fattening ration of high protein content for 6 months. At the end of that time the sheep were shorn and put on a maintenance ration for an additional 6 months. At the end of this period they were shorn again. At this time the sheep were subjected to a sub-maintenance ration limited in quantity to the point which caused them to gradually lose weight. This ration was continued for 6 months following which the sheep were shorn. It was shown that the fleeces on the fattening ration when compared to the fleeces grown by the same animals on a sub-maintenance ration were:

(a) Approximately 343 per cent heavier in average raw weight.
(b) Approximately 319 per cent heavier in clean weight.
(c) Superior in crimp.
(d) Approximately 172 per cent greater staple length.
(e) Approximately 206 per cent stronger.
(f) Coarser in fibre thickness.

Mitchell *et al.* (1928) fed a group of sheep on a sub-maintenance ration, and although the animals were in a negative energy and nitrogen balance for over

200 days, they stored energy and nitrogen in their wool at approximately normal rates.

Weber (1931) conducted experiments on the effect of the plane of nutrition on wool growth. Two groups of uniform Shropshire ewes sired by one ram were used. Lot II received a full feed of ground corn and alfalfa meal, and Lot I alfalfa meal in such amounts as to cause a slight loss in body weight. At the end of 6 months the sheep were shorn and the clean fleece weights determined. Lot I produced 1·4 lb of clean wool, while losing 7 lb in body weight. Lot II produced 3·3 lb of clean wool and gained 45 lb in weight. A second feeding period of 6 months produced similar results. It was concluded that the low plane of nutrition produced less clean wool because the wool fibres did not grow normally in either thickness or length.

Swart (1936) fed one group a ration estimated to be adequate in protein for maintenance and production; a second group received half this amount. His data indicated that the higher ration gave increased raw fleece weight, staple length, and fibre thickness. He concluded that Merino sheep tend to grow wool at the expense of body maintenance.

Bowstead and Larose (1938) reported that the feeding of sub-maintenance rations to pregnant ewes during the winter season seriously reduced wool growth. Maintenance of ewes in fair condition of flesh resulted in the production of fleeces that could be considered normal, and heavy feeding of ewes slightly increased the clean fleece weight. Average clean fleece weights produced during a 5-year period were 4·1, 4·5, and 4·8 lb for the low, medium, and high planes of nutrition, respectively.

Marston (1948) found that the weight of wool determined in seven sheep subjected to four levels of food intake, increased from 400 to 600 per cent from the lowest to the highest level of feeding, and this change was reflected in both the mean length and the mean thickness of fibres. The rations were approximately $\frac{1}{2}$, 1, $1\frac{1}{4}$, and 2 times the maintenance level. It is interesting to note that after a change in the maintenance level there was a time lag of 3 months or more between the change of feed level and the establishment of an equilibrium for wool growth.

Cystine and Wool Production

Although considerable controversy existed among early workers in this field as to the role of cystine in wool production it was due primarily to relatively poor methods of analyses in feedstuffs and herbage which led to erroneous interpretations and the postulation of several theories concerning cystine synthesis. In general, it was suggested that the cystine content of the forages fed was insufficient to meet the requirements of the animal for normal wool production. It remained so until about 1950 when a number of workers, including Hale and Garrigus (1953) confirmed the fact that sheep could synthesize cystine from elemental sulphur using ^{35}S as a tracer. There also is some indication that sulphate sulphur is better utilized than elemental sulphur for cystine synthesis.

SULPHUR AND WOOL PRODUCTION

Because sulphur is present in the keratin molecule and because some bacteria in the rumen have synthesizing powers, many workers have fed sulphur to sheep in the hope of increasing wool production. With one exception, no response has ever been obtained. Recent work with radioactive sulphur has indicated that inorganic sulphur can be utilized in the synthesis of sulphur amino acids in ruminants. Results from the Research Station, Lethbridge, have indicated that the sulphur requirements of the mature ewe for lamb and wool production do not exceed 0·1 per cent of the ration (Whiting *et al.*, 1954).

TRACE ELEMENTS AND WOOL PRODUCTION

Selenium

Published results have shown that selenium will produce toxic results when ingested in sufficient amounts by various classes of livestock.

Rosenfeld and Beath (1946) found that selenium poisoning in sheep caused a gradual decrease in vitamin A and total protein in the blood. A more rapid decrease in the ascorbic acid and an increase in non-protein nitrogen was noted in all sheep dying from selenium poisoning. On the other hand selenium has been found to be an important factor in the prevention of white muscle disease in lambs and calves, exudative diathesis in chicks, and necrotic liver degeneration in rats and mice.

Results obtained by Slen *et al.* (1961) have shown that the administration of sub-toxic levels of selenium produced a significant increase in clean fleece weight and fibre thickness. In this study the average increase in the selenium content of the treated animals when compared with the controls was approximately 45 per cent. It was also demonstrated by Demiruren and Slen (1963) that fibres from selenium treated sheep contracted less in both diameter and length than those from the controls when treated in 50 per cent LiBr solution at 100°C for 3 hr. Probably the molecular structure of the selenium treated fibres was modified by the formation of new cross-linkages between the peptide chains. The mechanical properties of the selenium-treated fibres such as dry tenacity, extension at break, and work load to break individual fibres were not affected when compared with controls.

Copper

Copper has long been known to cause a condition in sheep known as 'swing-back' or 'sway-back' exemplified by incoordination in the hind legs and paralysis, which has been diagnosed in Peru, Sweden, Great Britain, South Africa, United States, and Australia. Although the amount of copper required to maintain normal physiological functions is extremely small, it has certain vital functions to perform. Copper occurs as hemocuprein in blood cells and in certain enzyme systems. It is known that although iron may be absorbed and deposited in the liver, no normal blood regeneration occurs if copper is deficient. It also has a significant function in the growth and pigmentation of wool and hair.

When weaned lambs are subjected to a copper deficient ration, they produce abnormal wool and their general development is slow. In adult sheep the most noticeable lesion is in the wool and the changes which occur are quite definite. The normal character as indicated by the crimp of the wool is practically lost and in cases where the deficiency is borderline, large 'secondary waves' occur which appear to be superimposed on the crimp. If the deficiency is accentuated, the abnormality is more pronounced and in extreme cases of deficiency the wool becomes 'steely' or 'silky'. The wool now is almost entirely void of crimp, although there may be some indefinite waves and the wool has a typical glossy sheen. The wool feels soft and slippery and the general impression is that the wool lacks bulkiness. The affected fleeces usually lack strength and when the entire staple is affected there is no definite break but the wool tends to be weak or 'rotten'.

One interesting feature of this abnormality in wool is that it occurs only in that part of the staple grown during the time in which there was insufficient copper and after the liver stores have been exhausted. Typical 'silky' wool has been observed in previously normal-woolled sheep some time after their introduction into copper deficient areas. When these animals were removed, their wool again became normal.

Some observations have been made on black sheep grazing on land known to be copper deficient. These animals showed intermittent white bands along the length of their black staples and there was a strong correlation between the occurrence of these white bands and the appearance of changes typical of copper deficiency in the wool of white sheep. In addition, the black pigment returned rapidly to this wool when the animals were given copper drenches.

Research has shown that if an adequate amount of copper is administered during the gestation period, normal lambs can be produced. Bennetts and Beck (1942) suggested a daily intake of 10–20 mg of copper during the gestation period was effective. Investigations in South Australia indicated that a daily intake of 5 mg of copper was the minimum amount required by sheep on extremely deficient pastures. Drenching may have some value as a supplementary control measure. However, care must be taken to avoid overfeeding of copper since overdosing will kill animals in a relatively short time.

Cobalt

Only sheep and cattle are known to be affected by a cobalt deficiency and it has been shown that sheep are more susceptible than cattle. The recognition of cobalt as an essential nutrient developed from long-term studies of certain peculiar wasting diseases. Like copper deficiency, the disease was limited to certain areas and could be cured by moving the animals to healthy areas. The discovery of cobalt deficiency in animals resulted from the discovery of its lack in the soil and consequently in the forage. It is now recognized that cobalt is an essential part of the vitamin B_{12} molecule.

In cobalt-deficient sheep the wool growth is seriously retarded, the fibres are weak, and the wool has a 'dead' appearance. The skin gradually alters during the course of the disease from normal pink to a pale bloodless colour. In the latter stages of the disease, the skin becomes extremely fragile and breaks when the wool is parted.

Molybdenum

It has been shown that feeding an excess of molybdenum to ruminants brings on the physical symptoms of copper deficiency and interferes with copper metabolism. Dick (1954) found that when the molybdenum intake of sheep was raised to high levels, the concentration of copper in the blood increased if the diet also contained inorganic sulphate. He has suggested on the evidence of the 'steely wool' lesion associated with this rise in blood copper, that the circulating copper was not available to the enzyme systems concerned with fleece development, and that the higher intakes of molybdenum and sulphate produced a 'physiological' copper deficiency in the sheep. (For additional discussion on trace elements see Chapter 10.)

NUTRITION AND ITS RELATION TO GROWTH IN SHEEP

A comprehensive presentation on ruminant nutrition has been published by Blaxter (1954) which includes that of the young lamb. Although he has indicated the requirements in considerable detail, little has been said directly concerning its relation to growth. (See also Chapter 25.)

Body growth in Corriedale sheep was studied by Slen and Banky (1961). Body weights increased from 1 to 428 days of age, but after 300 days, approached constancy asymptotically. In general, the average body weight of single lambs was significantly greater than that of twins, and wethers were heavier than females at all ages. According to Brody (1945) body growth pattern is a function of time and in sheep may best be characterized by a sigmoid curve. In general, weight increases rapidly from birth until the point of inflection is reached at about two months of age and from then on growth rate decreases gradually.

Type of birth has a marked effect on both birth weights and gain in young lambs. De Baca *et al.* (1956) studied the effect of a number of environmental factors on the weight gains of crossbred lambs and found type of birth to be the most important factor. In this experiment single lambs were heavier at all ages up to thirteen weeks and gained at a faster rate. These workers suggested that the advantage of singles over twins was due partly to the higher birth weights of the single lambs. Donald and McLean (1935) and Phillips and Dawson (1940) also have shown that heavier lambs at birth tend to increase their weight more rapidly. On the other hand, Guyer and Dyer (1954) have attributed the higher rate of gain of single lambs compared with twins to the influence of better nutrition during the nursing period.

It is well known that the plane of nutrition at breeding has a marked influence on the type of birth and this will not be discussed here. However, since birth weight is related to subsequent growth, some discussion of the influence of nutrition during pregnancy is warranted. (See also Chapter 12.)

Protein Requirements for Growth

Jordan (1950) and Klosterman *et al.* (1951) have reported that increasing the digestible crude protein intake of pregnant ewes from approximately 0·12 to 0·33 lb daily, when receiving rations equivalent in total digestible nutrients, did not appreciably influence birth weights of lambs or their weight gains during the first 50 days of life. The latter workers also found that pregnant Shropshire ewes (weighing about 135 lb) remained in positive nitrogen balance, made satisfactory gains, and gave birth to thrifty lambs when fed an average of 0·10 lb digestible crude protein per ewe daily with a TDN intake of 1·8 lb. However, these results are not in agreement with those of Slen and Whiting (1952). In this work three groups of mature ewes were fed individually for one year isocaloric rations containing 7, 10, and 13 per cent total protein or 0·13, 0·23, and 0·29 lb of digestible crude protein. Table 1 demonstrates the influence of protein feeding on lamb weights.

Birth weights of the lambs from ewes receiving the low protein ration were significantly lighter than those from ewes receiving the higher protein rations. Birth weights were not influenced by increasing the total protein content of the ration from 10 to 13 per cent. It also was evident that lambs from the ewes receiving the two higher levels of protein gained more rapidly during the first 6 weeks after birth than those from ewes receiving the low levels of pro-

TABLE 1
The Weights of Lambs as Influenced by the Level of Protein in the Ration of Mature Ewes

	Lot I	Lot II	Lot III
Av. birth wt. (lb)			
Singles	9·5	11·4	11·0
Twins	7·5	8·6	9·0
Wt. per ewe lambed	11·9	14·6	15·2
Av. wt. at 6 weeks of age			
Born singles, raised singles	27·6	34·3	33·4
Born twins, raised singles	27·4	32·8	32·9
Born twins, raised twins		25·4	24·1
Weaning weight			
All lambs	59·8	67·5	65·8

(2·2 lb=1 kg)

tein. Some of the ewes on the low protein ration had insufficient milk to nurse single lambs and none had sufficient milk for twins. All ewes on the two higher levels of protein had sufficient milk for their lambs. It is now evident that a ration containing approximately 0·13 lb of DCP daily was adequate for the maintenance of body weight until the advanced stages of pregnancy but was not adequate for maximum lamb growth. Rations containing 0·23 and 0·29 lb DCP daily were adequate for maximum lamb production but feeding the higher level of protein did not result in heavier lambs at birth, 6 weeks of age, or at weaning.

Since a ration containing 0·13 lb DCP supplies sufficient protein for normal body gains until the latter stages of pregnancy and since it is during this latter stage that the body requirements for most nutrients are the highest, further studies were conducted to ascertain the influence of increased protein intake during the last 6 weeks of pregnancy on lamb production. It is now well established that raising the digestible crude protein content of the ration from 0·13 to 0·23 lb daily during the latter stages of pregnancy and lactation was sufficient to obtain maximum growth.

ENERGY IN RELATION TO GROWTH

It might be well to point out that plane of nutrition has been confounded with energy levels in a great deal of research work. Early studies showed that the birth weights of lambs were directly related to the plane of nutrition of the ewe during pregnancy. A diet which is restricted so that body weight is just maintained during this period will cause birth weights to be considerably lighter. An adequate diet on which ewes gain 20 to 30 lb will produce heavier lambs at birth and this superiority will be present until at least 20 weeks of age.

Results of experiments with sheep using digestibility and nitrogen retention as criteria of protein utilization, have shown that the addition of energy in the form of starch, cerelose, or corn oil to low protein rations will reduce the digestibility of the protein (Head, 1953; Pfander *et al.*, 1957; Woods *et al.*, 1956) and may reduce nitrogen retention (Fontenot *et al.*, 1955).

It is interesting to observe the results when protein is provided at constant levels of intake and the energy values are varied. In work by Whiting and Slen (1958) when the energy content of the ration was increased by approximately 10 per cent during advanced pregnancy and early lactation at any of three levels of protein (0·16, 0·20, 0·24 lb DCP daily), it had no influence on birth weights of single lambs. At the two lower levels of protein intake, feeding additional energy reduced the birth weights of twins significantly. In general, at the high level of protein intake, feeding additional energy will increase 6–8-week weights of single lambs but has little effect at the lower levels.

The results from numerous trials would indicate that the addition of starch or high energy feeds to a low protein ration reduces the apparent digestion of the protein. However, at protein intakes above minimum requirements this depression of protein utilization apparently does not occur. This indicates that the minimum protein requirements of a ruminant animal for any function at a given body weight, is not a fixed quantity but depends on the energy content of the ration, and consequently should be stated in relation to energy intake and the form of energy in the ration.

The following tabulations (Table 2) give some indication of the TDN and DCP requirements considered adequate by authorities in the field.

TABLE 2
The Daily DCP and TDN Requirements for Growing Ewe Lambs as Taken from the Published Literature (values in lb)

Wt. of lamb	Morrison		NRC		Blaxter and Mitchell	
	TDN	DCP	TDN	DCP	TDN	DCP
60	1·4	0·16	1·6	0·16	1·6	0·17
70	1·6	0·19	1·7	0·15	1·7	0·17
100	1·8	0·21	1·7	0·14	1·7	0·16

Taken from Whiting and Slen (1958)

Energy for Maintenance and Growth

An adequate supply of energy is essential for the proper development of an animal and maximum utilization of the protein ingested. Garrett *et al.* (1959)

have studied the energy requirements for sheep for maintenance and gain and obtained some interesting results, as presented in the following tables from their paper (Table 3).

TABLE 3
A Summary of the Energy Consumption and Weight Gains of Sheep

Level of feeding*	TDN (lb)	DE (kcal)	ME (kcal)	Mean body wt. (lb)	Mean daily gain (lb)
M	0·91	1924	1554	67·8	0·08
0·1	1·21	2565	2096	72·8	0·17
0·2	1·43	3050	2511	77·6	0·26
0·3	1·65	3527	2918	79·7	0·33

*M=Maintenance; other levels are the expected daily gains (lb)

On the basis of their work the estimated daily energy requirements of sheep for maintenance of energy equilibrium can be obtained in terms of the different measures of food energy by the following expressions:

$$\begin{aligned} \text{TDN} &= 0{\cdot}036W^{\frac{3}{4}} \\ \text{DE} &= 76W^{\frac{3}{4}} \\ \text{ME} &= 62W^{\frac{3}{4}} \\ \text{NE} &= 35W^{\frac{3}{4}} \end{aligned}$$

They found that the partial efficiency of food utilization was independent of body size and therefore one relationship between energy intake and energy gain will express the energy requirements for any rate of gain.

$$\text{DE} = 76W^{\frac{3}{4}}\,(1{\cdot}02)^{ge/W^{\frac{3}{4}}}$$

where *ge* is gain in kilocalories.

This latter formula has been expressed in terms of body weight gain and is most useful in calculating expected rates of gain for comparing different rations.

The relationship of protein and energy in a ration is most important and must be considered in relation to growth. Research has shown that increasing the amount of protein in a ration will produce an increase in the apparent digestibility of the protein but no change in the percentage of protein retained or in the digestibility of the gross energy. However, increasing the digestible energy content of a ration will reduce the digestibility and retention of protein. In an experiment conducted at Lethbridge (Whiting, Slen and Bezeau, 1959) it was observed that body weight gains of lambs fed rations containing wheat straw or oat hulls did not increase to so great an extent as those fed cornstarch. It is evident that energy is a limiting factor for maximum production in rations having a lower digestible energy content. On the basis of these results the average requirements of a ewe lamb weighing 85 lb was 1·3 lb TDN and 0·13 lb DCP.

Volatile Fatty Acids for Maintenance and Growth

In view of the previous material it would appear useful to discuss the role of volatile fatty acids in relation to maintenance and growth. The acids produced in the rumen are mainly acetic, propionic, and butyric and the manner in which they are utilized is now reasonably well understood. This is most important from the standpoint of the livestock producer since these acids are important sources of energy. The type of feed bears a direct relationship to the types of acids formed. High energy or starchy feeds tend to favor the production of a higher proportion of propionic acid while low energy feeds favor the production of acetic acid. Van Soest (1955) and Leffel (1960) have reported that an increased proportion of propionic acid in the rumen of animals 1·5 to 4·0 hr after feeding either ground or pelleted forage compared with chopped or long forage. Balch (1960), who has done a great deal of research on the subject, has summarized some of the data in the literature and these are presented in Table 4.

Table 4

Some Examples of the Mixtures of Volatile Fatty Acids (as molar per cent of Total Acids) Found in the Rumen with Different Diets (Balch, 1960)

Diet	Acetic	Propionic	Butyric
Cows			
70 lb arable silage	74	17	7
38–42 lb hay	70	20	8
18 lb hay	69	18	10
28–32 lb ground hay	68	20	9
Grazing old pasture, cow B	68	18	11
Grazing old pasture, cow O	65	19	12
20 lb hay, 20 lb concentrate	64	22	12
16 lb hay, 16 lb concentrate, cow C	64	19	12
16 lb hay, 16 lb concentrate, cow D	62	20	14
16 lb ground hay, 4 lb glucose	56	18	18
6 lb hay, 18 lb heated corn	54	23	16
18 lb ground hay, 18 lb concentrates, cow D	52	31	12
18 lb ground hay, 18 lb concentrates, cow C	49	32	12
6 lb ground hay, 18 lb heated corn	49	40	6
2 lb hay, 24 lb concentrate	41	36	11
Flaked corn, minerals	34	46	6
Sheep			
Grazing lush pasture	53–60	24–26	12–16
Grazing through year	50–62	21–30	12–17

The Hannah Dairy Research Institute has studied the utilization of energy from the lower volatile fatty acids, administered singly and in combination, by sheep on either a fasting or a fattening plane of nutrition. When the acids were given individually to fasting sheep, to supply approximately 700 kcal/day, it was found that the heat increment, as a proportion of the metabolizable energy, was 41 per cent for acetic acid, 13 per cent for propionic acid, and 16 per cent for butyric acid. A mixture of these acids in the molar ratio of 5:3:2 produced a heat increment of 17 per cent. Other mixtures, in which the proportion of acetic acid varied from 25 to 90 per cent gave heat increments ranging between 13 and 17 per cent. It is apparent that the animal has some difficulty in utilizing acetic acid rapidly, but under normal circumstances the other acids have a marked synergistic effect. The utilization of a mixture of acids, irrespective of their composition, was highly efficient in the fasting animal. However, when the sheep were placed on a fattening ration it was found that the acid utilization changed significantly. The administration of individual acids gave heat increments of 67 per cent for acetic acid, 44 per cent for propionic, and 38 for butyric. In addition, when the acids were combined and administered, the composition of the mixture appeared to have a marked effect on the size of the heat increment.

In feeding trials using pelleted rations, Rhodes and Woods (1962) obtained a lower acetate to propionate ratio when compared with a similar hay that was ground. However, their long hay rations unexpectedly produced lower acetate to propionate ratios than ground mixed rations. In general, they obtained a greater total quantity of volatile fatty acids by feeding pelleted rations than when feeding ground mixed rations. It may be concluded that volatile fatty acids are used more efficiently for gain when the ratio of acetic to propionic acid is narrow and this relationship is influenced markedly by the physical form of the rations.

Milk Production of Ewe in Relation to Growth

It is well known that birth weights are closely related to the nutritional status of the ewe *in vitro* and that this weight is reflected in weaning weight. Growth of the lamb is markedly influenced during the early stages of life by the milk production of its dam since its rumen is essentially unfunctional for approximately the first 3 weeks of its life. A number of studies have been conducted concerning milk production and its influence on lamb growth. McCance and Alexander (1959) have emphasized that the early period in a lamb's life is a most critical one nutritionally. They concluded that a delayed onset of lactation in the ewe reduced the chances of lamb survival. Burris and Baugus (1955) found that body gains of young lambs and milk production was highly correlated ($r = +0{\cdot}90$) but that the correlation declined as the lambs grew older. In general, there are significant breed differences (Neidig and Iddings, 1919; Slen *et al.*, 1963) in milk production which, in itself, suggests the reason for greater production and higher growth rates in certain breeds. In these tests the Suffolk and Hampshire, which are classed basically as mutton breeds, ranked as the heaviest milk producers. Neidig and Iddings found only small differences between the Cotswold, Shropshire, Rambouillet, Lincoln, and Southdown, while Slen and associates found no differences between the Rambouillet, Romnelet, or Corriedale.

In the latter study milk production was measured by weighing the lambs prior to and after nursing and crediting the gain in weight to milk produced. The ewes nursing twins produced significantly more milk than those ewes nursing singles. It is interesting to note here that the milk drawn by the twin lambs was considered to be a reasonably good measure of milk production by the ewe while that drawn by single lambs was probably a measure of the *ad lib.* consumption.

Single lambs have been found to gain significantly more than twin lambs but generally breed differences are not consistent. Gain in body weight varies with the amount of milk consumed and is influenced by differences in creep-feed consumption.

Significant correlation coefficients between milk production and body weight gain of lambs suggest that milk production is a major factor in determining weight gains in lambs. Generally, little differences have been obtained in the protein or fat content of the milk and it would appear that they are adequate and have little influence on weight gains in lambs. These data are shown in Tables 5 and 6.

TABLE 5
Average Milk Production, Milk Composition, and Gains Made by Lambs from Five Breeds of Sheep

	Suffolk		Hampshire		Rambouillet		Canadian Corriedale		Romnelet	
	Single	Twin	Single	Twin	Single	Twin	Single	Twin	Single	Twin
Milk production (lb/day)										
1955	3·2	3·8	2·9	3·3	2·7	4·3	2·5	4·4	3·1	4·0
1956	3·9	4·2	3·5	3·9	3·5	3·7	3·2	4·3	3·2	4·2
Gain per lamb in 8 weeks (lb)										
1955	35·5	24·6	33·8	21·6	31·0	25·0	26·8	23·1	31·6	23·9
1956	34·1	23·6	30·4	22·4	30·0	24·0	30·1	23·0	29·0	23·6
Protein content of milk (%)										
1955 only	5·4	5·2	5·4	5·4	5·3	5·2	5·4	5·0	5·0	4·8
Fat content of milk (%)										
1955 only	7·5	6·4	7·8	6·9	6·7	6·6	6·3	6·4	6·6	7·2

TABLE 6

Correlations of Lamb Gain with Average Daily Milk Production

	Suffolk		Hampshire		Rambouillet	
	r	95% confidence limits	*r*	95% confidence limits	*r*	95% confidence limits
Singles						
1955	−0·10	−0·28 to 0·45	−0·24	−0·57 to 0·17	0·48	0·11 to 0·71
1956	0·62	0·31 to 0·79	0·67	0·37 to 0·82	0·61	0·28 to 0·80
Twins						
1955, 1956 (combined)	0·36	0·07 to 0·59	0·54	0·34 to 0·69	0·28	0·02 to 0·48

	Canadian Corriedale		Romnelet	
	r	95 % confidence limits	*r*	95 % confidence limits
Singles				
1955	−0·04	−0·37 to 0·38	−0·05	−0·37 to 0·38
1956	0·82	0·63 to 0·86	0·57	0·25 to 0·76
Twins				
1955, 1956 (combined)	0·65	0·46 to 0·76	0·14	−0·16 to 0·40

VITAMIN REQUIREMENTS FOR LAMB GROWTH

It is well known that the ruminant animal is able to synthesize its requirements for the B complex vitamins under normal conditions. This is obtained through the action of the rumen microflora on the ingested food. In the case of the fat soluble vitamins this is not so with the exception of vitamin K.

Vitamin A. This vitamin has been known as a dietary essential for livestock for a long time. However, it has been found in the United States, Canada, and Australia that vitamin A stores of sheep are depleted very slowly. They appear to be able to store sufficient amounts of vitamin A during those periods when they are grazing green forage to carry them through the winter or dry season when no carotene is available. At the Research Station, Lethbridge, Alberta, an attempt was made to deplete lambs of the vitamin A reserves using a ration of wheat straw and barley. After 2 years on this ration the blood levels of vitamin A still remained at normal levels. Underwood and

Conochie (1941) and Pierce (1945), working in Australia, observed that only during extended drought periods did vitamin deficiency affect the general well being of sheep. It would appear that vitamin A is more useful in maintaining the reproductive performance of an ewe than in the growth of the lamb since growth occurs normally during periods when large supplies are available. Unless the ewes are practically depleted during pregnancy—in which case weak lambs could result—it is most unlikely that extra vitamin A will affect growth rate significantly.

Recent work on vitamin A in lambs by Nelson and Hironaka (1965) indicates that deficient lambs are more susceptible to sheep ked infestation and that the populations persist for longer periods of time. Since these parasites suck blood and cause weakened lambs, this could be an indirect effect of vitamin A deficiency.

Vitamin D. Vitamin D is essential for the normal utilization of calcium and phosphorus in bone development. Usually sheep receive sufficient vitamin D from the action of its ultraviolet rays of sunlight on the sterols in the animal body. However, sheep kept indoors or in regions of high latitude may develop a condition which will respond to vitamin D therapy. Such a condition, called 'bent leg', was discovered in Scotland by Elliot and Crichton (1926) which occurred in ram lambs, seven to twelve months of age, and was prevented by the administration of small doses of vitamin D in the form of cod liver oil. Since that time similar conditions have been described in New Zealand, South Africa, and the United States.

Lambs are born with sufficient vitamin D to last about 6 weeks and the milk furnishes sufficient calcium and phosphorus to maintain a normal balance as long as they are nursing. Under normal conditions following weaning there should be no setback in growth since vitamin D is available through the action of ultraviolet rays.

Vitamin E. White muscle disease is a condition which occurs in young lambs and can reach epidemic proportions. It is characterized by a white discoloration of the muscles involved and a degeneration of the fibre bundles in the skeletal and cardiac muscles. Culik *et al.* (1951) produced the condition experimentally in sheep by feeding rations low in vitamin E. Willman and associates (1945) at Cornell had also shown that the disease could be reduced significantly by feeding wheat germ oil or tocopherol acetate to lambs from ewes on a ration which caused muscular dystrophy. However, Whiting *et al.* (1949) found that feeding 80 mg of mixed tocopherols daily to ewes, starting 6 weeks before lambing, was not effective in controlling the disease. Work by Oldfield *et al.* (1960) indicated that ration supplemented with vitamin E and selenium was effective in preventing much muscle disease in lambs. The relationship of selenium and vitamin E to growth will be discussed further in the section under minerals.

Mineral Requirements for Lamb Growth

The relationship of certain essential minerals to wool growth has been discussed briefly in a previous section. These elements will now be discussed as they pertain to lamb growth. (See also Chapter 11.)

Copper. Copper-deficient soils are known to exist in certain areas of the world and to produce a muscular incoordination of the hindquarters in lambs born of ewes grazing such areas. The condition is known as enzootic ataxia or sway back; it is usually progressive, ending in paralysis and death. Histologically, the main lesions are extensive demyelinization of the cerebrum and secondary degeneration of the motor paths in the spinal cord. The copper in the liver of affected lambs is usually less than 10 $\mu g/g$. The condition can be controlled readily by top-dressing pastures with approximately five pounds of copper sulphate per acre. Cunningham (1954) found that 0·5 per cent copper sulphate in a salt lick was effective providing the ewes consumed 2 oz of salt per week. If amounts are consumed in excess of this amount, there is danger of copper poisoning. There must be a proper balance between copper molybdenum, and inorganic sulphate if young sheep are to grow normally. The literature would indicate that further work is necessary to clarify fully the problem of copper requirements and its effect on body growth.

Cobalt. Cobalt apparently serves no function in the animal body other than as a constituent of vitamin B_{12}, although certain tissues contain small amounts of this element. It is synthesized in the rumen by the rumen microorganisms and this is an essential step. Cobalt deficiency was first discovered in Australia and was called enzootic marasmus. This was similar to bush sickness in New Zealand and to conditions found in Africa, Great Britain, United States, and Canada. This disease is limited to those areas that have a very low cobalt content in the soil which is reflected in a deficient forage.

In mature sheep cobalt deficiency is characterized by a wasting condition, a depraved appetite, and anemia. Lambs from deficient ewes may be born dead or weak, or they may also appear healthy and grow normally for a month or so and then gradually lose their appetites, growth ceases, and body weight losses occur.

The cobalt requirement of the ruminant animal is very small in relation to its needs for many other elements. The cobalt content of forages and pastures varies greatly but normally lie within the range of 0·1–0·3 ppm on a dry matter basis with the animal requirements being of the order of 0·1 ppm. In areas of cobalt-deficient pastures the problem has been controlled by top dressing annually with approximately 5 oz of cobalt sulphate per acre, or through the use of cobalt pellets which remain in the rumen and provide a source of cobalt. In areas of borderline deficiency it appears sufficient to incorporate cobalt into the salt lick normally provided for livestock.

Molybdenum. Molybdenum is of interest nutritionally because of its relationship to copper. It has been shown that the symptoms of copper deficiency may occur even though normal concentrations of copper exist in the blood when the intakes of molybdenum and inorganic sulphates are high (Dick, 1956). Apparently molybdenum ties up the copper in a form which is not available for physiological functions. Ellis *et al.* (1958) have reported the results of 3 years' work which demonstrated the beneficial role of molybdenum in sheep nutrition. Lambs fed semi-purified diet containing 0·36 ppm of molybdenum and supplemented with 2·0 ppm of sodium molybdate made significantly greater gains than those on a similar ration with no added molybdenum. Digestibility of the cellulose was increased in the molybdenum supplemented ration when compared with the basal ration. The other proximate constituents of the ration were not affected by the addition. It is possible that the feeding of molybdenum may be useful when feeding low quality roughages to ruminants.

Iodine. Iodine has been known as an essential dietary nutrient for livestock for many years and is unique among trace elements in that it functions solely as an indispensable constituent of the thyroid hormones—thyroxine and triiodothyronine. These are produced by the thyroid gland and are generally considered to maintain control over the rate of energy metabolism and certain other effects which may be considered to be secondary manifestations of thyroid disfunction.

The normal functioning of the thyroid gland has a marked influence on lambs *in utero.* Marsh (1958) has indicated that only simple goitre occurs in sheep and is primarily confined to lambs. Lambs usually are born dead or die shortly after birth if they show an enlarged thyroid. Goitrous lambs have a markedly enlarged thyroid as evidenced by a swollen neck. The wool is *usually* scanty and the body edematous. Where the condition is borderline and the lambs survive, retardation occurs in general growth. In somewhat older sheep which are approaching sexual maturity estrus and libido may be markedly retarded or absent.

Iodine requirements for normal thyroid function are small though essential. An early estimate by Mitchell and McClure (1937) indicated that the iodine requirements for a sheep weighing 110 lb was 50 to 100 μg daily. The National Research Council (1957) now indicate that the use of stabilized iodine (KI) in salt mixed at the 0·01 per cent level has been found to be effective. The increased use of iodized salt in endemic and borderline areas has served to reduce goitre significantly.

It should be pointed out that goitrogenic substances occur in animal feeds which may play some role in endemic goitre. Underwood (1956) indicates that high intakes of arsenic and possibly fluorine may be significant factors in the production of goitre where intakes of iodine are low or marginal. He also indicated that cabbage was goitrogenic in sheep resulting in glands that were four to six times larger than normal. The Agricultural Research Council (1965) recommend that the iodine intake should probably be increased by about 1.2 mg/kg dry matter when the feeds contain goitrogens.

Calcium and phosphorus. These two elements are required for the proper development of bone tissue during the growing period of the lamb. A deficiency of either calcium or phosphorus, an imbalance between the two, or lack of vitamin D, or certain related factors of inhibition or resorption will cause rickets. This condition occurs in young sheep in which ossification does not develop normally and the bones become soft and distorted, i.e. 'bent-leg'. Rickets usually does not occur in young nursing lambs as they receive sufficient calcium and phosphorus from the ewe's milk. However, after nursing ceases, rickets may develop if the proper conditions exist. Lambs suffering from rickets will show the characteristic symptoms of stiffness and lameness. The joints are sore and painful as indicated by the fact that the lambs lie down a great deal. The appetite is reduced with the resulting loss in body weight gains. Cases of deficiency may be detected by assaying the blood plasma for phosphorus but in the case of calcium the condition is usually well advanced before there is any reduction in blood serum calcium. Marsh (1958) suggests that little significance be given to blood levels until they are below 4 mg per cent for inorganic phosphorus and 7 mg per cent for calcium. Since phosphorus deficiency appears to occur more often than calcium deficiency because phosphorus is more likely to be lacking in the soil, care should be taken to insure that the intake is adequate through proper supplementation. Lambs being fattened in the feedlot on a high grain ration should be given additional calcium to insure a proper balance between these two elements.

Selenium. Following on the work that has been done on the toxicity of selenium to livestock it is most interesting to find that the presence of selenium in the ration will protect lambs from white muscle disease (Oldfield *et al.*, 1960). Further work on this problem by Oldfield and associates involved treated and untreated lambs and ewes. Lambs given vitamin E supplements gained 17·3 lb during a 6-week period compared with 15·3 lb for an unsupplemented group which developed white muscle disease. Similar lambs given selenium gained 20.8 lb while the increase of lambs from selenium-fed ewes was 27·2 lb. These workers suggested that selenium fed to lambs in the post natal period, while correcting the white muscle disease, was not available early enough to allow for maximum growth which occurred in lambs from the selenium-treated ewes. It was further suggested that selenium may be an essential nutrient *per se* necessary for normal growth even in the absence of white muscle disease. A growth response in lambs due to selenium has been obtained by Slen *et al.* (1961) in Canada, and Andrews *et al.* (1964) in New Zealand.

Iron. Although iron supplementation is of great practical significance in the rearing of pigs it has virtually none in lamb production. Iron deficiency has never been demonstrated in sheep under natural conditions and since forages contain reasonably large amounts, it is most unlikely to develop. Lambs have good liver stores and hemoglobin levels at birth but are not adequate to prevent anemia within 6–8 weeks of age if they are restricted to a milk diet. Since lambs begin to consume feeds which contain iron at 2–3 weeks of age there is virtually no danger of an iron deficiency or simple anemia developing. However, it should be noted that copper poisoning, copper and cobalt deficiencies, and certain other conditions may be manifested in varying degrees of anemia.

Fluorine. Considerable interest has been shown in fluorine; firstly, because of its toxicity to livestock and secondly because of its relation to the incidence of dental caries. Although numerous experiments have been conducted in an attempt to demonstrate that fluorine is an essential element, no clear-cut evidence has been obtained. Harris *et al.* (1963) found that feeding fattening lambs rations containing 50 and 100 ppm of fluorine in the form of sodium fluoroxide did not improve their feedlot performance. When 200 ppm were fed in the ration there was a significant reduction in feed intake and average daily gains.

REFERENCES

AGRICULTURAL RESEARCH COUNCIL, London (1965) Requirements of Farm Livestock, No. 2 Ruminants.

ANDREWS, E. D., GRANT, A. B. and STEPHENSON, B. J. (1964) Weight responses of sheep to cobalt and selenium in relation to vitamin B_{12} and selenium concentrations in liver and kidney. *New Zealand J. Agr. Res.* **7**, 17–27.

ANONYMOUS (1936) *Tenth Annual Report of the Council for Scientific and Industrial Research*, pp. 39–40.

BALCH, C. C. (1960) Rumen digestion and herbage utilization. *Proc. 8th Int. Grassland Congress.*

BENNETTS, H. W. and BECK, A. B. (1942) Enzootic ataxia and copper deficiency of sheep in Western Australia. *C.S.I.R. Bull.* No. 47.

BLAXTER, K. L. (1954) Ruminant nutrition. *Progress in the Physiology of Farm Animals.* Vol. 1. Edited by Sir John Hammond. Butterworth Sci. Publications, London.

BOWSTEAD, J. E. and LAROSE, P. (1938) Wool growth and quality as affected by certain nutritional and climatic factors. III Results and Conclusions. *Can. J. Res. D.* **16**, 361–74.

BRODY, SAMUEL (1945) *Bioenergetics and Growth.* Reinhold, New York.

BURRIS, M. J. and BAUGUS, C. G. (1955) Milk consumption and growth of suckling lambs. *J. Animal Sci.* **14**, 186–99.

CULIK, R., BACIGALUGO, F. A., THORP, F., LUECKE, R. W. and NELSON, R. H. (1951) Vitamin E deficiency in the lamb. *J. Animal Sci.* **10**, 1006–16.

CUNNINGHAM, I. J. (1954) Copper deficiency in cattle and sheep. *New Zealand J. Agr.* **88**, 369–74.

DE BACA, R. C., BOGART, R., CALVIN, L. D. and NELSON, O. M. (1956) Factors affecting weaning weights of crossbred spring lambs. *J. Animal Sci.* **15**, 667–79.

DEMIRUREN, A. S. and SLEN, S. B. (1963) Effects of selenium on certain mechanical properties of the wool fibre. Can. *J. Animal Sci.* **43**, 233–6.

DICK, A. T. (1954) Preliminary observations on the effect of high intakes of molybdenum and of inorganic sulfate on blood copper and on fleece character in crossbred sheep. *Austral. Vet. J.* **30**, 196–202.

DICK, A. T. (1956) Molybdenum in animal nutrition. *Soil Sci.* **81**, 229–36.

DONALD, H. P. and MCLEAN, J. W. (1935) The growth rate of lambs in Canterbury. *New Zealand J. Sci. Technol.* **17**, 497–519.

DOWELL, A. A. and BOUSTEAD, J. E. (1923) Roughages for pregnant ewes. *Univ. Alta. Bull.* 1.

ELLIOT, W. and CRICHTON, A. (1926) A report on the addition of mineral salts to the ration of sheep. *J. Agr. Sci.* **16**, 65–77.

ELLIS, W. C., PFANDER, W. H., MUHRER, M. E. and PICKETT, E. E. (1958) Molybdenum as a dietary essential for lambs. *J. Animal Sci.* **17**, 180–8.

FONTENOT, J. P., GALLUP, W. D. and NELSON, A. B. (1955) Effect of added carbohydrate on the utilization by steers of nitrogen in wintering rations. *J. Animal Sci.* **14**, 807–17.

FRANKLIN, M. C. (1952) Maintenance rations for Merino sheep. 1. A comparative study of daily and weekly feeding on rations containing high proportions of wheat and several proportions of roughage to concentrate. *Austral. J. Agr. Res.* **3**, 168–186.

FRASER, A. H. H. (1931) Research on the influence of feeding on the weight and quality of the fleece. *Proc. Imp. Wool Res. Conf.* (1930), pp. 78–79.

GARRETT, W. N., MEYER, J. H. and LOFGREEN, G. P. (1959) The comparative energy requirements of sheep and cattle for maintenance and gain. *J. Animal Sci.* **18**, 528–47.

GUYER, P. O. and DYER, A. J. (1954) Study of factors affecting sheep production. *Missouri Agr. Exp. Sta. Bull.* 558.

HALE, W. H. and GARRIGUS, U. S. (1953) Synthesis of cystine in wool from elemental sulfur and sulfate sulfur. *J. Animal Sci.* **12**, 492–6.

HARRIS, L. E., RALEIGH, R. J., MADSEN, M. A., SHUPE, J. L., BUTCHER, J. E. and GREENWOOD, D. A. (1963) Effect of various levels of fluorine, stilbestrol, and oxytetracycline, in the fattening ration of lambs. *J. Animal Sci.* **22**, 5–55.

HEAD, M. J. (1953) The effect of quality and quantity of carbohydrate and protein in the ration of sheep on digestibility of cellulose and other constituents of the ration with a note on the effects of adding vitamins of the B-complex on the digestibility and retention of the nutrients of a hay ration. *J. Agr. Sci.* **43**, 281–93.

HENNEBERG, J. Cited by KNOBLICH, P. *Influence of increased protein feeding on wool growth.* Cited from translation file. Library, Univ. of Wyo. Original: Insug. Diss. Univ. Breslau, Muenchner Zeitung G.m.b.g. Munsterberg in Schles (1926).

HILL, J. A. (1921) Studies in the variation and correlation of fleeces from range sheep. *Wyo, Agr. Exp. Sta. Bull.* 137.

JORDAN, R. M. (1950) Rations for wintering breeding ewes. *So. Dakota Agr. Exp. Sta. Bull.* 339.

JOSEPH, A. E. (1926) Effect of feeding and management on the tensile strength and elasticity of wool. *J. Agr. Res.* **33**, 1073–89.

KLOSTERMAN, E. W., BUCHANAN, M. L., BOLIN, D. W. and BOLIN, F. M. (1951) Levels and sources of protein in rations for pregnant ewes. *J. Animal Sci.* **10**, 257–65.

LEFFEL, E. C. (1960) Diet and rumen volatile fatty acid production. *Proc. Univ. of Maryland Nutr. Conference* 1960:93.

MARSH, HADLEIGH (1958) *Newsom's Sheep Diseases.* Williams & Wilkins, Baltimore.

MARSTON, H. R. (1948) Nutritional factors involved in wool production by Merino sheep. 1 The influence of fodder intake on the rate of wool growth. *Austral. J. Sci. Res. B.* **1**, 362–7.

MCCANCE, I. and ALEXANDER, G. (1959) The onset of lactation in the Merino ewe and its modification by nutritional factors. *Austral. J. Agr. Res.* **10**, 699–719.

MITCHELL, H. H., KAMMLADE, W. G. and HAMILTON, T. S. (1928) Relative energy value of alfalfa, clover, and timothy hay for the maintenance of sheep. *Ill. Agr. Exp. Sta. Bull.* 217.

MITCHELL, H. H. and MCCLURE, F. J. (1937) Mineral nutrition of farm animals. *U.S. Nat. Res. Council Bull.* No. 99.

NATIONAL ACADEMY OF SCIENCES—National Res. Council (1957) Nutrient requirements of domestic animals No. V. Nutrient requirements of sheep. Publ. 504.

NEIDIG, R. E. and IDDINGS, E. J. (1919) Quantity and composition of ewe's milk: Its relation to growth in lambs. *J. Agr. Res.* **17**, 19–32.

NELSON, W. A. and HIRONAKA, R. (1965) Private communication.

OLDFIELD, J. E., MUTH, O. H. and SCHUBERT, J. R. (1960) Selenium and vitamin E as related to growth and white muscle disease in lambs. *Proc. Soc. Exptl. Biol. Med.* **103**, 799–800.

PFANDER, W. H. and VERMA, I. S. (1957) Physical factors that influence the response of sheep to added corn oil. *J. Animal Sci.* **16**, 1087.

PHILLIPS, R. W. and DAWSON, W. N. (1940) Some factors affecting survival, growth, and selection of lambs. U.S. Dept. of Agr. Circ. 538.

PIERCE, A. W. (1945) The effect of intake of carotene on the general health and on the concentration of carotene and of vitamin A in the blood and liver of sheep. *Austral. J. Expl. Biol. Med. Sci.* **23**, 295–303.

RHODES, R. W. and WOODS, W. (1962) Volatile fatty acid measurements on the rumen contents of lambs fed rations of various physical form. *J. Animal Sci.* **21**, 483–8.

ROSENFELD, I. and BEATH, O. A. (1946) The influence of protein diets on selenium poisoning. *Amer. J. Vet Res.* **7**, 52–56.

SLEN, S. B. and BANKY, E. C. (1961) Wool and body growth in lambs during the first fourteen months of life. *Can. J. Animal Sci.* **41**, 78–88.

SLEN, S. B., CLARK, R. D. and HIRONAKA, R. (1963) A comparison of milk production and its relation to lamb growth in five breeds of sheep. *Can. J. Animal Sci.* **43**, 16–21.

SLEN. S. B., DEMIRUREN, A. S. and SMITH, A. D. (1961) Note on the effects of selenium on wool growth and body gains in sheep. *Can. J. Animal Sci.* **41**, 263–5.

SLEN, S. B. and WHITING, F. (1952) Wool production as affected by the level of protein in the ration of the mature ewe. *J. Animal Sci.* **11**, 156–65.

SLEN, S. B. and WHITING, F. (1952) Further observations on the effect of protein in the ration of the mature ewe on wool and lamb production. *Sci. Agr.* **32**, 375–9.

SMOLIAK, S. and SLEN, S. B. (1958) Effects of certain supplements and methods of feeding during late pregnancy on lamb and wool production of mature ewes. *Can. J. Animal Sci.* **38**, 91–96.

SWART, J. C. (1936) The influence of the quantity of feed on the growth properties of wool. *Ztachr. Zuchtung B.* **36**, 197–226. (Cited from *Nut. Abst. and Rev.* **6**, 1136, 1936.)

UNDERWOOD, E. J. (1956) *Trace Elements in Human and Animal Nutrition.* Academic Press, New York.

UNDERWOOD, E. J. and CONOCHIE, J. (1941) Vitamin A in the nutrition of sheep in Western Australia. *Austral. Vet. J.* **17**, 202–11.

VAN SOEST, P. J. (1955) Interrelationships between composition and physical condition of feeds, rumen fermentation products, blood constituents, and milk fat in ruminants. Ph.D. Thesis. University of Wisconsin, Madison, Wisconsin.

WEBER, A. D. (1931) The effect of the plane of nutrition on wool growth. *Rec. Proc. Amer. Soc. Animal Prod.*, pp. 228–230.

WHITING, F. and SLEN, S. B. (1958) The influence of protein and energy content of the ration on lamb and wool production of range ewes. *Can. J. Animal Sci.* **38**, 208–16.

WHITING, F., SLEN, S. B. and BEZEAU, L. M. (1959) The influence of the amount of protein and energy in the ration of replacement ewe lambs on body weight gains and wool production. *Can. J. Animal Sci.* **39**, 64–70.

WHITING, F., SLEN, S. B., BEZEAU, L. M. and CLARK, R. D. (1954) The sulphur requirements of mature range ewes. *Can. J. Animal Sci.* **34**, 261–8.

WHITING, F., WILLMAN, J. P. and LOOSLI, J. K. (1949) Tocopherol (vitamin E) deficiency among sheep fed natural feeds. *J. Animal Sci.* **8**, 234–42.

WILLMAN, J. P., LOOSLI, J. K., ASDELL, S. A., MORRISON, F. B. and OLAFSON, P. (1945) Prevention and cure of muscular stiffness (stiff lamb disease) in lambs. *J. Animal Sci.* **4**, 128–32.

WILSON, J. F. (1931) The influence of the plane of nutrition upon various factors related to wool growth. *National Wool Growers* **21**, 23–28.

WOODS, W. R., THOMPSON, C. M. and GRAINGER, R. B. (1956) The effects of varying levels of protein and cerelose on the utilization of mature timothy hay by sheep. *J. Animal Sci.* **15**, 1141–6.

CHAPTER 23

Problems of the Nutrition of Ruminants in the Tropics

W. J. A. PAYNE

F.A.O./U.N.D.P., Dairy Training and Research Institute, Philippines

THE TROPICAL ENVIRONMENT

The term 'tropics' geographically defines the area between the Tropics of Cancer and Capricorn, but the climate in this area is very varied and from a bio-geographical point of view the limits of the tropics are hard to delineate.

1. Climate

Within the tropics can be found areas of dense rain forest, savannah, dry deciduous forest, thorn-bush and desert, so that the popular notion of a typical 'tropical' climate is meaningless and there are numerous tropical sub-climates. The climate in any one locality varies with unalterable factors such as latitude, altitude, distribution of land and water and changeable factors such as ocean current, winds, rainfall and vegetation.

In all their diversity the climates of the tropics have some characteristics in common. Daily and seasonal temperature variability is relatively small. It is least at the equator and gradually increases north and south, though of course there are local variations, particularly in the drier areas. The general trend is well demonstrated in the Philippines, where at Jolo (6°N) the mean monthly temperature range is 1·9°C (3·4°F) while at Laong (18°N), it is 3·8°C (6·8°F). Daylength is almost constant throughout the area, the variability being 30 min at the equator and 60+ min at the tropics. Thus differences within the tropics in the total hours of sunshine and the total solar radiation depend primarily on the degree of cloud cover.

The influence of wind on the climatic environment is considerable. The oceanic monsoons are of importance from 7° latitude north or south of the equator in the Indian sub-continent, South East Asia, northern Australia and West Africa. The trade winds that blow north and south of the equator create two distinct climatic environments on most mountainous oceanic islands, a wet humid climate on the windward side and a relatively dry climate on the leeward side. Typhoons are of local importance in the West Indies, the Philippines and the islands of the South Pacific, not only on account of the devastation that they cause, but also because they are a very important source of rain. It has been estimated that in the Philippines 25 to 33 per cent of all rain is due to typhoons (Huke, 1963).

The most important single factor, determining differences in climatic environment within the tropics, with the exception of altitude, is the total amount and the monthly distribution of rainfall. It is abundant and regular from the equator to approximately 7° latitude north or south and then progressively diminishes, except where other factors modify its distribution (Fig. 1*a*).

On account of the varied nature of the tropical climatic environment there is difficulty attendant on any attempt to create even a moderately satisfactory climate classification. De la Rue, Bourlière and Harroy (1957) have tried to do this by defining four zones, the equatorial, the tropical, the desert and the mountain, each with very different total rainfall and/or plant associations.

Equatorial. This climatic zone is characterized by relatively high and uniform temperatures, the extremes, being of the order of 18 and 35°C (64 and 95°F), normal daily temperatures varying between 22 and 30° C (72 and 86°F); an annual precipitation usually above 1905mm (75 in.) with two normally well-defined peak rainy seasons and no true dry season; and high humidities. It exists 6 to 7° latitude north and/or south of the equator in the centre of the Congo basin, in the Indian sub-continent, the Malayan peninsula, Indonesia, New Guinea, the southern Philippines and in part of the Amazon basin, but not in East Africa, the Ceará coast of Brazil, or the western coast of Ecuador and Peru around the gulf of Guayaquil. It is also found in some areas rather far from the equator such as part of the west coast of Columbia and the east coast of Madagascar.

Tropical. This climatic zone includes areas that are humid, semi-humid, semi-arid and arid; the climate varying with the total precipitation and the topography. Temperature extremes are wider than in the equatorial zone and are of the order of 10 to 45°C (50 to 113°F), the daily variation being as wide as 20 to 25°C (50 to 113°F), particularly in the semi-arid and arid areas. A typical feature of this climatic zone is that the rains are concentrated in one season and that there is a dry season of from 2 to 8 months' duration. Total precipitation is of the order of 305 to 2032 mm (12 to 80 in.) per annum and in general decreases with increasing distance from the equator. In the semi-arid and arid areas total annual rainfall is very variable and may occur as a small number of very heavy downpours. There are, however, many exceptions to these generalizations, particularly in the oceanic islands, where abnormally heavy rainfall may be experienced during the wet months on the windward side of the islands. For example on Kauai, an island situated at latitude 22°N in the Hawaiian group, at one station on the windward side the mean annual

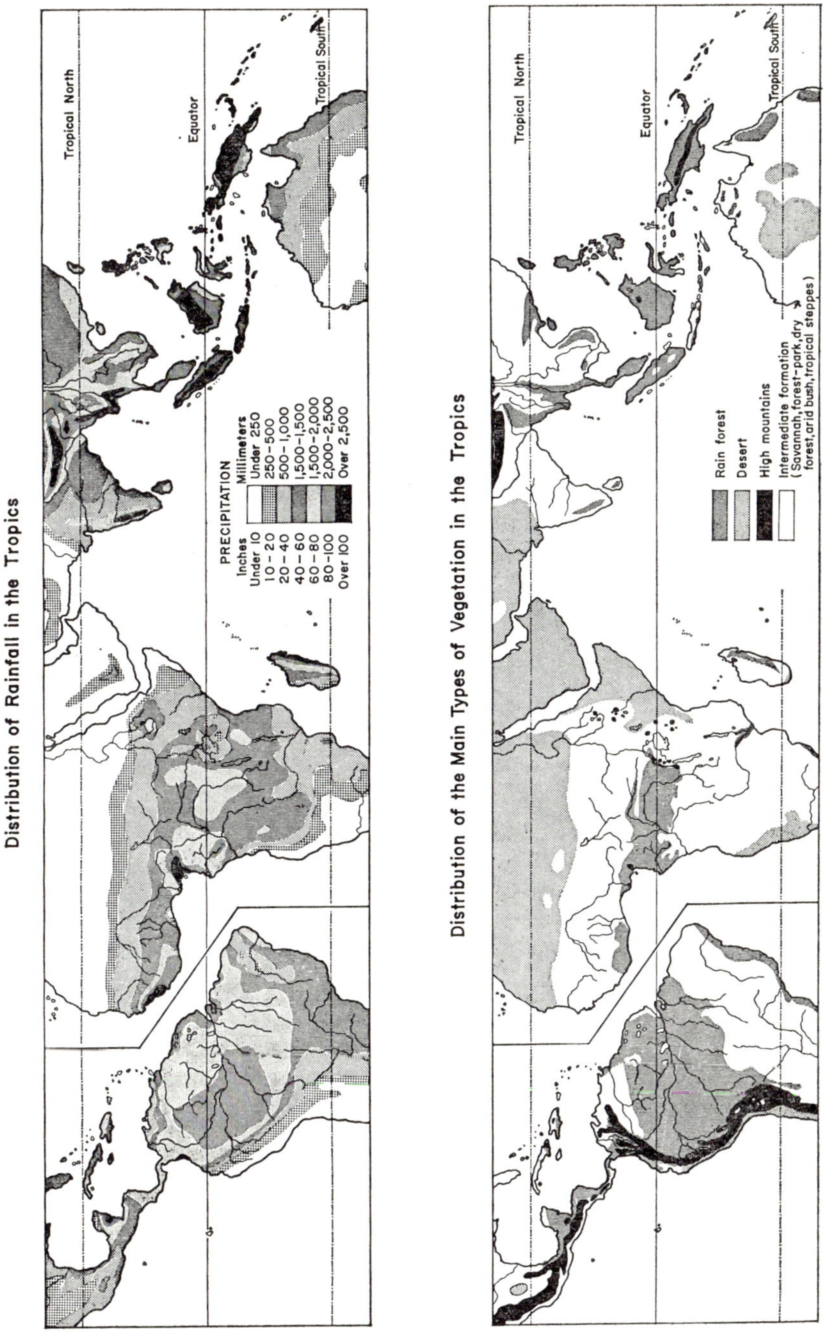

Distribution of Rainfall in the Tropics
Tropical North
Equator
Tropical South
PRECIPITATION
Inches Millimeters
Under 10 Under 250
10 – 20 250–500
20 – 40 500–1,000
40 – 60 1,500–1,500
60 – 80 1,500–2,000
80 –100 2,000–2,500
Over 100 Over 2,500
Distribution of the Main Types of Vegetation in the Tropics
Tropical North
Equator
Tropical South
Rain forest
Desert
High mountains
Intermediate formation
(Savannah, forest-park, dry forest, arid bush, tropical steppes)

a FIG. 1 *b*

rainfall is 12,014 mm (473 in.), while 24 km (14 miles) away on the leeward side, it is less than 508 mm (20 in.).

In Africa a humid tropical climate is found in the southern Sudan, northern Congo, parts of West Africa and in smaller areas in East Africa, while drier tropical climates predominate over most of East Africa, part of West Africa, Central Africa and Angola. Apart from a narrow coastal belt, drier tropical climates predominate in tropical Australia. The whole range of tropical climates are found in the Indian sub-continent, South East Asia and tropical America.

Desert. Most of the true deserts are located north and south of the Tropics, but there are relatively small areas within the tropics, particularly in the southern Sahara, south-west Arabia and the Pacific coast of northern Chile. This climatic zone is characterized by temperature extremes of the order of 0 to 50°C (32 to 122°F) and by the fact that there is no seasonal rhythm in the rainfall, which is insignificant in total amount. For example, in Aden the total annual rainfall is 51 mm (2 in.) while at Iquique, Chile, it is 5 mm (0·2 in.).

Mountain. Altitude strongly influences ambient temperature and to some extent total rainfall. In continental areas within the tropics there is a decrease of 1·8°C (3·3°F) for every 305m (1000 ft) increase in altitude and on oceanic islands the decrease is larger, being of the order of 2·3 to 2·9°C (4·2 to 5·3°F).

It is difficult to generalize, but in humid areas the rain forest zone becomes cooler with increasing altitude, and in the drier areas the open plains vegetation merges into a cool humid forest zone. Above the rain forest there may be bamboo forest, as in East Africa, or conifer forest, as in some areas of the Philippines, or open grasslands. An alpine vegetation is usually found above 3965 m (13,000 ft).

In general, areas at higher altitude in the tropics are characterized by low and uniform temperatures and by high solar radiation. For example at Quito in Ecuador, situated at 2745 m (9000 ft) the mean temperature is 14°C (57°F) while the variation in the monthly mean is 1°C (1·8°F). At medium altitudes temperatures are higher and not quite so uniform. At Kabete in Kenya situated at 1820 m (5971 ft) the extreme temperatures recorded are 5 to 31°C (41 to 87°F), but the variation in monthly mean is still only 4°C (7°F).

It is not always recognized that large areas within the tropics, particularly in East and Central Africa, Madagascar, and in the Americas from Mexico to Bolivia, are situated at medium altitude, and that not inconsiderable areas are at high altitude.

2. Soils

The exuberance of indigenous vegetative growth in the humid tropics has created a belief that the soils of these areas are inherently fertile. This is not necessarily so. Generally speaking, soils in the tropics are not as fertile as those found in the temperate regions, and they are often more easily destroyed.

Changes known as laterite formation, take place during the weathering of rocks and the process of soil formation in the humid tropics. Under the influence of constant heat and heavy rainfall, complete decomposition of the silicate minerals occurs, followed by a progressive leaching out of bases, so that the proportion of iron and aluminium hydroxides in the soil is dramatically increased. The lateritic soils that are produced in this way are often very deep and are inherently infertile. They may support a dense but shallow rooted rain forest vegetation, but if this is removed, the humus content of the top soil is drastically reduced, crop yields are low and the soil is very subject to erosion.

Not all tropical soils are lateritic, and some are very fertile. The latter include the alluvial soils found on the littoral and in the river valleys, volcanic soils of recent origin, and the soils of the cooler humid forests found at higher altitudes.

3. Vegetation and Animals

The vegetation of the tropics may be classified as: rain forest; dry forest, savannah and dry bush; desert steppe; and high mountain. The distribution of these four major vegetation zones in the tropical world is shown in Fig. 1*b*. The transitional forms are, however, so numerous, that exact divisions are difficult to establish.

Rain forest is generally found 10° latitude north and south of the equator in areas receiving at least 2032 mm (80 in.) of rainfall. It is a dense, evergreen forest, with the largest trees 45·7 m (150 ft) or more high, characterized by a multiplicity of species and full of epiphytes and lianas. At about 610 m (2000 ft) altitude it merges into mountain rain forest, moss covered and wrapped in clouds for most of the year. When total annual rainfall diminishes or becomes more seasonal, rain forest merges into semi-deciduous or monsoonal forest, this being somewhat more open, the majority of species losing their leaves during the dry season.

The littoral and the valley bottoms have been cultivated for centuries in many equatorial and monsoonal areas, the staple crop being rice in South-East Asia and various roots, bananas, plantains and maize in other parts of the world. Livestock husbandry was, and still is, ancillary to crop production. An important domestic ruminant in this region, especially in South-East Asia, is the water buffalo (*Bubalus bubalis*). The indigenous inhabitants of the forest areas were, in general, 'shifting' cultivators and hunters of forest game, rather than livestock raisers. Much of this forest is now being cut-over to remove the most valuable timber and is not being re-afforested, the less mountainous areas being suitable for arable or treecrop farming and/or animal husbandry. The exploitation of these areas using conventional domestic ruminant stock will present many new problems. It should also not be forgotten that some ruminant game species are indigenous to the rain forest, particularly in South-East Asia, and that these can also be utilized for meat production (Payne, 1965b).

As total annual rainfall progressively diminishes or becomes more seasonal, the monsoonal rain forest merges into dry forest, succeeded by savannah, and finally by dry bush. The dry forests providing cover for forage species, are particularly susceptible to fire destruction. Savannah or open grassland with trees and/or bushes studded around, presenting a park-like appearance, is found where there is adequate seasonal rainfall (762 to 1524 mm; 30 to 60 in.), and is characterized by the fact that the trees can tolerate a very long dry season. There are often wide expanses of open grasslands in this region, but it is generally thought that these are ecologically unstable fire-climax environments. As the climate becomes drier (254 to 762 mm; 10 to 30 in.), trees give way to low thorn bush or in tropical America to cactus, agaves and yuccas.

These regions, particularly in Africa, were in the past the habitat of vast herds of wild ruminants and other game. Even today, wild ruminants are numerous and could be utilized for meat production (Talbot *et al.*, 1965), but they are rapidly dwindling in numbers as the land is enclosed and cultivated and the herds of domestic ruminant livestock, often owned by semi-nomadic people, multiply. A major proportion of the domestic ruminant population of the tropics is to be found in this area, the majority of the cattle being of the *Bos indicus* or cross-bred *Bos indicus* type.

The dry bush merges into desert steppe where the total annual rainfall falls to 254 to 381 mm (10 to 15 in.). Vegetation in this region consists of widely spaced tufts of grass, scattered herbs and occasional thorn bushes, with bare stony soil between. The only productive use to which this area can be put is game farming or very extensive animal husbandry.

The vegetation of the high mountain regions varies according to latitude, altitude and rainfall and may consist of bamboo forest, conifer forest, tall grass prairies, heath lands or alpine-like meadows. Much of the area is suitable for animal husbandry and could be used for raising highly productive *Bos taurus* dairy and beef cattle and temperate-type sheep.

NUMBER AND DISTRIBUTION OF DOMESTIC RUMINANTS IN THE TROPICS

The total number and distribution of domestic ruminants, by continent, is shown in Table 1. Sheep are relatively unimportant in the tropics, camels are of local importance in parts of tropical Africa and the water buffalo is of major importance in tropical Asia. More than one-third of the world cattle population is found within the tropics, with the highest concentrations in the drier regions.

TABLE 1
The Number of Domestic Ruminants in the Tropics
F.A.O. (1963)

Continent	Cattle	Buffalo	Sheep	Goats	Camels
			(in thousands)		
Central America (1)	13,732	4	1231	2523	—
South America (2)	111,028	—	44,218	18,945	—
Asia (3)	110,812	42,821	31,813	38,700	152
Africa	91,931	—	62,602	81,067	6155
Oceania (4)	6539	1	52,593	69	3
Total No. in Tropics	334,042	42,826	192,457	141,304	6310
As percentage No. in world	35	42	19	40	61

Notes:
(1) Data for Mexico not included.
(2) Data for the whole of Brazil included.
(3) Data for China and Pakistan not included. One-half of India's ruminant livestock population included.
(4) One-third of Australia's ruminant livestock population included.

Estimates have been made of the total domestic ruminant livestock population expressed as ruminant units, the number of hectares available per ruminant unit, and the number of ruminant units per head of population (Table 2). The number of hectares available per ruminant unit is approximately the same in the tropical and the non-tropical world, while the number of ruminant units per head of population is higher in the tropics. Within the tropics the number of ruminant units per head of population is lowest in Central America and in Asia, where the climate is extensively humid, popula-

TABLE 2
Relation of Domestic Ruminant Units to Land Area and to Total Population
F.A.O. (1963)

Continent	No. ruminant units x 1000	Total area of land x 1000 ha.	Total population x 1000	No. ha. per ruminant unit	No. ruminant units per head of population
Central America	14,364	75,284	33,841	5·2	0·4
South America	121,555	1,400,973	120,342	11·5	1·0
Asia	187,023	623,987	475,559	3·3	0·4
Africa	125,109	2,180,229	194,050	17·4	0·6
Oceania	15,323	314,836	6792	20·5	2·3
Tropics	463,374	4,595,309	830,584	9·9	0·6
Remainder of world	882,942	8,933,691	2,283,416	10·1	0·4

Note: The ruminant unit used is the equivalent of one head of cattle.

tion growth explosive, and the area of land available per ruminant unit relatively small.

As total production of milk and meat is far lower in the tropics than in many other regions of the world, the data quoted above suggests that productivity per ruminant unit is very low and that the major problem is not to increase the numbers of ruminant livestock but to raise productivity per animal and per hectare. Improved nutrition will materially assist in the achievement of these aims.

PROBLEMS ASSOCIATED WITH THE CLIMATIC ENVIRONMENT

These problems are particularly complex as the climatic environment affects the ruminant animal directly and indirectly, while the level and quality of the nutritive intake and the method of feeding in turn affect the reaction of the animal to the climatic environment.

1. The Effect of the Climate on the Animal

As the climatic environment may exert a direct effect on the animal or an indirect effect on feed supplies, nutritional problems vary with the relative importance of these factors in different tropical climates.

In the equatorial and humid tropics, the direct effects of the climatic environment are probably more important, as forage growth is usually continuous throughout the year, though there are seasonal differences in yield. In the semi-arid and arid tropics the indirect effects assume a greater importance, as the feed supply progressively becomes more seasonal and limited.

1. *Direct effects*

There are profound differences between species (Findlay, 1954), between breeds or types within a species (Worstell and Brody, 1953) and between animals within a breed (Payne and Hancock, 1957) in ability to withstand the direct effects of climate, particularly heat stress.

(a) *On feed intake.* There is considerable experimental evidence available regarding the effect of the climatic environment on feed intake of cattle, but not on that of other ruminant species.

The data available suggests that high ambient temperatures depress the feed intake of all types of cattle, but that the feed intake of *Bos taurus* is depressed at lower ambient temperature than is that of *Bos indicus* breeds (Worstell and Brody, 1953; Johnson, Ragsdale and Yeck, 1958; Johnson *et al.*, 1960;

Johnston *et al.*, 1961a,b; Allen, Pan and Hayman, 1963; Randel and Rusoff, 1963).

The effect of very high ambient temperatures on feed intake is very pronounced, Ragsdale *et al.* (1948) showed that the feed consumption of Holstein and Jersey cattle virtually stopped when ambient temperature rose to 40·6°C (105°F), and Robinson and Klemm (1953) stated that rumination in Illawarra Shorthorn cows ceased at body temperatures above 40°C (104°F). More recently Aliev (1963) concluded that in a hot dry environment (28 to 44°C; 82·4 to 111·2°F) the contractions of smooth muscle in all parts of the digestive tract of buffaloes became weak and infrequent and that the passage of digesta was very much depressed.

In the equatorial and humid tropics ambient temperature and humidity are consistently high, but in the semi-arid and arid tropics humidity may be very low when ambient temperature is high. It is, therefore, important to have some evidence of the independent effect of humidity on feed intake. This problem has been studied at Missouri, where Ragsdale *et al.* (1953) reported that increasing humidity at ambient temperatures above 23·9°C (75°F) depressed the feed intake of Holstein, Jersey, Brown Swiss and Brahman cattle, while Johnson *et al.* (1963) have shown that when lactating Holsteins were managed at an ambient temperature of 32·2°C (90°F), feed intake was normal when relative humidity (r.h.) was 20 per cent, but when it rose to 40 per cent there was severe depression in both dry matter (DM) and total digestible nutrient (TDN) intake.

The effects of radiation on the intake of ruminants are more likely to be of importance in the semi-arid and arid tropics than in the equatorial or humid zones, where cloud cover is often dense. In the mountain tropics dense cloud cover alternates with clear skies, when radiation becomes intense. Brody, *et al.* (1954) studied the effect of radiation on the feed intake of lactating Holstein and Jersey and non-lactating Brahman cattle and found that when the animals were exposed to 5, 40, 90, 130 and 180 Btu/ft^2/hr of radiant energy while the ambient temperature was 21·1 to 26·9°C (70° to 80°F), the TDN intake of the Holsteins declined more than that of the Jerseys and that the intake of the Brahmans was unaffected.

Field data should reflect more accurately than climatic chamber data the overall effect of the climatic environment on the feed intake of ruminants. Unfortunately the data available is limited and inconclusive. Hancock and Payne (1955) and Payne and Hancock (1957) compared the performance of eight sets of *Bos taurus* identical twins fed hay and concentrates, one twin of each set being reared in a warm temperate climate in New Zealand and the other in a humid tropical climate in Fiji. They concluded that there was a wide variation in the intake of feed between individual twins reared in the same climatic environment, and that hay intake was significantly lower in the tropical environment although overall starch equivalent (SE) intake was approximately the same in the two environments. Johnston *et al.* (1957b) compared the DM intake of lactating Holstein cows in an air-conditioned chamber maintained at 21·1°C (70°F) with a matched group kept in an open barn and exposed to Louisiana summer temperatures. There were no significant differences in the DM intake of the two groups.

If voluntary feed intake of domestic ruminants falls with increasing ambient temperature and the beginning of this fall coincides with the first rise in body temperature (Worstell and Brody, 1953), then it appears that appetite, the immediate regulator of food intake, responds to thermal stimulation. Brobeck (1960) suggested that the hypothalamus acted as an integrator for regulating food intake and experimental evidence for this theory has been provided by Andersson and Larsson (1961). They warmed an area of the hypothalamus of goats with thermodes and caused a hungry animal to immediately stop eating.

(b) *On water intake.* The direct effect of climate on the water intake of ruminants is very complex, as water is required for at least two different purposes; first as an essential nutrient and component of the body and secondly to assist the animal lose heat by conductive or evaporative cooling.

Winchester and Morris (1956) formulated mathematical relationships between the DM and water intake of cattle and the ambient temperature. They found that between —12·2 and 4·4°C (10 and 40°F) the rate of water intake per unit of DM remained relatively constant, while above 4·4°C (40°F) it increased with the ambient temperature at an accelerating rate. However, the relationship between water intake and ambient temperature is not simple, as Reagan and Mead (1939), Ragsdale *et al.* (1949) and Ragsdale *et al.* (1951) have shown that the water consumption of Holstein, Brown Swiss and Jersey milking cows rose as ambient temperature increased from 4·4 to 26·7°C (40 to 80°F), but that above 29·4°C (85°F) it declined. Appleman and Delouche (1958) have also reported on a similar phenomena occurring with goats. When non-milking animals were studied, a decline in water intake did not occur until the ambient temperature further increased (Ragsdale *et al.*, 1950; Ragsdale *et al.*, 1951). It has been suggested that the decline in water intake above a certain ambient temperature can be attributed to a decline in feed intake and production, and to a rise in body temperature (Winchester, 1964).

Ambient temperature has a differential effect on the water intake of the different breeds of cattle (Johnson *et al.*, 1958; Johnson *et al.*, 1960; Johnson and Yeck, 1964) and it appears that acclimatised animals require less water than unacclimatized when managed at a high ambient temperature.

Harbin *et al.* (1958) demonstrated that there was no relationship between relative humidity and water intake if ambient temperature remained constant, but if the latter altered water intake was affected by humidity. Ragsdale *et al.* (1953) found that at ambient temperatures above 23·9°C (75°F) increasing humidity decreased water consumption, while the frequency of drinking increased. This may, of course, simply reflect decreased feed consumption and/or a decrease in the ability of the animal to utilize water for evaporative cooling purposes.

According to Brody *et al.* (1954), the effect of increased radiation intensity on lactating Holstein and Jersey and non-lactating Brahman cattle was to increase water consumption. Presumably this is due to the animal utilizing increasing amounts of water for evaporative cooling purposes, when subject to increased radiation stress.

Water intake in the field depends on many factors but the experimental evidence suggests that in the case of most domestic ruminants there is a direct relationship between climatic stress and water intake (Mullick, Murty and Kehar, 1952; Hancock and Payne, 1955; Rollinson, Harker and Taylor, 1955; Payne and Hancock, 1957; Harbin *et al.*, 1958; Macfarlane, 1958; Pagot and Delaine, 1958).

(c) *On efficiency of utilization of nutrients.* If climate can affect the level of nutrient intake, body temperature, respiration and pulse rates, and thus the general metabolic function of ruminants, it may also affect efficiency of utilization of feed. Experimental evidence of the overall effect of climatic environment on efficiency of utilization is limited, but what there is suggests that under controlled conditions increasing ambient temperature decreases efficiency, though under field conditions the difference may be insignificant (Hancock and Payne, 1955; Payne and Hancock, 1957; Armstrong *et al.*, 1958a; Wayman *et al.*, 1962).

As Graham *et al.* (1959) showed that digestibility was inversely related to the plane of nutrition in ruminants subject to controlled climatic conditions, it might be expected that as ambient temperature rises and feed intake declines, digestibility of ingested feed would increase. However, the experimental evidence is contradictory (Johnston *et al.*, 1957a; Johnston *et al.*, 1960; Davis and Merilan, 1960; Blaxter and Wainman, 1961; Johnston *et al.*, 1961b; Randel and Rusoff, 1963; Bailey, 1964; Kofb and Pfander, 1965).

It has been recognized for some time that the heat increment of feeding in ruminants is related to acetic acid production in the rumen (Armstrong and Blaxter, 1957a, b; Armstrong, Blaxter and Graham, 1957; Armstrong *et al.*, 1958b). More recently, Weldy, McDowell and Van Soest (1962) and Weldy *et al.* (1964) have shown that high ambient temperatures cause an insignificant decrease in the concentration of rumen VFA but a significant decline in the acetate: propionate (A: P) ratio, and in one experiment the VFA concentration was significantly correlated with rectal temperature; while Kelley, Martz and Johnson (1965), who maintained non-lactating cows at 1·6°C (34·8°F), 18·2°C (64·8°F) and 37·7°C (99·8°F) at 50 per cent r.h. on a constant feed intake, have demonstrated a marked decline in acetic, propionic and total VFA from values of 9·44, 3·76 and 15·31 to 4·67, 1·06 and 6·63 m-equiv./100 ml at 1·6° and 37·7°C, respectively.

(d) *On loss of nutrients by sweating and drooling.* With the exception of cattle, little experimental work has been conducted on the direct effect of climate on loss of nutrients by sweating and/or drooling. Cattle sweat glands are present in some number and are functional (Ferguson and Dowling, 1955), though some types of cattle possess more glands per unit area of skin than others (Dowling, 1955) and the volume of the sweat glands appear to be more important than the number per unit area of the skin (Pan, 1963). The subject has been extensively reviewed by Bianca (1965b).

Sweating behaviour, as ambient temperature rises, also differs between cattle types (Allen, 1962; Banerjee *et al.*, 1964). What experimental evidence is available suggests that generally loss of minerals through sweating and drooling is not of practical significance, though it is possible that unacclimatised temperate-type cattle that sweat profusely and drool saliva and mucus, may lose significant quantities, particularly in the arid tropics where mineral nutrient intake may be minimal. According to Bonsma (1940) losses of saliva and minerals in cattle through drooling can amount to 18 l. and 50 to 80 g per day, respectively.

2. *Indirect effects*

The major indirect effect of climate is on the quantity and quality of feed available for the ruminant animal. The most important climatic factors that limit plant growth, and hence the quantity of the feed available, are temperature, effective precipitation, length of daylight and the intensity of solar radiation. The quality of feed depends mainly on effective precipitation and on the intensity of solar radiation. The very real differences in climate that exist between the humid and the arid tropical areas thus present two broadly distinct nutritional problems, though there are many exceptions and the distinctions become blurred in the intermediate climatic zones.

(a) *In the humid tropics.* Generally forage growth is continuous and very rapid though still seasonal, and under 'high' farming conditions very heavy annual yields may be obtained. There are numerous papers detailing wet and DM yields of humid tropical forage and French (1957, 1959, 1961) has reviewed the data on composition and digestibility. Predictions of the productivity of humid tropical forage, in terms of milk production, have recently been reviewed by Hardison (1966).

(i) *On feed intake.* It is logical to expect that forage grown under conditions of abundant rainfall and high humidity should contain a high water content, and most studies show that this is so.

Payne (1962) suggested that the ability of the ruminant animal grazing humid tropical forage to obtain a sufficiently high DM intake, may often be impaired by the high water content of, or the free water on, the ingested fodder. The experimental evidence for this theory is inconclusive. In the temperate zone, MacLusky (1955), Hillman *et al.* (1958), Moore, Thomas and Sykes (1960), Campling and Balch (1961), Holmes and Lang (1963) and Mahapatro and Leffel (1964) have obtained data that suggests that the DM intake of cattle is unaffected by high water content or by free water on the forage, though Dodsworth and Campbell (1953) in Aberdeenshire and McMeekan (1958) in New Zealand have stated that excess water might limit the intake of silage and grass. Experimental evidence from the humid tropics is limited, but what is available suggests, that the high water content of, or the free water on forage, may affect the amount of feed consumed (Harrison, 1942; Butterworth, Groom and Wilson, 1961; Hardison, 1966).

(ii) *On the nutrient content of the forage.* Hardison (1966) has reviewed reports on the relationship between rainfall and the nutritional value of tropical forage. The evidence is conflicting, but in general it suggests that grasses are more nutritious in the wet than in the dry season. Most, but not all reports conclude that there is a positive correlation between rainfall and the crude fibre (CF), silica-free ash and nitrogen-free-extract content of the forage, and an inverse relationship between the rainfall and the CF content. It is possible that these conflicting reports may be due to an attempt to compare the behaviour of different forage species in what are really different climatic environments, and to the fact that although cutting intervals may be the same, the species may be at different stages of growth at different periods within the same season. Field observations suggest, that after the first 'flush' at the beginning of the rainy season, growth slows down due to continuous cloudy weather, but that it may increase very rapidly during sunny intervals.

It has generally been assumed that the nutritive value of humid tropical forage is severely limited by its reported low crude protein (CP) content, but Hardison (1966) concluded that the digestible crude protein (DCP) content of the herbage for which data was available was considerably more variable throughout the year than the TDN content and that the latter was likely to limit milk production for more months of the year than the level of DCP. This finding is similar to that of workers in other regions (Reid *et al.*, 1955; Milford and Minson, 1964).

The crude fibre (CF) content of humid tropical forage is not only inversely related to the amount of rainfall, but appears to be consistently higher than that of temperate forage at the same stage of growth (French, 1959). This is an additional disadvantage to animals that are already under heat stress and finding difficulty in eating a sufficient bulk of 'watery' forage.

(b) *In the arid and semi-arid tropics.* It is a characteristic of arid and semi-arid areas that the total rainfall is low, varying widely in amount from year to year and strongly seasonal and/or erratic in incidence. Thus it is usual for the rainy season 'flush' of highly nutritious forage to be followed by a long dry period, when growth ceases completely and the forage dries up. If a drought intervenes, the dry season may not be followed by a normal rainy season, but by a few scattered showers and an even longer dry period. Under these circumstances ruminant animals have to graze for the major part of the year on what is essentially standing hay, and in drought periods they may have to exist on this type of feed for very long periods, at a time when surface water resources are diminishing. In the truly arid areas forage growth is ephemeral, occurring only after the infrequent rains.

Thus, the major problems of nutrition of ruminants in the arid and semi-arid tropics are the intensely seasonal nature of the forage resources and the possibility of low nutrient intake and water deprivation during the dry season.

(i) *Effect of feed intake.* The DM content of forage in the arid and semi-arid tropics is high throughout most of the year and grazing animals have no difficulty in obtaining an adequate DM intake if ample amounts of forage are available. Thus, the crucial factors in the feeding of domestic ruminant livestock in these areas are keeping the stocking rate within the 'carrying capacity' of the dry season grazings and providing adequate supplies of free water throughout the year.

(ii) *Effect on water intake.* Ruminant species differ considerably in their demands for free water (Schmidt-Nielsen, 1964). Cattle are a species requiring access to some free water at all times though the water demands of different types of cattle vary and many workers have shown that *Bos indicus* apparently requires less free water than *Bos taurus* breeds, when managed in the same environment (Ittner, Kelly and Guilbert, 1951; French, 1956b; Phillips, 1960, 1961a; Horrocks and Phillips, 1961; Howes, Hentges and Davis, 1963; Payne, 1965a). Sheep may be better adapted to semi-arid conditions than cattle as Macfarlane, Morris and Howard (1963) concluded that the water turnover per unit of body weight is larger in cattle than in sheep.

When adequate free water is available, the water intake of domestic ruminant livestock rises during the dry season (Pagot and Delaine, 1958; Wilson, Barratt and Butterworth, 1962). If, however, the volume of free water available is restricted, then the animals will be water deprived for a longer or shorter period, and this often occurs in the arid and semi-arid tropics during the dry season.

Effect of water deprivation. This affects water and feed intake, metabolism and productivity.

On water intake. Water deprivation reduces the overall water consumption of sheep and cattle (Clark and Quin, 1949a; French, 1956a, b; Payne, 1965a;

Weeth and Lesperance, 1965). Cattle can acquire some degree of acclimatization to water deprivation (Payne, 1965a).

On feed intake. Experimental evidence from temperate and tropical regions suggests that cattle and sheep restrict DM intake when they are water deprived (Larsen, Hungerford and Bailey, 1917; Clark and Quin, 1949a, b; Balch *et al.*, 1953; French, 1956a; Lloyd *et al.*, 1962; Gordon, 1965) and that the decrease in feed intake can be as high as 46 per cent when sheep are water deprived for 96 hr.

The higher the ambient temperature the more depressing the effect of the same level of water deprivation on DM intake (Bianca, Findlay and McLean, 1965). If at the same time the animals are managed in such a way as to restrict their grazing time, and this is normal managerial practice in many tropical countries (Smith, 1965), this may also decrease feed intake. While the animal is water deprived, the nutritive content of available feed may decrease, as is normal in the semi-arid tropics with advancing dry season, and this may lead to a further decrease in voluntary DM intake (Crampton, 1957; Blaxter, Wainman and Wilson, 1961; Payne, 1965a). The cumulative effect of all these factors may mean that the nutritive intake of domestic ruminants in this environment is very inadequate.

Bos indicus water deprived in the same manner does not reduce its DM intake to the same extent as does *Bos taurus* (Bonsma, 1949; Phillips, 1960; Payne, 1965a).

On rumen function. As rumen fluid provides a source of water available to offset water losses during the initial stages of dehydration (Macfarlane *et al.*, 1961; Hecker, Budtz-Olsen and Ostwald, 1964), it could be expected that water deprivation would affect rumen function. Available experimental evidence suggests that this is so (Balch *et al.*, 1953; Phillips, 1961b; Gordon, 1965).

On digestibility. It has been reported from the temperate and tropical regions that water deprivation of the ruminant animal increases the digestibility of the feed and in particular the CF component (Larsen *et al.*, 1917; French, 1938; Balch *et al.*, 1953; French, 1956a, b; Phillips, 1961b).

The improvement in digestibility need not be attributed directly to water deprivation but may be due to an indirect effect such as reduced feed intake (Blaxter, Graham and Wainman, 1956) or to the excretion of additional saliva that favours the fermentation process in the rumen (Balch *et al.*, 1953; Phillips, 1961b).

On utilization of nutrients. Subjective field observations suggest that ruminant livestock thrive better than could be expected in the very severe nutritional stress environment experienced towards the end of the dry season in semi-arid tropical areas. One obvious reason for the better-than-expected performance is the animal's undoubted ability to select those parts of the forage plant that are usually of the highest nutritive value (Hardison *et al.*, 1954; Weir and Torell, 1959). However, towards the end of the dry season there is often no leaf material left on the grazing or browze species and the animal is forced to ingest feed of extremely low nutritive value. It is therefore likely that there are other factors that assist the survival of ruminant livestock under extreme nutritional stress conditions. Schmidt-Nielsen *et al.* (1957) investigated the metabolism of the camel managed under these environmental conditions and found it was able to reduce nitrogen output and particularly urea excretion in its urine to a very low level. Schmidt-Nielsen and Osaki (1958) then showed that sheep apparently possess the same facility, as do *Bos taurus* and *Bos indicus* cattle (Livingston, Payne and Friend, 1962). Payne (1965a) has since reported that when the CP intake of *Bos taurus* and *Bos indicus* cattle was so low that they were in negative nitrogen balance, severe water deprivation further reduced nitrogen output, so that the nitrogen balance of the cattle improved, even becoming positive in individual instances.

One explanation of these experimental results is that when nitrogen intake is low, nitrogen output is reduced and nitrogen is re-cycled in the body (Schmidt-Nielsen *et al.*, 1958; Houpt, 1959), possibly via the salivary glands (Somers, 1961) and/or other channels. Severe water deprivation reduces nitrogen output still further, thus enhancing the value of any re-cycling mechanism.

Rogerson (1963) using *Bos indicus* and *Bos taurus* cattle investigated the effect of declining nutrient intake and water deprivation on energy metabolism. The animals were maintained in a chamber at 20°C (68°F) ambient temperature and at 50 per cent r.h. As the CP content of the ration fed fell from 12 to 4 per cent, there was a decline of 60 per cent in digestible energy intake and a 20 per cent decrease in heat production. Rogerson concluded that cattle on a low protein roughage diet might be better able to maintain themselves in positive energy balance if they were subject to some degree of water restriction.

On mineral metabolism. Water deprivation directly affects mineral metabolism. French (1956b) concluded that when *Bos indicus* cattle were water deprived for 72-hr periods, there was a slight reduction in their sodium and chloride balances, their potassium balance was lowered and they went into negative magnesium balance. These results are interesting on account of the suggestion by Macfarlane *et al.* (1961) that the chances for survival of dehydrated sheep are improved if sodium excretion is high and potassium excretion low.

In many semi-arid regions the mineral content of the only free water available, progressively rises as the dry season advances, and may become so concentrated as to be unsuitable for drinking purposes. This is yet another hazard for ruminant animals living in these areas and Peirce (1960a) has suggested that tolerance to saline water is an important survival factor for sheep in some arid and semi-arid areas, as no doubt it also is for other species. Tolerance to highly mineralized water can be induced (Mulhearn, 1957), is a function of adaptation, and presumably happens in practice during the dry season in the semi-arid areas as the free water supplies contract.

There is considerable experimental data on the concentration of minerals that sheep and cattle will tolerate in their drinking water (French, 1956b; Embry *et al.*, 1959; Johnson, Harbers and Prescott, 1959; Peirce, 1959, 1960b; Weeth, Haverland and Cassard, 1960; Denton *et al.*, 1961; Cenni, 1963), the limits for well-adapted cattle and sheep being approximately 15,000 and 19,000 ppm of total solids, respectively.

On productivity. The short-term effect of water deprivation on the liveweight of ruminant animals can be very dramatic (Macfarlane, Morris and Howard, 1958; Payne, 1965a), but this may be mainly due to loss of body water. The long-term effects are not so well documented, but as water deprivation reduces feed intake it would be expected that it would also decrease liveweight gain. The experimental evidence available suggests that this is so. French (1938) stated that water deprivation for 48 hr reduced the liveweight gain of cattle by 50 per cent. Payne (1965a) reported that over a two-year experimental period intermittently water-deprived identical twins weighed 14·9 per cent less than their normally watered co-twins, but that the difference in liveweight gain between the two groups occurred during the first 6 months of the experiment. In this experiment, the effect of seasonal changes in the quantity and quality of the forage available was far more marked than the effect of water deprivation on liveweight gain.

There is one phenomena of liveweight loss, the 'green grass loss' experienced by cattle at the beginning of the 'rainy' season in the semi-arid tropics that is of special interest. After the onset of 'rains' and the growth of fresh, green grass, cattle continue to lose weight, Payne (1965a) reported that 60 per cent of the seasonal loss of bodyweight experienced by cattle in a semi-arid area of Tanzania took place during the first 3 weeks after the onset of the rains, and that it was 11 or 12 weeks before the liveweight of the cattle recovered to the end of dry season pre-rain level. French (1956b) also noted this phenomena and conducted an experiment to simulate what occurs out on the pasture at the beginning of the 'rains'. The daily liveweight gain of zebu cattle fed on dry hay was +0·39 kg (0·85 lb) but when the hay was wetted it fell dramatically to −0·43 kg (0·95 lb) although the total feed intake was only marginally lower.

(iii) *Effect on the nutrient content of the forage.* The ruminant animal in the arid and semi-arid regions has to exist for long periods on forage that is essentially mature 'standing hay' of low nutrient value (French, 1957;

quality is low as the animal then becomes more selective in grazing. Part of this extra grazing must take place at night if climatic stress is excessive during the middle of the day. Joblin (1960), working at Serere in Uganda, where the seasonal fluctuation in the quantity and quality of feed is wide, concluded that the availability of night grazing is critical for *Bos indicus* cattle during periods of marginal forage shortage; while under the best conditions the daylight period is sufficiently long for the cattle to obtain their needs, under very dry conditions no amount of grazing time will allow for adequate intake. Restriction of night grazing led, in the Serere environment, to a significant decline of 30 per cent in liveweight gain. These conclusions are supported by the observations of Harker, Taylor and Rollinson (1956).

No detailed observations have been made of the grazing behaviour of domestic ruminants managed under nomadic or semi-nomadic conditions in the semi-arid tropics. During the dry season the decreasing water content of the available forage increases the animal's demand for water at a time when surface water resources are dwindling, and the animal has to walk further and further to obtain adequate feed and water. Additional walking raises feed and water demand, as increased muscular activity requires additional feed and generates extra heat that has to be dissipated, further depleting the animal's water resources. At the same time the nutrient content of the available feed decreases as the dry season advances, the decreasing supplies of free water may become highly mineralized, and ambient temperatures may rise with a consequent further increase in the animal's water requirements. All these factors combine to subject partially nomadic ruminant livestock in the semi-arid tropics to very considerable physiological stress that may substantially reduce productivity.

The experimental evidence available suggests that within ill-defined limits, domestic ruminant livestock are able to adapt their nutritive needs to some of the direct and indirect stresses imposed by tropical climates, but that their productivity is affected to a greater or less degree.

II. The Effect of the Nutrition of the Ruminant Animal on Reaction to Climatic Stress

1. *Plane of Nutrition*

Most experimental evidence suggests that ruminant animals fed on a high plane are likely to experience more difficulty in disposing of body heat than those fed on a low plane, when ambient temperature is high (Clark and Quin, 1947; Robinson and Lee, 1947; Riek *et al.*, 1950; Quartermain, 1964).

Another possible effect of a high plane of nutrition would be to increase the subcutaneous deposition of fat and thus decrease 'heat tolerance' by causing interference with the mechanism of heat loss from the skin. Ledger

(1959) reported on an interesting difference between *Bos taurus* and *Bos indicus* cattle in this respect. He concluded that *Bos indicus* of equivalent fatness to *Bos taurus* cattle possess less subcutaneous fat, and that the distribution of fat within the body of these two types of cattle was different. This may be yet another characteristic that assists the survival of *Bos indicus* cattle in a semi-arid environment.

2. *Energy*

An explanation of the fact that the ruminant animal experiences more difficulty in disposing of body heat when fed on a high plane of nutrition at high ambient temperature comes from experimental work on energy metabolism. There appears to be an inverse relationship between the plane of nutrition and the range of ambient temperatures at which heat production is minimal. Armstrong *et al.* (1958a, b) maintained sheep at 8 to 38°C (46·4 to 100·4°F) ambient temperature and fed them 600, 1200 and 1800 g of dried grass daily. They found that the environmental temperatures at which heat production was minimal were 39 to 40°C (102·2 to 104°F), 33°C (91·4°F) and 24 to 27°C (75·2 to 80·6°F) for the sheep fed on low, medium and high level planes of nutrition, respectively. Above these ambient temperatures heat production increased exponentially with rectal temperature. Rogerson (1960) measured the heat production of steers, maintained on low and high planes of nutrition at 20°C (68°F), 30°C (86°F) and 40°C (104°F), and found that increasing ambient temperature had no effect on the heat production of the low-low plane but radically increased the heat production of the high plane animals. Blaxter and Wainman (1960; 1961) also showed that increasing ambient temperature from −5 to 35°C (23 to 95°F) had no effect on losses of heat per unit of body surface by convection, conduction or radiation, when steers were fed sub-maintenance or maintenance rations. More recently, Graham (1964b) has demonstrated that when Merino ewes were subjected to ambient temperatures of 10 to 35°C (50 to 95°F) minimal heat production was within the range 28 to 35°C (82·4 to 95°F), the critical temperature being lower when the sheep were fed on a higher plane of nutrition or during pregnancy.

3. *Fibre*

Most experimental evidence available suggests that low fibre rations produce a lower heat increment on digestion than those with high fibre and thus the former are most useful to feed to ruminants managed at high ambient temperature (Leighton and Rupel, 1956; Badr, 1957; Rupel and Leighton, 1957; Breidenstein *et al.*, 1960; Stott and Moody, 1960; Alim,

1963). However, Rogerson (1960) reported that the replacement of roughage by concentrates did not reduce heat production in his experiment and Wayman *et al.* (1962) concluded that a 2 per cent difference in the CF content of the rations had no significant effect on the amount or composition of the milk of Holsteins managed at 18·3 to 31·1°C (65 to 88°F) ambient temperature. It may be that the difference in CF content of rations has to be of the order of 6 to 10 per cent, as it was in some of the experimental work quoted above, before the increase in heat production on digestion becomes significant.

The products of the microbial digestion of fibre are fatty acids and in particular acetic acid, and Armstrong and Blaxter (1956) have suggested that this is an explanation for the rise in the heat increment of feeding with a rise in the fibre content of the ration, because heat production increases as the percentage of VFA, and especially acetic acid, rises in the rumen.

4. *Protein*

The experimental evidence is limited but what there is suggests that increasing the protein content of the diet has no significant effect on the ruminant's 'heat tolerance' (Robinson and Lee, 1947; Leighton and Rupel, 1960).

5. *Vitamins*

Experimental evidence of the effect of the vitamin content of the ration on the reaction of ruminant animals to climatic stress is very limited, but Page, Erwin and Nelms (1959) concluded that cattle deprived of vitamin A reserves were unable to withstand high summer temperatures.

6. *Water*

When ruminant animals are exposed to high ambient temperatures, they may utilize water in at least two ways to reduce their 'heat load'. Any water consumed in excess of metabolic needs, at a temperature lower than body temperature, and then excreted at body temperature as urine or in the faeces, directly reduces the 'heat load' of the animal, and may be used for evaporative cooling.

Thus, an adequate intake of free water is likely to assist the ruminant animal to improve its 'heat tolerance'. There is experimental evidence to support this conclusion. Payne and Hancock (1957) recorded a significant correlation between the water intake and body temperatures and respiration rates of lactating *Bos taurus* type twin heifers managed in two different climatic environments. Macfarlane, Morris and Howard (1958) showed that when sheep were managed at ambient temperatures varying from 4·4 to 46·1°C (40 to 115°F), there was a correlation between the water intake per kg of body weight and the respiration rate, and that mature sheep panted and drank less than growing sheep. Thompson, Worstell and Brody (1949) showed that as ambient temperature rose the urine output of cattle was positively correlated with the water intake, and that the greater the water intake the less the rise in body temperature. Kamal, Johnson and Ragsdale (1959) confirmed that as the ambient temperature and the water intake of cattle rose, so did urine excretion.

It is therefore reasonable to suppose that any measures that are taken to increase the volume of water intake or to lower the temperature of the ingested water should assist in reducing the 'heat load' of the ruminant animal. Rusoff *et al.* (1957, 1958) reported on an attempt to increase the volume of water intake of milking cows maintained at 35°C (95°F) by adding saccharin to the drinking water. The cows receiving saccharin treated water drank more and gained body weight, but there was no significant increase in milk production. The gain in body weight was probably due to the retention of additional water in the body. Kelly, Bond and Ittner (1955) attempted to raise the intake of water by feeding additional salt to cattle. Water consumption of the salt fed cattle increased by 60 per cent but there was no effect on liveweight gain.

Lowering the temperature of ingested water is known to have a more marked effect on the 'heat load' than increasing the volume ingested. Ruminal and rectal temperatures were shown to be depressed when cold water was ingested by sheep and cattle (Noffsinger, Otagaki and Furukawa, 1961; Bailey, Hironaka and Slen, 1962; Cunningham, Martz and Merilan, 1964), and the degree of depression depended on the volume and temperature of the ingested water. This effect can be of some practical significance as Noffsinger *et al.* (1961) reported that ingestion of water at 25°C (77°F) lowered the rumen and rectal temperature of sheep by 5·6°C (10°F) and 0·8°C (1·5°F), respectively. Bianca (1964) has also reported that steers drinking 50 l. of cold water at 14°C (57·2°F) after a period of water restriction experienced a precipitous fall in rectal, skin and subcutaneous temperature of about 1·7°C (3·1°F).

This effect has practical significance as Ittner *et al.* (1951), Ittner, Bond and Kelly (1954), Kelly *et al.* (1955) and Ittner, Bond and Kelly (1958) have shown that cooling drinking water increased the liveweight gain of beef cattle managed at high ambient temperatures in the Californian summer.

Ingestion of a large volume of cooled water could presumably affect the activity of the micro-organisms in the rumen and therefore digestibility. Bailey *et al.* (1962) concluded, however, that in sheep ingestion of cooled water has no significant effect on the digestibility of the feed.

If increasing water intake assists homeothermy, then water deprivation could be expected to have the opposite effect. Lee and Robinson (1941) reported that witholding water from sheep during exposure to very hot conditions did not produce any marked effect, but that a more drastic dehydration might have produced a reduction in 'heat tolerance'. More recent work suggests that this observation may be correct as Weeth, Hunter and Piper (1962) reported that dehydration, caused by adding sodium chloride to the drinking water of heifers, affected thermoregulation unfavourably, and Bianca (1965a) and Bianca *et al.* (1965) reported that when *Bos taurus* steers were water deprived for 4 days at an ambient temperature of 40°C (104°F), there was a decline in respiratory rate and a delay in the outbreak of sweating, with resulting increase in body temperature.

Leitch and Thomson (1944) reviewed the water requirements and metabolism of ruminant livestock and concluded that total loss of water by vaporization increased slowly to 15°C (59°F), then rapidly to 21°C (69·8°F) and at ambient temperatures above this more slowly. They concluded that other factors being equal, loss of water by vaporization rose with a rise in ambient temperature and with a rise in the plane of nutrition of the animal. Riek *et al.* (1950) concluded that in the case of sheep, increased evaporative loss was not well established until ambient temperature exceeded 35°C (95°F). Kibler and Brody (1953) reported that increasing humidity depressed the water vaporization rate from the skin of cattle.

There is experimental evidence that some types of cattle are more efficient than others at utilizing consumed water for evaporative cooling purposes (Rhoad, 1940; Findlay, 1950; Kibler and Yeck, 1959; Brown and Motasem, 1965). The reasons for this are not clear but some ruminants such as camels are apparently able to allow their body temperature to fluctuate and rise during the hottest period of the day (Schmidt-Nielsen, 1956), thus enabling them to conserve body water that otherwise would have to be used for evaporative cooling purposes in order to maintain a constant body temperature.

7. *Processing and Method of Feeding*

The method by which feed is processed and the manner in which it is fed affects the 'heat load' of the ruminant. Grinding and pelleting hay, and steaming and flaking corn, decrease rumen acetic concentration according to Shaw *et al.* (1960), and this decreases the heat increment. It can therefore be concluded that certain methods of processing feed will decrease the 'heat load' during feeding. Confirmation of this theory comes from the experimental work of Graham (1964a): who showed that at all intakes of metabolizable energy sheep fed long hay had higher heat production by 100 to

150 kcal per 24 hr and lower retention of energy by 400 to 600 kcal per hr than sheep fed pellets; and Leighton (1965) who compared the feeding of long and chopped hay to cattle during hot weather. The latter showed that the feeding of chopped hay reduced body temperature, pulse and respiration rates.

Brody (1945) once suggested that increasing the frequency of feeding should reduce heat production and more recently Rakes, Lister and Reid (1961) found that the computed heat production was significantly lower in frequently fed sheep that were rapidly gaining liveweight, but not in older animals. It is likely that the decline in the 'heat load' due to more frequent feeding may be due to a more uniform rate of absorption of nutrients and the spreading of the total 'heat increment' due to feeding over a much longer period of time and thus not overloading the animal's heat dissipation mechanisms at any one period.

8. *The Role of the Hormonal System*

The adaptation necessary for survival of highly productive ruminant livestock in a tropical climate may involve physiological changes that place a limit on productivity. If one effect of high ambient temperature on various breeds of cattle is to depress thyroid gland activity (Blincoe and Brody, 1955; Johnson and Kamal, 1958; Johnston *et al.*, 1959; Premachandra, Pipes and Turner, 1960; Johnson and Ragsdale, 1960; Johnston *et al.*, 1961a; Johnson and Kibler, 1963; and Thompson *et al.*, 1963), and if part of this depression can be directly attributed to high ambient temperature, as is suggested by the work of Lundgren and Johnson (1964) and Yousef and Johnson (1965). Then one direct effect of high ambient temperature will be a lower metabolic rate, thus reducing the 'load' on the animal's heat dissipation mechanisms, with a consequent reduction in productivity as expressed by growth and milk secretion. Thompson *et al.* (1963) suggested that under these circumstances adrenal cortical stimulation may increase plasma glucocorticoids, which would aid the stressed animal to maintain homeostasis but depress availability of some nutrients on account of the action of the glucocorticoids in deaminating amino acids. As support for this theory they showed that when Holstein heifers were exposed to high ambient temperature, thyroxine secretion decreased and plasma steroids increased.

Differential species and type response in lowered feed intake to high ambient temperature may be due in part to genetic variation in normal thyroid activity. This has been found to be higher in *Bos taurus* than in *Bos indicus* cattle (Howes, Feaster and Hentges, 1962; Post, 1965) twice as high in Jersey as in Holstein or Brown Swiss calves (Johnson and Ragsdale, 1960) and 67 per cent higher in dairy than in beef cattle (Pipes *et al.*, 1963).

OTHER PROBLEMS

Even when nutritional problems are understood and techniques to solve them evolved, the difficulties of implementing improved feeding of ruminant livestock in the tropics are very formidable. The existing limited animal industry in the wet tropics has generally evolved as an ancilliary activity within mono-cultural agricultural systems and in the dry tropics it is often not a normal farming activity but a 'way of life' for nomadic or semi-nomadic people.

The average livestock owner manages a small number of animals that he does not value as a source of income, but as motive power and/or a form of social capital. He has inherited a modicum of knowledge of traditional animal husbandry practices, that may have been very useful in the past but are inadequate in a rapidly changing environment. His formal education will be limited and his contacts with the outside world restricted. Markets for animal products, particularly highly priced products such as milk and meat, are generally underdeveloped or non-existent. Thus, ignorance, prejudice, conservatism and lack of incentives restrict the spread and widespread use of improved nutritional techniques.

Although highly productive ruminant livestock are increasingly fed on concentrates in the developed countries, this is not possible in many developing countries, where food production often lags behind population increase and there is always the possibility of man and livestock competing for inadequate feed supplies. This situation severely limits the use of concentrate feeds in the nutrition of highly productive milking cattle.

Another problem that is severe in the wet but not in the dry tropics is that of the interaction of gastro-intestinal parasites and nutrition. A wet tropical climatic environment is optimal for the development of many internal parasites. A low plane of nutrition, common in the wet tropics, encourages parasitism of the animal, and the presence of gastro-intestinal parasites in large numbers then affects the nutrition of the host by depressing appetite and decreasing the digestibility of certain nutrients (Gibson, 1963).

High productivity can only be achieved when the plane of nutrition of ruminant livestock is consistently high. In some tropical environments it is essential to store feed for some part of the year in order to supplement the animal's diet when feed supplies are minimal. Losses of stored feeds are particularly high in a wet tropical environment (Aylward, 1961) thus adding to the difficulty and expense of maintaining animals on a high plane of nutrition.

PRACTICAL IMPLICATIONS

In the wet tropics a major effect of constant high temperature and humidity on the ruminant animal is to depress feed intake. It is therefore of the utmost importance to ensure that available feed is of the highest nutritive value, and that it has a low fibre content. As pasture species in the wet tropics mature very rapidly, it is necessary to cut or graze them at frequent intervals.

The most suitable methods of managing the more productive ruminant animals such as high-grade or pure-bred *Bos taurus* cattle have not yet been finalized, but they will probably vary according to local circumstances. It can be reasoned that the adverse direct effects of the climatic environment can be ameliorated to some degree by managing and feeding the animals indoors. If this managerial system is adopted, feed and as cool a water supply as possible should be available throughout the 24 hr. Indoor management and feeding should reduce maintenance nutritive requirements, decrease the heat production resulting from grazing and walking, and provide roughage of a higher DM content, as it is likely to be wilted before it is consumed. However, indoor feeding will undoubtedly reduce the ability of the animal to select the most nutritious feed available and it will probably be advantageous to chop all roughage fed indoors. Animals managed outdoors should have access to high quality pasture and to as cool a water supply as possible throughout the 24 hr. Good shade should always be available in the day grazing paddocks and the most productive paddocks should be available for night grazing. Pastures should be grazed as rapidly as possible and at frequent intervals. A managerial system that will ensure this is one where the stocking rate is very heavy, so that a large number of cattle can be concentrated on a small area of pasture for a short time, and where the cattle are rotated around the pastures at frequent intervals. Concentrate feeds for animals fed indoors or outdoors should be ground and pelleted, of high nutritive value, and be fed at as frequent intervals as possible.

On account of the highly seasonal nature of forage growth, the low nutritive value of the forage during many months of the year, and the lack of free water, the dry tropics are most suited to extensive animal husbandry operations, and except under unusual circumstances, the utilization of indigenous ruminant livestock. In some favoured areas it may be economic to conserve forage from the wet season 'flush' for feeding during the dry season, or to use feed supplements of various types. The use of supplementary blocks containing minerals, molasses and urea is definitely gaining favour in some parts of the dry tropics. However, in general, stocking rates will inevitably be low and should be geared to the availability of feed during the dry season. Paddocks should be large, and shade and a piped supply of water are highly desirable. In some areas, paddocks are given a one-year rest from grazing every three or four years and this practice encourages the growth and retention in the pastures of the more desirable forage species. In other areas some reliance can be placed on the utilization of browze by the animals during the dry season. If free water is scarce, it should be rationed during the dry season, and under very severe drought conditions when feed is very scarce and the nutrient content of the available feed very low, ruminant livestock may benefit from deliberate water deprivation as long as the period between waterings is not too prolonged.

As ruminant species undoubtedly differ in their nutritive needs and in their reaction to the direct and indirect effects of the climatic environment, the problem of feeding ruminants in the tropics is in part ecological, or a question of utilizing the species most suited to the environment.

At the present stage of development it may be economically difficult or undesirable to produce and utilize animal feeds with a high nutrient content in some economies, even when climate is favourable for forage and other animal feed production, as it often is in the wet tropics. The solution in these areas is therefore to exploit ruminant livestock that utilize available feeds of low nurtitive value most efficiently. The water buffalo (*Bubalus bubalis*) appears to be a more efficient convertor of feeds of low nutritive value than either *Bos taurus* or *Bos indicus* cattle (Ichhponani *et al.*, 1962; Mullick, 1964) and the differences in efficiency of conversion appear to be wider, the lower the nutrient content of the feed, thus, in rice, maize and sugar-cane growing areas the water buffalo may continue to be very useful if it is used to convert low nutrient content straws and other by-products into meat and milk, in addition to being used as a work animal.

In the dry tropics, the experimental and practical experience available suggests that *Bos indicus* are better adapted in many ways than *Bos taurus* cattle to the stress caused by the direct effect of climatic environment, and are more effective in utilizing the low nutrient content roughage that is the only feed available for so many months of the year (Duckworth, 1946; French, 1940; Phillips, 1961a; Ashton, 1962). These differences may be partially due to differences in rumen fermentation rates (Hungate *et al.*, 1960; Phillips, Hungate, MacGregor and Hungate, 1960).

It has also been suggested that the DCP requirements for maintenance of some indigenous cattle, specifically Afrikander and Mashona cattle in Rhodesia, are lower than accepted standards and that these cattle may be adapted to nutritional stress and to a low CP intake (Elliott, 1963; Elliott and Topps, 1963a). One possible reason for the efficient utilization of nitrogen by the cattle used in these experiments was the high carbohydrate content of the feed. This explanation was tested experimentally (Elliott and Topps, 1963b) and found to be valid. When the supply of energy was adequate, requirements for protein of these breeds were found to be less than 60 per cent of accepted standards.

Undoubtedly *Bos indicus* and cross-bred cattle can adapt themselves more effectively to lack of free water than *Bos taurus* cattle, and it is likely that sheep and goats are more adaptable than cattle, and camels more adaptable than any other domestic livestock. The use of camels for meat production in the very dry areas may have considerable possibilities that have not yet been exploited.

The capacity of indigenous wild ruminants to adapt themselves to nutritional stress conditions is virtually unknown at present, but observations suggest that some of them may be more adaptable than any domestic livestock. It is feasible that it could be economic to utilize some areas in the tropics, and particularly those that are very humid or very dry for the 'farming' of wild rather than domestic ruminants. This subject is at last meriting serious consideration (Talbot *et al.*, 1965).

REFERENCES

ALIEV, A. A. (1963) *Fiziol. Z. SSSR. Secenova*, **49**, 1109. Abstracted from *Nutr. Abstr. Rev.* **34**, 95 (1964).
ALIM, K. A. (1963) *E. Afr. Agric. For. J.* **28**, 209.
ALLEN, T. E. (1962) *Aust. J. Agric. Res.* **13**, 165.
ALLEN, T. E., PAN, Y. S. and HAYMAN, R. H. (1963) *Aust. J. Agric. Res.* **14**, 580.
ANDERSSON, B. and LARSSON, B. (1961) *Acta Physiol. Scand.* **52**, 75.
APPLEMAN, R. D. and DELOUCHE, J. C. (1958) *J. Anim. Sci.* **17**, 326.
ARMSTRONG, D. G. and BLAXTER, K. L. (1956) *Nature, Lond.* **177**, 1183.
ARMSTRONG, D. G. and BLAXTER, K. L. (1957a, b) *Br. J. Nutr.* **11**, 247, 413.
ARMSTRONG, D. G., BLAXTER, K. L. and GRAHAM, N. McC. (1957) *Br. J. Nutr.* **11**, 392.
ARMSTRONG, D. G., BLAXTER, K. L., GRAHAM, N. McC. and WAINMAN, F. W. (1958a) *Proc. Nutr. Soc.* **17**, xxxii.
ARMSTRONG, D. G., BLAXTER, K. L., GRAHAM, N. McC. and WAINMAN, F. W. (1958b) *Br. J. Nutr.* **12**, 177.
ASHTON, G. C. (1962) *J. Agric. Sci., Camb.* **58**, 333.
AYLWARD, F. (1961) *Proc. Nutr. Soc.* **20**, 112.
BADR, M. F. (1957) *Indian J. Dairy Sci.* **10**, 176.
BAILEY, C. B. (1964) *Can. J. Anim. Sci.* **44**, 68.
BAILEY, C. B., HIRONAKA, R. and SLEN, S. B. (1962) *Can. J. Anim. Sci.* **42**, 1.
BALCH, C. C., BALCH, D. A., JOHNSON, V. W. and TURNER, J. (1953) *Br. J. Nutr.* **7**, 212.
BANERJEE, M. R., BRANTON, C., JOHNSTON, J. E., GUIDRY, A. J. and BREIDENSTEIN, C. P. (1964) *J. Dairy Sci.* **47**, 346.
BIANCA, W. (1964) *Res. Vet. Sci.* **5**, 75.
BIANCA, W. (1965a) *Res. Vet. Sci.* **6**, 33.
BIANCA, W. (1965b) *J. Dairy Res.* **32**, 291.
BIANCA, W., FINDLAY, J. D. and McLEAN, J. A. (1965) *Res. Vet. Sci.* **6**, 38.
BLAXTER, K. L., GRAHAM, N. McC. and WAINMAN, F. W. (1956) *Br. J. Nutr.* **10**, 69.
BLAXTER, K. L. and WAINMAN, F. W. (1960) *Proc. Nutr. Soc.* **19**, xvi.
BLAXTER, K. L. and WAINMAN, F. W. (1961) *J. Agric. Sci., Camb.* **56**, 81.
BLAXTER, K. L., WAINMAN, F. W. and WILSON, R. S. (1961) *Anim. Prod.* **3**, 51.
BLINCOE, C. and BRODY, S. (1955) *Res. Bull. Mo. Agric. Expt. Sta.*, No. 579.
BONSMA, J. C., SCHOLTZ, G. D. J. and BADENHORST, F. J. G. (1940) *Fmg. S. Afr.* **15**, 7.
BONSMA, J. C. (1940) *Fmg. S. Afr.* **15**, 373.
BONSMA, J. C. (1949) *J. Agric. Sci., Camb.* **39**, 204.
BREDON, R. M. and HORRELL, C. R. (1961) *Trop. Agric., Trin.* **38**, 297.
BREDON, R. M. and HORRELL, C. R. (1962) *Trop. Agric., Trin.* **39**, 13.
BREIDENSTEIN, C. P., JOHNSTON, J. E., HINDERY, G. A. and RUSOFF, L. L. (1960) *J. Dairy Sci.* **43**, 443.
BROBECK, J. R. (1960) *Recent Progress in Hormone Research*, **16**, 439. Academic Press, New York.
BRODY, S. (1945) *Bioenergetics and Growth.* Rheinhold, New York.
BRODY, S., RAGSDALE, A. C., THOMPSON, H. J. and WORSTELL, D. M. (1954) *Res. Bull. Mo. Agric. Exp. Sta.*, No. 556.

Brown, M. A. and Motasem, M. M. (1965) *J. Dairy Sci.* **48**, 788.
Butterworth, M. H., Groom, C. G. and Wilson, P. N. (1961) *J. Agric. Sci., Camb.* **56**, 407.
Campling, R. C. and Balch, C. C. (1961) *Br. J. Nutr.* **15**, 523.
Cenni, B. (1963) *Ann. Fac. Med. Vet., Pisa*, **16**, 61. Abstracted from *Nutr. Abstr. Rev.* **35**, 213.
Clark, R. and Quin, J. I. (1947) *Onderstepoort J. Vet. Sci. Anim. Ind.* **21**, 317.
Clark, R. and Quin, J. I. (1949a) *Onderstepoort J. Vet. Sci. Anim. Ind.* **22**, 335.
Clark, R. and Quin, J. I. (1949b) *Onderstepoort J. Vet. Sci. Anim. Ind.* **22**, 345.
Crampton, E. W. (1957) *J. Anim. Sci.* **16**, 546.
Cunningham, M. D., Martz, F. A. and Merilan, C. P. (1964) *J. Dairy Sci.* **47**, 382.
Duckworth, J. (1946) *Trop. Agric., Trin.* **23**, 4.
Davis, A. V. and Merilan, C. P. (1960) *J. Dairy Sci.* **43**, 871.
de la Rüe, E. A., Bourlière, F. and Harroy, J. (1957) *The Tropics.* A. A. Knoff, New York.
Denton, D. A., Goding, J. R., Sabine, R. and Wright, R. D. (1961) *Salinity problems in the arid zones.* Proc. Teheran Symposium, Arid Zone Research. UNESCO, Paris.
Dodsworth, T. L. and Campbell, W. H. McK. (1953) *J. Agric. Sci., Camb.* **43**, 166.
Dowling, D. F. (1955) *Aust. J. Agric. Res.* **6**, 645.
Elliott, R. C. (1963) *J. Agric. Sci., Camb.* **61**, 417.
Elliott, R. C. and Topps, J. H. (1963a) *Br. J. Nutr.* **17**, 539.
Elliott, R. C. and Topps, J. H. (1963b) *Br. J. Nutr.* **17**, 549.
Embry, L. B., Hoelscher, M. A., Wahlstrom, R. C., Carlson, C. W., Krista, L. M., Brosz, W. R., Gastler, G. F. and Olson, O. E. (1959) *Bull. S. Dak. Agric. Exp. Sta.*, No. 481.
F.A.O. (1963) *Production Yearbook*, **17**. F.A.O., Rome.
Ferguson, K. A. and Dowling, D. F. (1955) *Aust. J. Agric. Res.* **6**, 640.
Findlay, J. D. (1950) *Bull. Hannah Dairy Res. Inst.*, No. 9.
Findlay, J. D. (1954) *Met. Monogr.* **2**, 19.
French, M. H. (1938) *Tanganyika Dept. Vet. Sci. and Anim. Hus. Annual Report 1938*, Part II, p. 80.
French, M. H. (1940) *J. Agric. Sci., Camb.* **30**, 503.
French, M. H. (1956a) *Emp. J. Exp. Agric.* **24**, 128.
French, M. H. (1956b) *E. Afr. Agric. J.* **21**, 171.
French, M. H. (1957) *Herb. Abstr.* **27**, 1.
French, M. H. (1959) *Turrialba*, **9**, 127.
French, M. H. (1961) *Turrialba*, **11**, 78.
Gibson, T. E. (1963) *Proc. Nutr. Soc.* **22**, 15.
Goonesekera, G. C. M. (1954) *Trop. Agric. Mag. Ceylon Agric. Soc.* **110**, 25.
Gordon, J. G. (1965) *J. Agric. Sci., Camb.* **64**, 31.
Graham, N. McC., Wainman, F. W., Blaxter, K. L. and Armstrong, D. G. (1959) *J. Agric. Sci., Camb.* **52**, 13.
Graham, N. McC. (1964a) *Aust. J. Agric. Res.* **15**, 100.
Graham, N. McC. (1964b) *Aust. J. Agric. Res.* **15**, 982.
Hancock, J. and Payne, W. (1955) *Emp. J. Exp. Agric.* **23**, 55.
Harbin, R., Harbaugh, F. G., Neeley, K. L. and Fine, N. C. (1958) *J. Dairy Sci.* **41**, 1621.
Hardison, W. A., Reid, J. T., Martin, C. M. and Woolfolk, P. G. (1954) *J. Dairy Sci.* **37**, 89.
Hardison, W. A. (1966) *Dairy Train. Res. Inst., Philipp., Res. Bull.*, No. 1.
Harker, K. W., Taylor, J. I. and Rollinson, D. H. L. (1954) *J. Agric. Sci., Camb.* **44**, 193.
Harker, K. W., Taylor, J. I. and Rollinson, D. H. L. (1956) *J. Agric. Sci., Camb.* **47**, 44.
Harker, K. W., Rollinson, D. H. L., Taylor, J. I., Gourlay, R. N. and Nunn, W. R. (1961) *J. Agric. Sci., Camb.* **56**, 137.
Harrison, E. (1942) *Trop. Agric., Trin.* **19**, 147.
Hecker, J. F., Budtz-Olsen, O. E. and Ostwald, M. (1964) *Aust. J. Agric. Res.* **15**, 961.
Hillman, D., Lassiter, C. A., Huffman, C. F. and Duncan, C. W. (1958) *J. Dairy Sci.* **41**, 721.

HOLMES, J. C. and LANG, R. W. (1963) *Anim. Prod.* **5**, 17.
HORROCKS, D. and PHILLIPS, G. D. (1961) *J. Agric. Sci., Camb.* **56**, 379.
HOUPT, T. R. (1959) *Am. J. Physiol.* **197**, 115.
HOWES, J. R., FEASTER, J. P. and HENTGES, J. F. (Jr.) (1962) *J. Anim. Sci.* **21**, 210.
HOWES, J. R., HENTGES, J. F. (Jr.) and DAVIS, G. K. (1963) *J. Anim. Sci.* **22**, 22.
HUNGATE, R. E., PHILLIPS, G. D., HUNGATE, D. P. and MACGREGOR, A. (1960) *J. Agric. Sci., Camb.* **54**, 196.
HUKE, R. E. (1963) *Shadows on the Land. An Economic Geography of the Philippines.* Bookmark, Manila.
ICHHPONANI, J. S., MAKKAR, G. S., SIDHU, G. S. and MOXON, A. L. (1962) *J. Anim. Sci.* **21**, 1001.
ITTNER, N. R., BOND, T. E. and KELLY, C. F. (1954) *J. Anim. Sci.* **13**, 867.
ITTNER, N. R., BOND, T. E. and KELLY, C. F. (1958) *Bull. Calif. Agric. Exp. Sta.*, No. 761.
ITTNER, N. R., KELLY, C. F. and GUILBERT, H. R. (1951) *J. Anim. Sci.* **10**, 742.
JOBLIN, A. D. H. (1960) *J. Br. Grassld. Soc.* **15**, 212.
JOHNSON, C. E., HARBERS, L. H. and PRESCOTT, J. M. (1959) *J. Anim. Sci.* **18**, 599.
JOHNSON, H. D. and KAMAL, T. H. (1958) *J. Anim. Sci.* **17**, 1228.
JOHNSON, H. D. and KIBLER, H. H. (1963) *J. Appl. Physiol.* **18**, 73.
JOHNSON, H. D., RAGSDALE, A. C. and YECK, R. G. (1958) *Res. Bull. Mo. Agric. Exp. Sta.*, No. 683.
JOHNSON, H. D. and RAGSDALE, A. C. (1960) *Res. Bull. Mo. Agric. Exp. Sta.*, No. 709.
JOHNSON, H. D., RAGSDALE, A. C., YECK, R. G. and JONES, J. F. (1960) *Res. Bull. Mo. Agric. Exp. Sta.*, No. 786.
JOHNSON, H. D., RAGSDALE, A. C., BERRY, I. L., SHANKLIN, M. D. and MCLARNEY, S. (1963) *Res. Bull. Mo. Agric. Exp. Sta.*, No. 846.
JOHNSON, H. D. and YECK, R. G. (1964) *Res. Bull. Mo. Agric. Exp. Sta.*, No. 865.
JOHNSTON, J. E., HINDERY, G. A., HILL, D. H. and GUIDRY, A. J. (1961a) *J. Dairy Sci.* **44**, 976.
JOHNSTON, J. E., HINDERY, G. A., HILL, D. H. and GUIDRY, A. J. (1961b) *J. Dairy Sci.* **44**, 1191.
JOHNSTON, J. E., HINDERY, G. A., TURNIPSEED, T. and THOMPSON, D. (1960) *J. Dairy Sci.* **43**, 871.
JOHNSTON, J. E., MCDOWELL, R. E., SHRODE, R. R. and LEGATES, J. F. (1959) *Southern Co-operative Ser. Bull.*, No. 63.
JOHNSTON, J. E., PAYNE, W. J. A., ABADIE, J. W., SCHRADER, G. T., HAMBLIN, F. B. and FRYE, J. B. (Jr.) (1957a) *Louisiana State Uni., Dairy Dept., Pub.* No. 10, 21.
JOHNSTON, J. E., STONE, E. J., SMITH, J. W., SCHRADER, G. T. and FRYE, J. B. (Jr.) (1957b) *J. Dairy Sci.* **40**, 616.
KAMAL, T. H., JOHNSON, H. D. and RAGSDALE, A. C. (1959) *J. Dairy Sci.* **42**, 926.
KELLY, C. F., BOND, T. E. and ITTNER, N. R. (1955) *Agric. Engng., St. Joseph, Mich.* **35**, 173.
KELLEY, R. O., MARTZ, F. A. and JOHNSON, H. D. (1965) *J. Dairy Sci.* **48**, 819.
KIBLER, H. H. and BRODY, S. (1953) *Res. Bull. Mo. Agric. Exp. Sta.*, No. 522.
KIBLER, H. H. and YECK, R. G. (1959) *Res. Bull. Mo. Agric. Exp. Sta.*, No. 701.
KOFB, A. R. and PFANDER, W. H. (1965) *J. Anim. Sci.* **24**, 892.
LAMPKIN, G. H., QUARTERMAN, J. and KIDNER, M. (1958) *J. Agric. Sci., Camb.* **50**, 211.
LAMPKIN, G. H. and QUARTERMAN, J. (1962) *J. Agric. Sci., Camb.* **59**, 119.
LARKIN, R. M. (1954) *Qd. J. Agric. Sci.* **11**, 115.
LARKIN, R. M. (1956) *Qd. Agric. J.* **82**, 96.
LARSEN, C., HUNGERFORD, E. H. and BAILEY, D. E. (1917) *Bull. S. Dak. Agric. Exp. Sta.*, No. 175.
LEE, D. H. K. and ROBINSON, K. W. (1941) *Proc. R. Soc. Qd.* **53**, 189.
LEDGER, H. P. (1959) *Nature, Lond.* **184**, 1405.
LEIGHTON, R. E. and RUPEL, I. W. (1956) *Texas Agric. Exp. Sta. Prog. Rep.*, No. 1889.
LEIGHTON, R. E. and RUPEL, I. W. (1960) *J. Dairy Sci.* **43**, 443.
LEIGHTON, R. E. (1965) *J. Dairy Sci.* **48**, 796.
LEITCH, I. and THOMSON, J. S. (1944) *Nutr. Abstr. Rev.* **14**, 197.

LIVINGSTON, H. G., PAYNE, W. J. A. and FRIEND, M. T. (1962) *Nature, Lond.* **194**, 1057.
LLOYD, L. E., DONEFER, E., BOWMAN, A. L. and CRAMPTON, E. W. (1962) *J. Anim. Sci.* **21**, 1036.
LUNDGREN, R. G. and JOHNSON, H. D. (1964) *J. Anim. Sci.* **23**, 28.
MACFARLANE, W. V. (1958). *Climatology and Microclimatology*, p. 227. UNESCO, Paris.
MACFARLANE, W. V., MORRIS, R. J. H. and HOWARD, B. (1958) *Aust. J. Agric. Res.* **9**, 217.
MACFARLANE, W. V., MORRIS, R. J. H. and HOWARD, B. (1963) *Nature, Lond.* **197**, 270.
MACFARLANE, W. V., MORRIS, R. J. H., HOWARD, B., McDONALD, J. and BUDTZ-OLSEN, O. E. (1961) *Aust. J. Agric. Res.* **12**, 889.
MACLUSKY, D. S. (1955) *Proc. Br. Soc. Anim. Prod.*, p. 45.
MAHAPATRO, B. B. and LEFFEL, E. C. (1964) *J. Anim. Sci.* **23**, 883.
McMEEKAN, C. P. (1958) *Proc. N.Z. Soc. Anim. Prod.* **18**, 34.
MILFORD, R. (1960) *Aust. J. Agric. Res.* **11**, 138.
MILFORD, R. and MINSON, D. J. (1964) *CSIRO Division of Tropical Pastures. Ann. Rept.* 1963-64, p. 57.
MOORE, L. A., THOMAS, J. W. and SYKES, J. F. (1960) *Int. Grassl. Congr. VIII, Reading*, p. 701.
MULHEARN, C. J. (1957) *J. Dep. Agric. S. Aust.* **61**, 49.
MULLICK, D. N. (1964) *Ind. J. Dairy Sci.* **17**, 45.
MULLICK, D. N., MURTY, V. N. and KEHAR, N. D. (1952) *J. Anim. Sci.* **11**, 42.
NOFFSINGER, T. L., OTAGAKI, K. K. and FURUKAWA, C. T. (1961) *J. Anim. Sci.* **20**, 718.
OYENUGA, V. A. (1958) *Nutr. Abstr. Rev.* **28**, 985.
PAGE, H. M., ERWIN, E. S. and NELMS, G. E. (1959) *Am. J. Physiol.* **196**, 917.
PAGOT, J. and DELAINE, R. (1958) *Revue. Elev. Méd. vét. Pays trop.* **11**, 293.
PAN, Y. S. (1963) *Aust. J. Agric. Res.* **14**, 424.
PAYNE, W. J. A. (1962) *Span*, **5**, 3.
PAYNE, W. J. A. (1965a) *Qualitas Pl. Mater. veg.* **12**, 269.
PAYNE, W. J. A. (1965b) *Development of Animal Husbandry Resources in South East Asia.* Proc. Conf. Con. Nat. and Nat. Res. Trop. S.E. Asia, Bangkok, Nov. 29th–Dec. 4th, 1965 (in the Press).
PAYNE, W. J. A. and HANCOCK, J. (1957) *Emp. J. Exp. Agric.* **25**, 321.
PAYNE, W. J. A. and HUTCHISON, H. G. (1963) *J. Agric. Sci., Camb.* **61**, 255.
PAYNE, W. J. A., LAING, W. I. and RAIVOKA, E. N. (1951) *Nature, Lond.* **167**, 610.
PEIRCE, A. W. (1959) *Aust. J. Agric. Res.* **10**, 725.
PEIRCE, A. W. (1960a) *Arid Zone Technical Conference*, Washburton, Victoria. **1**. Paper No. 25.
PEIRCE, A. W. (1960b) *Aust. J. Agric. Res.* **11**, 548.
PHILLIPS, G. D. (1960) *J. Agric. Sci., Camb.* **54**, 231.
PHILLIPS, G. D. (1961a) *Res. Vet. Sci.* **2**, 202.
PHILLIPS, G. D. (1961b) *Res. Vet. Sci.* **2**, 209.
PHILLIPS, G. D., HUNGATE, R. E., MACGREGOR, A. and HUNGATE, D. P. (1960) *J. Agric. Sci., Camb.* **54**, 417.
PIPES, G. W., BAUMAN, T. R., BROOKS, J. R., COMFORT, J. E. and TURNER, C. W. (1963) *J. Anim. Sci.* **22**, 476.
POST, T. B. (1965) *Aust. J. Agric. Res.* **16**, 229.
PREMACHANDRA, B. N., PIPES, G. W. and TURNER, C. W. (1960) *Res. Bull. Mo. Agric. Exp. Sta.*, No. 727.
QUARTERMAIN, A. R. (1964) *J. Agric. Sci., Camb.* **62**, 333.
RAGSDALE, A. C., BRODY, S., THOMPSON, H. and WORSTELL, D. M. (1948) *Res. Bull. Mo. Agric. Exp. Sta.*, No. 425.
RAGSDALE, A. C., THOMPSON, H. J., WORSTELL, D. M. and BRODY, S. (1950) *Res. Bull. Mo. Agric. Exp. Sta.*, No. 460.
RAGSDALE, A. C., THOMPSON, H. J., WORSTELL, D. M. and BRODY, S. (1951) *Res. Bull. Mo. Agric. Exp. Sta.*, No. 471.
RAGSDALE, A. C., THOMPSON, H. J., WORSTELL, D. M. and BRODY, S. (1953) *Res. Bull. Mo. Agric. Exp. Sta.*, No. 521.

RAGSDALE, A. C., WORSTELL, D. M., THOMPSON, H. J. and BRODY, S. (1949) *Res. Bull. Mo. Agric. Exp. Sta.*, No. 449.
RAKES, A. H., LISTER, E. E. and REID, J. T. (1961) *J. Nutr.* **75**, 86.
RANDEL, P. F. and RUSOFF, L. L. (1963) *J. Dairy Sci.* **46**, 368.
REAGAN, W. M. and MEAD, S. W. (1939) *J. Dairy Sci.* **22**, 472.
REID, J. T., TURK, K. L., HARDISON, W. A., MARTIN, C. M. and WOOLFOLK, P. G. (1955) *J. Dairy Sci.* **38**, 20.
RHOAD, A. O. (1938) *Rec. Proc. Am. Soc. Anim. Prod.*, 31st Ann. Meeting, p. 284.
RHOAD, A. O. (1940) *Emp. J. Exp. Agric.* **8**, 190.
RIEK, R. F., HARDY, M. H., LEE, D. H. K. and CARTER, H. B. (1950) *Aust. J. Agric. Res.* **1**, 217.
ROBINSON, K. W. and KLEMM, G. H. (1953) *Aust. J. Agric. Res.* **4**, 224.
ROBINSON, K. W. and LEE, D. H. K. (1947) *J. Anim. Sci.* **6**, 182.
ROGERSON, A. (1960) *J. Agric. Sci., Camb.* **55**, 359.
ROGERSON, A. (1963) *Nature, Lond.* **197**, 1222.
ROLLINSON, D. H. L., HARKER, K. W. and TAYLOR, J. I. (1955) *J. Agric. Sci., Camb.* **46**, 123.
ROSE INNES, R. (1963) *Emp. J. Exp. Agric.* **31**, 1.
RUPEL, I. W. and LEIGHTON, R. E. (1957) *J. Dairy Sci.* **40**, 1391.
RUSOFF, L. L., PAYNE, W. J. A., STONE, E. J. and HINDERY, G. A. (1957) *Louisiana State Uni. Dairy Dept., Pub.* No. 10, p. 28.
RUSOFF, L. L., PAYNE, W. J. A., STONE, E. J. and HINDERY, G. A. (1958) Ann. Mtg. So. Div., ADSA, So. Agric. Workers, Little Rock, Arkansas, Feb. 3–5, 1958.
SCHMIDT-NIELSEN, K. (1956) *The Future of Arid Lands.* p. 368. American Association for the Advancement of Science, Washington.
SCHMIDT-NIELSEN, K. (1964) *Desert Animals.* O.U.P., Oxford.
SCHMIDT-NIELSEN, B. and OSAKI, H. (1958) *Am. J. Physiol.* **193**, 657.
SCHMIDT-NIELSEN, B., SCHMIDT-NIELSEN, K., HOUPT, T. R. and JARNUM, S. A. (1957) *Am. J. Physiol.* **188**, 477.
SCHMIDT-NIELSEN, B., OSAKI, H., MURDAUGH, H. V. (Jr.) and O'DELL, R. (1958) *Am. J. Physiol.* **194**, 221.
SHAW, J. C., ENSOR, W. L., TELLECHEA, H. F. and LEE, S. D. (1960) *J. Nutr.* **71**, 203.
SMITH, C. A. (1959) *J. Agric. Sci., Camb.* **52**, 369.
SMITH, C. A. (1965) *J. Agric. Sci., Camb.* **64**, 295.
SOMERS, M. (1961) *Aust. J. Exp. Biol. Med. Sci.* **39**, 145.
STOTT, G. H. and MOODY, E. G. (1960) *J. Dairy Sci.* **43**, 871.
TALBOT, L. M., PAYNE, W. J. A., LEDGER, H. P., VERDCOURT, L. D. and TALBOT, M. H. (1965) *Comm. Bur. Anim. Breed. Gen., Tech. Comm.*, No. 16.
THOMPSON, R. D., JOHNSTON, J. E., BREIDENSTEIN, C. P., GUIDRY, A. J., BANERJEE, M. R. and BURNETT, W. T. (1963) *J. Dairy Sci.* **46**, 227.
THOMPSON, H. J., WORSTELL, D. M. and BRODY, S. (1949) *Res. Bull. Mo. Agric. Exp. Sta.*, No. 436.
WAYMAN, O., JOHNSON, H. D., MERILAN, C. P. and BERRY, I. L. (1962) *J. Dairy Sci.* **45** 1472.
WEETH, H. J., HAVERLAND, L. H. and CASSARD, D. W. (1960) *J. Anim. Sci.* **19**, 845.
WEETH, H. J., HUNTER, J. E. and PIPER, E. L. (1962) *J. Anim. Sci.* **21**, 688.
WEETH, H. J. and LESPERANCE, A. L. (1965) *J. Anim. Sci.* **24**, 589.
WEIR, W. C. and TORELL, D. T. (1959) *J. Anim. Sci.* **18**, 641.
WELDY, J. R., MCDOWELL, R. E. and VAN SOEST, P. J. (1962) *J. Anim. Sci.* **21**, 1031.
WELDY, J. R., MCDOWELL, R. E., VAN SOEST, P. J. and BOND, J. (1964) *J. Anim. Sci.* **23**, 147.
WHYTE, R. O. (1962) *Trop. Agric., Trin.* **39**, 1.
WILSON, P. N. (1959) *J. Agric. Soc. Trin. Pap.*, No. 879.
WILSON, P. N. (1961a) *J. Agric. Sci., Camb.* **56**, 351.
WILSON, P. N. (1961b) *Turrialba*, **11**, 57.
WILSON, P. N., BARRATT, M. A. and BUTTERWORTH, M. H. (1962) *J. Agric. Sci., Camb.* **58**, 257.

WINCHESTER, C. F. (1964) *J. Anim. Sci.* **23**, 254.
WINCHESTER, C. F. and MORRIS, M. J. (1956) *J. Anim. Sci.* **15**, 722.
WORSTELL, D. M. and BRODY, S. (1953) *Res. Bull. Mo. Agric. Exp. Sta.*, No. 515.
YOUSEF, M. K. and JOHNSON, H. D. (1965) *J. Dairy Sci.* **48**, 813.

CHAPTER 24

The Nutrient Requirements of Farm Livestock

*2. Ruminants**

(SUMMARY OF ESTIMATED REQUIREMENTS)

FOREWORD

By SIR DAVID CUTHBERTSON, C.B.E., F.R.S.E.
Department of Pathological Biochemistry, Royal Infimary, Glasgow, C.4

Following representations that there was a need for a reappraisal of the accepted standards for the feeding of farm livestock, the Agricultural Research Council set up in 1959 a Technical Committee charged with the task of reviewing periodically the literature on the nutrient requirements of various classes of stock, and of making recommendations or alterations that seemed necessary in published feeding standards. To assist it in discharging its duties, the Committee established Working Parties to consider the requirements of ruminants, of poultry, and of pigs, as well as one to consider the nomenclature to be used in the Committee's reports. In giving guidance to the Working Parties dealing with different classes of livestock the Technical Committee realized that the approaches of these groups would not necessarily be the same.

The findings of the Working Party on Ruminants form the basis of the following series of *Summaries of Estimates of Nutrient Requirements*. In the course of its studies the Working Party has amassed much information of interest which could not be included in the Summaries; this information has been published in the form of detailed Technical Reviews.

It is the Committee's intention to keep under review future developments in the study of the nutrient requirements of farm livestock, and to publish from time to time revised summaries of estimates of requirements and, when the material warrants, further Technical Reviews. It would therefore be appreciated if persons possessing unpublished data pertinent to any aspect of the Committee's task would make them available, as they come to hand, to the Secretary of the Technical Committee.

*This chapter is reprinted from A.R.C. publication No. 2, H.M.S.O. 1965.

INTRODUCTION

The Working Party on the Nutrient Requirements of Ruminants was appointed in 1959 to study and report on the nutrient requirements of cattle and sheep, the word 'requirement' being defined as the amount of a nutrient that must be supplied in the diet to meet the needs of the normal healthy animal given a completely adequate diet in an environment compatible with good health.

The subject has been dealt with under the following main headings (1) Water, (2) Major mineral elements, (3) Minor or trace elements, (4) Vitamins, (5) Protein, (6) Energy and (7) The expression of nutrient requirements as dietary concentrations. A Technical Review and a Summary of the main findings in that review have been prepared for each of the first six of these seven sections. The aim has been to estimate the requirements of cattle and sheep for each nutrient for growth, maintenance, pregnancy and lactation, and, where applicable, for wool growth. The first step was to prepare a critical review of the relevant literature, and from the information so obtained to arrive at what appeared to be the best estimate of requirement.

The results of the Working Party's research and discussions are being published in two forms, one of which consists of Technical Reviews containing enough detail to enable all who are actively concerned with the more theoretical aspects of the subject to see for themselves the data on which the calculations of requirements have been based and the methods of calculation that have been adopted, and each Technical Review is accompanied by a Summary of the main findings. The volume containing the six Technical Reviews and their Summaries includes also the seventh section referred to in the list given above. The other form of publication is that presented here and consists of the Summaries of the first six sections together with an abridged form of the seventh section. It is intended for those who simply want to know the estimated requirements themselves without the details of the methods by which they have been derived. It is fully realized that as more research is done several of the requirements given in these Summaries may well need to be modified, and for that reason one of the aims of preparing the Technical Reviews in such detail has been to show where more knowledge is needed and to stimulate the research work necessary to provide that knowledge.

The methods of determining the best estimate of requirement differed for different nutrients and these different methods are dealt with in detail in the Technical Reviews.

The metric system has been used throughout since it was felt that the adoption of one uniform system would minimize the risk of errors in presenting the requirements and also in their application. All the terms used in the Summaries, whose meaning may not be clear, have been brought together in a glossary which will be found on p. 911.

The information given in these Summaries should be applicable in many parts of the world, but it must be emphasized that the Working Party has considered primarily the needs of ruminants under conditions that prevail in the United Kingdom. For countries with very different climatic conditions and in which different breeds and feeds and different systems of management exist, some of the requirements put forward here would almost certainly have to be modified.

J. A. B. SMITH, C.B.E., Chairman, Working Party on the Nutrient Requirements of Ruminants

SECTION 1. WATER REQUIREMENTS

The water requirements of ruminants are supplied in part from the water present in the food, in part from water consumed voluntarily and to a smaller but nevertheless significant extent from the water formed within the body as the result of oxidation in the tissues. The suggested water requirements given here relate only to the sum of the first two, i.e. the water in the food and the water consumed voluntarily.

It must be emphasized that there is considerable variation between individual animals and also that the nature of the food can affect water requirements. Thus high protein contents in foods and the addition of salt to the ration can increase the need for water.

Suggested water intakes to meet requirements are summarized below. The values given relate only to livestock maintained under environmental conditions pertaining in this country.

Class of stock	Environmental temperature (°C)	Water intake (kg/kg dry matter consumed)
Calves		
(first 5–6 weeks of life)	—	6·5
Cattle		
(above 100 kg, not pregnant and not lactating)	− 17 to + 10°	3·5
	10–15°	3·6
	15–21°	4·1
	21–27°	4·7
	above 27°	5·5

Pregnant cows. For the last 4 months of pregnancy the values given above for cattle should be multiplied by 1·5.

Lactating cows (indoor feeding). As for non-pregnant cattle (above) with an additional allowance of 0·87 kg water/kg milk produced.

Grazing cows. The allowance for non-pregnant cattle times 1·5, with the additional allowance of 0·87 kg water/kg milk produced.

Class of stock	Environmental temperature (°C)	Water intake (kg/kg dry matter consumed)
Sheep		
(growing and fattening)	Up to 15°	2·0
	15–20°	2·5
	Exceeding 20°	3·0

Pregnant ewes. During the third month of pregnancy the values given above for growing and fattening sheep should be multiplied by 1·5. In the fourth and fifth months the above values should be multiplied by 1·8 and 2·2 respectively.

Lactating ewes. During the first 8 weeks of lactation the values for growing and fattening sheep should be multiplied by 1·5 and during the next 8 weeks by 1·25.

SECTION 2. REQUIREMENTS FOR THE MAJOR MINERAL ELEMENTS, CALCIUM, PHOSPHORUS, MAGNESIUM, POTASSIUM, SODIUM AND CHLORINE

The requirements for the major mineral elements have been determined by a factorial method. This involves estimating the net requirement for the element, that is the sum of the amounts inevitably lost from the body (the endogenous loss) and the amounts stored in the body in growth or pregnancy or excreted in the milk. The net requirement divided by the availability of the element in the diet is an estimate of the amount that must be supplied by the diet. These factorial estimates have then been checked by reference to practical feeding trials in which the amount of the element in the diet has been varied and the performance of the stock assessed. In this checking process, which has involved consideration of some 50 experiments, no major discrepancies were found, and the requirements determined by the factorial method thus appear to conform to those obtained by more practical procedures. It must be pointed out that practical trials have a poor resolving power when used to establish a quantitative need. In most instances it is only possible to state from them whether or not a particular mineral intake was adequate. Such conclusions depend on the criteria of adequacy used, and it often happens that visible criteria of mineral deficiency appear only after serious depletion of body reserves.

Full details of the methods employed, the basic data examined and the approximations necessitated by paucity of information are given in the Technical Review. In this summary the estimates of dietary need are presented together with some discussion of their reliability and an indication of their agreement with existing standards.

Calcium, Phosphorus and Magnesium

Growth requirements. Analysis of the data on mineral deposition showed that the amount of calcium and to a lesser extent the amounts of phosphorus and magnesium deposited in each kg of weight gained during growth depend on the age and the previous nutritional history of the animal. This stems from the fact that the skeleton and soft tissues can and do grow at different rates. Quantitative estimates of this joint relationship were possible with cattle but with sheep, lack of data precluded such an analysis. In addition, there is considerable evidence that the availability of these three elements in the diet of cattle declines as the animal matures. Table 2.1 summarizes the requirements of cattle for calcium, phosphorus and magnesium when the cattle are growing at different rates.

The requirements of sheep are summarized in Table 2.2.

Pregnancy. Dietary requirements for pregnancy in cattle have been based on determinations of the amounts of mineral elements retained in the foetus and adnexa, these being added to the amounts required for maintenance. With pregnant heifers a small additional allowance has been included for their own growth. The requirements for mature cattle can be applied to heifers with little error simply by assuming that the requirements of a heifer are those of a mature animal weighing 100 kg more than the heifer. Table 2.3 gives the estimates for a mature cow weighing 500 kg; estimates for cows of other weights can be obtained from the Technical Review.

Table 2.1
Dietary Calcium, Phosphorus and Magnesium Requirements of Growing Cattle

Body-weight (kg)	Rate of gain 0·33 kg/day			Rate of gain 0·5 kg/day			Rate of gain 1·0 kg/day		
	Calcium (g/day)	Phosphorus (g/day)	Magnesium (g/day)	Calcium (g/day)	Phosphorus (g/day)	Magnesium (g/day)	Calcium (g/day)	Phosphorus (g/day)	Magnesium (g/day)
50	6·7	4·3	0·4	9·6	6·2	0·5	18	12	0·8
100	11	5·4	1·1	15	7·3	1·2	27	13	1·7
200	14	8·1	3·7	18	9·8	4·0	30	15	5·0
300	18	13	5·2	21	15	5·5	33	20	6·5
400	22	22	6·7	26	24	7·0	37	29	8·0
500	25	28	8·2	28	29	8·5	38	33	9·5

Table 2.2
Dietary Calcium, Phosphorus and Magnesium Requirements of Growing Sheep (30-50 kg body-weight)

Rate of growth (g/day)	Requirement (mg/kg body-weight per day) Calcium	Phosphorus	Magnesium
50	110	65	14
100	130	75	16
200	180	90	21

TABLE 2.3

Dietary Calcium, Phosphorus and Magnesium Requirements of Cattle during Pregnancy (mature cow weighing 500 kg)

Stage of pregnancy (months)	Calcium	Phosphorus (g/day)	Magnesium*
5–6	20	26	7·9
6–7	27	30	8·0
7–8	35	34	8·8
8–9	33	34	9·4
Non-pregnant	18	26	7·5

* It is recommended that this minimal allowance should be increased by 2·0 g to take into account variation in need from individual to individual.

The requirements of pregnant sheep obviously vary with the number of lambs the ewe carries. The difference in the dietary needs of ewes bearing single lambs and those bearing twins, however, is not large and is apparent only in the final stages of pregnancy. Table 2.4 gives the estimates which refer to a ewe weighing 50 kg.

TABLE 2.4

Dietary Calcium, Phosphorus and Magnesium Requirements of Pregnant Ewes (50-kg ewe)

Month of pregnancy	Ewes bearing single lambs			Ewes bearing twins		
	Calcium	Phosphorus (g/day)	Magnesium	Calcium	Phosphorus (g/day)	Magnesium
2	4·5	3·7	0·59	4·5	3·7	0·59
3	4·8	3·8	0·61	5·2	4·0	0·65
4	5·8	4·2	0·68	7·1	4·8	0·80
5	7·6	5·1	0·85	9·3	5·8	0·98

Lactation. The requirements of the lactating cow for calcium appear to vary with both the age of the cow and the fat content of her milk. Thus the calcium content of milk increases by 20 per cent when the fat content increases from 3 to 5 per cent, but at the same time the phosphorus content increases by only 8 per cent. Age rather than mature size of animal appears important in that there is evidence, not restricted to that from cows, that the availability of dietary calcium declines with advancing age. The calculated requirements for calcium, phosphorus and magnesium of Friesian cows (500 kg body-weight) and of Jersey cows (350 kg body-weight) for yields ranging from 5 to 30 kg milk/day are given in Table 2.5. It is recommended that the allowances for magnesium should be increased by 2·0 g/day, because the requirement we have calculated is an average estimate of minimal need, and individuals can well have slightly higher requirements.

TABLE 2.5

*Calculated Dietary Requirements of Milking Cows for Calcium, Phosphorus and Magnesium (g/day)**

Milk yield (kg)	Friesian cow, weight 500 kg 3·8% fat in milk			Jersey cow, weight 350 kg 6·0% fat in milk		
	Calcium	Phosphorus	Magnesium†	Calcium	Phosphorus	Magnesium†
5	32	34	11	29	23	8
10	46	43	14	44	32	12
15	60	51	17	60	40	15
20	74	60	20	76	49	18
25	88	68	23	92	57	21
30	102	77	26	108	66	24

* Based on dietary requirements as follows:

	Calcium	Phosphorus	Magnesium
Friesian maintenance (g/day)	17·8	25·5	7·5
Jersey maintenance (g/day)	12·4	14·9	5·3
Friesian milk (g/kg per day)	2·8	1·7	0·63
Jersey milk (g/kg per day)	3·2	1·7	0·63
Availability (%)	45	55	20

† It is recommended that these minimal allowances for magnesium should each be increased by 2·0 g to take into account variation in need from individual to individual.

With lactating sheep estimates have been made of the milk production likely under two different sets of circumstances. The results given in Table 2.6 apply to lowland breeds producing 2·5 kg/day at the peak of lactation and to hill breeds producing 1·8 kg milk/day at that time.

TABLE 2.6

Dietary Calcium, Phosphorus and Magnesium Requirements of Lactating Ewes (maintenance and lactation)

Month of lactation	Lowland breeds (50-kg ewe)			Hill breeds (35-kg ewe)		
	Calcium	Phosphorus (g/day)	Magnesium	Calcium	Phosphorus (g/day)	Magnesium
1	14	9·3	2·6	11	7·0	2·0
2	12	8·3	2·2	9·3	6·2	1·8
3	10	7·0	1·7	7·6	5·2	1·4
4	7·8	5·7	1·3	5·8	4·2	1·0

REQUIREMENTS FOR POTASSIUM, SODIUM AND CHLORINE

Although we have estimated requirements for potassium there appears little likelihood that ruminant diets would be deficient in this element. The only possibilities are those of the very young calf given an artificial milk diet and of older cattle on diets consisting almost entirely of cereals. For the 50-kg calf on an artificial milk diet an allowance of 2·6 g potassium/day would be adequate.

Estimates of the minimal sodium and chlorine requirements of growing stock are summarized in Table 2.7. These are lower than most current practical recommendations suggest, and in this regard it should be remembered that the values in Table 2.7 are minimal requirements and can be exceeded. Lactation in cows results in the minimal needs shown in Table 2.8. With ewes a peak lactational performance of 2·5 kg milk/day is met by an allowance of 1·9 g sodium and 4·1 g chlorine.

TABLE 2.7

Dietary Requirements of Sodium and Chlorine (g/day) of Cattle and Sheep of Different Body-weights and Growing at Different Rates

Body-weight (kg)	Cattle Growth rate per day					
	0·33 kg		0·5 kg		1·0 kg	
	Sodium	Chlorine	Sodium	Chlorine	Sodium	Chlorine
50	1·4	1·6	1·6	1·8	2·3	2·2
200	3·9	5·5	4·1	5·7	4·8	6·1
400	7·3	10·7	7·5	10·9	8·2	11·3

TABLE 2.7 (*continued*)

Body-weight (kg)	Sheep Growth rate per day					
	0·1 kg		0·2 kg		0·4 kg	
	Sodium	Chlorine	Sodium	Chlorine	Sodium	Chlorine
10	0·31	0·35	0·45	0·44	0·73	0·62
30	0·65	0·87	0·79	0·96	1·07	1·14
50	0·99	1·39	1·13	1·48	1·41	1·66

TABLE 2.8

Sodium and Chlorine Requirements of Milking Cows (maintenance and lactation)

Milk yield (kg)	Sodium (g/day)	Chlorine (g/day)
5	12	19
10	15	25
20	21	36
30	27	48

It should be noted that in the young animal the ratio of sodium to chlorine required is greater than the ratio of sodium to chlorine in sodium chloride. This stems in part from the fact that sodium is deposited in bones whereas chlorine is not. It follows, therefore, that in translating the estimated requirements in Tables 2.7 and 2.8 into practical terms of amounts of common salt, the sodium figure should always be used.

SECTION 3. REQUIREMENTS FOR TRACE MINERAL ELEMENTS

Published information on the trace mineral elements in relation to the nutrition of ruminants is less extensive than that on the major mineral elements. It has therefore not been possible, except for iron, to use the factorial method for the calculation of requirements. The suggested requirements are based instead on the published results of feeding trials, in which the performance of cattle or sheep has been assessed in relation to differing known dietary levels of the element, and are stated in terms of dietary concentration. In some regions, known to be subject to trace mineral deficiencies, the levels suggested here for individual minerals may be unsatisfactory because of interactions between a number of different minerals present in these areas either in insufficient or in excessive amounts. Until knowledge of these interactions is adequate to support further generalizations, the use of trace mineral supplements in such areas should be based on expert advice specific to the area, rather than on the levels suggested here for more normal conditions.

Iron. Factorial estimates of requirements can be made from the results of isotopic studies with milk-fed calves and from observations of the iron content of live-weight gain. It appears from these studies that a dietary concentration of 200 mg iron/kg whole milk dry matter should be adequate for milk-fed animals. Required concentrations for ruminating animals are likely to be substantially less because of the lower energy content of the feeds, and a concentration of 30 mg iron/kg dry matter should be adequate. This figure is based on assumed values for availability, but precision is relatively unimportant in this instance, since virtually all ruminant feeds will contain more than the required amounts of iron.

Copper. From the results of feeding trials it appears that dietary concentrations of 5 mg copper/kg dry matter for sheep and 10 mg/kg for cattle are unlikely to be associated with deficiency signs and these levels are therefore suggested here as the amounts that should be contained in the diets. In special circumstances it may be necessary to exceed these levels, but such instances must be treated individually and with some caution especially with sheep since they have shown toxicity signs when copper has been supplied at a concentration of 30 mg/kg dry matter, and young lambs are particularly susceptible to copper poisoning.

Molybdenum. The ruminant has a requirement for molybdenum; the amount is probably low but is not at present known. Many of the reports on molybdenum in ruminant nutrition come under the heading of toxicity and in this respect molybdenum appears to be intimately linked with the metabolism of sulphur and of copper. It is probable that molybdenum toxicity in fact induces a physiological copper deficiency. Molybdenum toxicity has been reported at levels of 3 mg molybdenum/kg of diet and the maximum safe level has been reported to be 0·5 mg/kg of diet.

Cobalt. It is suggested that cobalt should be present in the diet at a level of 0·1 mg/kg feed dry matter. Sheep have been shown to tolerate cobalt at the high level of 4 mg/kg body-weight per day. Cobalt deficiency tends to be associated with soils derived from acid igneous materials.

Zinc. Signs of deficiency have been noted in cattle given 18–36 mg zinc/kg feed dry matter and in calves given 12 mg zinc/kg of diet but not in similar calves given 50 mg/kg. It is therefore suggested that the diet should contain 50 mg zinc/kg dry matter.

Manganese. There have been a number of studies of the effects of manganese on growth, reproduction, milk yield, and blood haemoglobin levels. From the results of these studies it is suggested that manganese should be present in the diet at a level of 40 mg/kg dry matter.

Iodine. A level of 120 μg/kg dry matter probably meets requirements except during pregnancy and lactation when the level should be 800 μg/kg. If goitrogens are present, as, for example, in kale or white clover, these values should be increased by 1·2 mg/kg dry matter.

Selenium. The requirement for selenium and also its toxicity are referred to in relation to vitamin E in the section on vitamin requirements (p. 130 of Reviews).

Toxicity

Lead. Chronic lead poisoning due to consumption of contaminated herbage under natural conditions is unlikely; the main hazard is acute or chronic poisoning due to ingestion of paint containing lead.

Fluorine. The occurrence of fluorosis in the United Kingdom is mainly confined to relatively small and defined areas in the vicinity of particular factories and power stations. There is no practical solution to the problem when the fluorine in the feed in a locality reaches a certain concentration, other than importing feed into the area or sending the stock out of the area for at least a part of their life. Contamination of the diet by sodium fluoride or other soluble fluorides should not exceed 30–50 mg of fluorine/kg dry matter for dairy cattle and 70–100 mg/kg for sheep. Somewhat higher levels of 60–100 mg and 100–200 mg fluorine/kg dry matter of feed may be tolerated when the fluorine is derived from rock phosphate or phosphatic limestone. Fluorine is also present in drinking waters and since 1 mg fluorine/kg water may be equivalent to an additional 4 mg/kg of feed this source should be considered when assessing the total amount of fluorine ingested.

In normal circumstances any requirement that ruminants may have for fluorine will be adequately covered by the feed.

The various values arrived at here can be summarized as follows:

Element	*Dietary concentration (mg/kg dry matter)*
Iron (in milk)	200
Iron (in dry feed)	30
Cobalt	0·1
Copper	5*, 10†
Molybdenum	Unknown
Zinc	50
Manganese	40
Iodine‡ (non-pregnant animals)	0·12
Iodine‡ (pregnant and lactating animals)	0·80

* Sheep † Cattle
‡ If feed contains goitrogens these levels should probably be increased by 1·2 mg/kg dry matter.

SECTION 4. REQUIREMENTS FOR VITAMINS

Only those vitamins have been considered whose presence in the diet is, or may be, essential for the well-being of ruminants.

Requirements for vitamin A

Units: 1 i.u. vitamin A = 1 USP* unit
= 0·300 μg vitamin A alcohol
= 0·344 μg vitamin A acetate.

The requirements for the vitamin are given in terms of vitamin A alcohol or of its major precursor β-carotene. Although the relation of the biological efficiency of the two is not constant, consideration of the experimental evidence available has led to the decision to regard 5 μg β-carotene (carotene) as being equivalent to 1 μg vitamin A alcohol (vitamin A).

In assessing minimal requirements for the vitamin or its precursor, attention has been paid to choosing levels which, it is considered, will not only ensure freedom from the commonly recognized syndromes of the deficiency state but also maintain hepatic stores of the vitamin at least at a low level.

The carotene requirements of cattle and sheep for maintenance and growth, pregnancy and lactation are given in Table 4.1. The requirements given for cattle cover all breeds. The possibility of a higher requirement for the Channel Island breeds was considered but not accepted on the evidence available. The requirement for pregnancy in cattle is that which should prove sufficient to avoid deficiency in the cow or abnormality in the new-born calf. The requirement for lactation is one sufficient to protect the cow from a deficiency of the vitamin and enable her to produce that yield of milk which her inherent ability

* United States Pharmacopoeia

and environment will permit: it is not necessarily that needed to ensure levels of vitamin A and of carotene in the milk adequate for the well-being of the milk-fed calf. This is dealt with separately below.

TABLE 4.1
Daily Requirements of Cattle and Sheep for β-carotene

	β-carotene required (μg/kg live weight)		
	Maintenance (including growth if any)	Maintenance and pregnancy	Maintenance and lactation
Cattle	80	150	130
Sheep	50	50	100

Few data exist on which to establish the requirement for pregnant and lactating sheep, but there is some evidence that pregnancy and lactation will continue normally in Merino ewes with daily intakes of carotene of 50 μg/kg live weight. Nevertheless it is suggested that for lactating ewes the value be raised to 100 μg/kg to ensure adequate amounts of vitamin A in the milk for the well-being of the suckling lamb.

The supply of vitamin A for the milk-fed calf. On the basis of a requirement of 80 μg β-carotene/kg live weight and with a 5:1 ratio for the relative efficiency of carotene and vitamin A, the daily basic requirement of the calf for preformed vitamin A is 16 μg/kg live weight. Consideration has been given to the extent to which this requirement of the new-born calf can be met under normal farm practice. Of the three possible sources of the vitamin available to the new-born calf, namely the liver reserve of the vitamin at birth and the amounts of the vitamin or its precursor present in the colostrum or milk, only the last two are of any practical significance. Table 4.2 shows the number of days for which a 40-kg calf would obtain sufficient of the vitamin from colostrum of two potencies. The assumption has been made that the total endowment from colostrum is equivalent to twice that supplied during the first day

TABLE 4.2
Number of Days for which a 40-kg calf would obtain sufficient Vitamin A from Colostrum

Colostral fat consumed during first 24 h (g)	Potency of colostral fat (1st day after parturition)	
	40 μg/g fat (normal winter diet pre-partum)	60 μg/g fat (normal summer diet pre-partum)
50	6	9
100	12	18
150	18	(27)*

* Possible over-estimate—see text.

TABLE 4.3

The Age in days of the Milk-fed Calf from which, with no Other Source of Vitamin A, it is likely to require Supplementary Vitamin A at the rate per day of 16 μg vitamin A/kg live weight†

Dam's pre-partum diet	Dam's post-partum diet	Calf receiving no colostrum				Calf receiving colostrum (supplying 200 g colostral fat)			
		Milk consumed daily (kg)				Milk consumed daily (kg)			
		2	4	6	8	2	4	6	8
		Days of age after which a supplement is likely to be required							
Normal summer diet (Colostral fat 50 μg vitamin A/g)	Normal summer diet (Milk fat 10μg vitamin A/g)	*	*	*	*	*	*	*	*
	Normal winter diet (Milk fat 7 μg vitamin A/g)	6	*	*	*	*	*	*	*
	Low carotene diet (Milk fat 4 μg vitamin A/g)	3	12	*	*	30	*	*	*
Normal winter diet (Colostral fat 30 μg vitamin A/g)	Normal summer diet (Milk fat 10 μg vitamin A/g)	*	*	*	*	*	*	*	*
	Normal winter diet (Milk fat 7 μg vitamin A/g)	3	*	*	*	*	*	*	*
	Low carotene diet (Milk fat 4 μg vitamin A/g)	1	6	*	*	18	*	*	*
Low carotene diet (Colostral fat 20 μg vitamin A/g)	Normal summer diet (Milk fat 10 μg vitamin A/g)	*	*	*	*	*	*	*	*
	Normal winter diet (Milk fat 7 μg vitamin A/g)	From birth	*	*	*	26	*	*	*
	Low carotene diet (Milk fat 4 μg vitamin A/g)	From birth	From birth	*	*	11	37	*	*

† In calculating the values given in the table the following assumptions have been made. Weight of calf 40 kg; fat content of milk 3·5 per cent, liver reserves for a normal summer diet 1000 μg; for a normal winter diet 500 μg, and for a low carotene diet 0 μg.
* No supplement should be required.

after parturition. There is evidence to suggest that on this assumption over-estimates of protection are given where the period exceeds 2–3 weeks.

In considering the calf's supply of the vitamin or its precursor from milk, variation in the intake and vitamin potency of the milk make it difficult to be precise. Nevertheless for a calf entirely dependent on milk for its vitamin A the daily intake of carotene by the cow should approximate to 400 μg/kg live weight in the early days of the calf's life. As the calf's consumption of milk increases, the cow's intake of carotene could be allowed to fall to 200 μg/kg live weight daily.

The data given in Table 4.3 are intended to act as a guide to the need for giving supplementary vitamin A to milk-fed calves. The variables include two situations commonly met in calf-rearing practice in the United Kingdom, and a third, the unlikely situation of a very low carotene intake.

Requirements for vitamin D. Units: 1 i.u. vitamin D = 1 USP unit = the antirachitic activity of 0·025 μg crystalline vitamin D_3.

Ergocalciferol (vitamin D_2) and cholecalciferol (vitamin D_3) are considered to be equally effective for ruminants. The minimum requirements given here are those considered to be necessary to prevent the development of signs of vitamin D deficiency in animals receiving adequate dietary levels of calcium and phosphorus. In the presence of inadequate amounts of calcium the requirement for vitamin D is increased. The same probably applies to phosphorus but the evidence is not so clear. It appears also that the vitamin D requirement of young stock increases with growth rate.

The suggested requirements for vitamin D are listed in Table 4.4. It must be emphasized that those referring to growing cattle and to pregnant and lactating cattle are based on very limited information.

TABLE 4.4

Daily Requirements for Vitamin D

	Vitamin D (i.u./kg live weight)
Calf	4
Growing cattle	2·5
Cows—pregnancy and lactation	10
Sheep (all categories)	5

No valid assessment of the extent to which particular requirements for vitamin D can be met by the exposure of the animals to sunlight is possible with the information available at present.

The evidence on the possibility of improvement in bovine fertility by dosing with large amounts of vitamin D before mating is conflicting and much of it is unsatisfactory. Nevertheless it seems noteworthy that sometimes when an improvement in fertility has been noted the amounts used were by no means much above the suggested minimum daily requirement given in Table 4.4.

The use of massive oral doses of vitamin D given pre-partum to dairy cows is undoubtedly effective in reducing the incidence of milk fever. Nevertheless, there is need to predict to within a day or two the date of calving if the treatment is to be successful.

The toxicity to sheep and cattle of massive doses of vitamin D, given daily over an extended period, has been established and it is recommended that such treatment should be given only under veterinary supervision.

Requirements for vitamin E and selenium. The metabolic functions of vitamin E are closely related to those of the trace element selenium and thus requirements for these micronutrients have been considered jointly. The international unit for vitamin E is defined as the specific activity of 1 mg synthetic DL-α-tocopheryl acetate. The defined potency of DL-α-tocopherol is 1·1 i.u./mg and of D-α-tocopherol 1·49 i.u./mg.

The enzootic muscular dystrophies of cattle and sheep are prevented by the administration of either minute amounts of selenium or amounts of vitamin E which are very large relative to the presumptive needs of vitamin E of the animal. Until more is known concerning the inter-relationship between these two micronutrients it is not possible to state requirements for either the vitamin or selenium. Nevertheless, as a guide it appears that whereas in the presence of minute amounts of selenium and on normal diets, the daily requirements for vitamin E of growing lambs and calves lie in the range 0·1–0·3 i.u./kg live weight, on diets low in selenium, the minimum daily prophylactic dose of vitamin E at least for calves is probably in the range 0·7–0·9 i.u./kg. The vitamin E requirements of adult stock appear to be very small and should be adequately met from the content of tocopherols in natural feeds.

The addition of considerable quantities of glycerides of unsaturated fatty acids to feeds creates an additional requirement for vitamin E. A daily dose of 200 i.u. vitamin E has been found to be effective in preventing the muscular dystrophy found in calves receiving a diet supplemented daily with 30 ml cod-liver oil.

In certain regions of the world the addition of minute amounts of selenium to the diets of farm livestock has resulted in beneficial effects including improved liveweight gains. In parts of Scotland, where from geochemical considerations responses to selenium might be expected to be larger than in the rest of Britain, increases in gain after selenium injections have proved to be small. Selenium is a very toxic element and owing to the tendency of some plants to concentrate selenium from the soil, the margin between even this small benefit and disaster to the livestock industry is too narrow to warrant its indiscriminate use. It is recommended that on soils known to be selenium-deficient, and where the crops are thus likely to induce muscular dystrophy among farm livestock, prophylactic treatment for calves and lambs should normally consist of oral administration of vitamin E.

Requirements for the vitamin B complex. It is known that ruminants have a requirement for riboflavine, nicotinic acid, thiamine, pyridoxine, folic acid and vitamin B_{12}, but only with the very young ruminant need they be present in the diet. Less certain is the proof of a requirement for biotin and pantothenic acid.

After the young ruminant reaches the stage where rumination is established and solid feeds are eaten, it becomes largely independent of an exogenous supply of the vitamins of the B complex, although with reference to B_{12} it is essential that the animal receives an adequate supply of dietary cobalt.

From the information available on the contents of the vitamins of the B complex in cow's and ewe's milk, it seems likely that the milk-fed calf or lamb will normally receive adequate supplies of thiamine, riboflavine, pyridoxine, pantothenic acid and nicotinic acid, but that in some circumstances the supply of vitamin B_{12} may be barely adequate.

SECTION 5. REQUIREMENTS FOR PROTEIN

The General Approach

A factorial method has been chosen for estimating the minimum protein requirements of ruminants. The requirements have been compiled from estimates of the inevitable losses of protein from the body which have to be replaced, the retention of protein during growth, pregnancy and lactation, the output of protein in milk during lactation, and the biological value of dietary protein. It should be understood that the factorial estimate provides a minimum requirement and not a recommended level of feeding. Whether the latter should differ from the minimum, and by how much, will depend on the class of stock, the system of management, and on economic factors. On this question only general guidance can be given, based on comparisons of the factorial estimates with feeding trial results.

Estimates of the Factors Contributing to Protein Requirements

Endogenous urinary excretion. Estimates of this factor are given in Table 5.1. They are based on a study of published estimates obtained from experiments in which cattle and sheep were kept for some time on diets containing little or no protein but adequate in other respects.

Loss of nitrogen by cattle in hair and scurf shed from the skin. This loss is estimated to be ($0{\cdot}02W^{0{\cdot}73}$) g nitrogen/day, where W is the live weight in kg.

TABLE 5.1

Accepted Estimates of Endogenous Urinary Nitrogen Excretion

Cattle			Sheep		
Live weight		Endogenous excretion g/day per $kg^{0.73}$	Live weight		Endogenous excretion g/day per $kg^{0.73}$
kg	$kg^{0.73}$		kg	$kg^{0.73}$	
50	17·4	0·20	5	3·24	0·17
75	23·4	0·18	10	5·37	0·15
100	28·8	0·17	15	7·22	0·14
125	33·9	0·15	20	8·91	0·12
150	38·8	0·14	25	10·5	0·10
175	43·4	0·13	30	12·0	0·09
200	47·8	0·12	Over 30		0·09
Over 200		0·12			

Retention of nitrogen for fleece growth of adult sheep. From average clip weights, this retention is estimated to vary from an average of 0·6 g nitrogen/day for hill breeds of sheep to one of 1·5 g nitrogen/day for long-wooled lowland breeds.

Metabolic faecal excretion of nitrogen. This is estimated as 5 g nitrogen/kg dry matter intake for sheep and cattle, except for calves on liquid diets for which the estimate is 2·5 g nitrogen/kg dry matter intake.

Retention of nitrogen for liveweight growth. From carcass-composition data it is estimated that for growing cattle the gain in body nitrogen, as a percentage of the corresponding liveweight gain, is 2·4 per cent for live weights up to 600 kg. The estimates for growing sheep are 2·5 per cent up to 40 kg live weight and 2·4 per cent from 40 kg up to maturity.

The additional retention of nitrogen during pregnancy. From observations relating to cows producing calves of about 45 kg at birth, the total daily retention of nitrogen in the growing foetus, placenta, foetal fluids and maternal uterus is estimated to be 1·7 g nitrogen/day in the 5th and 6th months of pregnancy, 5·1 g/day in the 7th month, 12·0 g/day in the 8th month and 29·0 g/day in the 9th month. Corresponding rates for ewes producing single lambs of 5·9 kg or twin lambs of 10 kg are estimated to be 0·18 or 0·24 g/day in the 2nd month, 0·34 or 0·96 g/day in the 3rd month, 1·45 or 3·07 g/day in the 4th month, and 4·96 or 7·40 g/day in the 5th month.

Output of nitrogen in milk. This is estimated as 0·97 g, 0·61 g and 0·53 g/100 g milk for ewes, Channel Island cows and other cows respectively.

Biological value (BV) of dietary protein for ruminants. A large number of estimates of BV were obtained from a survey of nitrogen balance studies, and they were found to display considerable variation. Some of the differences could be attributed to differences in the protein concentration or in the level of feeding and some to genuine differences in BV between comparable diets. Although no single estimate could be reconciled with all the observations it was not considered that multiple estimates could be justified at present. A value of 65 per cent is proposed for sheep and one of 70 per cent for cattle, except for calves on liquid diets based on milk for which 80 per cent is proposed. It is highly probable that these estimates, in particular that of 70 per cent, will have to be revised as more knowledge is obtained, and quite likely that they will have to be replaced by separate estimates appropriate to different feeds and feeding systems.

Combination of the factors to provide estimates of protein requirements. The estimates described in the previous section have been combined to produce the estimates of dietary protein requirement given in Tables 5.2–5.8. The estimates are of available protein requirement, which does not include provision for the replacement of metabolic faecal losses. This is convenient for tabulation purposes since the estimate is thus independent of dry matter intake (D kg/day). For practical use, the estimates can be easily converted to estimates of apparently digestible crude protein requirement by adding the amounts 16·8D g/day (for sheep), or 13·4D g/day (for cattle), or 3·9D g/day (for calves on liquid diets), whenever D can be estimated.

TABLE 5.2

The Minimum Requirements of Available Protein for Maintenance and Growth of Cattle

Live weight (kg)	Rate of gain in live weight (kg/day)						
	0·0	0·2	0·4	0·6	0·8	1·0*	1·2*
	Available protein (g/day)						
50	35 (30)†	75 (65)	120 (105)	165 (145)	210 (180)	250 (220)	290 (250)
75	40 (35)	85 (75)	130 (115)	170 (150)	210 (185)	260 (220)	300 (260)
100	50 (45)	90 (80)	135 (120)	175 (155)	220 (195)	260 (230)	310 (270)
125	50 (45)	95 (85)	135 (120)	180 (160)	220 (195)	270 (230)	310 (270)
150	55	100	140	185	230	270	310
200	60	105	145	190	230	270	320
250	70	110	155	200	240	280	330
300	80	120	165	210	250	290	340
350	90	130	175	220	260	300	350
400	100	140	185	230	270	310	360
450	110	150	190	235	280	320	370
500	115	160	200	245	290	330	370
550	125	170	210	255	300	340	380
600	135	175	220	260	300	350	390
650	140	185	225	270	310	360	400
700	150	190	235	280	320	360	410

* Estimates for gains of 1·0 and 1·2 kg/day are of more doubtful validity than for lower rates of gain because of extrapolation from the data.

‡ Values in brackets refer to calves on liquid milk-based diets.

TABLE 5.3

The Minimum Requirements of Available Protein for Maintenance and Growth of Sheep

Live weight (kg)	Rate of gain in live weight (kg/day)				
	0·05	0·10	0·20	0·30	0·40
	Available protein (g/day)				
5	17	30	55	75	100
10	20	30	55	80	105
15	20	35	60	80	105
20	20	35	60	80	105
25	20	35	60	80	105
30	20	35	60	80	105
40	25	35	60	85	110
50	25	40	60	85	110
60	30	40	65	85	110
70	30	40	65	90	110

TABLE 5.4

The Minimum Requirements of Sheep for Available Protein for Maintenance and Wool Production

Live weight (kg)	Hill breeds	Lowland breeds Short wools	Lowland breeds Long wools	Crosses
	Available protein (g/day)			
30	16	19	25	20
40	19	21	27	22
50	21	24	30	25
60	23	26	32	27
70	25	28	34	29

TABLE 5.5

Supplementary Requirements of Available Protein of Pregnant Cows for Growth of Foetus and Foetal Tissues

Stage of pregnancy (months)	Birth-weight of calf (kg)			
	30	35	40	45
	Available protein (g/day)			
5th-6th	10	12	13	15
7th	30	35	40	45
8th	70	85	95	110
9th	175	200	230	260

TABLE 5.6

Supplementary Requirements of Available Protein of Pregnant Ewes for Growth of Foetus and Foetal Tissues

Stage of pregnancy (months)	Single lamb (6 kg)	Twin lambs (10 kg)
	Available protein (g/day)	
3rd	3	9
4th	14	30
5th	50	70

TABLE 5.7

Supplementary Requirements of Available Protein of Lactating Cows for Milk Production

Milk yield (kg/day)	Guernsey and Jersey cows*	Other breeds†
	Available protein (g/day)	
5·0	270	240
7·5	410	350
10·0	540	470
12·5	680	590
15·0	810	710
17·5	950	830
20	1090	950
25	1360	1180
30	1630	1420
35	1910	1660

* Assumed contents, fat 5·0 per cent, protein 3·9 per cent, nitrogen 0·61 per cent.
† Assumed contents, fat 3·7 per cent, protein 3·35 per cent, nitrogen 0·53 per cent.

TABLE 5.8

Supplementary Requirements of Available Protein of Lactating Ewes for Milk Production

Stage of lactation (months)	Estimated total yield in 16-week lactation (kg)					
	80	100	120	140	160	180
	Available protein (g/day)					
1st	95	115	140	165	185	210
2nd	75	95	115	135	155	175
3rd	55	70	80	95	110	120
4th	35	40	50	60	65	75

COMPARISON OF FACTORIAL ESTIMATES WITH RESULTS OF FEEDING TRIALS

The estimates given in Tables 5.2–5.8 have been compared with published results of feeding trials and protein standards based on feeding trials. The factorial estimates were found to agree with the practical standards for calves and lambs up to about 4 months, growing at moderate rates, and for pregnant and lactating cows, but to be substantially lower than the practical standards for older or more rapidly growing calves and lambs and to be higher than the practical standards for pregnant and lactating ewes.

Although the discrepancies could well arise in part from errors in the factorial estimates the main explanation is believed to be that the practical standards are not minimum amounts. There are several possible reasons for this, but the most important is that an acceptable ration, made up from the farm feedingstuffs commonly used in temperate climates and providing adequate energy for a particular rate of growth, will very often contain more than the factorial estimate of minimum protein requirement. The discrepancy in the estimates for pregnant and lactating ewes is in the opposite direction, but here the factorial estimate provides the minimum that might enable the ewe to maintain her condition whereas the practical standard probably allows the ewe to decrease considerably in weight, particularly in early lactation, as is common farm practice.

Conclusions

The factorial estimates given in Tables 5.2–5.8 are put forward as the amounts of available protein that will suffice to meet minimum protein requirements provided that they are consumed in a ration containing adequate energy. It does not follow that these are the best amounts of available protein to allow. It appears that larger amounts are advisable, under British conditions, for most classes of growing cattle and sheep, particularly when rapid rates of growth are sought. Exact estimates of dietary allowances cannot be attempted here, since they will depend very much on variable economic factors, and also on more detailed information than is available on the effects of dietary manipulations on voluntary intake and on energy values. One general suggestion that can be made is that the allowance should exceed the factorial estimate where the latter would result in less than about 9 per cent crude protein in the dry matter of the diet, since voluntary intake of feed is liable to be reduced at lower levels of protein. Some approximate allowances for growing cattle and sheep, based partly on this suggestion and partly on published conclusions from feeding trials, are set out in Tables 5.9–5.11. They are again given as available protein and can be converted into terms of apparently digestible crude protein as already described provided that dry matter intake can be estimated. The amounts given in Tables 5.9–5.11 are not to be regarded as recommendations; they are merely illustrative of protein allowances that have been suggested on an empirical basis.

TABLE 5.9

Suggested Approximate Allowances of Protein for Growing-fattening Cattle on Traditional High-roughage Diets

Live weight (kg)	Rate of gain (kg/day) 0·50	1·0
	Available protein (g/day)	
50	130	210
100	180	260
200	280	380
300	340	480
400	360	550
500	370	630

TABLE 5.10

Suggested Approximate Allowances of Protein for Dual-purpose and Dairy Types of Cattle fed ad libitum *on High-concentrate Diets*

(Expected gain about 1 kg/day)

Live weight (kg)	Available protein (g/day)
50	230
100	380
200	510
300	630
400	720
500	760

TABLE 5.11

Suggested Approximate Allowances of Protein for Growing Sheep

Live weight (kg)	Rate of gain (g/day) 150	400
	Available protein (g/day)	
5	50	100
10	55	120
20	60	140
30	65	160
40	70	170
50	75	180
60	80	190
70	85	200

SECTION 6. REQUIREMENTS FOR ENERGY

There is growing evidence that the Starch Equivalent System of assessing the energy needs of ruminants is not fully in accord with experimental evidence about their utilization of the energy of food. Somewhat similar criticisms can be levied against the Total Digestible Nutrient System of assessing energy needs. In particular, the efficiencies with which the metabolizable energy of different diets is used to meet energy expenditures or to promote energy retentions in growth and fattening or energy secretion as milk are not simple multiples one of another.

In the Technical Review of which this is a summary, the literature, particularly that referring to recent experimental work in Britain, has been reviewed. The review shows that the efficiency with which the metabolizable energy of diets is used to maintain animals, increases slightly as the 'concentration' of the diet increases, the term 'concentration' meaning the content of metabolizable energy in the feed. Experimentally it has been found that efficiency of utilization of metabolizable energy for maintenance (k_m) increases according to the equation

$$k_m = 54{\cdot}6 + 6{\cdot}75 \frac{M}{D} \tag{1}$$

where $\frac{M}{D}$ is Mcal metabolizable energy of the diet per kg dry matter. This equation shows that with good roughage rations ($\frac{M}{D} = 2{\cdot}0$ Mcal/kg) efficiency of energy utilization for maintenance is 68 per cent, and with diets consisting entirely of good quality concentrated foods ($\frac{M}{D} = 3{\cdot}0$ Mcal/kg) this rises to 75 per cent. Similar efficiencies apply to muscular work, except for the muscular activity of shivering when metabolizable energy is used with an efficiency of 100 per cent to keep the animal warm.

In fattening, certainly in the later stages of growth and possibly in post-weaning growth, the efficiency of utilization of metabolizable energy, varies much more widely with the concentration of the diet than it does for maintenance. The relationship between the efficiency of utilization of metabolizable energy for fattening (k_f) and diet concentration is

$$k_f = 3{\cdot}0 + 18{\cdot}4 \frac{M}{D} \tag{2}$$

This equation shows that efficiency of utilization for fattening when good roughages provide the sole source of energy is about 40 per cent and when medium quality concentrates provide the sole source this value rises to 58 per cent.

For lactation there is a paucity of calorimetric information available on which to base estimates of efficiency, but it appears that efficiency of utilization of metabolizable energy varies very little with the concentration of the diet. Within the range of diet concentration exemplified by a diet consisting entirely of good quality roughage and one consisting solely of medium quality concentrates, efficiency of utilization of metabolizable energy for lactation varies from 65 to 70 per cent. These efficiencies refer specifically to utilization of energy for lactation: they imply that body reserves of energy undergo no change. In fact, the practical problem of providing optimal diets for cows necessarily involves considerations of the effects of diet on the partition of metabolizable energy surplus to maintenance needs as between milk secretion and deposition of body fat and protein. Calorimetric data provide very little information on this important aspect which, however, has been discussed in the Technical Review.

The efficiencies of utilization of the metabolizable energy of a diet for maintaining animals, for fattening animals and for supporting lactation clearly differ. Furthermore the relative values of the efficiencies for one type of diet differ from those for another. These facts are illustrated by the following figures which are quoted merely as examples.

	Efficiency of utilization of metabolizable energy		Relative values as energy sources when maintenance = 100	
$\frac{M}{D}$ (Mcal/kg)	All roughage diet 2·0	All concentrate diet 3·0	All roughage diet 2·0	All concentrate diet 3·0
Maintenance	68	75	100	100
Fattening	40	58	59	77
Lactation	66	68	97	90

There is evidence that in the ruminant, as the level of feeding increases, the metabolizable energy of unit weight of diet declines, the decline being greater for the poorer quality diet. The relationship can be expressed quantitatively by stating that when the intake of a good roughage diet is increased from an amount sufficient to maintain the animal to twice that amount, the metabolizable energy of the total diet falls by 10 per cent. With a diet consisting of medium quality concentrates a similar doubling of intake reduces its metabolizable energy value by 3 per cent only. This reduction in the metabolizable energy of diets with increase in the amount given appears to be a major factor accounting for the apparent decline in animal productivity with increased feed intake. The efficiency of utilization of metabolizable energy for fattening is not affected by feeding level.

The New Approach

From the above observations a new feeding system has been devised which is based on metabolizable energy. The metabolizable energy values ascribed to individual feeds are those determined at the maintenance level of feeding, and when diets are calculated, the metabolizable energy to be expected at the particular feeding level adopted is calculated from the relationship between feeding level and the metabolizable energy of the diet as a whole. The production of weight gain or of milk yield from a particular diet is assessed by using the best available measure of the efficiency of utilization of the metabolizable energy of the diet as a whole. At present the best measures are those embodied in equations (1) and (2) and in the general relationship described above for lactation.

These methods enable estimates to be made of the amounts of metabolizable energy supplied by a diet as computed from tabulated values of metabolizable energy of individual feeds per unit weight that are required to produce a given amount of weight-gain or milk production. These estimates necessarily include a classification of *diets* by concentration, thus recognizing that the type of diet has a major effect in determining its nutritive value, and that individual feeds can be assessed only as components of diets. It is recognized that the productive value of a feed cannot be given an absolute expression, a fact which is pertinent to problems concerned with assessing the economics of substitution of one feed by another.

The Technical Review gives in detail the results of investigations which deal with the basal energy expenditure of cattle and sheep, with energy expenditures in muscular work, with the calorific value of gains of body-weight and energy retention in pregnancy and with the determination of the metabolizable energy of feeds. Details of the methods of calculation of energy requirements are also given and the results of a number of such calculations summarized. In these tables two criteria of a ration are used, firstly its concentration, that is the number of Mcal metabolizable energy/kg dry matter of diet, determined at the maintenance level of feeding (M/D), and secondly the total number of Mcal of metabolizable energy supplied. The former designates the quality of the diet as a whole, low values of M/D being characteristic of poor roughages and high values being characteristic of diets containing liberal amounts of cereals and other more digestible feeds. Tables 6.1 and 6.2 give estimates of the metabolizable energy requirements for lactation and pregnancy in cattle. The standards for lactation in Table 6.1 provide less feed at low levels of lactation and more feed at high levels of lactation than do the existing British Starch Equivalent standards, a provision which is in accord with the results of practical feeding trials. At moderate to high levels of milk production the

TABLE 6.1

Daily Requirements of a Mature Cow weighing 500 kg, giving milk of 4 per cent fat content and walking 1 mile

(Mcal metabolizable energy/day)

Concentration of diet (Mcal metabolizable energy/kg dry matter)	Milk yield (kg/day)					
	5	10	15	20	25	30
1·8	19·1	28·6	—	—	—	—
2·2	17·7	24·9	33·3	—	—	—
2·6	16·8	23·0	29·8	37·2	45·4	—
3·0	16·3	22·4	28·7	35·2	42·1	49·3
3·4	16·5	22·9	29·5	36·1	43·0	49·9

TABLE 6.2

Requirements of a Cow, weighing 500 kg, taking exercise equivalent to walking 1 mile/day before it became pregnant, and during pregnancy

(Mcal metabolizable energy/day)

Concentration of diet (Mcal/kg dry matter)	Non-pregnant weight 500 kg	Weight 8–4 weeks from term 510 kg	Weight 4–2 weeks from term 530 kg	Weight 2–0 weeks from term 560 kg
1·8	11·7	15·8	20·2	22·4
2·2	11·3	14·6	18·4	20·1
2·6	10·9	13·5	17·0	18·5
3·0	10·5	12·5	15·9	17·3

new system provides about the same amount of feed as the American TDN system.

In pregnancy the energy requirements increase owing to an increase in maternal heat production and the allowances given, which allow for no maternal gain in weight, agree with practical opinion that dairy cows should be given additional food before calving, that is steaming-up allowances.

With growing cattle, the requirements given in Table 6.3 provide slightly less feed than does the British Starch Equivalent System. This difference may well be more apparent than real because of the arbitrary upward correction of the starch values for hays made in the British Starch Equivalent System. No estimates have been made for calves weighing less than 200 kg.

TABLE 6.3

Estimates of the Energy required by Growing and Fattening Cattle

(Mcal metabolizable energy/day)

Live weight (kg)	Ration concentration (Mcal/kg dry matter)	Rate of gain (kg/day)						
		0	0·25	0·50	0·75	1·00	1·25	1·50
200	1·8	8·6	10·5	12·9	15·8	—	—	—
	2·2	8·3	9·8	11·5	13·6	16·3	—	—
	2·6	8·0	9·2	10·6	12·3	14·3	16·7	19·7
	3·0	7·7	8·7	9·9	11·3	13·0	14·9	17·3
	3·4	7·5	8·4	9·4	10·6	12·0	13·8	15·7
300	1·8	10·5	12·8	15·3	18·8	—	—	—
	2·2	10·2	11·9	13·9	16·3	19·5	—	—
	2·6	9·8	11·2	12·8	14·7	17·2	20·3	—
	3·0	9·4	10·6	12·1	13·6	15·6	18·0	21·0
	3·4	9·1	10·2	11·4	12·8	14·6	16·5	19·1
400	1·8	12·4	15·1	18·0	22·1	—	—	—
	2·2	12·0	14·1	16·4	19·3	22·9	—	—
	2·6	11·5	13·2	15·0	17·4	20·2	23·9	—
	3·0	11·1	12·6	14·1	16·0	18·3	21·3	25·0
	3·4	10·8	12·1	13·4	15·0	17·0	19·5	22·7

Table 6.4 gives estimates of the energy requirements of growing sheep. These provide less energy than does the British Starch Equivalent standard largely because of the considerable over-estimation of maintenance needs by the latter standard.

Values for lactation in sheep are given in Table 6.5.

TABLE 6.4
Estimates of the Energy required by Fattening Lambs
(Mcal metabolizable energy/day)

Live weight (kg)	Ration concentration (Mcal/kg dry matter)	Daily gain (g/day)				
		0	50	100	200	300
20	1·8	1·00	1·58	2·35	—	—
	2·2	0·97	1·42	1·88	3·42	—
	2·6	0·93	1·29	1·69	2·70	4·20
	3·0	0·87	1·20	1·52	2·32	3·37
30	1·8	1·31	1·91	2·79	—	—
	2·2	1·27	1·76	2·32	3·88	—
	2·6	1·22	1·60	2·06	3·15	4·63
	3·0	1·17	1·51	1·89	2·74	3·82
40	1·8	1·59	2·30	3·20	—	—
	2·2	1·54	2·14	2·71	4·25	—
	2·6	1·48	1·92	2·41	3·54	5·00
	3·0	1·42	1·80	2·21	3·11	4·22

TABLE 6.5
Energy Requirements of a Ewe (weighing 70 kg and walking 2 miles/day) when neither lactating nor pregnant and at four different stages of its lactation
(Mcal metabolizable energy/day)

Concentration of diet (Mcal/kg dry matter)	Neither lactating nor pregnant	Month of lactation			
		1	2	3	4
1·8	2·02	—	—	5·10	3·80
2·2	1·95	—	6·04	4·52	3·46
2·6	1·87	6·01	5·37	4·18	3·25
3·0	1·80	5·72	5·15	4·06	3·16

In conclusion, the new system has been developed largely because a more versatile method of computing diets and assessing the nutritive value of feeds is required than that provided by the Starch Equivalent System. The latter ascribes to individual feeds constant nutritive values which are independent of the other ingredients of the diet, the amount of the diet which is given or the type of production the diet is designed to support. The assumptions involved in doing this cannot be justified in the light of modern knowledge. The new system provides an overall evaluation of diets, and takes into account the effects of level of feeding. Furthermore it is capable of considerable expansion to take into account future developments in this field. In many practical circumstances little error may be incurred by using the Starch Equivalent System in assessing the energy needs of ruminants, but it is suggested that the new system should gradually replace the Starch Equivalent System in the practical formulation of diets for ruminants.

SECTION 7.* THE EXPRESSION OF NUTRIENT REQUIREMENTS AS DIETARY CONCENTRATIONS

In the tables in Section 6 which give requirements in terms of metabolizable energy, the weight of dry food required to meet energy needs can be obtained by dividing the values in the table by the appropriate dietary concentration. Having obtained the amount of dry matter needed to meet energy requirements calculations can then be made of the amounts of other nutrients that must be present in that food to ensure that the requirements of these other nutrients are also met. The results of such calculations are recorded in Tables 7.1–7.4.

General consideration of these tables shows that as the energy concentration of a diet is increased, the concentration of nutrients in a diet designed to support a particular level of production also increases but at a greater rate. More of a nutrient is required per Mcal metabolizable energy at high energy concentrations than at low energy concentrations since the metabolizable energy of diets with a high energy concentration is more efficiently utilized than that of diets with a low energy concentration.

For a given energy concentration of the diet, as the intensity of production increases and more of the same energy source is given to an animal, concentrations of nutrients in the dietary dry matter also tend to rise. With both cattle and sheep the dietary concentration of digestible crude protein at the highest levels of milk production is approximately twice that at maintenance (Table 7.1), and at the highest rates of liveweight gain it is about 50 per cent greater than at maintenance (Table 7.2, factorial estimates of requirements of digestible crude protein).

The trends, however, are less consistent with minerals. While the calcium concentration required in the dry diet rises with rate of gain in cattle and remains almost constant with increase in milk yield, it falls slightly with increased rates of growth and lactation in sheep. The phosphorus and magnesium requirements as dietary concentrations decline with increased level of

* An abridged form of Section 7 as published in the volume containing the Technical Reviews.

production in all classes of stock because our assessment of maintenance needs shows them to be high relative to those for production.

Standards for ad libitum *feeding*. If appetite limits could be estimated with precision it would be possible to formulate complete diets adequate in all nutrients including energy and capable of promoting a wide range of production from ruminants under conditions of *ad libitum* feeding.

The best approach to this problem is to state that a suitable *ad libitum* diet to attain a given level of production is one which is of such an energy concentration that the amount of it that must be consumed to satisfy energy needs is also the amount which satisfies appetite.

Tables 7.1–7.4 have been adjusted to exclude estimates of dry matter requirements thought likely to exceed normal appetite, and in preparing these tables the ability of ruminants to eat larger amounts of high quality than of low quality diets has been taken into account. The interpretation of these tables can be illustrated by the following example. It can be estimated from Table 7.1 that a dry non-pregnant cow could probably be satisfied by an *ad libitum* diet containing 1·8 Mcal/kg. This diet should contain at least 3·1 per cent digestible crude protein (or 5 per cent to avoid appetite limitation), 0·28 per cent calcium, 0·39 per cent phosphorus and 0·12 per cent magnesium. From the same table a cow of 500 kg yielding 30 kg milk of 4 per cent fat would not be able to satisfy its needs on a diet containing 1·8 or 2·2 Mcal/kg and might well require a diet containing 3·0 Mcal/kg with concentrations of 10·7 per cent digestible crude protein, 0·62 per cent calcium, 0·48 per cent phosphorus and 0·17 per cent magnesium in the dry matter.

GLOSSARY

This glossary gives a list of terms that apply particularly to this volume of Summaries. The more comprehensive glossary of terms used in the Technical Reviews is published in the volume comprising these Reviews.

Apparent digestibility. The amount of dry matter or of a nutrient ingested minus the amount excreted in the faeces whether a residue from the feed or of endogenous origin, expressed as a percentage of the intake.

$$\text{Apparent digestibility} = 100 \times \frac{\text{Amount in feed} - \text{amount in faeces}}{\text{Amount in feed}}$$

Available protein. The amount of truly digestible crude protein in the feed less the amount of that protein required to make good the loss of nitrogen as metabolic faecal nitrogen that the ingestion of the feed necessarily entails. Available protein (%) = Apparently digestible crude protein (%) − 6·25MFN $\left(\frac{100}{\text{BV}} - 1\right)$

TABLE 7.1

The Nutrient Requirements of an Adult Dairy Cow weighing 500 kg and walking 1 mile/day, expressed in terms of amounts of dry feed needed each day, and dietary concentrations of nutrients in the dry matter

Level of production	Requirement	Energy concentration of diet (Mcal/kg dry matter) 1·8	2·2	2·6	3·0
Dry non-pregnant (maintenance)	Dry matter (kg/day)	6·5	5·1	4·2	3·4
	Concentration in dry matter (%)				
	Digestible crude protein	3·1	3·6	4·1	4·6
	Calcium	0·28	0·36	0·44	0·53
	Phosphorus	0·39	0·50	0·61	0·73
	Magnesium	0·12	0·15	0·18	0·21
Pregnant, 4–2 weeks from term carrying a 40-kg calf	Dry matter (kg/day)	11·2	8·4	6·5	5·3
	Concentration in dry matter (%)				
	Digestible crude protein	4·5	5·5	6·7	8·0
	Calcium	0·30	0·40	0·51	0·63
	Phosphorus	0·30	0·40	0·52	0·63
	Magnesium	0·10	0·14	0·18	0·22
Producing 10 kg milk/day	Dry matter (kg/day)	15·9	11·3	8·8	7·5
	Concentration in dry matter (%)				
	Digestible crude protein	5·0	6·5	8·0	9·1
	Calcium	0·29	0·41	0·52	0·61
	Phosphorus	0·27	0·38	0·49	0·57
	Magnesium	0·10	0·14	0·18	0·21
Producing 20 kg milk/day	Dry matter (kg/day)	—	—	14·3	11·7
	Concentration in dry matter (%)				
	Digestible crude protein	—	—	8·8	10·4
	Calcium	—	—	0·52	0·63
	Phosphorus	—	—	0·42	0·51
	Magnesium	—	—	0·15	0·19
Producing 30 kg milk/day	Dry matter (kg/day)	—	—	20·1*	16·4
	Concentration in dry matter				
	Digestible crude protein	—	—	9·0*	10·7
	Calcium	—	—	0·50*	0·62
	Phosphorus	—	—	0·38*	0·48
	Magnesium	—	—	0·14*	0·17

* This will be in excess of maximal appetite with most cows.

TABLE 7.2

The Nutrient Requirements of Beef Cattle weighing 400 kg expressed in terms of amounts of dry feed needed each day, and dietary concentrations of nutrients in the dry matter

Level of production	Requirement	Energy concentration of diet (Mcal/kg dry matter) 1·8	2·2	2·6	3·0	3·4
Maintenance	Dry matter (kg/day)	6·9	5·5	4·4	3·7	3·2
	Concentration in dry matter (%)					
	Digestible crude protein (Factorial)*	2·8	3·1	3·6	4·0	4·5
	Digestible crude protein (Practical)*	3·8	4·4	5·2	5·9	6·6
	Calcium	0·21	0·26	0·33	0·39	0·45
	Phosphorus	0·27	0·34	0·43	0·51	0·59
	Magnesium	0·072	0·090	0·112	0·134	0·155
0·25 kg gain/day	Dry matter (kg/day)	8·4	6·4	5·1	4·2	3·6
	Concentration in dry matter (%)					
	Digestible crude protein (Factorial)	3·1	3·7	4·3	4·9	5·5
	Digestible crude prote (Practical)	4·5	5·5	6·5	7·7	8·7
	Calcium	0·24	0·31	0·39	0·48	0·56
	Phosphorus	0·25	0·33	0·42	0·51	0·59
	Magnesium	0·076	0·101	0·127	0·154	0·179
0·75 kg gain/day	Dry matter (kg/day)	12·3	8·8	6·7	5·3	4·4
	Concentration in dry matter (%)					
	Digestible crude protein (Factorial)	3·5	4·3	5·2	6·3	7·3
	Digestible crude protein (Practical)	5·0	6·5	8·1	9·9	11·7
	Calcium	0·26	0·36	0·47	0·60	0·72
	Phosphorus	0·21	0·30	0·38	0·49	0·60
	Magnesium	0·061	0·085	0·111	0·140	0·169
1·25 kg gain/day	Dry matter (kg/day)	—	—	9·2	7·1	5·7
	Concentration in dry matter (%)					
	Digestible crude protein (Factorial)	—	—	5·4	6·6	7·9
	Digestible crude protein (Practical)	—	—	9·2	11·5	14·0
	Calcium	—	—	0·47	0·61	0·76
	Phosphorus	—	—	0·34	0·44	0·55
	Magnesium	—	—	0·092	0·119	0·148

* See p. 902

TABLE 7.3

The Nutrient Requirements of Adult Ewes weighing 70 kg and walking 2 miles/day, expressed in terms of amounts of dry feed needed each day, and dietary concentrations of nutrients in the dry matter

Level of production	Requirement	Energy concentration of diet (Mcal/kg dry matter) 1·8	2·2	2·6	3·0
Dry non-pregnant (maintenance)	Dry matter (kg/day)	1·1	0·9	0·7	0·6
	Concentration in dry matter (%)				
	Digestible crude protein	4·3	4·9	5·8	6·5
	Calcium	0·57	0·69	0·89	1·04
	Phosphorus	0·46	0·56	0·73	0·85
	Magnesium	0·074	0·090	0·116	0·135
Pregnant		Not available			
Producing 0·65 kg milk/day	Dry matter (kg/day)	2·11	1·57	1·25	1·05
	Concentration in dry matter (%)				
	Digestible crude protein	6·0	7·5	9·0	10·4
	Calcium	0·43	0·57	0·72	0·85
	Phosphorus	0·32	0·43	0·53	0·64
	Magnesium	0·065	0·089	0·110	0·131
Producing 1·05 kg milk/day	Dry matter (kg/day)	2·83	2·05	1·61	1·35
	Concentration in dry matter (%)				
	Digestible crude protein	6·3	8·1	9·8	11·4
	Calcium	0·38	0·52	0·66	0·79
	Phosphorus	0·27	0·38	0·48	0·57
	Magnesium	0·061	0·085	0·108	0·129
Producing 1·50 kg milk/day	Dry matter (kg/day)	—	2·75	2·06	1·72
	Concentration in dry matter (%)				
	Digestible crude protein	—	8·0	10·1	11·8
	Calcium	—	0·46	0·61	0·73
	Phosphorus	—	0·32	0·43	0·51
	Magnesium	—	0·078	0·104	0·124
Producing 1·80 kg milk/day	Dry matter (kg/day)	—	—	2·31	1·90
	Concentration in dry matter (%)				
	Digestible crude protein	—	—	10·5	12·4
	Calcium	—	—	0·60	0·73
	Phosphorus	—	—	0·42	0·50
	Magnesium	—	—	0·104	0·126

TABLE 7.4

The Nutrient Requirements of Fattening Sheep weighing 40 kg, expressed in terms of amounts of dry feed needed each day, and dietary concentrations of nutrients in the dry matter

Level of production	Requirement	Energy concentration of diet (Mcal/kg dry matter) 1·8	2·2	2·6	3·0
Maintenance	Dry matter (kg/day)	0·88	0·70	0·57	0·47
	Concentration in dry matter (%)				
	Digestible crude protein (Factorial)*	3·4	3·8	4·3	4·9
	Digestible crude protein (Practical)*	—	—	—	—
	Calcium	0·36	0·46	0·56	0·68
	Phosphorus	0·25	0·31	0·39	0·47
	Magnesium	0·052	0·066	0·081	0·098
100 g gain/day	Dry matter (kg/day)	1·78	1·23	0·93	0·74
	Concentration of dry matter (%)				
	Digestible crude protein (Factorial)	3·7	4·5	5·4	6·4
	Digestible crude protein (Practical)	4·5	5·8	7·1	8·4
	Calcium	0·28	0·41	0·43	0·68
	Phosphorus	0·16	0·23	0·30	0·38
	Magnesium	0·036	0·052	0·069	0·086
200 g gain/day	Dry matter (kg/day)	—	1·93	1·36	1·04
	Concentration in dry matter (%)				
	Digestible crude protein (Factorial)	—	4·8	6·1	7·5
	Digestible crude protein (Practical)	—	6·3	8·3	10·3
	Calcium	—	0·35	0·50	0·65
	Phosphorus	—	0·18	0·25	0·33
	Magnesium	—	0·042	0·060	0·079
300 g gain/day	Dry matter (kg/day)	—	—	1·92	1·41
	Concentration in dry matter (%)				
	Digestible crude protein (Factorial)	—	—	6·1	7·7
	Digestible crude protein (Practical)	—	—	8·5	10·9
	Calcium	—	—	0·44	0·60
	Phosphorus	—	—	0·21	0·28
	Magnesium	—	—	0·052	0·071

* See p. 902

Where MFN = g of metabolic faecal nitrogen/100 g dry feed,
and BV = the biological value of the feed protein.

Availability. The availability of a mineral element is the percentage of the element supplied by the food that can be used by the animal to make good endogenous losses and to meet the demands of growth, production and pregnancy. With most mineral elements the urinary loss is small so the availability can be estimated as :—

$$\text{Availability} = 100 \times \frac{\text{Amount in feed} - \text{amount in faeces} + \text{endogenous faecal loss}}{\text{Amount in feed}}$$

The estimate is usually made when the amount of the element in the diet is low, since with some elements availability falls as the amount of the element given is increased.

Biological value (BV). The percentage of the truly digestible nitrogen of the feed protein that is retained by the animal and used for maintenance, growth, production and pregnancy.

$$\text{BV} = 100 \times \frac{\text{I} - (\text{F} - \text{MFN}) - (\text{U} - \text{EUN})}{\text{I} - (\text{F} - \text{MFN})}$$

$$= 100 \times \frac{\text{NR} + \text{MFN} + \text{EUN}}{\text{I} - (\text{F} - \text{MFN})}$$

Where I = dietary intake of nitrogen, F = total faecal nitrogen,
MFN = metabolic faecal nitrogen, U = total urinary nitrogen,
EUN = endogenous urinary nitrogen, and NR = nitrogen retention, all in g/day.

Calorie. The amount of heat required to raise the temperature of 1g water through 1° C from 14·5 to 15·5° C. By international agreement a calorie (abbreviation 'cal') is now defined as being equal to 4·1852 International Joules.

Crude protein. Since proteins, on average, contain 16 per cent nitrogen, crude protein is usually taken as being total nitrogen multiplied by 6·25. For milk, however, the factor generally adopted is 6·38.

Efficiency of utilization of metabolizable energy (k). This is defined as the increase in energy retention resulting from an increase of 100 kcal in the intake of metabolizable energy.

$$\text{k} = 100 \times \frac{\text{Increment in energy retention}}{\text{Increment in intake of metabolizable energy}}$$

If the two measurements of energy retention on which the estimate of the increment of retention is based are both negative or if one is negative and the other zero, the efficiency measured is *the efficiency of utilization of metabolizable energy for maintenance* (k_m).

If both measurements are positive or if one is positive and the other zero and the animal is growing or fattening, the efficiency measured is *the efficiency of utilization of metabolizable energy for growth or for fattening* (k_f).

The efficiency of utilization of metabolizable energy for lactation (k_l) is measured as the increase in the amount of milk energy secreted resulting from an increase of 100 kcal in the intake of metabolizable energy. The efficiency of utilization of metabolizable energy for lactation when there is no gain or loss of body substance is termed *the efficiency of utilization of metabolizable energy for lactation at zero body retention* (k_{lo}).

Endogenous loss or excretion. Endogenous loss of a nutrient is the loss originating from the tissues even when the nutrient is absent from the diet.

Kilocalorie (kcal). One thousand calories.

Maintenance. At the maintenance level of feeding, the animal's requirement for a nutrient or for energy for the normal continuance of vital processes within the body, including the replacement of endogenous losses in faeces and urine and from the skin, is just met so that the net gain or loss of the nutrient from the tissues of the animal as a whole is zero.

Megacalorie (Mcal). One million calories or 1000 kcal.

Metabolic faecal nitrogen (MFN or MF). The nitrogen, other than the undigested feed nitrogen, which is excreted in the faeces. It arises from the unabsorbed constituents of the digestive juices, from cellular material and mucus derived from the gastric and intestinal mucosa.

Metabolizable energy. The metabolizable energy of a feed (ME, or M in formulae and equations) is the heat of combustion of the feed minus the heats of combustion of the faeces, urine and gaseous products of digestion.

Nitrogen retention (NR). This can be positive or negative and is the net gain or loss of nitrogen by the animal. It is the intake of nitrogen minus the nitrogen lost in the faeces and urine and from or through any of the external surfaces of the animal.

Requirement. The requirement for any given nutrient is the amount of that nutrient which must be supplied in the diet to meet the needs of the normal healthy animal given an otherwise completely adequate diet in an environment compatible with good health.

Starch equivalent (SE). The number of pounds of starch that would promote in the adult steer the storage of the same amount of fat as 100 lb of the feedingstuff under consideration.

Total digestible nutrients (TDN). This is an assessment of nutritive value used in the U.S.A. The TDN value is calculated from the percentages of apparently digestible nutrients in the following way:

TDN per 100 lb of feed = % apparently digestible crude protein
+ % apparently digestible nitrogen-free extract
+ % apparently digestible crude fibre
+ 2·25 (% apparently digestible crude fat).

The analysis of the feed for this purpose is done by the Weende scheme.

True digestibility. The amount of a nutrient in the feed minus the amount in the residue of the feed excreted in the faeces, expressed as a percentage of intake.

True digestibility =

$$100 \times \frac{\text{Amount in feed} - (\text{amount in faeces} - \text{metabolic faecal loss})}{\text{Amount in feed}}$$

Weende scheme of analysis. A scheme of analysis in which the concentrations of moisture, ash, crude protein, crude fibre and fat are determined by standard procedures, and the remainder, calculated by difference, is termed the 'nitrogen-free extract'.

Weights and volumes

1 kilogram (kg) = 1000 grammes (g) = 2·205 pounds (lb).
1 lb = 453·6 g.
1 oz = 28·35 g.
1 gallon (gal) = 10 lb = 4·536 kg = 4·536 litres*.

To convert:	*Multiply by:*
lb to kg	0·4536
kg to lb	2·205
gal to kg	4·536
kg to gal	0·2205

* This applies where the specific gravity is 1·000. A typical value for the specific gravity of cow's milk is 1·032.

TECHNICAL COMMITTEE ON THE NUTRIENT REQUIREMENTS OF FARM LIVESTOCK

Chairman	SIR DAVID CUTHBERTSON, C.B.E., M.D., D.SC., LL.D., F.R.C.P.E., F.R.S.E.	Rowett Research Institute
	PROFESSOR R. G. BASKETT, O.B.E., D.SC., F.R.I.C.	National Institute for Research in Dairying
	*K. L. BLAXTER, Ph.D., D.SC., N.D.A., F.R.S.E.	Hannah Dairy Research Institute
	A. W. BOYNE, B.SC.	Rowett Research Institute
	R. BRAUDE, Ph.D., D.SC.	National Institute for Research in Dairying
	A. EDEN, M.A., Ph.D., F.R.I.C.	National Agricultural Advisory Service
	PROFESSOR S. K. KON, C.B.E., Ph.D., D.SC., F.R.I.C.	National Institute for Research in Dairying
	F. H. MOULT, M.SC.	Agricultural Research Council
	PROFESSOR A. T. PHILLIPSON, M.A., Ph.D., D.V.SC., M.R.C.V.S., F.R.S.E.	School of Veterinary Medicine, Cambridge
	PROFESSOR ALEX. ROBERTSON, C.B.E., M.A., B.SC., Ph.D., M.R.C.V.S., F.R.I.C., F.R.S.H., F.R.S.E.	Royal (Dick) School of Veterinary Studies
	J. A. B. SMITH, C.B.E., Ph.D., D.SC., F.R.I.C., F.R.S.E.	Hannah Dairy Research Institute
	D. B. WALLACE, M.A.	School of Agriculture, Cambridge
Technical Secretaries	PROFESSOR I. A. M. LUCAS, B.SC., M.S. (resigned September, 1961)	Rowett Research Institute
	J. L. CORBETT, B.SC., M.Agr.SC. (resigned December, 1962)	Rowett Research Institute
	C. F. MILLS, M.SC., Ph.D., F.R.I.C.	Rowett Research Institute

* *Director, Rowett Research Institute, with effect from 1.10.65*

WORKING PARTY ON NUTRIENT REQUIREMENTS OF RUMINANTS

Chairman	J. A. B. SMITH, C.B.E., Ph.D., D.SC., F.R.I.C., F.R.S.E.	Hannah Dairy Research Institute
	*D. G. ARMSTRONG, M.SC., Ph.D., F.R.I.C.	University of Newcastle upon Tyne
	C. C. BALCH, Ph.D., D.SC.	National Institute for Research in Dairying
	K. L. BLAXTER, Ph.D., D.SC., N.D.A., F.R.S.E.	Hannah Dairy Research Institute
	PROFESSOR W. O. BROWN, M.Agr., Ph.D., D.SC., F.R.I.C.	Queen's University of Belfast
	T. L. DODSWORTH, J.P., B.SC., Ph.D.	North of Scotland College of Agriculture
	A. EDEN, M.A., Ph.D., F.R.I.C.	National Agricultural Advisory Service
	A. S. FOOT, M.SC.	National Institute for Research in Dairying
	PROFESSOR W. HOLMES, B.SC., Ph.D., N.D.A., N.D.D.	Wye College, University of London
	I. MCDONALD, M.A.	Rowett Research Institute
	F. H. MOULT, M.SC.	Agricultural Research Council
	PROFESSOR J. E. NICHOLS, M.SC., Ph.D., F.R.S.E.	University College of Wales, Aberystwyth
	R. J. PERKINS, B.A., M.A.(ECON).	University of Newcastle upon Tyne
	T. R. PRESTON, B.SC., Ph.D. (resigned September, 1964)	Rowett Research Institute
	PROFESSOR ALEX. ROBERTSON, C.B.E., M.A., B.SC., Ph.D., M.R.C.V.S., F.R.I.C., F.R.S.H., F.R.S.E.	Royal (Dick) School of Veterinary Studies
	W. J. THOMAS, M.SC. (resigned September, 1964)	University of Manchester

* Dr. Armstrong acted also as Technical Secretary of the Working Party.

CHAPTER 25

The Nutrition of the Horse

N. O. OLSSON

Veberöd, Sweden

INTRODUCTION

The rapid advance of mechanization during the last 50 years has progressively reduced the use of horses for traction and for riding. The result has been a considerable reduction of the number of horses in many countries. The reduction in some European countries, the United States and Canada between 1912 and 1963 is shown in Table 1 (Institut International d'Agriculture, 1911–12, 1922, 1931–2. F.A.O. 1947; 1952. *Statistical Yearbook, United Nations*, 1955–64).

TABLE 1
Number of Horses in some European Countries, the United States and Canada between 1912 and 1963 (in thousands)

Year	Country					
	Denmark	France	Sweden	Great Britain and Ireland	United States	Canada
1912	535	3222	588	2059	23,016	2596
1920	602	2635	728	2213	19,766*	3400
1930	494*	2924	653	1670	13,165*	3295
1940	575*	2115†	617	1509	10,193*	2780*
1951	465	2397	415	845	4993*	1304*
1954/55	309	2215	312	570	2973*	835*
1955/56	282	2161	276	509	2710*	787*
1959/60	171	1825	209	381	2130*	574*
1962/63	81	1526	151	—	—	451*

* Horses in towns not included. † Army horses not included.

By 1960 the number of horses in the United States had fallen to about 9·3 per cent of the number in 1912. During the same time the number of horses in Canada was reduced to about 22·1 per cent of the number in 1912, and in Great Britain and Ireland about 18·5 per cent.

Table 2 gives data for the estimated number of horses in the whole world in 1939–63 (F.A.O., 1947, 1952. *Statistical Yearbook, United Nations*, 1956–64).

TABLE 2
Number of Horses in the Whole World

Year	Millions
1939	74·4
1950/1	60·6
1955/6	73·0
1960/1	66·0
1962/3	64·4

In spite of the great decrease in the developed countries, there are still 64·4 million horses in the world. Even if agriculture and forestry are further mechanized so that their need for horses is further reduced, the horse will for long have some importance for traction, and it is for that reason that its correct feeding is still important.

For a discussion on the physiology of digestion of the horse reference should be made to Alexander (1954).

Of published reports on the feeding of horses only a few are based on accurate experiments. Most have arisen from practical experience or from revision of older experiments.

DIGESTION IN HORSES AND RUMINANTS

Although the digestive system of the horse is unlike that of the ruminant in anatomical structure, both live on the same feeding stuffs and there is little difference in their capacity to digest them. The horse has a single stomach and does not ruminate; cattle have chambered stomachs and ruminate. Cattle also have longer and roomier digestive canals, as appears from the following comparison (Crasemann, 1945):

Digestive Canal

Length (m)		Volume (l.)		Distribution of volume (l.)		
Total	Ratio to body length	Total	Per 100 kg liveweight	Stomach	Intestines total	Caecum
Horse 30	20 : 1	200	40	10	190	30
Cattle 50	30 : 1	310	50	200	110	10

The capacity of both to digest some of the feeds which appear most often in their diets is shown by the digestibility coefficients in Table 3 (Crasemann, 1945).

TABLE 3

Comparison of Digestion in Horses and Cattle

Feed	Animal	Digestibility coefficient				
		Organic matter	Crude protein	Ether extract	N-free extract	Fibre
Meadow hay, early (22 per cent fibre)	Horse	58	63	22	65	48
	Cattle	67	65	57	68	63
Wheat straw (41 per cent fibre)	Horse	21	28	—	28	18
	Cattle	42	4	31	37	50
Oats (10 per cent fibre)	Horse	69	80	71	75	29
	Cattle	70	76	80	76	28
Maize (2 per cent fibre)	Horse	89	76	61	92	40
	Cattle	90	72	89	95	58

The organic matter in feeding stuffs of low fibre content, such as cereal grains, is equally well digested by horse and cattle, but the organic matter of feedingstuffs of high fibre content is less well digested by horses. The same is true of nitrogen-free extract. Protein is about equally digested by horses and cattle. Digestibility coefficients for crude fat and fibre are almost always less when a feed is used for horses than when it is used for cattle.

With rising fibre content in a ration, digestibility of organic matter is less for both horses and cattle, but the reduction is greater for horses. The relation between fibre content (x) and coefficient of digestibility of organic matter (y) is shown in the following equations, derived from a summary of 71 feeding experiments on horses with mixed rations and from an analysis of Kellner's digestibility experiments with cattle (Axelsson, 1940a, 1941):

$$\text{For horses } y = 97{\cdot}0 - 1{\cdot}26\,x,$$

$$\text{For cattle } y = 86{\cdot}0 - 0{\cdot}66\,x.$$

The digestibility of organic matter has been studied also with feeds differing in fibre content by Olsson (Larsson *et al.*, 1951). The results are shown in Table 4.

TABLE 4
Digestibility Coefficients of Organic Matter in Horses and Cattle

Feed	Approximate fibre content (per cent)	Digestibility coefficient	
		Horse	Cattle
Wheat straw	39	35	40
Clover-grass hay	30	50	63
Meadow hay	30	52	59
Pasture	18	64	81
Oats	10	78	72
Fodder cellulose	68	81	80

The discrepancy is smallest for fodder cellulose; but Hvidsten (1945b) records even higher values for ruminants, and Edin, Helleday and Nordfeldt (1941) give values of from 63 to 92 per cent for horses, depending on the nature of the product tested. From the results of roughly 1000 digestion experiments, Olsson (1949) has fitted an equation of the third degree to express the relation of digestibility of dry matter to fibre content (see p. 946).

NUTRITIONAL REQUIREMENTS OF HORSES

Requirements for maintenance, work, reproduction and lactation, and growth must be considered separately, and with reference to energy supply and the essential components of food, namely protein, minerals and vitamins, etc. Knowledge of metabolism, and especially metabolism of the individual components of feed, is incomplete and in some respects negligible.

A number of rationing systems have been used. Those best known are based on starch equivalent (SE), the Scandinavian feed unit (ScFU), the feed unit for ruminants (FUr) based on metabolizable energy according to Axelsson, the Russian oat feed unit, and total digestible nutrients (TDN). Computations in terms of the Scandinavian unit are based on the net energy value of the feed; the unit is the value of 1 kg average barley. Olsson (1951) claims that this unit gives a reliable measure of the net energy requirements of the horse. (1 ScFU = about 2820 kcal metabolizable energy).

REQUIREMENTS FOR MAINTENANCE

Energy

The energy requirement of the horse for maintenance has been computed by several workers. The mean results are shown in Table 5.

According to some writers (Ehrenberg, 1932; Hansson, 1938) there is a direct relation between the horse's energy requirement for maintenance and its liveweight; others have found a somewhat lower requirement of energy per unit of weight as liveweight rises (Axelsson, 1943; Crasemann, 1945;

Jespersen, 1949; Larsson *et al.*, 1951). In these calculations it is assumed that the requirement is proportional to the 5/8 power of liveweight (Axelsson, 1943; Larsson *et al*, 1951). It has been confirmed in investigations in the United States (Procter *et al.*, 1934) and elsewhere that maintenance energy requirement per unit liveweight decreases as weight increases.

TABLE 5
Daily Energy Requirement for Maintenance

Authority	Unit of measurement	Liveweight (kg)					
		300	400	500	600	700	800
Kellner and Fingerling (1924)	SE (kg)	2·35	2·8	3·3	3·7	4·1	—
Hansson (1938)	ScFU	2·7	3·6	4·5	5·4	6·3	7·2
Jespersen (1949)*	ScFU	2·9	3·6	4·3	5·0	5·7	6·3
Crasemann (1945)†	SE (kg)	2·1	2·6	3·0	3·4	3·75	4·1
Ehrenberg (1932)‡	SE (kg)	—	—	—	3·0	3·5	4·0
Axelsson (1943)	FUr	4·0	4·8	5·5	6·2	6·8	7·4
Popov (1946)	Russian (oat) FU	4·2	4·8	5·5	6·2	—	—
Morrison (1937)	TDN	—	—	3·8	4·3	4·9	5·5
Larsson *et al.* (1951) §	ScFU	—	*3·9*	*4·5*	*5·0*	*5·5*	*6·0*

* Computed from the formula: requirement in ScFU = 0·03 $W^{4/5}$, where W is the liveweight in kg.
† Computed from the formula: requirement in kcal metabolizable energy = 179·06 $W^{2/3}$, where W is liveweight in kg; 1 kg. SE = 3760 kcal metabolizable energy.
‡ In the computation it is assumed that the energy requirement for maintenance of heavier horses is 5·0 kg SE per 1000 kg liveweight. § Norms recommended by the author.

Table 6 compares the standards in Table 5, all in ScFU. The comparison is based on the assumption that the fodder used is grass hay harvested before flowering.

TABLE 6
Energy Requirement for Maintenance in Scandinavian Feed Units

Authority	Liveweight (kg)					
	300	400	500	600	700	800
Kellner and Fingerling (1924)	3·2	3·9	4·5	5·1	6·0	—
Hansson (1938)	2·7	3·6	4·5	5·4	6·3	7·2
Jespersen (1949)	2·9	3·6	4·3	5·0	5·7	6·3
Crasemann (1945)	2·9	3·5	4·1	4·6	5·1	5·6
Ehrenberg (1932)	—	—	—	4·1	4·7	5·4
Axelsson (1943)	3·0	3·6	4·1	4·6	5·1	5·5
Popov (1946)	3·5	4·0	4·6	5·2	—	—
Morrison (1937)	—	—	3·5	4·0	4·5	5·1
Larsson *et al.* (1951)*	—	3·9	4·5	5·0	5·5	6·0

* Norms recommended by the author.

The highest estimate is that of Popov. Kellner and Fingerling's estimates for light horses and Hansson's for heavy are high. Morrison's standard is relatively low.

Protein

Table 7 shows estimates by different writers for average protein requirements for maintenance.

TABLE 7
Protein Requirement for Maintenance per 100 kg Liveweight

Authority	Digestible protein (g)	Digestible true protein (g)
Hansson (1938)	—	65–70
Jespersen (1949)	—	about 52
Crasemann* (1945)	—	62–80
Nitsche (1939)	53–68	48–59
Ehrenberg (1932) (heavy horses)	—	60
Popov (1946)	—	55–65
Morrison (1937)	60–80	—
Axelsson (1943)	about 77–100	—
Larsson *et al.* (1951)	—	55–70
Norms recommended by the author	65–80	

* Computed from the formula: digestible true protein requirement in g = $5 \cdot 8\ W^{2/3}$ for horses of 400 to 800 kg liveweight, where W = liveweight in kg.

According to most writers, protein requirement is directly proportional to the animal's liveweight. Crasemann (1945), on the other hand, computes that the horse's requirement of true protein for maintenance rises as the 2/3 power of liveweight, but does not regard the point as sufficiently well established. Most authors give the horse's requirement of protein for maintenance in terms of digestible true protein and the standards vary between 48 and 80 g per 100 kg liveweight. Axelsson (1943) and Morrison (1937) express the requirement in terms of digestible protein. Axelsson's standard is higher than Morrison's. The expression of the requirement as digestible protein involves the assumption that amides are at least partly utilized.

REQUIREMENTS FOR WORK

Energy

The energy consumption and output by working horses are calculated by Procter *et al.* (1934). However, the energy requirement of working horses depends on several modifying circumstances, namely the kind of work, the incline of the ground to be worked, the speed of work, the liveweight of the horse, its training and its individuality. In addition, energy requirement may be affected by handling, including harnessing, the care of the hooves and the degree of tiredness, and by climate and weather. Zuntz and Hagemann (1898) studied energy expenditure of horses at work with reference to modifying circumstances. But it is impossible in practice to take account of all such circumstances, especially in such variable work as agriculture. Therefore, energy requirement has been computed for a few classes of work.

Kellner and Fingerling (1924) allocated work to three grades, assuming a horse of 500 kg weight, and including the cost of moving the body:

Light work	in a day about	700,000 kg-m
Medium (normal) work	„ „	1,300,000 „
Heavy work	„ „	2,250,000 „

Other authors (Crasemann, 1945; Hansson, 1938; Høie and Tilrem, 1951) have introduced a fourth grade, very heavy work. Wrede (1932) suggests the following grades of performance for a working horse weighing 500 kg.:

Grade	Mean daily work, including movement of the body (million kg-m)	
Light work	0·0–1·0	mean 0·5
Normal work	1·0–2·0	„ 1·5
Heavy work	2·0–3·0	„ 2·5
Very heavy work	over 3·0	

Ehrenberg (1932) and Jespersen (1949) divide work performance of horses into five grades: very light, light, medium, heavy and very heavy work.

The energy requirement given by different workers for the standard 500 kg horse is shown in Table 8, in which 3 or 4 grades of work are assumed.

The standards given in Table 8, re-computed to Scandinavian feed units on the assumption that the horse is fed on oats of normal composition, are shown in Table 9.

TABLE 8

Energy Requirement above Maintenance for Work, Computed for a Horse of 500 kg Liveweight

Authority	Unit	Grade of work			
		Light	Normal	Heavy	Very heavy
Kellner and Fingerling (1924)	SE (kg)	1·5	2·4	4·2	—
Hansson (1938)	ScFU	2·0	3·3	4·8	Over 5·5
Crasemann (1945)	SE (kg)	1·3	2·0	3·25	4·15
Axelsson (1943)	FUr	1·7	3·4	5·1	
Popov (1946)	Russian (oat) FU	1·9	3·6	6·0	
Morrison (1937)	TDN (kg)	0·9	4·4	7·3	
Larsson *et al.* (1951)*	ScFU	1·6	3·2	4·9	

* Norms recommended by the author.

TABLE 9
Energy Requirement for Work in ScFU Computed for a Horse of 500 kg Liveweight

Authority	Grade of work			
	Light	Normal	Heavy	Very heavy
Kellner and Fingerling (1924)	2·2	3·6	6·3	—
Hansson (1938)	2·0	3·3	4·8	Over 5·5
Crasemann (1945)	1·9	3·0	4·8	6·1
Axelsson (1943)	1·7	3·4	5·1	—
Popov (1946)	1·7	3·3	5·5	—
Morrison (1937)	1·3	2·6	4·3	—
Larsson *et al.* (1951)*	1·6	3·2	4·9	—

* Norms recommended by the author.

It appears from Table 9 that Kellner and Fingerling's standard is highest and Morrison's lowest.

Jespersen (1941a) has given energy standards for work, by grade and per hour of work (Table 10).

TABLE 10

Grade	Addition per hour of work (ScFU)
Very light	0·2
Light	0·3
Medium	0·5
Heavy	0·7
Very heavy	1·0

Recently Tatsumi *et al.* (1958a) have studied the energy expense of horses in different kinds of agriculture work and (1958b) of horses walking in water.

Protein

According to Nitsche (1939) the true protein requirement of the horse, computed per unit of work (0·1 million kg-m), falls as work output increases. He gives the allowances shown in Table 11.

TABLE 11

Million kg-m of work	Digestible true protein (g per 0·1 million kg-m)
0·5–0·6	80
0·6–0·7	81
0·7–0·8	67
0·8–0·9	65
0·9–1·0	59
1·0–1·1	53
1·1–1·2	—
1·2–1·3	50

Meyer (1944), on the other hand, holds that increase in intensity of work has no effect on protein requirement. Morrison (1937) says that the older protein standards are too high. On the other hand, he believes (p. 68) that if the protein allowance is too low, the digestibilities of other nutrients may be reduced (see also p. 12). He suggests that the microorganisms in the digestive tract which normally break down cellulose will attack sugars or starch if the ration is rich in these more readily available carbohydrates. The best ratio of nitrogen-containing to nitrogen-free nutrients for working horses is thought to be 1:9–11 and for non-working horses 1:10–12. Axelsson (1943) takes for work the same protein requirement as for maintenance, namely 75–83 g digestible protein per FUr. Hansson (1938) considered 65–70 g digestible true protein per ScFU sufficient for working horses.

For different grades of work, Ehrenberg (1932) proposes the following supplements of digestible true protein per 1000 kg. liveweight daily:

Grade of work	Digestible true protein (g daily)
Very light	200
Light	300
Medium	600
Heavy	900
Very heavy	1250

Jespersen (1949) gives the true protein requirement of working horses as 60 g per ScFU with variations between 56 and 60 g on the assumption that the energy allowed is not greater than is needed for the work. Larsson *et al.* (1951) assess the working horse's requirement of digestible true protein as 70–75 g per ScFU.

TABLE 12

Protein Requirement in Grams for Different Grades of Work, Computed for a Horse of 500 kg Liveweight

Authority	Grade of work			
	Light	Normal	Heavy	Very heavy
Kellner and Fingerling (1924)*†	500	700	1000	—
Hansson (1938)*	145	220	320	Over 370
Axelsson (1943)‡§	134	269	403	—
Popov (1946)*§	180	335	500	—
Morrison (1937)‡§	113	204	295	—
Larsson *et al.* (1951)*	115	235	355	—
Norms recommended by the author‡	125	260	390	—

* Digestible true protein. † Including digestible true protein for maintenance.
‡ Digestible protein. § Computed values.

Table 12 gives the mean protein standards recommended by different authors for 3 or 4 grades of work. It is evident that agreement is not complete on the requirements for work. For work alone Popov's are highest and Morrison's lowest.

Objections have been made to the energy and protein standards of Kellner and of Hansson on the basis of German experiments under Ehrenberg's guidance. In most of the objections it is argued that both standards are too high (Asam, 1924; Bartsch, 1926; Klose, 1929; Temper, 1927; Teschner, 1927; Wrede, 1932). One of these (Klose, 1929) maintains that Hansson's protein standards of 1926 are too low.

Requirements for Reproduction

Most statements on requirements for reproduction are based on practical experience and judgement, not on the results of exact experiment.

Pregnant Mares

Energy. During the first half of pregnancy no increase in energy allowance above normal is considered necessary. Additional feed for growth of the foetus is not considered necessary until the seventh month.

Some authors (Axelsson, 1943; Popov, 1946) believe that the energy requirement for foetal growth is related to the liveweight of the mare. Others (Larsson *et al.*, 1951) think that no distinction should be made between mares of different weights and that their requirements for growth of the foetus are almost equal (see Table 13). Popov (1946) states the energy requirement for the foetus in terms of the weight increase of the mare in the last months of pregnancy. According to him, the mare puts on 10 to 12 per cent of weight during pregnancy. He considers that per kilogram liveweight the mare needs 4 to 4·5 Russian (oat) FU (3·2–3·6 ScFU). On that basis the requirements for the foetus according to Popov are twice those proposed by Axelsson and by Larsson *et al.* (1951) (see Table 13). Jespersen (1949) thinks the energy allowance of the pregnant mare should be increased by ½ to 2 ScFU daily after 6 months. Dobrynin (1950) found that the plane of nutrition affected the duration of pregnancy. With correct feeding it was on the average 333 days, and with restricted diet 352 days.

Protein. As pregnancy advances the protein requirement of the mare increases. Axelsson (1943) considers that the supplementary feed given to the pregnant mare should have 160 g digestible protein per FUr. Jespersen (1949) recommends 80–100 g digestible true protein per ScFU of total feed. Popov (1946) suggests 90–100 g digestible true protein per Russian (oat) FU as a suitable proportion in the total feed. That corresponds to 112–125 g digestible true protein per ScFU. Olsson (Larsson *et al.*, 1951) recommends from the seventh month of pregnancy 80–85 g digestible true protein per ScFU of total feed, equivalent to 90–95 g digestible protein per ScFU.

The protein content of the ration may affect the mare's fertility. Jespersen (1949) asserts that the mare is less likely to become pregnant if she is given a low-protein diet. He recommends at least 80 g digestible true protein per ScFU for the non-pregnant mare.

TABLE 13

Energy Requirement for Growth of the Foetus

Month of pregnancy	Liveweight of the mare (kg)						Larsson *et al.*‡ (1951)
	500		600		700		
	Additional feed in ScFU daily according to						
	Axelsson* (1943)	Popov† (1946)	Axelsson*	Popov†	Axelsson*	Popov†	
6–7	0·2		0·2		0·2		0·2
8–9	0·5	1·5	0·6	1·8	0·7	2·1	0·5
10–11	0·8		1·0		1·2		0·9
Additional feed required for the whole pregnancy	90·0	180·0	108·0	220·0	126·0	252·0	96·0

* On the assumption that the additional feed consists of the usual concentrates, 1 FUr = 1 ScFU.

† Computed, 1 Russian (oat) FU taken as 0·8 ScFU.

‡ Norms recommended by the author.

Lactating Mares

Energy and protein. The requirements of the mare during lactation depend on the milk yield and the composition of the milk. On the assumption that mare's milk contains on the average (per cent): protein 1·98, fat 1·03 and sugar 6·19, and that the energy value of the fat is 9·11, of the protein 5·86 and of the sugar 3·95 kcal/g, the energy value of mare's milk is about 450 kcal/kg (cow's milk with 4 per cent fat has an energy value of about 750 kcal/kg). That implies that the requirement of energy for production of milk of the above composition is 0·25 ScFU/kg if the mare's efficiency is comparable with that of the dairy cow. The protein requirement for milk production may, in the same way, be about 37 g digestible true protein or 45 g digestible protein per kg milk.

According to Axelsson (1943) 0·28 FUr and 45 g digestible protein are required to produce 1 kg mare's milk. Popov (1946) computes the energy requirement to be 0·33 Russian FU (about 0·26 ScFU) with 30–35 g digestible true protein per kg milk. Ehrenberg (1932) and Prowosnik (1931) consider that 200 g starch equivalent (about 0·28 ScFU) and 40 g digestible true protein are required to produce 1 kg mare's milk. Prowosnik recommends for the lactating mare doing moderate work about 16·5 kg SE, 2·9 kg digestible true protein and 35 kg dry matter daily, per 1000 kg liveweight. On the basis of his own experiments, Ehrenberg (1932) proposed the standards shown in Table 14 for the breeding mare, assuming that 200 g SE and 40 g digestible true protein are required per kg milk.

TABLE 14

Milk production daily (l.)	Requirements per 1000 kg liveweight		
	Digestible true** protein (kg)	Starch equivalent (kg)	ScFU*
1. With light work			
5	1·0	7·5	10·7
10	1·2	8·5	12·2
15	1·4	9·5	13·5
20	1·6	10·5	15·0
25	1·8	11·5	16·4
2. With moderate work			
5	1·4	9·2	13·2
10	1·6	10·2	14·6
15	1·8	11·2	16·0
20	2·0	12·2	17·4
25	2·2	13·2	18·9

* Conversion: 1 kg SE = 1·43 ScFU.
** Conversion: 1 kg digestible true protein = about 1·1 kg digestible (crude) protein.

Stallions

Energy and protein. The services required of the stud stallion may vary widely and that affects the requirements. Axelsson (1943) considers that stud stallions, in addition to the energy for maintenance and work, should have 0·5–1·0 FUr daily. During the breeding season, and depending on the demands made, he recommends the following additional allowances:

Light service	0·5–1·5 FUr	per head daily.
Moderate service	1·5–2·5	,, ,,
Heavy service	2·5–3·5	,, ,,

The total feed should provide 100–125 g digestible protein per FUr and the supplementary feed 200–230 g per FUr. According to Jespersen (1949) a stallion that serves 70–100 times in 4 or 5 months requires about as much energy as when working. The protein requirement for a breeding stallion serving once daily he estimates at 700–800 g and for a stallion serving 3 or 4 times daily 1000 g. Popov (1946) recommends 0·95–1·2 kg digestible true protein and 9·5–10 Russian FU (7·6–8·0 ScFU) daily for stallions of light breeds, with a liveweight of 450–500 kg, and regards these amounts as adequate even for intensive breeding for 3 to 5 months. Ehrenberg (1932) proposes for breeding stallions an increase of the daily allowance by 200 g digestible true protein per 1000 kg liveweight but only a small increase in energy supply. Seidler and Zolkiewski (1957) reported that increased supply of energy and protein to stallions could perform increased numbers of services. For this purpose the authors (1959) recommend use of lucerne, fresh or as hay, oats, beans, peas, meatmeal and skimmed milk. Olsson (Larsson *et al.*, 1951) estimates for breeding stallions with moderate intensity of breeding an increase of 0·5–1·0 ScFU daily, with 100 to 200 g digestible true protein (= 110–220 g digestible protein) per ScFU.

Requirements for Growth

Hansson and Müller (1929) made feeding experiments during the years 1923–8, and recorded rate of growth and feed consumption of foals of light breed. Olsson (1952) has published the results of a study in 1945–50 of the rearing of Ardennes horses. The results of their experiments provide a good guide for the computation of feeding standards for foals of these breeds under Swedish conditions. Studies on the growth rate of the horse are reviewed by Green (1961).

Table 15

*Daily Energy Requirement for Growing Foal in ScFU**

Liveweight (kg)	Hansson and Müller (1929)	Axelsson (1943) Liveweight at 3 years of age		Morrison (1937)	Olsson (1952)† Liveweight at 3 years of age	
		600 kg	800 kg		600 kg	800 kg
250–300	5·8	6·3	7·8	4·4	*4·8*	*5·4*
300–350	6·3	6·5	7·9	4·9	*5·1*	*5·7*
350–400	6·8	6·7	8·0	5·5	*5·4*	*6·0*
400–450	—	6·9	8·2	6·1	*5·7*	*6·3*
450–500	7·2	6·9	8·3	6·6	*6·1*	*6·6*
500–550	7·8	7·0	8·4	7·0	*6·4*	*6·9*
550–600	—	6·9	8·5		*6·5*	*7·2*
600–650	8·4		8·6		*6·6*	*7·5*
650–700			8·7			*7·7*
700–750			8·7			*7·9*
750–800			8·7			

* Provided that the requirement is met by 65 per cent grass hay cut at the bud stage and 35 per cent oats or other cereal. † Norms recommended by the author.

TABLE 19

	Requirements for different rates of weight increase (g daily)		
	1 kg	0·8 kg	0·5 kg
Digestible protein	950–1000	850–900	750–800
Fat	350–370	300–320	250–270
Sugar	350–400	350–400	350–400
Calcium	65–67	60–65	55–60
Phosphorus	32–35	30–33	25–30

FATTENING HORSES FOR MEAT

Recently experiments with fattening horses for meat were started in certain districts of east Europe and Asia (Venjarskij, 1959; Barmincev and Necaev, 1959; Venjarskij, 1959–60; Kulaev, 1961; Tjulegenov, 1964; Borkum, 1964). The experiments concerned foals as well as older horses up to 5 years old or more.

Venjarskij (1959) reported a mean daily weight gain of about 600 g. The yearlings increased their weight by about 43 per cent, and the horses aged 5 years or older by 23 per cent. In-winter fattening of horses of Kirgiz breed Tjulegenov (1964) noted a mean daily gain in weight from 800 to 895 g. The young horses used 6·5 to 7·0 and the adult horses 7·2 to 7·5 FU (Russian) per kg gained. In an experiment with foals of several breeds Borkum (1964) reported a gain from birth to weaning of on an average 810 g daily and the weaning weight at 5–7 months was 210 kg. They used 5·8 FU (Russian) per kg gained.

MINERALS

Horses as a rule require more sodium chloride and more of some other inorganic salts than are present in their common feeds.

Salt (*sodium chloride*)

Requirements for sodium and chlorine depend to some extent on the work done, the temperature and the humidity of the air and other circumstances. The need to add salt depends also on how much is already present in the feed. According to experiments in Russia (Popov, 1946) excretion of NaCl by horses increases with the intensity of work. During a pull of 50 km, 50–60 g NaCl was excreted in sweat and 35 g in the urine, or 85–95 g daily. The diet of 5 kg hay and 5 kg oats provided only 30–35 g NaCl and part of that was lost in the faeces. Consequently the salt balance was negative.

Kellner and Fingerling (1924) recommend 15–25 g salt daily for grown horses, to be mixed with the feed. They regard salt blocks as unsuitable, since horses take too much. According to Liebscher (1934), sudden death as the result of heavy work with excessive sweating is due to loss of chloride. Such a catastrophe may be avoided if the horse has free access to a salt block or is given 30–70 g salt after work. Ehrenberg (1932) suggests 40 g salt per 1000 kg liveweight as a suitable allowance, and adds that an extra supplement of salt should be given when molasses is used in the feed. He, too, does not approve of salt blocks. Dobrynin, Davydova and Skachkov (1949) give the following daily salt allowance for horses:

	Young horses (g)	Grown horses (g)
Stud horses of light breeds	15–30	30–50
Stud horses of heavy breeds and working horses	20–40	40–70

According to Jespersen (1949) horses need 10–15 g salt daily. Olsson (Larsson *et al.*, 1951) recommends the following daily allowances of salt (in g):

Young horses	15–40
Working horses	20–40
Pregnant mares	25–45
Lactating mares	30–45
Stallions in the breeding season	30–45

These standards correspond in general with Axelsson's (1943).

Calcium (Ca) and Phosphorus (P)

Svanberg and Johansson (1936) maintain that deficiency of calcium or phosphorus or both, if large enough, will cause disorders of bone, even in fully grown horses. Where there is plenty of lime, horses are relatively less sensitive than ruminants to lack of phosphorus, but these authors believe that horses' rations are often deficient in phosphorus.

On the basis of experiments by Scheunert, Schattke and Wiese (1923) and Bang (1925), Breirem (1938) computed that daily rations of 12–15 g Ca and 12–15 g P are sufficient for full-grown horses of 500 kg liveweight. He maintains that the Ca:P ratio should not be less than 0·7. Axelsson (1943), on the other hand, suggests 30–40 g Ca daily for full-grown horses. He considers that the Ca:P ratio for working horses may be less than that recommended for young horses, 1·5–1·7.

iron in horses is not known. However, it is supposed that horses ordinarily fed with green fodder, hay and grain, in general, are supplied with 80 ppm or more iron in the dry matter, which is in excess of their requirements.

Copper (*Cu*). It is proved that copper is a necessary element for the animals. Thus it is found that copper and iron together are needed for the haemoglobin synthesis. Cupps and Howell (1949) found that the copper requirement of growing foals is less than 8 ppm of the feed. It is supposed that 5–8 ppm Cu in the feed for horses will be adequate.

Cobalt (*Co*). The requirements for cobalt in horses is seen to be very low, less than 0·05 ppm. It is found, that horses without any trouble have grazed pastures so low in cobalt that cattle and sheep have shown signs of cobalt deficiency. In spite of those observations, it is probable that a cobalt deficiency in the ration of horses may be possible (Cascin and Bogdanov, 1958).

Manganese (*Mn*). The ordinary rations used in the feeding of horses, contain adequate amounts of manganese. Therefore manganese deficiency in horses has never been shown. No estimates of manganese requirements of horses have been carried out.

Zinc (*Zn*). Zinc deficiency in horses has not been shown, and the requirements for zinc in horses are not estimated. The roughages used in the rations of horses in general contain abundant amounts of zinc.

Fluorine (*F*). The metabolism of and the requirements for cobalt and fluorine in horses were studied by Magidov (1962). The effects of fluorine on the dental development are known, but not the effects of excess fluorine in the rations. The highest fluorine content allowed in rations for cattle, sheep and swine to prevent poisoning effects is 0·01 per cent of the dry matter of the ration.

Molybdenum (*Mo*). The effect of different molybdenum contents in the rations for horses are not known. However, it is shown that large amounts of molybdenum in forages does not appear to affect horses with disturbances as it does in cattle and sheep.

Selenium (*Se*). Selenium, except in the minutest traces, is a poisoning element and is commonly not recognised to be an essential nutrient for horses. As a result of studies, particularly in the United States, it was found that soil areas containing more than 5 ppm of selenium may cause poisoning effects, of which 'alkali disease' develops in horses. The symptoms of the affection include loss of hair from the mane and tail and the occurrence of deformed hooves, and in terminal stages of the disease, even occurrence of blindness and paralysis (Moxon, 1937; Walsh *et al.*, 1951; Knott, McCray and Hall, 1958). Recently, Stewart (1960) found that a muscle disorder in thoroughbreds, known as 'being tied up' occurred when large amounts of oats were given, obtained from an area where selenium deficiency had been diagnosed.

VITAMINS

Fat-soluble Vitamins

Little is known of the vitamin needs of horses, but there are some statements in the literature.

Vitamin A. Klemola (1933) showed that army horses which had been fed for a long time on hay, oat straw and out grain without fresh green forage had deformed hooves. The horny layer was unevenly laid down and the horn was brittle. Extra rations of fresh grass, AIV silage or cod-liver oil prevented the abnormality, which was regarded as due to deficiency of vitamin A. Scheunert and Schieblich (1934) and Howell, Hart and Ittner (1941) also have shown that horses are sensitive to lack of vitamin A or carotene.

Edwards (1937–8) discusses disorders in horses caused by deficiency of vitamin A. He assesses the requirement of carotene at about 30 μg per kg liveweight daily.

Guilbert, Howell and Hart (1940) fed horses on fodder of low vitamin A content and from the results estimated the requirement of carotene or vitamin A. The first sign of deficiency was night blindness. If vitamin A was not given then, the signs increased gradually for 3 or 4 months. The coat became rough, the cornea became opaque and daylight vision also was impaired. Instead of xerophthalmia, there was lachrymation. The animal was restless and resentful of handling. Within a month of treatment with vitamin A most of the signs had disappeared. The horse's requirement of vitamin A is estimated at 4·2–5·3 μg per kg liveweight daily or 20–30 μg carotene, for instance, in dried alfalfa. These quantities correspond to 14–18 i.u. vitamin A (alcohol) or 33–50 i.u. carotene and are considered sufficient for growth and to prevent deficiency, but not enough to give any storage in the body. Further, 3 or 4 times as much was thought to be necessary for reproduction and lactation.

Crasemann (1945), on the basis of the requirements of other domestic animals, computed the requirement of a 500 kg horse as 10·0–12·5 mg carotene daily. In the same way, he computes the vitamin D requirement as 0·04–0·08 mg daily.

According to Popov (1946), deficiency of vitamins A and D causes low fertility in mares and abortion of premature birth of weakly foals. Young mares are believed to be more sensitive than older ones to deficiency of vitamins.

Ocsag and Sréter (1955) found that abortions were especially common in mares with serum carotinoid values below 15 μg per cent and highest on farms where the value was 15–20 μg per cent during the winter.

Later, Gondos-Cromicov, Romanovschi and Maxim (1958) reported that high intake of carotene improved fertility, raised the levels of carotene and vitamin A in the milk and reduced morbidity and mortality of foals up to 6 months of age.

The studies of Galichnikova and Kulikova (1951) showed that deficiency of vitamin A produced keratinization of the mucosa of the digestive canal and so reduced glandular secretions. In addition the hydrochloric acid of gastric juice was abnormally low, so that digestion could not proceed normally.

Sokolov and Sekulov (1951) found that a supplement of 50 mg carotene dissolved in 100 ml oil improved the digestibility of a carotene-free ration and the balances of nitrogen, sodium, calcium and phosphorus.

Olsson (Larsson *et al.*, 1951) gives the following allowances for vitamin A:

	Per 100 kg liveweight daily	
	Vitamin A (i.u.)	Carotene (mg)
Foals and young horses	5000–7500 *or*	12–15
Working horses		10–12
Pregnant mares and stud stallions		15–20
Lactating mares		100–150

Vitamin D. The requirements for vitamin D in adult horses are not known. In general, the horses are supplied with sun-cured hay containing large amounts of vitamin D. In addition the horses are supplied with vitamin D obtained by the action of sunlight upon their skin. However, in foals rachitic symptoms have been observed which possibly may depend on vitamin D deficiency. For rachitic foals Morrison (1937) recommends about 30 g fish liver oil daily or an equivalent amount of some other source of vitamin D. Probably adult horses confined with restricted exposure to sunlight may have a need for vitamin D. Under such circumstances the National Academy of Sciences (1961) recommends 300 i.u. of vitamin D per 100 lb (about 660 i.u. per 100 kg) body weight.

Vitamin E. Only a few experiments have been carried out on the vitamin E requirements of horses. Rjabov (1946) gave vitamin E experimentally as wheat germ, 0·5 kg per head daily for 3 or 4 days. The result was oestrus in barren mares and better quality of semen in stallions.

Darlington and Chassels (1956a, b) reported good effects of large doses of α-tocopherol to brood mares and racing horses. In New Zealand Dodd *et al.* (1960) studied muscle degeneration and yellow fat disease in foals. Some affected foals were supplied with up to 2 g of α-tocopheryl acetate daily for several days. Because there was no control it is difficult to assess the value of the treatment. However, judging from the results, the treatment with α-tocopheryl acetate were seen to have had a favourable effect.

'Dry coat' in horses was studied by Marsh (1961). He found this condition in thoroughbred horses, especially imported. These horses were seen to be liable in hot, wet climates to develop failure to sweat, with dullness of the coat and loss of hair. The trouble was stated to be prevented in about 90 per cent of ninety-seven horses treated with 1000 to 3000 'units' of vitamin E daily by mouth for one month, alone or combined with one or other change of environment.

Water-soluble Vitamins

Ascorbic acid. Very few observations are published about the reaction for ascorbic acid (vitamin C) in horses. Though there are no evidences for a requirement of vitamin C in horses, they probably have a need for this vitamin.

Pearson, Sheybani and Schmidt (1943) found that when horses were supplied orally with 6 g of ascorbic acid the ascorbic acid value in the plasma rose by about 50 per cent and that there was a significant increase in urine excretion of ascorbic acid during the following 24 hr, though the total amount excreted in that time was only a small proportion of the amount ingested.

Davis and Cole (1943) reported that ascorbic acid administered parenterally or orally to a single Belgian stallion, consistently promoted an increase in the number and motility of the sperm. In mares the values for plasma ascorbic acid tended to be higher in good than in inferior breeders.

Forenbacher, Maltar and Marzan (1957) found that myoglobinuria in horses could be prevented when the horses were supplied with 2–4 g of vitamin C daily.

Thiamine. It is not certain to what extent microorganisms in the intestine of the horse can synthesize B vitamins as they do in cattle and other ruminants. That deficiency disease could not be demonstrated in horses fed on forage of low vitamin B content under certain circumstances suggests that intestinal synthesis is possible. Under other circumstances deficiency of thiamine has been demonstrated, but that does not prove that there is no possibility of intestinal synthesis in the horse.

Carlström and Hjärre (1939) have shown that deficiency of thiamine can occur in horses fed on grain and hay, both of poor quality. The deficiency was characterized by emaciation in spite of good appetite; the coat became dull and there was anaemia and disturbance of heart action. The emaciation was thought to arise from faulty carbohydrate breakdown, resulting from thiamine deficiency. Health was restored when a supplement of 250 g dried yeast was given daily.

Reports of bracken (*Pteris aquilina*) poisoning (due to a thiaminase) in the horse responding to injections of thiamine have appeared (Roberts, Evans and Evans, 1949; Carpenter, Phillipson and Thomson, 1950; Evans, Evans and Roberts, 1951; Forenbacher, 1953).

Westerlund (1938, 1941) insists that it is important that horses have plenty of thiamine and nicotinic acid. Kjos-Hansen (1944) found that a foal that had convulsions recovered after a week's intravenous injection of thiamine and simultaneous administration by mouth of 'vitamin B'.

Thiamine deficiency in horses also have been reported by Forenbacher (1953) when fed on acid hay containing from 20 to 50 per cent of *Equisetum arvense* or *E. palustre*. Lott (1951) reported successful treatment of horses from Equisetum poisoning with 2500 mg thiamine hydrochloride given intravenously.

Mackay (1961) found that 1000 mg of thiamine given to racehorses by intramuscular or intravenous injection slowed the pulse rate and had a tranquilizing effect. However, from 500 to 2000 mg given by mouth had no observable effect.

According to Olsson (Larsson *et al.*, 1951) thiamine deficiency in horses will not occur if 40–50 g dried yeast is given daily in their feed.

Riboflavin. Riboflavin is a dietary essential for horses. Symptoms of riboflavin deficiency begin with signs of catarrhal conjunctivitis in the eyes, accompanied by photophobia and lachrymation, which can lead to impaired vision and even blindness. Pearson, Sheybani and Schmidt (1944) have found that the riboflavin requirement of horses is met by 44 μg per kg liveweight daily.

Niacin. Pearson and Luecke (1945) reported that nicotinic acid is synthesized by the horse and that special supplies of this vitamin are not needed. Of interest in this connection is, that the horse can convert dietary tryptophan to niacin (Schweigert, Pearson and Wilkening, 1947).

Pantothenic acid. Pearson and Schmidt (1948) supplied twenty-two male Shetland ponies, 9 months old, for periods of 6-12 months with pantothenic acid in doses which varied from 38 to 150 μg per kg body weight. They found that 38 μg per kg body weight was adequate for these animals. The common rations used in the nutrition of the horse in general will contain sufficient quantities of pantothenic acid to fill the requirement.

Vitamin B_{12}. Zeller (1954) found that anaemia, with large numbers of inclusion bodies in the erythrocytes in horses could be cured by treatment with vitamin B_{12}.

Clifford, Henderson and Wilkins (1956) found that 15 μg vitamin B_{12} daily for 8 weeks to mature debilitated horses was followed by a considerable improvement whether the horses previously were supplied with 1 g procaine penicillin daily or not.

The studies by Kureeva (1958, 1960) as regards the vitamin B_{12} metabolism in horses and the influence of nutritional factors and season on the vitamin B_{12} content of blood and milk of horses have given further evidences for the importance of sufficient vitamin B_{12} content in the rations for horses. However, at present it is impossible to give reliable norms for the supply of vitamin B_{12} to horses.

ANTIBIOTICS

The value of antibiotics in the nutrition of horses is for the present not known. Clifford, Henderson and Wilkins (1956) supplied mature debilitated horses with 1 g procain benzyl penicillin daily without any apparent effect. Nor was Henricson (1960) able to prove any effect of a daily supplement of 200 mg of oxytetracycline to north Swedish colts less than a year old. The colts were given a daily ration of oats and bran 2·5 kg hay 5 kg and straw 3 kg.

DIGESTIBILITY OF FEED

The number of digestion experiments made on horses is much less than on ruminants, and still too low to afford a firm basis for the evaluation of feeding stuffs for horses. They provide information on the digestibility of some single feeding stuffs. Table 24 shows the averages for chemical composition and coefficients of digestibility of nutrients in 577 published digestion experiments with horses in which single feeds were studied.

CIRCUMSTANCES WHICH AFFECT DIGESTION IN THE HORSE

Individuality

The individuality of the horse to some extent determines the digestibility of feed. Edin and Eriksson (1929) and Olsson (1943a, b) found that individual differences in digestion of a particular feed were large. In addition, several outside influences affect digestion (Maynard and Loosli, 1956).

Chemical Composition of Feed

It is known that the digestion of feed depends partly on its chemical composition. At the National Animal Experiment Station, Sweden, the results of 1094 published digestion experiments have been summarized (Olsson, Kihlén and Cagell, 1949). From these it appears (Olsson and Ruudvere, 1955) that the relation of the digestibility coefficient for organic matter (y) to percentage of fibre in dry matter (x) of the feeding stuff may be expressed by a curve with the equation

$$y=99{\cdot}71-3{\cdot}066x+0{\cdot}0679x^2-0{\cdot}00056x^3$$

The result is illustrated by a diagram in Fig. 1.

With increase of fibre in dry matter the digestibility of other nutrients is reduced. In the feeds represented, fibre content ranged from 2 to 48 per cent of dry matter. No account was taken of differences in chemical composition of fibre.

From this collection of experimental data it appeared also that digestibility coefficients for protein rose with increase of protein in the feed to about 20 per cent and then fell (cf. pp.954-55). With rising protein content, the digestibility coefficient of N-free extract rose and that for ether extract fell. As the percentage of crude fat in dry matter rose, the digestibility coefficient rose also, as did the coefficients for protein and N-free extract.

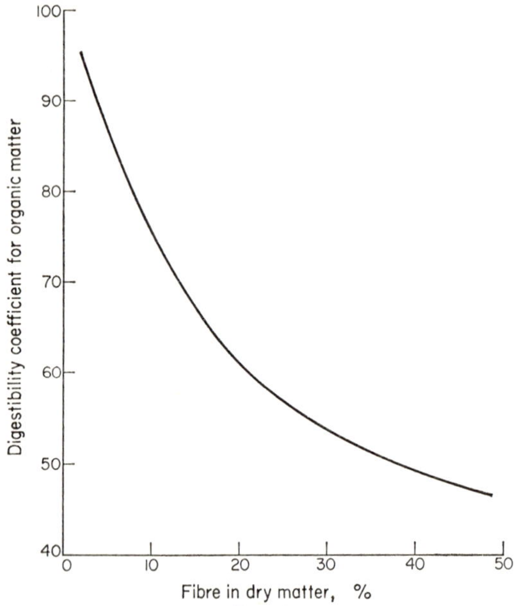

FIG. 1

Digestibility of organic matter, like that of the several nutrients, depends not only on the gross fibre content of the feed, but in large degree on the composition of the fibre, as was shown by Nordfeldt (1954) in experiments on pigs. It is not known to what extent this is true for horses, but there is much evidence in favour of it.

Axelsson (1943) recommends the limits shown in Table 21 to fibre in dry matter for growing horses of different ages, breeding stock and working horses.

Feed Capacity

The digestion of feed depends partly on the amount of feed given. In experiments on heavy-work horses, Hansson (1924) found that hay was better digested when only 6–8 kg per head was given daily. Edin and Eriksson (1929) showed that the amount of feed dry matter has no definite effect on digestion until it exceeds 2 kg per 100 kg liveweight daily.

Rate of Working

From experiments made it appears that digestion may be affected in horses by work or exercise in such a way that it is improved by light exercise and may

TABLE 21

	Fibre in dry matter (%)
Young horses 6–12 months	20–23
12–18 months	22–25
18–24 months	24–27
24–30 months	25–29
30–36 months	25–30
Breeding stock	
Mares in late pregnancy (light work)	25–28
Lactating mares (light work)	18–21
Stud stallions	18–22
Working horses	
Maintenance	27–31
Light work	26–29
Normal work	22–25
Heavy work	18–21

be inhibited by heavy work. That is true, for instance, of the studies of Grandeau and Le Clerc (1884). With digestibility at rest as 100, relative values for different amounts of work were:

Rest	100
Walking without load	106
Walking with load	101
Trotting without load	98
Trotting with load	97

Kellner (1880), on the other hand, could find no difference in the digestion of organic matter by horses doing light and heavy work; but Edin and Eriksson (1929) found that the digestibility of organic matter was less when horses moved very quickly.

Grinding, Water Content and Time of Transit through the Digestive Tract

Grinding feed (crushing or milling) under certain circumstances improves digestion. Morrison (1937) considers that the nutritive value of oats for horses, young and old, is improved by milling. The chopping of roughage seems to have no effect on digestibility (Axelsson, 1940b) or on nutritive value (Hofman-Bang, 1910). According to Ehrenberg (1932) and Jespersen (1949), fodder should not be chopped shorter than 3 cm or it may cause disturbances of digestion.

The water content of feed may affect digestion. Taubert (1934) showed that horses digest feed less well if the water content is high. Wittig (1938) found that much energy is lost in sweating if the intake of water is much in excess of what is normally eliminated in urine and faeces.

Time of transit through the digestive tract has been studied by Edin and Eriksson (1929) and at the National Animal Experiment Station (Olsson, Kihlén and Cagell, 1949). The latter study was made with meadow grass, hay and oat straw with crushed oats as supplement; chromic oxide was used as indicator. The results are shown in Table 22 and Fig. 2.

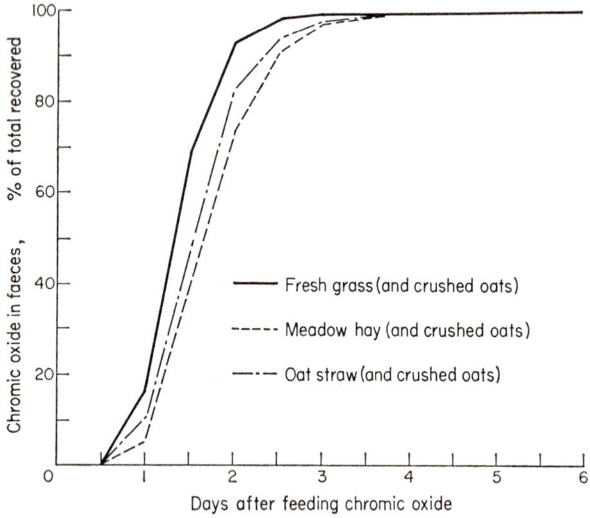

FIG. 2

TABLE 22

Time interval (days)	Percentage—chromic oxide recovered in faeces		
	Fresh grass and oats	Hay and oats	Straw and oats
0–0·5	0·16	0·06	· 0·21
0·5–1·0	16·61	6·47	10·39
1·0–1·5	51·34	32·34	38·14
1·5–2·0	24·73	34·35	33·85
2·0–2·5	5·32	17·84	11·41
2·5–3·0	1·07	6·13	3·79
3·0–3·5	0·27	1·86	1·23
3·5–4·0	0·12	0·59	0·47
4·0–4·5	0·16	0·22	0·26
4·5–5·0	0·11	0·08	0·19
5·0–5·5	0·08	0·04	0·06
5·5–6·0	0·03	0·02	0·00
	100·00	100·00	100·00

From this it appears that freshly cut grass passes through the digestive tract of the horse more rapidly than hay of the same cut. The difference depends, probably, on the difference in water content. Straw moves at approximately the same rate as hay. Hill (1952) confirmed the great variation in time of transit of feed by X-ray with a barium feed.

Bulk

This is determined chiefly by the amount of fodder in the ration and particularly by the fibre present. The more fibre there is in the dry matter of a ration, the greater the bulk.

A new method for digestibility studies of carbohydrates, proteins and fats in animals with fistulas was described by Protasenja (1956).

IMPORTANCE OF DIFFERENT FEEDING STUFFS

The views of different writers about the importance and use of different feeding stuffs in rations for horses differ widely.

Coarse Fodder

Kellner and Fingerling (1924) give the daily allowance of roughage for horses doing light work as 10–15 kg per 1000 kg liveweight. Hansson (1938) recommends 7–8 kg roughage (straw and hay) per horse daily.

For horses doing an average amount of work Simms and Williams (1931) compared the following rations of hay: 3·6, 5·4 and 7·3 kg per head daily, with 5–7 kg oats or oats and ground maize. The weights of the horses were between 540 and 770 kg. From the results, 3·6 kg hay, corresponding to 0·5–0·7 per cent of the liveweight of the horses, appeared to be the minimum desirable amount per day.

Harvey (1935) investigated the reaction of horses to different amounts of roughage in the ration, and found that 0·2 kg hay per 100 kg liveweight was insufficient for horses doing moderate amounts of work, but that 0·6 kg was suitable when the horses had, at the same time, 0·6–1·2 kg oats per 100 kg liveweight daily. More recently other writers (Bünger, 1944; Crasemann, 1945; Hvidsten, 1945a; Karlsen, 1946; Schürch, 1943) have experimented with varying and sometimes very large daily rations of roughage for working horses. The results are not sufficiently clear to indicate the best amount of roughage for horses. According to Earle, Ellis and Greer (1943), the minimum amount of roughage (hay) a horse should have daily is 0·4 kg per 100 kg liveweight.

On the basis of extensive studies on Ardennes working horses, Olsson (1951) found that when they were doing no work or light work they were not satisfied with 5–6 kg roughage (hay and straw); but when they did moderate to heavy work and had more concentrates they were satisfied with that amount of roughage, which corresponds to 0·6–0·7 kg per 100 kg liveweight. Olsson recommends not less than 7 kg for rest or light work.

Pasture and Green Fodder

According to Blechschmidt (1933) a 2-year-old horse of heavy breed in dry weather can eat 40 kg pasture grass. Nietsch (1935) says that a year-old foal on the average will eat 21–38 kg grass daily. But the amount eaten on free grazing may vary within wide limits, depending on air temperature and other climatic differences. According to Schmidt, Patov and Kleisch (1945), working horses will eat 80–110 kg green fodder harvested at an early stage of growth, or 50–60 kg at a later stage of development.

Zakrzewska (1957) reported that foals reared in the open all the time with 1 ha pasture per animal (for 26 months) were larger and healthier and consumed less feed than foals kept in stables and during the grazing season on pasture during the day. Both the groups were given the same feed supplements. In rearing experiments by Dusek (1960) the cost of feeding was higher in winter than in summer by 28 per cent for yearling foals and by 70 per cent for 2- and 3-year-olds.

Changes in the blood of horses under different conditions of pasturing have been studied by Turlin (1961).

Silage

Morrison (1937) says that grass silage can with advantage replace a third or a half of the ration of hay. Popov (1946), on the basis of Russian experiments, states that 6–12 kg is a suitable amount of grass silage for working horses, and 20 kg of maize silage. It is stated that maize silage could be more widely used for horses (Poplavskij, 1958). Axelsson (1943) thinks that the amount of silage should be restricted to 5–6 kg daily; Jespersen (1949) recommends 10–15 kg and Chaljuk (1946) 12–20 kg.

The use of maize silage for pure-bred mares was studied by Caljuk (1964) and the value for horses of silage of maize cobs at the milk-vax stage by Magidov (1956).

Potatoes and Roots

These may be of great importance for feeding horses. The amount will obviously depend on the total bulk and on the work to be done. Kellner and Fingerling (1924) recommend 1·5–2·5 kg raw potatoes per head daily for horses doing intensive and heavy work. For horses doing lighter work and moving slowly, and for 3- to 4-year-old horses, 12 kg potatoes per 1000 kg liveweight daily is a suitable ration. Hansson (1938) gives 5–7 kg potatoes as suitable. On the other hand, according to some writers (Galichnikova and Kusnetsova, 1951; Pärn, 1949) both grown and half-grown horses may be given up to 16 kg per head daily. More may be given of cooked potatoes; Hansson recommends 12–15 kg boiled potatoes daily and Wrede (1932) 40 kg steamed ensiled potatoes per 1000 kg liveweight daily.

The most suitable roots are those relatively high in dry matter. Löwe (1943) uses roots as a substitute for cereal (oat) meal from weaning to 3 years of age without any unfavourable result. Bünger (1944) recommends 20–25 kg roots as the daily maximum for working horses. According to Jespersen (1949), half the ration for horses doing heavy work may be sugar-beet, provided the protein content is 50 g per ScFU. From other Danish experiments (Jespersen, 1939) the conclusion has been drawn that the amount of fodder beet per head daily should be restricted to 25 kg. Hansson (1923) believes that the feeding value of sugar-beet for horses is greater when the daily allowance is limited to 10–12 kg. Popov (1946) is more cautious with roots and proposes for heavy horses 6–8 kg carrots or other roots, or at most 15–20 kg. Working horses doing light to moderate work, according to Asam (1924) may be given 20 kg fodder beet or more. He reckons that 5 kg beet can replace 1 kg oats.

Root Tops

These, too, have been given to working horses. Bartsch (1926) says that up to 20 kg sugar-beet leaves can be given to horses doing strenuous work. If the protein requirement is met, 5 kg sugar-beet leaves can replace 1 kg oats. Temper (1927) came to the same conclusion regarding the use of tops and leaves of roots.

Molasses

Molasses is a valuable feed for horses, especially when it is given with straw as roughage. Popov (1946) approves of the use of molasses, apart from the risk of colic. Views differ on suitable daily allowances. Axelsson (1943) recommends 1·5–2 kg and says that larger amounts cause increased sweating and are laxative. According to Danish experiments (Hofman-Bang, 1917) up to 2·5 kg molasses can be given to working horses, and 4 parts by weight of molasses can replace 5 of oats. In Germany distiller's potato wash has sometimes been used to feed horses; according to Müller (1931), 40 l. may be given daily.

Cereals

Oats is the concentrate most used for horses and is more important in the feeding of horses than any other feeding stuffs. But barley also and other cereals may be used. Jespersen (1949) says the prices of cereals should decide which will be used. Rye should perhaps be excepted because it causes colic. According to Kellner and Fingerling (1924), rye may be used to feed horses but should be mixed with 2 or 3 parts of oats, and Morrison (1937) agrees. On the other hand, Teschner (1927) records that rye in experiments caused no trouble, even when 4·5–6 kg was used for working horses. In his view, rye is a valuable feeding stuff of which 1 kg can replace 1·20–1·25 kg oats.

Sprouted oats or other germinated grains, at least 1 kg daily, are believed to have a good effect on stud stallions (Skatkin, 1950).

Magidov (1955) reported that maize is a valuable feed for horses. However, it should be accompanied by fodder rich in protein, such as legume hay, bran and oats. Magidov says that maize should not be moistened or steamed before feeding, for if this is done it ferments rapidly in the stomach of the horse.

Turancic and Bunjevac (1958) state that it is necessary to crack maize for feeding horses. Caskin and Tjulegenov (1964) has studied the use of maize in rations of horses at race courses.

Yeast

Sokolov, Kostin and Gololobov (1953) found that hydrolyzed yeast is a valuable ingredient in rations for horses to fill their protein requirements. And Magidov *et al.* (1956) reported that trotter horses given yeast from a sulphite-alcohol factory, during hippodrome training gave better performance than the control horses, not supplied with yeast.

Oilseeds

Views on the use of oilseeds in rations for horses vary. Jespersen (1949) advises against rapid change to oilseeds since they may cause digestive disturbances. Sunflower-seed cake is thought to be the best for horses, but if too much is given it may cause disorders. Mustard-seed cake and cottonseed cake are considered unsuitable for horses and so is hempseed cake. But there are other views: Grigorjev and Feodorov (1946) found that 40 per cent of the concentrate ration might be of cottonseed cake without ill effect. Of the concentrate mixture in question, 6-10 kg were given daily for 6 months.

Seidler and Zolkiewski (1955) reported that 0·2 kg of horse beans can replace 1 kg oats in rations for working horses.

Fodder Cellulose

In many countries fodder cellulose was used to feed horses during the Second World War. From the studies made it appears that cellulose is an excellent feed and may be used in relatively large amounts, provided the requirements for protein, vitamins and minerals are met. Digestibility coefficients for crude fibre in fodder cellulose, varying from 88·1 to 93·6, are considerably higher than in untreated cellulose (Olsson, 1943a). Edin (1918) found that 0·95 kg fodder cellulose could replace 1 kg oats. According to Jespersen (1941b), 3 kg fodder cellulose is a suitable daily ration, but Olsson (1943b) found in extensive experiments that medium-weight working horses could without harm be given 4·7 kg fodder cellulose daily, and Hvidsten

(1945a) states that working horses of 600 kg liveweight were given up to 5 kg fodder cellulose daily.

Fishmeal

Caljuk and Runjanceva (1958) found that 200-250 g of fishmeal in the rations for stallions resulted in an increase of semen activity and a considerable increase of survival spermatozoa. The authors pointed out that unless fishmeal for horses is fat-free with little fishy taint, it is necessary to start with small amounts, about 5 g daily.

MAXIMUM DAILY ALLOWANCES OF DIFFERENT FEEDING STUFFS

Of different feeding stuffs, Popov (1946) gives (Table 23) the maximum daily quantities for heavy working horses, on the assumption that the feeds are of good quality.

Table 23

	Kg		Kg
Oats	8·0 or more	Wheat bran	4·0
Maize, barley	6·0	Malt culms	1
Rye, wheat	4·0	Brewer's draff, dried	3
Sorghum	4·0	Distiller's wash, dried	2
Vetches	2·0	Potatoes	8
Linseed and sunflower-seed		Fodder beet	12
cake	3·5		
Hempseed cake	2·5	Carrots	10
Cottonseed cake	1·5	Draff, fresh	12 l.
		Silage	8 kg

A similar table has been compiled by Davydova (1946).

PELLETED FEEDS FOR HORSES

Today the use of complete feed mixtures for horses is very common in certain countries. In general those feed mixtures are pelleted. Sevkovic *et al.* (1959) believe that pelleted feeds would be especially useful for army horses. They tested two types of pellets representing complete rations and one containing concentrates and fed together with hay to army horses. One of the complete rations was found to give the best result. This ration contained (in per cent): meadow hay 47, lucerne meal 5, oats 14, maize 18, bran 5, molasses 5, maize gluten 4, dried yeast 1·2, bonemeal 0·5, and salt 0·3. Riding horses were fed 7 kg pellets daily and draught horses 7·5. The pellets were best eaten dry and not broken.

TABLE 24

Summary of the Results of Digestion Trials with Horses

Feeding stuff	No. of experiments	Dry matter (%)	Chemical composition of dry matter (%)						Coefficient of digestibility				
			Organic matter	Crude protein	Crude fat	N-free extract	Fibre	Ash	Organic matter	Crude protein	Crude fat	N-free extract	Fibre
Green fodder	40	21·0	89·8	17·5	2·6	40·4	29·3	10·2	64·1	77·9	−3·6	77·1	47·2
Hay													
Meadow hay	141	85·3	92·1	10·1	2·6	51·4	28·0	7·9	50·1	52·2	13·5	54·9	42·6
Timothy hay	12	88·8	94·9	6·2	2·5	52·0	34·2	5·1	47·2	40·6	10·2	52·2	42·5
Clover hay	5	89·9	91·6	13·7	2·9	35·9	39·1	8·4	51·5	56·0	28·6	63·7	37·7
Alfalfa and alfalfa mixture hay	28	86·7	91·2	17·4	1·9	41·1	30·8	8·8	57·3	73·0	−14·1	67·0	39·8
Cereal hay	28	90·4	93·2	6·4	58·2		28·6	6·8	52·9	58·9	60·8		34·7
Straw													
Oat straw	20	88·0	92·1	4·7	2·6	43·3	41·5	7·9	43·2	24·6	23·9	41·2	47·7
Wheat straw and chaff	15	84·3	92·2	3·9	1·6	42·8	43·9	7·8	32·0	21·5	79·2	29·8	33·0
Rye straw	6	85·5	92·4	2·5	2·2	42·4	45·3	7·6	47·6	−23·8	78·8	42·6	52·2
Potatoes and roots													
Potatoes	11	22·0	95·1	9·0	0·5	82·7	2·9	4·9	90·9	59·0	—	96·4	48·3
Fodder sugar-beet and sugar-beet	7	18·6	93·0	7·1	0·3	78·7	6·9	7·0	83·8	74·1	—	90·7	11·6
Turnips	6	8·4	91·3	16·5	1·9	61·2	11·7	8·7	74·5	112·7	−85·1	45·7	116·0
Carrots	6	14·8	92·5	9·1	1·4	71·0	11·0	7·5	97·4	72·5	33·8	82·7	102·8
Industrial residues													
Beet pulp	7	84·5	92·0	12·2	0·5	65·5	13·8	8·0	72·6	54·7	—	87·9	30·7
Potato pulp, dried	8	86·3	94·9	4·3	0·5	69·0	21·1	5·1	82·1	−8·5	103·8	92·7	65·8
Sulphate and sulphite cellulose	14	91·9	99·5	0·1	0·6	21·6	77·2	0·5	81·1	—	—	73·1	90·5

TABLE 24 (*continued*)
Summary of the Results of Digestion Trials with Horses

Feeding stuff	No. of experi-ments	Dry matter (%)	Chemical composition of dry matter (%)						Coefficient of digestibility				
			Organic matter	Crude protein	Crude fat	N-free extract	Fibre	Ash	Organic matter	Crude protein	Crude fat	N-free extract	Fibre
Cereals													
Oats	132	86·5	96·1	13·2	5·5	65·1	12·3	3·9	70·0	81·0	70·5	75·5	24·4
Barley	7	85·3	97·0	12·5	2·2	76·9	5·4	3·0	82·7	80·8	34·7	88·3	31·0
Mixed oats and barley	18	85·1	96·7	11·9	3·0	76·0	5·8	3·3	79·8	79·1	39·1	87·8	3·5
Maize	16	85·4	98·4	11·1	4·5	79·9	2·8	1·6	87·7	71·2	62·2	91·3	−48·4
Legumes (beans, sweet lupins, peas)	13	87·3	96·1	35·1	2·1	46·7	12·7	3·9	86·0	87·0	17·6	90·4	67·5
Milling offal													
Wheat bran and wheat feed	15	88·4	92·8	16·9	4·9	60·6	10·4	7·2	54·6	75·3	31·4	60·8	−7·0
Oat bran, oat chaff and oat feed	8	91·4	93·5	4·9	1·8	56·7	30·1	6·5	35·0	77·9	68·9	30·6	34·6
Barley feed	4	89·8	95·2	15·5	3·8	64·7	11·2	4·8	57·8	80·0	−15·1	73·6	−39·8
Oilseeds													
Cottonseed cake meal	4	91·8	93·1	39·4	7·9	29·6	16·2	6·9	65·2	85·8	93·9	56·4	36·3
Linseed cake meal	4	89·7	94·1	36·7	6·8	41·9	8·7	5·9	51·5	84·5	34·8	53·0	−65·7
Soya bean meal	2	87·4	93·7	50·9	0·6	36·7	5·5	6·3	85·8	96·2	—	76·1	84·8

WATER

The water metabolism and water requirements of horses were studied by Caljuk (1961). It was found that mares in the ninth month of pregnancy consumed on an average about 40 kg of water daily, of which about one-sixth was with the feed. Immediately before foaling the intake was about 42·7 kg and in the first month of lactation 57 kg. Fattening horses in average condition consumed about 27 and when they were fat 25 kg of water daily.

Working horses consumed 68 kg water daily when the air temperature was 13·2°C and the relative humidity 72 per cent, 79 kg when the temperature rose to 21·1°C and the relative humidity fell to 58 per cent. Horses not working consumed 37 kg of water daily and increased their intake to 76 kg with moderate and 86 kg with heavy work.

References to Table 24

Green fodder: Ehrenberg, Nieschling and Lipinski (1935); Müntz and Girard (1898); Olsson, Kihlén and Cagell (1949); Taubert (1934); Winkler (1941b); Wittig (1938). *Meadow hay*: Engler (1933); Grandeau and Le Clerc (1886); Grandeau and Alekan (1904); Hagemann (1911a, b); Hofmeister (1866); Hötzel (1931); Lindsey, Beals and Archibald (1926); Müntz (1882); Müntz and Girard (1882); Olsson, Kihlén and Cagell (1949); Tangl (1902, 1905); Tangl and Weiser (1906); Weiser (1906); Wolff (1877, 1878, 1879, 1881, 1884, 1887); Wolff and Kreutzhage (1895). *Timothy hay*: Lindsey, Beals and Archibald (1926); Patterson (1879). *Clover hay*: Wolff (1881, 1884). *Alfalfa and alfalfa mixture hay*: Lindsey, Beals and Archibald (1926); Müntz and Girard (1898); Winkler (1940, 1941a); Wolff (1878, 1884). *Cereal hay*: Perkins *et al.* (1914). *Oat straw*: Dunbar (1925); Olsson, Kihlén and Cagell (1949); Wolff and Kreutzhage (1895). *Wheat straw and chaff*: Müntz (1882); Olsson, Kihlén and Cagell (1949); Wolff (1877, 1879). *Rye straw*: Hötzel (1931); Weiser and Zaitschek (1921). *Potatoes*: Engler (1933); Naumann (1940); Nieschling (1934); Olsson, Kihlén and Cagell (1949); Wolff (1884); Wrede (1932). *Fodder sugar-beet and sugar-beet*: Olsson, Kihlén and Cagell (1949); Scholz (1933). *Turnips*: Günther (1934); Olsson, Kihlén and Cagell (1949). *Carrots*: Müntz and Girard (1883); Nieschling (1935); Olsson, Kihlén and Cagell (1949); Wolff (1884). *Beet pulp*: Olsson, Kihlén and Cagell (1949). *Potato pulp dried*: Bizer (1940); Ehrenberg and Lachmann (1942); Newerla and Müller (1943). *Sulphate and sulphite cellulose*: Olsson (1943b). *Oats*: Grandeau and Le Clerc (1888); Lindsey, Beals and Archibald (1926); Müntz (1882); Müntz and Girard (1883); Nitsche (1939); Olsson, Kihlén and Cagell (1949); Patterson (1879); Tangl (1902, 1905); Wolff (1877, 1879, 1884, 1887); Wolff and Kreutzhage (1895). *Barley*: Lindsey, Beals and Archibald (1926); Müntz and Girard (1884); Patterson (1879). *Mixed oats and barley*: Olsson, Kihlén and Cagell (1949). *Maize*: Lindsey, Beals and Archibald (1926); Müntz (1882); Patterson (1879); Winkler (1940); Wolff (1879, 1887). *Legumes (beans, sweet lupins, peas)*: Müntz (1882); Müntz and Girard (1883); Toepsch (1937); Wolff (1879, 1881). *Wheat bran and wheat feed*: Hagemann (1911a); Lindsey, Beals and Archibald (1926); Müntz (1882); Olsson, Kihlén and Cagell (1949); *Oat bran, oat chaff and oat feed*: Lahtorp and Bohstedt (1938); Lindsey, Beals and Archibald (1926). *Barley feed*: Hötzel and Müller (1933). *Cottonseed cake meal*: Lindsey, Beals and Archibald (1926). *Linseed cake meal*: Lindsey, Beals and Archibald (1926). *Soya bean meal*: Olsson, Kihlén and Cagell (1949).

REFERENCES

Alexander, F. (1954) *Brit. Vet. J.* **110**, 146, 196.
Asam, M. (1924) *J. Landwirtsch.* **71**, 16.
Axelsson, J. (1940a) *Biedermanns Zentralbl. (B) Tierernährung* **12**, 414.

Axelsson, J. (1940b) *Djurnäringens grunder* Vol. 2. Nordisk Rotogravyr, Stockholm.
Axelsson, J. (1941) *Biedermanns Zentralbl. (B) Tierernährung* **13**, 399.
Axelsson, J. (1943) *Hästarnas utfodring och skötsel.* Nordisk Rotogravyr, Stockholm.
Bang, O. (1925) *Kgl. vet. Landbohöjsk. Aarsskr.* 383.
Barmincev, Ju. and Necaev, I. (1959) *Konevodstvo*, **29**, No. 9, 21.
Bartsch, E. (1926) *Landwirtsch. Jahrb.* **63**, 157.
Bizer, E. (1940) *Biedermanns Zentralbl. (B) Tierernährung* **12**, 217.
Blechschmidt, H. (1933) *Landwirtsch. Jahrb.* **77**, 463.
Borkum, V. (1964) *Konevodstvo*, No. 10, 11.
Breirem, K. (1938) *Vitaminer og mineralstoffer i husdjurens ernaering.* Gröndahl & Sons, Oslo.
Bünger, H. (1944) *Mitt. Landwirtsch.* **59**, 115.
Caljuk, E. A. (1961) *Trudy Vses Inst. Konevodstvo* **23**, 295.
Caljuk, E. (1964) *Konevodstvo*, No. 1, 13.
Caljuk, E. and Runjanceva, V. (1958) *Konevodstvo*, 28, No. 6, 42.
Carlström, B. and Hjärre, A. (1939) *Biedermanns Zentralbl. (B) Tierernährung* **11**, 121.
Carpenter, K. J., Phillipson, A. T. and Thomson, W. (1950) *Brit. Vet. J.* **106**, 292.
Caskin, I. and Bogdanov, P. (1958) *Konevodstvo*, 28, No. 5, 22.
Caskin, I. and Tjulegenov, D. (1964) *Konevodstvo*, No. 6, 29.
Chaljuk, E. (1946) *Konevodstvo*, No. 10/12, 31.
Clifford, R. J., Henderson, G. N. and Wilkins, J. H. (1956) *Vet. Rec.* **68**, 48.
Crasemann, E. (1945) *Landwirtsch. Jahrb. Schweiz* **59**, 504.
Cupps, P. T. and Howell, C. E. (1949) *J. Animal Sci.* **8**, 286.
Darlington, F. G. and Chassels, J. B. (1956a) *The Summary* **8**, 1, 10. Nat. Stud Farm, Oshawa, Canada.
Darlington, F. G. and Chassels, J. B. (1956b) *The Summary* **8**, 55, 71. Nat. Stud Farm, Oshawa, Canada.
Davis, G. K. and Cole, C. L. (1943) *J. Animal Sci.* **2**, 53.
Davydova, L. P. (1946) *Konevodstvo*, No. 10/12, 21.
Derevlev, A. (1946) *Konevodstvo*, No. 4/5, 20.
Dobrynin, V. P. (1950) *Konevodstvo*, No. 3, 23.
Dobrynin, V. P., Davydova, L. P. and Skachkov, M. E. (1949) *Konevodstvo*, No. 6, 19.
Dodd, D. C., Blakely, A. A., Thornbury, R. S. and Dewes, H. F. (1960) *N.Z. Vet. J.* **8**, 45.
Dunbar, B. A. (1925) *S. Dakota State Coll. Agric. Exp. Stat. Bull.*, No. 212.
Dusek, J. (1960) *Zivocisna Vyroba* **5**, 527.
Earle, I. P., Ellis, N. R. and Greer, C. S. (1943) *Army Vet. Bull. Washington*, No. 37, 95.
Edin, H. (1918) *Centralanst. Försöksväs. Jordbruksområd. Medd.*, No. 168, 11.
Edin, H. and Eriksson, K. (1929) *Centralanst. Försöksväs. Jordbruksområd. Medd.*, No. 365, 39.
Edin, H., Helleday, T. and Nordfeldt, S. (1941) *Lantbrukshögsk. Husdjursförsöksanst. Medd.*, No. 6, 42.
Edwards, F. F. (1937–8) *J. Roy. Army Vet. Corps* **9**, 60.
Ehrenberg, P. (1932) *Arb. deutsch. Gesellsch. Züchtungskunde*, Heft 52.
Ehrenberg, P. and Lachmann, G. (1942) *Biedermanns Zentralbl. (B) Tierernährung* **14**, 112.
Ehrenberg, P., Nieschling, H. and Lipinski, G. (1935) *Züchtungskunde* **10**, 321.
Engler, G. (1933) *Landwirtsch. Jahrb.* **78**, 421.
Evans, E. T. R., Evans, W. C. and Roberts, H. E. (1951) *British Vet. J.* **107**, 364.
Food and Agriculture Organization of the United Nations (FAO) (1947, 1952) *Yearbook of Food and Agriculture Statistics.*
Forenbacher, S. (1953) *Bull. Sci. Yugoslav.* **1**, 87.
Forenbacher, S., Maltar, J. and Marzan, B. (1957) *Vet. Arhiv.* **27**, 1.
Galichnikova, L. A. and Kulikova, Y. V. (1951) *Konevodstvo*, No. 7, 21.
Galichnikova, L. A. and Kusnetsova, V. M. (1951) *Konevodstvo*, No. 3, 34.
Gondos-Gromicov, M., Romanovschi, S. and Maxim, V. (1958) *Lucr. Stiint. Inst. Cercet. Zooteh. Bukarest*, **16**, 517.

GRANDEAU, L. and ALEKAN, A. (1904) *Ann. Sci. Agronom. Paris*, 2 série, 9 année, **1**, 30, 330.
GRANDEAU, L. and LE CLERC, A. (1884) *Ann. Sci. Agronom. Paris*, 1 année, **2**, 325.
GRANDEAU, L. and LE CLERC, A. (1886) *Ann. Sci. Agronom. Paris*, 3 année, **2**, 351.
GRANDEAU, L. and LE CLERC, A. (1888) *Ann. Sci. Agronom. Paris*, 5 année, **2**, 211.
GREEN, D. A. (1961) *Brit. Vet. J.* **117**, 181.
GRIGORJEV, P. E. and FEODOROV, W. (1946) *Konevodstvo*, No. 3, 18.
GUILBERT, H. R., HOWELL, C. E. and HART, G. H. (1940) *J. Nutrit.* **19**, 91.
GUNTHER, H. (1934) *Landwirtsch. Jahrb.* **80**, 723.
HAGEMANN, O. (1911a) *Landwirtsch. Jahrb. 41*, Ergänzungsband, **1**, 129.
HAGEMANN, O. (1911b) *Pflügers Arch.* **137**, 571.
HANSSON, N. (1923) *Centralanst. Försöksväs. Jordbruksområd. Medd.*, No. 251, 30.
HANSSON, N. (1924) *Centralanst. Försöksväs. Jordbruksområd. Medd.*, No. 253, 36.
HANSSON, N. (1938) *Husdjurslära*, Vol. 2. C. E. Fritzes Förlag, Stockholm.
HANSSON, N. and MÜLLER, W. (1929) *Centralanst. Försöksväs. Jordbruksområd. Medd.*, No. 351.
HARVEY, A. L. (1935) *Amer. Soc. Animal Prod., Proc. 27th Annual Meeting* 97.
HARVEY, A. L., THOMAS, B. L., CULBERTSON, C. C. and COLLINS, E. V. (1943) *J. Animal Sci.* **2**, 103.
HENRICSON, B. (1960) *Vet. Rec.* **72**, 515. Dept. Animal Nutrit. Genetics Hyg., Royal Vet. Coll. Stockholm.
HILL, H. (1952) *Arch. Tierernährung, Beihafte*, No. 3, 64.
HOFMAN-BANG, N. O. (1917) *Forsogslab. Kobenhavn Beretn.*, No. 96, 31.
HOFMAN-BANG, N. O. (1910) *Forsogslab. Kobenhavn Beretn.*, No. 72, 39.
HOFMEISTER, V. (1866) *Landwirtsch. Versuchsstat.* **8**, 99.
HÖIE, J. and TILREM, H. (1951) *Husdjurslaere*. Grondahl & Sons, Oslo.
HÖTZEL, K. (1931) *Landwirtsch. Jahrb.* **73**, 93.
HÜTZEL, K. and MÜLLER, H. (1933) *Biedermanns Zentralbl. (B) Tierernährung* **5**, 398.
HOWELL, C. E., HART, G. H. and ITTNER, N. R. (1941) *Amer. J. Vet. Res.* **2**, 60.
HVIDSTEN, H. (1945a) *Norges Landbrukshögsk. Foringsforsök. Beretn.*, No. 59, 41.
HVIDSTEN, H. (1945b) *Norges Landbrukshögsk. Foringsforsök. Beretn.*, No. 60, 38.
INSTITUT, INTERNATIONAL D'AGRICULTURE. *Annuaire internat. Statistique Agric.* (1911-12, 1922 1931-2).
JESPERSEN, J. (1939) *Forsögslab. Köbenhavn Beretn.*, No. 183, 17.
JESPERSEN, J. (1941a) *Maanedsskr. Dyrlaeger* **53**, 409.
JESPERSEN, J. (1941b) *Forsogslab. Kobenhavn Beretn.*, No. 195, 18.
JESPERSEN, J. (1949) *Hestens avl og fodring*. Kgl. Vet. Landbohöjskole, Copenhagen.
KARLSEN, G. (1946) *Nutrit. Abst. Rev.* **16**, 223.
KELLNER, O. (1880) *Landwirtsch. Jahrb.* **9**, 651.
KELLNER, O. and FINGERLING, G. (1924) *Die Ernährung der landwirtschaftlichen Nutztiere*, 10th ed. Paul Parey, Berlin.
KJOS-HANSEN, J. (1944) *Norsk. Vet. Tidsskr.* **56**, 215.
KLEMOLA, V. (1933) *Biedermanns Zentralbl. (B) Tierernährung* **5**, 657.
KLOSE, K. H. (1929) *Landwirtsch. Jahrb.* **69**, 177.
KNOTT, S. G., MCCRAY, C. W. R. and HALL, W. T. K. (1958) *Queensland J. Agric. Sci.* **15**, 43.
KULAEV, V. (1961) *Konevodstvo*, No. 8, 6.
KUREEVA, N. (1960) *Konevodstvo*, No. 1, 28.
KUREEVA, N. (1958) *Konevodstvo*, 28, No. 6, 44.
LAHTORP, A. W. and BOHSTEDT, G. (1938) *Univ. Wisconsin Agric. Exp. Stat. Res. Bull.*, No. 135.
LARSSON, S., OLSSON, N., JARL, F. and OLOFSSON, N. E. (1951) *Husdjurslära*, Vol. 2. L. T.:s Förlag, Stockholm.
LIEBSCHER, W. (1934) *Wien. Landwirtsch. Ztg.* **84**, 210.
LINDSEY, J. B., BEALS, C. L. and ARCHIBALD, J. G. (1926) *J. Agric. Res.* **32**, 569.
LOTT, D. G. (1951) *Canad. J. Comp. Med.* **15**, 274.
L¨WE, H. (1943) *Ztschr. Züchtung (B) Tierzücht. u. Züchtungsbiol.* 55, 171.

MACKAY, A. (1961) *N.Z. Vet. J.* **9**, 129.
MAGIDOV, G. A. (1962) *Sel'hozizdat.* Moscow, 142.
MAGIDOV, G. (1955) *Konevodstvo* **25**, No. 5, 31.
MAGIDOV, G. (1956) *Konevodstvo* **62**, No. 1, 24.
MAGIDOV, G., TUMANOVA, E., KUREEVA, N. and STAROVA, V. (1956) *Konevodstvo* **26**, No. 5, 24.
MARSH, J. H. (1961) *Vet. Rec.* **73**, 1124, 1127.
MAYNARD, L. A. and LOOSLI, J. K. (1956) *Animal Nutr.* McGraw-Hill Book Co. Inc.
MEYER, L. (1944) *Mitt. Landwirtsch.* **36**, 808.
MITCHELL, H. H. and MCCLURE, F. J. (1939) *Bull. Nat. Res. Counc. U.S.A.*, No. 99.
MORRISON, F. B. (1937) *Feeds and Feeding.* Handbook for the Student and Stockman, 20th ed. Ithaca, N.Y.
MOXON, A. L. (1937) *S.D. Agr. Expt. Sta. Bull.* **311**, 91.
MÜLLER, H. G. (1931) *Landwirtsch. Jahrb.* **73**, 169.
MÜNTZ, A. (1882) *Ann. Inst. nat. agronom.*, No. 5, 5 année, 195.
MÜNTZ, A. and GIRARD, A. C. (1882) *Ann. Inst. nat. agronom.*, No. 5, 4 année, 229.
MÜNTZ, A. and GIRARD, A. C. (1883) *Ann. Inst. nat. agronom.*, No. 8, 7 année, 161, 183.
MÜNTZ, A. and GIRARD, A. C. (1884) *Ann. Inst. nat. agronom.*, No. 9, 8 année, 91.
MÜNTZ, A. and GIRARD, A. C. (1898) *Ann. agronom.*, **24**, 5.
NATIONAL ACADEMY OF SCIENCES—NATIONAL RESEARCH COUNCIL, COMMITTEE ON ANIMAL NUTRITION (1961) *Nutrient Requirements of Domestic Animals* 6.
NAUMANN, K. (1940) *Ztschr. Tierernährung Futtermittelk.* **3**, 193.
NEWERLA, H. and MÜLLER, H. (1943) *Biedermanns Zentralbl. (B) Tierernährung* **15**, 193.
NIESCHLING, H. (1934) *Ztschr. Spiritusindust.* **57**, 213, 222, 228, 231.
NIESCHLING, H. (1935) *Ztschr. Züchtung (B) Tierzücht. u. Züchtungsbiol.* **32**, 331.
NIETSCH, H. (1935) *Landwirtsch. Jahrb.* **81**, 525.
NITSCHE, H. (1939) *Biedermanns Zentralbl. (B) Tierernährung* **11**, 214.
NORDFELDT, S. (1954) *Kgl. Lantbrukshögsk. Ann.* **21**, 1.
OCSAG, I. and SRETER, F. (1955) *Rep. Fac. Animal Husb. Hung. Univ. Agric. Sci.* **1**, No. 2, 3.
OLSSON, N. (1943a) *Biedermanns Zentralbl. (B) Tierernährung* **15**, 421.
OLSSON, N. (1943b) *Lantbrukshögsk. Husdjursförsöksanst. Medd.*, No. 12.
OLSSON, N. (1949) *Kgl. Lantbrukshögsk. Ann.* **16**, 644.
OLSSON, N. (1951) *Kgl. Lantbrukshögsk. Statens Husdjursförsök. Medd.*, No. 47.
OLSSON, N. (1952) *Kgl. Lantbrukshögsk. Statens Husdjursförsök. Medd.*, No. 49.
OLSSON, N., KIHLEN, G. and CAGELL, W. (1949) *Kgl. Lantbrukshögsk. Husdjursförsöksanst. Medd.*, No. 36.
OLSSON, N. and RUUDVERE, A. (1955) *Nutrition Abst. and Revs.* **25**, 1.
PÄRN, M. A. (1949) *Konevodstvo*, No. 5, 46.
PATTERSON, H. J. (1879) *Maryland Agric. Exp. Stat. Bull.*, No. 51.
PEARSON, P. B. (1947) *J. Dairy Sci.* **30**, 73.
PEARSON, P. B. and LUECKE, R. W. (1945) *Arch. Biochem.* **6**, 63.
PEARSON, P. B. and SCHMIDT, H. (1948) *J. Animal Sci.* **7**, 78.
PEARSON, P. B., SHEYBANI, M. K. and SCHMIDT, H. (1943) *J. Animal Sci.* **2**, 175.
PEARSON, P. B., SHEYBANI, M. K. and SCHMIDT, H. (1944) *Arch. Biochem.* **3**, 467.
PERKINS, A. J., PHILLIPS, J. H., SPAFFERD, W. J. and MAY, N. S. (1914) *Dept. Agric. S. Austral. Bull.*, No. 82.
POPLAVSKIJ, I. (1958) *Konevodstvo*, **28**, No. 1, 36.
POPOV, I. S. (1946) *Kormlenie sel'skohozjaistvennyh ziwotnyh.* Seljhozgiz, Moscow.
PROCTER, R. C., BRODY, S., JONES, M. W. and CHITTENDEN, D. W. (1934) *Missouri Agric. Exp. Stat. Res. Bull.*, No. 209.
PROTASENJA, T. P. (1956) *Fiziol. Z. SSSR Secenova* **42**, 430.
PROWOSNIK, L. (1931) *Biedermanns Zentralbl. (B) Tierernährung* **3**, 35.
RJABOV, M. H. (1946) *Konevodstvo*, No. 3, 16.
ROBERTS, H. E., EVANS, E. T. R. and EVANS, C. W. (1949) *Vet. Rec.* **61**, 549.
SCHECKER, H. (1943) *Berl. Münch. tierärztl. Wochenschr.* **54**, 397.
SCHEUNERT, A., SCHATTKE, A. and WIESE, M. (1923) *Biochem. Ztschr.* **139**, 1.
SCHEUNERT, A. and SCHIEBLICH, M. (1934) *Biedermanns Zentralbl. (B) Tierernährung* **6**, 133.

SCHMIDT, J., PATOV, C. and KLEISCH, J. (1945) *Züchtung Ernährung und Haltung der landwirtschaftlichen Haustiere*. Bes. Teil. Paul Parey, Berlin.
SCHOLZ, H. A. (1933) *J. Landwirtsch.* **81**, 304.
SCHWEIGERT, B. S., PEARSON, P. B. and WILKENING, M. C. (1947) *Arch. Biochem.* **12**, 139.
SCHÜRCH, A. (1943) *Tech. Indust. u. schweiz. Chem. Ztg.*, No. 1/2.
SEIDLER, S. and ZOLKIEWSKI, A. (1955) *Rocz. Nauk rol. (B)* **69**, 381.
SEIDLER, S. and ZOLKIEWSKI, A. (1957) *Rocz. Nauk rol. (B)*, **71**, 689.
SEIDLER, S. and ZOLKIEWSKI, A. (1959) *Rocz. Nauk rol. (B)*, **73**, 789.
SEVKOVIC, N., DJUKIC, R., STAPAR, N. and BLANKA, I. (1959) *Vet. Glasnik* **13**, No. 1, 9.
SIMMS, J. A. and WILLIAMS, J. O. (1931) *Storrs Agric. Exp. Stat. Bull.*, No. 173, 28.
SKATKIN, P. H. (1950) *Konevodstvo*, No. 1, 22.
SOKOLOV, A. W. and SEKULOV, J. (1951) *Konevodstvo*, No. 9, 19.
SOKOLOV, A. V., KOSTIN, A. A. and GOLOLOBOV, A. G. (1953) *Konevodstvo*, No. 6, 20.
STEWART, J. M. (1960) *N.Z. Vet. J.* **8**, 82.
SVANBERG, O. and JOHANSSON, I. (1936) *Mineralämnen och vitaminer i husdjurens utfodring*. Part 1. Nordisk Rotogravyr, Stockholm.
TANGL, F. (1902) *Landwirtsch. Versuchsstat.* **57**, 329-381.
TANGL, F. (1905) *Landwirtsch. Jahrb.* **34**, 1.
TANGL, F. and WEISER, S. (1906) *Landwirtsch. Jahrb.* **35**, 159.
TATSUMI, H., CHIBA, H., KATO, M., UENO, K., OKABE, T., SETO, Y. and KANO, Y. (1958a) *Bull. Nat. Inst. Agric. Sci. Japan*, No. 14, 81.
TATSUMI, H., CHIBA, H. and UENO, K. (1958b) *Bull. Nat. Inst. Agric. Sci. Japan*, No. 15, 27
TAUBERT, F. (1934) *Biedermanns Zentralbl. (B) Tierernährung* **6**, 173.
TEMPER, K. (1927) *Landwirtsch. Jahrb.* **66**, 297.
TESCHNER, M. (1927) *Landwirtsch. Jahrb.* **66**, 361.
TJULEGENOV, D. (1964) *Konevodstvo*, No. 1, 7.
TOEPSCH, H. (1937) *Landwirtsch. Jahrb.* **84**, 493.
TURANCIC, V. and BUNJEVAC, S. (1958) *Vet. Glasnik* **12**, 163.
TURLIN, A. A. (1961) *Vestn. sel'skohoz. Nauki*, No. 2, 66.
UNITED NATIONS (1955-64) *Statistical Yearbook*.
VENJARSKIJ, A. D. (1959-60) *Sborn. nauc. Trud. Semipalatinsk. zooteh.-vet. Inst.*, **2**, 108.
VENJARSKIJ, D. O. (1959) *Konevodstvo*, No. 7, 16.
WALSH, T., FLEMING, G. A., O'CONNOR, R. and SWEENEY, A. (1951) *Nature* **168**, 881.
WEISER, S. (1906) *Landwirtsch. Jahrb.* **35**, 224.
WEISER, S. and ZAITSCHEK, A. (1921) *Landwirtsch. Versuchsstat.* **97**, 57.
WESTERLUND, A. (1938) *Lantbrukarnas fick-kalender*, 7*:de årg*. 146. Gullviks Fabriks AB, Malmö.
WESTERLUND, A. (1941) *Lantbrukarnas fick-kalender*, 10*:de årg*. 199. Gullviks Fabriks AB, Malmö.
WINKLER, G. (1940) *Biedermanns Zentralbl. (B) Tierernährung* **12**, 22.
WINKLER, G. (1941a) *Biedermanns Zentralbl. (B) Tierernährung* **13**, 124.
WINKLER, G. (1941b) *Ztschr. Tierernährung Futtermittelk.* **5**, 113.
WITTIG, H. (1938) *Biedermanns Zentralbl. (B) Tierernährung* **10**, 496.
WOLFF, E. (1877) *Landwirtsch. Versuchsstat.* **20**, 125.
WOLFF, E. (1878) *Landwirtsch. Versuchsstat.* **21**, 19.
WOLFF, E. (1879) *Landwirtsch. Jahrb.* **8**, Suppl. I, 6, 34, 73.
WOLFF, E. (1881) *Landwirtsch. Jahrb.* **10**, 559, 585, 594.
WOLFF, E. (1884) *Landwirtsch. Jahrb.* **13**, 245, 257, 271.
WOLFF, E. (1887) *Landwirtsch. Jahrb.* **16**, Suppl. III, 1, 49.
WOLFF, E. and KREUTZHAGE, C. (1895) *Landwirtsch. Jahrb.* **24**, 125.
WREDE, W. (1932) *Landwirtsch. Jahrb.* **75**, 869.
ZAKRZEWSKA, G. (1957) *Rocz. Nauk rol. (B)*, **71**, 715.
ZELLER, R. (1954) *Berl. Münch. Tierärztl. Wochenschr.* **67**, 270.
ZUNTZ, N. and HAGEMANN, O. (1898) *Landwirtsch. Jahrb.* 27, Ergänzungsband **3**, 1.

CHAPTER 26

Rearing Young Pigs

A. S. JONES

Rowett Research Institute, Bucksburn, Aberdeen

INTRODUCTION

When pigs are allowed to suckle and are not removed at birth, the suckling period can vary from 1 to 70 days. In current farm practice weaning takes place at 56 days. The artificial rearing of pigs from birth is rarely done. It has, however, been used as part of a disease irradication programme in which it is essential to sever the contact between generations. In this chapter the problems of rearing pigs both naturally and artificially are considered. Consideration is given to the growth of the young piglet following different methods of rearing. Because the deposition of fatty tissues generally begins to exceed the deposition of lean tissue at a weight of about 65 kg the effects of rearing methods on growth and fattening subsequent to the body weight of 65 kg have not been considered.

THE ARTIFICIAL REARING OF PIGS

There is very little transfer of antibodies through the placenta in the sow[1] so that the ingestion of colostrum is important for the conference of passive immunity. The gut wall of the neonatal pig is a permeable membrane through which large protein, and other molecules are absorbed by pinocytosis, and this activity of the gut ceases after colostrum has been ingested. The cessation of pinocytosis is referred to here as 'closure'.

Closure and the Nature of Absorption

A number of studies have been undertaken to determine ways in which large molecules are absorbed from the gut and the factors responsible for the duration of the period during which such absorption occurs. Young and Underdahl (1949)[2] determined the haemaglutinin inhibitor titer of piglets suckling sows immunized to swine influenza and concluded that after about 8 hours some physiological change had occurred in the intestinal epithelium that retarded further absorption of the haemaglutinin inhibitor substances.

Other studies have shown that very little absorption takes place after periods of up to 48 hours.[3-7]

Lecce and Morgan (1962)[8] gave newborn pigs cow's colostrum at various times after birth, those being given colostrum at a late time having received water only. They found that the feeding of between 300 and 400 ml of cow's colostrum caused closure in 24 hours, but that when only water was given absorption could still take place after 86 hours. Pigs given only water have been shown to have marked absorption of gamma globulin even after 106 hours.[9]

The absorption is non-selective in that a range of molecules, for example ovalbumin or polyvinyl pyrrolidone can be demonstrated in sera 6 hours after the introduction of these molecules into the stomach.[8,10,11] Recently it has been shown[11] that boiled colostral whey, devoid of detectable amounts of fat but containing traces of protein, could cause closure. Small amounts of colostrum cause closure, but when piglets were given simple salt solutions

$$Na, K, PO_4, Cl \text{ ions}$$

$$\text{or}$$

$$Na^+, K^+, PO_4^{\equiv}, Cl$$

alone, or containing 1·5 per cent porcine serum protein—either globulin or albumin—or 8 per cent of egg white protein closure had not occurred at 24 hours. A dialyzate from cow's colostrum or from skin milk caused rapid closure. It thus appears that protein, carbohydrates, and minerals are not by themselves responsible for closure. A factor present in colostrum which is heat stable and has a molecular weight between 10^4 and 2×10^4 has been postulated as responsible for closure. Recently,[12] however, it has been claimed that glucose alone will cause closure and the problem is by no means solved.

While the gut is permeable to large molecules, it is probably also permeable to bacteria. This is undoubtedly the reason why heavy losses associated with *E. coli* are encountered amongst piglets deprived of colostrum and artificially reared. Indeed it has been suggested that the development of antibodies in such piglets, in which *E. coli* has been established in the intestine, arises from the massive immunizing dose of *E. coli* they received when the microbe penetrated into the tissues.[13]

Changes in Serum Protein Composition

The concentrations of the various protein fractions in the blood of the neonate are very different to those in the adult pig. The change toward the adult pattern is termed maturation of the serum proteins.

Many studies have been made of the composition of serum proteins and of the changes that occur with age. At birth total serum protein is low (2·2 g/100 ml) but it rises rapidly to 6.0 g/100 ml within 36 hours when suckling occurs and then approaches the normal values of 7 g/100 ml within three months.[14] The main components of serum proteins and their relative proportions are shown in Fig. 1. Albumin and gamma globulin, both low at birth, rapidly increase at the expense of alpha globulin. Although a number of workers have been unable to detect gamma globulin in the serum of pigs at birth, Olsson (1959)[6] found small amounts of gamma globulin in the protein of serum taken from the cords of baby pigs, and using modern techniques other workers have also demonstrated low concentrations.[11,15,16] As far as immuno globulins are concerned, after suckling steadily increasing amounts of the IgG, IgM and IgA fractions have been demonstrated.[1,17]

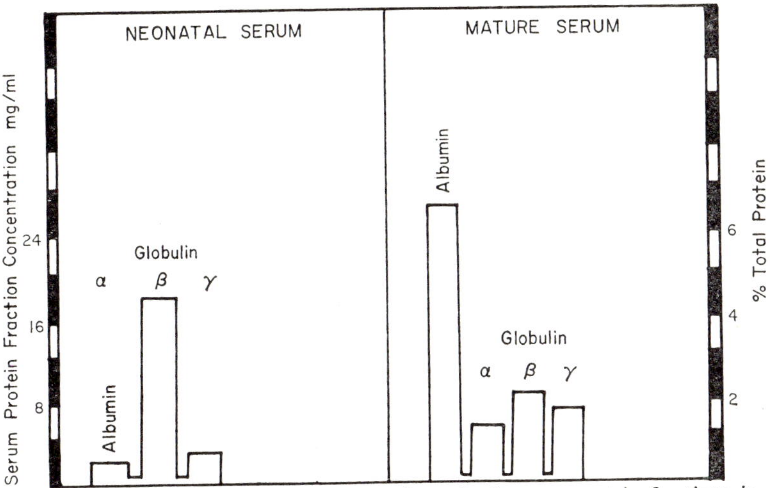

FIG. 1. The concentration of total protein and of serum protein fractions in neonatal and mature serum.

The rate of maturation of serum proteins is affected by the diet given and the biggest differences in serum protein composition are found between piglets which have suckled and have thus had colostrum, and those which have been artificially reared without colostrum. Lecce and co-workers[10] concluded that the sow through her mammary secretions appears to influence serum protein changes in two ways, firstly by providing for immediate increases in serum proteins and secondly by providing for continuing and uninterrupted changes in serum protein composition.[1] After a number of studies[18,19,20] they concluded that rapid maturation of serum proteins was associated with health, vigour and rapid growth, while arrested and immature profiles were associated with bacteraemia and morbidity.[21] Thus the effect of nutrition on development of serum protein composition is paramount.

When piglets are deprived of colostrum the change, both in total serum protein concentration and its composition are dependent on the diet given.[18,19,21] Whereas the feeding of cow's colostrum will produce changes in serum protein profile quantitatively similar to those found in pigs given sow's colostrum but smaller in magnitude, the feeding of cow's milk or artificial milk[18,19,21] is associated with slow increases in albumin and gamma globulin and thus such diets can be regarded as causing a delay in the maturation of the protein pattern. It is of interest that piglets given an 'amino acid milk' diet—similar in composition to the artificial milk referred to above[18,19] except that hydrolysed casein and lactalbumin replaced non hydrolysed forms of these proteins—did not show maturation of serum protein even after 14 days, and the initial increases in gamma globulin and albumin concentrations brought about by suckling for 24 hours were reduced.

The ability of artificial diets to produce changes in serum protein concentration depends also on the age of weaning. Piglets weaned to cow's milk after four days of suckling show no delay in protein profile development, whereas those weaned at one day do show a delay; and piglets given an amino acid milk diet and weaned at one or four days show considerable delay. However, if weaned at 14 days, the composition of the already mature serum will be maintained.

Although diet affects the rate of changes in serum protein profile, the appearance or disappearance of specific serum proteins as illustrated by immuno diffusion techniques is dependent on time and independent of diet.

The production of antibodies is important particularly in piglets deprived of colostrum. Early work[22,23] indicated that antibodies were not produced until 8–10 weeks but more recently it has been shown that in response to antigenic stimulus they are produced as early as 14 days of age.

It is difficult to rear pigs from birth which have been deprived of colostrum.[24,25] The piglets develop diarrhoea[20,26] and even when they are produced by hysterectomy or caught in sterile towels and then reared in stainless steel cages in an isolation room, they develop diarrhoea at about three days of age. Reports[20,26] suggest that it is not unusual for mortality to be 50 per cent by two weeks of age and frequently *E. coli* have been isolated from the blood and liver of morbid pigs. Piglets with bacteraemia usually die within four days of the isolation of the first positive blood culture of *E. coli* and the majority of deaths in colostrum deprived pigs occur within the first week of life.[26,27]

In one particular study, *E. coli* type–08 was the organism most frequently isolated from the blood and liver of dead pigs and this strain accounted for 67 per cent of the typable *E. coli* isolated. In the faeces taken at random from healthy neonatal pigs or those with diarrhoea 24 per cent was type–08, and 85 per cent of the bacteraemia and death associated with *E. coli* occurred in piglets devoid of gamma globulin.[26] After two weeks of age there was little trouble associated with *E. coli* even in pigs given 10^9 *E. coli* –08 and which at that time were antibody and gamma globulin free. Other workers[28] found serotype O and K (including O—2, 26, 141, 138 and K—60, 85 and 88) in scouring pigs which had been healthy to 10–11 weeks of age and whose serum contained no gamma globulin detectable by electrophoresis using the cellulose acetate technique.

The administration of porcine gamma globulin improved the survival rate of colostrum deprived pigs.[29,30] Parenteral administration or oral administration for one day was not effective in the provision of passive immunity, but there appear to be some immunity when gamma globulin was given orally for five days; serum gamma globulin increased in proportion to the amount of globulin given.[30]

It is possible to reduce mortality by strict control of hygiene. Young and Underdahl[31] showed that given a proper environment, colostrum deprived pigs could be reared satisfactorily. They described an isolation unit for this purpose in which, in 22 experiments, they successfully reared 51 out of 62 pigs or 82 per cent. Eight of those which died were believed to have died because the isolation procedure was poor. These workers stress the importance of immediate and adequate isolation of the newborn pig if it is to be deprived of colostrum. They found that piglets dropped in their dam's environment and then isolated, died in 3–4 days if they had been deprived of colostrum. Reber[32] concluded that newborn pigs deprived of colostrum must be isolated for 4–8 weeks because of their high susceptibility to disease due to their inability to synthesize antibodies adequately in this time.

Several workers have described units suitable for rearing colostrum deprived piglets.[33,36] The survival rate of pigs reared to 14 days in an isolation unit and then introduced to clean premises was significantly better than that for pigs reared naturally from the same herd (Table 1). It is of interest that in one study the greatest mortality of the colostrum deprived pigs occurred at 11 weeks of age at a time when serum gamma globulin was still low. When subsequently, further groups of piglets were put through the same unit and premises but after they had been sterilised, mortality was less severe and serum gamma globulin and total serum protein were higher at an earlier age[27] (Table 2).

TABLE 1
Mortality of Naturally Reared Pigs and Colostrum deprived Pigs reared in Isolation

	No. Born	Mortality at		
		3 weeks (%)	6 weeks (%)	14 weeks (%)
Colostrum deprived reared in isolation	126	5·5	13·5	31·5
Naturally reared	1061	28·5	30·8	—

Diets

Very little is known about the nutritional requirements of pigs during the first week of life and, as described above, studies with colostrum deprived pigs are complicated by disease. A number of workers have attempted to rear colostrum deprived piglets on proprietary food prepared for human infants, but have not achieved good results.[27,34,35] Cow's colostrum is an effective substitute for sow's colostrum, but when colostrum is not given, diets based on cow's milk rather than on synthetic ingredients are the most satisfactory.[18,37] It is customary to give supplements of vitamins A and K—either by injection or orally—and the minerals Fe, Mn and Cu. Cow's milk is lower in protein and dry matter than sow's milk and hen egg is often added. This increases dietary protein concentration and also increases the proportion of the essential fatty acids. Reduced growth rates have been reported when egg was omitted.[27,37]

TABLE 2

Total Serum Protein and γ–Globulin of Colostrum deprived Pigs reared in isolation to 14 days then introduced to a clean environment at intervals

	Born		Age			
			10	14	20	26
Group 1	June 1966	Total serum protein g %	3·3	—	5·4	6·3
		γ–globulin mg/ml	None	—	5·8	9·2
Group 2	Aug. 1966	Total serum protein g %	—	—	6·0	—
		γ–globulin mg/ml	—	—	8·2	—
Group 3	Sept. 1966	Total serum protein g %	4·2	5·5	—	—
		γ–globulin mg/ml	6·8	9·5	—	—
Group 4	Nov. 1966	Total serum protein g %	4·1	—	—	—
		γ–globulin mg/ml	7·4	—	—	—

Table 3 shows the composition of a typical basal diet, but various workers have used different mineral and vitamin supplements.[31,34,35]

It has been shown that the growth rate of piglets given a diet based on cow's milk is inferior to that of pigs on sow's milk[36] (see Table 4). Probably this is due to the fact that cow's milk is lower in dry matter (12 per cent compared with 20 per cent) than sow's milk, and has a lower energy content (820 kcal/litre compared with 1110 kcal/litre).[39] Various trials have been made with fortified cow's milk. When the energy content of a cow's milk diet is increased the growth rates achieved on it are comparable with those for

TABLE 3
The Composition of a Diet suitable for Colostrum-deprived Piglets to 14 days of age

Ingredient	Amt./litre cow's milk
Cow's milk	—
Hen's egg	1
Minerals:	
Iron (mg)	100
Copper (mg)	3
Manganese (mg)	7
Zinc (mg)	35
Vitamins:	
Nicotinic acid (mg)	3
Pantothenic acid (mg)	1
Vitamin B_{12} (mg)	1·5
Vitamin D (iu)	1200
Vitamin K (μg)	60

TABLE 4
The Weight of Pigs reared on the Sow or given a Diet based on Cow's Milk

	Weight at 14 days (kg)
Sow reared	3·53
Cow's milk	2·89

naturally reared piglets and in some cases have exceeded those of the latter.[34] Young and Underdahl increased the energy content of their basal egg/milk diet by the addition of cereal cream, whereas other workers have achieved the increase merely by reducing the dietary moisture content. The way in which the energy content is increased is important since the pig's ability to digest carbohydrate at this early age depends on the form of carbohydrate and on the sufficiency of enzymatic secretions.[25] The activity of most enzymes increases to about five weeks of age[40,41] with the exception of lactase which tends to decrease. The activity of tributyrinase, amylase and proteinase in pancreatic tissue is significantly lower in weaned, compared with suckled pigs.[40] Schneider and Sarett (1966)[34] gave diets to young pigs with higher energy contents than sow's milk, but with lower protein content (3·8 g cf 5·2 g/litre). The lower protein apparently produced poorer growth rates.

Desquamation of skin has been reported in colostrum deprived pigs given a diet based on cow's milk and supplemented with egg.[27,42] The condition is similar to that produced by linoleic acid deficiency and although linoleic acid is low in the cow's milk diet compared with sow's milk, the addition of maize oil, containing 53 per cent linoleic acid, to a cow's milk diet did not prevent desquamation.[27] The addition of ascorbic acid has been shown to improve this condition.[42]

The amount of milk given to young pigs reared artificially is generally increased daily as they grow. Overfeeding is frequently associated with diarrhoea, and improved performance with less diarrhoea has been obtained when the frequency of feeding has been increased.[27,37] Seven meals per day gave better growth rate than three.[37] Near *ad libitum* feeding has been satisfactorily achieved by Schneider and Sarett,[43] but it is easier to control diarrhoea by feeding to a predetermined scale, giving an amount of a diet containing 20 per cent dry matter. The nutrients supplied daily by a diet and a particular level of feeding which have proved satisfactory are compared with estimates of those supplied by the sow together with the respective growth rates of the pigs and are given in Fig. 2. After 14 days it is customary to introduce solid feed and diets designed for early weaned pigs are suitable for this purpose.

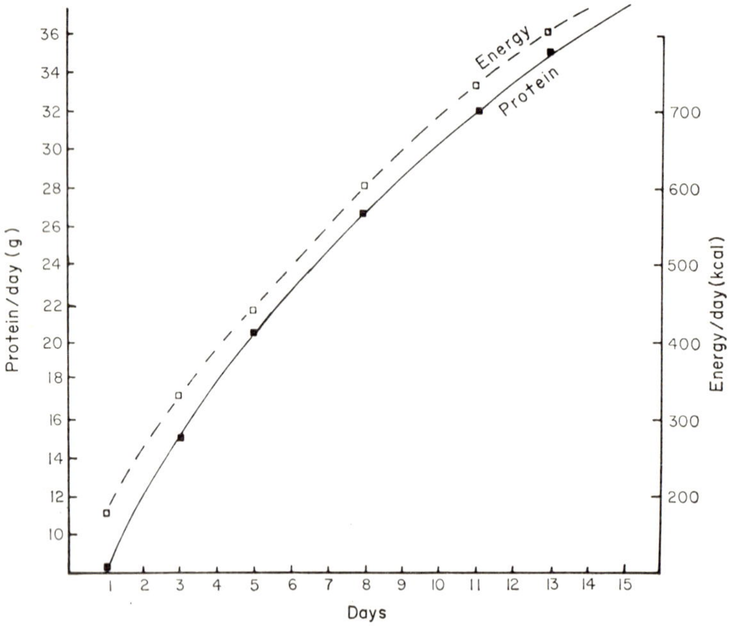

FIG. 2. The energy and protein supply to baby pigs given an artificial diet.

The use of sterilized diets apparently has not increased the survival rate of colostrum deprived pigs; whereas some workers[33] reared on average 87 per cent of pigs given sterilized diets compared with only 70 per cent on non-sterilized diets, others have reported good survival rates when pigs were given non-sterilized diets together with suitable protection from infection and an adequate environment[27,31] 0·1 per cent formalin has been used satisfactorily to control bacterial growth in non-sterilized diets.[27]

Early Weaned Pigs

Although it is usual to wean pigs at 56 days of age they can be successfully weaned as early as 10 days when they weigh 4 or 5 kg.[44,50]

As lactation in the sow proceeds there is a decrease in the efficiency of the utilization of sow's milk[51,52] by the piglets and since it is reported[53] that 50 per cent of the energy intake of a litter is recovered as carcase energy at weaning while only 25 per cent of the energy given to the sow is recovered in this form, it should be more efficient to wean early. Salmon-Legagneur and Jouandet[53] have shown that when weaning was at five weeks the pigs grew 8 per cent faster and the efficiency of the utilization of feed energy by the sow during lactation was increased by 40 per cent compared with weaning at eight weeks. In addition, the length of the reproductive cycle was reduced by 16 per cent in the sows of the early weaned group. By weaning at 20 instead of 56 days it has been estimated[54] that there is a saving of at least 21 per cent of feed energy per unit gain of litter weight and that the energy recovered in the litter per 100 Mcal feed energy given to the sow increased from 18 Mcal for pigs weaned at 56 days to 23·5 Mcal for those weaned at 20 days.

As discussed earlier, young pigs which receive colostrum acquire passive immunity through the globulin which the colostrum provides, and the difficulties of rearing such pigs are therefore relatively less than for pigs

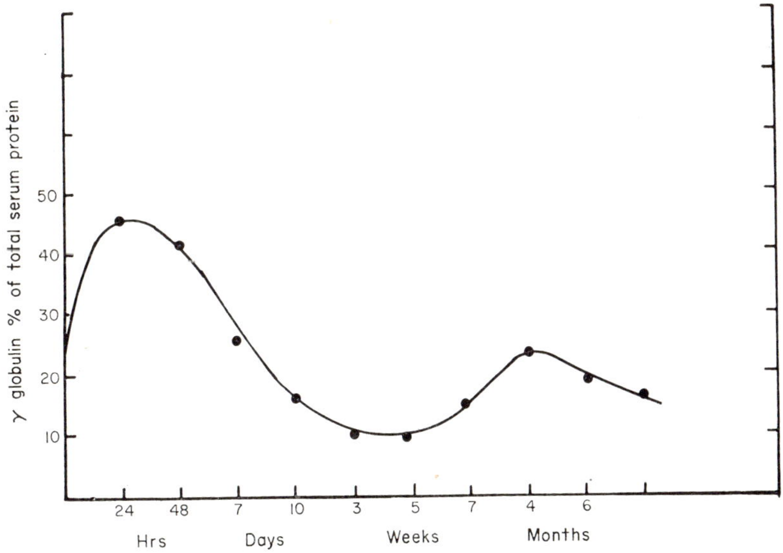

FIG. 3. Changes in serum γ-globulin with age (after Miller *et al*, 1960).

deprived of colostrum. Nevertheless, there are periods when diarrhoea, often associated with *E. coli*, is particularly troublesome. In pigs weaned at 56 days, these periods occur at 2–3 days, at three weeks of age, and post weaning—usually at about nine weeks of age. The changes that occur in serum gamma globulin as the pig ages are shown in Fig. 3. The early scouring may be the result of invasion of the gut wall by organisms in the lumen of the gut before 'closure' has occurred and at a time when passive immunity is low. The scouring at three weeks of age may result from changes in the gut flora brought about by the introduction of solid feed, and possibly from changes in the composition and supply of sow's milk due to the onset of oestrus. Both these changes occur when both passive and acquired immunity are low. Post-weaning scouring is frequently associated with abrupt changes in diet and environment. A change of environment brings the piglet into contact with new strains of bacteria, and abrupt changes in diet are believed to cause digestive upsets which in turn provide favourable substrates for the proliferation of these bacteria.

Lucas and Lodge[55] have described briefly the techniques for early weaning pigs and they have given estimates of nutrient requirements for young pigs at different liveweights.

Growth and Body Composition

Smith[56] demonstrated that pigs could be weaned at ten days of age, but in the period of 7–14 days following weaning growth was slow. This check in growth was followed by a period during which a compensatory growth took place.

There are differences between the rates of growth and the composition of gains of early weaned pigs and those reared on the sow[57,60] and there are three factors which have been claimed to affect the growth and composition of early weaned pigs. These are the effects of the method of rearing, of different rates of growth in early life, and of checks in growth rate.

The re-evaluation of the data published by McMeekan[61,62] and Palsson[63] by other workers[64,65,66] showed that the proportions of body components, other than fat, could be predicted from the total fat-free weight of the pig, and this re-evaluation has helped in the elucidation of the effects of rate of energy intake at different stages of growth on body composition. A reduction in the rate of metabolizable energy intake in the early stages should cause a uniform retardation in the growth of the components of the fat-free body. When the rate of metabolizable energy intake is increased beyond that required for the maximum growth of the fat-free body, the proportion of fatty tissue will be increased. However, from a review of the experimental work there appears to be confusion over the effects of rate of metabolizable energy intake on growth rate and body composition.

Many comparisons have been made between the growth rate and feed conversion efficiencies of early weaned and suckled pigs.[25,58,67,72] There are as many which report that early weaned pigs perform better than suckled pigs as *vice-versa* and the differences probably reflect the differences in rearing techniques, in the diets used and in the milking capacities of the different sows. It is possible to achieve better growth rates from early weaned pigs than from suckled ones[70,73,74] and although the former have a check in growth immediately after weaning they grow faster in the later part of the early growth period.[60,73]

When Lucas *et al.* weaned pigs at 4 kg and gave them different amounts of feed so that they grew at different rates for 56 days, no differences in the efficiency of food utilization were found. However, early weaned pigs given feed *ad libitum* and which weighed 23 kg at 56 days, grew 7–8 per cent more slowly and utilized food 7–9 per cent less efficiently than pigs which were reared on the sow to 23 kg liveweight at 56 days of age. The early weaned pigs also produced carcases at 90 kg liveweight which contained less lean and more fat than the normally reared control pigs. Similarly, when pigs were weaned at 21 or 56 days and after 20 kg allowed feed *ad libitum* or in restricted amounts, those weaned at three weeks were considerably fatter at 90 kg irrespective of the level of feeding after 20 kg.[75] These observations about carcass composition at 90 kg are not supported by those made by Mandel[76] or by Elsley[77] who found no differences in carcass composition of early weaned or suckled pigs at 90 kg liveweight. The difference between these latter experiments and Lucas' may reside in the rate of energy intake provided from weaning to 90 kg. However, Fowler[65] showed that differences in the composition of pigs at 23 kg remained until 90 kg liveweight both when the level of feeding between 23 and 90 kg liveweight was high or when it was low.

Turning now to the question of the effect of different growth rates before the pig is 56 days of age on carcass composition, it has been shown that when pigs are weaned at 4–5 kg, and by control of food intake, subsequently reach weights of 13, 18 or 23 kg at 56 days of age, the amount of lean tissue is greater and the amount of fat tissue is larger in the slower growing animals.[59] A low weight at 56 days of age was also associated with an increase in growth rate and feed conversion from 22–90 kg and to significant reductions in the fat content of the pigs at 90 kg although the proportionate weights of joints and organs were not affected.[77] In another experiment of the same type Frape *et al.* (1959)[58] varied the amount of feed given to early weaned pigs so that a range of weight was achieved at 56 days. Although 85 per cent of the variation in the time taken to reach 90 kg liveweight could be attributed to the growth rate before 54 kg there were no differences in carcass quality as between the various groups. Pigs which weighed 1·0 kg less than the average at 54 days of age took a day longer to reach bacon weight.

It has been suggested that a check in growth in early weaned pigs may be responsible for the greater proportion of fat in the carcass at 56 days of age[59] and at 90 kg.[60] However, severe restriction of growth from 3–21 days produced no deleterious effects[78] on carcass quality. Compared with early weaned pigs which had suffered but a small check in growth, those checked for ten days by underfeeding so that they gained only 908 g in 21 days did not have more fat in their carcasses at 90 kg liveweight.[79] Even when pigs did not grow for three weeks at 23, 45 or 68 kg[80] or for five weeks at 23 or 45 kg[79] no differences could be found in carcass quality at 90 kg liveweight. The same was true when the growth check was imposed either by underfeeding or by the restriction of protein rather than energy intake.

Protein

Much work has been done in an attempt to establish the protein requirements of young pigs. A feeding trial with baby pigs is not a good method to use in this respect because the requirement for protein decreases very rapidly early in life[81] and the weight range over which trials must be carried out to establish differences in growth is large. The length of the trial is also important because during this early growth period the enzyme secretions of the pig are changing and the digestibility of some proteins alter with age. For example, the digestibility of the protein of soyabean increases from 78 per cent at 14 days of age to 82 per cent at 35 days.[82-85] The digestibility of the protein of skim milk, however, does not change with age. Furthermore, there are differences between experiments in the type of protein concentrate used. It is not usually possible in retrospect to allow for these differences in the absence of complete analytical data.

The requirements for protein and for the essential amino acids by young pigs were estimated by Lucas and Lodge[55] in 1961.

Since that time several comparisons of the value of protein concentrates in diets for young pigs have been made.[86-93] In general skim milk promotes good growth and feed conversion and soyabean meal appears to be a better source of protein than fish meal.[83,95-99] A combination of equal parts of fish meal and dried skim milk have given better results than diets containing greater proportions of dried skim milk.[100,101] For pigs up to 12 kg liveweight, replacement of half of the 40 per cent skim milk in a diet with fish meal improved the rate of gain by 4 per cent but total replacement reduced rate of gain by 6 per cent.[60] The amount of protein required by pigs up to about 11 kg liveweight is 22–29 per cent crude protein, the amount within this range depending on the source of the protein. Thereafter diets containing 18 per cent crude protein are satisfactory.

There is evidence that the dietary concentration of protein required is related to the calorie density of the diet[99,102-106] and for very young pigs, Manners and McCrea suggested that with diets low in fat 25 per cent crude protein is required and with diets high in fat 30 per cent crude protein is needed. For older pigs of five weeks of age an increase in dietary energy from 4–4·6 Mcal/kg increased the concentration of dietary protein required from 17·5 to 20·6 per cent.[103] Similarly there is an interaction between protein quality and dietary energy concentration.[60,100]

Lucas and Lodge[55] summarized the results of experiments which reported the amino acid content of satisfactory diets, the amino acid content of the bodies of young pigs and the effects of amino acid supplementation of diets. They estimated from this evidence the amino acid requirements of young pigs. In a recent re-evaluation of relevant literature[107] a similar approach was made but data from experiments have been disregarded if the diets used contained unusually high or unusually low protein concentrations. The new estimates of the essential amino acid requirements are shown in Table 5 together with estimates of requirements for crude protein. These estimates are based on experimental work and there is a lack of work on pigs weighing between 4·5 and 9 kg.

TABLE 5

Estimates of the Requirements of Pigs for Crude Protein and the Essential Amino Acids

Amino acid % of L-isomer in dry matter	Liveweight of pig (kg)		
	Birth–4·5	4·5–9	9–20
Lysine	—	1·1	1·1
Methionine + cystine	0·9	0·8	0·7
Tryptophan	0·19	—	0·15
Histidine	—	0·2	0·2
Tyrosine + phenylalanine	—	—	0·5
Threonine	0·95	—	0·5
Leucine	1·32	—	0·7
Isoleucine	—	—	0·7
Valine	—	—	0·5
Crude protein % in dry matter	32	26	20

TYPES OF DIETS FOR EARLY WEANED PIGS

Generally, for pigs weaned before ten days of age liquid diets have been more satisfactory than solid diets, but for pigs weaned after this age, solid diets have produced better results. Most diets for early weaned pigs are based on cow's milk supplemented so as to simulate sow's milk, and the composition of sow's milk has been reported by Braude[108] and by Perrin.[39] Sow's colostrum, however, is thought to contain a trypsin inhibitor which prevents the digestion of protein in very young pigs[109,110] so that it is not possible to simulate mammary secretion exactly.

Sow's milk contains approximately 20 per cent dry matter but for milk substitute diets the best results were obtained when 15 per cent dry matter was used.[111,112] The composition of a typical solid diet for early weaned pigs is given in Table 6. Many other formulations have been reported.[113,118]

TABLE 6
The Composition of a Diet for Pigs from 4 to 11 kg liveweight

Composition	%
Rolled oat groats	46·5
Whitefish meal	13·5
Roller-dried skim-milk	20·0
Sugar (sucrose)	10·0
Dried yeast (75% unextracted)	5·0
Ground limestone	1·0
Salt (NaCl)	0·5
Cod-liver oil	2·0
Trace mineral supplement*	1·0
Antibiotic supplement†	0·5
Vitamin B_{12} supplement‡	+

* 1 lb trace mineral supplement contained $FeSO_4{\cdot}7H_2O$ 18 g; $MgSO_4{\cdot}7H_2O$ 23 g; $CuSO_4{\cdot}5H_2O$ 3 g; $MnSO_4{\cdot}4H_2O$ 7 g; $ZnSO_4{\cdot}7H_2O$ 2 g; KIO_3 0·076 g; $CoSO_4{\cdot}7H_2O$ 0·428 g; maize starch 200 g; ground flaked maize 200 g.

† 'Aurofac 2A' (Cyanamid of Great Britain Ltd.) containing 3·6 g chlortetracycline hydrochloride per lb.

‡ 50 g of a vitamin B_{12} supplement containing 7 mg B_{12} per lb.

SUCKLED PIGS

The peak of milk yield occurs in the sow at about three weeks after farrowing[119] and it is therefore usual to introduce supplementary or "creep" feed at this time so that as lactation declines creep feed intake by the piglet increases. This practice increases the weaning weight at 56 days.[119-122] There is evidence[122] that when the piglets are given feed the milk yield of the sow declines more rapidly than when she is the sole source of sustenance for her young. Creep feeding, however, increases the overall energetic efficiency of the sow in producing her litter by about 7 per cent.

Several observations have been made [123-126] which suggest that the heavy weaners at 56 days have the better subsequent performance than lighter ones, but other work[127,128] indicates that the rate of growth before 56 days has no effect on subsequent growth rate or carcass quality. It was shown that when pigs have a genetic potential for fast growth weaning weight does not affect subsequent carcass quality but a high energy intake by piglets with a low growth potential was detrimental to carcass quality.[129]

The variations in 56-day weaning weight of pigs are due to difference in birth weight, to differences in the total yield of nutrient in the milk of the sow and to the differences in the intake of creep feed. Milk yield and composition are to some extent affected by the nutrition of the sow.[130,131] For example, the addition of fat to sow diets has been shown to increase the weaning weights of the piglets.[132] Litter size affects the weaning weight directly due to the competition for feed between individuals as litter size increases and also because birth weight, which is correlated with weaning weight, is related to litter size.[133-136] The biggest factor affecting weaning weight, however, is the amount of creep feed consumed, and the variation between litters due to differences in consumption of creep feed has been estimated by different workers to be 77,[137] 36,[138] 57,[52] and 20 per cent for individual pigs.[139]

High concentrations of crude protein, such as are required by early weaned pigs are not required in creep feeds. In one study[140] raising the concentration of protein above 17 per cent did not improve growth rate and in another[141] there was no difference in performance to 56 days between pigs given either 13 or 18 per cent crude protein in diets containing either 3·8 or 4·5 Mcal/kg. When creep feeds containing 14, 18 or 22 per cent crude protein were given from 21 to 56 days of age there were no differences in performance to 56 days or in carcass composition at 56 days. These results are not surprising because

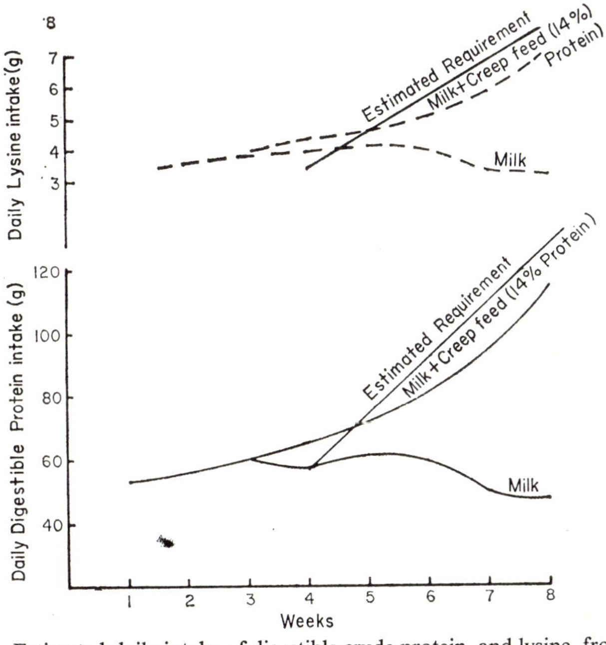

FIG. 4. Estimated daily intake of digestible crude protein, and lysine, from sow's milk and a creep feed containing 14 per cent crude protein. (Digestibility of milk protein assumed to be 97 per cent, and digestibility of crude protein in feed, 80 per cent; lysine in protein of creep feed, 4.4 per cent.)

of the large contribution to daily crude protein requirements made by the sow's milk which in addition has a high digestibility and biological value.[142] In Fig. 4 the estimated contribution made by sow's milk and by a 'creep' feed which contains 14 per cent crude protein to the daily digestibility protein supply of a piglet is compared with the estimates of the animal's requirement of digestible crude protein given by Lucas and Lodge.[55] The figure also shows a similar comparison of lysine supply and need. Lysine is probably the limiting amino acid for the young pig.

The relationship between total feed consumption and liveweight gain from birth to eight weeks of age has been shown to be linear.[143] The milk yields reported by Lodge[119] and the mean weekly feed consumptions reported by Witt (1957)[144] have been used to plot Fig. 5 which shows that liveweight gain is more closely related to digestible energy intake than to feed intake.

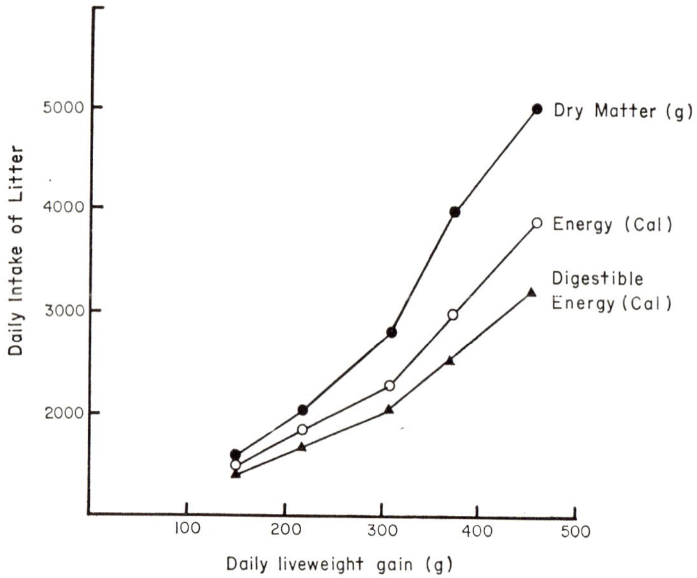

FIG. 5. The relationship between liveweight gain of young pigs and dry matter of energy intake from sow's milk and creep feed.

Little is known about the importance of protein quality in 'creep' feeds. In those experiments in which protein concentrates of different quality have been used[92,141,145] the diets differed in other respects so that the effects of palatability and protein quality have been confounded. When skim milk is given in addition to 'creep' feed but instead of water, the piglets eat less 'creep' feed but have equal intakes of dry matter which suggests that questions of food preference are important.[146]

Several workers have studied the palatability of 'creep' feeds and of solid complete diets for early weaned pigs. The use of lard or beef tallow increased feed consumption.[147,148] Cane sugar also increased feed consumption. Many reports have confirmed this thinking and in only two reports has no response to 10 per cent sugar been noted.[44,148] In some cases saccharin has been shown to increase feed consumption[149] but this is not always so.[118,148,150] Sugar beet has also been shown to improve performance.[151] Attempts have also been made to improve feed consumption directly by the use of flavouring agents, such as aniseed[139] and some of these have been successful. Other attempts have been to increase feed consumption indirectly notably by inducing hyperglycaemia.[152,153] Although this condition can be induced, it does not appear to affect feed consumption. The physical form of the diet does not in general seem to be important since there are as many reports which indicate that pigs have a preference for pelleted feed as there are those which show that pelleting does not improve feed consumption.

THE NUTRITION OF PIGS AFTER 56 DAYS

Energy Metabolism

Changes in Body Composition with Age

Several workers have studied the chemical composition of pigs, and the partition of metabolizable energy (ME) of the feed as between maintenance, body protein and body lipid deposition. Wide variations in the chemical composition of pigs exist because there are genetical differences in lean growth potential as between strains of pigs.

There are two ways in which the composition of carcasses have been expressed. Firstly in terms of chemical composition—lipid, N $\times$ 6·25 and so on—and secondly in terms of lean tissue, fat and bone. The two represent different things because lean tissue contains some lipid and water, and the chemical composition of lean tissue and indeed most tissues is not constant but alters with advancing maturity.[61] There are also, for example, significant differences in the chemical composition of individual muscles at any given time.[154,155]

TABLE 7
The Chemical Composition of Young Pigs

Empty liveweight	Protein	Lipid	Water	Ash
1·0 (Birth)	11·0	1·4	81·5	3·5
5·3	14·4	14·2	68·6	2·8
16·2	14·6	14·6	67·8	3·1
24·2	15·1	15·0	66·6	3·4

As shown in Table 7 in the early stages of growth marked changes in chemical composition occur, characterized by rapid increases in lipid and a decrease in water and ash content. At birth lipid is very low[156,158,159] and the amounts of carbohydrate are relatively large (up to 15·5 per cent of dry matter).[156,159,160] Studies with older pigs[161,162,75] show that there are significant differences in composition between breeds and strains.

The rate at which energy intake is partitioned into energy of protein or energy of lipid is shown in Fig. 6. The data used to plot this figure were taken from work reported by Oslage and Fliegel[163] and demonstrate that the energy value of the gains of body protein is always less than the energy value of the

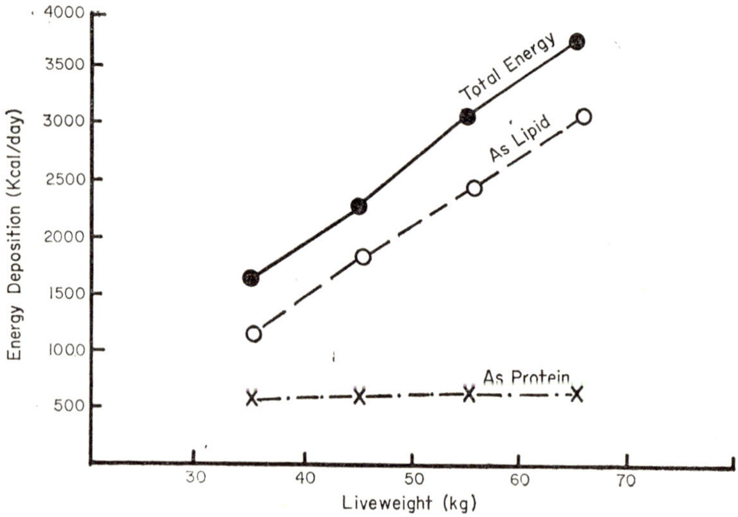

FIG. 6. The distribution of the total energy deposited between lipid and protein tissue.

gains of body lipid. The proportion of energy deposited as protein depends to some extent on the biological value of dietary protein. If the heats of combustion of fat and protein are taken to be 9·46 and 5·79 kcal/g respectively.[164] Oslage and Fliegel's results show that 9–10 g of protein are deposited daily at weights between 20 and 30 kg. This daily gain is low since it would not be unreasonable to expect N–retention by pigs of this weight to be in the region of 14 g/day.[165,166] Thus the values found by the Hannover workers cannot be taken as absolute. They will vary between strains of pig and with the rate of energy intake.

Although the percentage of protein deposited declines as the pig grows, the absolute amounts of nitrogen retained does not fall till after 120 kg liveweight.[163]

As far as lean and fat tissues are concerned since the calorific values of these tissues are in the ratio of 2:7, then actual deposition of lean tissue is not superseded by that of fat tissue till late in the pig's life. There have been many reports of the composition of pigs in terms of lean and fat tissue,[167-170] but it is difficult to compare the different data because of the differences in the techniques used for dissection. Most data, however, show that fatty tissue deposition does not exceed that of lean tissue until about 65 kg liveweight although some Danish data[169] suggest that this may not occur till about 100 kg liveweight.

The Effect of Energy Intake on Body Composition

Very little has been done on the effect of energy intake on the partition of retained energy as protein and lipid. The only data available are those reported by Oslage[170,171] from experiments in which he studied the effect of energy restriction from 60–110 kg liveweight. His results are presented in Table 8 and indicate that restriction of energy intake by 10 or 20 per cent during the period from 60–90 kg liveweight reduced the deposition of lipid and increased the deposition of protein, but from 90–110 kg liveweight this effect was reversed. Oslage also suggests[171] that restriction in the early stages of growth, before 60 kg, reduces the deposition of both protein and lipid.

TABLE 8
The Effect of Restriction on the Composition of Energy Weight Gain in Pigs

Liveweight (kg)	60–90			90–110		
Degree of feed restriction	Control (O)	10%	20%	Control (O)	10%	20%
Water (%)	39·9	42·7	50·0	38·8	39·7	37·7
N x 6·25 (%)	13·0	14·0	16·8	14·3	13·8	13·5
Fat (%)	45·3	39·3	30·0	42·8	43·9	48·0

Extensive tissue dissections were undertaken by McMeekan[61,62,133] to determine the effect of rate of energy intake at different stages of growth on tissue development. He suggested that the order in which the tissues developed was firstly bone, then muscle, then fat, each overlapping to a degree determined by the rate of growth. The rate of energy intake had a differential effect on development, early restriction limiting bone development and late restriction limiting fat development. As mentioned earlier, other workers,[64,66] using McMeekan's data, showed that on a fat-free basis restriction of energy intake did not affect the development of body tissues differentially and that excess energy intake at any time resulted in fat deposition.

Kielanowski calculated the energy cost of protein and fat deposition to be 7·51 kcal ME/g and 11·65 kcal ME/g respectively. Based on calorific values for protein and fat of 5·79 and 9·46 kcal/g, the efficiencies of utilization of metabolizable energy above maintenance are 77 and 80 per cent respectively.

There is a suggestion[163] that the efficiency of utilization metabolizable energy falls with age from about 78 per cent at 35 kg liveweight to 63 per cent at 160 kg liveweight, but the efficiency does not decline with increase in energy intake above maintenance,[172] although it does vary with the type of feed.[173]

Effect of Feed Restriction on Feed Utilization and Carcass Quality

It is to be expected that as feed intake is increased above the amount required for maintenance feed conversion (kg feed/kg liveweight gain) would improve since as the rate of energy intake and growth rate are increased, the proportion of energy spent on maintenance becomes less. At very high intakes of energy, however, the proportion of energy deposited as fat tissue will increase and since this tissue has a high calorific value, body weight gain would be less, and the feed efficiency could well decline. Thus whether or not the efficiency of feed utilization will improve with increase in energy intake will depend on the amount of feed energy given over maintenance and the pigs potential for lean tissue development.

In a review by Lucas and Calder in 1956[174] on the effects of feed restriction on the performance of pigs it was shown that although feed restriction generally improved carcass quality it did not always improve feed conversion efficiency (kg feed/kg liveweight gain). Vanschoubroek *et al.*[175] in an examination of data in the literature for pigs up to 90 kg computed that an average restriction of 15·8 per cent in feed intake caused daily weight gain to decrease by 0·78 per cent. Backfat thickness decreased by 0·53 per cent for each 1 per cent restriction while feed conversion improved 0·31 per cent for each 1 per cent restriction. As feed intake was reduced, the efficiency of feed conversion improved at first but eventually decreased as the restriction became more severe than about 25 per cent.

In most of the more recent experiments feed conversion efficiency from weaning to slaughter did not change by more than an average of 2–4 per cent over a wide range of energy intakes.[107]

The effect of feed restriction is different in different stages of growth. A restriction of 10 per cent gave an improvement in the efficiency of feed conversion up to 45 kg liveweight, but not in the period beyond that weight,[176] and a 25 per cent restriction gave an improvement up to 36 kg liveweight, but not up to 73 kg liveweight.[177] Such effects are to be expected since in the period after 50 kg, the rate of deposition of fat tissue overtakes the rate of deposition of lean tissue.

The results of feeding trials show that feed conversion is not always worse on self-feeding than on restricted feeding, but it has been estimated that self feeding in the period between 20–95 kg is likely to make efficiency of feed conversion worse by about 5 per cent. Killing-out percentage may increase by 1·5 units; backfat thickness by 3–5 mm, and body fat by 1·5 per cent. If self feeding is restricted to the period up to 55 kg liveweight, these effects might be reduced by about 50 per cent.[107]

The feed intakes of pigs which produced satisfactory carcass qualities and the corresponding growth rates are shown in Fig. 7. The data are for Landrace and Large White gilts and castrates which were reared at the British progeny testing stations. The growth rates for the two breeds were similar. Castrates ate 1–3 per cent more feed than gilts of similar weights up to 50 kg and 4–6 per cent more thereafter.[107]

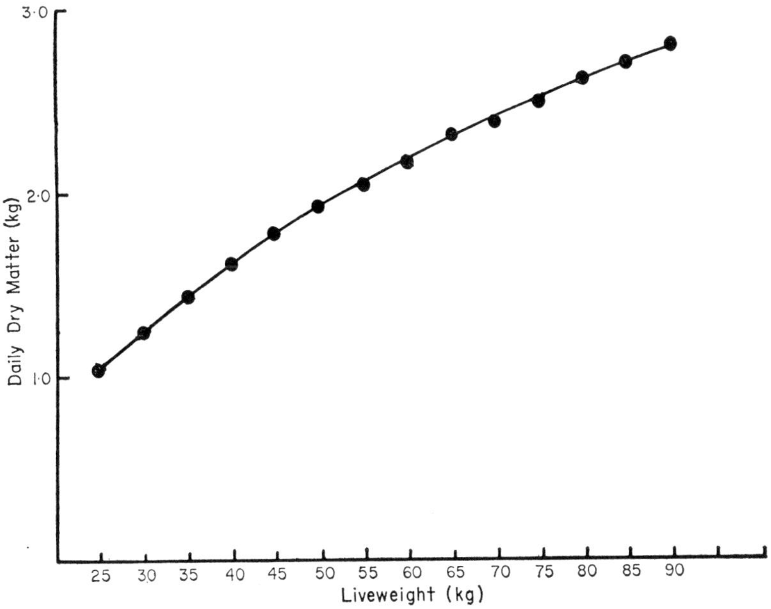

FIG. 7. The dry matter intake of pigs given feed twice daily to appetite in relation to liveweight.

The Utilization of Fat

Diets for young pigs based on barley and supplemented with fish meal or extracted vegetable protein contain about 2 per cent fat but pigs may be reared on diets containing considerably less than this amount.[179,180] The possibility of using fat to increase the energy concentration of diets has been considered by many. When fats are incorporated in diets it is necessary to use antioxidants to preserve their keeping quality.

The digestibility of a fat depends on its molecular weight and digestibility varies from 71 for those containing long chain saturated fatty acids, to 96 for short chain acids or highly unsaturated ones. For example, beef tallow is about 86 per cent digestible and soyabean oil about 93 per cent.[181] If the heat of combustion of fat is taken as 9·4 kcal/g then on average the digestible energy of a diet would change by 500 kcal for each 1 per cent addition of fat

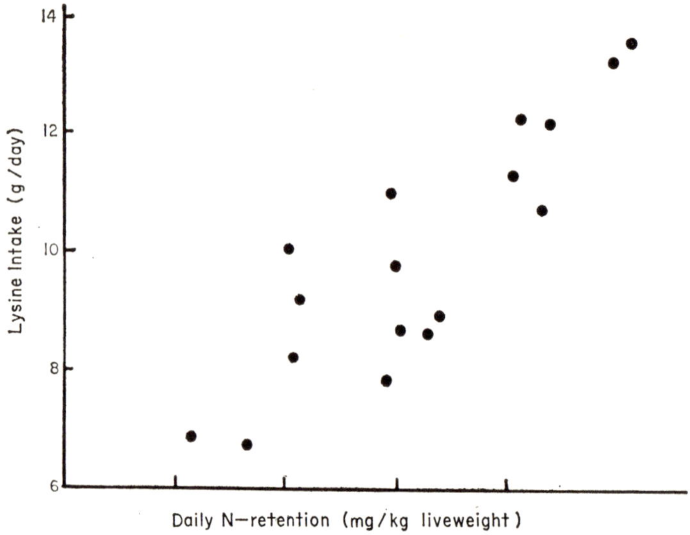

FIG. 9. The effect of lysine intake on the nitrogen retention of the young pig.

The large variation that exists in the amino acid content of protein concentrates means that the nitrogen (N) requirement of the young pig cannot be accurately defined in terms of crude, or digestible protein. Data from Woodman and Evans[216,217,218] have been used to plot Figs. 8 and 9 and data for the same pigs were used in both figures. The figures indicate that total lysine intake, lysine being the first limiting amino acid in the diets, was more closely related to N–retention than was the total protein intake.

The protein requirements of the young pig from 56 days to about 50 kg liveweight have been given as 16–17 per cent of crude protein in air dry diet and the estimates of the requirements for the various essential amino acids are given in Table 9.[107] As far as British type rations are concerned—and these are normally based on barley, with either fish meal, soyabean, groundnut or meat and bone meal—lysine is usually the first limiting amino acid, and methionine (or methionine + cystine) the second. A statistical examination of the data from N–balance experiments reported in the literature showed that the rate of liveweight gain, feed conversion and the N–content of liveweight gain of pigs up to 40 kg liveweight were affected by the lysine concentration in the diets used. An increase in 0·1 per cent lysine in diets for pigs up to 27 kg, for example, was associated with an increase in daily gain of 18 ± 3·6 g. The criteria mentioned were not affected by the dietary concentration of methionine + cystine, and total dietary lysine was a more important determinant than either the dietary crude protein concentration, or the dietary concentration of total digestible nutrients (TDN).[107]

In Fig. 10 the ability of various diets containing 16–17 per cent crude protein to supply lysine and methionine + cystine needs of the young pig is

TABLE 9

The Crude Protein and Essential Amino Requirements of Pigs from 56 days of age to 45 kg liveweight

	% of DM in diet
Crude protein	18·5–20
L–lysine	0·9–0·95
DL–methionine + cystine	0·6–0·7
L–tryptophan	0·15–0·2
L–threonine	0·5–0·6
L–isoleucine	0·75

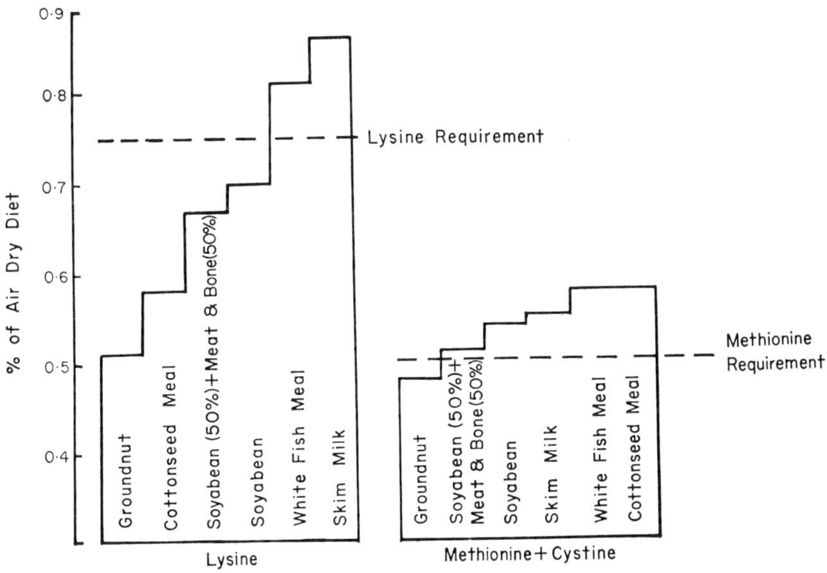

FIG. 10. The ability of diets containing 16.5 per cent crude protein to supply the lysine and methionine+cystine requirements of the young pig.

shown, and further demonstrates the limitation of crude protein in defining protein needs.

The dietary concentration of total protein supplied to the pig can be reduced if the quality of the protein is increased.[219,220] Since lysine is usually the first limiting amino acid the quality of the protein depends to a large extent on the dietary concentration of lysine, the response of pigs to dietary supplements of lysine either synthetic or natural, has been variable. The effects of dietary additions of lysine depend of course on the main source of protein in the diet and on the dietary protein concentration. The addition of 0·1 per cent lysine produces highly significant improvements in the rate of liveweight gain when added to a diet containing 12 per cent crude protein but not when the diet contained 14 per cent crude protein, the reason being that the extra 2 per cent crude protein did itself provide some lysine.[112] When lysine alone or with methionine have been added to maize-soyabean diets or to diets containing rye and vegetable protein, there have been considerable improvements in rate of liveweight gain.[302,211,40,222]

Generally, if diets contain 16 per cent crude protein and 65 per cent TDN, then supplements of lysine will not give improvements in growth rate and N–retention when the diets contain fish meal as the main source of protein. If the main source of protein is groundnut, soyabean, or some other vegetable protein, then significant improvements can be expected. This of course is a very general statement because there are differences from diet to diet in the availability of the essential amino acids, and between strains and sexes of pigs in their ability to produce lean tissue (see page 21).

Possibly the biggest improvements in growth and N–retention are observed when synthetic amino acids are added to all cereal diets. Satisfactory performance in pigs have been achieved by feeding all cereal diets (mainly barley) supplemented with amino acids[223] but a number of other essential amino acids are also required to ensure optimal performance.[224,225]

The addition of synthetic DL–methionine to diets containing 14–17 per cent crude protein has not generally improved growth rate in young pigs.[226,227,165,228]

The variation in response to amino acid supplements may be due either to the differences in the energy content of the diets or to the differences in the availability of the amino acids in the diets, or to interaction between total protein concentration and amino acid concentration.[229] It has already been mentioned that with baby pigs there is an interaction between the energy content of a diet and the amino acid requirement expressed as a concentration in the diet, and similar interactions have been observed with older pigs.[186,230,231,187] For example, the addition of 0·1 per cent lysine caused a growth depression in young pigs given a diet containing either 14 or 18 per cent crude protein, but the magnitude of the depression was reduced as the caloric content of the diet was increased.[187]

There is also an interaction between dietary protein concentration and the concentration of lysine in protein. For example, when pigs were given either 14 or 18 per cent dietary crude protein with either 4 or 4·6 per cent total lysine in protein, N–retention improved as the lysine in the protein increased at the low, 14 per cent dietary protein concentration but significantly decreased at 18 per cent dietary protein, and when slaughtered at 90 kg liveweight there were significant differences in the linear measure of carcass fat.[229] Although it has been shown the high B–vitamin supplementation is necessary to obtain maximum N–retention,[232] the interaction observed above also occurred where the experiment was repeated and large amounts of the B–vitamin were given. More recent work[233] has indicated that the conversion efficiency, in terms of meal consumed per kg of lean meat produced, falls as the lysine concentration is lowered at a high dietary protein concentration, but not at a low dietary protein concentration, and that the same efficiency measure fell as the energy intake increased at either 12, 14, 16 or 18 per cent protein concentration.

Part of the variation in the response by pigs to different concentrations of dietary lysine is due to the difference in the availability of amino acids, in particular protein concentrates. The extent of these variations has been reported recently.[234] As protein concentrates are heated it is thought that a Maillard reaction takes place between pentose sugars and free α-amino groups[235] and this change is not apparent from assay of amino acid hydrolysates. The availability of lysine in animal by-products—such as fish meal for example—may be estimated by the method of Carpenter[236] which is based on the reaction between free ε-NH_2 groups and fluoro–2:4–dinitrobenzene. With groundnut protein there is evidence that heating causes change in the arachin and conarachin fraction, and these changes have been associated with differences in the performance of chicks[237] and with cottonseed meal the lysine has a low availability because of its reaction with gossypol.[238,239] Heating white fish meal for 36 hours at 150°C reduced the available lysine content as measured by Carpenter's method by approximately 50 per cent and when young pigs were given this damaged meal so that it supplied 66 per cent of their dietary protein, there were significant reductions in growth rate, feed conversion and nitrogen retention.[240,241] The addition of lysine improved these measures, but growth and N–retention were not as good as when the unheated meal was given. The difference between the heated meal with lysine and the unheated meal may have been due to the lowering, by over-heating, of the availability of the other amino acids—for example of methionine—but in addition the heat treatment also significantly reduced the digestibility of dry matter. Not all of this reduction was due to the change in the digestibility of the protein. Other workers[242,243] have measured differences in the availability of lysine and methionine in fish meal and have associated these with differences in the performance of pigs. It is reasonable to suggest that at least for those diets containing protein concentrates in which the availability of the lysine is low because of over-heating, that the ability of such diets to meet the N requirements of the pig should be defined in terms of 'available lysine' rather in terms of crude protein or total lysine. In a recent experiment[244] young pigs were given diets containing different protein concentrates and which contained different concentrations of total crude protein and total lysine, but the same concentration of available lysine. Similar growth rates were obtained and the carcass qualities at 90 kg as indicated by linear fat measurements were not significantly different.

It has been suggested that boars have a higher protein requirement than castrates and in one study the slower growth rate of boars compared with castrates was attributed to the boars being unable to satisfy their need for protein on a diet which was adequate for castrates.[245] Sex differences between pigs in the response to protien are well documented,[246,247] and it is reasonable to assume that the protein requirement of boars, which have consistently been shown to have significantly leaner carcasses than castrates at equal slaughter weights, will be higher, but experimental evidence on this is lacking.

Lameness in Young Pigs

There are many reports of lameness in pigs but in most cases the lameness has been cured by adjusting dietary concentrations of calcium, phosphorus, or vitamin D. Reports recently, however, suggest that manipulation of dietary mineral and/or vitamins is without effect on some forms of lameness or leg weakness. A high incidence of leg weakness was reported by the Pig Industry Development Authority in boars at the National testing stations[248] and there is evidence that 1st and 2nd generation pigs produced by hysterectomy are prone to leg abnormalities.[249,250] The Agricultural Research Council (1967)[107] estimated the mineral and vitamin requirement for pigs and the leg weaknesses referred to above have developed in pigs given diets adequate in respect of these nutrients.

Pigs affected with leg weakness stand on their toes, at least in the early stage of the deformity, and they appear to have difficulty in walking; there is flexion of the carpal, metacarpo-phalangeal and inter-phalangeal joints and this is often accompanied by lateral deflection of the leg bones which produces a 'knockneed' or 'bow legged' condition. The ulnae and radii are frequently distorted about their longitudinal axis. In the later stage of deformity there is considerable extension of the metacarpo-phalangeal joints which produces 'dropped' pasterns. As far as the hind legs are concerned, the pigs place the feet well forward under the belly during walking and this gives a swinging gait not unlike that of pigs with pantothenic acid deficiency. Plates 1 and 2 show cases of leg weakness.

Lameness often occurs in pigs growing rapidly[251,252,253] and it is clinically at its worst when the pigs weigh between 45–65 kg liveweight. Growth rate is only reduced when the condition becomes so severe that the pigs find difficulty in walking to their feeders. The incidences are not confined to any particular sex.[254,255] Unevenness of claws was observed in a high proportion of the pigs[254] and frequently there are swellings on the lower part of the legs but these later have been shown to be chronic bursitis[256] and not necessarily associated with the leg condition. Similarly, unevenness of claws is not thought to be part of the syndrome, but is hereditary.[257]

Penny *et al.*[257] have studied the conditions of trotters of slaughterhouse pigs. They describe various types of foot rot which could cause lameness, and which collectively accounted for a high proportion of the pigs they examined at slaughter weight. Other workers[255] have suggested that leg weakness may be due to infection at some stage with *Erysipelothrixerhusiopathiae*, but

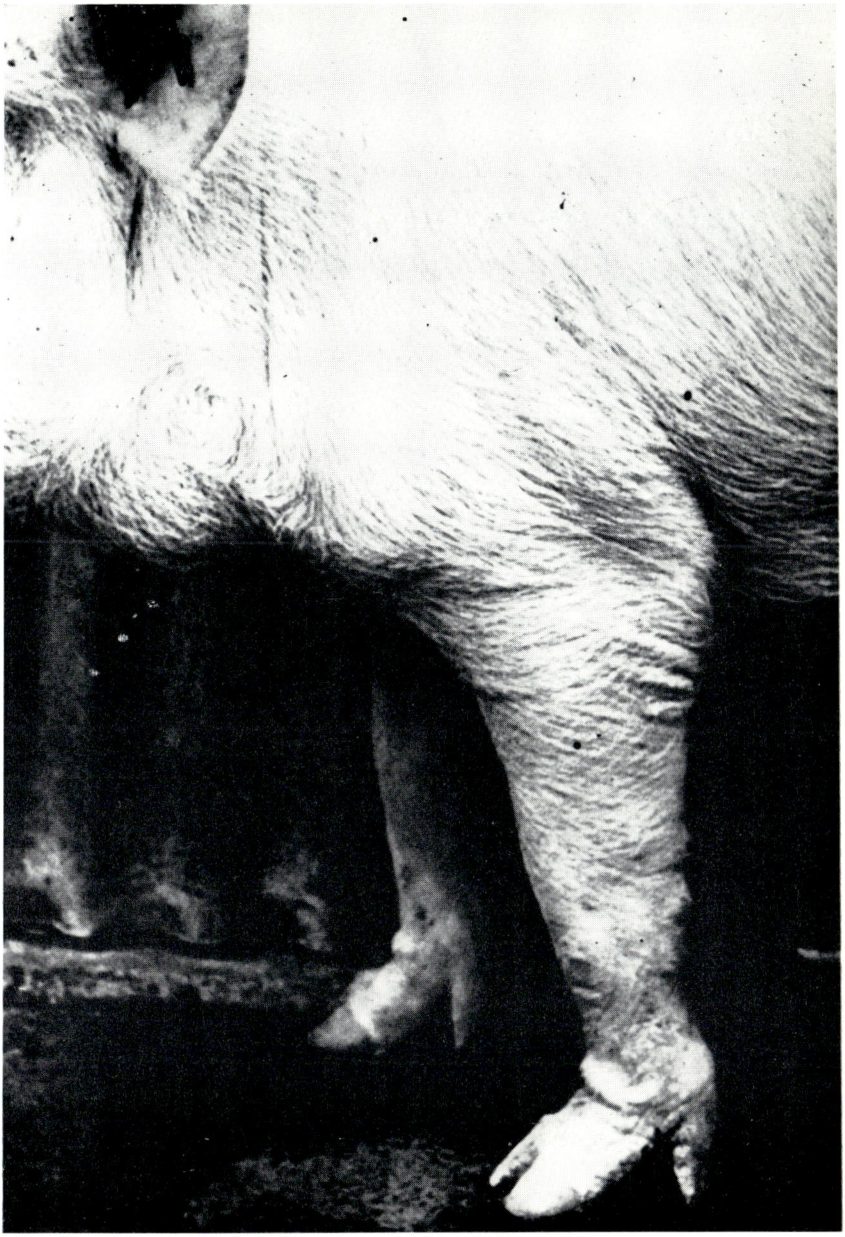

Plate 1. Boar with dropped pasterns.

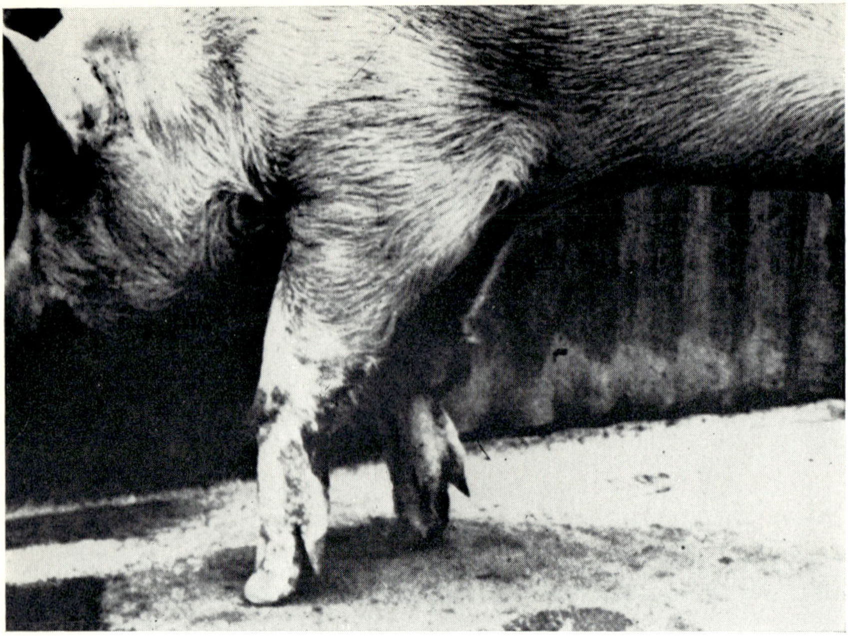

PLATE 2. Incipient stage of leg weakness—boar walking on toes.

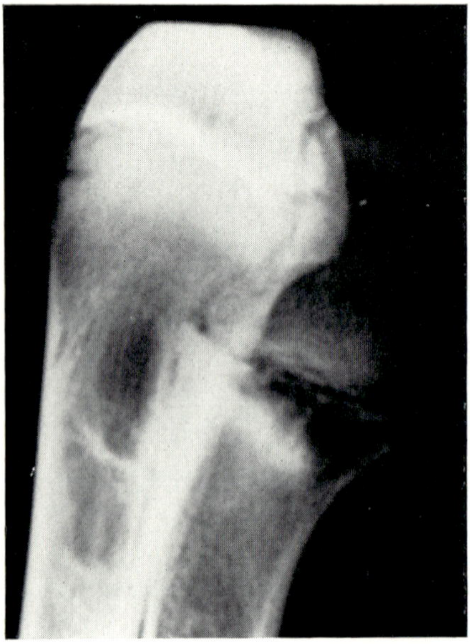

PLATE 3. Rarefaction of the distal ulna.

other workers do not substantiate this.[257,254] The extent of the erosion of the articular surfaces of cartilages was not correlated with clinical observations of leg weakness.[258]

Lesions or radio-translucent areas in the diaphysis of the distal ulna (see Plate 3) have been associated with leg weakness,[259] and studies on the effect of growth rate on the development of these lesions indicated that the extent of the lesions was reduced as growth rate and energy intake were reduced,[254,259] although the extent of the lesions was not necessarily associated with the severity of the clinical syndrome. Histological and radiographical examinations suggested that the primary lesion of the ulna was necrosis of the epiphyseal cartilage, the lesion occurring almost exclusively on the proximal surface of the epiphyseal plate. The bone lesion which followed was due to fracture of greatly elongated columns of calcified cartilaginous lattice, and there was evidence of recovery and repair.[260] Although histologically the condition is similar to skeletal scurvy, oral or intravenous administration of ascorbic acid to pigs of 20 kg liveweight or heavier, did not prevent the condition.[254] Similarly in other studies, where the lameness was characterized by loosening of the articular or epiphyseal cartilage of the femoral head, ascorbic acid supplements were without effect.[261] The ascorbic acid concentration of plasma in affected pigs appears normal,[254] although in young minimal disease pigs and in early weaned pigs which are both prone to leg weakness, plasma ascorbic acid concentrations are lower than in suckled pigs.[262]

The syndrome does not appear to be due to deficiencies of the major minerals or vitamin D. There were no significant differences in the ash content of ulnae taken from boars with leg weakness, and the manipulation of dietary concentrations of calcium and phosphorus within the range 0·8–1·6 per cent of air dry diet, or the addition of vitamin D, had no influence on the incidence of leg weakness even though some of the dietary treatments produced significant improvements in the retention of calcium.[254] Dietary supplements of manganese did not improve the condition[254] although similar lesions have been produced by feeding a manganese deficient diet.[263]

A study [264] of the various pressures exerted on leg bones, and of the changes in these pressures brought about by unevenness of claws illustrates the magnitude of these pressure loads. The severity of lesions in the ulna is increased by increasing the weight borne by the fore limb, or reduced by protecting it from stress.[260] The lesions found in the ulna are similar to those reported in rabbits and which were produced by experimental compression of the joints.[265] Thus stress appears to be a factor in the development of ulnar lesions, perhaps by compression of the vascular bed and there is support for this view from other work.[266,267,268] It is of interest, however, that necrosis was common[260] regardless of previous history and that this was exaggerated by stress. Most of the studies to date have been made with pigs of 20 kg liveweight or more, many of which could have had mild necrotic lesions at that weight, and there is a need to study the effect of nutritional and stress factors earlier in life.

Leg deformities were reported by Blair[269] in early weaned pigs and the lameness occurred even on diets which provided adequate amounts of minerals and vitamins according to estimated requirements. However, there are considerable difficulties in the assessment of mineral requirements. For example, optimal retention of these minerals within the body, or optimal breaking strength of bones are not necessarily indicative of 'requirement' and it would appear that the concentrations of calcium or phosphorus suggested by the Agricultural Research Council[107] are inadequate for early weaned pigs. Certainly the retentions of calcium and phosphorus increase when graded amounts of these minerals are given up to 1·6 per cent of the air dry diet,[270] and in another trial pigs given diets containing 1 or 0·8 per cent calcium and phosphorus showed signs of calcium deficiency with leg weakness.[271] It has been suggested that early weaned pigs should be given diets containing 1·3 per cent calcium and 1 per cent phosphorus.[272]

REFERENCES

1. Karlsson, B. W. *Acta Path. Microbiol. Scand.* **67**, 237 (1966).
2. Young, G. A., and Underdahl, N. R. *Cornell Vet.* **39**, 120 (1949).
3. Barrick, E. R., Matrone, G., and Osborne, J. *Proc. Soc. Exp. Biol. Med.* **87**, 92 (1954).
4. Hoerlein, A. B. *Amer. J. Vet. Res.* **13**, 67 (1952).
5. Speer, V. C., Brown, H., Quinn, L., and Catron, D. V. *J. Immunol.* **83**, 632 (1959).
6. Olsson, B. *Nord. Vet-Med.* **11**, 41 (1959).
7. Nelson, J. B. *J. Exp. Med.* **56**, 835 (1932).
8. Lecce, J. G. and Morgan, D. O. *J. Nutr.* **78**, 263 (1962).
9. Payne, L. C. and Marsh, C. L. *J. Nutr.* **76**, 151 (1962).
10. Lecce, J. G., Matrone, G. and Morgan, D. O. *J. Nutr.* **73**, 158 (1961).
11. Lecce, J. G., Morgan, D. O. and Matrone, G. *J. Nutr.* **84**, 43 (1964).
12. Lecce, J. G. and Morgan, D. O. *J. Nutr.* **90**, 3 (1967).
13. Sterzl, J., Kostka, J., Riha, I. and Mandel, L. *Folia Microbiol.* **5**, 29 (1960).
14. Miller, E. R., Ullrey, D. E., Inge Ackerman, Schmidt, D. A., Hoefer, J. A. and Luecke, R. W. *J. Anim. Sci.* **20**, 31 (1961).
15. Foster, J. F., Friedell, R. W., Catron, D. and Diekmann, M. R. *Iowa State College J. Sci.* **24**, 421 (1950).
16. Nordbring, F. and Olsson, B. *Acta Soc. Med. Upsalien.* **62**, 193 (1957).
17. Karlsson, B. W. *Acta Path. Microbiol. Scand.* **67**, 83 (1966).
18. Lecce, J. G. and Matrone, G. *J. Nutr.* **70**, 13 (1960).
19. Lecce, J. G. and Matrone, G. *J. Nutr.* **73**, 167 (1961).
20. Lecce, J. G. and Matrone, G. *J. Nutr.* **70**, 33 (1960).
21. Lecce, J. G., Matrone, G. and Morgan, D. O. *Ann. N.Y. Acad. Sci.* **94**, 250 (1961).
22. Staub, H. and Boguth, W. *Zentralbl. Vet.-Med.* **3**, 653 (1956).
23. Hoerlein, A. B. *J. Immunol.* **73**, 112 (1957).
24. Bustard, L. K., Ham, W. E. and Cunha, T. J. *Arch. Biochem.* **17**, 249 (1948).
25. Weybrew, J. A., Stewart, H. A., Matrone, G. and Peterson, W. J. *J. Anim. Sci.* **8**, 207 (1949).
26. Lecce, J. G. and Reap, B. R. *J. Exp. Med.* **115**, 491 (1962).
27. Alexander, V. A. W. and Jones, A. S. Personal communication.
28. Sharman, G. A. M. Personal communication.

29. Pond, W. G., Roberts, S. J., Dunn, J. A., King, J. M., Maner, J. H., Lowrey, R. S. and Olafson, P. *J. Anim. Sci.* **20**, 88 (1961).
30. Owen, B. D., Bell, J. M., Williams, C. M. and Oakes, R. G. *Can. J. Anim. Sci.* **41**, 236 (1961).
31. Young, J. A., Jr. and Underdahl, N. R. *Arch. Biochem. Biophys.* **32**, 449 (1951).
32. Reber, E. F. *Vet. Med.* **57**, 23 (1962).
33. Betts, A. O., Lamont, P. H. and Littlewort, M. C. G. *Vet. Rec.* **72**, 461 (1960).
34. Schneider, D. L. and Sarett, H. B. *J. Nutr.* **89**, 158 (1966).
35. Meyer, R. C., Bohl, E. H., Henthorne, R. D., Thrap, V. L. and Baldwin, D. E. *Lab. Anim. Care* **13**, 655 (1963).
36. Whitehair, C. K. and Waxler, G. L. *Lab. Anim. Care* **13**, 665 (1963).
37. Berry, T. H., Jenson, A. H., Becker, D. E. and Woods, J. T. *J. Amer. Vet. Med. Ass.* **141**, 689 (1962).
38. Thorbek, G. and Ludvigson, J. B. *Proc. 8th Int. Congr. Anim. Husb.* **2**, 95 (1961).
39. Perrin, D. R. *J. Dairy Res.* **21**, 55 (1954).
40. Hartman, P. A., Hays, V. W., Baker, R. O., Neagle, L. H. and Catron, D. V. *J. Anim. Sci.* **20**, 114 (1961).
41. Bailey, C. B., Kitts, W. D. and Wood, A. J. *Canad. J. Agric. Sci.* **36**, 51 (1956).
42. Johnson, W. S. *Vet. Rec.* **79**, 363 (1966).
43. Schneider, D. L. and Sarett, H. B. *J. Nutr.* **89**, 43 (1966).
44. Smith, H. and Lucas, I. A. M. *J. Agric. Sci. Camb.* **48**, 220 (1956).
45. Smith, H. and Lucas, I. A. M. *J. Agric. Sci. Camb.* **49**, 405 (1957).
46. Smith, H. and Lucas, I. A. M. *J. Agric. Sci. Camb.* **49**, 409 (1957).
47. Crane, F. M. *J. Anim. Sci.* **12**, 912, Proc. (1953).
48. Crampton, E. W. *J. Anim. Sci.* **12**, 912, Proc. (1953).
49. Crampton, E. W. and Ness, O. M. *J. Anim. Sci.* **13**, 357 (1954).
50. Speer, V. C., Ashton, G. C., Diaz, F. and Catron, D. V. *Iowa Farm Sci.* **8**, 10 (1954).
51. Salmon-Legagneur, E. and Aumaitre, A. *Ann. Zootech.* **4**, 181 (1962).
52. Aumaitre, A. and Salmon-Legagneur, E. *Ann. Zootech.* **10**, 127 (1961).
53. Salmon-Legagneur, E. and Jouandet, C. *Ann. Zootech.* **11**, 311 (1962).
54. Smith, D. M. *N.Z. J. Agric. Res.* **3**, 896 (1960).
55. Lucas, I. A. M. and Lodge, G. A. The nutrition of the young pig, Tech. Comm. No. 22, Commonw. Agric. Bureaux, Commonw. Bureau of Anim. Nutr., Rowett Inst. (1961).
56. Smith, H. Ph.D. Thesis, Univ. Aberdeen (1956).
57. Walker-Love, J. *Fmrs. Wkly.*, Nov. 1959.
58. Frape, D. L., Hays, V. W., Speer, V. C., Jones, J. D. and Catron, D. V. *J. Anim. Sci. Proc.* **18**, 1492 (1959).
59. Elsley, F. W. H. *J. Agric. Sci.* **61**, 233 (1963).
60. Lucas, I. A. M., Calder, A. F. C. and Smith, H. *J. Agric. Sci.* **53**, 136 (1959).
61. McMeekan, C. P. *J. Agric. Sci. Camb.* **30**, 276 (1940).
62. McMeekan, C. P. *J. Agric. Sci. Camb.* **30**, 387 (1940).
63. Pálsson, H. *Progress in the Physiology of Farm Animals*, ed. J. Hammond, Butterworths Scientific Publications, London, 1955.
64. Wallace, L. R. *J. Agric. Sci. Camb.* **38**, 93 (1948).
65. Fowler, V. R. Ph.D. Thesis, Univ. Leeds (1964).
66. Elsley, F. W. H., McDonald, I. and Fowler, V. R. *Anim. Prod.* **6**, 141 (1964).
67. Grummer, R. H. and Self, H. L. *J. Anim. Sci.* **14**, 1241 Abs. (1955).
68. Self, H. L. and Grummer, R. H. *J. Anim. Sci.* **17**, 862 (1958).
69. Bowland, J. P. *35th Ann. Feeders Day, Univ. Alberta* **41**, 10 (1956).
70. Sewell, R. F. and Manner, J. H. *J. Anim. Sci.* **19**, 643 (1960).
71. Vukavic, D., Zivkovic, S. and Sreckovic, A. *Stokarstvo* **12**, 519 (1958).
72. Lehrer, W. P., Jr., Moore, P. R., Wiese, A. C. and Pahnish, O. F. *J. Anim. Sci.* **8**, 107 (1949).
73. Hartman, P. A. Hays, V. W., Baker, R. O., Neagle, L. H. and Catron, D. V. *J. Anim. Sci.* **20**, 114 (1961).

74. TOMME, M. F. *Svinovodstvo* (5), 17 (1961).
75. CLAUSEN, H. Leaflet by Trouw & Co., Amsterdam, Feb. (1963).
76. MANDEL, L., TRAVNICEK, J., KVAPIL, O., SIMEK, L. and SYRINEK, F. *Sborn. cal. Akad. Zemed. Ved. Ziv. vyroba.* **6**, 413 (1961).
77. ELSLEY, F. W. H. *J. Agric. Sci.* **61**, 243 (1963).
78. CLAUSEN, H. *Cornell Nutr. Feed Mfr. Proc.* p. 113 (1959).
79. LUCAS, I. A. M., LIVINGSTONE, R. M. and MCDONALD, I. *Proc. 6th Int. Congr., Edin.*, 568 (1964).
80. LUCAS, I. A. M., LIVINGSTONE, R. M. and MCDONALD, I. *Proc. 8th Int. Congr. Anim. Prod., Hamburg. General Reports*, 209 (1961).
81. PEO, E. R., Jr., ASHTON, G. C., SPEER, V. C. and CATRON, D. V. *J. Anim. Sci.* **16**, 885 (1957).
82. HAYS, V. W., SPEER, V. C., HARTMAN, P. A. and CARLTON, D. V. *J. Nutr.* **69**, 179 (1959).
83. COMBS, G. E., OSEGUEDA, F. L., WALLACE, H. D. and AMMERMAN, C. B. *J. Anim. Sci.* **20**, 929 (1961).
84. LLOYD, L. E. and CRAMPTON, E. W. *Can. J. Anim. Sci.* **38**, 65 (1958).
85. LLOYD, L. E., CRAMPTON, E. W. and MACKAY, V. G. *J. Anim. Sci.* **16**, 383 (1957).
86. BAKER, J. P., BERRY, T. H., BECKER, D. E., JENSEN, A. H. and NORTON, H. W. *J. Anim. Sci.* **21**, 516 (1962).
87. LOWREY, R. S., POND, W. G., LOOSLI, J. K. and BARNES, R. H. *J. Anim. Sci.* **20**, 941 (1961).
88. LOWREY, R. S., POND, W. G., LOOSLI, J. K. and BARNES, R. H. *J. Anim. Sci.* **22**, 109 (1963).
89. DANIELSON, D. M., PEO, E. R., Jr. and HUDMAN, D. B. *J. Anim. Sci.* **19**, 1055 (1960).
90. LOWREY, R. S., POND, W. G., LOOSLI, J. K. and MANER, J. H. *J. Anim. Sci.* **21**, 746 (1962).
91. PEO, E. R., Jr. and HUDMAN, D. B. *J. Anim. Sci.* **19**, 477 (1960).
92. THRASHER, D. M. and RODRIGUEZ, P. S. *J. Anim. Sci.* **21**, 378 Abs. (1962).
93. WILBUR, R. D., KELLOGG, T. F., QUINN, L. Y., SPEER, V. C., HAYS, V. W. and CATRON, D. V. *J. Anim. Sci.* **18**, 1514 (1959).
94. PUCHAL, F., HAYS, V. W., SPEER, V. C., JONES, J. D. and CATRON, D. V. *J. Nutr.* **76**, 11 (1962).
95. AQUIRRE, A., WALLACE, H. D. and COMBS, G. E. *J. Anim. Sci.* **19**, 1246 (1960).
96. COMBS, G. E. and WALLACE, H. D. *J. Anim. Sci.* **21**, 95 (1962).
97. MANER, J. H., POND, W. G. and LOOSLI, J. K. *J. Anim. Sci.* **20**, 641 (1961).
98. MANER, J. H., POND, W. G. and LOWREY, R. S. *J. Anim. Sci.* **18**, 1487 (1959).
99. SEWELL, R. F., SHEFFY, B. E., EGGERT, R. G. and LOOSLI, J. K. *J. Anim. Sci.* **12**, 597 (1953).
100. DELIC, I., PALUSEK, I., PUACA, V. and BRKOVIC, M. *Stocarstvo* **15**, 197 (1961).
101. HARVEY, D. *Tech. Commun. Commonw. Bur. Anim. Nutr.* No. 19 (1956).
102. SEWELL, R. F., THOMAS, M. C. and PRICE, D. *J. Anim. Sci.* **20**, 820 (1961).
103. AUMAITRE, A., JOUANDET, C. and SALMON-LEGAGNEUR, E. *Ann. Zootech.* **13**, 241 (1964).
104. BECKER, D. E. and TERRILL, S. W. *Arch. Biochem. Biophys.* **50**, 399 (1954).
105. REBER, E. F., WHITEHAIR, C. K. and MACVICAR, R. *J. Nutr.* **50**, 451 (1953).
106. MANNERS, M. J. and MCCREA, M. R. *Brit. J. Nutr.* **16**, 475 (1962).
107. AGRICULTURAL RESEARCH COUNCIL. *The Nutrient Requirements of Farm Livestock*, No. 3, London, 1967.
108. BRAUDE, R. *Problems in Pig Rearing*, European Association for Animal Production, Rome, 1964.
109. HEDBOM, A. Personal communication quoted by Nordbring and Olsson, *Acta Soc. Med. upsal* **63**, 25 (1958).
110. LASKOWSKI, M., KASSELL, B. and HAGERTY, G. *Biochem. Biophys. Acta* **24**, 300 (1957).
111. NELSON, L. F., ASHTON, G. C., MADDOCK, H. M. and CATRON, D. V. *J. Anim. Sci.* **11**, 771 Abs. (1952).

112. CATRON, D. V., ACKER, D. B., ASHTON, G. C., MADDOCK, H. M. and SPEER, V. C. *J. Anim. Sci.* **12**, 910 (1953).
113. TRIBE, D. E. *Vet. Rec.* **66**, 862 (1954).
114. LLOYD, L. E. and CRAMPTON, E. W. *Canad. J. Anim. Sci.* **38**, 65 (1958).
115. LLOYD, L. E. and CRAMPTON, E. W. *J. Anim. Sci.* **20**, 172 (1961).
116. MORIMOTO, H., TAKAHASKI, A. and TAKAHASKI, S. *Bull. Nat. Inst. Agric. Sci., Japan*, Series G. (20), 61 (1961).
117. SCHLEGEL, W. *Wien. tierärztl. Mschr.* **49**, 293 (1962).
118. BLAIR, R. *Brit. J. Nutr.* **17**, 1 (1963).
119. LODGE, G. A. Ph.D. Thesis, Univ. Aberdeen (1957).
120. COMBS, G. E., HAINES, C. E. and WALLACE, H. D. *Surg. Molec.* **11**, 18 (1960).
121. FINKE, K. J. Thesis, Rheinisch Friedrich Wilhelms-Univ. Bonn (1960).
122. SMITH, D. M., *N.Z.J. Agric. Res.* **3**, 792 (1960).
123. CRAY, A. S. *J. Frmrs' Cl. London* (3) (1955).
124. OSBALDISTON, G. W. *Aust. Vet. J.* **35**, 217 (1959).
125. MINAAR, G. F. *Fmg. in S. Afr.* **35**, 16 (1960).
126. WARD, H. K., REMPEL, W. E. and ENFIELD, F. D. *J. Anim. Sci.* **21**, 976 Abs. (1962).
127. VOGT, D. W., COMSTOCK, R. E. and REMPEL, W. E. *J. Anim. Sci.* **22**, 214 (1963).
128. BOAZ, T. G. and ELSLEY, F. W. H. *Anim. Prod.* **4**, 13 (1962).
129. NIELSEN, H. *N.J.F. Conf., Helsinki* (1963).
130. LODGE, G. A. *J. Agric. Sci. Camb.* **53**, 172 (1959).
131. LODGE, G. A. *J. Agric. Sci. Camb.* **53**, 177 (1959).
132. SALMON-LEGAGNEUR, E. *Ann. Zootech*, **11**, 295 (1962).
133. MCMEEKAN, C. P. *N.Z.J. Agric.* **52**, 278 (1936).
134. BUCHANAN SMITH, A. D. and DONALD, H. P. *J. Agric. Sci.* **27**, 485 (1937).
135. COEY, W. E. *Res. Exp. Rec. Minist. Agric. N. Ire.* **3**, 106 (1955).
136. FREDEEN, H. T. and PLANK, R. N. *Can. J. Anim. Sci.* **43**, 135 (1963).
137. LODGE, G. A. and MCDONALD, I. *Anim. Prod.* **1**, 139 (1959).
138. LODGE, G. A., MCDONALD, I. and MACPHERSON, R. M. *Anim. Prod.* **3**, 261 (1961).
139. SIBBALD, I. R. and BOWLAND, J. P. *Can. J. Agric. Sci.* **36**, 19 (1956).
140. TERRILL, S. W., MEADE, R. J., NELSON, T. S. and BECKER, D. E. *J. Anim. Sci.* **11**, 777 Abs. (1952).
141. AUMAITRE, A. and SALMON-LEGAGNEUR, E. *Ann. Zootech.* **13**, 237 (1964).
142. LIVINGSTONE, R. M., LODGE, G. A., LUCAS, I. A. M., MACPHERSON, R. M. and MACPHERSON, S. J. Unpublished data, Rowett Research Institute (1959).
143. HEADLEY, V. E., MILLER, E. R. and HOEFER, J. A. *J. Anim. Sci.* **19**, 1266 Abs.(1960).
144. WITT, M., ANDREAE, V. and SCHRÖDER, J. *Züchtungskunde* **29**, 142 (1957).
145. WILLIAMS, L. and DIGGS, B. G. *J. Anim. Sci.* **20**, 391 (1961).
146. BRAUDE, R. Unpublished data (1962).
147. MCMILLAN, F. A. and WALLACE, H. D. *J. Anim. Sci.* **13**, 993 Abs. (1954).
148. TEAGUE, H. S. and RUTLEDGE, E. A. *J. Anim. Sci.* **18**, 1486 Abs. (1959).
149. ALDINGER, S. M., SPEER, V. C., HAYS, V. W. and CATRON, D. V. *J. Anim. Sci.* **20**, 249 (1961).
150. RICHTER, K. and ANTONI, J. *Schweinazucht u. Schweinemast*, **9**, 82 (1961).
151. SALMON-LEGAGNEUR, E. and FEVRIER, R. *Ann. Zootech*, **5**, 73 (1956).
152. PEKAS, J. C., COMBS, G. E., VANDEPOPULIERE, J. M. and WALLACE, H. D. *J. Anim. Sci.* **17**, 1221 Abs. (1958).
153. PEKAS, J. C., COMBS, G. E., VANDEPOPULIERE, J. M. and WALLACE, H. D. *J. Anim. Sci.* **18**, 1282 (1959).
154. ELSON, C. E., FULLER, W. A., KILNE, E. A. and HAZEL, L. N. *J. Anim. Sci.* **22**, 946 (1963).
155. GRUHN, K. *Nahrung*, **9**, 325 (1965).
156. ELSLEY, F. W. H. *Ann. Zootech*, **13**, 75 (1964).
157. MCMEEKAN, C. P. *J. Agric. Sci. Camb.* **30**, 276 (1940).
158. WOOD, A. J. and GROVES, T. D. D. *Can. J. Anim. Sci.* **45**, 8 (1965).
159. MITCHELL, H. H., CARROLL, W. E. HAMILTON, T. S. and HUNT, G. E. *Univ. Ill. Agric. Exp. Sta. Bull.* No. 375 (1931).

160. BERGE, S. and INDREBØ, T. *Meld. Norg. Landbruhogsk*, **34**, 481 (1954).
161. HORNICKE, H. *Ztschr. Tierphys. Tierernah. Futtermittelk* **16**, 344 (1961).
162. WOOD, T. B. *J. Agric. Sci. Camb.* **16**, 425 (1926).
163. OSLAGE, H. J. and FLIEGEL, H. *Energy Metabolism*, ed. K. L. Blaxter, Academic Press, London, 1965.
164. KIELANOWSKI, J. *Energy Metabolism*, ed. K. L. Blaxter, Academic Press, London, 1965.
165. JONES, A. S. The nitrogen metabolism of the growing pig. Ph.D. Thesis, Univ. Aberdeen (1960).
166. LUDVIGSEN, J. and THORBEK, G. *Beretn. Forsøgslab, Kobenhavn* No. 320 (1960).
167. CUTHBERTSON, A. and POMEROY, R. W. *J. Agric. Sci. Camb.* **59**, 215 (1962).
168. BROOKS, C. C., THOMAS, H. R., KELLEY, R. F., GRAHAM, P. P. and ALLEN, L. B. *Virginia Agric. Exp. Sta. Tech. Bull.* No. 176, 15 (1964).
169. CLAUSEN, H. George Scott Robertson Memorial Lecture, Queen's Univ., Belfast (1953).
170. OSLAGE, H. J. *Ztschr. Tierphysiol. Tierernahrung Futtermittelk* **17**, 357 (1962).
171. OSLAGE, H. J. *Ztschr. Tierphysiol. Tierernahrung Futtermittelk* **18**, 14 (1963)
172. NEHRING, K., SCHIEMANN, R., HOFFMANN, L. and KLIPPEL, W. *Archiv. fur Tierernährung* **10**, 275 (1960).
173. NEHRING, K., HOFFMANN, L. and SCHIEMANN, R. *Archiv. fur Tierernährung* **13**, 147 (1963).
174. LUCAS, I. A. M. and CALDER, A. F. C. *J. Agric. Sci. Camb.* **47**, 287 (1956).
175. VANSCHOUBROEK, F., WILDE, R. and LAMPO, P. L. *Anim. Prod.* **9**, 67 (1967).
176. LODGE, G. A. and MACPHERSON, R. M.
177. ROBINSON, D. W. *Anim. Prod.* **6**, 227 (1964).
178. GARTON, G. A., DUNCAN, W. R. H., MADSEN, K. A., SHANKS, P. L. and BEATTIE, I. S. *Brit. J. Nutr.* **12**, 97 (1958).
179. LEAT, W. M. F. *Proc. Nutr. Soc.* **18**, xxxi (1959).
180. BAYLEY, H. S. and LEWIS, D. *J. Agric. Sci. Camb.* **64**, 373 (1965).
181. THRASHER, D. M., BROWN, R. E., MULLINS, A. H., HANSARD, S. L. and BROWN, P. B. *J. Anim. Sci.* **18**, 1494 (1959).
182. KENNINGTON, M. H., PERRY, T. W. and BEESON, W. M. *J. Anim. Sci.* **17**, 1166 (1958).
183. RUPNOW, E. H. and ENSMINGER, M. E. *J. Anim. Sci.* **20**, 683 (1961).
184. BARNES, R. H., KWONG, E., POND, W., LOWRY, R. and LOOSLI, J. K. *J. Nutr.* **69**, 269 (1959).
185. CLAWSON, A. J., BLUMER, T. N., SMART, W. W. G., Jr. and BARRICK, E. R. *J. Anim. Sci.* **21**, 62 (1962).
186. NOLAND, P. R. and SCOTT, K. W. *J. Anim. Sci.* **19**, 67 (1960).
187. ABERNATHY, R. P., SEWELL, R. F. and TARPLEY, R. L. *J. Anim. Sci.* **17**, 635 (1958).
188. BAYLEY, H. S. and LEWIS, D. *J. Agric. Sci. Camb.* **61**, 121 (1963).
189. CHRISTENSEN, K. D. *Acta Agric. Scand.* **13**, 249 (1963).
190. CORNELISSEN, J. P., BOOGAERDT, J. and BLOM, B. *Landbouwk. Tijdschr. 's-Grav.* **71**, 791 (1959).
191. LEAT, W. M. F., CUTHBERTSON, A., HOWARD, A. N. and GRESHAM, G. A. *J. Agric. Sci. Camb.* **63**, 311 (1964).
192. NELSON, E. C., DEHORITY, B. A., TEAGUE, H. S., SANGER, M. and POUNDEN, W. D. *J. Nutr.* **76**, 325 (1962).
193. LANNEK, N., LINDBERG, P., NILSSON, G., NORDSTRÖM, G. and ORSTRADIUS, K. *Res. Vet. Sci.* **2**, 67 (1961).
194. AXELSSON.
195. FORBES, R. M. and HAMILTON, T. S. *J. Anim. Sci.* **11**, 480 (1952).
196. WOODMAN, H. E. and EVANS, R. E. *J. Agric. Sci. Camb.* **37**, 202 (1947).
197. POIJÄRVI, I. *Suom. Maat l. Sevr. Julk.* **57**, No. 1 (1944).
198. FISHWICK, V. C. *Agric. Prog.* **20**, 11 (1945).
199. NORDFELDT, S. *Meddn. Lantbrhögsk. Husdjursförs Anst. Norrtälje* No. 23 (1946).
200. NORDFELDT, S. *K. Lantbrhogsk. Annlr.* **21**, 1 (1954).
201. WOODMAN, H. E. and EVANS, R. E. *J. Agric. Sci. Camb.* **37**, 211 (1947).

202. Woodman, H. E., Evans, R. E. and Menzies Kitchin, A. W. *J. Agric. Sci. Camb.* **22**, 657 (1932).
203. Friend, D. W., Cunningham, H. M. and Nicholson, J. W. G. *Can. J. Anim. Sci.* **42**, 156 (1963).
204. Friend, D. W., Cunningham, H. M. and Nicholson, J. W. G. *Can. J. Anim. Sci.* **42**, 55 (1962).
205. Lloyd, L. E. and Crampton, E. W. *J. Anim. Sci.* **14**, 693 (1955).
206. Cunningham, H. M., Friend, D. W. and Nicholson, J. W. G. *Can. J. Anim. Sci.* **42**, 167 (1962).
207. Lucas, I. A. M. M.Sc. Thesis, McGill University (1949).
208. Mitchell, H. H. *Comparative Nutrition of Man and Domestic Animals*, Vol. 1, Academic Press, New York, London, 1962.
209. Barber, R. S., Braude, R. and Mitchell, K. G. Unpublished data, N.I.R.D. (1960).
210. Bohstedt, G. and Fargo, J. M. *Bull. Wis. Agric. Exp. Sta.* No. 425, 114 (1933).
211. Dinusson, W. E., Bolin, D. W. and McIlroy, D. L. *J. Anim. Sci.* **20**, 930 (1961).
212. Jensen, A. H., Becker, D. E. and Terrill, S. W. *J. Anim. Sci.* **18**, 701 (1959).
213. Crampton, E. W., Ashton, G. C. and Lloyd, L. E. *J. Anim. Sci.* **13**, 327 (1954).
214. Jensen, A. H., Becker, D. W. and Terrill, S. W. *J. Anim. Sci.* **18**, 1356 (1959).
215. Hale, O. M., Beardsley, D. W. and Southwell, B. L. *J. Anim. Sci.* **21**, 377 (1962).
216. Woodman, H. E. and Evans, R. E. *J. Agric. Sci. Camb.* **38**, 51 (1948).
217. Woodman, H. E. and Evans, R. E. *J. Agric. Sci. Camb.* **38**, 354 (1948).
218. Woodman, H. E. and Evans, R. E. *J. Agric. Sci. Camb.* **41**, 102 (1951).
219. Kropf, D. H., Bray, R. W., Phillips, P. H. and Grummer, R. H. *J. Anim. Sci.* **18**, 755 (1959).
220. Jones, A. S. *J. Sci. Fd Agric.* **12**, 353 (1961).
221. Pfander, W. H. and Tribble, L. F. *J. Anim. Sci.* **12**, 927 (1953).
222. Whitehair, C. K. and MacVicar, R. *Misc. Publs. Okla. Agric. Exp. Sta. M.P.* **27**, 34 (1952).
223. Bellis, D. B. *Anim. Prod.* **3**, 331 (1961).
224. Ericson, L. E., Larsson, S. and Ostholm, C. O. *Acta Agric. Scand.* **12**, 157 (1962).
225. Robinson, D. W. and Lewis, D. *J. Sci. Fd Agric.* **14**, 806 (1963).
226. Pfander, W. H. and Tribble, L. F. *J. Anim. Sci.* **14**, 545 (1955).
227. Curtin, L. V., Loosli, J. K., Willman, J. P. and Williams, H. H. *J. Anim. Sci.* **11**, 459 (1952).
228. Jones, A. S., Hepburn, W. R. and Boyne, A. W. *J. Sci. Fd Agric.* **12**, 353 (1961).
229. Jones, A. S., Livingston, D. M. S. and Livingstone, R. M. *Anim. Prod.* **5**, 222 (1963).
230. Bowland, J. P. and Berg, R. T. *Can. J. Anim. Sci.* **39**, 102 (1959).
231. Jones, A. S., Livingston, D. M. S. and Cadenhead, A. *Anim. Prod.* **6**, 259 (1964).
232. Clausen, H. *Wld. Rev. Anim. Prod.* **1** (No. 1), 28 (1965).
233. Raeburn, J. R. and Blair, R. Report to PIDA Conference, Brighton (1966).
234. Boyne, A. W., Carpenter, K. J. and Woodham, A. A. *J. Sci. Fd Agric.* **12**, 832 (1961).
235. Lea, C. H. and Hannan, R. S. *Biochem. Biophys. Acta* **4**, 518 (1950)
236. Carpenter, K. J. *Biochem. J.* **77**, 604 (1960).
237. Dawson, R. Personal communication (1967).
238. Clawson, A. J., Smith, F. H., Osborne, J. C. and Barrick, E. R. *J. Anim. Sci.* **20**, 547 (1961).
239. Hale, F. and Lyman, C. M. *J. Anim. Sci.* **20**, 734 (1961).
240. Jones, A. S. and Cadenhead, A. *Anim. Prod.* **4**, 297 (1962).
241. Jones, A. S. and Cadenhead, A. *J. Sci. Fd Agric.* **16**, 38 (1965).
242. Barber, R. S., Braude, R., Chamberlain, A. G., Hoskin, Z. D. and Mitchell, K. G. *Brit. J. Nutr.* **18**, 545 (1964).
243. Bunyan, J. and Woodham, A. A. *Brit. J. Nutr.* **184**, 537 (1964).
244. Jones, A. S., Livingstone, R. M. and Cadenhead, A. *Anim. Prod.* **7**, 286 (1965).
245. Prescott, J. H. D. and Lamming, G. E. *J. Agric. Sci.* **63**, 341 (1964).
246. Robinson, D. W. *Anim. Prod.* **6**, 227 (1964).
247. Robinson, D. W. and Lewis, D. *J. Agric. Sci.* **63**, 185 (1964).
248. Johnson, R. Personal communication.

249. Frape, D. Personal communication.
250. Jones, A. S. Personal communication (1967).
251. Lodge, G. A. and MacPherson, R. M. *Anim. Prod.* **3**, 19 (1961).
252. Jetes, D. L., Conrad, J. H., Plumlee, M. P. and Beeson, W. H. *J. Anim. Sci.* **19**, 226 (1960).
253. Goodwin, R. F. W. *J. Comp. Path.* **72**, 214 (1962).
254. Walker, T. Ph.D. Thesis, Univ. Aberdeen (1966).
255. Duthie, I. F. and Lancaster, M. C. *Vet. Rec.* **76**, 263 (1964).
256. Fell, B. F. Personal communication.
257. Penny, R. H. C., Osborne, A. D. and Wright, A. I. *Vet. Rec.* **75**, 1225 (1963).
258. Done, J. T., Melrose, R. and Jones, A. S. Personal communication.
259. Walker, T. and Jones, A. S. *Anim. Prod.* **4**, 297 (1962).
260. Walker, T., Fell, B. F., Jones, A. S., Boyne, R. and Elliot, M. *Vet. Rec.* **79**, 472 (1966).
261. Hupka, E. Uber die Begattungaispotens der Jungeber (Epyphysiolysis), *Dtsch. Tierärztl. Wochenschr.* **66**, 201 (1959).
262. Jones, A. S. and Ball, S. E. Personal communication.
263. Neher, G. M., Doyle, L. P., Thrasher, D. M. and Plumlee, M. P. *Amer. J. Vet. Res.* **17**, 121 (1956).
264. Gray, J. *J. Expt. Biol.* **20**, 88 (1944).
265. Trueta, J. and Trias, A. *J. Bone and Joint Surg.* **43-B**, 800 (1961).
266. Freese, H. H. *Arch. Tierernahrung* **8**, 330 (1958).
267. Vaughan, L. C. PIDA Conference on Leg Weakness in Boars (1962).
268. Trueta, J. and Amato, V. P. *J. Bone and Joint Surg.* **42-B**, 57 (1960).
269. Blair, R. *J. Agric. Sci. Camb.* **57**, 373 (1961).
270. Livingstone, R. M., Lucas, I. A. M. and McDonald, I. *Anim. Prod.* **4**, 297 (1962).
271. Dudley, W. A. *Dissertation Abst. XXI*, **5**, 1088 (1960).
272. Blair, R. and Benzie, D. *Brit. J. Nutr.* **19**, 91 (1964).

CHAPTER 27

The Production of Meat from Pigs

K. L. ROBINSON

Agricultural Chemistry Department, The Queen's University of Belfast

INTRODUCTION

PIGS are kept solely for the production of meat and it might be thought that this was a simple and uncomplicated objective. In fact, however, pigs are produced for two main markets: (i) Wiltshire bacon, which requires animals of about 90 kg at slaughter. For this the carcasses must conform to certain specific measurements emphasizing leanness because the processing method involves brine-curing of whole untrimmed sides, which are then sliced to provide bacon rashers of acceptable size and proportions of lean and fat, (ii) Pork and by-products, which require pigs varying from 45 kg to 130 kg at slaughter, depending upon local demand. For these the uncured sides are cut and varying proportions sold as fresh uncured pork, or manufactured as cured products such as bacon and ham, or as sausages, pies and processed meats. Carcass measurements are not of major importance, for providing there is a good development of lean meat, excess fat can be trimmed off and used in other products.

The requirements of these contrasting methods of disposal have had marked effects on pig production. Thus in countries such as Denmark, Canada and the United Kingdom in which the Wiltshire bacon trade is prominent, breeding and management policies give major emphasis to producing lean pigs, whereas in the U.S.A. and in some European countries the predominance of the cutter trade allows greater attention to factors such as maximum growth rate. At the same time an increasing consumer preference for lean meat in all pig-producing countries is having recognizable effects on production methods. With farm animals, in general, the primary means of bringing about desirable patterns of production is by the transmission of appropriate characteristics through selective breeding. The changes in productive performance thus achieved can best be appreciated by first considering the growth pattern of the species.

GROWTH AND BODY CONFORMATION

In its simplest expression the growth of pigs, like that of other mammals, has three main phases. In the first the skeleton grows more rapidly than in any subsequent period. This phase is followed by a period in which muscular development predominates. In the third phase the rate of fat deposition reaches a maximum. Fat is first deposited in the subcutaneous tissues and around the kidneys, stomach and intestines; as maturity approaches, the back fat increases in thickness and fat is deposited in and between the muscles.[1]

During these phases, as the pig becomes older, there is a regular change in the proportions of the body. An anterior to posterior sequence of development takes place, starting at the head and neck, then passing to the shoulders and lastly to the loins. As a consequence the head and neck are proportionately large in relation to the loin in the young animal, but by the time slaughter weight is reached the position is reversed. While this description illustrates the changes which take place in pigs which have been bred to meet consumers' requirements, it must be realized that without continual selection for this quality of change in body proportions, as the animal grows up, improvement of the pig for market requirements would cease.[2] For example, the unselected wild pig grows up with but little change in body proportions and puts on its weight in the early developing, low-priced, neck and shoulders while the desirable loin remains poorly developed. More detailed anatomical changes brought about by crossing the Danish Landrace with the European Wild pig have been shown by Clausen[3]; some of these are given in Table 1.

These figures indicate the slow growth rate of the wild pigs and correspondingly greater food requirements and suggest that selection for rapid growth and efficient utilization of food, in commercial terms, involves an unwitting selection for increasing alimentary capacity. Also evident is the greater length of body, with increased number of ribs and the reduction in thickness of peripheral fat in the Landrace. A subjective visual scoring showed that the Wild pig had very heavy shoulders and poorly developed hams.

Original selection of pigs for meat production was undoubtedly based on visual characters, but a direct relation should not be assumed between the alterations in body conformation thereby recognized and assessed and the superior growth rates of the modern breeds. The striking improvements in the productive performance of pigs in some countries have been secured by officially sponsored schemes of performance and progeny testing. These will be referred to again in a later section.

Conformation scores and body measurements have been used in attempts to classify morphological variation in farm animals but have not been efficient as predictors of growth rates and at present the relationships of the qualitative aspects of body form to the genetic variation in rates and extents of growth lack definition. It appears unlikely that improved growth rate in animals can be brought about by selection for a particular body conformation, but the rate at which an animal approaches maturity may be related to body

TABLE 1

The Results of Crossing the Danish Landrace with European Wild Pig
(From data of Clausen [3])

Generation	Wild pig	Landrace	Age in days at liveweights of		Daily weight gain (g)	F.U.* per kg liveweight gain	Length of intestine (m)			Diameter of small intestine (mm)	Weight of stomach (empty) (g)
			20 kg	90 kg			Small	Colon + Rectum	Total		
	Per cent	Per cent									
Backcross	75	25	142	388	286	5·79	13·81	4·14	17·95	24·6	424
F1	50	50	107	256	469	4·44	16·50	4·18	20·48	25·0	484
F2	25	75	85	217	528	3·74	17·55	4·49	22·04	27·9	509
F3	12·5	87·5	85	198	592	3·57	17·69	4·44	22·13	28·7	556
F4	6·3	93·8	78	193	621	3·43	18·85	4·57	23·42	27·4	552
F5	3·1	96·9	80	181	695	3·12	21·38	4·64	26·02	29·7	616
Landrace 1951–52	—	100	76	180	674	3·06	21·32	4·68	26·00	31·7	620

* One F.U. (Feed Unit) corresponds to the feeding value of 1 kg barley.

structure.[4] Some aspects of the genetic basis of size and conformation, the problems of altering these characters, and the extent to which others may be affected in the process, have recently been discussed by Taylor.[5]

GROWTH AND NUTRITION

The efficiency of feed utilization *when expressed as units of food/1 unit liveweight gain* generally diminishes with increasing age and body size. There is nevertheless sufficient variation in the relationships to indicate that ability to utilize nutrients efficiently is a heritable characteristic. Kunkel[4] gives the following estimates of heritability of growth characteristics of pigs:

	Range (per cent)	Average (per cent)
Post-weaning growth	14–58	29
Efficiency of feed utilization	8–72	31

An example of the exploitation of the genetic potential for improvement in food utilization is seen in the results of the Danish Progeny Testing Stations[3] where during the period 1907–53 the feed required for 1 kg liveweight gain fell from 3·77 kg to 3·06 kg.

Hammond's theory of differential or heterogenic growth[7] offers an explanation of the general principles of growth and development in farm animals in relation to nutrition. Various organs and tissues are considered to differ in the order, in time, of their maximum rate of growth, the order of development being, (i) brain and nervous tissue, (ii) bone, (iii) muscle, (iv) fat. Hammond[8] took the view that various components of the body therefore constitute a series of priorities in competing for available nutrients for growth. When the plane of nutrition is sufficiently high all the demands will be met; if the plane is progressively lowered a stage is reached when fat ceases to be put on, the other tissues continuing to grow. At a still lower plane bone and brain continue to grow, but at a much reduced rate, muscle growth ceases and fat depots are depleted to supply energy. This hypothesis, originally developed from observations made on sheep, became the basis of extensive investigations at Cambridge by Hammond's associates which have contributed much to knowledge of the patterns of growth in pigs. [6,10,11]

McMeekan[6] showed that a high level of feeding, which allowed fast growth, was associated with increased fat deposition particularly if feeding was stepped up from a low to a high plane at about 45 kg liveweight. However, a re-evaluation of McMeekan's data by Elsley *et al.*[12] has shown that if growth is expressed in terms of fat-free body mass (the skeletal bone and muscle) the relative growth of bone and muscle conforms to the allometric equation[13,14], $y = ax^b$, where y = weight of muscle and x = weight of muscle and bone. This relationship appears to be unaffected by differences in plane of nutrition.

This implies that at any stage of growth food restriction will limit fat deposition and if it retards skeletal development this will occur uniformly throughout the body and not in relation to the order of maturity of the different parts. On the other hand, food given in excess of that required for maintenance and maximum development of muscle and bone is deposited as fat. Although this can occur at any stage in growth the likelihood of excess food being consumed becomes greater with age because of the progressive diminution of growth rate as the animal approaches maturity.

This latter point should perhaps not be over-emphasized, however, because a pig at slaughter weight of 90 kg may be only about 40 per cent of its mature weight[15] and still on the rapidly ascending part of its growth curve (Fig. 1). The reasons for frequently observed differences in fatness between animals receiving uniform treatment are by no means clear, for example to what extent are the variations the result of genetically differing mature weights, genetically differing metabolic patterns for individual parameters of growth—muscle, bone or fat—or of individual responses to behavioural and environmental stresses.

At all events it is clear that the expression, or suppression, of the animal's growth potential is greatly affected by nutrition. In an experiment with Large White pigs Holme[16] imposed daily gains of 668 g or 450 g by high or low levels of feeding from 18 kg liveweight to 55 kg, both followed by a high level allowing daily gains of 818 g to slaughter at 90 kg. The animals on the high plane throughout had the fatter carcasses, suggesting that the level of feeding in the early stages provided nutrients in excess of those required for fat-free body growth. The general conception of the role of depot fat in the animal economy has in the past been that of an inert reservoir of energy, to be augmented or depleted as the occasion demanded. Newer knowledge is making it increasingly evident that adipose tissue is more than a repository for fat and considerable evidence has accumulated that it functions metabolically in the synthesis of lipids, the storage of triglycerides and the release of unsaturated fatty acids to the blood. It has been suggested[18] that the large absolute amount of adipose tissue, coupled with its lipogenic ability, is responsible for most of the lipogenic syntheses in the body. Although this knowledge derives mainly from work with rats, its possible significance in relation to the pig should not be overlooked. For example, Lucas *et al.*[19] and Boaz and Elsley[20] found that with early weaned pigs rapid growth up to about 23 kg as compared with restricted growth, both followed by uniform treatment to slaughter at 90 kg, was associated with fatter carcasses. A similar effect was obtained by Nielsen[78] and, with suckled pigs, by Vanschoubroek.[22] In this respect the position in Denmark where improvement in food conversion and carcass quality has been associated with relatively lighter weaner weights,[78] is perhaps of some significance.

It seems possible that a pig which has acquired a depot of adipose tissue might henceforth metabolize its nutrients differently from a leaner animal.

COMPENSATORY GROWTH

Many animals show capacity to recover from periods of underfeeding by making better than normal rates of growth when restored to normal nutrition. This phenomenon has been called compensatory growth and is the physiological basis of the production of 'store' animals. Although it is well established that compensatory growth can occur with pigs the effects of this on feed conversion efficiency and carcass conformation are less certain. McMeekan[6] found that low–high planes of nutrition produced fatter pigs, in contrast to the findings of Winters *et al.*[25]; Brugman[26] and Lucas *et al.*[27] It is probable, however, that factors such as sex and age of animal, diet and duration of underfeeding will affect the ultimate response to nutritional recuperation. The effects of moderate pre-weaning restriction have already been considered. Those during the post-weaning stage have been recently studied by Robinson[28] in an experiment in which low levels of feeding were applied to different groups of pigs at different stages through the weight range 20 kg to 90 kg. By the use of the instantaneous growth rate equation of Brody[29],

$$R = \frac{dw/dt}{w}$$

where R is a function of growth already made and w and t are bodyweight and time, comparisons were made between treatment groups over equal weight intervals. In all cases the relative (but not the actual) growth rates of the restricted groups were greater than the controls in the last stage of growth, when all the animals were on the control plane of nutrition. Growth immediately following a lifting of food restriction was, on the whole, directly related to the duration of restriction and the response of gilts was greater than that of castrates. In no case did the period of rapid growth following the lift of restriction fully compensate for the previous retardation but there were indications that equality might have been achieved if growth had continued to about 115 kg. Although at some stages the efficiencies in feed conversion showed improvement following restriction, the overall figures showed little difference. No estimates of carcass lean meat were obtained but the restricted groups had superior commercial grading, suggesting, as was found by Winters *et al.*[25], Brugman[26] and Lucas *et al.*[27] that the pigs were less fat.

Although not likely to happen under commercial conditions a consideration of the effects of severe under-nutrition of young pigs is of interest. McCance[30,31] has shown that pigs can be held to a weight of less than 6 kg for a period of a year by giving them very little of a nutritionally adequate and acceptable diet. They acquired an abnormal appearance and characteristics but these were abolished when the animals were allowed to eat freely and made rapid weight gains. The animals in many cases grew to almost, if not quite, normal adult size, with quite normal body proportions. The similarity of the shape of the growth curves (Fig. 1)[30] emphasizes the usefulness of the Brody[29] growth equation

$$\frac{dw}{dt} = -k_2\,(A-W)$$

(where A is the mature weight of the animal)

in defining the parameters of growth. The equation states that the ultimate limiting factor to growth is the animal's inherent mature body size, regardless of how growth may be controlled. Growth rate is thus represented as a condition of the growing organism in a particular environment and not a direct function of time or age.

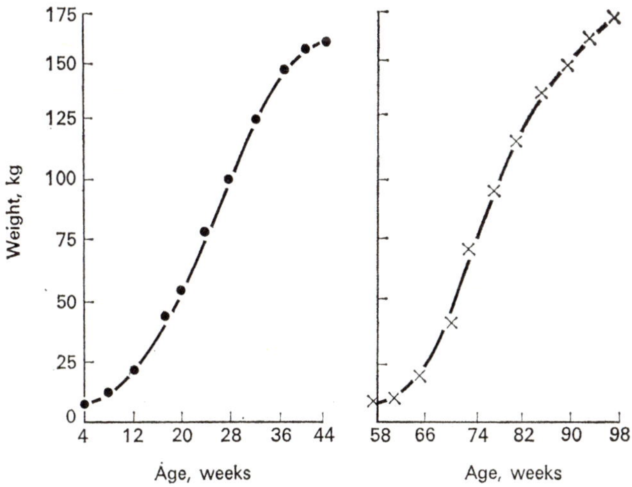

FIG. 1

Not all of McCance's animals had an uneventful growth stasis and rehabilitation; for example some developed recognizable infections and some had abnormalities in their bones. The author[31] points out, however, that if a low plane of nutrition does curtail the attainment of a full genetic stature the adjuvant role of infections will have to be defined. In other words the permanent effects of a low plane of nutrition on stature may not be metabolic at all; under-nutrition may merely pave the way for pathological invasions. These views are of particular significance in the light of the possibilities of nutritional and physiological studies with germ-free or specific-pathogen-free pigs.

GROWTH AND BODY/CARCASS COMPOSITION

Some matters of definition must first be considered. Studies of the composition of the body of the pig and of its carcass (the edible part of the body) are made for various purposes, and consequently the results may be expressed in a number of different ways. Where the experiment is concerned with the retention of nutrients the whole body may be chemically analysed and the results expressed as weights and percentages of water, protein, fat, and mineral matter (ash). In other investigations the body or carcass may be dissected anatomically into its specific organs, regions, muscles and bones, the data for the carcass being summarized in terms of muscle tissue, fatty tissue and bone. In commercial or economic studies attention may be confined to the separation of the carcass into marketable joints or cuts and the manual dissection of these as 'separable' lean, fat and bone or as 'usable meat' (boneless lean and fat); the extensive literature on evaluation techniques and procedures has been reviewed by Harrington,[32] with particular reference to contemporary British practice, and by Hankins *et al.*[33] mainly in relation to U.S.A. practice.

Complete anatomical carcass dissection is extremely laborious—Cuthbertson and Pomeroy[11] give a figure of 110 man-hours per half carcass—and in many experiments reliance has to be placed on the partial dissections possible during factory processing, and on the correlative support of various linear carcass measurements and weights of joints and bones taken at various times between killing and processing. The predictive value for lean-meat content of some of these indices has been studied recently by Buck *et al.*[34] Their measurements included thickness of backfat at various joints from shoulder to rump, area and depth (the 'B' measurement) of eye-muscle—the cross-sectional area and depth of the longissimus dorsi, exposed by a transverse cut at the last rib—and the thickness of fat over the eye-muscle (the 'C' and 'K' measurements), Fig. 2.

The results confirm that sex and breed are important determinants of leanness and show that in using conventional backfat measurements as predictors of carcass leanness on individual animals high residual standard errors (up to 2 per cent of side-weight) are involved. However, it should be borne in mind that the errors are much less if the appropriate regression equations are used simply to place carcasses in a number of grades or classes according to the calculated percentage leanness.

It has been found by Buck *et al.*[34] that better predictions may be obtained by using the 'internal' carcass measurements—the B, C and K measurements—and they suggest that the possibility of making these on Wiltshire bacon carcass, without the transverse cutting of the carcass, must be considered. They consider that the ultrasonic technique[35], previously used by Price *et al.*[36] to map the eye-muscle boundary in live pigs, could, with the development of

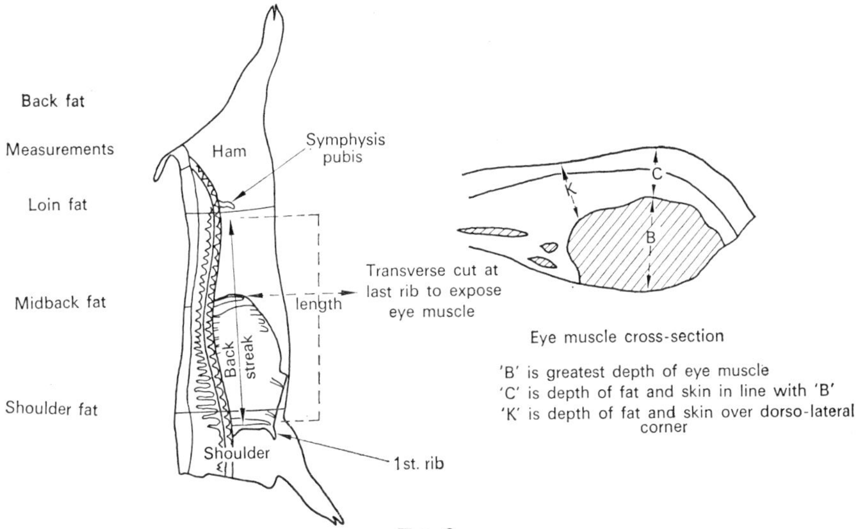

FIG. 2

suitable apparatus, be used for the 'B' measurement. The present ultrasonic apparatus and the Danish optical probe are both considered suitable for the 'C' and 'K' measurements.

Yet another prediction index of leanness is that of carcass density, determined by weighing the sides in air and in water.[37] The method is based on the low density of fat relative to that of the fat-free tissues (muscle and bone). Bias may be introduced if allowances are not made for sex and breed, which bring about variations in muscle/bone ratio,[34] and errors may arise, such as the incomplete removal of the chine bone—the central portion of the vertebral column—during factory handling.[11] Unlike most of the simple indices of carcass composition, however, density determinations can be carried out on individual cuts or joints as well as on the side, thus providing information about the distribution of fatty tissue. Accuracy is also unaffected by carcass deformations during handling and hanging.

KILLING-OUT PERCENTAGE

Before considering some body and carcass analyses in more detail, attention must be drawn to different applications of the terms carcass and carcass weight. In British commercial practice[60] the pig producer is paid on the weight of the bled, scraped and eviscerated body, i.e. the body less the blood, hair, internal organs and the alimentary tract and contents but including the kidneys and associated fat. The weighing is made whilst the carcass is still warm but a standard deduction is made for loss of weight on cooling. The experimenter is more likely to base his results on a headless, de-chinned chilled carcass, as is generally the case in the determination of carcass densities.

As a rule the commercial 'hot' carcass is about 70 to 75 per cent of the liveweight of the animal; clearly this killing-out or dressing percentage will be affected by the weight of the gut contents and therefore by the time and amount of the last meal before slaughter and it is also affected by the weight of the animal. As will be seen from the data in Table 2 from Clausen[3] as a pig approaches slaughter weight the gain in edible meat and fat is much greater

TABLE 2
Composition of the Danish Landrace Pig during Growth
(From data of Clausen[3])

Liveweight (kg)	Net* weight (kg)	Blood (kg)	Bristle and hoofs (kg)	Offal: Edible (kg)	Offal: Non-edible (kg)	Fat† (kg)	Meat (kg)	Bones (kg)
20	17·70	0·97	0·09	1·19	1·46	3·31	8·04	2·64
50	44·50	2·14	0·18	2·52	3·05	11·27	19·95	5·39
90	81·59	3·48	0·28	3·95	4·66	25·60	35·35	8·27
120	109·73	4·32	0·33	4·76	5·56	38·51	46·10	10·15
90–120	28·14	0·84	0·05	0·81	0·90	12·91	10·75	1·88

* The pig without stomach content, intestines and other losses (evaporation, etc.)
† Including leaf fat and intestinal fat and skin.

than that of the gut, internal organs (offal) and blood, so that as the pig becomes heavier the killing-out percentage rises.

The relationship between liveweight and carcass weight increments of pigs from about 13 kg to 105 kg liveweight has been studied by Smith.[38] He found the regression of hot or cold carcass weight on liveweight to be linear ($Y - 0{\cdot}84x - 4{\cdot}4 \pm 3{\cdot}1$ where Y = chilled carcass weight and x = liveweight). The corresponding relationship between dressing percentage and liveweight is shown in Fig. 3[38].

From what has already been said it will be evident that information about the effect of weight on composition is required under several headings: (i) the composition of the body at different weights, in terms of the chemical constituents, water, protein, fat and ash; (ii) the composition of the carcass at different weights in terms of the skeleton, muscle or lean meat and fat; (iii) the composition of the weight gains as muscle, fat and bone at different body weights; (iv) the composition of specific tissues, as chemical constituents, at different body weights.

(i) CHEMICAL COMPOSITION OF THE BODY

Some of the earliest sets of data for the chemical composition of the empty body—liveweight less the contents of the alimentary tract—are those of Ellis

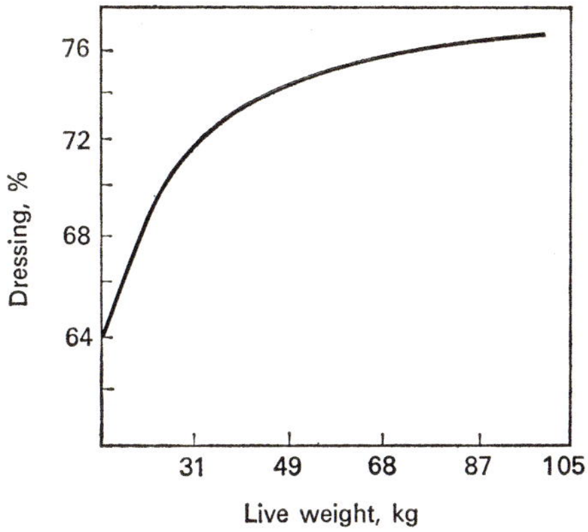

FIG. 3

and Hankins,[39] derived from dissection and analysis, ranging from 6 kg up to 100 kg at slaughter and fed *ad lib*. Table 3 and Fig. 4 give means of 4–6 and 2–5 animals respectively, at each stage of growth, the breeds including Hampshires, Chester Whites, Durocs and Berkshires. Where the liveweights correspond, the figures for protein, water, fat and ash are in general agreement with those of Weninger,[40] with those of Wood and Groves[41] for young pigs, and with the trend which can be deduced from the data of Wood,[42] McMeekan[9] and Cuthbertson and Pomeroy.[11] Although the individual results show that the compositional changes associated with growth take place at different rates for each animal a general pattern is discernible (Fig. 4[39]).

TABLE 3
Composition of Pigs at Certain Stages of Growth
(From data of Ellis and Hankins[39])

Stage of growth	1	2	3	4	5	6
No. of pigs	5	4	4	4	4	6
Liveweight at slaughter, kg	6·3	12·8	30·3	49·4	73·2	100·0
Total weight of parts analysed, kg	6·0	11·5	28·0	46·2	69·5	95·6
Percentage of analysed weight*						
Water	64·12	69·84	66·64	57·40	50·54	43·74
Protein	14·64	15·33	16·14	15·36	14·01	12·32
Fat	18·25	11·29	14·25	24·07	32·83	41·78
Ash	3·15	3·66	3·08	2·96	2·65	2·27
Weight of fat by periods, kg	—	0·2	2·7	7·2	11·7	17·2
Percentage fat gained by periods	—	3·16	13·19	36·06	46·15	61·88

* Based on the combined results obtained from the separate analyses of blood, organs, meat, bone and skin.

The body water declines steadily from about 60 per cent in the 18 kg animal to 44 per cent in the 90 kg animal and the fat content rises from about 18 per cent to about 40 per cent. Protein shows a small rise from an initial 14 per cent to 16 per cent at about 20 kg weight, and then slowly declines to 12 per cent at 90 kg. In a parallel manner the ash content rises slightly and then declines to around a figure of 3 per cent.

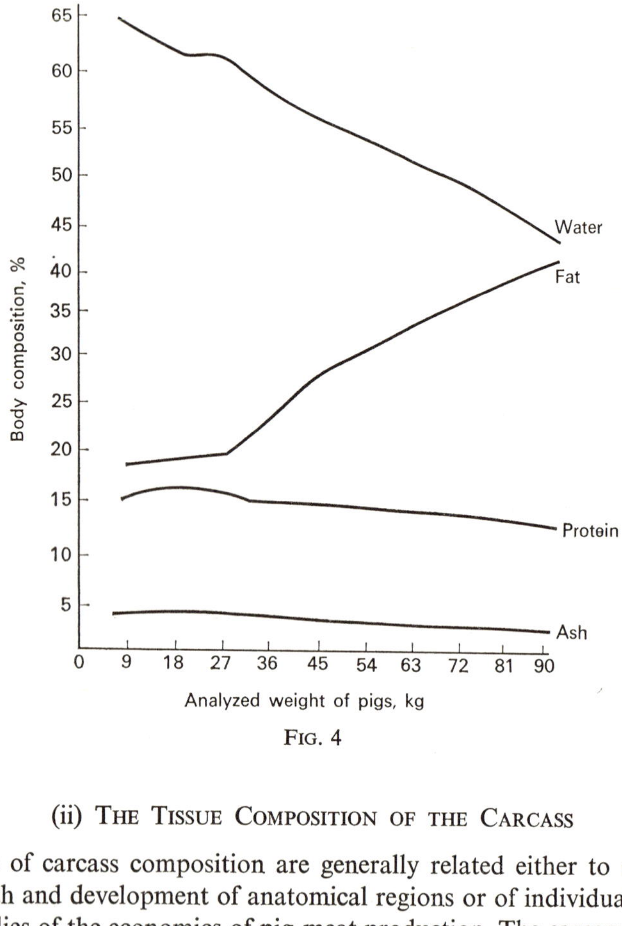

FIG. 4

(ii) The Tissue Composition of the Carcass

Studies of carcass composition are generally related either to studies of the growth and development of anatomical regions or of individual muscles, or to studies of the economics of pig meat production. The carcass examination is usually confined to the manual dissection and separation of tissues into specific categories of muscles, fat and skeleton, or more simply as separable lean meat, fat and bone. Chemical data appears to be seldom obtained and therefore some caution must be exercised in making deductions about chemical composition from carcass dissection data; it must be borne in mind that muscle, or lean meat, contains some fat and mineral matter as well as its main constituents of protein and water, and separable fat contains small amounts of the other three substances. The dissection technique is also subjective and arbitrary procedures have to be adopted, for example it may be difficult to decide whether a particular item of fat should be considered intermuscular or subcutaneous. Nevertheless, the agreement between separable and analytically determined fat in the carcasses of 90 kg pigs has been found to be sufficiently close[37] to justify transposing such data where a wide range in fat contents is being examined.

The general pattern of carcass composition is indicated by the figures in

TABLE 4
Carcass Composition of Danish Landrace Pigs
1945 (From data of Clausen[3])

Liveweight (kg)	Carcass weight (kg)	Proportion of carcass (%)		
		Meat	Fat	Bone
20	14	57	24	19
50	37	54	31	15
90	70	51	37	12
120	95	49	41	10
1964 (data of Clausen[168])				
90	65·3	56·2	32·8	11·0

Table 4 which have been recalculated to a carcass weight basis from Clausen's data for Landrace pigs (Table 2). As is to be expected, the percentage of fat shows the same trend as in Ellis and Hankins whole-body data (Table 3), but the percentage of meat falls appreciably with increasing liveweight because it includes a large proportion of the markedly declining amount of body water.

Cuthbertson and Pomeroy[11] carried out quantitative anatomical studies of the composition of the Large White pig fed semi-*ad lib.*, as defined on p. 1018, to 68 kg, 91 kg or 118 kg liveweight, corresponding to carcass weights of 50, 68 and 92 kg. The rounded-off figures for the composition of the side (de-chinned, head-off half carcass) are given in Table 5 and follow the same pattern

TABLE 5
*Carcass Composition of Large White Hogs**
(From data of Cuthbertson and Pomeroy[11])

Liveweight (kg)	Carcass weight (kg)	Proportions of one side (%)			
		Muscle	Fat	Skeleton	Remainder
68	50	50	31	11	8
91	68	48	35	9	8
118	92	44	41	8	7

*Hog = male castrate.

of those of Clausen (Table 4). They are also in line with the detailed dissection and analytical data of Kropf *et al.*[43]

(iii) Composition of Weight Gains at Different Liveweights

The data of Cuthbertson and Pomeroy also show the composition of the daily gains during three periods of growth. These are given as tissue percentages of the daily gain in Table 6, and as g/day gain of tissues in Fig. 5.[11]

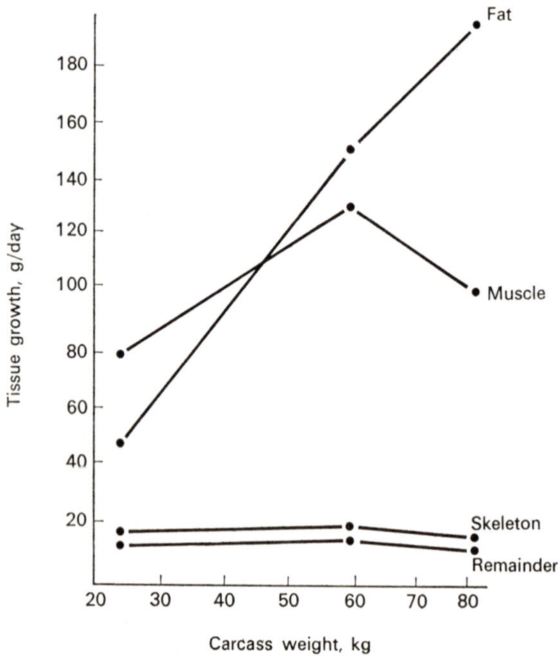

FIG. 5

The authors suggest that the positions of equi-point of muscle and fat production and peak of muscle production may not be very accurate in view of the wide intervals of time between stages at which the pigs were dissected.

TABLE 6
Rate of Growth of Tissues in the Carcass
(From data of Cuthbertson and Pomeroy[11]*)*

Carcass Weight (kg)	Daily growth of carcass (%)			
	Muscle	Fat	Skeleton	Remainder
1–50	50	31	10	9
50–68	41	46	7	6
68–92	32	59	5	4

The figures in Table 6 indicate that in the later stages of finishing to bacon weight, with these pigs fed semi-*ad lib.*, the weight gains are about 50 per cent fat.

At this stage it should be noted that body composition, and particularly the composition of serial weight gains, may be expressed in terms of energy deposition. This terminology may be applied to the results of carcass dissections by the use of conversion factors (e.g. 1 g fat = 9·4 kcal; 1 g protein = 5·75 kcal) or by carrying out simultaneous nitrogen and carbon balances[45] in a respiration chamber.[46, 47] The determinations of nitrogen and carbon deposition in the body by simultaneous balances has the great advantage that serial results can be obtained on each animal in the experiment, but for the assessment of carcass composition the technique has certain limitations, which will be considered in the next section.

The results of Osinska[48] are an example of energy deposition figures obtained by calculation in an experiment with Large White pigs slaughtered at about 90 kg weight and with empty body weights of about 86 kg. The mean values of protein, chemical fat and energy in the edible carcass, carcass and empty body are given in Table 7.

TABLE 7

Composition of Empty Body, Whole Carcass and Edible Parts of Carcasses of Pigs as Protein, Chemical Fat and Energy (From data of Osinska[48])

	Protein (N×6·25)	Chemical fat	Energy
	(kg)	(kg)	(Mcal)
Edible part	7·4	21·9	248
Carcass	10·4	24·0	285
Empty body	12·7	25·4	312
Edible part as per cent of empty body	58·5	85·0	80

The results indicate that only about 58 per cent of the empty body protein is found in the edible carcass as compared with 85 per cent of the fat. If, however, the results are expressed on an energy basis, 80 per cent of the energy of the body is found in the edible carcass. Oslage and Fliegel[49] carried out simultaneous serial N and C balances on pigs fed *ad lib.*, from 20 kg up to 160 kg liveweight and from these calculated the deposition of protein, fat and total energy. The results suggest that even at the beginning of the trial pigs of 30–40 kg weight were depositing as fat almost 70 per cent of their total energy retention. With increasing energy intake this proportion increased very rapidly so that in pigs of 50–60 kg it was 80 per cent and rose still further to 87 per cent. The relationship between protein and fat synthesis in these pigs can be seen in Table 8. Although the protein is only about 30 per cent

TABLE 8
Protein and Fat Synthesis in Pigs
(From data of Oslage and Fliegel[49] and Oslage[167])

Liveweight kg	Protein deposition g/day	Fat deposition g/day	N intake dig. N g/day	N retention % of intake
20–30	99	—	30	52
30–40	104	148	39	43
40–50	111	218	48	37
50–60	111	264	43	41
60–70	115	369	51	36
70–80	108	412	43	40
80–90	114	386	46	40
90–100	107	385	47	37
100–110	105	409	48	35
110–120	107	405	52	33
120–130	117	395	54	35
130–140	92	355	53	28
140–150	93	357	55	27
150–160	87	345	54	26

of the total energy deposition (as kcal) at the 30–40 kg stage, and about 13 per cent at the 150–160 kg stage, it is clearly of greater magnitude on a weight-basis. It is therefore evident that the various ways in which protein deposition can be considered: as crude protein ($N \times 6{\cdot}25$), as a fat-free lean tissue (hydrated protein) and as a component of energy retention, all affect the interpretations which may be placed on food utilization and the efficiency of the process and this point must always be kept in mind.

It is perhaps worth reiterating that even with these quite young growing pigs fat deposition rather than protein deposition dominated energy retention. As Oslage and Fliegel[49] point out this situation makes it very difficult to discern a growth (protein synthesis) phase from a fattening phase. They also suggest that although the energetic efficiency of protein formation may be higher than that of fat, the dominant role of fat synthesis in the young pig makes it not surprising that the efficiency of energy utilization for the growth process as a whole is frequently not substantially above the energy efficiency in fat formation by adult animals.

(iv) THE COMPOSITION OF SPECIFIC TISSUES AT DIFFERENT BODY WEIGHTS

It has long been known that the chemical composition of the body fat changes with increasing liveweight. Typically Ellis and Hankins found that the iodine value of the fat fell from 72 in the 6 kg pig to 59 at the 100 kg stage. This was associated with an almost constant value for oleic acid of about 50 per cent and a fall in linoleic acid from 11 to 6 per cent; the saturated acids rose from about 31 per cent to 36 per cent. The explanation generally proposed is that oleic acid and the saturated acids may be synthesized from carbohydrate but linoleic and the other more unsaturated acids, which promote soft fat of high iodine value, are of dietary origin. Therefore, in a situation of rapid fat synthesis the body fat is increasingly derived from carbohydrate and becomes correspondingly harder. Hence it is that, given a standard diet, younger pigs tend to have soft backfats of different chemical composition from that of older pigs. Equally it follows that at any weight stage there is an association between the thickness of the back fat and its chemical composition and physical texture. For example, severe food restriction, by limiting fat synthesis, will bring about a thin, soft backfat[50]. Diets high in unsaturated fatty acids, for example diets high in maize, weatings, or oily fish or meat meals, may also contribute to soft unsaturated backfats, which are commercially undesirable.[1]

Less well recognized are the physical or physico-chemical changes which may take place in the lean and fatty tissue of the body with increasing maturity. From the data of McMeekan[6] (Table 9), for the backfat and the longissimus dorsi muscle, it is evident that in both the fatty and muscular tissue there is a decline in water content and an increase in fat content with increasing age/weight. Since in the fat-free muscle the water content falls and the dry-matter content rises it would seem that the requirement of nitrogen per unit of fat-free muscle increases with increasing liveweight.

If these figures are accepted it is clearly not valid to assume that pig muscle has a constant chemical composition and it would seem desirable to attempt to relate nitrogen retention to the stage of growth of the animal. It may well be that similar considerations apply to fat synthesis and that the energy expenditure for this purpose at any one time is dependent upon the composition of the pre-existing depots.

If, therefore, protein and fat depositions as determined by N and C balances cannot be converted to weight gain in the growing pig by assuming fixed tissue compositions it is desirable to have information about the other major component, body water, both in terms of overall retention and in its distribution within tissues at different liveweights. There would seem to be a great need for the development of *in vivo* methods of determining body water, in conjunction with more detailed studies of tissue composition by dissection and analysis. The methods of *in vivo* water determination in pigs include air displacement, helium dilution, underwater weighing, antipyrine dilution, tritium oxide dilution, deuterium oxide dilution and urinary creatinine, all of which are described in the literature.[51] The lean body mass may also be determined by gamma ray measurements of 40-potassium[51]. Biopsy and core techniques for estimating body composition, as described by Bray[75] and Everitt[24] appear to hold considerable promise. The degree of precision required and the value of various techniques in this respect has been con-

TABLE 9

Age Changes in the Chemical Composition of Tissues of Pigs
(From data of McMeekan[6])

Empty veweight (kg)	Age (weeks)	Back fat†						Muscle (longissimus dorsi)			
		Outer layer			Inner layer						
		Fat	Fat-free tissue	Water	Fat	Fat-free tissue	Water	Fat	Muscle tissue	Water	Water, fat-free basis
		%	%	%	%	%	%	%	%	%	%
1·3*	birth	6·2	8·9	84·9	6·2	8·9	84·9	1·9	16·6	81·5	83·1
6	4	75·4	5·1	19·5	76·8	4·7	18·5	4·3	19·9	75·8	79·2
13	8	76·7	6·7	16·7	76·9	8·0	15·1	4·7	19·0	76·2	80·0
36	16	84·4	6·4	9·1	87·8	4·8	7·4	3·4	21·0	75·7	78·3
52	20	85·9	3·9	10·2	90·7	2·7	6·6	4·0	21·6	74·4	77·5
100	28	92·4	2·6	4·9	94·3	1·9	3·8	5·6	22·6	71·8	76·0

* Inner and outer layers of fat bulked owing to small size of samples.
† Subcutaneous strip running from head to tail.

sidered by Panaretto[52] who points out that methods applicable to one species or age may be unsuitable in other circumstances. Kay and Jones[53] found that a tritium oxide dilution method considerably overestimated body water in pigs of 55 and 90 kg but that *N*-acetyl 4-amino antipyrene (NAAP) injected intravenously gave a difference between predicted and actual body water of $-$ 1·5 per cent $\pm$ 2·06 per cent. They[54] found poor relationships between body composition and body density, determined by a combined technique of water displacement and helium dilution. Recently Groves and Wood have reported encouraging results for the *in vivo* determination of total body water in the suckling pig by a deuterium oxide dilution technique,[55] and consider that the relative simplicity of the technique and equipment recommend them for more extensive use in sequential studies of body composition of growing animals.

It is disappointing to have to conclude at this point that *in vivo* methods of determining body composition are still very much in the experimental stage. Undoubtedly the search for suitable methods is well worth pursuing for the sequential determination of body composition could contribute greatly to the interpretation of growth and balance studies.

THE FOOD OF THE PIG AND THE MANNER OF FEEDING

Many sources of information are available about the nature and composition of the major components of pig feeds and the formulation of standard diets.[56,57,58,59,60] Detailed descriptions of feeding stuffs will therefore not be attempted and accounts of diets will be largely in relation to specific topics.

It has been a matter of comment by Eden[61] and Crampton,[62] that the major dietary components of mixed feeds—cereals, cereal by-products, animal and vegetable protein foods—are pretty much as they were 20 or 30 years ago. There has been much improvement, however, in the processing of these materials, for example in the heat treatment of proteins, to avoid the destruction of essential nutrients. Today there is a greater knowledge of the nutritional requirements of the pig and of the nutrient content of feeding stuffs, and hence an extensive development in the application of scientific principles of feed formulation. It should be noted, however, that according to Crampton[62] the trend has been towards keeping down costs and his survey of U.S.A. and Canadian practice suggests little improvement in recent years in items such as growth rate and feed conversion. But most important of all has been the growing recognition of the importance of those micro-dietary supplements and 'additives'—vitamins, minerals, antibiotics, hormones, organo-arsenicals—some of which safeguard the available nutrient levels in the feed and others the well-being of the animal against non-specific infections. Hence it is that modern pig feed formulation is adaptable to the electronic computation of linear programmed least-cost rations meeting minimum nutrient requirements, or allowing specified margins of safety, say, in protein, lysine, methionine, calcium, phosphorus, riboflavin, pantothenic acid, and vitamins B_{12}, A and D.[62,63]

The advent of feed additives has made farm mixing of rations increasingly difficult for the individual farmer and he tends to rely on ready-mixed feeds or buys a supplement (the so-called balancer) containing the appropriate proteins, vitamins and minerals for supplementation of his home-grown cereals. As Breirem[64] has pointed out, at the present state of knowledge of nutrition, feed rations can easily be balanced to prevent gross deficiencies and the progressive farmer need not be too worried in this respect. Energy and protein being quantitatively the decisive factors in nutritive requirements, it follows that variations, first in energy intake and secondly in protein, exert the major effects in controlling growth. In the case of suitable balanced rations growth rate in pigs seems to be largely determined by energy intake[65, 66] and it is a matter of dispute whether the balancing protein is best given on a basis of a daily intake, or as a percentage of the energy. In view of the relative protein and fat depositions in full-fed pigs noted earlier,[49] it may be that ideally a constantly changing protein/calorie ratio is required but this is not always practicable on the farm. In British practice for bacon production two rations are generally used: a 'grower' feed, for pigs up to about 60 kg, and a 'finisher' feed thereafter. However, the particular objectives of pork cutter or heavy cutter production may modify the weight ranges over which these diets are fed. Because the percentage retention of dietary protein falls as growth proceeds, and protein has a diminishing influence on growth rate, the finisher ration is made up to contain a lower level of protein. It is open to question whether these relative levels are in keeping with present carcass requirements and this point will be considered later.

To restate these nutritional factors in model terms, it is probable that when a pig is fed a normally balanced ration without stint it will show a relatively more rapid rate of liveweight gain, but with a higher proportion of fat in the weight increment, than if it had been given a lesser amount of food. The fatter carcass is the less acceptable product and therefore the various interrelated factors involved in the production of the leaner carcass merit detailed consideration, for even if the market requirements for lean meat are met by reducing the slaughter weights, or by increasing the slaughter weights and manually trimming off surplus fat tissue, the basic need for lean-meat production still remains. It will help to simplify the position if it can be accepted that the pig, by reason of its subcutaneous backfat depots, is physiologically destined to produce a carcass which, relative to other animals, is very fat. For example, Meyer and Nelson[67] full-fed rats, chicks and pigs the same diets to give liveweight gains characteristic of the species. The gains of the rats contained 10 per cent fat and 20 per cent protein, those of the chicks 6 per cent fat and 30 per cent protein, and those of the pigs 50 per cent fat and 12 per cent protein. These figures give some indication of the inherent differences between broilers and pigs in relation to their nutrition. The extent to which breeding and selection towards leaner types of pigs may be fruitful—or is even justifiable in view of world food situations[23]—therefore has to be considered. It would seem improbable that any dramatic changes can result from the application of new nutritional knowledge, unless perhaps from the manipulation of the endocrine system, and any improvement is most likely to arise from a better understanding of the specific effects of each of a number of interrelated nutritional and environmental factors.

Much research work on pigs has necessarily been towards immediately practical objectives by following the overall growth of the animal in a given set of conditions and less progress has been made in the analysis of the basic mechanisms of physiological responses to the individual nutritional and environmental factors which determine growth. Under more stable and less exacting conditions of production and marketing than exist today, empirical observations and conclusions could have a wide application, but this approach is increasingly inadequate to meet the variable and constantly changing systems of feeding, housing and management of the present age. The value of the fundamental approach is that it attempts to establish a pattern of interrelationship between the factors which collectively determine growth and thus to enable the prediction of the effects of variation in any of them. In the light of what has just been said a consideration of the methods of feeding pigs is of particular importance.

METHODS OF FEEDING PIGS

Some of the developments in the practical methods of feeding pigs have been introduced without much regard for their possible nutritional effects. Although the proportionate cost of the feed is high—as much as 80 per cent of the total in Britain—the desire to reduce labour requirements, by automation or other devices, is evident. This has led to a variety of current feeding methods which differ to varying degrees in factors such as the physical condition and acceptability of the food, the method of dispensing, the distribution of the meal-times, the amount of food made available and the amount of competition between pigs for food. Thus trough-feeding wet or dry meal, off-the-floor feeding of dry meal or cubes, self-operated dry meal feeding and trough or bowl feeding of pumped slurries of meal with water, whey, or separated milk each may impart gradations in nutrient intake the possible effects of which are ignored or unrecognized. For example, the results of Braude *et al.*[68] suggest that under particular conditions the method of providing water *per se* may influence appetite, irrespective of the amount of water consumed. Undoubtedly some of the current feeding systems used experimentally[44] have been adopted with the intention of regulating food intake but even so it should not be assumed that the comparative effects obtained are solely the result of quantitative differences in food intake. In relation to methods of limited feeding Clausen[44] mentions the following generally accepted procedures; feeding *ad libitum* (*ad lib.*) = feeding to appetite from self-feeders; feeding semi-*ad libitum* (semi-*ad lib.*) = feeding by hand two or three times daily as much as the pigs will eat in 20 to 30 min; limited (restricted) feeding = feeding by hand once or twice daily an amount of food according to a fixed scale based on liveweight. While the absolute intake of nutrients is probably the main difference between the various systems, the possible effects of the distribution of meal-times should not be overlooked. Unfortunately in most comparisons of restricted and *ad lib.* feeding of pigs the effects of frequency of feeding are confounded with those of magnitude of intake.

It should also be noted that in comparisons of restricted and *ad lib.* feeding it is implicit that the method of presentation of the food has no effect on its acceptability and its apparent digestibility. Some consideration must therefore be given to these questions before attempting to summarize the effects of food restriction.

APPETITE IN THE PIG

Some caution must be exercised in the interpretation of appetite. What is implied here is the expression of the animal's voluntary intake of food. There is little evidence about any food preferences, or indications that particular foods or ingredients can increase overall food intake by organoleptic means in the growing pig.

The physiological factors involved in voluntary food intake have been reviewed by Kleiber.[69] Blaxter[70,71] points to the considerable evidence that appetite for food is determined by both the physical distention which the food causes in the gut and by homeostatic mechanisms whereby the animal equates its intake of energy to its need. The physiological effects of ingested food which invoke homeostasis are postulated as those of rise of heat production, rise of blood sugar or rise in the rate of fat synthesis. The particular stage at which the pig on *ad lib.* feed stops eating will depend on the digestible energy concentration of the diet. In theory, therefore, if the digestible energy level of the diet be lowered a pig will eat correspondingly more of the less concentrated diet to readjust its energy intake, so that growth rate would be independent of the energy concentration of the diet. However, if the energy value of the diet is progressively lowered a stage is reached where the animal is physically incapable of raising its food intake sufficiently to compensate for the decline in energy value.

It should thus be possible to control the rate of gain of pigs on *ad lib.* feeding by lowering the energy concentration of the diet to below the point where compensation for energy intake is possible. From the considerable literature on the subject, which has been reviewed by Troelson and Bell,[72] it is clear that the effect of increasing the crude fibre level of a basal diet of about 4 per cent crude fibre and 3000 kcal DE/kg, even by quite small percentages, is nearly always to reduce the daily liveweight gains of pigs fed *ad lib.* In some experiments pigs have been found to eat more of a lower energy diet, in others not, differences due possibly to the various kinds of fibrous diluents used, but even where food intake rises it does not do so sufficiently to offset the reduced energy level of the diet. Typical figures are those of Robinson[73] who found that the *ad lib.* intakes of pigs, over the range 20 kg–90 kg, on three fishmeal/cereal diets of: (i) 3 per cent crude fibre and 3500 kcal DE/kg; (ii) 6 per cent crude fibre and 3150 kcal DE/kg; and (iii) 9 per cent crude fibre and 2750 kcal DE/kg averaged 2·5, 2·5 and 2·6 kg/day respectively. Meyer and Nelson[67] fed pigs *ad lib.* at two levels of crude fibre and energy: (i) 7·6 of fibre and 3600 kcal DE/kg; (ii) 13·6 per cent and 3000 kcal DE/day (the diets were fat-supplemented) starting at 35 kg weight and continuing until those on the high energy diets had reached 90 kg (a period of 90 days). The mean daily food intakes were 2·5 and 2·2 kg respectively. Similar results were obtained by Cunningham *et al.*[74]

The question arises as to what may be regarded as the pattern of voluntary food intake of growing pigs on a representative high energy diet. Guilbert and Loosli[15] give figures of 1·4, 2·4, 3·1, 3·4 and 3·8 kg feed/day for pigs of 23, 45, 68, 90 and 113 kg body weight on a diet of 3300 kcal DE/kg. They consider these figures for total feed capacity bear the same general relationship with the metabolic body size (liveweight$^{0.75}$) as do basal heat production and maintenance requirement. It is of interest that the mean food intake of pigs from 23 to 90 kg weight (2·5 kg/day) on Guilbert and Loosli's estimate is similar to that found by Robinson[73] and by Meyer and Nelson[67] over a similar range.

It is evident that the growing-finishing pig is very sensitive to quite small percentage alterations in crude fibre level, and to the consequent changes in digestible energy, which are not far removed from the range of normal variation in conventional feeding stuffs. This underlines the necessity for particular care in the obtaining and in interpreting experimental data in this connection.[100]

Just as the consensus of experimental results is that increase in dietary fibre reduces voluntary feed intake and rate of liveweight gain, it is equally conclusive that the slower growth is accompanied by a reduction in carcass fatness as indicated by thinner backfat and other carcass indices. Troelson *et al.*[72] in their experiments with Large White pigs, using corn cobs, cellulose, alfalfa, oathulls or wheatbran, fed diets of mean energy levels of 3000, 2710 and 2620 kcal DE/kg and obtained shoulder fat thicknesses of 44·5, 42·5 and 41·3 mm respectively. The eye-muscle areas on the two lower energy levels were both greater than on the highest level. The authors conclude as a result of these and other experiments that while it is possible to regulate energy intakes in a predictable manner by the addition of bulky fibrous feeds to high energy rations it is difficult to cite an ideal rate of gain in the finishing period in view of the wide difference in inherent carcass quality characteristics among breeds and strains of swine. A further difficulty in the way of prediction is that plant fibre includes a number of structural cellulosic materials and the variable behaviour of these[76] may be further modified by differences in fineness of grinding, by cubing, and by the age of the animal.[77] The increase in digestive ability with age may be thought to be due to a corresponding increase in the capacity of the large intestine, since the fibre is believed to be digested by bacterial action in the caecum and colon, but several workers[74] were unable to find any change in the empty weights of stomach and intestines of pigs on high fibre diets compared with those on low fibre diets. Although the digestibility of fibre varies considerably the products of the cellulose degradation have been found to be well utilized; with fibre from several sources—lucerne, woodflock (wood cellulose), wheat straw and oats husks in decreasing order of digestibility—the metabolizable energy was a fairly constant proportion of digestible energy, ranging from 93·2 to 98·6 per cent[79]. Nevertheless Carpenter[21] points to uncertainty about the efficiency of utilization of gross energy from cellulose which disappears from the gut and to the need for calorimetric studies.

Again, the change in digestible energy with change in level of crude fibre is not determined solely by the digestibility of the fibre; for example added fibre reduces the apparent digestibility of the dietary protein and there is a high degree of correlation between the output of metabolic faecal nitrogen and faecal dry matter. Zelter *et al.*[81] found the regression to be $Y = 0{\cdot}07X + 0{\cdot}850$ ($r = +\,0{\cdot}900$) where $Y =$ metabolic faecal nitrogen and $X =$ faecal dry matter. To generalize, it would appear from an examination of a large number of published results[77] that in pig diets crude fibre has a digestibility of about 25 per cent and that with diets having a crude fibre content of 3–6 per cent, a rise of 1 percentage unit of fibre reduces the apparent digestibility of organic matter and of the crude protein by about 1·5 percentage units.

It will be apparent that limitation of energy intake on *ad lib.* feeding, by dilution with fibre, is difficult in practice. The choice between this method, as for example practised in the U.S.A. and Canada, and that of restricting food intake by one of the systems of hand feeling, as is customary in Scandinavia and Britain, may be determined by relative costs of feeding stuffs and of labour. As Clausen[44] has stated the case, limited feeding requires more labour and reduces the yearly turnover in the piggery, but it decreases feed consumption per unit of gain and improves carcass quality. When fibre-fillers are used in quantities sufficiently large to reduce backfat thickness down to the level achieved with limited feeding, feed consumption per unit of gain increases to such an extent that payment for the filler is negligible. Great interest must therefore be directed towards self-feeders which can weigh out daily rations with sufficient accuracy to meet the fixed scale recommended by limited feeding.

One further economic deterrent to high-fibre diets is the reduction in killing-out percentage, amounting to several units, which they bring about.[74,82] In effect, the liveweight gains are exaggerated by increased weights of digesta, due apparently to their higher water-holding capacity.

RESTRICTED FEEDING

The practice of restricted feeding of bacon pigs includes a number of applications. In one of these the pigs are fed *ad lib.* from a weight of, say, 20 kg to 50 kg and thence restricted to slaughter, at 90 kg. The underlying assumption here seems to be that the pig first passes through a phase of maximum muscle deposition which can be enhanced by liberal feeding; in the succeeding phase of relatively high fat synthesis deposition of fatty tissue is limited by restricted feeding. Although a peak of muscle synthesis has been shown to exist,[3, 11] it is questionable whether it can be favourably affected by *ad lib.* feeding. As has been suggested earlier, there is the alternative hypothesis that a high energy intake in the early stages may encourage metabolic habits of fat synthesis more difficult to limit later on. In other variants of restricted feeding food is given on a basis of body weight according to some particular scale, of which there are a number, such as, for example, allowances of 0·8 or 1 kg, food at 20 kg weight, rising to a maximum of about 2·5 kg, reached at 70 kg weight or to 3 kg, reached at 80 kg weight. Superimposed on these may be once or twice daily feeding. Finally there is the semi-*ad lib.* system of feed to appetite twice daily.

It has been pointed out earlier on that differences may exist between these systems other than from food intake. The work of Cohn and Joseph[83] has shown that rats force-fed twice daily an amount of food equivalent to that eaten by animals feeding continuously throughout the 24 hr deposited more fat and less protein in equal weight gains. This change in lipogenesis has been associated with corresponding alterations in respiratory quotients, nitrogen excretions, and in certain tissue enzyme patterns. Changes in lipid metabolism[84] and nitrogen metabolism[85] in man have also been associated with feeding frequency. However, Friend and Cunningham[86] found no appreciable differences in weight gains or carcass conformation with pigs fed either once or five times daily to a restricted intake, nor did Braude *et al.*[87] with pigs fed once or twice daily, and Crampton[88] found that hand-fed pigs with a limited number of meal-times but eating the same amount of food as self-fed pigs had the leaner carcasses even though they grew 9 per cent faster.

With so many variables, comparisons and particularly economic comparisons, between the effects of different levels of food intake, or planes of nutrition, in experiments on growth rate and carcass composition are greatly improved by large-scale trials with large numbers of animals and replicates. In this respect the co-ordinated trials conducted by Braude on behalf of the Agricultural Research Council of the United Kingdom have provided valuable information. In the present context the results of Experiment 2 of Braude *et al.*[89] are instructive. In a comparison of three degrees of restriction (treatments 2, 3 and 5) against *ad lib.* feeding (treatment 1) it was found that increasing restriction correspondingly reduced growth rate and shoulder-fat thickness, reductions of 2·6, 4·1 and 5·5 mm from the shoulder-fat thickness of 51·3 mm on the *ad lib.* treatment being associated with increases of 4, 15 and 25 days in growing from 30 to 90 kg weight; the effects on eye-muscle dimensions were only slight. The system of feeding on treatment 2 was *ad libitum* to 55 kg, followed by about 3 kg food/day, that on treatment 3 was a widely used scale of feeding, 5 per cent body weight at 20 kg liveweight dropping gradually to 4 per cent at about 75 kg, while on treatment 5 the animals received food at the rate of 3·8 per cent body weight until about 80 kg with no increase thereafter. The reductions in backfat thickness were, as was to be expected, accompanied by corresponding improvements in commercial carcass grading.

Lucas and Calder[50] have reviewed many experiments comparing the effects on the carcass of varying the food intake of hand-fed pigs to 90 kg slaughter weight. They found that, in general, restriction of food intake had little effect on carcass length and killing out percentage, and that moderate restriction, causing delays of up to 30 days to slaughter, reduced average backfat thickness by up to 5 mm but only increased the eye-muscle area (cm^2) by 1·0–1·5 units; the iodine value of the fat was higher by up to 5 units. The authors conclude, however, that the reduced carcass fatness associated with moderate food restriction and growth rate may be no greater than that associated with variations in the genetic potential of pigs. So here again, although the overall effect of calorie restriction, by direct limitation of food, is to reduce carcass fatness there are found to be wide variations in responses within the general pattern which make it difficult to predict results with any degree of precision. A further complication lies in the absence of any clear standards for attainment apart from those involved in the commercial grading and value of the carcass. It is therefore difficult to integrate parameters such as rate of growth, food consumption and carcass composition except in terms of a detailed economic analysis such as that carried out by Braude *et al.*[89] An assessment of this kind is, however, markedly affected by the prevailing price structure.

In another context, however, Braude[90] has listed some targets for efficient pig production (Table 10). Whatever the feasibility of these they have the

TABLE 10

Targets for Efficient Pig Production
(From data of Braude[90]*)*

	1962	To be attained within 10 years
Weaning weight (56 day) of suckled pig, kg	16	20
Time from weaning to 90 kg, days	124	94
Daily liveweight gain, weaning to 90 kg, g	600	750
Food conversion ratio, 18–90 kg, kg standard meal/1 kg liveweight gain	3·5	below 3
Carcass length, pelvis–1st rib, mm	810	825
Backfat thickness, shoulder, mm	45–50	35–40
loin, mm	25–30	20–25
Eye-muscle area, cm^2	27–28	28–30

very considerable merit of introducing a measure of objectivity for comparisons of present and future performances. From the previous discussion it is evident that progress to more rapid liveweight gains or to appreciably thinner backfats cannot be made by manipulations of energy intake alone; the effects of dietary variations in protein still have to be examined. However, studies of the genetic parameters of British Large White and Landrace bacon pigs[91, 92] suggest that prospects of improvement by breeding and selection are encouraging. It is all the more important therefore that every effort be made towards a better understanding of the nutritional and physiological factors in pig growth and development, so that breeding programmes can be put on the best possible basis.

It is instructive to examine in rather more detail Braude's requirement of a reduction in shoulder fat thickness of 10 mm. Using McMeekan's[9] regression equation $Y = 636{\cdot}0\,X - 2775$ [where Y is the weight (g) of fat in the carcass and X (mm) is the thickness of shoulder fat] the backfat thickness of 50 mm corresponds to a carcass fatness of 42 per cent, and that of 40 mm to one of 33 per cent. It is well known[93] that in relation to sex the order of fatness at a given weight is (leanest) boars $<$ gilts $<$ castrates and the data of Carroll *et al.*[94] show boar carcasses to have about 7 percentage units less fat than castrated males. It is thus evident that a discontinuance of the practice of castration would in itself result in a decisive improvement in lean meat production and merits consideration.

Any study of economic pig production is in the main concerned with the animals' ability to convert food into economic products. In this context we have already examined the overall effect of the factors of appetite and food digestibility on the intake of food. We now have to consider the animals' efficiency in converting the absorbed nutrients into products of marketable value.

EFFICIENCY OF FOOD CONVERSION

The efficiency of food conversion or food utilization in farm animals is, striclty, the ratio of intake of food to the output of animal product. In the case of the meat-producing pig the unit of product is usually a unit of liveweight gain and the ratio will be then 'kg food per 1 kg liveweight gain'. Because of the highly variable composition of increments of liveweight gain, especially in terms of protein and calorie content, the ratio is an imprecise reflection of food conversion efficiency. None the less it is the most generally used single figure index for comparisons of diets, feed additives, management practices and of individual animals in breeding programmes. This popularity is largely because the high proportionate cost of food makes the index a simple measure of profitability, magnified by the fact that a carcass of a high protein content, and therefore of high market grade, is of a correspondingly low calorie content, requiring less food to produce than a less valuable fatter carcass.

Where the composition of the ultimate product is not markedly affected by level of food intake the correlation between rate of gain and efficiency of feed conversion will be high, as is the case with broiler poultry. Stated in physiological terms, the amounts of nutrients absorbed by an animal depend on the level of intake and the efficiency of digestion; the rate of gain depends on the partition of the absorbed nutrients between maintenance and production. The more quickly an animal reaches a given weight the less of its energy intake over the period it will have used in maintenance and the more for growth; therefore the higher will be its efficiency of food conversion. If the level of food intake can be raised without markedly affecting either food digestibility or the composition of the final product, then, as has already been indicated, the correlation between growth rate and efficiency of food utilization will be very high. If, however, the increase in food leads to an increased deposition of fat rather than protein, this will result in a reduced weight increment per unit of food increment and the efficiency of food conversion will fall; efficiency may also be lowered by a reduction in food digestibility at high intakes, but this will probably be slight. On the other hand, the temperament of the animal, and hence its energy requirement, may be affected by level of food intake; Kleiber[69] has pointed out that food is the most widely used of tranquillizers. Little is known about the specific effects of food satiety on food utilization for in the pig they are confounded with changes in body composition. It should be noted that the term 'satiation' has a specific meaning in pig husbandry[95] and refers to a recognized condition of loss of appetite, with growth stasis, brought on by *ad lib.* feeding of *wet* food (slop) but not encountered on *ad lib.* dry feeding.

It is clear from the results of performances of large numbers of pigs, individually fed semi-*ad lib.* in progeny testing stations[92] that there is a marked negative correlation between daily liveweight gain and efficiency of food utilization. Apparently the higher the intake of food the greater the proportion of fat deposition, with a corresponding decline in food conversion efficiency because of the higher energy requirement per unit of weight. It will be appreciated that quite apart from a lower energy requirement for the deposition of a unit of protein as compared with fat (a calculation from the data of Kielanowski[96] puts the amounts at approximately 8·0 and 12·4 kcal DE/g), the protein in the bacon carcass, is combined with 3 to 3·5 times its weight of water in the hydrated muscular tissue, while the fatty tissue contains 85–90 per cent fat. At any rate the net effect appears to be that any saving in energy of maintenance, resulting from increase in growth rate of liberally fed pigs is outweighed by the higher energy cost of the increased fat

TABLE 11

Effect of Increasing Daily Feed Intake on Carcass Composition of Pigs (From data of Clausen and Ludvigsen[44])

S.F.U./pig/day*	1·68	1·89	2·18	2·48	2·60
Daily liveweight gain, g	536	621	699	762	790
Growing period, 20 to 90 kg, days	131	113	100	92	89
S.F.U./kg liveweight gain	3·14	3·06	3·13	3·27	3·31
Thickness of back fat, mm	31·8	32·4	33·9	36·9	37·1
Per cent lean meat in carcass	58·1	56·0	54·1	51·8	51·8

* 1 S.F.U. = 1 Scandinavian feed unit = feeding value of 1 kg barley $\simeq$ 3333 kcal DE.

synthesis which liberal energy intakes promote. Whether the variations in energy are imposed by variations in the amount of food fed (Table 11) or by giving, on a liveweight scale, foods of differing energy content (Table 12) the association of increased energy intake with reduced efficiency of food

TABLE 12

Food Conversions with Two Diets, One of Low and One of High Energy Content, during Three Periods of Growth, expressed as Weight/Weight and Energy/Weight Ratios (From data of Costain and Morgan[165])

Energy level of diet	Period 1*	Period 2	Period 3
	kg food/1 kg liveweight gain		
Low	2·9	3·8	4·0
High	3·0	3·6	3·7
	Mcal DE/1 kg liveweight gain		
Low	8·6	11·3	11·8
High	10·3	12·3	12·7

*Period 1, 25–45 kg; Period 2, 45–68 kg; Period 3, 68–90 kg liveweight.

conversion, or of digestible energy conversion can be seen. Two points should be noted from Table 12. First, that the efficiency is declining more rapidly on the low energy diet; this is presumably because the high energy diet is better able to meet the increasing energy demands for fat synthesis. Secondly, that the mode of presentation of food conversion data, as kg food/kg weight gain gives no indication of how much dietary energy is used per unit of liveweight gain; it is evident that this method of stating the results may conceal important aspects of the utilization of energy. However, before considering this further, it is worth looking at some recent carcass composition data in relation to age, and growth rate.

Mention has already been made of the different objectives in various dissection studies, and the consequent difficulty in relating results expressed in terms of chemical analysis of sides (half carcasses) and as tissue composition of sides, anatomical regions or commercial cuts. To these complications must now be added those of feeding régimes. Thus in some experiments the pigs were fed semi-*ad lib.* throughout, some according to a liveweight scale, and some in which an initial *ad lib.* feeding was followed by restriction. However, perhaps because the pig's metabolism is so obviously directed towards fat synthesis, any specific effects of these nutritional sequences as applied to the 20–90 kg pig do not seem to mask the main compositional trends in the carcass. In another important respect details are generally lacking, that of environment of the piggery. Some attention is given to this topic in a later section, but for the present it should be borne in mind that conditions of temperature, humidity and air movements can have important effects on the caloric economy of the pig and comparisons of results assume comparable environments.

The studies of Buck[97] provide information from the British Progeny Testing Stations about the effects of growth rates and food intake on carcass composition of Large White pigs at three stages of growth (68, 90 and 118 kg) corresponding to pork cutters, baconers and heavy cutters individually fed semi-*ad lib.* The author concludes from the results that food consumed up to 68 kg weight is not predictive of the food consumed between 68 and 90 kg. A similar position was found with liveweight gain. This is an important finding, since it suggests that there can be no short cut in progeny testing. The author considers that a carcass evaluation for the 68 kg pig could be achieved by an ultrasonic technique since the regression equations for the B, C and K measurements on the carcasses at all weights enabled their lean meat contents to be estimated with a standard error of 2 per cent of the carcass weight. The weight of the different commercial cuts as a percentage of the side weight does not seem to change much between slaughter weights. The carcass becomes less lean as the slaughter weight increases and this difference becomes more observable between the 90 to 118 kg weights than between 68 and 90 kg. In these ranges at least as much lean as fat is put on the shoulder and ham but more fat than lean on the back and streak. For all cuts and for both sexes the percentage of lean meat added in the range

90 to 118 kg liveweight is less than that for the range 68 to 90 kg. This difference is more marked for hogs (castrated males), especially in the back cut, than for gilts (unbred females) which are leaner at all stages and in all respects. Some of the results are given in Tables 13, 14 and 15.

The data of Hill and O'Carroll[98] provide information about the chemical composition of carcasses of Large White × Landrace and Large White pigs slaughtered at pork, 56 kg, bacon, 90 kg, and manufacturing, 120 kg liveweight. Tables 16 and 17 indicate the percentage composition of the right side, and its composition on a fat-free and on a fat and bone-free basis.

TABLE 13
Carcase Composition of Pigs at Three Different Weights
(From data of Buck[97])

Slaughter Weight (kg)	Lean as a percentage of side		Fat and skin as a percentage of side		Bone as a percentage of side		Shoulder fat thickness (mm)	
	Hogs†	Gilts‡	Hogs	Gilts	Hogs	Gilts	Hogs	Gilts
68	54·7	57·5	33·6	30·7	11·7	11·8	36·1	34·4
90	51·7	54·6	37·8	34·7	10·5	10·7	44·4	40·8
118	48·0	51·9	42·6	38·5	9·4	9·6	51·2	48·2

† Hog = male castrate. ‡ Gilt = unbred female.

TABLE 14
Composition of Four Cuts from Hogs and Gilts at Three Different Weights
(From data of Buck[97])

Slaughter weight (kg)	HOGS				GILTS			
	Wt. of cut as % of side	Fat and skin as % of cut	Lean as % of cut	Bone as % of cut	Wt. of cut as % of side	Fat and skin as % of cut	Lean as % of cut	Bone as % of cut
Shoulder cut								
68	32·3	30·3	55·1	14·5	32·3	27·8	57·5	14·7
90	31·3	33·1	53·7	13·3	31·4	30·3	56·0	13·6
118	30·5	36·6	51·1	12·2	30·6	33·5	54·2	12·4
Back cut								
68	28·5	40·7	47·1	12·2	28·1	35·4	52·2	12·5
90	30·4	46·0	43·5	10·5	29·3	41·0	48·1	10·9
118	31·7	52·0	39·3	8·7	30·5	45·3	45·4	9·3
Ham cut								
68	22·0	22·1	66·0	11·9	22·4	20·5	67·8	11·6
90	21·3	25·2	63·5	11·3	21·8	23·3	65·7	11·0
118	20·8	29·1	60·5	10·3	21·3	26·7	63·3	10·0
Streak cut								
68	17·2	42·7	51·8	5·5	17·3	41·3	53·4	5·4
90	17·0	47·1	48·0	4·9	17·5	46·1	49·0	4·9
118	17·0	52·0	43·7	4·3	17·6	49·9	45·7	4·3

TABLE 15
Carcass Composition of Pigs at Three Slaughter Weights (68, 90 and 118 kg) with Estimated Changes in Composition and in Efficiency Factors between these Weights
(*From data of Buck*[97])

Liveweight ranges (kg)	Lean (kg)	Fat and skin (kg)	Bone (kg)	Fat as a percentage of lean	Food consumed (kg)	Food conversion: Per kg liveweight (kg)	Food conversion: Per kg carcass weight (kg)	kg food to produce 1 kg lean	Days on test	Daily gain (g)
22–68	22·9	13·1	4·8	57·2 *	147·8	3·10	4·17	6·4	74	650
22–90	29·9	20·4	6·0	68·2	234·1	3·34	4·40	7·8	104	690
22–118	37·9	30·7	7·2	80·9	353·6	3·64	4·66	9·3	140	700
68–90	7·0	7·6	1·2	104·2 †	86·4	3·88	4·87	12·3	30	760
68–118	15·0	17·6	2·4	117·0	205·9	4·12	5·09	13·7	66	760
90–118	8·0	10·3	1·2	128·3	119·5	4·31	5·26	14·9	36	760

* Carcass composition at slaughter weights of 68, 90 and 118 kg.
† Carcass weight increments from 68–90, 68–118 and 90–118 kg liveweights.

TABLE 16
*Composition of Pig carcasses at Three Slaughter Weights**
(From data of Hill and O'Carroll[98])

Slaughter weight (kg)		Percentage composition					
		Moisture	Fat	Skin	Protein	Ash	Bone
56	Hogs	47·97	26·53	—	13·22	0·71	11·57
	Gilts	49·46	24·43	—	14·25	0·74	11·13
90	Hogs	38·28	37·27	3·22	11·37	0·57	9·29
	Gilts	42·26	31·73	4·09	12·53	0·66	8·73
120	Hogs	35·85	42·11	3·66	10·19	0·51	7·67
	Gilts	37·77	39·19	3·13	11·40	0·56	7·96

*Moisture, fat, protein and ash from chemical analysis of the boneless meat.

TABLE 17
Composition of Carcasses on a Fat-free and Fat and Bone-free Basis
(From data of Hill and O'Carroll[98])

Slaughter weight (kg)		Percentage on fat-free basis				Percentage on fat and bone-free basis		
		Moisture	Protein	Ash	Bone	Moisture	Protein	Ash
56	Hogs	65·29	18·00	0·97	15·74	77·50	21·36	1·15
	Gilts	65·45	18·85	0·98	14·72	76·75	22·11	1·15
90	Hogs	64·32	19·10	0·97	15·60	76·21	22·64	1·14
	Gilts	65·86	19·51	1·03	13·61	76·23	22·58	1·19
120	Hogs	66·08	18·80	0·94	14·16	76·98	21·91	1·09
	Gilts	65·47	19·76	0·97	13·79	75·95	22·92	1·12

Tables 18 and 19 show the composition of the additional carcass weight gain between two successive stages of slaughter and of the effect of slaughter weight on the composition of the subcutaneous fatty tissue; all the information is for both hogs and gilts and again illustrates the considerable difference in performance between the sexes. The methods of feeding to the three slaughter weights, *ad lib.* throughout, *ad lib.* to 65 kg followed by restriction to 90 kg and *ad lib.* throughout, make the treatments not strictly comparable but they represent commercial practice for the three types of production.

Data (not given here) for the chemical composition, and lean meat of the shoulder, back, belly (streak) and leg (ham) dissected at specific anatomical points have also been obtained. The data has been used by the authors to study the error involved in making assumptions about the constancy of body water content, on a fat-free basis, a widely used standard of reference in biological measurements which has been treated theoretically by Wedgwood.[99] They conclude that assumptions of this kind, although valid for some purposes, can only be general ones and that separate determinations are necessary where accuracy is required. This finding has important implications in relation to the proposed usage of the body water figure to calculate the percentage of fat, as for example by the dilution methods previously discussed, and suggests that much work is required to establish sequential compositional changes in the growing and finishing pig before these techniques can be generally applied.

TABLE 18

Composition of Additional Carcass Weight Obtained at Increasing Slaughter Weights
(From data of Hill and O'Carroll[98])

Stage of growth (kg) liveweight	Additional carcass weight (kg)	Percentage composition of additional carcass weight					
		Moisture	Chemical fat	Skin	Protein	Ash	Bone
56–90	27·3	29·9	53·7	*	9·7	0·5	6·1
90–120	20	25·9	60·0	2·6	7·1	0·3	4·2

* Included with fat.

TABLE 19

Percentage Fat in Subcutaneous Fatty Tissue Samples, showing Effect of Slaughter Weight and Litter
(From data of Hill and O'Carroll[98])

Slaughter weight (kg)	Litter A	Litter B	Litter C	Combined
90	84·0	85·4	87·0	85·5
120	86·8	87·0	88·8	87·6

CALORIFIC ASPECTS OF FOOD CONVERSION BY PIGS

The two components, the diet and the carcass, pose rather different problems of terminology and will therefore be considered separately. A calorie conversion ratio of the same form as that for food conversion would indicate the number of food calories required for the deposition of one kg of body gain. In making experimental comparisons this presentation would obscure differences in utilization due to any major differences in the digestibility of different diets. It is therefore usually desirable to express the food component of the ratio in some term of utilizable energy. The procedure adopted may depend on the nature of the comparison, and the values for Total Digestible Nutrients (TDN), Digestible Energy (DE) and Metabolisable Energy (ME), usually as kcal/kg food, all have their particular advantages. Nevertheless, there is much to be said for a standardized procedure, and the most recent edition of the NRC recommendations[57] uses Digestible Energy (DE), defined in the NRC glossary of Energy Terms[101] as apparent digestible energy or ADE (the gross energy of the food intake minus the energy of the faecal output). Not much data for directly determined DE values are yet available, [102, 103] and meanwhile values are obtained from the published TDN values without much loss of accuracy,[104] by assuming that 1 kg TDN $\simeq$ 4500 kcal DE. The data in Table 12 in terms of DE have been calculated in this way to give expression to the higher calorie content of the high energy diet which, when compared on this basis, is less efficient than the low calorie diet. Since there were no marked differences in growth rates, it would appear that the high energy diet promoted greater fat synthesis, as was indicated by the carcass measurements.

The expression of body weight gain as calories is obtained by direct calorimetry on tissue samples or is calculated on the basis of standard equivalents for protein and fat; Kleiber[69] gives 1 g protein = 5·7 kcal and 1 g fat = 9·5 kcal. This estimation, of course, requires data for body gain in terms of protein and fat and these are conspicuously lacking in the literature. It is extremely difficult to find experimental results from feeding trials which include details of all the required components—the diet and its digestibility, food intake, rate of liveweight gain and the composition of the weight increments either as calories or as protein and fat contents. Many partial citations are to be found such as composition of weight increments without food and growth data, and weight increments in terms of tissues or meat but without chemical analysis, but from the vast amount of food/growth-rate investigations on pigs only a small amount of calorimetric information can be derived. The reasons for this are not hard to appreciate. For input/output studies on the basis of the chemical composition of weight increments it is necessary to use the comparative slaughter trial, at the start of which a representative group of animals is slaughtered to give an estimate of the initial body composition.[105, 106] This, together with the need to sample and analyse the whole body adds greatly to expense and labour and there is consequently little incentive to incorporate detailed studies of body composition into trials which have other primary objectives. Comparative slaughter trials may have the advantage of a greater number of observations with animals kept under normal conditions and making more acceptable growth rates than in the alternative nitrogen/carbon balance or calorimetric trials. The results of Oslage and Fleigel,[49] however, indicate that good growth rates can be obtained over long periods with the balance technique which in this case has the great advantage of giving the sequence of metabolic events within short time spans without the serial slaughter of animals. Carcass dissection and analysis is desirable at the end of carbon/nitrogen balance trials if the results are to be interpreted in terms of meat production. However, in the main they have been concerned with fundamental studies of energy partition.[107]

Oslage and Fleigel[49] found that the daily energy retention of protein plus fat fell from about 53 kcal/kg liveweight with pigs at 40/50 kg liveweight to 35 kcal/kg liveweight at 120–130 kg liveweight. This sequence was because of a rise in daily fat deposition from 4·2 g/kg liveweight at 40–50 kg liveweight to a maximum of 5·3 g/kg/day at 70–80 kg weight followed by steady fall thereafter, associated with a steady fall in protein retention from 2·3 g/kg/day to 0·9 g over the weight range. The effect of these events is that the efficiency of energy utilization, expressed in the form of total energy deposition as per cent of the ME available for production (intake ME – maintenance ME), fell from about 80 per cent in the weight range 40–80 kg to 68 per cent at 130 kg. The figures cited by Leroy[108] show a range of values from 28 to 52 for energy of gain expressed as a percentage of metabolizable energy of intake; when put in the form of kcal ME for 1 kcal gain the ratios range from 2 to 3·5 for pigs about 24 to 90 kg liveweight. As these pigs were on *ad lib.* feeding and making rapid rates of gain there would appear to be scope, and certainly there is the need, to study the effect of change of energy intake on both the rate of liveweight gain and body composition. For this the carbon/nitrogen balance technique seems well fitted and despite the expense of establishment and the difficulties of running, it is to be hoped that this kind of facility will come to be regarded as indispensable to the work of any research centre concerned with the nutrition of the pig.

THE UTILIZATION OF PROTEIN

Although protein cannot readily be given a separate expression in an efficiency ratio unless by virtue of nitrogen balances or carcass analyses, its effects on the overall food conversion ratio and associated rate of liveweight gain, and more recently on carcass composition, have been widely studied. It will be apparent that in so far as utilization is a function of intake and requirement, efficiency will show wide variations caused by the physiological differences of age and sex and the innate synthetic ability of the individual. This in the main is genetically determined but conceivably modified by the stress of husbandry and environment. The quality of protein in terms of the amounts and availability of its constituent amino acids is also a major factor in determining utilization. The absence of a generally acceptable method of carcass evaluation, short of complete dissection and analysis, makes it difficult to collate the results of different workers, and a solution to this problem is urgently needed. Although predictions of lean content solely from linear carcass measurements may be adequate for breeding programmes they are less likely to be so for discriminating between nutritional treatments. In view of these complexities the pattern of protein utilization is only slowly emerging, a process which may not be helped by the constantly changing genetic material available for study.

It is certain that the growing pig cannot be satisfactorily taken to bacon weight on a cereal diet without a protein supplement,[109,110] and the improvement in growth rate, food conversion and reduction in carcass fatness brought about by even a small supplement of fish meal sufficient to raise the level of crude protein from 10 per cent to 12·6 per cent[110] is quite dramatic. In many experiments the effectiveness of the level of protein supplement has been assessed largely by the effects on liveweight gain and food conversion. This work has been reviewed by Holme[111] and Morgan and Lewis.[112]

The data from a large number of papers have been summarized graphically by Holme[111] to illustrate the connection between the level of crude protein (N x 6·25) in the diet and rate of gain and efficiency of feed conversion from weaning to 45 kg liveweight. Increasing levels of protein are seen to have been accompanied by increased rates of gain up to a level of 16–17 per cent with a levelling off beyond this stage. The contribution of American results to this effect is considerable and well defined and has been found with both fast- and slow-growing pigs. Similarly, results for efficiency of feed conversion suggest a general improvement with increasing protein content until a level of 16 per cent protein is reached. This relationship appears to be broadly similar for experiments with good and experiments with poor conversions. For the period from 45 to 90 kg liveweight the small amount of data available suggest that in relation to growth rate there is little advantage in more than 12 per cent protein in the diet. Results are very divergent and it is clear that the protein nutrition of pigs of more than 45 kg can only be usefully considered in relation to carcass composition. Most British experiments have shown only small responses to changes in protein level but it should be noted that, in contrast to those in the U.S.A., restricted feeding was usually employed and that the levels of energy in the diets, and the rates of liveweight gain, were lower.

Since additions of protein to a cereal diet increase food intake this will in the first place, on *ad lib.* feeding, lead to faster growth and effects on efficiency of feed conversion will be secondary. With restricted feed all the animals at the outset receive the same amount of food, based on body weight, and any relative increase in growth, by pigs on any particular treatment, can only be achieved in the first instance by an improvement in efficiency of food conversion, thereby achieving a greater weight increment and qualifying for relatively more food on the liveweight scale. The potential differences between various levels of protein may for this reason tend to be masked, particularly when, as is customary, the level of protein is reduced at about 50 kg liveweight by various amounts or sequences. Feeding practices have come, therefore, to be rather empirically determined by the economics of feeding stuffs and labour with, for example, the U.S.A. leaning to high protein/high energy soya bean/maize diets, which with *ad lib.* feeding impose no restrictions on protein or energy, and some European countries, with a need for the lean bacon carcass relying on the restricted feeding of diets lower in protein and energy and based on the sparing use of fish meal, meat meal and skim milk together with barley, oats and wheat products.

It has long been recognized that protein-rich foods of animal origin are superior to plant protein concentrates for supplementing cereals and the relative values of fish meal, meat meal, earthnut meal and soya bean meal have been well established.[113] From these and kindred studies[114, 109] has come the realization of the ultimate need to study the nitrogen nutrition of the pig in terms of amino acids, and, as a first approach, of the most limiting amino acid lysine. Coupled with this development has been the use of the nitrogen balance technique as a means of establishing the relative values in efficiencies of protein in terms of per cent retention of the intake of digestible or crude protein and of the supplemental value in this respect of various amino acids. Using this technique, Jones *et al.*[115] found large increases in N retention in young pigs with fish meal increments of up to 15 per cent, and with groundnut meal increments when supplemented with L-lysine, in diets providing between 12 and 21 per cent crude protein. However, the maximum N retention for each diet, as g/kg liveweight was obtained when the diet contained 18 per cent crude protein.

In a later restricted-feeding experiment[116] it was found, in line with the earlier results of Kropf[43] with *ad lib.* feeding, that a diet of 12 per cent protein, with added lysine, gave as good liveweight gain as one of 18 per cent protein with no added lysine. These results, taken along with the effectiveness of small additions of lysine-rich protein supplements—fish meal; separated milk—in improving rate of liveweight gain suggest that lysine level may have a particular effect in promoting liveweight gain whilst protein level may have more specific effect in promoting nitrogen retention. This is an important possibility which requires further investigation. The increased N retentions as g/kg liveweight obtained by Jones *et al.*[115], as compared with those of Woodman and Evans[117] have been attributed to higher dietary energy levels. However, the results of Robinson *et al.*[118] for pigs up to 53 kg do not lend much support to this view and suggest that protein and energy exert their effects independently of one another except perhaps at low energy levels. It should be mentioned that very little nitrogen-balance data appears to exist for pigs of over 50 kg weight.

The successful use of nitrogen balances in determining the supplemental value of widely dissimilar proteins, using young responsive animals for the tests, has led to attempts to relate N balance data from less contrasting diets to the carcass composition of the test animals at conventional slaughter weights. Here the technique has been less successful. In carcass composition studies N balance data are generally presented as: (i) N retention as per cent of digested or ingested N; (ii) N retention g/day; (iii) N retention g/kg liveweight/day or g/kg $W^{0.75}$/day; (iv) N retention (NR) as per cent of weight gain. Only the last of these, termed NR per cent, which is calculated as

$$\frac{\text{N retention during period of balance} \times 100}{\text{Liveweight gain during period of balance}}$$

gives an indication of the body composition which may be taking place during the balance. Its usefulness is limited in that it gives no indication of the distribution of the N of the weight gain between edible and non-edible tissue, whether the mass of a tissue is increasing without change in per cent N or whether the per cent N is increasing without change in mass. Nevertheless, providing two major experimental difficulties are overcome, the determination of NR per cent could be a useful indicator of the effect of specific treatments on body composition. Examples of this are given by Cunningham *et al.*[45] The accuracy of NR per cent depends upon the accuracy with which the liveweight increment can be determined. This calls for long balance periods, of about three weeks, or perhaps continuous balances. Metabolism crates which can be wheeled on to a weigh-bridge are very desirable. Appreciable errors in determining N retention can occur—food spillage, loss of volatiles, inadequate mixing of faeces, and, probably most of all, inadequate sampling of the food. These errors are considered by Graham *et al.*[119] and Rasch *et al.*[120] Hedin and Adachi[121] have suggested that there may also be a loss of elemental N from the food in the digestive tract. Lassiter *et al.*[122] have considered the precision of long and short balance periods and their results suggest possible advantages in a series of spaced shorter balance periods with larger numbers of animals and with graphical assessment of the results, as an alternative to longer balances with fewer animals. The establishment of response curves to dietary N, in parameters which will reflect carcass composition is much to be desired, and to this end the nitrogen balance method clearly merits careful study.

THE ENVIRONMENT OF THE PIG

The environment includes climatic conditions of temperature, light, humidity and air movement, physical conditions of living space, management practices, particularly those relating to feeding, and exposure to risk of disease. Although some may be impossible to control in practice, it is important to understand the factors concerned in the optimum physiological response of the pig to its surroundings. What constitutes the measure of optimum response is perhaps a matter of opinion but for the present it will be regarded as efficient food utilization with the concomitant production of lean meat. The environmental factor with the biggest effect on pig productivity is temperature but it has not always been clear how important are changes in air temperature, radiant temperature and air movement (draughts) in relation to one another in affecting heat lost from the animal.[123] The extensive experiments of Heitman and his associates and of Moustgaard and Sørensen to study the effects of hot and cold environments on growth rate and food conversion have been reviewed by Sørensen.[124] More recently Mount and Ingram[125] have studied the effects of ambient temperature and air movement on heat loss and Mount has discussed[126] the relationship of the pig to its physical environment. Below thermal neutrality heat loss takes place mainly by radiation and convection but sometimes considerable loss can occur by conduction to the floor where these are not insulated. Mount[126] suggests that this may be a vital channel for shedding the heat burden in hot humid piggeries[127] where the pig's skin temperature is close to air temperature and where the vapour pressure gradient from the respiratory tract to the surroundings is small. Because the pig does not sweat, variations in humidity have little effect on its metabolic rate and heat loss.[128] Both the thermoneutral range and the rate of increase in metabolic rate with fall in ambient temperature depend partly on the size of the pig. This is shown in Fig. 6[126], for new-born pigs, singly and in groups, and for pigs of 90 kg. Given the requisite

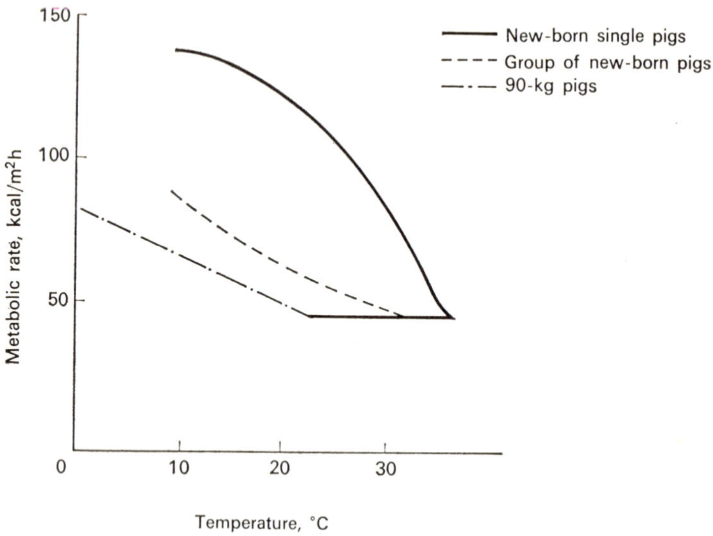

FIG. 6

attention to matters of building construction such as insulation and ventilation, a diagram of this kind permits the selection of an appropriate ambient temperature corresponding to a metabolic rate just above the minimum. This allows the economical use of food and at the same time allows a margin of animal activity which does not induce hyperthermia due to inability to lose heat.[126] The diagram also illustrates the effect of huddling—the pigs' behavioural response to low temperatures[129]—in which close grouping reduces surface area, heat loss and metabolic rate. The increase in rate of heat loss as the ambient temperature falls from 35°C to 0°C, in still air, was 4·6 kcal/m²/hr per °C fall in ambient temperature for 20 kg pigs, and 3·0 kcal for 60 kg pigs.[125] The smaller loss in the 60 kg pig is due to the insulation of its greater depth of subcutaneous fat. Mount[123] estimates that for a temperature of about 6°C below the optimum level during 130 days in the fattening house a pig would require about 12 per cent more food. This estimate is in line with the data of Heitman and Hughes[130] who found the best weight gains and feed conversions at 24°C for pigs of about 30 kg to 65 kg weight and 17°C for pigs of 75 kg to 120 kg weight. In practice temperatures within the range 16–21°C prove most suitable for pigs penned in groups where huddling is possible. Where pigs are penned individually, as may be required in an experimental house, an ambient temperature in the range 21°C to 23°C is necessary for recently weaned pigs of about 20 kg.[131]

Because of its negligible coat insulation the pig is responsive to small air movements, in the low range of 8–35 cm/sec., impinging on the insulating boundary layer of air surrounding the body. Mount and Ingram[125] found that heat loss increases with increased air movement and that the rate of increase tends to be greater with younger pigs and at lower ambient temperatures, being much greater below the critical temperature, 20°C for the 90 kg pig and 25°C for the 18 kg pig. For example, 12-weeks-old pigs in still air at 15°C lose 115 kcal/m²/hr from their backs; raising the temperature to 20°C without air movement would reduce the loss to 95 kcal/m²/hr. If, however, the temperature is raised by blowing in warm air, air movement will increase and if it rises to more than 30 cm/sec at pig level the net effect will be increased heat loss. This points to the necessity for careful baffling of fans.

Sørensen[124] points to the necessity for distinguishing between reactions against acute exposure to a sudden change in environmental conditions and acclimatization over a long period. This is illustrated by the disturbance in thyroid function[132] caused by abruptly changing pigs from an ambient temperature of 3°C to one of 18°C. After 3 weeks the thyroid secretion rate was still much higher than that of a control group kept at 18°C throughout, indicating that some time is required for the abatement of the elevated secretion appropriate to the cold environment (Table 20). It would appear that during this period of adjustment the animal must be under greater physiological stress than that imposed by the low temperature, with consequent deterioration in productive performance. Considerations of this kind are of obvious importance too when assembling pigs for experiments.

The physiological sequence following a sudden drop in ambient temperature has been considered by Sørensen[124]. The first reaction is stated to be an

TABLE 20
Thyroid Secretion Rate in Pigs during Exposure to 3°C, Period 1, and after a change of Air Temperature to 18°C, Period 2
(*From data of Moustgaard and Sørensen*[132])

	Liveweight (kg)	Temperature °C	Thyroxine secretion rate	
			mg/day	mg/day/100 kg liveweight*
Period 1 (6 weeks)				
Experimental group	26	3	0·9	2·4
Control group	29	18	0·3	0·8
Period 2 (3 weeks)				
Experimental group	52	18	0·8	1·2
Control group	59	18	0·5	0·7

* Corrected linearly with liveweight.$^{0 \cdot 7}$

increased adrenaline secretion associated with increased nitrogen excretion and decreased protein utilization. On the other hand, Cunningham[133] describes experiments in which adrenaline has increased nitrogen retention in pigs, and in which the feeding of nicotine at 5 mg/kg of diet resulted in a decrease of 10 per cent in carcass fat and an increase of 5·3 per cent in protein content. It is considered that the primary effect of the nicotine supplement was to stimulate the release of adrenaline from the adrenal medulla, although it has not been possible to demonstrate this, after the manner of adrenaline injections, by elevation of blood reducing sugar and free fatty acids. This disparity between the effects of an endogenous hormone secreted under the stimulation of particular environmental conditions, and those of an exogenous source functioning under quite different external conditions indicates the need for biochemical criteria to assess responses of this kind.

It seems to be generally assumed that the concentration in numbers of pigs in buildings, and the high rate of throughput, which are associated with intensive farming bring problems of disease risk. However, following the advent of antibiotics, which initially were used in the treatment of obvious malaise, it became apparent that penicillin and the tetracyclines could promote improved rates of liveweight gain and food conversion when fed to pigs considered to be in normal health. That this response to antibiotics was not specifically associated with housed pigs was apparent from the survey of Braude *et al.*[134] wherein pasture-fed pigs, in the U.S.A., were shown to be as responsive as those in floored pens. The extensive use of antibiotics in pig feeding from 1952 onwards, reviewed by Robinson[135], and arising from it the concept of sub-clinical infections controllable by fortifying the diet, has been followed by the introduction of other antibiotics, and of additives such as copper sulphate[136] and organic arsenicals,[137] all of which, in the absence of alternative evidence, are assumed to act by modifying the activities of the enteric flora. Almost all these antibacterials are subject to legislation which varies from country to country and perhaps for this reason no rationale appears to have emerged for the systematic use, in conjunction with one another, of these valuable adjuncts to productive efficiency.

Although the number of pigs in a pen or in a piggery must be an important determinant of productive performance, population pressures may be dictated in the first instance by other considerations, such as the methods of feeding and watering, the segregation of the areas for feeding, resting and excreting and the need to train the pigs to use these as intended. However, the effect of space allotment has been considered in some studies of growing pigs and estimates per pig for optimum performance vary from 7–14 ft^2.[138,139]. An optimum is difficult to achieve in practice because the increasing size of the pigs cannot be readily accommodated by removal and remixing, owing to the behavioural pattern of pigs in rejecting strangers. Inevitably, therefore, a pen tends to be under-populated at the beginning of the feeding period and over-populated at the end. Nofziger,[139] in studying the possible stresses which might arise, found no differences in rate of gain or food efficiency when pigs of eight in a pen were allotted 8, 12 or 24 ft^2 each and fed *ad lib.* from weaning until 90 kg weight. The pigs allowed 8 ft^2, however, had rather leaner carcasses.

An important development in the search for ways of alleviating the disease burden of pigs has been that of the production of specific pathogen-free (SPF) pigs by the use of hysterectomy and subsequent segregation from other pigs or contacts. This technique, described by Betts,[140] has obvious advantages in providing animals for various aspects of pig research. It has also had some commercial application in the U.S.A. and U.K. Apart from the difficulties imposed by the rigid husbandry requirements of segregation and isolation the attractiveness of the proposition at present is lessened in that the diseases eradicated are respiratory infections, and do not apparently include the enteric infections which cause the all-too-prevalent 'scours' in pigs.[166]

BREEDING AND SELECTION

Since in the main areas of production the present-day pig produces fat in excess of market requirements, efficiency in food conversion depends on the ability of farmers to provide that integration of nutrition and environment in which their pigs can achieve maximum growth of muscle. When this stage is reached further progress must be sought in the selection of tested breeding stock. The studies of Smith *et al.*[91] and Smith and Ross[92] indicate that a large part of the important performance traits are of genetic origin and amenable to improvement by selection. The breeding herds are largely in the hands of farmers themselves and breeding policy therefore depends on their judgements, in which they may be assisted by information about the productive capacity of their stock from progeny and performance-testing stations. However, only a small proportion of operational breeding stock can be tested and in this situation much depends on success in the development of artificial insemination, which could increase the use of *élite* boars 20-fold. At present A.I. is limited by the fall in viability of semen during storage and the difficulty in discerning the correct time for insemination, but there appear to be good prospects of these difficulties being solved. Presumably A.I. would also enable a wider exploration of the value of cross-breeding[141] in improving productivity.

A more fundamental problem in breeding is the choice of criteria on which to base selection. Smith *et al.*[91] point out that although a genetic analysis makes a closer approach to underlying relationships it leaves a major problem unsolved, namely, in what manner the principal components revealed by statistical analysis can be related to biological mechanisms. In their analyses the first principal component could be interpreted as some function of fatness, but the authors conclude that until there are physiological measures of fat metabolism available, with which to correlate the fatness factor, the delineation of factors would be misleading. It is by no means clear to what extent changes brought about by breeding are changes in age at maturity, and to what extent new physiological characteristics, for example improved absorption of nutrients, have been introduced. The propositions involved in growth towards maturity have recently been considered by Taylor.[17]

Nalbandov[142] has analysed the results of a study carried out with his co-workers[143] in which the hypophyses of two strains of pigs, selected over eight generations for rapid or slow growth were assayed for somatotrophin content. At all ages at which comparisons were made the glands of the faster-growing strain contained greater amounts of hormone per unit of gland tissue. It would appear that the factor being selected for was the gene-controlled rate of secretion of the somatotrophic hormone by the adenohypophysis.

The concept that metabolic differences associated with variable growth and body composition might be indicated by the level of metabolites has been the basis of many chemical studies involving determinations such as serum proteins and amino acids, non-protein nitrogen, creatine and creatinine, glucose, and sulphydryl compounds notably glutathione. Few of these studies appear to have been concerned with the pig and significant relationships generally have not been established. A difficulty here may be that the only basis of comparison is that afforded through the sporadic variations in performance of experimental animals drawn from a closed herd, variations which may not all be due to the same cause. It would be much more satisfactory if metabolic studies could be applied to different selected strains of pigs whose differential performances, as with those of Baird *et al.*[143] are predictable. In this context an experiment recently reported[144] is of particular interest. Here the main intention is to study the use of carcass density as a criterion of selection and to this end four herds have been established, two of which are being selected on the basis of high density and two for low density. If successful this work could provide valuable material for physiological studies of growth and carcass composition.

Recent innovations in the study of the inheritance of commercially valuable characters in pigs seek to make use of the existence of gene-controlled biochemical polymorphism.[145] The parameters under investigation are principally red cell factors, and serum protein and plasma esterase variants. The study of red cell factors uses reagent antibodies and progress has been facilitated by international co-operation between Blood Group Laboratories in the standardization of these reagents. Starch-gel electrophoresis is the basis of the protein and enzyme studies. A list of recent published world research on blood grouping in pigs can be seen in the Index of Current Research on Pigs.[146] Blood typing by red cell groupings can help to identify parentage, for example in identifying the progeny of each boar used in a double mating. The extent to which blood groups may be associated with useful production factors is a matter of active investigation. A total of about thirty heritable characters can now be identified by blood group techniques. The facilities needed for this work are obviously elaborate and call for specialized Units or Departments. Apart from the purely laboratory aspects a reference panel of pigs of known blood type is required and association with a progeny testing station seems to be necessary in order to ensure a broad basis of genetic material for study.

MEAT QUALITY

The main criterion of carcass quality in the pig is the proportion of lean to fat, which does not include any assessment of the characteristics of the lean meat itself. The quality of lean is usually judged from two aspects, mainly by subjective tests. Firstly, factors contributing to appearance, including colour, firmness and texture, and secondly factors contributing to palatability, such as tenderness, juiciness, flavour and aroma.[147] These aspects of meat quality, on which a number of texts are available,[33,80,148] will not be considered in detail but attention will be given to some problems of pig production, of a physiological nature, which seem to have arisen out of the breeding and selection of pigs for leaner carcasses.

It is well known that the appearance of fresh pork can vary from the normal uniform red colour to a pale grey. In some cases all the skeletal muscles are discoloured, in others some differentiation may occur, giving rise to a characteristic two-toned appearance. The greyish colours are also associathe with a change from a firm elastic structure to a loose and inelastic musculature with watery infiltrations between the bundles and of unattractive appearance and impaired palatibility. This abnormal condition has been variously termed muscular degeneration or MD,[149] watery pork[150] and two-toned pork.[151] According to Ludvigsen[149] all pigs from the Danish Progeny Stations are graded, 24 hr *post mortem*, on the basis of colour and structure of the cross-section exposure at the last rib of the longissimus dorsi muscle. The following scores are used:

Score	Specification
0·5	Grey, same colour as boiled meat; very moist surface.
1·0	Very pale, pinkish; moist.
1·5	Pale pink; slightly moist.
2·0	Slightly paler than desirable; almost dry surface.
2·5–3·0	Ideal red colour; dry firm surface.
3·5–4·0	Slightly darker than desirable; dry firm structure.
4·5–5·0	Very dark.

Thus scores of 0·5 to 1·5 include various degrees of pale watery muscle, whereas scores above 2·0 represent red, firm, dry muscle. About 15 per cent of pig carcasses from Danish Progeny Testing Stations in 1958/9[149] had scores of 1·5 or less, 22 per cent scores of 3·0 or higher, 24 per cent scored 2·0 and 38 per cent 2·5 but, as will be discussed later, the sequence of events leading to slaughter may affect the incidences. Ludvigsen states that in the Wild Pig × Landrace crosses (Table 1) the seven generations from 75 per cent Wild pig to 100 per cent Landrace had scores of 4·2, 3·8, 2·7, 2·4, 2·1, 1·6 and 1·8 and from this, and because it has been possible to establish an inbred strain of pigs showing a very high incidence of MD, and also because there seems to be some association with certain breeds, he suggests that the disorder may be due to a shift in homeostasis to a borderline insufficiency in some pituitary hormone systems as a result of continued selection for leanness. It is thought that watery muscle is not common in the U.K.[152] According to Clausen[44] the pH level, at the time when post mortem lactic acid formation from glycogen has ceased, is a main factor in determining colour and structure in muscle. The precise conditions are a rapid fall in pH and a rapid onset of rigor mortis in 30 to 40 min after slaughter. The pH drops to around 5·4–5·7, compared with normal range of 6·1–6·5. The closer the pH to the iso-electric point of the muscle proteins—about pH 5·4—the more pronounced is their decreased water-holding capacity, with a consequent outflow of water in the extracellular space, resulting in the pale, watery appearance of the meat. Muscle pH differences are less pronounced 12 to 24 hr after slaughter and pale colours may develop in the course of 24 hr post mortem, especially if glycogen depots are high. It will be clear, therefore, that conditions immediately prior to slaughter, and particularly muscular activity, have a big influence on pork quality. This was recognized and investigated by Callow[153] who also demonstrated that transport fatigue was associated with a high muscular pH, which was directly related to a greater tendency to bacterial spoilage. Callow[153] sought to define the variations in muscle structure in terms of electrical resistance but, on the whole, objective routine methods have not been successful for this purpose. In apparent conflict with Callow's findings, Ludvigsen[149] found that immediate slaughter after transport was conducive to low muscle pH and watery muscle. The differences are apparently determined by the state of muscle glycogen reserves. If these are low after prolonged fatigue extensive lactic acid formation cannot take place. On the other hand, factors involved in transport can trigger off rapid lactic acid development where glycogen reserves exist. Therefore Ludvigsen[149] postulates excitement or stress rather than exhaustion or fatigue as an operative factor in MD and has investigated the role of hormonal regulation in its incidence. He has found that dietary methyl thiouracil promotes and iodinated casein diminishes the incidence of MD and from this it has been suggested that the seasonal variations are to some extent correlated with the corresponding seasonal variations in thyroid activity. Of even greater interest is the finding that treatment with cortical steroids prior to enforced muscular activity prevents acute symptoms of stress and that the lactic acid produced by the activity passes from the musculature into the blood stream. In the control animals lactic acid accumulates, and only limited amounts pass into the blood stream because of a marked vasoconstriction of the muscles. In some experiments watery muscle has been produced by treatment with somatotrophic hormone. Ludvigsen[149] therefore suggests that selection for leanness may be correlated to selection of pigs for increased production of growth hormone and that this, and a reciprocally lower corticotrophin and thyrotrophin, accounts for the high incidence of MD in lean-type pigs. In their recent review of the subject, Bendall and Lawrie[150] suggest that long-term research should be directed to deciding whether intensive genetic selection is likely to exaggerate the physiological peculiarities which give rise to watery muscle, particularly the activity of the phosphorylase systems which are concerned in the breakdown of glycogen to lactic acid.

Reference has earlier been made to the fact that boars are appreciably leaner than castrated males of the same weights. Since this holds out the prospect that they may also be more efficient, a number of investigators have studied this question, particularly in the U.S.A. More recently the development of boar performance testing at progeny testing stations is providing much additional data for study. From a review of the literature, Blair and English[154] derive an indication that boars grow faster, utilize food more efficiently and are less fat at slaughter weights than castrated males. The results of their own experiments support this conclusion (Table 21) and suggest that the production of boar meat is worthy of serious consideration on the score of leanness and cost of production. An obvious difficulty in practice is the management of large numbers of boars but experience at

TABLE 21
Performance of Boars and Hogs
(*From data of Blair and English* [154])

	Boars	Hogs
Number of animals	24	24
Daily liveweight gain, g (20–90 kg liveweight)	655	605
Food conversion, kg food/1 kg gain	3·06	3·45
Shoulder fat thickness, mm	39	43
Eye-muscle area, cm^2	24·8	21·7

progeny testing stations suggests that this is not serious when the boars are penned separately. Another danger is that of the possible use of some of these animals for breeding, which would run counter to the animal licensing regulations in some areas.

The most serious drawback, however, is that of boar taint, an objectionable combination of odour and flavour which may appear when boar meat is cooked. Dutt *et al.*[155] showed that this sex odour is due to a secretion from the preputial gland which is fat soluble and Craig *et al.*[156] have found it to be generally distributed in the unsaponifiable fraction of the body fat at slaughter. The incidence of boar taint is variable as judged by organoleptic tests. Some investigators[157] finding it high while others[94, 158] were unable to detect it. The need for further work on the physiology and chemistry of the taint, and for analytical methods for its assessment is evident. Feeding or implanting stilboestrol has been studied as a means of limiting boar taint[159] with some success, but this reopens the question of feeding or implanting castrates to improve carcass characters. Some years ago a number of workers studied the effects of stilboestrol and testosterone on pig performance[160] but attention was mainly directed to growth rates, which were found to be little affected. More recently it has been found[161, 162] that carcass fatness in castrates is reduced without much alteration in growth rate, by steroid hormone treatments.

As has been underlined by Lamond[163] the understanding of the effects of the four categories of growth—plane of nutrition, impetus to grow, size and age[164]—on body composition has remained relatively static for some years. The roles of growth hormone and thyroid, gonadal and adrenal hormones in simple situations such as change in body weight have not been elucidated. Perhaps the next major contribution in the field of meat production will come through studying the relations between the endocrine system and growth.

REFERENCES

1. CALLOW, E. H. Carcass quality of the pig in relation to growth and diet. *Emp. J. Exp. Agric.* **3**, 80 (1935).
2. HAMMOND, J. Growth and conformation in the pig. *Pig Breeders' Annu.*, 1927–8 (1928).
3. CLAUSEN, H. *The Improvement of Pigs.* G. S. Robertson Memorial Lecture, Queen's University, Belfast (1953).

4. KUNKEL, H. *Biochemical and Fundamental Physiological Bases of Genetically Variable Growth of Animals.* MP–499. Texas, A. and M. Coll. (1961).
5. TAYLOR, ST. C. S. Pathways to maturity. *Anim. Breeding Res. Org. Rep.*, pp. 25–31. London: H.M.S.O. (1965).
6. MCMEEKAN, C. P. Growth and development in the pig with special reference to carcass quality characters, I, II and III. *J. Agric. Sci.* **30**, 276, 387 and 511 (1940).
7. PALSSON, H. Conformation and Body Composition. In: *Physiology of Farm Animals*, Vol. 2. London: Butterworths (1955).
8. HAMMOND, J. Pigs for pork and pigs for bacon. *J. R. Agric. Soc. Engl.* **93**, 131 (1932).
9. MCMEEKAN, C. P. Growth and development in the pig with special reference to carcass quality characters. IV. *J. Agric. Sci.* **31**, 1 (1941).
10. POMEROY, R. W. The effect of a submaintenance diet on the composition of the pig. *J. Agric. Sci.* **31**, 50 (1941).
11. CUTHBERTSON, A. and POMEROY, R. W. Quantitative anatomical studies of the composition of the pig at 50, 68 and 92 kg carcass weight. II. Gross composition and skeletal composition. *J. Agric. Sci.* **59**, 215 (1962).
12. ELSLEY, F. W. H., MCDONALD, I. and FOWLER, V. R. The effect of plane of nutrition on the carcasses of pigs and lambs when variations in fat content are excluded. *Anim. Prod.* **6**, 141 (1964).
13. HUXLEY, J. S. Constant differential growth and ratios and their significance. *Nature* **114**, 895 (1924).
14. MITCHELL, H. H. *Comparative Nutrition of Man and Domestic Animals*, Vol. 1. London: Academic Press (1963).
15. GUILBERT, H. R. and LOOSLI, J. K. Comparative nutrition of farm animals. *J. Anim. Sci.* **10**, 22 (1951).
16. HOLME, D. W. Effect of rates of gain on carcass quality in the bacon pig. *Anim. Prod.* **5**, 220 (1963).
17. TAYLOR, ST. C. S. A relation between mature weight and the time taken to mature in mammals. *Anim. Prod.* **7**, 203 (1965).
18. FRITZ, I. B. Factors influencing the rates of long chain fatty acid oxidation and synthesis. *Physiol. Rev.* **41**, 52 (1961).
19. LUCAS, I. A. M., CALDER, A. F. C. and SMITH, H. The early weaning of pigs. IV. Comparisons of level of antibiotic and sources of protein in diets for pigs weaned at 9 lb. liveweight. *J. Agric. Sci.* **53**, 125 (1959).
20. BOAZ, T. G. and ELSLEY, F. W. H. The growth and carcass quality of bacon pigs reared to different weights at 56 days old. *Anim. Prod.* **4**, 13 (1962).
21. CARPENTER, K. J. The evaluation of dietary energy. In: *Nutrition of Pigs and Poultry.* London: Butterworths (1962).
22. VANSCHOUBROEK, F. X., DE WILDE, R. O. and VAN SPAENDONCK, R. L. The influence of the level of feeding of suckled pigs on subsequent performance during fattening. *Anim. Prod.* **7**, 111 (1965).
23. BOAZ, T. G. Progress in pig production. *Agric. Progr.* **39**, 123 (1964).
24. EVERITT, G. C. Component analysis of meat production using biopsy techniques. *Proc. Tech. Conf. Carcase Comp. Aust. Anim. Prod. Commit.* Melbourne: C.S.I.R.O. (1963).
25. WINTERS, L. M., SIERK, C. F. and CUMMINGS, J. N. The effect of plane of nutrition on the economy of production and carcase quality of swine. *J. Anim. Sci.* **8**, 132 (1949).
26. BRUGMAN, H. H. The effect of the plane of nutrition on the carcass quality of a line of swine based on a Cheshire White and Danish Landrace cross. *J. Anim. Sci.* **9**, 602 (1950).
27. LUCAS, I. A. M., LIVINGSTONE, R. M. and MCDONALD, I. Further observations on the effect of growth checks in pigs of 50 lb liveweight and over. *Anim. Prod.* **4**, 195 (1962).
28. ROBINSON, D. W. The plane of nutrition and compensatory growth in pigs. *Anim. Prod.* **6**, 227 (1964).
29. BRODY, S. *Bioenergetics and Growth.* New York: Reinhold Publishing Co. (1945).
30. MCCANCE, R. A. Severe undernutrition in growing and adult animals. I. Production and general effects. *Brit. J. Nutr.* **14**, 59 (1960).

31. McCance, R. A. The bearing of early nutrition on later development. *Proc. 6th Int. Congr. Nutr.* Edinburgh: E. and S. Livingstone & Co. (1963).
32. Harrington, G. Pig Carcass Evaluation. Tech. Commun. 12, Commonw. Bur. Anim. Breeding and Genet. Edinburgh (1958).
33. Hankins, O. G., Gaddis, A. M. and Sulzbacher, W. L. Meat Research Techniques pertinent to Animal Production Research. In: *Techniques and Procedures in Animal Production Research.* Amer. Soc. Anim. Prod. Albany, N.Y. (1959).
34. Buck, S. F., Harrington, G. and Johnson, R. F. The prediction of lean percentage of pigs of bacon weight from carcass measurements. *Anim. Prod.* **4**, 25 (1962).
35. East, E., Taylor, J., Miller, I. T. and Widdowson, R. W. Measurements of backfat thickness on live pigs by ultrasonics. *Anim. Prod.* **1**, 129 (1959).
36. Price, J. F., Pearson, A. M. and Emerson, J. A. Measurement of the cross-sectional area of the loin eye muscle in live swine by ultrasonic reflections. *J. Anim. Sci.* **19**, 785 (1960).
37. Holme, D. W., Coey, W. E. and Robinson, K. L. The prediction of pig carcass composition from measurements of carcass density. *J. Agric. Sci.* **61**, 9 (1963).
38. Smith, D. M. The relationship between liveweight and carcass–weight increments of pigs. *N.Z. J. Sci. and Technol.* **38**, 803 (1957).
39. Ellis, N. R. and Hankins, O. G. Soft pork studies. I. Formation of fat in the pig on a ration moderately low in fat. *J. Biol. Chem.* **66**, 101 (1925).
40. Weninger, J. H. Changes in percentage chemical composition of pigs from four breeds. *Arch. Tierernahrung.* **4**, 293 (1955) cited by Braude, R., p. 174, *Scientific Principles of Feeding Farm Livestock, 1959.* London: Farmer and Stockbreeder, Publications Ltd.
41. Wood, A. J. and Groves, T. D. D. Body composition studies on the suckling pig. I. Moisture, chemical fat, total protein and total ash in relation to age and body weight. *Canad. J. Anim. Sci.* **45**, 8 (1965).
42. Wood, T. B. Studies of the nutrition of young animals. I. Energy exchanges in the growing pig. *J. Agric. Sci.* **16**, 425 (1926).
43. Kropf, D. H., Bray, R. W., Phillips, P. H. and Grummer, R. H. Effect of protein level and quality in swine rations upon growth and carcass development. *J. Anim. Sci.* **18**, 755 (1959).
44. Clausen, H. and Ludvigsen, J. Quantity and quality of final product—pigs. *Proc. 5th Int. Congr. Nutr., Fed. Proc.* **20** (1) Supplement (1961).
45. Cunningham, H. M., Friend, D. W. and Nicholson, J. W. G. Efficiency of conversion of protein, energy and carbon in pigs restricted in the late fattening period. *Canad. J. Anim. Sci.* **42**, 176 (1962).
46. Thorbek, G. Description of an open-circuit respiration apparatus for pigs. *Proc. 2nd Sympt. Energ. Metab.* Publ. 10, Europ. Assoc. Anim. Prod.: Rome (1961).
47. Thorbek, G. and Neergaard, L. Some remarks about instruments and methods applied to an open-circuit respiration apparatus for pigs. *Proc. 3rd Symp. Energ. Metab.* Energy Metabolism, Publ. 11, Europ. Assoc. Anim. Prod. London: Academic Press (1964).
48. Osinska, Z. Estimation of protein, chemical fat and energy content in pigs. *Anim. Prod.* **4**, 391 (1962).
49. Oslage, H. J. and Fliegel, H. Nitrogen and energy metabolism of growing-fattening pigs with an approximately maximal feed intake. *Proc. 3rd Symp. Energ. Metab.* Energy Metabolism, Publ. 11, Europ. Assoc. Anim. Prod. London: Academic Press (1964).
50. Lucas, I. A. M., McDonald, I. and Calder, A. F. C. Some further observations upon the effects of varying the plane of feeding for pigs between weaning and bacon weight. *J. Agric. Sci.* **54,** 81 (1960).
51. Symposium on body composition. *Ann. N.Y., Acad. Sci.* **110**, 1–1018 (1963).
52. Panaretto, B. A. Estimation of body composition in living animals. *Proc. Tech. Conf. Carcass Comp. Aust. Anim. Prod. Commit.* Melbourne: C.S.I.R.O. (1963).
53. Kay, M. and Jones, A. S. Estimating body water in live pigs using tritium or antipyrene substances. *Anim. Prod.* **5**, 220 (1963).

54. KAY, M. and JONES, A. S. The relationship between body density, body fat and body water in living pigs. *Anim Prod.* **4**, 296 (1962).
55. GROVES, T. D. D. and WOOD, A. J. Body composition studies on the suckling pig. II. The *in vivo* determination of total body water. *Canad. J. Anim. Sci.* **45**, 14 (1965).
56. SCHNEIDER, B. H. *Feeds of the World—their digestibility and composition.* West Virginia Univ., Morganstown (1947).
57. COMMITTEE ON ANIMAL NUTRITION. *Nutrient Requirements of Swine*, Publ. 1192, Nat. Acad. Sci.–Nat. Res. Coun., Washington D.C. (1964).
58. COMMITTEE ON ANIMAL NUTRITION. *Tables of Feed Composition* (*Nutrition data for U.S.A. and Canadian Feeds*), Publ. 659, Nat. Acad. Sci.–Nat. Res. Coun., Washington D.C. (1959).
59. MINISTRY OF AGRICULTURE, FISHERIES AND FOOD. *Rations for Livestock*, Bull. 48. London: H.M.S.O. (1960).
60. MINISTRY OF AGRICULTURE, FISHERIES AND FOOD. *Pig Feeding and Management*, Bull. 193. London: H.M.S.O. (1963).
61. EDEN, A. Additives in feeding stuffs. *Agric. Progr.* **35**, 54 (1960).
62. CRAMPTON, E. W. *The Potential for Increasing Efficiency of Feed Utilization through Newer Knowledge of Nutrition. Topic 5 (D), The Pig.* World Conf. Anim. Prod., Vol. 1, Europ. Assoc. Anim. Prod.: Rome (1963).
63. BOWLAND, J. P. Electronic computation of growing and finishing rations for pigs to meet U.S. N.R.C. (1959) nutrient requirements with and without a margin of safety. *Canad. J. Anim. Sci.* **42**, 191 (1962).
64. BREIREM, K., EKERN, A. and HOMB, T. Relation of nutrition of the young animal to subsequent fertility and lactation. *Proc. 5th Int. Congr. Nutr., Fed. Proc.* **20** (1) Supplement 7 (1961).
65. CRAMPTON, E. W. *Applied Animal Nutrition.* W. H. Freeman & Co., San Francisco (1956).
66. ROBERTSON, A. *The Potential for Increasing Efficiency of Feed Utilization through Newer Knowledge of Nutrition. Topic 3, Influence of Breeding on Production Efficiency.* World Conf. Anim. Prod., Vol. 1. Rome: Europ. Assoc. Anim. Prod.: Rome (1963).
67. MEYER, J. H. and NELSON, A. D. Efficiency of feed utilization by various animal species fed similar rations. *J. Nutr.* **80**, 343 (1963).
68. BARBER, R. S., BRAUDE, R. and MITCHELL, K. G. Further studies on the water requirements of the growing pig. *Anim. Prod.* **5**, 277 (1963).
69. KLEIBER, M. *The Fire of Life—an introduction to animal energetics.* London: John Wiley & Sons, Inc. (1961).
70. BLAXTER, K. L. *The Scientific Principles of Feeding Farm Livestock.* London: Farmer and Stockbreeder Publications Ltd. (1959).
71. BLAXTER, K. L. Efficiency of feed conversion by different classes of livestock in relation to food production. *Proc. 5th Int. Cong. Nutr., Fed. Proc.* **20** (1) Supplement 7 (1961).
72. TROELSEN, J. E. and BELL, J. M. Ingredient and processing interrelationships in swine feeds. IV. Effects of various levels and kinds of fibrous diluents in finisher rations, fed as meal or pellets on performance and carcass quality in swine. *Canad. J. Anim. Sci.* **42**, 63 (1962).
73. ROBINSON, K. L. How much fibre in the ration? *Pig Farming* **6**, 32 (1958).
74. CUNNINGHAM, H. M., FRIEND, D. W. and NICHOLSON, J. W. G. The effect of a purified source of cellulose on the growth and body composition of growing pigs. *Canad. J. Anim. Sci.* **41**, 120 (1961).
75. BRAY, R. W. Biopsy and core techniques for estimating composition. *Ann. N.Y. Acad. Sci.* **110**, 302 (1963).
76. COOPER, P. H. and TYLER, C. Some effects of bran and cellulose on the water relationships in the digesta and faeces of the pig. I. The effects of including bran and two forms of cellulose in otherwise normal rations. *J. Agric. Sci.* **52**, 332 (1959).
77. NORDFELDT, S. Digestibility experiments with pigs. *Ann. R. Agric. Coll. Sweden* **21**, 1 (1954).
78. NIELSEN, H. E. Growth rate of piglets as related to their later performance. In: *Physiologie Nutritionnelle et Sevrage des Porcelets. Ann. Zootech.* **13**. Suppl. (1964).

79. FORBES, R. M. and HAMILTON, T. S. The utilization of certain cellulosic materials by swine. *J. Anim. Sci.* **11**, 480 (1952).
80. LAWRIE, R. A. *Meat Science.* Oxford: Pergamon Press (1966).
81. ZELTER, S. Z. and CHARLET-LERY, G. The efficiency of some alimentary protids in the pig. 1. The excretion of faecal metabolic nitrogen and endogenous urinary nitrogen; effect of age, body weight, quantity of dry matter. *Ann. Biol. Anim. Bioch. Biophys.* **1**, 29 (1961).
82. COEY, W. E. and ROBINSON, K. L. Some effects of dietary crude fibre on liveweight gains and carcass conformation of pigs. *J. Agric. Sci.* **45**, 41 (1954).
83. COHN, C. and JOSEPH, D. Effects on metabolism produced by the rate of ingestion of the diet. *Amer. J. Clin. Nutr.* **8**, 682 (1960).
84. GWINUP, G., BYRON, R. C., ROUSH, W. H., KRUGER, F. A. and HAMWI, J. Effect of nibbling versus gorging on serum lipids in man. *Amer. J. Clin. Nutr.* **13**, 209 (1963).
85. CUTHBERTSON, D. P. and MUNRO, H. N. The relationship of carbohydrate metabolism to protein metabolism. 1. Roles of total dietary carbohydrate and of surfeit carbohydrate in protein metabolism. *Biochem. J.* **33**, 128 (1939).
86. FRIEND, D. W. and CUNNINGHAM, H. M. Effects of feeding frequency on metabolism, rate and efficiency of gain and on carcass quality of pigs. *J. Nutr.* **83**, 251 (1964).
87. BRAUDE, R., TOWNSEND, M. J., HARRINGTON, G. and ROWELL, J. G. A comparison of feeding growing pigs once or twice daily. *J. Agric. Sci.* **60**, 389 (1963).
88. CRAMPTON, E. W. Self versus hand feeding for market hogs with particular reference to effects on carcass measurements and quality. *Sci. Agric.* **17**, 529 (1937).
89. BRAUDE, R., TOWNSEND, M. J., HARRINGTON, G. and ROWELL, J. G. A large-scale test of the effects of food restriction on the performance of fattening pigs. *J. Agric. Sci.* **51**, 208 (1958).
90. BRAUDE, R. Some concepts of the nutrition of the pig. In: *Nutrition of Pigs and Poultry.* London: Butterworths (1962).
91. SMITH, C., KING, J. W. B. and GILBERT, N. Genetic parameters of British Large White bacon pigs. *Anim. Prod.* **4**, 128 (1962).
92. SMITH, C. and ROSS, G. J. S. Genetic parameters of British Landrace bacon pigs. *Anim. Prod.* **7**, 291 (1965).
93. HAMMOND, J. and MURRAY, G. N. The body proportions of different breeds of bacon pigs. *J. Agric. Sci.* **27**, 394 (1937).
94. CARROLL, M. A., HILL, F. and O'DONOVAN, P. B. Some effects of castration on pig carcasses of pork and bacon weights. *Irish J. Agric. Res.* **2**, 177 (1963).
95. DAVIDSON, H. R. *The Production and Marketing of Pigs.* London: Longman, Green & Co. (1948).
96. KIELANOWSKI, J. Estimates of the energy cost of protein deposition in growing animals. *Proc. 3rd Symp. Energ. Metab.* Energy Metabolism, Publ. 11, Europ. Assoc. Anim. Prod. London: Academic Press (1964).
97. BUCK, S. F. A comparison of pigs slaughtered at three different weights. 1. Carcass quality and performance. *J. Agric. Sci.* **60**, 19 (1963).
98. HILL, F. and O'CARROLL, F. M. The chemical composition of pig carcasses at pork, bacon and manufacturing weights. *Irish J. Agric. Res.* **1**, 115 (1962).
99. WEDGWOOD, R. J. Inconstancy of the lean body mass. *Ann. N.Y. Acad. Sci.* **110**, 141 (1963).
100. MITCHELL, H. H. The measurement of digestibility. In: *The Comparative Nutrition of Man and Domestic Animals*, Vol. 2. London: Academic Press (1964)
101. COMMITTEE ON ANIMAL NUTRITION. *Glossary of Energy Terms*, Publ. 1040. Nat. Acad. Sci.–Nat. Res. Coun. Washington, D.C. (1962).
102. ROBINSON, D. W., PRESCOTT, J. H. D. and LEWIS, D. The protein and energy nutrition of the bacon pig. IV. Digestible energy values of cereals in pig diets. *J. Agric. Sci.* **64**, 59 (1965).
103. DIGGS, B. C., BECKER, D. E., TERRILL, S. W. and JENSEN, A. H. The energy values of various feedstuffs for the young pig. *J. Anim. Sci.* **18**, 1492 (1959).
104. SWIFT, R. W. The calorie value of TDN. *J. Anim. Sci.* **16**, 753 (1957).

105. THOMPSON, D. G. Energy retention in lambs as measured by the comparative slaughter technique. *Proc. 3rd Symp. Energ. Metab.* Energy Metabolism, Publ. 11, Europ. Assoc. Anim. Prod. London: Academic Press (1964).
106. DAVIDSON, J. The efficiency of conversion of dietary metabolizable energy to tissue energy in young chickens as measured by body analysis. *Proc. 3rd Symp. Energ. Metab., Energy Metabolism*, Publ. 11, Europ. Assoc. Anim. Prod. London: Academic Press (1964).
107. MITCHELL, H. H. and KELLY, M. A. R. Energy requirements of swine and estimates of heat production and gaseous exchange for use in planning the ventilation of hog houses. *J. Agric. Res.* **56**, 811 (1938).
108. LEROY, A. M. Analysis of the origin of the heat loss of pigs fed normally. *Proc. 3rd Symp. Energ. Metab.*, Energy Metabolism, Publ. 11, Europ. Assoc. Anim. Prod. London: Academic Press (1964).
109. MADSEN, A. The influence of protein, and especially lysine and methionine on the carcass quality of pigs. *Proc. Pfizer Europ. Agric. Res. Conf. Lucerne* (1960).
110. HOLME, D. W., COEY, W. E. and ROBINSON, K. L. The effect of dietary protein on the carcass composition of bacon pigs. *Anim. Prod.* **7**, 363 (1965).
111. HOLME, D. W. Carcass quality in the bacon pig. Ph.D. Thesis, Queen's University, Belfast (1961).
112. MORGAN, J. T. and ROBINSON, D. W. Dietary factors and the performance of growing finishing pigs. In: *Nutrition of Pigs and Poultry*. London: Butterworths (1962).
113. EVANS, R. E. The minimum amount of white fish meal required to supplement the proteins in some vegetable-protein concentrates. *J. Agric. Sci.* **53**, 230 (1959).
114. HUTCHINSON, H. D., JENSEN, A. H., TERRILL, S. W. and BECKER, D. E. The lysine requirement of the weaning pig. *J. Anim. Sci.* **16**, 558 (1957).
115. JONES, A. S., HEPBURN, W. R. and BOYNE, A. W. The nitrogen metabolism of the young pig. I. Supplemental value of certain amino-acids when added to groundnut meal rations. *J. Sci. Fd. and Agric.* **12**, 353 (1961).
116. JONES, A. S., HEPBURN, W. R., CADENHEAD, A. and BOYNE, A. W. The effect of variations in protein quality and protein level in diets on the performance of young pigs. *Anim. Prod.* **4**, 185 (1962).
117. WOODMAN, H. E. and EVANS, R. E. Nutrition of the bacon pig. XIV. The determination of the relative supplemental values of vegetable protein (extr. decortic. groundnut meal) and animal protein (white fish meal). *J. Agric. Sci.* **41**, 102 (1951).
118. ROBINSON, D. W., MORGAN, J. T. and LEWIS, D. Protein and energy nutrition of the bacon pig. I. Effect of varying the protein and energy content of grower rations. *J. Agric. Sci.* **62**, 369 (1964).
119. GRAHAM, N. MCC., BLAXTER, K. L. and ARMSTRONG, D. G. Analytical and other techniques used in respiration calorimetry and their errors. *Proc. 1st Symp. Energ. Metab.*, Publ. 8, Europ. Assoc. Anim. Prod. Rome (1958).
120. RASCH, G., LUDVIGSEN, J. and THORBEK, G. Evaluation of errors in determination of N, P and Ca in feeds, faeces and urine. *Proc. 1st Symp. Energ. Metab.*, Publ. 8, Europ. Assoc. Anim. Prod. Rome (1958).
121. HEDIN, D. A. and ADACHI, R. A. Effect of diet and time of feeding on gastrointestinal gas production in rats. *J. Nutr.* **77**, 229 (1962).
122. LASSITER, J. W., TERRILL, S. W., BECKER, D. E. and NORTON, H. N. Protein levels for pigs as studied by nitrogen balance. *J. Anim. Sci.* **15**, 392 (1956).
123. MOUNT, L. E. Food, Meat and Heat Conservation. *Proc. Conf. on Efficient Pig Feeding*. Pig Indust. Devel. Authority: London (1963).
124. SØRENSEN, P. H. Influence of climatic environment on pig performances. In: *Nutrition of Pigs and Poultry*. London: Butterworths (1962).
125. MOUNT, L. E. and INGRAM, D. L. The effects of ambient temperature and air movement on localized sensible heat-loss from the pig. *Res. Vet. Sci.* **6**, 84 (1965).
126. MOUNT, L. E. The young pig and its physical environment. *Proc. 3rd Symp. Energ. Metab.* Energy Metabolism, Publ. 11, Europ. Assoc. Anim. Prod. London: Academic Press (1964).

127. GORDON, W. A. M. Environmental studies in pig-housing. 1. Air velocity. *Brit. Vet. J.* **118**, 171 (1962).
128. INGRAM, D. L. The effect of humidity on temperature regulation and cutaneous water loss in the young pig. *Res. Vet. Sci.* **6**, 9(1965).
129. HAFEZ, E. S. E. The behaviour of swine. In: *The Behaviour of Domestic Animals.* London: Baillière, Tindall & Cox Ltd. (1963).
130. HEITMAN, H. and HUGHES, E. H. The effects of air temperature and relative humidity on the physiological well-being of swine. *J. Anim. Sci.* **8**, 171 (1949).
131. HALE, R. W. and COEY, W. E. *Two Levels of Humidity compared in an Experimental Pig Fattening House.* Final Rep. 8th Int. Congr., Anim. Prod., Europ. Assoc. Anim. Prod.: Rome (1962).
132. MOUSTGAARD, J. and SØRENSEN, P. H. Influences of environmental factors including nutrition on thyroid function. *Proc. 6th Int. Congr. Nutr.* Edinburgh: E. and S. Livingstone & Co. (1963).
133. CUNNINGHAM, H. H. Effect of epinephrine and nicotine on protein and fat metabolism of pigs. *Proc. 3rd Symp. Energ. Metab.* Energy Metabolism, Publ. 11, Europ. Assoc. Anim. Prod. London: Academic Press. (1964).
134. BRAUDE, R., WALLACE, H. D. and CUNHA, T. J. The value of antibiotics in the nutrition of swine: a review. *Antibiot. and Chemother.* **3**, 271 (1953).
135. ROBINSON, K. L. The value of antibiotics for growth of pigs. In: *Antibiotics in Agriculture.* London: Butterworths (1962).
136. BRAUDE, R. Copper as a stimulant in pig feeding. *Proc. Symp. Cuprum pro vita Vienna.* Int. Copper Devel. Coun. p. 55-66 (1965).
137. HANSON, L. E. Feed additives—Rationale for medicated feeds: in Symposium on feed additives and significance of their residues in animal tissues. *J. Agric. and Fd. Chem.* **11**, 365 (1963).
138. HEITMAN, H. (JR.), HAHN, L., KELLY, C. F. and BOND, T. E. Space allotment and performance of growing-fattening swine raised in confinement. *J. Anim. Sci.* **20**, 543 (1961).
139. NOFZIGER, J. C. Effect of some environmental stresses on swine. *Dissert. Abs.* **22**, 955 (1961).
140. BETTS, A. O. 'Pathogen-free' pigs for research and the practical control of pig diseases. *Vet. Rec.* **73**, 1349 (1961).
141. KING, J. W. B. Crossbreeding pigs. *Anim. Breeding. Res. Orgn. Rep.*, pp. 32–36. London: H.M.S.O. (1965).
142. NALBANDOV, A. V. Symposium on Growth: Endocrine causes of growth and growth stasis. *J, Anim. Sci.* **22**, 558 (1963).
143. BAIRD, D. M., NALBANDOV, A. V. and NORTON, H. W. Some physiological causes of genetically different rates of growth in swine. *J. Anim. Sci.* **11**, 292 (1952).
144. HOLME, D. W. Archimedes and the pig. *Ann. Rep. Agric. Res. Inst., N.I.*, 1964–5 (1965).
145. OGDEN, A. L. Biochemical polymorphism in farm animals. *Anim. Breed. Abstr.* **29**, 127 (1961).
146. AGRICULTURAL RESEARCH COUNCIL. *Index of Current Research on Pigs, XI.*, Tech. Commit. Pig. Res., N.I.R.D., Shinfield, Reading (1965).
147. COOK, C. F. Techniques for measuring some quality characteristics of meat. *Proc. Tech. Conf. Carcass Comp., Aust. Anim. Prod. Comm.* Melbourne: C.S.I.R.O. (1963).
148. AMERICAN MEAT INSTITUTE FOUNDATION. *The Science of Meat and Meat Products.* San Francisco: W. H. FREEMAN (1960).
149. LUDVIGSEN, J. B. Maladaptation syndromes in pigs. *Proc. 2nd Int. Anim. Nutr. Conf., U.S.D.A./U.S. Fd. Grains/Soyabean Coun., Madrid,* 1960 (1960).
150. BENDALL, J. R. and LAWRIE, R. A. Watery Pork. *Anim. Breed. Abstr.* **32**, 1 (1964).
151. BRISKEY, E. J., BRAY, R. W., HOEKSTRA, W. G., PHILLIPS, P. H. and GRUMMER, R. H. The effect of exhaustive exercise and high sucrose regimen on certain chemical and physical pork ham muscle characteristics. *J. Anim. Sci.* **18**, 173 (1959).
152. INGRAM, M. Feeding meat animals before slaughter. *Vet. Rec.* **76**, 1305 (1964).

153. CALLOW, E. H. Ultimate pH of muscle tissue: Electrical resistance and microstructure: Transport by rail and its after effects on pigs (with T. G. BOAZ). *Rep. Fd. Invest. Bd.*, 1937, pp. 41, 46 and 56. London: H.M.S.O. (1938).
154. BLAIR, R. and ENGLISH, P. R. The effect of sex on growth and carcass quality in the bacon pig. *J. Agric. Sci.* **64**, 169 (1965).
155. DUTT, R. H., SIMPSON, E. C., CHRISTIAN, J. C. and BORNHART, C. E. Identification of preputial glands as the site of production sexual odour in the boar. *J. Anim. Sci.* **18**, 1557 (1959).
156. CRAIG, H. B., PEARSON, A. M. and WEBB, N. B. Fractionation of the component(s) responsible for sex odour/flavour in pork. *J. Fd. Sci.* **27**, 29 (1962).
157. WILLIAMS, L. D., PEARSON, A. M. and WEBB, N. D. Incidence of sex odour in boars, sows, farrows and gilts. *J. Anim. Sci.* **22**, 166 (1963).
158. PRESCOTT, J. H. D. and LAMMING, G. E. The effects of castration on meat production in cattle, sheep and pigs. *J. Agric. Sci.* **63**, 341 (1964).
159. PLIMPTON, R. F. (JR.), OCKERMAN, H. W., CAHILL, V. R. and TEAGUE, H. S. Palatibility and quality of pork from stilboestrol implanted boars. *J. Anim. Sci.* **24**, 867 (1965).
160. BRAUDE, R. The use of hormones and additives as growth promoters in pigs. *Vet. Rec.* **69**, 178 (1957).
161. DAY, B. N., ZOBRISKY, S. E., TRIBBLE, L. F. and LASLEY, J. F. Effects of stilboestrol and combination of progesterone and estradiol on growing finishing swine. *J. Anim. Sci.* **19**, 898 (1960).
162 BEACOM, S. E. The effect of diethylstilboestrol and estradiol-testosterone implants on rate and efficiency of gain and on carcass quality of market pigs fed different finishing diets. *Canad. J. Anim. Sci.* **43**, 374 (1963).
163. LAMOND, D. R. The endocrine basis of growth in relation to the design of experiments influencing body composition. *Proc. Tech. Conf. Carcass Comp., Aust. Anim. Prod., Comm.* Melbourne: C.S.I.R.O. (1963).
164. MCCANCE, R. A. and WIDDOWSON, E. M. Nutrition and Growth: in Symposium on biochemistry and nutrition. *Proc. Roy. Soc. B*, **153**, 326 (1962).
165. COSTAIN, R. A. and MORGAN, J. T. The effect of variations in the energy and protein status of the diet upon productive capacity in the pig. *Anim. Prod.* **3**, 199 (1961).
166. GOODWIN, R. F. W. The possible role of hysterectomy and related procedures for the eradication and control of pig diseases. *Vet. Rec.* **77**, 1070 (1965).
167. OSLAGE, H. J. Untersuchungen zum Stoff-und Energieumsatz wachsender Mastschweine. Landbauforsch. Volken & ode, **15**, 107 (1965).
168. CLAUSEN, H. Private communication (1966).

CHAPTER 28

Nutrition of the Sow

G. A. LODGE

*University of Nottingham School of Agriculture, Sutton Bonington, Loughborough**

THE first essential in attempting to assess nutrient requirements of livestock kept for economic purposes is to establish criteria of adequacy based on the performance characteristics of major economic significance. In the case of the sow, whose sole economic function is the production of weaned pigs, the ultimate measure of dietary sufficiency must be the cost of producing healthy weaned pigs of suitable weight and conformation.

TABLE 1
*Variation in Cost of Producing Weaned Pigs (PIDA, 1963a)**

Feed cost per pig reared (shillings)	Proportion of herds (%)
less than 50	9
50–60	20
60–70	25
70–80	20
80–90	9
90–100	6
100 and over	11

* Feed is estimated to account, on average, for 70 per cent of total costs.
PIDA = Pig Industry Development Authority.

The cost of producing weaned pigs varies widely in practice (Table 1) and for this there seem to be two main reasons: first is the wide variation in productivity of sows, and second is the equally wide variation in overhead costs, of which feed is by far the greatest single item accounting for some 70 per cent of the total (PIDA, 1963a).

SOW PRODUCTIVITY

Sow productivity, in terms of number or weight of pigs produced per unit time, is the end result of several individual processes, and there is no single, generally accepted measurement which expresses adequately the relative importance of these.

* Present address: Canada Department of Agriculture, Animal Research Institute, Ottawa.

A common measure in comparative assessments of sow productivity is weight of piglets produced per annum or during the reproductive life. In Table 2 are presented performance data of recorded sows in the United Kingdom which, if converted into a single measure, indicate an average production of 657 lb of weaned pig per annum (2·2 lb = 1 kg). It is evident, however, from published data on average milk yields of sows and the efficiency of conversion of milk into weight gain (Lucas and Lodge, 1961) that, on average, only some 28 lb of the 8-week weight of a pig can be attributed to the sow and the remainder must be attributed to feed consumed by the pigs themselves. Of a total weaning weight of 657 lb per annum, therefore, only some 460 lb may be attributed directly to the sow.

Even when this allowance is made for the contribution of supplementary feed, it is questionable whether total litter weight is a good measure of sow productivity. In the above example, weaning at the traditional age of 8 weeks resulted in the annual output of 460 lb of weaned pigs per sow; if those pigs had been weaned at 3 weeks of age, when they averaged 13·0 lb per head, then 2·5 litters per sow per annum would have been possible, but the total weight of pigs produced per sow would have been only 289 lb.

It is debatable which of the above sows is the more productive; that which produces 460 lb weight in 16·5 pigs per annum or that which produces only 289 lb weight but in 22·25 pigs per annum. The answer to this question depends upon nutritional and economic aspects of rearing young pigs, which are the subjects of Chapters 26 and 34, but the view may be stated here that as the sow has no substitute for the production of piglets, while there are alternative methods of rearing other than on mother's milk, sow productivity may be better assessed in terms of number of pigs weaned rather than their total weight.

Actual and Potential Sow Productivity

The average performance of recorded sows in the United Kingdom is presented in Table 2.

Table 2
*Average Performance Data of Litter-Recorded Sows** (*PIDA*, 1963*b*)

Age at first parturition (days)	403
Litters per sow per annum	1·9
Number per litter born alive	10·5
Number per litter at 3 weeks	8·9
Average weight at 3 weeks (lb)	13·0
Number per litter at 8 weeks	8·7
Average weight at 8 weeks (lb)	39·8
Total weight of weaned pigs per annum (lb)	657

* Based on approximately 53,000 litters in 1700 herds between April 1962 and March 1963 of the following breeds: Large White, 51%; Landrace, 33%; Welsh, 4%; Wessex Saddleback, 3%; other pure breeds, 1%; crossbreds, 8%.

It may be assumed that as these *were* recorded sows, their selection and management would be above the general average and that their performance also would be above the general average for the country; even so, it falls far short of the potential expressed in terms of number of pigs weaned per annum.

Breeding Age

The first factor which influences the overall productivity of the sow is age at first parturition. The average age of 403 days given in Table 2 means that the average age at first effective service was about 290 days. As puberty is generally between 175 and 200 days of age (Phillips and Zeller, 1943; Robertson *et al.*, 1951a; Burger, 1952; Schmidt and Bretschneider, 1954; Lodge and Macpherson, 1961), the above gilts were mated, on average, at the fifth heat after puberty. It has been reported that ovulation rate increased up to the second post-pubertal heat (Warnick *et al.*, 1951; Robertson *et al.*, 1951a), so, on average, some 7–10 weeks were being wasted through failure to serve gilts early.

Optimal economic age for first parturition would appear to be about 330 days, and the nutrition of gilts during the rearing period to first service should be designed with this end in view. Not only does mating at an early age and light weight reduce the initial cost of breeding stock—and so the overhead cost which each weaned pig must carry—but it may also allow subsequent savings in feed costs through reducing the requirements of the animal for maintenance.

Litter Frequency

The second important factor determining sow productivity is number of litters produced per annum, a measure commonly called the 'farrowing index' and expressed as the interval in days between one parturition and the next.

The average interval between farrowings in recorded herds (Table 2) is 193 days. With an average pregnancy duration of 114 days and interval between weaning and subsequent oestrus of 5 days, the lowest possible farrowing index with 8-week weaning is 175 days. So even discounting the fact that some of the breeders contributing to the average of a 193-day farrowing index wean at earlier ages than 8 weeks, and few if any will exceed this, the inference is that the sows were mated, on average, not at the first but at the second heat after weaning. In practice the low average farrowing index must be due partly to sows which returned to service subsequent to mating at first oestrus.

By weaning at ages earlier than 8 weeks the average farrowing index may be reduced to the minimum of about 130 days which is possible with weaning at under 1 week old (Dyrendahl *et al.*, 1958). Weaning at such an early age is impracticable, however, under farm conditions, and the lowest realistic farrowing index is about 140 days with weaning at 3 weeks. Even 3-week weaning has its problems and necessitates the use of relatively costly diets (see Chapter 26), and weaning at 5 weeks of age with a farrowing index of about 155 days could be regarded as an acceptable target at the present time, thus gaining some 40 days with each litter over the current average.

As daily requirements for all nutrients are appreciably higher during lactation than during pregnancy (Lodge and Lucas, 1959), weaning age has considerable influence on the nutrient requirements of sows.

Numbers Born

The third major factor affecting sow productivity is number of pigs born alive per litter. The average of 10·5 shown in Table 2 conceals considerable variation between breeds, and reference to the original publication (PIDA, 1963b) shows that average numbers born alive range from 10·9 and 10·8 for the Welsh and Large White breeds respectively, through 10·1 for Landrace and Wessex Saddleback and 9·8 for Essex, to 9·5 for the small number of other breeds represented; crossbred sows averaged 10·3 live pigs born.

Number of pigs born alive is dependent upon ovulation rate, number of ova fertilized, incidence of embryo and foetal mortality and loss at parturition. If all the ova shed survived through parturition, the number of pigs born alive would be about double the present average (Hammond, 1914, 1921; Corner, 1923; Squiers, Dickerson and Mayer, 1950; Robertson *et al.*, 1951a; Burger, 1952; Perry, 1955; Rathnasabapathy, Lasley and Mayer, 1956).

Although this 50 per cent loss through prenatal mortality in the sow has been recognized for over 50 years, its cause has never been established. The high proportion of the loss which occurs in early pregnancy, when nutritional requirements of the embryos must be negligible and there could be no competition for space, suggests that a major cause must be internal defects of the germ cells and embryos. Nevertheless, as is discussed later, nutrition of the sow is known to influence both ovulation rate and embryo survival.

In view of the inability of most sows to rear successfully more than twelve pigs, the problem of foetal mortality is largely academic and likely to remain so until an effective and economical means is developed for the artificial rearing of piglets from birth. The problem in practice is not so much a low average number born but rather the great variation which occurs, from two or three at one end of the scale to over twenty at the other. Both extremes are undesirable as the former inevitably reduces the number weaned per sow while the latter increases the incidence of postnatal mortality.

The ideal litter size at birth would appear to be twelve live pigs (Menzies-Kitchen, 1937; Stahl, 1930; Olbrycht, 1943; Longwill, 1952a, b), which is only 1·5 pigs more than the current average, but if the range could be reduced to 10–12, this could result in an appreciable increase in the number weaned.

Numbers Weaned

Number of pigs weaned per litter is the fourth major factor determining sow productivity, and while it is basically dependent on number born it is influenced also by the other sow effects of birth weight and viability of the pigs and yield and composition of milk, as well as by the other environmental factors of housing and disease.

It can be seen from Table 2 that nearly two pigs per litter are lost between parturition and 8 weeks of age and that most of this loss occurs during the first 3 weeks of life. More detailed surveys of piglet mortality show that most of the loss, in fact, occurs during the first 3 days and represents some 25 per cent of all pigs born (Veterinary Investigation Service, 1959).

There has been considerable speculation as to the basic causes of this high mortality, and while that most commonly cited, namely, crushing by the sow, is undoubtedly important, the high correlation frequently found between number born and neonatal mortality (Menzies-Kitchen, 1937; Stahl, 1930; Olbrycht, 1943; Korkman, 1947; Longwill, 1952a, b; Lodge, McDonald and Macpherson, 1961; PIDA, 1963b) suggests that competition for an inadequate number of functioning teats is a major cause of early loss.

A negative relationship has been clearly established between birth weight and mortality, and in any population there is a critical weight below which the chances of survival are small (Husby, 1933; Jespersen and Olsen, 1939; Meyer, 1940; Olofsson, 1950; Holder, 1959; Pomeroy, 1960; Lodge, McDonald and Macpherson, 1961). The critical weight itself will vary with environmental circumstances, and whereas Pomeroy (1960) considered it to be 1·9 for a herd of inbred Large Whites, Lodge and Pratt (1963) found that when the range in weight within litters was reduced by reallocating piglets to sows on the basis of birth weight, pigs of as little as 1·1 lb at birth were reared successfully to 8 weeks of age.

Because of the close correlation which has been found between birth weight and survival, it is widely believed that an increase in average birth weight would result in a marked reduction in the incidence of piglet mortality. However, the indication that much of the mortality in the larger litters, where the incidence is highest, is due to competition for functioning teats implies that this loss would occur regardless of birth weight; the lightest pigs would be penalized whether they weighed 1 or 3 lb.

The implication of this in relation to sow feeding is that although high birth weights may be desirable for subsequent ease of rearing (unless abnormally high and causing parturition difficulties), attempts to increase the average birth weight through high levels of nutrition, even if successful, may be of little value unless accompanied by a reduction in variation within litters and in number born to below twelve. In any event, the economic value of increasing birth weights through costly feeding regimes would require careful assessment.

A factor of extreme importance in the early survival of piglets is the conference of passive immunity through the colostrum of the sow. This is a subject more appropriate to Chapter 26, which deals with the nutrition of the young pig, but it merits some mention in the context of sow nutrition because of any influence which the feeding of the sow may have on the quantity and composition of her colostrum.

Nothing seems to be known about the influence of diet on the quantity of colostrum, or on the significance of the amount of colostrum consumed by each piglet on its subsequent resistance to infection. On a qualitative basis, however, it is known that the composition of colostrum may be influenced by the diet of the sow.

At parturition the colostrum of the sow has a high dry-matter content and a protein (N $\times$ 6·38) content which may be as high as 70 per cent of the dry matter; of this protein over 50 per cent is albumin and globulin (Salmon-Legagneur and Guegen, 1962). During the first 3 days *post-partum*, the composition of colostrum changes rapidly; the percentages of ash, lactose and fat rise while the percentage of protein falls. The transition to normal milk is generally complete by the seventh day of lactation (Perrin, 1955; Travnicek, 1960; Salmon-Legagneur and Guegen, 1962).

The high levels of albumin and globulin and the relatively low level of casein, by comparison with normal milk, are of considerable significance in the conferring of passive immunity to the newborn pigs, but there is no evidence that the level or composition of colostral protein can be influenced by the feeding of the sow.

Also of possible significance in the resistance of piglets to infection is the intake of vitamin A. The vitamin A level of colostrum is some three times that of normal milk (Braude *et al.*, 1947; Heidebrecht *et al.*, 1951; Bowland *et al.*, 1949a, b, 1951), and while the level in milk may be greatly influenced by the intake of the sow (Schofield, Ingle and McFarlane, 1941; Braude *et al.*, 1947; Thomas, Loosli and William, 1947; Bowland *et al.*, 1949a, b, 1951), it is less evident that the level in colostrum can be appreciably increased by the administration of high levels to the sow, or that higher levels than those generally present are of benefit to the pigs.

The vitamin C level in colostrum of the sow is about double that in normal milk (Braude *et al.*, 1947; Bowland *et al.*, 1949a; Heidebrecht *et al.*, 1951; Gregory, Kon and Thompson, 1952), and while it may be influenced by the diet (Bowland *et al.*, 1949a) this would seem of little significance in view of the ability of the pig to synthesize its requirements (Braude, Kon and Porter, 1950).

There is clear evidence that the levels of B complex vitamins in sows' milk are greatly influenced by dietary levels (Owen and Bowland, 1952; Collins *et al.*, 1951; Davis *et al.*, 1951; Braude *et al.*, 1947; Catron, Culbertson and Shearer, 1948), with the apparent exception of vitamin B_{12} (Barber *et al.*, 1960), but as the levels of at least three of the B complex, namely thiamine, nicotinic acid and pantothenic acid, are lower in colostrum than in normal milk, the value of increasing colostral levels seems doubtful.

In general, dietary influences on the yield and composition of post-colostral milk and the effect of these on the growth of piglets would seem to be of more practical significance than any probable effects on colostrum.

Postnatal Growth

Although the number of pigs may be of greater economic significance than weight of pigs produced per sow per annum, individual weight of pigs is nevertheless important through its influence on the age at which they may be effectively weaned.

The three major determinants of weight for age in piglets are birth weight, milk consumption and the consumption of supplementary feed. The relative importance of these will vary with age of piglet under consideration; the effect of birth weight becomes progressively less and that of supplementary feed consumption progressively greater with increase in age. Thus the correlation between weight at birth and weight at 8 weeks is only of the order $r = +0{\cdot}4$ to $0{\cdot}5$ (Blunn, Warwick and Wiley, 1954; Combs *et al.*, 1959; Lodge and McDonald, 1959; Lodge, McDonald and MacPherson, 1961), which indicates that of the variation between pigs in 8-week weight only some 20 per cent may be attributed to differences in birth weight, whereas some 40 per cent of the variation in weight at 3 weeks may be attributed to this cause (Lodge, McDonald and MacPherson, 1961).

Of the 8-week weight of a pig only a small proportion is represented by the initial birth weight; the remainder must be attributed to intake of milk or other food. Generally, the greater proportion of the weight gain is attributable to milk (although the greater part of the *variation* between pigs may be attributable to differences in intake of supplementary feed) and only in the case of pigs in excess of 50 lb at 8 weeks is most of the total weight likely to be attributable to supplementary feed (Lucas and Lodge, 1961).

Milk secretion by the sow is, therefore, of considerable significance in the growth of young pigs, and the earlier they are weaned the more influential does milk secretion become.

Yield and Composition of Milk

Many estimates of average milk yields of sows have been published and some of the more recent are presented in Table 3.

TABLE 3
Some Published Estimates of Milk Yields of Sows

Number of lactations	Average daily yield (lb)	Average size of litter	Average daily yield per pig (lb)	Reference
9	9·3	6·9	1·3	(129)
15	14·4	7·4	1·9	(176)
61	12·1	7·1	1·7	(177)
28	16·5	8·5	1·9	(16)
3	13·7	10·7	1·3	(15)
6	11·4	8·0	1·4	(57)
71	11·7	8·2	1·4	(160)
24	17·8	8·8	2·0	(102)
63	12·5	7·3	1·7	(178, 179, 180, 181)
6	13·6	9·2	1·5	(73)
112	12·0	8·5	1·5	(162, 165)
39	15·7	8·7	1·8	(196)

The considerable variation between values indicates the extent to which individuals may differ, both within and between breeds, under different environmental conditions. An overall average yield would appear to be about 1·75 lb milk per day per piglet suckled but even within the same sow this would vary with stage of lactation and number of pigs in the litter.

Peak yield is normally reached between the third and fifth weeks (Niwa *et al.*, 1951; Smith, 1952a, b, 1959a, b, 1960a, b; Berge and Indrebø, 1953; Barber, Braude and Mitchell, 1955; Gill and Thomson, 1956; Salmon-Legagneur, 1956; Lodge, 1959a), but the shape of the lactation curve appears to vary with level of feeding. Under circumstances which allow the sow to lose appreciable weight during lactation a peak occurs, whereas when sows are maintained at relatively constant weight the result is more of a plateau in milk yield between weeks 3 and 6 (Lodge, 1959a).

For a sow fed to maintain weight during lactation and give an average daily yield of 1·75 lb per piglet, the actual daily yields during successive weeks of lactation have been estimated to be as follows (Lodge, 1962):

TABLE 4
Variation in Milk Yield with Stage of Lactation

	Week of lactation								
	1	2	3	4	5	6	7	8	Mean
Mean daily yield/ sow (lb)	11·5	15	16	16	16	15	12	10·5	14
Mean daily yield/ pig (lb)	1·4	1·9	2·0	2·0	2·0	1·9	1·5	1·3	1·75

A close correlation has been found between litter size and milk yield ($r = +0·8$, Berge and Indrebø, 1953; $r = +0·72$, Salmon-Legagneur, 1958), due to the fact that each piglet usually sucks only one teat (Allen and Lasley, 1958).

Berge and Indrebø (1953) found that within the range of 5 to 10 pigs per litter, daily milk yield increased almost linearly with increase in litter size; only in litters with fewer than five pigs or more than 10, did each pig receive respectively more or less.

It is evident that the two major factors influencing the nutritional requirements of the lactating sow are stage of lactation and litter size. A third important consideration is the composition of the milk.

Some estimates of the average composition of sows' milk are presented in Table 5.

TABLE 5
Some Published Estimates of Average Composition of Sow's Milk

Fat (%)	Protein (Nx 6·38) (%)	Lactose (%)	Ash (%)	Calcium (%)	Phosphorus (%)	Energy* (kcal/g)	Reference
8·2	5·6	4·8	0·94	0·25	0·17	1·23	(22)
6·8	7·3	5·1	0·99	—	—	1·19	(19)
8·0	6·0	4·7	1·03	—	—	1·22	(76)
10·0	5·3	4·2	0·91	—	—	1·35	(129)
7·7	6·2	5·0	—	—	—	1·22	(176)
9·6	6·1	4·6	0·92	—	—	1·37	(137)
—	—	—	—	0·25	0·15	—	(138)
6·2	5·8	5·4	0·91	—	—	1·08	(161)
6·6	5·7	4·7	0·89	—	—	1·26	(101)
—	—	—	—	0·21	0·18	—	(92)
—	—	—	—	0·21	0·15	—	(63)
6·8	5·4	5·3	0·84	—	—	1·10	(191)
7·8	6·0	5·4	0·97	—	—	1·23	(162)
7·2	5·6	6·4†	0·82	—	—	1·20	(146)

* By calculation from percentages fat, protein and lactose with Perrin's (1958) formula.
† By difference.

These are overall averages and the composition does not remain constant but changes with stage of lactation. Following the transition from colostrum to normal milk, during the first week of lactation, the percentages of protein and ash tend to rise while percentages of lactose and fat tend to fall, although there is considerable fluctuation in fat percentage throughout lactation (Perrin, 1954). Average changes in composition of sows' milk during the course of lactation, based on two sets of experimental data (Lodge, 1959a, b; Guegen and Salmon-Legagneur, 1959) are presented in Table 6.

Of particular relevance to the feeding of sows are the high fat content of sows' milk, necessitating a high energy intake, and the rise in protein percentage with advance in lactation, which means that protein requirement remains high even though milk yield is declining.

TABLE 6

*Average Changes in Milk Composition During Lactation**

Week of lactation	1	2	3	4	5	6	7	8	Mean
Protein (N x 6·38) (%)	5·5	5·0	5·3	5·4	5·7	5·9	6·2	6·6	5·7
Lactose (%)	4·9	5·0	4·9	4·8	4·7	4·6	4·6	4·6	4·8
Fat (%)	9·0	8·6	9·7	8·9	8·8	8·2	7·9	7·7	8·6
Ash (%)	0·8	0·7	0·8	0·9	0·9	1·0	1·0	1·1	0·9
Ca† (%)	0·18	0·18	0·19	0·20	0·22	0·23	0·25	0·23	0·21
P† (%)	0·14	0·13	0·15	0·15	0·16	0·16	0·16	0·17	0·15
K† (%)	0·12	0·10	0·10	0·11	0·09	0·10	0·10	0·10	0·10
Na† (%)	0·04	0·03	0·03	0·03	0·04	0·03	0·04	0·04	0·03
Mg† (%)	0·02	0·02	0·02	0·02	0·02	0·02	0·02	0·02	0·02
Gross energy kcal/g‡	1·29	1·24	1·35	1·28	1·28	1·23	1·22	1·22	1·26

* Means of twenty-four lactations of eight sows (Lodge, 1959a, b).
† Guegen and Salmon-Legagneur, 1959.
‡ Calculated with Perrin's (1958) formula from percentages of fat, protein and lactose.

The diet of the sow may influence both yield and composition of milk but, because of her considerable ability to utilize body reserves for milk secretion, the effects of dietary deficiency may not be immediately apparent. On a dietary regime which allows consistent weight loss from parturition there may be no reduction in milk yield for 2–3 weeks, and over the whole lactation (by comparison with sows fed to maintain weight) the reduction in milk yield will be less than the reduction in feed (Lodge, 1959a; Smith, 1959b).

Changes in milk composition are associated with reduction in yield brought about by inadequate feeding. Reduction in the plane of feeding during lactation, causing a reduction in intake of all nutrients, has resulted in increases in the percentages of fat and protein in the milk, thus counteracting to some extent the effect of reduced yield on intake of nutrients by the pigs (Lodge, 1959a; Smith, 1959b).

Level of feeding during pregnancy has been found also to affect milk composition; sows fed *ad lib.* during pregnancy produced milk of higher fat content than did sows restricted to half the *ad lib.* level regardless of level of feeding during lactation (Salmon-Legagneur and Jacquot, 1961b). It appeared from the results of this experiment that milk yield was mainly influenced by level of feeding during lactation while milk composition was affected more by level of feeding during the preceding pregnancy.

Milk composition is influenced also by the composition of the sow's diet. Mention has been made already of the effect which intake of vitamins may have on the levels in milk; there is little evidence that the mineral content may be influenced in a like manner and this is of significance in the prevention of piglet anaemia. Of the major constituents of milk, the nitrogen fraction does not appear to be influenced by protein intake during lactation (Salmon-Legagneur, 1964a) but the fat content may be influenced by intake of fat during pregnancy or lactation (Willet and Maruyama, 1946; Salmon-Legagneur, 1962, 1963, 1964b).

The significance of changes in the yield and composition of sows' milk on the growth of young pigs is not always apparent. A high milk yield might be considered to be always beneficial, but in spite of a 19 per cent higher milk yield from sows fed to a high-plane, in contrast to others allowed to lose appreciable weight, Lodge (1959a) found no significant difference in 8-week weight between the litters; partly because of the counteracting effects on milk composition and partly because most of the difference in yield occurred after the pigs were 3 weeks old and those of the 'low-plane' sows ate more solid food.

The implication may be that maximum milk yield is not desirable for the most economical growth in piglets, at least beyond the stage when they can take solid food. The value of a high milk yield in the early life of the pigs, in order to give high weights at 3 weeks, is indicated by the positive correlation found between 3-week weight and subsequent creep-feed consumption (Teter and Hanson, 1959; Teague and Wilson, 1957).

Of the major milk constituents, fat is the most readily influenced by diet, but evidence on the value of milk of high fat content in stimulating growth in piglets is confusing. Smith (1960b) increased the fat content of sows' milk by restricting feed intake during lactation and found that, with similar energy intakes, pigs consuming the milk of high fat content grew more slowly and less efficiently than pigs of sows fed to a higher level and producing milk of lower fat content; this depression in growth rate with milk of high fat content was attributed to widening of the energy/protein ratio beyond the apparent optimum.

Salmon-Legagneur (1962), on the other hand, increased the fat content of sows' milk by 15 per cent by giving diets containing 17 per cent lard and, as a result, obtained significantly faster growth in the piglets. This result may not necessarily conflict with Smith's, however, since in both experiments the sows which were losing most weight and, certainly in those of Salmon-Legagneur, utilizing subcutaneous fat, weaned the lightest litters.

The implication drawn by Smith was that to allow sows to build up body reserves during pregnancy to be catabolized during lactation is not only inefficient in terms of energy utilization by the sow but it may also be detrimental to the performance of the litter through its effects on milk composition.

The potential for growth in the early life of pigs is impossible to assess because the limits imposed by milk supply, on the one hand, and the physiology of the pig, on the other, are not known. Thus pigs of little over 1 lb at birth trebled their weight during the first week of life, whereas those of over 4 lb at birth barely doubled their weight during the same period (Lodge and Pratt, 1963), but it is not known to what extent this difference in relative growth was due to an insufficient diet for the larger pigs and to what extent it was due to reduction in growth potential in relation to body weight with an increase in size.

Individual 3-week weights in excess of 20 lb are not uncommon, and an individual milk yield of 514 lb in the 3 weeks of lactation has been recorded for a sow suckling 10 pigs (Lodge, 1958). If a sow with 12 pigs were to provide 50 lb of milk per head during the first 3 weeks of lactation, this could result in an average 3-week weight of 16 lb, and so a total litter weight of some 190 lb. The actual average for recorded sows (from Table 2) is 116 lb.

As already indicated, 8-week weight is less a reflection of the ability of the sow and more of the consumption of supplementary feed and the efficiency of the pigs themselves. Without supplementary feed an average 8-week weight of some 28 lb could be expected; with supplementary feed 8-week weights in excess of 60 lb are possible but have little to do with nutrition of the sow.

NUTRIENT REQUIREMENTS

Energy

The Rearing Period

Considerably more work has been done on the effects of total feed intake of gilts prior to first conception than of feed affects during subsequent pregnancies and lactations. Little has been done on energy effects *per se*, but as there is close agreement between results obtained from variation in total feed and those obtained from variation only in energy, it may be acceptable to regard all results as indicative of energy effects.

Fishwick (1944, 1945) appears to have been the first to suggest that level of feeding at a very early age, when gilts were between 8 and 12 weeks old, could influence their subsequent reproductive performance. Of thirteen gilts fed on a low-plane for 4 weeks' post-weaning and then changed to a higher plane, three subsequently failed to breed, whereas all thirteen fed to the higher plane throughout proved fertile and produced larger litters than their one-time restricted sisters.

The statistical significance of Fishwick's result is not known, but Lodge and MacPherson (1961) were unable to affect the reproductive performance of sows by varying the plane of feeding in the immediate post-weaning period. Three groups of twelve gilts were fed to scales rising to 5, 6·5 or 8 lb meal per day at 200 lb liveweight, and a fourth group were changed from the low to the medium-plane at 16 weeks of age. All were fed to the same scale after 200 lb until service at first oestrus after 300 lb and, where possible, taken through three reproductive cycles. There were no significant differences between groups in age at puberty or in subsequent reproductive performance; a suggestion that the incidence of infertility was increased by low-plane feeding was not confirmed by a second experiment in which sixteen gilts were reared on each of the medium- and low-plane treatments (Lodge and MacPherson, 1963).

Since Fishwick several other workers, mainly in the United States, have studied the effect of level of feeding on gilts before and immediately after first service. The results may be summarized into effects on age at puberty, effects on ovulation rate and effects on embryo survival; few workers have taken gilts through a complete pregnancy and lactation.

Effects on age at puberty have been very variable. Some workers found restriction of feed intake to 60–70 per cent of the self-feed level to have no effect (Robertson *et al.*, 1951b), others found that it hastened puberty (Christian and Nofziger, 1952; Self, Grummer and Casida, 1955; Hafez, 1958; Baker, 1959; Gossett and Sørensen, 1959), while others found that it delayed puberty (Goode, Warnick and Wallace, 1960; Zimmerman *et al.*, 1960a).

More severe restriction of feed intake to some 50 per cent of full-feed level, so that gilts weighed only 100 lb at 28 weeks of age, was found by Burger (1952) to cause considerable delay in puberty; 'low-plane' gilts at first oestrus were, on average, 118 lb and 235 days of age, while 'high-plane' gilts were, on average, 194 lb and 189 days of age. In contrast, Sørensen, Thomas and Gossett (1961) found no significant difference in age at puberty between gilts fed *ad lib.* a diet with either 93 or 55 Mcal productive energy per 100 lb. Average ages at puberty were 208 and 210 days for high- and low-energy groups respectively, but it should be noted that the weight difference was only 26 lb in contrast to the weight difference of 76 lb between Burger's high- and low-plane groups at puberty.

It seems from the above results that age is the main determinant of puberty in gilts but that a weight limit occurs below which puberty is delayed; the data of Burger (1952) and Lodge and MacPherson (1961) suggest that the minimum weight is about 115 lb.

Effects of level of feeding during the rearing period on ovulation at first and second oestrus have been variable also, but the majority finding has been an increase in ovulation rate with increase in feed or energy intake (Robertson *et al.*, 1951b; Christian and Nofziger, 1952; Self, Grummer and Casida, 1955; Hafez, 1958; Baker, Miller and Ulberg, 1958; Baker, 1959; Sørensen, Thomas and Gossett, 1961). However, in two experiments (Tribble *et al.*, 1956; Goode, Warnick and Wallace, 1960) gilts which had been subject to some degree of feed or energy restriction were found to have more corpora lutea than those fed to United States National Research Council standards.

In most of the experiments listed above, the levels of energy intake which influenced ovulation rate were given for several weeks before puberty, but it has been found that ovulation may be stimulated by increasing the energy intake for as little as 14, 10 or even 6 days before oestrus by reducing fibre level (Zimmerman, Self and Casida, 1957; Baker, 1959), by adding sugar (Zimmerman *et al.*, 1958) or by adding fat (Zimmerman *et al.*, 1960b; Rigor *et al.*, 1963). Conversely, ovulation may be reduced by the reverse process of starvation for some 10 days prior to oestrus (McGillivray *et al.*, 1962).

On the effect of level of feeding during the rearing period on subsequent embryo survival, agreement is unanimous that feed restriction is conducive to a high survival rate (Robertson *et al.*, 1951b; Christian and Nofziger, 1952; Self, Grummer and Casida, 1955; Casida, 1956; Tribble *et al.*, 1956; Hafez, 1958; Sørensen, Thomas and Gossett, 1961). Baker, Miller and Ulberg (1958) and McGillivray *et al* (1962) found more live embryos in their high-plane than in their low-plane gilts, but they imposed their different energy levels only from 10 days prior to service.

By testing various combinations of full- and restricted-feeding before puberty, between puberty and service at second oestrus, and for the first 25 days of pregnancy, Self, Grummer and Casida (1955) found that the greatest number of normal embryos at the twenty-fifth day resulted from restricted feeding to puberty, full feeding from puberty to service, then restricted feeding again following conception.

In the few experiments in which gilts have been allowed to farrow following different planes of feeding to puberty and service, there is generally no clear distinction between pre- and post-conception effects because the 'levels' have been continued through to parturition. In one case larger litters were obtained from the 'restricted' gilts (Christian and Nofziger, 1952), in another the self-fed gilts produced larger litters (Hanson, Ferrin and Annan, 1953), while in others restriction of feed intake to 50 or even 30 per cent of the *ad lib.* level had no significant effect on litter size at birth (Haines, Warnick and Wallace, 1959; Self *et al.*, 1960).

Some of the results quoted above appear to be conflicting but there is, perhaps, sufficient general agreement on most points to enable some conclusions to be drawn.

First, it would appear that puberty is determined mainly by age and normally occurs between 175 and 200 days. The optimal weight at puberty determines the required plane of feeding during the rearing period and must itself satisfy at least three conditions. First of these is the minimum weight for puberty itself, which appears to be about 110–120 lb. Second is the minimum size of gilt to allow effective natural service, until such time as artificial insemination is a satisfactory practical alternative; this will depend on the size of the boar, but is probably not less than 180 lb. The third condition to be satisfied is maximum efficiency of feed conversion; as feed intake approaches the maintenance level the efficiency of feed conversion must be depressed, and it appears from the results of Lodge and MacPherson (1961) that overall efficiency may be reduced when mean daily gain from weaning to puberty is below 1·0 lb per day. With a 56–day weight of 40 lb this would mean a weight at puberty between 160 and 190 lb, which is sufficient also to satisfy condition 2.

The second conclusion to be drawn is that ovulation rate may be increased by a high level of feeding, or energy intake, and that this is necessary only for as little as 7–10 days prior to service.

The third conclusion is that high energy intakes have an adverse effect on embryo survival unless for only a short period prior to conception.

There need be no conflict, therefore, between the three aims of: relatively slow but efficient rate of weight gain to puberty, a high ovulation rate and high embryo survival. These may be achieved by restricting growth between weaning and some 30 days after puberty, then increasing feed (energy intake) by some 30–40 per cent until service at the following oestrus and then reducing feed intake again to the desired pregnancy level.

The translation of the above generalization into terms of energy, or even feed, intake is extremely difficult, however, because of lack of information on daily feed intakes in the majority of experiments from the results of which the conclusions are drawn. In the experiment of Lodge and MacPherson (1961) all pigs were fed according to predetermined levels but all showed relatively poor efficiency of feed conversion and, therefore, growth rates; pigs of leaner genetic strain may be expected to tolerate higher levels of feed intake—and so rates of growth—before exhibiting any decline in efficiency of feed conversion.

In the absence of precise information on optimum weight at first service, it may be suggested that a scale of feeding suitable for Wiltshire bacon pigs would be suitable also for potential breeding stock, i.e. with a meal mixture of about 1350 kcal DE per lb, an allowance increasing from about 2 lb per head per day at 40 lb liveweight to about 6 lb per head per day at 200 lb liveweight. However, where such pigs are reaching 200 lb liveweight in 168 days or less, then the gilts may be unnecessarily heavy by the time they reach third oestrus and are ready for service; as leaner and more efficient strains of pig are developed it may prove desirable to restrict gilts for breeding more severely than those intended for slaughter.

Pregnancy

Several estimates of the energy requirements of the pregnant sow have been made by adding together separate estimates of requirement for maintenance, body gain and the development of the products of conception (Crowther, 1938; Mitchell and Kelly, 1938; Lodge and Lucas, 1959; Moustgaard, 1962; Vanschoubroek, 1963), but when such estimates are applied it is found that the resulting gain in net body weight greatly exceeds that which was anticipated. Thus, the estimates of Lodge and Lucas (1959) allowed for a body gain of 0·5 lb per day, but when applied the actual average gain was 0·7 lb per day or 40 per cent higher (Lodge, 1959a); similarly, Vanschoubroek (1963) calculated energy requirements to allow a daily net body gain of 285 g, but the average gain which resulted when the allowance was applied was 379 g or 33 per cent higher.

The most probable reason for the consistent discrepancy between anticipated and resulting body gains by pregnant sows is overestimation of the maintenance requirement. In all the estimates of requirement the allowance for maintenance has been that of a non-pregnant sow of the same body weight, but it has been shown clearly by Salmon-Legagneur and his colleagues (Salmon-Legagneur and Jacquot, 1961b; Salmon-Legagneur and Rerat, 1962) that a maintenance allowance for a non-pregnant sow will allow a pregnant sow of the same initial weight to produce a normal litter and also gain appreciable body weight. Thus, on an allowance of 2·0 kg meal per day non-pregnant sows gained about 4·0 kg in 4 months, whereas pregnant sows on the same allowance over the same period of time gained some 18 kg body weight in addition to producing a litter. At higher levels of feeding the same difference existed; where non-pregnant sows gained 39 kg pregnant sows of the same initial weight gained 54 kg in addition to the weight of litter, placenta and uterine fluids. Similarly, Heap and Lodge (1967) found that where non-pregnant sows given 4·5 lb meal per day gained 4 lb in 110 days, pregnant sows gained 36 lb in the same period of time.

The means by which this increased weight gain on a given feed intake is brought about in the pregnant sow has not been established, but it could be either a real improvement in efficiency of feed utilization through a reduction in the basal metabolic rate and the requirement for maintenance, or it could be merely an apparent improvement in efficiency caused by transfer of energy from body tissues of high caloric value to tissues of lower caloric value, so that weight would increase without any improvement in energy utilization.

There is little information on the nature of the body gain in the pregnant sow, and the little there is appears to be conflicting. Salmon-Legagneur and Rerat (1962) present data which indicate that most of the difference in body gain between pregnant and non-pregnant sows is accountable in the carcass while the total weight of internal organs remains much the same; furthermore, on the basis of internal- and back-fat determinations, pregnant sows had more muscle and less fat in the carcass than had non-pregnant sows of the same weight. On the other hand, Heap and Lodge (1967) found by carcass dissection that pregnant sows had more muscle and more fat than had non-pregnant sows after 110 days on the same feeding regime.

It should be noted, however, that in the latter of the two above experiments the sows were of the same *initial* weight and were given the same feed, whereas in the former they were of the same *final carcass* weight and, therefore, it must be presumed that the non-pregnant group received more feed than the pregnant.

The picture is further complicated by the finding of Elsley, Anderson and MacPherson (1965) that pregnant gilts also gained appreciably more weight than did non-pregnant gilts on the same level of feeding but that the carcass weights were similar in both groups and the difference in weight gain was accounted for by difference in weight of blood and internal organs. It would appear, therefore, that there is a difference in the effect of pregnancy on body composition between the young growing animal and the animal which has reached maturity.

Whatever the mode of action, however, the fact remains that the pregnant sow can achieve any desired weight gain on less feed than can a non-pregnant sow of the same weight and, therefore, the basing of energy requirements on values determined with non-pregnant animals results in excess weight gain. The problem which still remains is to decide what the weight gain during pregnancy should be.

The gross weight gain during pregnancy appears to be in three parts; that of the direct products of conception, which is obviously essential; that which results from intake of energy in excess of requirements for reproduction, which is, presumably, non-essential; and a third part, pregnancy anabolism, which is the gain in tissue weight over and above that of the direct products of conception, achieved on what would constitute a maintenance diet in the non-pregnant state. The significance of this pregnancy anabolism is still uncertain, but limited experimental data indicate that it is an essential part of normal pregnancy and if restricted the sow is, although still gaining weight, catabolizing body tissues with a resulting adverse effect on the developing embryos.

Thus it was found (Lodge, Elsley and MacPherson, 1966a) that gilts and sows making an average net gain of 42 lb during pregnancy produced litters as heavy as those of sows with an average net gain of 97 lb but a third group with an average net gain of only 9 lb produced litters significantly lighter at birth. The gain of 42 lb was achieved on what was calculated to be a maintenance ration, 4 lb meal per day, and so may be regarded as being close to the pregnancy anabolism gain.

Other workers have obtained similar results. Salmon-Legagneur and Jacquot (1961b) found no difference in litter weight between sows given 2 kg meal per day during pregnancy and those given 3·5 kg meal per day, and, at the lower end of the scale, American workers (Clawson *et al.*, 1963) found also that sows given 3 lb meal per day during pregnancy produced significantly lighter litters at parturition than did those given 6 lb meal per day.

It would appear from the above and other published data (Dean and Tribble, 1960; Eyles, 1959; Smith, 1960a, b) that a gain in body weight—over and above that of the parturition products—of about 45–50 lb is necessary for normal reproductive performance but that greater weight gains are without effect on the development of the litter *in utero*. As gains of this order appear to be obtainable on what would constitute a maintenance ration for the non-pregnant sow, it may be assumed that whatever the particular sow requires for maintenance prior to conception will be adequate also during pregnancy. What this means in terms of energy will depend on the size of sow and the environment in which she is kept, but suggested daily allowances for sows of different body weight are presented in Table 7.

TABLE 7
Estimated Energy Requirements for Pregnancy

Initial body weight (lb)	Daily requirement for digestible energy (Mcal)		Daily meal* allowance (lb)	
	First 3 months	Last month	First 3 months	Last month
300 or less	5·1	5·9	3·9	4·5
350	5·7	6·5	4·4	5·0
400	6·3	7·1	4·8	5·5
450	6·8	7·6	5·2	5·8

* Meal with DE value of 1·35 Mcal per lb.

Two further complications present themselves; one is the effect which stage of pregnancy has on energy requirement, the other is the effect of increasing maturity.

It is probable that energy requirements are greater at some stages of pregnancy than at others; the more critical periods which have been suggested are the first and last few weeks. Evidence for an increased energy requirement in early pregnancy is slight (Ballinger, 1940), but is supported by the reasoning that prior to implantation at around the twelfth day *post coitus* the embryos will be very vulnerable and that anything which inhibits their supply of nutrients could markedly affect embryo survival. On the other hand, the nutrient requirements of embryos at the blastocyst stage must be extremely small, and a direct effect of level of feeding of the sow on the nutrition of the embryos seems unlikely; any effect which the nutritional status of the sow might have on the nutrition of the embryos in early pregnancy would be more likely to occur indirectly through endocrine control of uterine secretions or introduction of a nutrient antagonist.

The results of recent feeding trials with sows do not indicate that litter size at birth is greatly affected by energy intake during early pregnancy. Thus Lodge, Elsley and MacPherson (1966a) found that sows given 3 lb meal per day for the first 76 days of pregnancy followed by 6 lb until parturition, produced litters larger on average than those of sows given 6 lb meal throughout; individual birth weights were significantly less by 1·8 lb, on average, but total litter weights were as great. Salmon-Legagneur (1962) compared a constant daily feed allowance of 4·6 lb per day with others in which 15 per cent more was allowed during the first half and 15 per cent less during the second, and vice versa; there were no significant effects of pattern of feed allowance on either numbers born or birth weights.

More recently (Heap, Lodge and Lamming, 1967) sows were slaughtered at the twenty-eighth day of pregnancy following the feeding from service of either 3, 6 or 9 lb feed per day; there were no significant effects of feeding on the growth or survival of embryos but there was a significant inverse relationship between level of feeding and placenta weight.

Although in general the level of feeding in early pregnancy would not appear to affect greatly reproductive performance, it is possible that a higher-than-average feed allowance may be of benefit during, perhaps, the first month in cases where considerable weight loss has occurred during the preceding lactation and the production of uterine secretions may be in competition with the replacement of greatly depleted body tissues.

Evidence for a higher-than-average energy requirement in late pregnancy is more convincing. On theoretical grounds the energy requirement of the sow will increase towards term because the embryos double their weight in the last month of pregnancy (Mitchell *et al.*, 1931; Gortner, 1945; Moustgaard, 1962); however, because the gross energy value of the products of conception is only some 5 per cent of the estimated net energy requirement for maintenance of a 200 lb gilt (Crowther, 1938), the increased requirement of the embryos during late pregnancy may be satisfied by only a small increase of not more than 10 per cent over the energy requirement in mid-pregnancy.

The value of increasing the feed allowance of sows during the last month or so of pregnancy has not been ascertained by the few feeding trials which have examined this point. Ballinger (1940) found no benefit from increasing the nutritional plane during the last 6 weeks of pregnancy by supplementing pasture with 5 lb meal per day, and Salmon-Legagneur (1962) reported no significant difference between litters from sows some of which received an increase in feed allowance during the second half of pregnancy while others suffered an equivalent reduction, although the litters of the former group were heavier on average at birth.

On the other hand, Lodge, Elsley and MacPherson (1966a) obtained significantly heavier litters from gilts and sows whose daily feed allowance was increased from 3 to 6 lb at the seventy-sixth day than from litter sisters continued on 3 lb per day through to parturition.

Accepting the fact that the energy requirements of the sow increase a little towards the end of pregnancy to meet the increasing needs of the rapidly growing embryos, whether or not there is any response to an increased feed allowance will depend upon the level of feeding up to the time of increase; thus, in the experiment of Lodge *et al.* (1966a) a beneficial response was obtained by increasing from 3 to 6 lb feed per day from the seventy-sixth day, but this merely indicates that 3 lb per day was insufficient over at least part of the last one-third of pregnancy and it does not follow that increasing from 4 or 5 to 6 lb per day would have been equally beneficial.

It must remain true, however, that some increase in feed allowance is justifiable during the last month or so of pregnancy, because if no response is obtained from an increase it implies that an excess is being given during the earlier period. More experiments are needed, however, to establish the optimal economic levels of feeding at different stages of pregnancy and the degree of increase which may be justifiable during the latter part.

Level of energy intake during pregnancy may affect not only the performance of the sow up to parturition but also during the subsequent lactation. Salmon-Legagneur, Gomez and Jacquot (1960) found that sows given 6·0 kg feed per day during the last month of pregnancy produced 15 per cent more milk than sows given only 3·9 kg feed per day. Such a difference might be expected to be reflected in a difference in growth in the litters, although Lodge (1959a) found that a 19 per cent difference in milk yield of sows did not affect litter growth, mainly because of difference in consumption of supplementary feed but also because of difference in milk composition. In the sow maximum milk production, if it means high levels of feeding, may not be the best economic proposition.

The influence of increasing maturity on energy requirement is also uncertain as the increasing needs for maintenance will be counterbalanced to some degree by the reducing need for growth. Thus, when gilts and sows are given the same daily feed allowance during successive pregnancies the net weight gain falls progressively as the initial weight increases (Lodge, McDonald and MacPherson, 1961). It may be presumed also that with increasing maturity the energy value of each unit of weight gain will increase and this, as much as change in maintenance requirement, could contribute to the above result.

With sows of equal age, however, difference in weight can be expected to affect energy requirement, and so allowance is made for this in the estimates of requirement presented in Table 7.

Lactation

Energy requirements for milk production can be calculated from information on milk yield, milk composition and the efficiency of energy conversion from feed to milk, but these will be also the *dietary* requirements only in the absence of tissue storage during pregnancy. As some net body gain is desirable during pregnancy to allow expression of the maximum reproductive potential, the full energy requirements for milk production will not normally be given and some weight loss will occur.

It is evident from simple calculations that the energy requirement for lactation is appreciably greater than for pregnancy. The pregnant sow may be expected to produce a total weight gain in the litter and other essential products of conception of some 50 lb in 114 days; the same sow with ten pigs during lactation is expected to produce a total weight gain of some 250 lb in 56 days. The pregnant sow would also produce some 50 lb of net body gain but this is used to subsidize subsequent lactation and does not reduce the actual energy requirement for milk synthesis.

Energy requirements for milk production may be calculated from the data on average yields and composition given in Tables 4 and 6, together with the information that the gross efficiency of energy conversion for milk production in the sow is some 45 per cent of the DE (digestible energy) of feed (Lodge, 1957; Smith, 1959a, b). Estimates calculated in this way, for sows with different sizes of litter and at different stages of lactation, are presented in Table 8.

Stage of lactation and size of litter are the two major variables which affect milk yield and so energy requirement, but a third variable is weight of sow. Difference in weight between sows will influence the gross efficiency of energy conversion, in that the higher the maintenance requirement the lower will be the efficiency value. In practice, the major difference will probably be between gilts and older sows, but as the improvement in efficiency through reduced maintenance requirement of the gilt will be counterbalanced, to some degree, by the reduction in efficiency resulting from the lower milk yield, coupled with the fact that the gilt is still growing even during lactation, no differentiation is made in estimating energy requirements for lactation between sows of different weight. A three-dimensional feeding scale is unwieldly to operate, and because the requirement for milk production varies more widely than the requirement for maintenance, the ignoring of weight difference introduces less inaccuracy than would the ignoring of litter size or stage of lactation.

In practice, stage of lactation might be ignored also and the sows fed to an average level which overfeeds in early and late lactation and underfeeds at the peak.

The energy requirements for lactation, if met entirely during lactation, necessitate the giving of feed levels considerably above those in common practice. Whether or not it is necessary to give the full requirement during lactation will depend on whether or not the sow laid down reserves during pregnancy and, if so, to what degree. If the pregnancy allowances presented in Table 7 are given, then some 50 lb net body gain would be expected and the full energy requirement for lactation need not be given during lactation. However, because of lack of experimental evidence on the actual net energy value of weight loss in the lactating sow and its equivalent in digestible energy in the diet, it is not possible to estimate inadequate energy allowances which would allow milk production and a predetermined weight loss. In Table 8, therefore, the full energy requirements for lactation with no allowance for weight loss are presented. If it is not desired that the sows should gain a full 50 lb during each reproductive cycle, then levels some degree lower than these

TABLE 8

Estimated Energy Requirements for Lactation

Litter size	Daily digestible energy requirement for each week of lactation (Mcal)*								Daily meal allowance for each week of lactation† (lb)							
	1	2	3	4	5	6	7	8	1	2	3	4	5	6	7	8
5	9·5	12·3	13·2	13·2	13·2	12·3	10·0	8·7	7·5	9·8	10·5	10·5	10·5	9·8	7·9	6·9
6	11·1	14·4	15·4	15·4	15·4	14·4	11·6	10·1	8·8	11·4	12·2	12·2	12·2	11·4	9·2	8·0
7	12·7	16·6	17·6	17·6	17·6	16·6	13·4	11·6	10·1	13·2	14·0	14·0	14·0	13·2	10·6	9·2
8	14·5	19·0	20·2	20·2	20·2	18·9	15·1	13·2	11·5	15·0	16·0	16·0	16·0	15·0	12·0	10·5
9	15·4	20·0	21·4	21·4	21·4	20·0	16·1	14·1	12·2	15·9	17·0	17·0	17·0	15·9	12·8	11·2
10	16·0	20·9	22·3	22·3	22·3	20·9	16·8	14·6	12·7	16·6	17·7	17·7	17·7	16·6	13·3	11·6
11	16·4	21·3	22·7	22·7	22·7	21·3	17·1	15·0	13·0	16·9	18·0	18·0	18·0	16·9	13·6	11·9
12	16·5	21·5	22·9	22·9	22·9	21·5	17·4	15·1	13·1	17·1	18·2	18·2	18·2	17·1	13·8	12·0

* From data on milk yields and composition and assuming 45 per cent gross efficiency of energy conversion (Lodge, 1957; Smith, 1959a,b, 1960b).

† Meal with DE value of 1·35 Mcal per lb.

would be adequate; how much lower is uncertain in the light of existing information.

The suggestion of feed levels for lactating sows in excess of 12 lb meal per day raises the question of whether they are within the capacity of the sow; the answer depends on the level of feeding given during the preceding pregnancy as it is evident that the voluntary feed intake after parturition is greatly influenced by the weight gained before. Sows fed *ad lib.* during pregnancy lost 44 kg body weight when fed *ad lib.* during the following lactation, whereas sows restricted during pregnancy to half the *ad lib.* level of feeding lost only 4 kg body weight when fed *ad lib.* during lactation (Salmon-Legagneur and Jacquot, 1961b).

The implication of these results was that the sows which were restricted in feed intake during pregnancy had greater appetites during lactation, but there is evidence to suggest that, while this is no doubt true, feed utilization is also more efficient following a low-feeding regime than when following a high. Over three pregnancies and lactations sows given 6 lb meal per day during pregnancy lost, on average, 38 lb body weight during lactation on a total feed intake of 562 lb, whereas sows given only 3 lb meal per day during pregnancy gained, on average, 7 lb during lactation on a total feed intake of 529 lb (Lodge, Elsley and MacPherson, 1966b).

It is evident that, basically, sows may be fed in one of two different ways; either given continually a similar amount so that reserves are accumulated when requirement is low for use when requirement is high, or fed according to current requirement in order to reduce, or even eliminate, fluctuations in body weight. The limited evidence available suggests that the former may be the easier method but the latter is the more efficient.

In terms of efficiency of energy utilization for litter growth, it has been calculated that the best feeding regime is one which allows only limited weight gain during pregnancy and prevents appreciable weight loss during lactation (Smith, 1960a, b; Salmon-Legagneur and Rerat, 1962). In a practical test of this hypothesis, Lodge (1966) found that sows given approximately one-third of their total feed during pregnancy and two-thirds during lactation utilized their feed about 15 per cent more efficiently than did sows whose feed was divided almost equally between the two periods.

In the above experiment the litters were weaned at 8 weeks of age and the improvement in efficiency from giving less feed during pregnancy and more during lactation would be expected to increase the earlier were the pigs weaned; such a feeding pattern is a logical corollary of earlier weaning.

PROTEIN

The Rearing Period

There is very little experimental evidence on which to base estimates of the protein requirements of young gilts for breeding, and the little there is merely indicates that they do not differ materially from the requirements for optimum growth in bacon pigs (Davidson, 1930; Robertson *et al.*, 1951b; Adams *et al.*, 1960).

However, the basic difference between gilts reared for slaughter and gilts reared for breeding purposes is that maximum carcass leanness is of less importance in the latter, except as a basis for selection, and that, therefore, protein levels which allow high efficiency of feed utilization may be acceptable even though they do not allow maximum muscle development.

Pregnancy

The requirement for protein is increased by pregnancy much more than is the requirement for energy; there is also a greater increase in the requirement for protein as pregnancy advances.

Estimates of protein requirement may be calculated in much the same way as for energy, namely by adding together individual estimates of requirement for maintenance, increase in body weight, and development of the products of conception. Thus Lodge and Lucas (1959) calculated from metabolism data of Mitchell *et al* (1931) that the protein requirement of a 400 lb sow carrying twelve pigs averaged 235 g DCP (digestible crude protein) per day over the whole pregnancy and was 294 g DCP at its maximum before parturition. Moustgaard (1962), from a consideration of similar data, estimated the average daily requirement to be 300 g DCP and the actual requirement to increase from 265 g for the first 90 days, to 350 g from day 90 to day 105 and to 400 g DCP for the final 10 days. No allowance was made for difference in body weight as it was concluded that increase in maintenance requirement with increase in weight would be offset by reduction in requirement for growth.

Data from nitrogen balance trials are another source of information on protein requirements for pregnancy. Evans (1929) found that sows given 80 g nitrogen per day retained, on average, 12·5 g per day and, in agreement with Mitchell *et al.* (1931), there was no consistent increase with advance in pregnancy; the N content of the litters at parturition was estimated to be only one-sixth of the total retained. Similarly, Lenkeit and co-workers (Lenkeit and Gutte, 1955; Lenkeit, Gutte and Streuter-Petermoller, 1955; Lenkeit *et al.*, 1955; Lenkeit *et al.*, 1956) found that sows on a high protein intake retained five times as much N in extrauterine body tissues as in the products of conception, but it was found also that while on a low daily intake of 193 g CP there was no increase in retention with advance in pregnancy, on a high daily intake of 662 g CP daily retention of N increased after the seventieth day of pregnancy. Over the first two-thirds of pregnancy retention was at a low level and N given in excess of this requirement was excreted in the urine.

Elsley and MacPherson (1964) also found that N retention increased in late pregnancy with gilts given 468 g CP per day but not with those given 287 or 176 g CP per day; all were in positive N balance throughout pregnancy, as were the gilts and sows of Heap (1966) given either 182 or 272 g CP per day. Even lower levels of crude protein have been given to gilts on balance trial during the last 45 days of pregnancy by Rippel and co-workers (Rippel *et al.*, 1965b) who found that a mean positive balance of 1·7 g N per day was obtained on a daily CP intake as low as 55 g and only on a N-free diet was the gilt in negative balance. Retention increased with advance in pregnancy with daily CP intakes of 164 g or more, but not with those of 109 g or less; maximum percentage retention was obtained with a daily intake of 221 g N and it was concluded that 230 g CP per day—from maize and soyabean meal—was sufficient even during the last one-third of pregnancy to meet the gilt's needs for amino acids and nitrogen.

The protein requirement of the pregnant sow does not appear to be influenced by energy intake as N retention is little different whether the daily intake is provided by a small amount of a high protein feed or a greater amount of a lower protein feed; the main determinant is daily intake of nitrogen (Elsley and MacPherson, 1964; Heap, 1966).

The significance of the increased N retention which may occur in late pregnancy is not clear since at all times, except on impracticably low intakes, retention greatly exceeds the uptake of the foetuses. Furthermore, the increased retention appears to be counterbalanced immediately after parturition by considerable loss of urinary nitrogen (Lenkeit *et al.*, 1956; Lenkeit *et al.*, 1962). If maximum N retention during late pregnancy is necessary for optimum reproductive performance, then it would appear that an increase in dietary protein sufficient to allow an increase in retention of some 15 g N would be required; if the protein of a typical sow feed is considered to have a biological value of 60 (Moustgaard, 1962) then this would necessitate an extra 185 g CP per day.

The implication of the results of the above metabolism trials with pregnant sows is that the protein requirement for the first two-thirds of pregnancy

TABLE 9

Estimated Protein Requirements for Pregnancy

Stage of pregnancy (weeks)	Daily requirement for crude protein (g)	Daily meal* allowance (lb)	Crude protein in feed (%)
1–11	250	4·4	12·5
12–14	325	5·0	14·3
15–16	400†	5·0	17·6†

* Meal allowance for sow of 300 lb initial weight (Table 7).

† CP requirement based on N-retention data; no practical evidence to indicate whether such a level is of benefit (see text).

would appear to be satisfied adequately by a daily intake of 250 g CP; this may be increased to 400 g CP per day to allow for an increase in retention during the last month of pregnancy (Table 9).

The few feeding trials which appear to have been carried out on protein allowances for pregnant sows lend support to these theoretical estimates. Boaz (1962) compared daily intakes ranging from 280 to 493 g CP throughout three successive pregnancies, provided by CP levels ranging from 10 to 18 per cent of air-dry feed at a daily intake of 6–7 lb, and found that the only significant effect of protein intake was on weight gain of the sow. There was a suggestion, however, that fertility might have been affected by protein intake in that the average intervals between weaning and effective re-service were 11·8, 11·4 and 16·7 days on high-, medium- and low-protein diets respectively. It should be noted, however, that in these experiments the low-protein diets were of cereals and cereal byproducts, so that any adverse effects might have been due to specific amino acid deficiencies as much as to lack of protein *per se*.

Other workers have compared lower levels than those of Boaz and found equally little effect on reproductive performance. Clawson *et al.* (1963) used semi-synthetic diets in which soyabean meal was the sole protein source and found that the only significant difference between gilts and sows given either 136 or 545 g CP per day was in body-weight gain, there were no effects on number or weight of pigs born. Rippel *et al.* (1965a) compared daily intakes as low as 91 and 272 g CP from maize and soyabean meal, from the sixtieth day of pregnancy until the fourteenth day *post-partum*, and found no effects on numbers born or birth weights. The only published report which suggests any benefit from increasing protein allowance above conventional levels appears to be that of Stevenson and Ellis (1957) who concluded that postnatal survival was improved by increasing the crude protein level of the pregnant sow's diet from 15 to 17·5 per cent but that no further benefit was obtained by increasing to 20 per cent; 15 per cent CP was thought to be adequate for lactation. This experiment seems to be the origin of the singular recommendation of the National Research Council of America (NRC, 1964) that the pregnant sow should be given a diet higher in protein than that of the lactating sow. The weight of evidence would appear to be against this suggestion.

It should be emphasized, however, that all the reported experiments on protein requirements of pregnant sows have been of a relatively short-term nature, the longest being over three reproductive cycles (Boaz, 1962). If low protein intakes result in deficiency of a particular amino acid for foetal development, it is possible that the adverse effects of this could be concealed for several reproductive cycles by the sow depleting body reserves. The subject of amino acid requirements and the meaning of 'protein quality' in relation to the pregnant sow is obviously of considerable significance in this respect but, unfortunately, little is known about it. Comparisons between proteins from different sources indicate that where differences in value do exist they are reflected in the weight gains of the sows and not in the litters (Rippel *et al.*, 1962, 1965a; Mayrose *et al.*, 1962; Salmon-Legagneur, 1964c).

Requirements for specific amino acids appear to have been studied only by Rippel and co-workers (Rippel *et al.*, 1965c, d) who, as a result of N-balance trials with gilts in late pregnancy, concluded that requirements of the pregnant gilt were similar to those of the 'finishing' pig for isoleucine, lysine, cystine and methionine but greater for threonine and valine. Their estimates of requirement as percentages of diet were: arginine, not more than 0·38; histidine, not more than 0·17; isoleucine, 0·37; leucine, not more than 0·56; lysine, 0·42; methionine, 0·19; cystine, 0·09; phenylalanine, not more than 0·30; tyrosine, 0·33; threonine, 0·34; tryptophan, not more than 0·07; and valine, 0·46.

There is little information on the relationship between the protein intake of the sow during pregnancy and the body composition of her piglets at birth. Livingstone (1962) found that sows given a diet with 10 per cent CP produced piglets with muscle fibres of significantly smaller diameter than those from sows given a 21 per cent CP diet. This effect did not persist, however, and there were no measurable differences by the time the pigs were slaughtered.

The translation of the above recommendations into practical terms of percentages crude protein in the feed and sources of protein is made difficult by the fact that changes in requirement for energy and protein change at different rates. Thus, if the daily CP allowance be increased from 250 to 400 g in the last month of pregnancy, to allow increased retention, this would necessitate an equivalent 60 per cent increase in energy intake if the composition of the diet were to remain unaltered; there is no evidence that the increased energy requirement is nearly as great as this and it would be satisfied probably by an increase in the order of 15 per cent.

The possibilities that are open are to overfeed with energy, underfeed with protein or introduce a second, high-protein diet for the last month of pregnancy. As the evidence for a high-protein requirement in late pregnancy is slight and satisfactory performance has been obtained on very low protein levels throughout, in the light of existing information the expense incurred by materially raising the protein percentage of the diet in late pregnancy would not seem to be justified.

Lack of clear evidence on the significance of protein quality in the feeding of pregnant sows also makes practical interpretation difficult. Thus a daily allowance of 250 g CP could be provided by 5·5 lb of an all-cereal diet, but it remains uncertain whether this would be satisfactory in the long term and so a higher protein allowance may have to be given in order to include a source of 'higher quality' protein. General practice in the United Kingdom is to include fish meal in the diets of pregnant sows, but American practice indicates that this may be replaced, if required, by soyabean meal without adverse effect.

Lactation

As with energy, the protein requirements for lactation are much greater than for pregnancy; a sow with an average milk yield of 14 lb per day may secrete at the peak of lactation as much protein in 3 days as is stored in the direct products of conception during the whole of pregnancy.

From the data on average milk yields and composition in Tables 3 and 5 it would appear that the average daily output of protein in milk is some 360 g; if the gross efficiency of conversion of dietary protein into milk protein is taken as 33 per cent (Lodge, 1959b), then the average daily requirement for CP is 1080 g. Gutte and Lenkeit (1960) calculated from metabolism data the protein requirements of a 200 kg sow suckling eight pigs and arrived at almost the same figure, namely 965 g CP per day. The difference between these two estimates is accounted for entirely by the different protein values adopted for sow's milk; 5·8 and 5·1 per cent respectively.

It would appear that the average protein requirement of a 400 lb sow suckling eight pigs is about 1000 g per day with appropriate variation for stage of lactation (Table 10). With a daily feed allowance of 14 lb the diet would be required to contain 17·5 per cent CP.

TABLE 10
Estimated Protein Requirements for Lactation

Stage of lactation (week)	Daily requirement* for crude protein (g)	Daily meal allowance (lb)	Crude protein† in feed (%)
1	858	11·5	16·4
2	1050	15·0	15·4
3	1161	16·0	16·0
4	1182	16·0	16·3
5	1248	16·0	17·2
6	1239	15·0	18·2
7	1023	12·0	18·8
8	951	10·5	20·0
Mean	1089	14·0	17·4

* For sow with litter of eight pigs; from data on yields and protein content of sow's milk and assuming 33 per cent conversion of dietary protein to milk protein (Lodge, 1959b).

† The increase in protein level with advance in lactation is caused by the increase in milk protein percentage and reduction in milk energy value. It has not been proved necessary by experiment and the problem does not arise with 6-week or earlier weaning.

There appear to have been no feeding trials in which diets of different protein level have been given to lactating sows, so the above theoretical estimates of requirement must go unsupported. Lower levels are commonly given in practice but it must be presumed that they are inadequate and result in excessive weight loss in the sow.

A similar complication arises in meeting the theoretical protein requirements for lactation as arose for pregnancy in that changes in requirement for energy and protein do not run parallel. With advance in lactation the energy requirement declines more rapidly than does the requirement for protein because the fat content of the milk tends to fall while the protein content rises. However, any practical advantage from increasing the protein content of the diet in late lactation has yet to be established and the problem would not arise to any marked degree if pigs were weaned at 5 or 6 weeks of age.

Just as total protein requirement is appreciably greater during lactation than during pregnancy, so the quality of that protein is more critical. Salmon-Legagneur (1964c) found that whereas groundnut meal and fishmeal were of similar value for maintenance and for pregnancy, fishmeal was measurably superior for lactation. To avoid excessive weight loss during lactation a high quality protein source would appear to be necessary in the diet.

MINERALS

The Rearing Period

There seems to be no evidence to suggest that the mineral requirements of future breeding stock differ in any way from those for normal growth.

Pregnancy

As with other classes of pig, most information on mineral requirements of pregnant sows relates to calcium and phosphorus; there is relatively little known about quantitative requirements for trace elements.

Calcium. There seems to have been considerable interest in calcium for sows in the late 1920s and most of the data on which present-day estimates are based, date from that time. Gross weakness, broken bones and failure to lactate in gilts receiving only 0·6 g calcium per day were reported by Missouri workers (Mumford, Hogan and McKenzie, 1927) and, subsequently, Hogan (1932) suggested that the minimum calcium requirement for the breeding sow was 0·4 per cent of the diet, or some 12 g per day.

Evans (1929, 1932) found from balance trials that calcium deficient sows utilized their own reserves of the element to allow normal calcification of the foetuses and estimated the requirement to be 13 g Ca per day. His conclusion that deficiency of calcium during pregnancy affected the dam rather than the progeny requires some modification in view of Davidson's (1930) finding that gilts given a calcium deficient diet produced normal litters at first but later litters showed progressive deterioration, with increase in number born dead and failure to lactate.

Also through balance trials, Mitchell *et al.* (1931) found the average daily retention of calcium to be 4·4 g, which represented only 32 per cent of intake; of that retained only 20 per cent was located in the uterus. Later, Mitchell and McClure (1937) estimated the maximum calcium requirement during pregnancy to be 11·1 g per day for a growing gilt of 250 lb.

More recently, Lenkeit and co-workers (Lenkeit *et al.*, 1959) found from balance trials that there was a marked increase in the excretion of calcium and phosphorus immediately after parturition which could not be prevented and which caused a negative balance for these elements during that time. Lenkeit (1959) estimated the calcium requirement to be 20 g per day during pregnancy; Moustgaard (1962), however, considers this to be the maximum requirement only during the final 10 days of pregnancy and 9 g per day to be adequate for the first 90 days.

The majority opinion, therefore, is that 12–14 g Ca per day is adequate as an average allowance for the pregnant sow. If Moustgaard's estimates be accepted, as they allow for the evident increase in retention during late pregnancy, the calcium requirements as percentages of feed would be as follows:

Stage of pregnancy (days)	Daily feed (lb)	Ca requirement	
		g per day	% of feed
0–90	4	9	0·5
91–105	6	15	0·6
106–115	6	20	0·7

It may be considered, therefore, that a calcium level in the diet of 0·6 per cent would be adequate at all stages of pregnancy except, perhaps, the final 10 days, but there is no evidence to support any practical need to increase the level of calcium in the diet for this short period.

Phosphorus. Mitchell *et al.* (1931) found that pregnant gilts retained, on average, 1·3 g P daily, which represented only 16 per cent of the intake and of which only 40 per cent was located within the uterus. Mitchell and McClure (1937) calculated the phosphorus requirement of the pregnant gilt to be not more than 7·3 g per day, or 0·3 per cent of the diet.

A much higher estimate of 16 g per day is made by Lenkeit (1959) who, as already mentioned with relation to calcium, found a high excretion rate after parturition; other recent estimates, however, support that of Mitchell and McClure (Dudley, 1960; Teter, 1961).

Moustgaard presents estimates for different stages of pregnancy, starting with 8 g per day for the first 90 days and increasing to 12 g from 91 to 105 days and 15 g from then until parturition. Expressed as percentages of feed these allowances would be as follows:

Stage of pregnancy (days)	Daily feed (lb)	P requirement	
		g per day	% of feed
0–90	4	8	0·5
91–105	6	12	0·5
106–115	6	15	0·6

It may be concluded, therefore, that a phosphorus allowance of 0·5 per cent of the feed would be adequate at all stages of pregnancy except, perhaps, the last few days but, as with calcium, there is no evidence to suggest any practical advantage from increasing the allowance for this short period.

It should be emphasized that with calcium and, especially, phosphorus there appears to be wide variability in availability of the elements from different sources of supply, and the estimates given by Moustgaard are based on availabilities of 50 and 40 per cent respectively. Several workers have found widely different values with other sources of phosphorus (Chapman, 1957; Plumlee *et al.*, 1958; Arrington, Ammerman and Davis, 1959; Dudley *et al.*, 1961; Dudley, 1960), but there is general agreement that di-calcium phosphate is highly available and would allow an appreciable safety margin with Moustgaard's estimates of requirement.

Other elements. Data on requirements of sows for other elements are fragmentary, with most on sodium chloride and manganese.

Evans (1932) considered that 5·5 g NaCl per day was sufficient during pregnancy, and Lenkeit (1959) places the estimate of requirement during pregnancy and lactation as 5–10 g NaCl per day. The highest estimate, 10 g, would be provided by 4 lb feed containing 0·5 per cent NaCl.

Deficiency of manganese in the diets of sows has resulted in irregular oestrus, resorption of foetuses, birth of small, weak pigs and failure to lactate (Johnson, 1943; Plumlee *et al.*, 1956), but while it is clear that a level of 0·5 ppm (parts per million air-dried feed) in the diet will result in these undesirable effects, it is not clear what level is adequate to prevent them; no single experiment appears to have employed several levels, and comparisons have been between 0·5 and 6 ppm (Johnson, 1943), 0·5 and 40 ppm (Plumlee *et al.*, 1956), and 3 and 100 ppm (Outler, Wallace and Davis, 1955). Grummer and co-workers (Grummer *et al.*, 1950) concluded that 12 ppm allowed for normal reproduction, but the performance of all their sows was poor; Johnson (1943) obtained satisfactory reproduction from sows given a diet containing 8·6 ppm over two generations. The inference to be drawn from this published data seems to be that 40 ppm are clearly adequate but that the true requirement is considerably less than this and probably in the region of 6–12 ppm. With this lower estimate there would normally be a considerable safety margin over requirement because, as pointed out by Plumlee *et al* (1956), pig diets generally contain over 20 ppm from natural sources.

Iodine was one of the earliest elements to create interest in relation to sows through its association with congenital goitre and the birth of 'hairless pigs' (Smith, 1917); this condition was prevented by giving potassium iodide at the rate of 45 g per 100 kg diet during pregnancy (Hart and Steenbock, 1918) or, more recently (Andrews *et al.*, 1948) by including in the diet 0·5 per cent common salt containing 70 ppm I, which would provide 0·35 ppm I. Information is not available on which to base an estimate of requirement.

Because of the perennial interest in piglet anaemia, some attention has been directed towards the iron intake of sows during pregnancy with the object of increasing the iron reserves of the young at birth. This has met with variable success, and the overall conclusion must be that even where the iron reserves of the piglets are increased through administration of high levels to the sow during pregnancy, the increase may delay but is still insufficient to prevent the development of anaemia (Hart *et al.*, 1930; Foot and Thompson, 1938; Urbanyi, 1940; Venn, McCance and Widdowson, 1947; Buchanan, Lasley and Bolan, 1949; Neseni, 1953; Rydberg *et al.*, 1959).

There is wide variation between estimates of iron uptake by the foetuses; Lintzel and Radeff (1931) put the total iron content of the newborn piglet at some 30 mg, Venn, McCance and Widdowson (1947) proposed a value of 50 mg and Moustgaard (1962) states that the deposition of iron in the uterus rises progressively to a maximum of 30 mg per day by the end of pregnancy. The daily requirement of the pregnant sow for iron remains unknown but it may be mentioned that Mitchell *et al.* (1931) found the daily intake of iron by the sow on normal feeding to be nearly as great as the amount stored in the products of conception during the whole of pregnancy.

The degree to which the copper reserves of the newborn piglet may be influenced by the copper intake of the dam during pregnancy has been investigated also, with varying results. On the one hand, Hart *et al.* (1930) gave gilts from weaning copper levels ranging from 2 to 50 mg per day but found no relationship between intake and liver copper levels of the first litters at birth, while Moustgaard and Olsen (1951) found that litters from sows given supplementary copper had liver copper levels double those of litters from unsupplemented sows. The copper requirement of the pregnant sow remains unknown.

Virtually nothing is known of magnesium requirements for pregnancy, other than that the daily deposition in a single foetus increases from about 0·1 mg during the first 2 months to 1·0 mg during the third month and to 12 mg during the final month. Thus, a sow with twelve foetuses consuming 6 lb feed per day would require 50 ppm Mg in the diet during the final quarter of pregnancy for foetal development alone.

The only information on zinc requirement of the pregnant sow appears to be that 30 ppm of the diet is insufficient for maximum reproductive performance (Roberts, Hoekstra and Grummer, 1961).

There is no published information on requirements for other elements, but the toxic effect of selenium has been studied. A level of 10 ppm was found to lower conception rate and increase the incidence of weak and stillborn pigs (Wahlstrom and Olsen, 1959a, b).

Lactation

Information on mineral requirements for lactation are equally as sparse as for pregnancy, but for some elements it is possible to calculate approximate requirements from data on milk composition and efficiency of utilization for milk secretion.

Calcium. Balance trials indicate that the sow utilizes about 50 per cent of her calcium intake for milk production (Lenkeit *et al.*, 1959) and as the average calcium content of the milk appears to be about 0·22 per cent (Hughes and Hart, 1935; Braude *et al.*, 1947; Perrin, 1955; Lenkeit, Gimmler and Seick, 1959; Guegen and Salmon-Legagneur, 1959) it can be estimated that the sow yielding an average of 14 lb milk per day is secreting some 14 g Ca per day and would require about 28 g Ca in the diet to provide this. However, as the calcium percentage rises in milk with advance in lactation (Guegen and Salmon-Legagneur, 1959; Lenkeit *et al.*, 1962), to ensure adequacy throughout a higher value of 0·25 per cent for milk calcium should be used and this would result in an estimated dietary requirement of some 32 g Ca per day. This estimate compares with 27 g per day suggested by Evans (1932) and 40–45 g per day suggested by Lenkeit *et al.* (1959).

Expressed in terms of percentage of feed, a sow with ten pigs consuming 12 lb feed per day would require 0·6 per cent Ca in the diet; a value close to that in common practice but little more than half the singularly high figure of 1·0 per cent proposed by Bryushinin (1958). The estimate of Lenkeit *et al.* (1959) would necessitate 0·7 per cent Ca in the diet.

Where relatively low planes of feeding are employed during lactation then, presumably, it would seem necessary that the calcium requirement as a percentage of feed be increased but it may be argued that if the sow is expected to utilize reserves of energy and protein to maintain milk secretion, then a corresponding use of reserves might apply for calcium and other minerals.

Phosphorus. In the same way as for calcium, it may be calculated that the phosphorus requirement to allow for a progressive increase in the percentage in milk with advance in lactation should be some 22 g per day for a sow yielding 14 lb milk. This would necessitate a dietary level of 0·4 per cent for a sow with eight pigs receiving 12 lb feed per day. Other estimates in the literature range from 0·3 to 0·35 per cent P in the diet (Dudley, 1960; Teter, 1961) to 0·55 per cent (Bryushinin, 1958) and 30–36 g P per day, equivalent to 0·55 per cent P for a sow receiving 12 lb feed per day (Lenkeit *et al.*, 1959).

Other elements. The majority of the few publications on trace mineral requirements of the sow make no differentiation between pregnancy and lactation. The only data specific to lactation appear to be those of Bryushinin (1958) who, from balance trials, estimated the daily requirements of sodium and potassium for lactation to be provided by 0·35 and 0·30 per cent of the diet respectively.

Specific requirements for other minerals have not been studied and only iron appears to have received some attention, through the interest in piglet anaemia, with the object of increasing the iron levels of milk. This, however, has met with even less success than the attempts to increase the iron reserves of the pig at birth (Venn, McCance and Widdowson, 1947; Habersang, 1952; Neseni, 1953; Koves, 1958; Anders and Schipper, 1959).

Toxic effects of high levels of fluorine have been reported (Kick *et al.*, 1935). Levels in excess of 190 ppm of the diet reduced milk production in sows in addition to affecting bone strength and teeth.

Estimates of requirement for minerals for pregnancy and lactation are presented in Table 11.

TABLE 11
*Estimated Mineral Requirements for Pregnancy and Lactation**

	Proportion in diet	
	Pregnancy	*Lactation*
Ca (%)	0·6	0·6
P (%)	0·5	0·5
NaCl (%)	0·5	
K (%)	—	0·3
Mg	—	—
Mn (ppm)	12	
Fe	—	—
Cu	—	—
Zn (ppm)	<30	—
I	—	—

* No or insufficient information for those marked —.

VITAMINS

The Rearing Period

As with minerals, there is no evidence to suggest that vitamin requirements of growing gilts for breeding differ from those for normal growth.

Pregnancy and Lactation

It is not possible with vitamins to distinguish between requirements for pregnancy and requirements for lactation, partly because no differentiation is made in the majority of experiments and partly because, in the case of fat-soluble vitamins, tissue storage makes it impossible to draw a distinction.

Estimates of requirement for vitamins are presented in Table 12.

TABLE 12
*Estimated Vitamin Requirements for Pregnancy and Lactation**

	Proportion in diet during pregnancy and lactation
Vitamin A (i.u./lb)	1200
Vitamin D_3 (i.u./lb)	100†
Tocopherol (ppm)	<22
Thiamine (mg/lb)	—
Riboflavine (mg/lb)	1·25
Nicotinic acid (mg/lb)	—
Pantothenic acid (mg/lb)	5·0
Pyridoxine (mg/lb)	<0·45
Vitamin B_{12} (μg/lb)	5·0
Vitamin C	No requirement

* No information for those marked —.
† Assumed to be same as estimated requirement for growing pigs.

Vitamin A. The vitamin A requirements of the young pig are provided partly by way of placental transfer and partly through the colostrum, and both of these sources of supply are directly influenced by the intake of the sow (Hjarde *et al.*, 1961). However, as Hjarde *et al.* demonstrated, transfer of vitamin A to the foetuses represents only a small fraction of the sow's normal intake, and as she has the capacity to store reserves adequate for long periods of deficiency there may be no direct effect of dietary deficiency upon reproductive performance. Thus a sow deprived of vitamin A produced three normal litters before deficiency symptoms appeared in the fourth (Braude *et al.*, 1951).

There is more information available on requirements of the sow for vitamin A than for any other vitamin. Nordfelt (1944–5) established that 10 i.u. per day of the vitamin was inadequate, in that pigs appeared to be normal at birth but few survived to weaning; other sows given approximately 3500 i.u. vitamin A in the form of carotene from lucerne meal produced and reared normal litters. Gilts given either 6500 or 7100 i.u. vitamin A per day (Parrish *et al.*, 1951) produced normal litters with liver levels equal to those from gilts given free access to lucerne hay; other gilts given carotene equivalent to 2000 or 2200 i.u. vitamin A produced normal, healthy litters but with reduced liver deposits and colostrum and milk had lower vitamin A levels also. Heaney *et al* (1959) studied a range of levels by giving depleted gilts either 7·5, 15 or 48 i.u. vitamin A per kg liveweight per day; equivalent for a 450 lb sow to 1500, 3000 and 9600 i.u. per day. The level of vitamin A had no effect on litter size or birth weight; liver stores occurred with all levels but were highest with the highest level of intake and, if this be taken as the criterion of adequacy, suggests that the true requirement lies between 3000 and 9600 i.u. per day.

More recently, an extensive experiment was reported (Selke, Barnhart and Chaney, 1964) in which a vitamin-A-free diet was supplemented with vitamin A palmitate to provide levels of 0, 666, 1333, 2666, 5332 and 10664 i.u. per lb feed and given to sows for two pregnancies and lactations. There was no significant effect on numbers born, birth weights or weights at 14 days of age, but there were significant increases in vitamin A levels of milk and piglet livers on dietary levels above 1333 i.u. per lb.

It may be concluded from the above summary of the evidence that the vitamin A requirement of the pregnant sow would be fully satisfied by a daily intake of 8000 i.u. and that no obvious defects would occur with appreciably lower levels.

As the average output of vitamin A in milk would increase the daily requirement of the sow by only some 700 i.u., it follows that a diet containing sufficient of the vitamin for pregnancy would provide considerable excess for lactation if total feed was at the recommended level. In fact it might be suggested that in the case of a nutrient like vitamin A, for which the sow has considerable storage capacity, the dietary level should be based on the average feed intake over the whole reproductive cycle; with the levels of feeding here suggested, the overall requirement would then be satisfied by a dietary level of 1200 i.u. vitamin A per lb feed.

Vitamin D. No experiments have been reported on vitamin D for sows so neither deficiency effects nor requirements are known. It is usual to assume that sows require the vitamin and to base the allowance on an estimate for growing pigs, e.g. 100 i.u. per lb feed (NRC, 1964).

Vitamin E. Information on dietary requirements of sows for vitamin E is sparse. There have been reports of improved reproductive performance following the addition of wheat germ meal or wheat germ oil to the diet (Aubel, Hughes and Lienhardt, 1929; Crampton, 1941), but data specific to vitamin E are limited to the report of Adamstone *et al.* (1949) that a diet with only 3 ppm total tocopherols caused muscular dystrophy in young pigs and embryo mortality in pregnant sows; these abnormalities did not occur with a total tocopherol level of 22 ppm.

There is no information other than this on which to base an estimate of requirement, but in practice deficiency is not likely to occur unless the diet contains a high level of unsaturated fat.

Vitamin B complex. The effects on sows of a deficiency have been established for most of the B complex vitamins (Ensminger *et al.*, 1947; Ensminger *et al.*, 1951), but for thiamine, nicotinic acid, folic acid and choline (commonly included in this group) there is insufficient information on which to base estimates of requirement. No requirements have been established for biotin and inositol (Ensminger *et al.*, 1951).

The requirement of sows for riboflavine was investigated by Miller and co-workers (Stevenson *et al.*, 1952; Miller *et al.*, 1953) who found that a level established as adequate for growing pigs, namely 0·83 mg per lb (Miller and Ellis, 1951) was inadequate for normal reproductive performance and that the deficiency was rectified with a level of 1·25 mg per lb of feed; a higher level of 1·65 mg per lb feed gave no further improvement.

More data than the above are available on the pantothenic acid requirement of sows. Owen and Bowland (1952) found no signs of deficiency with a basal diet containing 2·5 mg per lb, and the only effects of supplements of 3, 6 or 12 mg per lb calcium pantothenate were increases in the levels in blood and milk. However, Ullrey and co-workers (Ullrey *et al.*, 1955) compared levels of 0·7, 2·7, 5·7 and 8·7 mg per lb and found that 5·7 mg per lb was necessary to prevent the birth of abnormal piglets; similarly, Davey and Stephenson (1953), who compared a range of levels, concluded that 5·4 mg per lb was required in the diet of sows. It may be concluded, therefore, that the requirement of the sow for pantothenic acid is satisfied by some 5·0 mg per lb of feed.

Information on the requirement for pyridoxine appears to be limited to the finding of Ritchie and co-workers (Ritchie *et al.*, 1960) that supplementing a diet which already contained 0·45 mg per lb total pyridoxine gave no improvement in reproductive performance.

A requirement for vitamin B_{12} has been established, but it is evident that the sow has appreciable storage capacity for the vitamin (Anderson and Hogan, 1950a, b; Kralovanszky, Eori and Kallai, 1954; Frederick and Brisson, 1961). There is little evidence on which to base an estimate of requirement, but Anderson and Hogan (1950a) estimated it to be 5 μg vitamin B_{12} per lb feed.

Vitamin C. Although vitamin C occurs at a higher concentration in sow's milk than in that of any other domestic animal (Braude *et al.*, 1947; Bowland *et al.*, 1949b; Heidebrecht *et al.*, 1951), a dietary requirement for the vitamin has not been established (Orr and Crichton, 1924; Hughes, Aubel and Lienhardt, 1928; Braude *et al.*, 1950; Travnicek, Simck and Mandel, 1961; Barber, Braude and Cooke, 1962).

REFERENCES

ADAMS, C. R., BECKER, D. E., TERRILL, S. W., NORTON, H. W. and JENSEN, A. H. (1960) *J. Anim. Sci.* **19**, 1245.

ADAMSTONE, F. B., KRIDER, J. L. and JAMES, M. F. (1949) *Ann. N.Y. Acad. Sci.* **52**, 260.

ALLEN, A. D. and LASLEY, J. F. (1958) *J. Anim. Sci.* **17**, 1223.

ANDERS, R. J. and SCHIPPER, J. A. (1959) *Bi-m. Bull. M. Dak. Agric. Exp. Stn.* **20**, 9.

ANDERSON, G. C. and HOGAN, A. G. (1950a) *J. Anim. Sci.* **9**, 163.

ANDERSON, G. C. and HOGAN, A. G. (1950b) *J. Anim. Sci.* **9**, 646.
ANDREWS, F. N., SHREWSBURY, C. L., HARPER, C., VESTAL, C. M. and DOYLE, L. P. (1948) *J. Anim. Sci.* **7**, 298.
ARRINGTON, L. E., AMMERMAN, C. B. and DAVIS, G. K. (1959) *J. Anim. Sci.* **18**, 1503.
AUBEL, C. E., HUGHES, J. S. and LIENHARDT, H. F. (1929) *Rec. Proc. Am. Soc. Anim. Prod.* 1930, 133.
BAKER, B. (1959) *J. Anim. Sci.* **18**, 1160.
BAKER, B., MILLER, W. W. and ULBERG, L. C. (1958) *Proc. Ass. 5th Agric. Wkrs. C.*
BALLINGER, C. E. (1940) *N.Z. J. Agric.* **60**, 351.
BARBER, R. S., BRAUDE, R. and COOKE, P. (1962) *Proc. Nutr. Soc.* **21**, XXVII.
BARBER, R. S., BRAUDE, R., FORD, J. E., GREGORY, M. E., MITCHELL, K. G. and PORTER, J. W. G. (1960) *Br. J. Nutr.* **14**, 43.
BARBER, R. S., BRAUDE, R. and MITCHELL, K. G. (1955) *J. Agric. Sci. Camb.* **46**, 97.
BERGE, S. and INDREBØ, T. (1953) *Meld. Norg. Landbr. Hoisk.* **33**, 389.
BLUNN, C. T., WARWICK, E. J. and WILLEY, J. R. (1954) *J. Anim. Sci.* **13**, 383.
BOAZ, T. G. (1962) *Proc. 80th Annual Congress of Br. Vet. Ass.*
BOWLAND, J. P., GRUMMER, R. H., PHILLIPS, P. H. and BOHSTEDT, G. (1949a) *J. Anim. Sci.* **8**, 199.
BOWLAND, J. P., GRUMMER, R. H., PHILLIPS, P. H. and BOHSTEDT, G. (1949b) *J. Anim. Sci.* **8**, 98.
BOWLAND, J. P., GRUMMER, R. H., PHILLIPS, P. H. and BOHSTEDT, G. (1951) *J. Anim. Sci.* **10**, 533.
BRAUDE, R., COATES, M. E., HENRY, K. M., KON, S. K., ROWLAND, S. J., THOMPSON, S. Y. and WALKER, D. M. (1947) *Br. J. Nutr.* **1**, 64.
BRAUDE, R., KON, S. K., MITCHELL, K. G. and THOMPSON, S. Y. (1951) *Vet. Rec.* **63**, 671.
BRAUDE, R., KON, S. K. and PORTER, J. W. G. (1950) *Br. J. Nutr.* **4**, 186.
BRODY, S., PROCTOR, R. C. and ASHWORTH, U. S. (1934) *Res. Bull. Mo. Agric. Exp. Stn.*, No. 220.
BRYUSHININ, I. G. (1958) *Trudy Naucho. – issled Inst. Svinov.* No. 20, 189.
BUCHANAN, M. L., LASLEY, E. and BOLAN, D. W. (1949) *Bi-m. Bull. N. Dak. Agric. Exp. Stn.* **11**, 106.
BURGER, J. F. (1952) *Onderstepoort, J. Vet. Res.*, Suppl. 2.
CASIDA, L. E. (1956) *Proc. III Int. Congr. Anim. Reprod., Plenary Papers*, 19.
CATRON, D., CULBERTSON, C. C. and SHEARER, P. S. (1948) *J. Anim. Sci.* **7**, 522.
CHAPMAN, H. L. (1957) *Iowa St. Coll. J. Sci.* **31**, 375.
CHRISTIAN, R. E. and NOFZIGER, J. C. (1952) *J. Anim. Sci.* **11**, 789.
CLAWSON, A. J., RICHARDS, H. L., MATRONE, G. and BARRICK, E. R. (1963) *J. Anim. Sci.* **22**, 662.
COLLINS, R. A., HARPER, A. E., SCHREIBER, M. and ELVEHJEM, C. A. (1951) *J. Nutr.* **43**, 313.
COMBS, G. E., WALLACE, H. D., ALSMEYER, W. L. and KOGER, M. (1959) *J. Anim. Sci.* **18**, 361.
CORNER, G. W. (1923) *Am. J. Anat.* **31**, 523.
CRAMPTON, E. W. (1941) *Scient. Agric.* **21**, 750.
CROWTHER, C. (1938) *J. Minist. Agric. Fish.* **45**, 797.
DAVEY, R. J. and STEPHENSON, J. W. (1953) *J. Anim. Sci.* **12**, 912.
DAVIDSON, H. R. (1930) *J. Agric. Sci. Camb.* **20**, 233.
DAVIS, V. E., HEIDEBRECHT, A. A., MACVICAR, R., ROSS, O. B. and WHITEHAIR, C. K. (1951) *J. Nutr.* **44**, 17.
DEAN, B. T. and TRIBBLE, L. F. (1960) *J. Anim. Sci.* **19**, 1257.
DUDLEY, W. A. (1960) *Diss. Abstr. XXI*, 1008.
DUDLEY, W. A., BECKER, D. E., NORTON, H. W., JENSEN, A. H. and TERRILL, S. W. (1961) *J. Anim. Sci.* **18**, 1505.
DYRENDAHL, S., OLSSON, B., BJORCK, G. and EHLERS, T. (1958) *Acta Agric. Scand.* **3**, 334.

ELSLEY, F. W. H., ANDERSON, D. M. and MACPHERSON, R. M. (1965) *Anim. Prod.* **7**, 286.
ELSLEY, F. W. H. and MACPHERSON, R. M. (1964) *Anim. Prod.* **6**, 259.
ENSMINGER, M. E., BOWLAND, J. P. and CUNHA, T. J. (1947) *J. Anim. Sci.* **6**, 409.
ENSMINGER, M. E., COLBY, R. W. and CUNHA, T. J. (1951) *Circ. Agric. Exp. Stn. Wash. St. Inst. Agric. Sci.* 134.
EVANS, R. E. (1929) *J. Agric. Sci. Camb.* **19**, 752.
EVANS, R. E. (1932) *J. Minist. Agric. Fish.* **39**, 544.
EYLES, D. E. (1959) *Anim. Prod.* **1**, 41.
FISHWICK, V. C. (1944) *Agriculture, London* **51**, 41.
FISHWICK, V. C. (1945) *Agric. Prog.* **20**, 11.
FOOT, A. S. and THOMPSON, S. Y. (1938) *J. Minist. Agric. Fish.* **45**, 452.
FREDERICK, G. L. and BRISSON, G. J. (1961) *Can. J. Anim. Sci.* **41**, 212.
GILL, J. C. and THOMSON, W. (1956) *J. Agric. Sci. Camb.* **47**, 324.
GOODE, L., WARNICK, A. C. and WALLACE, H. D. (1960) *J. Anim. Sci.* **19**, 643.
GORTNER, W. A. (1945) *J. Biol. Chem.* **159**, 135.
GOSSETT, J. W. and SORENSEN, A. M. (1959) *J. Anim. Sci.* **18**, 40.
GREGORY, M. E., KON, S. K. and THOMPSON, S. Y. (1952) *Rep. Natn. Inst. Res. Dairying*, p. 88.
GRUMMER, R. H., BENTLY, O. G., PHILLIPS, P. H. and BOHSTEDT, G. (1950) *J. Anim. Sci.* **9**, 170.
GUEGEN, L. and SALMON-LEGAGNEUR, E. (1959) *C.R. Hebd. Seanc. Acad. Sci. Paris*, **249**, 784.
GUTTE, J. O. and LENKEIT, W. (1960) *Tierphysiol. Tierernahr Futtermittelk.***15**, 165.
HABERSANG, M. (1952) *Vet. Med. Diss.*, Berlin.
HAFEZ, E. S. E. (1958) *J. Anim. Sci.* **17**, 1212.
HAINES, C. E., WARNICK, A. C. and WALLACE, H. D. (1959) *J. Anim. Sci.* **18**, 347.
HAMMOND, J. (1914) *J. Agric. Sci. Camb.* **6**, 263.
HAMMOND, J. (1921) *J. Agric. Sci. Camb.* **11**, 337.
HANSON, L. E., FERRIN, E. F. and ANNAN, W. J. (1953) *J. Anim. Sci.* **12**, 919.
HART, E. B., ELVEHJEM, C. A., STEENBOCK, H., KEMMERER, A. R., BOHSTEDT, G. and FARGO, J. M. (1930) *J. Nutr.* **2**, 277.
HART, E. B. and STEENBOCK, H. (1918) *J. Biol. Chem.* **33**, 313.
HARTMAN, D. A., LUDWICK, T. M. and WILSON, R. F. (1962) *J. Anim. Sci.* **21**, 882.
HEANEY, D. P., ULLREY, D. E., MILLER, E. R. and HOEFER, J. A. (1959) *J. Anim. Sci.* **18**, 1494.
HEAP, F. C. (1966). Ph.D. Thesis, Univ. Nottingham.
HEAP, F. C. and LODGE, G. A. (1967) *Anim. Prod.* **9**, 237.
HEAP, F. C., LODGE, G. A. and LAMMING, G. E. (1967) *J. Reprod. & Fert.* **13**, 269.
HEIDEBRECHT, A. A., MACVICAR, R., ROSS, O. B. and WHITEHAIR, C. K. (1951) *J. Nutr.* **44**, 43.
HJARDE, W., NEIMANN-SORENSEN, A., PALLUDAN, B. and HAVSKOV SORENSEN, P. (1961) *Acta Agric. Scand.* **XI**, 13.
HOGAN, A. G. (1932) *Res. Bull. Mo. Agric. Exp. Stn.*, No. 167.
HOLDER, J. M. (1959) *Aust. Vet. J.* **35**, 296.
HUGHES, J. S., AUBEL, C. E. and LIENHARDT, H. F. (1928) *Tech. Bull. Kans. Agric. Exp. Stn.*, No. 23.
HUGHES, E. H. and HART, H. G. (1935) *J. Nutr.* **9**, 311.
HUSBY, M. (1933) *Norg. Landbr. Hoisk. Beretn. Foringsfors.* **35**.
JESPERSEN, J. and OLSEN, H. M. (1939) *Kobenhavn Beretn. Forsogslab.* **186**.
JOHNSON, S. R. (1943) *J. Anim. Sci.* **2**, 14.
KICK, G. H., BETHKE, R. M., ERDINGTON, B. H., WILDER, O. H. M., RECORD, P. R., WILDER, W., HILL, T. J. and CHASE, S. W. (1935) *Bull. Ohio Agric. Exp. Stn.*, No. 558.
KORKMAN, N. (1947) *Acta Agric. Suec.* **2**, 253.
KOVES, J. (1958) *Magy. Allatorv. Lap.* **13**, 212.
KRALOVANSZKY, U. P., EORI, E. and KALLAI, L. (1954) *Allattenyesztes* **4**, 331.
LENKEIT, W. (1959) *Zuchtungskunde* **31**, 410.

LENKEIT, W., GIMMLER, W. and SEICK, K. H. (1959) *Arch. Tierernahr.* **2/3**, 166.

LENKEIT, W. and GUTTE, J. O. (1955) *Tierernahr. Futtermittelk.* **10**, 94.

LENKEIT, W., GUTTE, J. O., KIRCHOFF, W. and SOEHNGEN (1959) *Tierphysiol. Tierernahr. Futtermittelk.* **14**, 3.

LENKEIT, W., GUTTE, J. O., KIRCHOFF, W., SOEHNGEN, F. K. and FARRIES, E. (1956) *Tierernahr. Futtermittelk.* **11**, 337.

LENKEIT, W., GUTTE, J. O. and STREUTER-PETERMOLLER, A. (1955) *Tierernahr. Futtermittelk.* **10**, 228.

LENKEIT, W., GUTTE, J. O., WARNACKE, W. and KIRCHOFF, W. (1955) *Tierernahr. Futtermittelk.* **10**, 351.

LENKEIT, W., MOLNAR, S., LANTZECH, H. J. and GUTTE, J. O. (1962) *Tierphysiol. Tierernahr. Futtermittelk.* **17**, 65.

LINTZEL, W. and RADEFF, T. (1931) *Arch. Tierernahr. Tierz.* **6**, 313.

LIVINGSTONE, D. M. S. (1962) *Anim. Prod.* **4**, 296.

LODGE, G. A. (1957) *J. Agric. Sci. Camb.* **49**, 200.

LODGE, G. A. (1958) *Ph.D. Thesis, Univ. Aberdeen.*

LODGE, G. A. (1959a) *J. Agric. Sci. Camb.* **53**, 177.

LODGE, G. A. (1959b) *J. Agric. Sci. Camb.* **53**, 172.

LODGE, G. A. (1962) In *Nutrition of Pigs and Poultry*, Proc. Univ. Nottingham 8th Easter School, p. 224, MORGAN, J. T. and LEWIS, D. (Ed.), Butterworths, London.

LODGE, G. A. (1966) Unpublished data.

LODGE, G. A., ELSLEY, F. W. H. and MACPHERSON, R. M. (1966a) *Anim. Prod.* **8**, 29.

LODGE, G. A., ELSLEY, F. W. H. and MACPHERSON, R. M. (1966b) *Anim. Prod.* **8**, 499.

LODGE, G. A. and LUCAS, I. A. M. (1959) In *Scientific Principles of Feeding Farm Livestock*, p. 125, Farmer and Stockbreeder Publs., London.

LODGE, G. A. and MACPHERSON, R. M. (1961) *Anim. Prod.* **3**, 19.

LODGE, G. A. and MACPHERSON, R. M. (1963) *Anim. Prod.* **5**, 44.

LODGE, G. A. and MCDONALD, I. (1959) *Anim. Prod.* **1**, 139.

LODGE, G. A., MCDONALD, I. and MACPHERSON, R. M. (1961) *Anim. Prod.* **3**, 261.

LODGE, G. A. and PRATT, P. D. (1963) *Anim. Prod.* **5**, 225.

LONGWILL, A. (1952a) *N.Z. J. Sci. Technol.* (A) **34**, 294.

LONGWILL, A. (1952b) *N.Z. J. Agric.* **84**, 97.

LUCAS, I. A. M. and LODGE, G. A. (1961) *The Nutrition of the Young Pig*, T.C.22, Commonw. Bur. Anim. Nutr., Aberdeen.

MAYROSE, V. B., SPEER, V. C., MCCALL, J. T. and HAYS, V. W. (1962) *J. Anim. Sci.* **21**, 1005.

MCGILLIVRAY, J. J., NALBANDOV, A. V., JENSEN, A. H., NORTON, H. W. and BECKER, D. E. (1962) *J. Anim. Sci.* **21**, 1005.

MENZIES-KITCHEN, A. W. (1937) *J. Agric. Sci. Camb.* **27**, 611.

MEYER, W. (1940) *J. Landwirt* **87**, 302.

MILLER, C. O. and ELLIS, N. R. (1951) *J. Anim., Sci.* **10**, 807.

MILLER, C. O., ELLIS, N. R., STEVENSON, J. W. and DAVEY, R. (1953) *J. Nutr.* **51**, 163.

MITCHELL, H. H., CARROLL, W. E., HAMILTON, T. S. and HUNT, G. E. (1931) *Bull. Ill. Agric. Exp. Stn.*, No. 375.

MITCHELL, H. H. and KELLEY, M. A. R. (1938) *J. Agric. Res.* **56**, 811.

MITCHELL, H. H. and MCCLURE, F. J. (1937) *Bull. Ntn. Res. Coun. Wash.*, No. 99.

MOUSTGAARD, J. (1962) In *Nutrition of Pigs and Poultry*, Proc. Univ. Nottingham 8th Easter School, p. 189, MORGAN, J. T. and LEWIS, D., (Ed.), Butterworths, London.

MOUSTGAARD, J. and OLSEN, N. J. H. (1951) *Nord. Vet. Med.* **3**, 763.

MUMFORD, F. G., HOGAN, A. G. and MCKENZIE, F. F. (1927) *Rec. Proc. Am. Soc. Anim. Prod.* 1925, 85.

N.R.C. (1964) *Nutrient Requirements for Swine*, National Research Council (U.S.A.) Publ. No. 295, Washington, D.C.

NESENI, R. (1953) *Mh. Vet. Med.* **8**, 139.

NIWA, T., ITO, S., YOKOYAMA, H. and OTSUKA, M. (1951) *Bull. Natn. Agric. Sci. Tokyo*, Ser. G, No. 1, 135.

NORDFELT, S. (1944–5) *LantbrHogsk, Annlr.* **12**, 204.

OLBRYCHT, T. M. (1943) *J. Agric. Sci. Camb.* **33**, 28, 74.
OLOFSSON, N. E. (1950) *LantbrHogsk. Medda.*, No. 42.
ORR, J. B. and CRICHTON, A. (1924) *J. Agric. Sci. Camb.* **14**, 114.
OUTLER, J. C., WALLACE, H. D. and DAVIS, G. K. (1955) *J. Anim. Sci.* **14**, 1219.
OWEN, B. D. and BOWLAND, J. P. (1952) *J. Nutr.* **48**, 317.
PARRISH, D. B., AUBEL, C. E., HUGHES, J. S. and WHEAT, J. D. (1951) *J. Anim. Sci.* **10**, 551.
PERRIN, D. R. (1954) *J. Dairy Res.* **21**, 55.
PERRIN, D. R. (1955) *J. Dairy Res.* **22**, 103.
PERRIN, D. R. (1958) *J. Dairy Res.* *25*, 215.
PERRY, J. S. (1955) *J. Embryol. Exp. Morph.* **2**, 308.
PHILLIPS, R. W. and ZELLER, J. H. (1943) *Anat. Rec.* **85**, 387.
PLUMLEE, M. P., JORDON, C. E., KENNINGTON, M. H. and BEESON, W. M. (1958) *J. Anim. Sci.* **17**, 73.
PLUMLEE, M. P., THRASHER, D. M., BEESON, W. M., ANDREWS, F. N. and PARKER, H. E. (1956) *J. Anim. Sci.* **15**, 354.
P.I.D.A. (1963a) *Feed Recording*, Pig Industry Development Authority Publs., London.
P.I.D.A. (1963b) *Litter Recording*, Pig Industry Development Authority Publs., London.
POMEROY, R. W. (1960) *J. Agric. Sci. Camb.* **54**, 31.
POND, W. G., VANVLECH, L. D. and HARTMAN, D. A. (1962) *J. Anim. Sci.* **21**, 293.
RATHNASABAPATHY, V., LASLEY, J. F. and MAYER, D. T. (1956) *Res. Bull. Mo. Agric. Exp. Stn.*, No. 615.
RIGOR, E. M., MEYER, R. K., FIRST, N. L. and CASIDA, L. E. (1963) *J. Anim. Sci.* **22**, 43.
RIPPEL, R. H., RASMUSSEN, O. G., NORTON, H. W., BECKER, D. E. and JENSEN, A. H. (1962) *J. Anim. Sci.* **10**, 841.
RIPPEL, R. H., HARMON, B. G., JENSEN, A. H., NORTON, H. W. and BECKER, D. E. (1965b) *J. Anim. Sci.* **24**, 209.
RIPPEL, R. H., HARMON, B. G., JENSEN, A. H., NORTON, H. W. and BECKER, D. E. (1965c) *J. Anim. Sci.* **24**, 373.
RIPPEL, R. H., HARMON, B. G., JENSEN, A. H., NORTON, H. W. and BECKER, D. E. (1965d) *J. Anim. Sci.* **24**, 378.
RIPPEL, R. H., RASMUSSEN, O. G., JENSEN, A. H., NORTON, H. W. and BECKER, D. E. (1965a) *J. Anim. Sci.* **24**, 203.
RITCHIE, H. D., MILLER, E. R., ULLREY, D. E., HOEFER, J. A. and LUECKE, R. W. (1960) *J. Nutr.* **70**, 491.
ROBERTS, H. F., HOEKSTRA, W. G. and GRUMMER, R. H. (1961) *J. Anim. Sci.* **21**, 1011.
ROBERTSON, G. L., GRUMMER, R. H., CASIDA, L. E. and CHAPMAN, A. B. (1951a) *J. Anim. Sci.* **10**, 647.
ROBERTSON, G. L., CASIDA, L. E., GRUMMER, R. H. and CHAPMAN, A. B. (1951b) *J. Anim. Sci.* **10**, 841.
RYDBERG, M. E., SELF, H. L., KOWALCZYK, T. and GRUMMER, R. H. (1959) *J. Anim. Sci.* **18**, 415.
SALMON-LEGAGNEUR, E. (1956) *Annls. Zootech.* **5**, 95.
SALMON-LEGAGNEUR, E. (1958) *Annls. Zootech.* **7**, 143.
SALMON-LEGAGNEUR, E. (1959) *Annls. Zootech.* **8**, 93.
SALMON-LEGAGNEUR, E. (1962) *Annls. Zootech.* **11**, 295.
SALMON-LEGAGNEUR, E. (1963) *C. R. Hebd. Seanc. Acad. Sci. Paris* **256**, 1372.
SALMON-LEGAGNEUR, E. (1964a) *Annls. Biol. Anim. Biochim. Biophys.* **4**, 49.
SALMON-LEGAGNEUR, E. (1964b) *Annls. Biol. Anim. Biochim. Biophys.* **4**, 141.
SALMON-LEGAGNEUR, E. (1964c) *Annls. Zootech.* **13**, 51.
SALMON-LEGAGNEUR, E. and GUEGEN, L. (1962) *Ann. Biol. Anim. Biochim. Biophys.* **2**, 299.
SALMON-LEGAGNEUR, E., GOMEZ, G. and JACQUOT, R. (1960) *C. R. Hebd. Seanc. Acad. Agric. Fr.* **46**, 445.

SALMON-LEGAGNEUR, E. and JACQUOT, R. (1961a) *C. R. Hebd. Seanc. Acad. Sci. Paris* **253**, 1497.

SALMON-LEGAGNEUR, E. and JACQUOT, R. (1961b) *C. R. Hebd. Seanc. Acad. Sci. Paris* **253**, 1618.

SALMON-LEGAGNEUR, E. and RERAT, A. (1962) In *Nutrition of Pigs and Poultry*, Proc. Univ. Nottingham 8th Easter School, p. 207, MORGAN, J. T. and LEWIS, D. (Ed.), Butterworths, London.

SCHMIDT, K. and BRETSCHNEIDER, W. (1954) *Tierzucht.* **8**, 119.

SCHOFIELD, F. W., INGLE, R. T. and MCFARLANE, W. D. (1941) *Ontario Vet. Coll.*, Rep. No. 29, p. 16.

SELF, H. L., GRUMMER, R. H. and CASIDA, L. E. (1955) *J. Anim. Sci.* **14**, 573.

SELF, H. L., GRUMMER, R. H., HAYS, O. E. and SPIES, H. G. (1960) *J. Anim. Sci.* **19**, 274.

SELKE, M. R., BARNHART, C. E. and CHANEY, C. H. (1964) *J. Anim. Sci.* **23**, 891.

SMITH, D. M. (1952a) *N.Z. J. Sci. Technol.*

SMITH, D. M. (1952b) *Proc. N.Z. Soc. Anim. Prod.* **12**, 102.

SMITH, D. M. (1959a) *N.Z. J. Agric. Res.* **2**, 1057.

SMITH, D. M. (1959b) *N.Z. J. Agric. Res.* **2**, 1071.

SMITH, D. M. (1960a) *N.Z. J. Agric. Res.* **3**, 598.

SMITH, D. M. (1960b) *N.Z. J. Agric. Res.* **3**, 745.

SMITH, G. E. (1917) *J. Biol. Chem.* **29**, 215.

SORENSEN, A. M., THOMAS, W. B. and GOSSETT, J. W. (1961) *J. Anim. Sci.* **20**, 347.

SQUIERS, C. D., DICKERSON, G. E. and MAYER, D. T. (1950) *J. Anim. Sci.* **9**, 683.

STAHL, — (1930) *Schweinez* **37**, 423.

STEVENSON, J. W. and ELLIS, N. R. (1957) *J. Anim. Sci.* **16**, 877.

STEVENSON, J. W., MILLER, C. O. and ELLIS, N. R. (1952) *J. Anim. Sci.* **11**, 766.

TEAGUE, H. S. and WILSON, R. F. (1957) *Res. Circ. Ohio. Agric. Exp. Stn.*, No. 46.

TETER, W. S. (1961) *Diss. Abstr. XXI*, 1008.

TETER, W. S. and HANSON, L. E. (1959) *J. Anim. Sci.* **18**, 104.

THOMAS, J. W., LOOSLI, J. K. and WILLIAM, J. P. (1947) *J. Anim. Sci.* **6**, 141.

TRAVNICEK, J. (1960) *Sborn. cls. Akad. Zemed. Ved. Zivocisna vyroba* **5**, 497.

TRAVNICEK, J., SIMEK, L. and MANDEL, L. (1961) *Sborn. csl. Akad. zemed. Ved. Zivocisna vyroba* **6**, 421.

TRIBBLE, L. B., PFANDER, W. H., LASLEY, J. F., ZOBRISKY, S. E. and BRADY, D. E. (1956) *Res. Bull. Mo. Agric. Exp. Stn.*, No. 609.

ULLREY, D. E., BECKER, D. E., TERRILL, S. W. and NOTZOLD, R. A. (1955) *J. Nutr.* **57**, 401.

URBANYI, L. (1940) *Mezogazd. Kutat.* **13**, 157.

VANSCHOUBROEK, F. (1963) *Studien en Onderzoekingen over de Voeding van Varkens.* Rijksuniversitiet Veearysenijschool Laboratorium voor Zootechnie, Gent.

VETERINARY INVESTIGATION SERVICE (1959) *Vet. Rec.* **71**, 777.

VENN, J. A. J., MCCANCE, R. A. and WIDDOWSON, E. M. (1947) *J. Comp. Path.* **57**, 314.

WAHLSTROM, R. C. and OLSEN, O. E. (1959a) *J. Anim. Sci.* **18**, 141.

WAHLSTROM, R. C. and OLSEN, O. E. (1959b) *J. Anim. Sci.* **18**, 578.

WARNICK, A. C., WIGGINS, E. L., CASIDA, L. E., GRUMMER, R. H. and CHAPMAN, A. B. (1951) *J. Anim. Sci.* **10**, 479.

WILLET, E. L. and MARUYAMA, C. (1946) *J. Anim. Sci.* **5**, 365.

ZIMMERMAN, D. R., SELF, H. L. and CASIDA, L. E. (1957) *J. Anim. Sci.* **16**, 1099.

ZIMMERMAN, D. R., SPIES, H. G., SELF, H. L. and CASIDA, L. E. (1958) *J. Anim. Sci.* **17**, 1212.

ZIMMERMAN, D. R., SPIES, H. G., RIGOR, E. M., SELF, H. L. and CASIDA, L. E. (1960a) *J. Anim. Sci.* **19**, 687.

ZIMMERMAN, D. R., SPIES, H. G., SELF, H. L. and CASIDA, L. E. (1960b) *J. Anim. Sci* **19**, 295.

CHAPTER 29

The Nutrient Requirements of Farm Livestock

*3. Pigs**

FOREWORD

By SIR DAVID CUTHBERTSON, C.B.E., F.R.S.E.

Chairman of the Agricultural Research Council Technical Committee on the Nutrient Requirements of Farm Livestock

Following representations that there was a need for a reappraisal of the accepted standards for the feeding of farm livestock, the Agricultural Research Council set up in 1959 a Technical Committee charged with the task of reviewing periodically the literature on the nutrient requirements of various classes of stock, and of making recommendations or alterations that seemed necessary in published feeding standards. To assist it in discharging its duties, the Committee established Working Parties to consider the requirements of ruminants, of poultry, and of pigs, as well as one to consider the nomenclature to be used in the Committee's reports. In giving guidance to the Working Parties dealing with different classes of livestock the Technical Committee realized that the approaches of these groups would not necessarily be the same.

The findings of the Working Party on Pigs form the basis of the following series of *Summaries of Estimates of Nutrient Requirements*. In the course of its studies the Working Party has amassed much information of interest which could not be included in the Summaries; this information has been published in the form of detailed Technical Reviews.

It is the Committee's intention to keep under review future developments in the study of the nutrient requirements of farm livestock, and to publish from time to time revised summaries of estimates of requirements and, when the material warrants, further Technical Reviews. It would therefore be appreciated if persons possessing unpublished data pertinent to any aspect of the Committee's task would make them available, as they come to hand, to the Secretary of the Technical Committee.

* *This chapter is reprinted from A.R.C. publication No. 3, H.M.S.O. 1966*

INTRODUCTION

By Sir Ronald Baskett, o.b.e., d.sc., f.r.i.c.
Chairman, Working Party on the Nutrient Requirements of Pigs

The Working Party on the Nutrient Requirements of Pigs was appointed in 1959 to study and report on the nutrient requirements of pigs, the word 'requirement' being defined as the amount of a nutrient that must be supplied in the diet to meet the needs of the normal healthy animal given an otherwise completely adequate diet in an environment compatible with good health.

The subject has been dealt with under the following main headings: (1) Energy, (2) Protein, (3) Minerals, (4) Vitamins, and (5) Water. Technical Reviews have been prepared dealing with the nutrient requirements of young pigs and growing pigs, pregnant and lactating sows, and boars. In addition, concise summaries of the main findings of each review have been prepared, and are published here as a separate outline as well as in the volume containing the Technical Reviews.

No account has been taken of possible differences in requirements between breeds of pigs, but it should be mentioned that experimental work carried out in the United Kingdom refers mainly to the Large White breed. Very little information was found concerning the nutrient requirements of growing pigs of live weights above approximately 90 kg, or of boars.

These estimates of requirements do not include margins of safety, as they make no allowance for variations between animals, for variations in availability of nutrients in different samples of feeding stuffs or for possible losses during manufacture and storage of feed.

Present-day emphasis on a high proportion of lean meat in the carcass means that the results of some of the earlier research are inapplicable today. In more recent research, diets have often been supplemented with feed additives such as antibiotics, high levels of copper, hormones, or tranquillizers; the value of such additives and their effect on nutrient requirements has not been taken into consideration in arriving at the final estimates of requirements for specific nutrients.

The metric system has been adopted throughout and biological units are expressed in international units. Nutrient requirements have been expressed in terms of nutrient per unit weight of dry matter in the diet.

The recommendations of nutrient requirements are based on information available approximately to the middle of 1964. In certain cases, however, later information has been included. The recommendations can only reflect the information available, and members of the Working Party are conscious that their conclusions will be subject to modification as more accurate information becomes available.

All the terms used in the Summaries, whose meaning may not be clear, have been brought together in a glossary which will be found on p. 1128.

SECTION 1

REQUIREMENTS FOR ENERGY

Growing Pigs from Birth to Slaughter

Growth

Growth does not progress at an equal rate throughout all tissues and parts of the body. Judged by percentage increments over weights of tissue at birth, skeleton develops and grows faster than muscle in early life, and later growth is characterized by a considerable increase in size of the fat depots. Changes in overall composition with age imply greater changes in the proportions of tissues in the daily weight gain. Although muscle continued to increase substantially in weight, more fat than muscle was deposited each day in one group of Large Whites from 60–70 kg live weight onwards, while energy retained as lipid has exceeded that retained as protein in pigs as light as 15–20 kg. As age increases, the lipid content of the fat depots increases and water content decreases, while in muscle, protein and lipid contents increase and water content decreases. These changes could cause inaccuracies in estimates of energy values of weight gains from carbon and nitrogen balance data where water retention is not known.

Nutrition substantially influences the pattern of growth. Considering pigs of up to 25 kg live weight, high 8-week weight was associated in survey data with reduced fatness at slaughter. But variation in growth rate of sucking pigs, *due to imposed variation in feed intake*, leading to weights at 8 weeks within the range 14 to 23 kg, is unlikely to affect carcass composition greatly at 90 kg live weight. In pigs weaned at about 10 days of age, fast growth sometimes increased fatness, whereas reduced growth rate up to 8 weeks of age reduced fatness both at 8 weeks and at 90 kg live weight.

With pigs weighing 25–90 kg, high feed intake encourages fat deposition. Currently there is controversy on the interpretation of observations relating intake to body development. One interpretation is that *at any stage of growth* feed given in excess of the requirements for maintenance and fat-free body growth is converted into body lipid, while a restricted feed intake reduces the rate of lipid deposition and causes a more or less uniform retardation of development of bone and muscle throughout the body. At the same time, the probability of giving excess feed and producing over-fat carcasses is greater with pigs of over 50 kg live weight. The precise pattern of response to changes in intake depends on genotype, physical environment and perhaps other variables.

Level of Feed Intake and Feed Conversion Efficiency

In older experiments with controlled feeding, reduction of intake for pigs fed twice daily sometimes led to savings of up to 16 per cent in feed requirements, but in others it increased requirements by up to 12 per cent. In most recent experiments feed conversion efficiency from weaning to slaughter did not change by more than 2–4 per cent over quite a wide range of planes of feeding.

With twice daily feeding, one afternoon feed may be omitted each week without detriment provided the total weekly allowance is unaltered and divided evenly between the remaining 13 feeds. Alternatively, performance is as good in pigs fed only once daily. But considerable unevenness in the daily division of the weekly feed ration might make feed conversion efficiency worse.

Compared with controlled feeding, feed given *ad lib.* from hoppers (self feeding) sometimes has no effect on feed conversion efficiency, but in United Kingdom experience it is more likely to make it worse by about 5 per cent. Inefficient hopper design can lead to feed wastage and then the apparent deterioration in efficiency is greater. The dressing percentage of pigs slaughtered at a live weight of 90 kg may be increased by up to 1·5 units, and backfat measurements increased by up to 3–5 mm, and there may be 1·5 per cent more fat but only 0·8 per cent less muscle in the carcasses. If self feeding is discontinued at 55 kg live weight, these differences are reduced by about one half.

Devising Feeding Standards or Recommendations

It was decided that digestible energy (DE) would be used to describe energy intakes. But if more evidence accumulates on the metabolizable energy (ME) of diets given to pigs it will be desirable to evaluate more closely ME against DE values.

In general, energy intakes of sucking pigs cannot be controlled closely, although broad differences can be imposed by allowing or denying access to creep feed, while early weaned pigs are usually fed *ad lib.* Seven sets of data were used to describe growth outputs achieved with various levels of energy input in young pigs. From these, estimates of the smoothed mean performance of sucking and early weaned pigs were made (Table 1.1). Variations from this pattern are known to occur. There is frequently low intake and growth during the week after early weaning, and sucking pigs might have lower intakes at about 2–4 weeks of age due to delay in starting to eat creep feed. Assuming average energy intake from sow's milk, a sucking pig would have to eat daily averages of about 40, 110, 175, 225 and 380 g dry matter from creep feed during weeks 3 to 7 to conform to the pattern of weekly energy intake shown in Table 1.1.

TABLE 1.1
Smoothed mean performance of sucking and early weaned pigs

Week	Sucking pigs			Early weaned pigs		
	Live weight at start of week	Weekly energy intake	Weekly gain	Live weight at start of week	Weekly energy intake	Weekly gain
	kg	Mcal DE	kg	kg	Mcal DE	kg
1	1·5	5·4	1·1			
2	2·6	7·1	1·4	2·5	6·5	1·5
3	4·0	8·7	1·7	4·0	9·1	1·9
4	5·7	10·5	1·9	5·9	12·3	2·4
5	7·6	12·0	2·2	8·3	15·9	3·0
6	9·8	13·5	2·4	11·3	19·5	3·5
7	12·2	15·1	2·6	14·9		
8	14·8					

The recommendation on energy intake with controlled feeding for pigs of 20–90 kg live weight should be based on a system giving optimum feed conversion efficiency and carcass quality. Although various scales of feed intake have been used in British experiments there is no reason why any one should be given preference. However, three sets of data were obtained from the National Pig Progeny Testing Stations, and the average intakes and gains derived from one of these (Table 1.2), for Large Whites and Landrace shown by dissection to have had high lean contents, are proposed as a base line for recommendations on energy nutrition. They have the advantage that the pattern of intake is that under which boars are selected by breeders using the National testing scheme; pigs are given twice daily a quantity of feed, with water added, which is subsequently increased if the troughs are found to be clean 20 minutes after feeding.

Although a joint regression of weekly weight gain with initial weekly live weight and feed intake in Progeny Tested pigs was significant, the variability in weekly gain accounted for was 12–40 per cent, and the regression was not considered to be of great value in predicting the effect of changes in intake upon gains. This was probably due to variability between pigs in feed conversion efficiency.

With self feeding, energy intake is not controlled, except perhaps through changes in energy value of the feed, and therefore specific recommendations are not made. There is good agreement between experiments on the pattern of intake with self feeding up to 70 kg live weight, but from 70 to 114 kg there is greater variation. At 30, 50, 70 and 90 kg live weight the average feed intake of self fed pigs is about 114, 117, 120 and 117 per cent respectively of the feed intakes shown in Table 1.2 of pigs fed twice daily. Rate of weight gain varies with the feed conversion efficiency, but it is noted as a rough guide that if self

TABLE 1.2
Average daily feed intakes and growth rates of pigs fed twice daily
(Based on data from the National Pig Progeny Testing Stations)

Live weight	For the following week			
	Expected daily live weight gain	Daily feed intake		
kg	g	kg diet	kg dry matter[1]	Mcal DE[2]
20[3]	500[4]	1·00[4]	0·87	3·00
25	550	1·20	1·04	3·60
30	625	1·45	1·26	4·35
35	690	1·65	1·44	4·95
40	750	1·85	1·61	5·55
45	775	2·05	1·78	6·15
50	790	2·20	1·91	6·60
55	790	2·35	2·04	7·05
60	790	2·50	2·17	7·50
65	790	2·65	2·31	7·95
70	790	2·75	2·39	8·25
75	790	2·85	2·48	8·55
80	790	3·00	2·61	9·00
85	790	3·10	2·70	9·30
90[3]	790	3·20	2·78	9·60

[1] Assuming 87% dry matter in the diet.
[2] Air-dry diets calculated to have, on average, 68 per cent Total Digestible Nutrients or 3·00 Mcal DE/kg.
[3] Figures obtained by extrapolation.
[4] Pigs of some 20 kg live weight often have a check, for example at weaning; consequently, they might eat less than is shown in the table, and grow more slowly.

feeding *per se* leads to a reduction of about 5 per cent in feed conversion efficiency, an increase of 17 per cent in daily feed intake with self feeding would be associated with 10–12 per cent faster weight gain.

QUANTITATIVE EFFECTS OF VARIATIONS IN ENERGY INTAKE

Most recent experiments indicate that over quite a wide range of levels of feeding, but excluding self feeding, overall feed conversion efficiency from weaning to slaughter is unlikely to vary by more than 2–4 per cent. Therefore it is unlikely that the error would be large in assuming that efficiency is not affected by imposing feed intakes within +5 per cent to −15 per cent of those in Table 1.2. A known percentage increase or decrease in daily energy intake would then lead to the same percentage change in rate of live weight gain.

There is no direct evidence on the effects on carcass composition of changing energy intakes from those in Table 1.2. However, using as a base line a different series of feed intakes rising to 10 Mcal DE/pig daily at 90 kg live weight, it was estimated that for each 10 per cent change in daily intake from 20–90 kg live weight, shoulder fat, mid-back fat and fat at 'C' over the eye muscle would change by 2·3, 1·4 and 1·5 mm. Also, for each change of 10 per cent in average daily intake from 45 to 90 kg live weight, these fat measurements would change by 1·4, 1·2 and 1·2 mm respectively. All measurements are for carcasses of pigs killed at 90 kg live weight, and increased energy intake is associated with increased fatness and *vice versa.* It was further estimated that a 10 per cent change in daily energy intake might alter the percentage fat in the carcass by 1·2 or 1·0 per cent, depending on whether the change was introduced at 20 or 45 kg live weight. It is stressed that these estimates are all tentative.

No data uncomplicated by dietary changes were found on the quantitative effects on growth rate and carcass conformation of variations in intake of self fed pigs. But the literature on increasing dietary fat or fibre, and so increasing or decreasing dietary DE values, is extensive.

There is considerable variation between experiments in the effects of fat and fibre. This is not all readily explained by known causes of variation such as the inverse relation between the mean molecular weights of fatty acids in fats and oils and their digestibilities, or the reduced digestibility of fibre associated with increased lignification.

Pigs tend to eat high-fat diets on an energy basis, and the addition of 5, 10 or 15 per cent fat to the diet leads to feed intakes reduced by about 2–8 per cent, 5–15 per cent and 10–20 per cent respectively. Increased dietary DE with 5 or 10 per cent added fat results in improvements of about 7 per cent and 16–17 per cent in feed conversion efficiency. This is frequently a little more than sufficient to counterbalance the effects of reduced feed intake, and increased growth rate (5–10 per cent with 10 per cent added fat) occurs. With low protein levels, decrease in feed intake when fat is added to the diet is greater than with high protein levels. Likewise, there is some evidence that low protein levels may exaggerate the increase in fatness usually observed in pigs given high-fat diets.

It was estimated that 5 and 10 per cent increases in crude fibre in the diet lead to reductions of 11 and 22 units in DE expressed as a percentage of gross energy (GE). Furthermore, additions of 2·5, 5·0, 10·0 and 15·0 per cent crude fibre would require that daily feed intakes be raised by 7, 14, 34 and 61 per cent to give a DE intake equal to a control diet with 2 per cent crude fibre, assuming no change in the GE of 3·6 Mcal/kg. Feeding trials show that the daily feed intake of self fed pigs changes by −5 to +10 per cent with 1–5 per cent extra crude fibre in the diet, and by +10 to +20 per cent with 5–10 per cent extra fibre. However, increased intake is not usually sufficient to compensate entirely for reduced dietary DE, and added fibre in diets given *ad lib.* causes some reduction in rate of weight gain. It also causes reduced carcass fat measurements, increased carcass lean and reduced dressing percentage.

Reduction in carcass weight associated with the latter can confuse the interpretation of performance with high-fibre diets.

Energy Balance Studies

Feed having one unit net energy for fattening in pigs [NE(FP)] has 1·13 units NE for maintenance in pigs [NE(MP)], and concentrates having one unit NE for fattening in cattle [NE(FC)] have 1·29 units NE(FP) and 1·46 units NE(MP).

At present the most satisfactory estimate of daily maintenance requirement is kcal NE(MP) $= 196 \times$ (live weight in kg)$^{0 \cdot 56}$. To establish total daily requirements, three methods of estimating energy needs for growth were used.

Firstly, daily energy retention as protein and lipid follows a reasonably consistent pattern if expressed per kg live weight. Therefore to the energy of maintenance was added 55, 52, 49, 46 and 43 kcal/kg live weight at 20, 40, 60, 80 and 100 kg live weight.

Secondly, estimates of the energy values per 100 g live weight gain (220, 280, 350, 410, 480 kcal at 20, 40, 60, 80 and 100 kg live weight) were coupled with the daily weight gains shown in Table 1.2 to give daily energy retentions; these were added to the energy of maintenance. To convert requirements for energy into DE of feed it was assumed that ME is 95 per cent of DE and that DE is 77 per cent and 64 per cent utilized for maintenance and growth. As ratios of energy of maintenance to energy of growth increased from 48:52 to 37:63 with increase in live weight, the percentage utilizations of DE were combined in these ratios and found to range only from 70·2 to 68·8. Therefore, a single value for requirement for energy as 70 per cent of DE or, conversely for DE as 143 per cent of requirement, was used in estimating total DE requirements.

Digestible energy requirements determined by these two methods ranged from 87 per cent to 102 per cent of those based on the performance of Progeny Tested pigs over the range 20–80 kg live weight; agreement was closer for pigs of 20–40 kg than at higher weights.

The third method of calculating energy requirements was based on the estimate that 7·51 and 11·65 kcal ME are required/g protein and lipid deposited, respectively; these figures suggest greater efficiency of utilization of ME for growth (79 per cent) than was assumed in the calculations made above (67 per cent). From the daily weight gains shown in Table 1.2 and estimates of percentages of protein and lipid in live weight gains, the ME needs for growth were calculated and added to those for maintenance. Daily ME requirement was then converted to DE, giving figures 80–96 per cent of those in Table 1.2.

Unfortunately none of these calculations estimates the effect of change in energy intake on changes in both rate of weight gain and body composition, which are relations of primary importance in considering practical recommendations. Also, additional energy requirements at environmental temperatures below the critical should be included in a complete model, although such conditions should not prevail in modern feeding practice.

Sows

Growing period to first conception. Much has been published on effects of total feed intake during the pre-pubertal period on age at puberty, subsequent ovulation rate and embryo survival, but very little on specific effects of energy. But as there is considerable measure of agreement between results based on 'planes of feeding' and those specifically from energy differences, the results from all may be taken as pertaining to energy requirements.

Effects on age at puberty have been variable; some workers found no effect from restriction of feed intake to 60–70 per cent of full feed level, others found that such restriction hastened puberty, while yet others obtained a delay in puberty with restriction of energy intake. More severe restriction of energy or total feed intake to 50 per cent of full feed level caused considerable delay in onset of puberty.

Ovulation rate has been generally increased with increase in feed intake, but in two experiments there were more *corpora lutea* found in gilts which had been subject to some degree of energy restriction than in those which had not. Ovulation rate has been increased also by maintaining level of feeding but increasing the energy intake for only 6–14 days before oestrus.

With only two exceptions, all experiments have shown more embryos alive in gilts subject to feed restriction than in those fed *ad lib.* The greatest number of embryos at the 25th day of pregnancy resulted from restricted feeding to puberty, full feeding from puberty to service at second oestrus, then restricted feeding again after conception.

There is little information on the effects of pre-conception nutrition beyond the 25th day of pregnancy, or on subsequent pregnancies and lactations, and such results as we have are confusing; some report larger litters from gilts on restricted feed intakes, some indicate that larger litters were produced by self-fed gilts, while others found no significant difference in size of litter between gilts fed *ad lib.* and those restricted to 50 per cent or even to 33 per cent of *ad lib.* level.

An adverse effect on subsequent fertility and reproductive performance from restricting the feed intake of gilts between the 8th and 12th weeks of age has been reported, but this was not confirmed by the results of another experiment in which rate of growth in early life had no significant effect on reproductive performance.

Pregnancy. Theoretical estimates of energy requirements for pregnancy have been made from metabolism trial data on energy intakes and energy values of products of conception but the validity of such estimates is limited by lack of information on (a) optimum weight gain during pregnancy, and (b) the energy requirement for maintenance in the pregnant animal.

A net weight gain of about 20 kg in addition to the products of conception, appears to be adequate to allow normal reproductive performance, and it is evident that such a gain can be achieved with a lower energy intake in the pregnant than in the non-pregnant sow. A maintenance ration for a non-pregnant sow will allow a pregnant animal of the same initial weight to produce a normal litter and in addition gain some 18 kg live weight.

It appears, therefore, that even at its maximum towards the end of gestation the energy requirement of the pregnant sow is less than 20 per cent above the maintenance need of the non-pregnant animal. Estimates are presented in Table 1.3. No addition is allowed immediately following service as the possible value of this would appear to depend upon the condition of the sow following lactation, but the daily requirement is increased slightly towards term to allow for the greater retention of nutrients by the sow and foetuses during this time.

TABLE 1.3
Estimated daily energy requirements of pregnant sows[1]

Initial live weight	DE for main-tenance	DE for products of conception		Total DE requirement		Diet to supply daily DE requirement[2]	
		weeks 1–12	weeks 13–16	weeks 1–12	weeks 13–16	weeks 1–12	weeks 13–16
kg	Mcal	Mcal	Mcal	Mcal	Mcal	kg dry matter	kg dry matter
135 and under	5·07	0·35	0·80	5·42	5·87	1·6	1·8
160	5·69	0·35	0·80	6·03	6·49	1·8	2·0
180	6·27	0·35	0·80	6·62	7·07	2·0	2·1
205	6·82	0·35	0·80	7·17	7·62	2·2	2·3
230	7·38	0·35	0·80	7·73	8·18	2·3	2·5

[1] No allowance is made for live weight gain as pregnant sows gain appreciable weight on what is a maintenance ration for the non-pregnant animal.
[2] Diet with 3·3 Mcal DE/kg dry matter.

Lactation. The energy requirement of the sow for milk production depends solely on the amount and composition of milk and the gross efficiency with which it is produced. The requirements for dietary energy, however, will depend also on the level of body reserves accumulated during the preceding pregnancy, so it is not possible to state categorical requirements for dietary energy during lactation without reference to the level of feeding employed during pregnancy. Absolute energy requirements for milk production can be calculated (Table 1.4) but these will be also the dietary energy requirement only in the absence of tissue storage during pregnancy.

Voluntary feed consumption and weight change during lactation are greatly influenced by level of feeding and weight gain during pregnancy. The energy requirements presented in Table 1.4 may be regarded as the absolute requirements of sows which have gained little live weight during pregnancy. Sows fed to the level for pregnancy suggested in Table 1.3 would probably make a net gain of some 20 kg live weight and have body reserves to draw upon, and hence would probably fail to consume the full lactation requirement and would lose some weight in consequence. Higher levels than those given in Table 1.4 could be consumed by sows giving above-average quantities of milk, although the desirability of this may be questioned, or by sows which had gained little or no weight during pregnancy and so may be expected to gain some weight during lactation.

The same results in terms of number and size of pigs weaned may be achieved from higher levels of feeding during pregnancy and lower levels during lactation than those suggested, but experimental evidence suggests that the pattern of feed distribution proposed here, in which about one-third of the

TABLE 1.4
Estimated mean daily energy requirements of lactating sows[1]

Litter size	Daily requirement for DE according to live weight[2]			Diet to supply DE requirement according to live weight[3]		
	Live weight			Live weight		
	135 kg	180 kg	225 kg	135 kg	180 kg	225 kg
	Mcal	Mcal	Mcal	kg dry matter	kg dry matter	kg dry matter
3	11·8	13·3	14·8	3·6	4·0	4·5
4	12·9	14·4	15·9	3·9	4·3	4·8
5	14·0	15·5	17·0	4·2	4·6	5·1
6	15·1	16·6	18·1	4·6	5·0	5·5
7	16·2	17·7	19·2	5·0	5·4	5·9
8	17·2	18·7	20·2	5·2	5·6	6·1
9	18·0	19·5	21·0	5·5	5·9	6·4
10	18·3	19·8	21·3	5·6	6·0	6·5
11	18·6	20·1	21·6	5·6	6·0	6·5
12	18·7	20·2	21·7	5·7	6·1	6·6

[1] The figures represent mean requirements of sows which have gained little or no weight during pregnancy. Actual requirements would follow the lactation curve and would be about 20 per cent lower in the first and eighth weeks of lactation and 15 per cent higher in the third, fourth and fifth weeks.

[2] Based on an average daily milk yield of 6·4 kg for a sow suckling 8 pigs, a mean gross energy value for sow's milk of 1·25 kcal/g, a gross efficiency of energy conversion of 45 per cent, and a conventional requirement for maintenance.

[3] Diet with 3·3 Mcal DE/kg dry matter.

total consumption is given during pregnancy and about two-thirds during an 8-week lactation, allows the more efficient use of feed energy. This pattern of feeding would fit particularly well into any system in which the pigs were weaned earlier than at 8 weeks, as reduction in fertility due to obesity may be a problem where considerable weight gain is allowed during pregnancy and where the lactation period is curtailed.

Boars

Little is known of the effect of energy intake on the development and reproductive function of boars. Retardation of growth by a low level of feeding has delayed development of testes and germinal epithelium and the first appearance of active spermatozoa but, on the other hand, a high level of feeding may be associated with the development of leg weakness.

Between boars reared on the recommended feed allowance of the (USA) National Research Council, or 70 or 50 per cent of that level, there was a significant difference in weight, but not in age, at puberty, and a reduction in semen volume per ejaculate but not in fertility. Similarly, no differences in fertility were obtained between mature boars fed for 15 months so as to gain either 136 or 39 kg live weight or to lose 60 kg, but those on the lowest feed intake (2 kg feed/day reduced to 1·5 kg after six months) eventually became lethargic and incapable of service.

SECTION 2

REQUIREMENTS FOR PROTEIN

The requirements of growing pigs from birth to slaughter and of breeding sows and boars have been estimated following a critical survey of the literature, but for pigs from 8 weeks of age to 90 kg live weight an attempt was also made to assess requirements from an analysis of nitrogen balance data. This study emphasized the importance of essential amino acids as well as protein, but the data were inadequate to permit reliable estimates of requirements to be made. Furthermore there is little consistent evidence of the extent to which maximum nitrogen retention at any stage of growth is related to carcass quality at slaughter.

The Young Pig from Birth to 8 Weeks of Age

As there is very little transfer of antibodies through the placenta, the newborn pig requires dietary protein not only as a source of essential amino acids but also for the conference of passive immunity through the globulins which it contains. The absorption of antibodies by pigs that have received feed from birth, normally takes place during the first 24–36 hours of life, although some absorption may occur up to about 3 days of age. As a result of protein absorption from colostrum the serum protein pattern of the pig changes rapidly during early life towards that of the mature animal.

The ability of the young pig to synthesize its own antibodies develops some time between 3 and 6 weeks of age, so a critical period of susceptibility to infection may occur during this period when colostral acquired immunity is declining and the active production of antibodies is not fully developed. The best alternatives to sow's colostrum for normal maturation of the serum profile appear to be bovine colostrum or equine serum; parenteral administration of gamma globulins is without value, although oral administration over an extended period has given an improvement in survival rate.

Complete proteins appear to be also of nutritional significance in the early life of the pig, in that benefit has been derived from natural milk in preference to synthetic milk of equivalent amino acid value. The difference may be explained on the grounds of the ability of the complete proteins to clot in the stomach, so increasing the efficiency of their utilization through a reduced rate of passage.

Estimates of the protein requirements of young pigs vary widely, due mainly to differences between test diets in digestibility, energy value and content of amino acids.

The digestibilities of milk protein and fish meal are high, over 90 per cent, and change little with increase in age of pig, but the digestibilities of less digestible proteins, from cereals and soyabean meal, increase between 2 and 8 weeks of age.

The energy value of the diet influences response to protein level and the addition of fat increases the level of protein required in the diet. Protein quality is also more critical at high energy intakes.

Quality of protein has considerable influence on the performance of young pigs and milk proteins appear to be the best single source; however, a combination of milk and fish meal proteins has given better results than either alone. Because of the wide differences between proteins in their value to the young pig, protein requirements cannot be expressed with any accuracy without reference to the source. However, as most diets for young pigs contain milk products, fishmeal and/or soyabean meal as major protein sources, estimates of requirement based on diets of this type may be of practical value and are presented in Tables 2.1, 2.2 and 2.3.

Estimates of amino acid requirements are also given in the tables; they have been derived from numerous sources and some, due to lack of published data on a particular amino acid for a particular weight of pig, have been derived by calculation.

The protein content of creep feeds presents a special case in that these are not sole diets but merely supplements to sow's milk. As in the second half of an 8-week lactation the protein in the dry matter of sow's milk exceeds the estimated requirements of the pig, the optimal protein level in a creep feed fed during this time is less than that of an early-weaning diet, perhaps 17 per cent of the dry matter, and protein quality will be less critical.

TABLE 2.1
Estimates of requirement of pigs weighing 1·4–4·5 kg for crude protein and essential amino acids

Amino acid (as the L-isomer)	Estimated		Calculated[1]	
	Per Mcal DE, g	In the dry matter in the diet, per cent	Per Mcal DE, g	In the dry matter in the diet, per cent[2]
Lysine	—	—	4·8	2·2
Methionine + cystine	2·2[3]	0·87	2·9	1·3
Tryptophan[4]	0·5	0·19	0·6	0·3
Histidine	—	—	1·0	0·5
Tyrosine + phenylalanine	—	—	2·2	1·0
Threonine	1·9	0·95	1·9	0·9
Leucine	2·6	1·32	3·0	1·4
Isoleucine	—	—	2·9	1·3
Valine	—	—	2·1	1·0
Crude protein	63	32	—	—

[1] Requirements were calculated by increasing the estimates of requirement (per Mcal DE) of 9–20 kg pigs by 60 per cent.
[2] Dry matter 95 per cent; DE 4580 kcal/kg dry matter in the diet.
[3] Level in a satisfactory diet in which methionine + cystine were probably limiting amino acids.
[4] With diets containing adequate nicotinic acid.
NOTE: For methionine and possibly also for phenylalanine, the D-isomer is believed to be utilized as efficiently as the L-form; the D-isomer of tryptophan appears to have only a part of the value of the L-form, while for the other amino acids shown in the table it is assumed that the D-forms are without value for the pig.

PIGS FROM 8 WEEKS OF AGE TO 90 KG LIVE WEIGHT

An examination of the literature relevant to the protein requirement of the pig revealed that growth rate and feed conversion efficiency were the most widely used criteria. In general terms carcass quality was frequently assessed but a great variety of methods of assessment were used such as the composition of selected cuts, specific gravity, and linear measurement of the depth of fat at selected points. Many papers also reported the results of nitrogen balance experiments carried out at various stages in the growth of the pig.

Almost all experiments directed towards assessing the protein requirements of the pig have been aimed at determining a difference in response to different concentrations of protein. Since maximal response appeared to be approached

TABLE 2.2

Estimates of requirement of pigs weighing 4·5–9·0 kg for crude protein and essential amino acids

Amino acid (as the L-isomer)	Estimated		Calculated[1]	
	Per Mcal DE, g	In the dry matter in the diet, per cent	Per Mcal DE, g	In the dry matter in the diet, per cent[2]
Lysine	3·0	1·11	4·2	1·5
Methionine + cystine	2·0	0·78	2·5	0·9
Tryptophan[3]	—	—	0·6	0·2
Histidine	0·5	0·21	0·8	0·3
Tyrosine + phenylalanine	—	—	2·0	0·7
Threonine	—	—	1·7	0·6
Leucine	—	—	2·7	1·0
Isoleucine	—	—	2·5	0·9
Valine	—	—	1·8	0·7
Crude protein	66	26	—	—

[1] Requirements were calculated by increasing the estimates of requirement (per Mcal DE) of 9–20 kg pigs by 40 per cent.
[2] Dry matter 90 per cent; DE 3612 kcal kg dry matter in the diet.
[3] With diets containing adequate nicotinic acid.
See note at foot of Table 2.1.

TABLE 2.3

Estimates of requirement of pigs weighing 9–20 kg for crude protein and essential amino acids

Amino acid (as the L-isomer)	Per Mcal DE, g	In the dry matter in the diet per cent[1]
Lysine	3·0	1·1
Methionine + cystine	1·8	0·7
Tryptophan[2]	0·4	0·15
Histidine	0·6	0·2
Tyrosine + phenylalanine	1·4	0·5
Threonine	1·2	0·5
Leucine	1·9	0·7
Isoleucine	1·8	0·7
Valine	1·3	0·5
Crude protein	52	20

[1] Dry matter 87 per cent; DE 3832 kcal/kg dry matter in the diet.
[2] With diets containing adequate nicotinic acid.
See note at foot of Table 2.1.

asymptotically, it was more difficult to distinguish between responses, the nearer the protein concentrations were to the 'requirement', i.e. the concentration associated with maximal response. Two further limitations were imposed by the data, the first that it was not possible to take breed and sex differences into account, the second that in most experiments changes in protein concentration were inevitably accompanied by changes in protein quality. These and similar limitations described in the Technical Review should be borne in mind when the estimates of requirements are considered.

From a consideration of the literature, it appears that as far as rate of gain and feed conversion efficiency are concerned, the requirement for crude protein up to about 40–50 kg live weight is about 18·5–20 per cent of the dry matter, and from this weight to slaughter it is about 15–16·5 per cent of the dry matter. There is also evidence that the energy content of the diet affects the response at different dietary protein concentrations, but it was not possible from the data to resolve the interaction between energy and protein in the effects on gain or feed conversion efficiency. The importance of protein quality, in particular the lysine concentration, was evident in several papers. Only with protein concentrations of between about 20–23·5 per cent of the dry matter of the diet were there consistent improvements in carcass quality with increase in dietary protein concentration within this range.

The nitrogen balance data emphasized the importance of protein quality, particularly with regard to its lysine content, but the experiments reported were not suitable for the examination of interactions between different nutrients, as for example, between the dietary concentrations of lysine and of total protein, in their effects on the responses measured.

A further section of the Technical Review reports a survey of papers dealing with the response of the pig to essential amino acids. Insufficient information was found on which to estimate the requirements of the pig for arginine,

TABLE 2.4
Estimates of protein and essential amino acid requirements of pigs between approximately 8 weeks of age and 90 kg live weight

Amino acid (as the L-isomer)	Requirement as a percentage of the dry matter in the diet	
	Up to 45–50 kg live weight	From 50–90 kg live weight
Lysine	0·9–0·95	0·7–0·75
Methionine + cystine	0·6–0·7	[1]
Tryptophan	0·15–0·2	[1]
Threonine	0·5–0·6	[1]
Isoleucine	0·75	[1]
Crude protein	18·5–20	15–16·5

[1] Although it is likely that for some amino acids the requirements may be less at the later stage of growth, there is insufficient evidence on which to specify requirements lower than those given for pigs up to 45–50 kg live weight.

See note at foot of Table 2.1.

histidine, phenylalanine, leucine or valine. The estimated requirements for the other essential amino acids are given in Table 2.4, along with those for crude protein.

In conclusion, it is pointed out that there is a need for more experiments to examine the interactions between different nutrients in their effects on the growth, feed conversion efficiency and carcass composition of the pig, and also for experiments designed to determine response curves, rather than to estimate those concentrations of protein which provide maximal response. The reasons for these suggestions are that maximal response is very ill-defined in practice, that the response to any particular nutrient is likely to depend on the concentrations of others in the diet and that only response curves of this type can provide the basis for the type of information which is required at the practical level.

Sows

The growing period to first conception

There is no evidence to indicate that the protein requirements are any greater than those of the bacon pig, and since carcass leanness, within limits, is of less significance in future breeding stock, the requirements may be based entirely on growth rate and feed conversion efficiency.

Pregnancy

Balance trial data suggest that a daily intake of 250 g crude protein/day

TABLE 2.5
Estimated protein requirements of pregnant sows[1]

Stage of pregnancy	Daily crude protein requirement[2]
week	g
1–11	250
12–14	325
15 and 16	400

[1] Based on the assumption that an increase in nitrogen retention is desirable in the last third of pregnancy.

[2] No differentiation is made between gilts and sows for daily protein requirement as the growth requirement of the former is offset by the increased maintenance requirement of the latter.

In applying the figures in this table, the proviso is made that the diet should contain not less than 14 per cent and not more than 21 per cent crude protein in the dry matter.

is adequate over the greater part of pregnancy but that nitrogen retention increases markedly during the last month if up to 400 g crude protein/day is given. The value of this increased retention is not clear and although sows on feeding trials have produced normal litters on intakes of less than 200 g/day through the gestation period, the long-term effects of such low protein intakes have not been assessed. The significance of amino acid balance for pregnant sows is not known but intakes of 250 g crude protein and upwards have allowed positive nitrogen balance and the production of normal litters even when derived from poor quality sources, such as cereals.

Lactation

A sow of average milking ability suckling 8 pigs appears to require on average about 1000 g crude protein/day. For maximum milk production throughout lactation, which may or may not be desirable, the actual requirement would be some 20 per cent less in the first and eighth weeks and some 20 per cent more in the fifth and sixth weeks. Variations in requirement due

TABLE 2.6
Estimated protein requirements of lactating sows

Week of lactation	Average daily milk yield[1]	Average protein content of milk	Output of milk protein	Daily crude protein requirement[2]
	kg	per cent	g/day	g/day
1	5·2	5·5	286	858
2	7·0	5·0	350	1050
3	7·3	5·3	387	1161
4	7·3	5·4	394	1182
5	7·3	5·7	416	1248
6	7·0	5·9	413	1239
7	5·5	6·2	341	1023
8	4·8	6·6	317	951
Average	6·4	5·7	363	1089

[1] For a sow suckling 8 pigs.

[2] Assuming a gross conversion efficiency of 33 per cent for dietary crude protein into milk protein.

As a guide, the *average* requirements of a sow based on the above figures and given the feed intakes indicated in Table 1.4 in the Energy section would be met by a diet containing approximately 18·5 per cent crude protein in the dry matter.

to differences in litter size and weight of sow may be met by altering the quantity of feed and maintaining crude protein content constant, but variation in requirement due to advance in lactation cannot be allowed for directly in this way as the requirement for protein declines less rapidly than the requirement for energy. There is insufficient evidence, however, to justify increasing the protein content of the feed, or feeding at above the level determined by energy requirement, in order to allow for the rise in milk protein content with advance in lactation.

Estimated protein requirements of pregnant and lactating sows are given in Tables 2.5 and 2.6 respectively.

Boars

There appears to be no published information on the effect of protein intake on the reproductive function of boars. There is some indication, however, that the protein requirements of the boar for maximum growth rate, which may or may not be desirable, is greater than that of gilts or castrates.

SECTION 3

REQUIREMENTS FOR MINERALS

The many experiments in which the mineral needs of pigs have been explored, though extensive, are far from comprehensive. Considerable controversy exists on all aspects of mineral nutrition and it is difficult to find information on which specific recommendations might be based. Results are very often complicated by differences in experimental conditions. Thus, differences in the composition of basal diets, breed, environmental effects and the criteria used to assess results may all confuse the interpretation of different experiments. The subject is further complicated by lack of knowledge of the interactions between various minerals and other constituents of the diet. Further difficulties arise because the standards of expected performance have been continually rising, thus making some of the earlier recommendations possibly obsolete.

The requirements of pigs for minerals are given in Table 3.1. The estimates were arrived at after critical assessment of information found in the literature which included, in some cases, data on the storage of minerals in the body, on endogenous losses and on availability of minerals from different sources. Where possible, the material was sub-divided to cover different classes of pigs. Hardly any information was found on the mineral requirements of boars and no estimates of requirement for this class of stock could be made.

Calcium

The availability of calcium is known to vary with the source, and there is evidence to show that the percentage calcium absorbed from any particular

TABLE 3.1

Summary of the estimates of requirements for minerals expressed as a concentration of the dry matter in the diet and derived from estimates of requirements in the literature

Mineral	Live weight or class of pig[1]	Estimated requirement	Range of individual estimates of requirement in the literature[2]	Toxic level
Calcium	Up to 20 kg	0·8%	0·2–1·3%	1·2% when zinc content of diet is less than 50 mg/kg dry matter[3]
	20–55 kg	0·8%	0·3–0·8%	
	55–90 kg	0·6%	0·3–0·8%	
	Sows	0·6%	0·4–1·0%	
Phosphorus	Up to 20 kg	0·7%	0·4–1·1%	
	20–55 kg	0·6%	0·3–0·6%	
	55–90 kg	0·5%	0·3–0·6%	
	Sows	0·5%	0·3–0·6%	
Chlorine	20–35 kg	0·15%		2·0% sodium chloride
Sodium	20–35 kg	0·10%		
Potassium	Up to 45 kg	0·25%	0·08–0·28%	
Magnesium	Up to 55 kg	400 mg/kg	300–2000 mg/kg	
Iron	Up to 20 kg	60 mg/kg	60–125 mg/kg	5 g/kg
Zinc	Up to 90 kg	50 mg/kg	14–80 mg/kg	2 g/kg
Manganese	Up to 90 kg	Evidence for a dietary requirement, but insufficient data on which to base an estimate		1 g/kg
	Sows	10 mg/kg	6–12 mg/kg	
Copper	Up to 90 kg	4 mg/kg		500 mg/kg
Cobalt	Essential as a component of vitamin B_{12}, but no other dietary requirement established			
Iodine	Evidence for a dietary requirement, but insufficient data on which to base an estimate			
Selenium	No dictary requirement established			7 mg/kg
Fluorine	No dietary requirement established			250 mg/kg

[1] For live weights or class of pig outside those stated, there were insufficient data on which to base an estimate of requirement. It may be reasonably assumed, however, that the requirements are not higher than those given for the lower weights of growing pigs.
[2] Percentage in air-dry diet.
[3] Not a *toxic* level but may interfere with the metabolism of zinc and other minerals.

diet falls progressively with increase in the age of the pig. The amount of the element required by the pig also depends upon the presence in the feed of other dietary factors, notably phosphorus and vitamin D, so adequate amounts of these nutrients must be offered. The daily requirement to ensure maximum growth rate is considerably below that needed to ensure good bone formation.

The recommendations for young pigs apply to diets in which the availability is about 80 per cent, those for growing pigs assume a value of between 70 per cent at 20 kg live weight and 45 per cent at 90 kg live weight. Comparatively little is known about the requirements of the sow for calcium, though satisfactory results are obtained by feeding diets of the type suitable for pigs between 45 and 90 kg live weight. These diets lead to good bone formation in the progeny and ensure satisfactory lactation without imposing too great a strain on the sow's own mineral reserves.

An excess of calcium is undesirable, as it can interfere with the metabolism of other minerals, e.g. zinc and iron (see below).

Phosphorus

Phosphorus requirements are closely associated with those for calcium. Availability is known to depend on the chemical form of the phosphorus in the diet; of particular importance is the availability of phytate phosphorus. Indications are that it is somewhat less available than most inorganic sources of phosphorus. An excessive amount of phytic acid should be avoided as this substance can reduce zinc absorption (see below). The amounts of phosphorus required for good bone quality, reproductive performance and lactation are higher than those needed to obtain maximum growth rate.

An average availability of 70 per cent has been assumed for phosphorus in early weaning diets for young pigs and a value of 50 per cent in practical cereal-based feeds as supplied to growing pigs, sows and boars.

Sodium and Chlorine

Lack of data makes it extremely difficult to estimate the requirement with confidence. However, the levels suggested are believed to be adequate for young pigs. Both elements appear to be highly available to pigs of about 25 kg live weight, for which values of about 90 per cent have been reported; availability to heavier pigs has not been adequately established.

There are indications that signs of salt toxicity may arise if diets with more than 2 per cent sodium chloride are fed, particularly if water intake is restricted.

Potassium

It is unlikely that a pig receiving diets compounded from conventional feeding stuffs will ever become depleted of potassium, as practical diets contain adequate amounts of this element. Again it is difficult to assess the requirements of the pig for potassium, as there is very little exact information on which to base them, but evidence gained from experiments on pigs of up to about 25 kg live weight indicates that the feed should contain about 0·25 per cent potassium. There is some evidence that the requirement may be relatively lower for heavier pigs, but it is insufficient to warrant a lower recommendation.

Magnesium

It has been shown that about 400 mg magnesium/kg dry matter in the diet is required for young growing pigs to ensure adequacy when the availability of the element is about 80 per cent.

Iron

Current information indicates that about 60 mg iron/kg dry matter in the diet is adequate for young pigs to ensure good growth and satisfactory haemoglobin level in blood. It is important that the small amount of copper necessary to ensure the efficient utilization of iron is included (see below). This estimated requirement is intended to ensure haemoglobin levels of approximately 8 g/100 ml blood, but will not make good the iron stores of anaemic pigs. If a higher haemoglobin level than 8 g/100 ml blood is desired, the amount of iron required will be proportionally greater.

There is good evidence to show that the availability of iron for young pigs depends upon the chemical form in which it is supplied, and the presence of excessive calcium or manganese can reduce the amount absorbed. An average value of 60 per cent has been assumed for the level of supplementation recommended.

While it is probable that the level could be reduced for growing pigs and sows, there is no direct evidence indicating the requirements of these classes of pig for iron.

Zinc

Factors affecting the metabolism of zinc include calcium, phytic acid and vitamin D levels of the diet, and possibly protein source. An increase in the first three factors reduces the availability of zinc. 50 mg zinc/kg dry matter in the diet of all classes of pig should provide protection against parakeratosis.

The toxic effects of zinc have received some consideration, and the toxic level is placed at 2 g/kg dry matter in the diet.

Manganese

It seems that as little as 0·5 mg manganese/kg dry matter in the diet may be sufficient to maintain good growth and feed conversion efficiencies in pigs from weaning to 90 kg live weight. However, this low level has in some cases led to abnormal carcass development. Also, low levels of manganese fed to gilts from weaning through to farrowing have caused poor farrowing performance. These facts indicate that a higher level than 0·5 mg manganese/kg dry matter in the diet should be fed. A level of 40 mg/kg dry matter in the diet has been shown to be satisfactory in all respects, but it is probable that this is an over-generous amount. Unfortunately, there is no clear indication from published papers as to where within the range 0·5 to 40 mg manganese/kg dry matter in the diet the requirement lies. For this reason no estimate of requirement has been made for growing pigs.

Diets for sows containing less than 3 mg manganese/kg have caused a reduction in the number of pigs born alive, lowered birth weights, and depressed the sow's milk yield compared to diets containing supplementary manganese. There are two reports indicating that 6 or 12 mg/kg diet is adequate for breeding sows, and the suggested requirement for this class of stock is 10 mg manganese/kg dry matter in the diet.

Excessive amounts of manganese have toxic effects. Growth rate has been depressed in some instances and it has been shown that manganese can interfere with haemoglobin synthesis. It is proposed that the concentration of manganese in the dry matter of diets should not exceed 0·1 per cent.

Copper

The relation between copper, iron and nutritional anaemia is well established. The level recommended is intended to satisfy the requirements of the pig for copper as a trace mineral; the growth-promoting effects of high-level feeding are not considered here. There is some information on copper availability; copper sulphate, copper carbonate and copper oxide appear to be equally available to young pigs.

The level at which toxic effects become apparent is hard to place. In this country levels of 250 mg copper/kg diet are fed routinely without ill effect; it would be unwise to exceed 500 mg copper/kg dry matter in the diet.

Cobalt

Considerable doubt exists about the function of cobalt in pig nutrition. While there can be no question of the animal's requirement for the cobalt of vitamin B_{12}, no other requirement for cobalt has been established.

There are, however, several reports in the literature suggesting that under certain conditions, often when the animal protein content of the diet is low, there may be an advantage from adding cobalt to the diet. It seems likely that such diets are low in vitamin B_{12}, and one hypothesis to explain the observed effects is that the addition of cobalt increases synthesis of this vitamin by intestinal microorganisms, and that growth is improved as the supply of the vitamin increases.

Iodine

The iodine requirements of the pig have not been measured with any accuracy, so no estimate of these is made.

Selenium

The relation between selenium and vitamin E has not been fully elucidated, so no definite requirement can be suggested. At levels above 7 mg/kg dry matter in the diet, selenium has been shown to have toxic effects.

Fluorine

Studies have been directed to an investigation of the toxicity of this element. Though the results are inconclusive, it would appear unwise to exceed 250 mg fluorine (as fluoride)/kg dry matter in the diet. There is no direct evidence that the element is essential for pigs.

SECTION 4

REQUIREMENTS FOR VITAMINS

The estimates of the requirements for vitamins have been based on available information in the literature. In common with other nutrients, the survey disclosed considerable gaps in knowledge on vitamin requirements and serious shortcomings in some of the experimental data available. It is essential that such limitations, which have been considered in detail in the Technical Review, should be borne in mind when considering the summarized estimates of requirement that follow. Hardly any information was found on the vitamin requirements of boars and no estimates of requirement for this class of stock could be made.

FAT-SOLUBLE VITAMINS

The estimates of requirements for vitamins A, D, E and K are given in Table 4.1, as a concentration of the dry matter in the diet. In the case of vitamins A and K, requirements are also given per unit of live weight.

Vitamin A

The requirement for vitamin A may be met by supplying either vitamin A or one of the provitamins, for example, β-carotene. The apparent efficiency of conversion of β-carotene to vitamin A depends on the criterion used to judge the adequacy of vitamin A status in the pig, and it decreases as increments are given above the amount which the pig actually requires. The conversion appears to be most efficient when growth is used as the criterion, but the adequacy of vitamin A status can only be judged correctly by reference to liver and blood stores and the efficiency of the conversion of β-carotene to vitamin A for the maintenance of blood and liver stores is much lower. Hence, for calculating the vitamin A potency of β-carotene, values of 6 μg carotene=1 μg vitamin A have been assumed. Thus, 1 i.u. vitamin A alcohol=1·8 μg β-carotene (see Glossary).

In general the vitamin A requirement has been assessed using the depletion-repletion technique and it has been assumed that requirement is proportional to live weight. The values therefore depend to a large extent on the length of the repletion period, and the net change in live weight over the assay.

For growing pigs up to 90 kg live weight, experiments designed to determine the daily requirement for β-carotene show that approximately 30 μg β-carotene/kg live weight are required while those designed to determine the daily requirement of vitamin A indicate that 18 μg vitamin A/kg live weight (60 i.u. vitamin A/kg live weight) are needed; thus the conversion of β-carotene to vitamin A on a weight basis appears to be better than 2:1. In these experiments, weight gain was the main criterion for which, as mentioned above, the efficiency of conversion of carotene to vitamin A is high. In some experiments liver storage was examined, but the number of animals used was small. The requirement has, therefore, been given as 60 i.u. vitamin A/kg live weight daily and the values for β-carotene shown in Table 4.1 have been calculated using the conversion ratio of 1 i.u. vitamin A = 1·8 μg carotene.

The data available on the requirements of the breeding sow are limited, but it is suggested that a daily intake of 40 i.u. vitamin A/kg live weight would cover requirements without depleting liver reserves of pregnant sows. The requirement for lactating sows has been calculated by increasing the requirement for pregnancy by 4 i.u./kg live weight daily, which would make good the loss in milk assuming the efficiency of utilization of dietary vitamin A for milk vitamin A is 100 per cent.

Vitamin D

The requirements for vitamin D have generally been determined by the addition of graded levels of vitamins D_2 and D_3, which have the same activity per international unit, to fixed amounts of diet. A requirement for growing

TABLE 4.1

Summary of the estimates of requirements for the fat-soluble vitamins

Vitamin	Live weight or class of pig	Estimated requirement: Per kg dry matter in the diet[1]		Estimated requirement: Per kg live weight daily	
		As vitamin A	As β-carotene[2]	As vitamin A	As β-carotene[2]
A	Up to 20 kg	1400 i.u.	2·5 mg	60 i.u.	108 μg
	20–55 kg	1600 i.u.	2·9 mg		
	55–90 kg	1900 i.u.	3·4 mg		
	Pregnant sows:				
	Up to 160 kg	3600 i.u.	6·5 mg	40 i.u.[3]	72 μg
	160–230 kg	4000 i.u.	7·2 mg		
D	Up to 90 kg	230 i.u. vitamin D			
	Sows	No data available to use as a basis for estimate			
E	Up to 90 kg	1·5 mg total tocopherols			
	Sows	Insufficient data on which to base an estimate			
K	0–5 kg	62 μg vitamin K		5 μg	
	Older pigs and sows	No data available to use as a basis for estimate			

[1] Converted where necessary from the amount/kg live weight daily using the daily dry matter intakes given in Table 1.2 for growing pigs and in Table 1.3 for pregnant sows.

[2] Calculated assuming that 1 i.u. vitamin A alcohol = 1·8 μg β-carotene.

[3] For lactating sows the estimated requirement/kg live weight daily is 44 i.u. vitamin A but no increase in the concentration of the vitamin in the diet is necessary as the increased levels of feed given during lactation would amply cover this increase in daily requirement.

pigs up to 90 kg live weight of 230 i.u. vitamin D/kg dry matter in the diet is given. There is some evidence that pigs with continued access to sunlight have a reduced dietary requirement. Growth rate, and therefore energy intake, influence requirement and it is important that the diet should contain the correct concentration of calcium and phosphorus. There were insufficient data on which to base an estimate of requirement of breeding sows.

Vitamin E

There is a wide variation in the vitamin E activity of the various tocopherols and selenium. Muscular dystrophy induced by feeding 'toxic' grain and reported in field outbreaks has been prevented by feeding tocopherol or selenium. Cod liver oil was also shown to induce muscular dystrophy.

From the limited evidence available, an estimate of requirement for growing pigs up to 90 kg live weight of 1·5 mg total tocopherols/kg dry matter in the diet is given as being sufficient to prevent signs of vitamin E deficiency. It is suggested that 8 mg total tocopherols/kg dry matter in the diet would be required to ensure freedom from muscular dystrophy in the presence of unsaturated fatty acids such as would be provided by 5 per cent cod liver oil, and a level of 46 mg total tocopherols/kg dry matter in the diet would be required to ensure complete protection against oxidation of body fat after slaughter.

Insufficient data are available on which to base an estimate of requirement for sows, but it would appear that their requirements are higher than those of growing pigs.

Diets used in practice for growing pigs are unlikely to be deficient in vitamin E unless fat which is rich in unsaturated fatty acids is included, but it cannot be said with certainty that dietary supplements of tocopherols are not required for sows.

Vitamin K

The estimated requirement for newly born pigs up to 5 kg live weight of 5 μg vitamin K/kg live weight is based on the results of one experiment. No work on the requirements of older pigs or of sows has been published.

WATER-SOLUBLE VITAMINS

The estimates of requirements for the eleven water-soluble vitamins as a concentration of the dry matter in the diet are given in Table 4.2, together with the range of individual estimates found in the literature.

In a large majority of the experiments dealing with water-soluble vitamin requirements, the criteria of growth rate, feed conversion efficiency and absence of clinical deficiency signs have been used to assess requirements. Due attention has been paid to those cases where other criteria have been considered, particularly when the requirement suggested by such criteria differed markedly from that indicated as needed for optimum growth.

Thiamine

The estimated requirement for growing pigs up to 90 kg live weight of 1·5 mg thiamine/kg dry matter in the diet is the mean of the range of 1·0 to 2·0 mg/kg dry matter suggested by the results of those experiments considered to be the most reliable. On high-fat diets the requirement for thiamine may be

TABLE 4.2

Summary of the estimates of requirements for the water-soluble vitamins (per kg dry matter in the diet)

Vitamin	Live weight or class of pig	Estimated requirement	Range of individual estimates of requirement in the literature
Thiamine	Up to 90 kg	1·5 mg	0·55 to 4·5 mg
	Sows	No data available to use as a basis for estimate	
Riboflavine	Up to 90 kg	2·5 mg	0·52 to 5·3 mg
	Sows	3·0 mg	2·1 to 3·2 mg
Nicotinic acid ("free")	Up to 20 kg	20·0 mg[1]	15·8 to 26·3 mg
	20–90 kg	12·0 mg[2]	2·2 to 13·9 mg
	Sows	No data available to use as a basis for estimate	
Pantothenic acid	Up to 90 kg	10·0 mg	4·0 to 36·7 mg
	Sows	10·0 mg	5·7 to 13·7 mg
Pyridoxine	Up to 90 kg	2·5 mg	0·72 to 5·5 mg
	Sows	1·0 mg	1·1 mg
Vitamin B_{12}	Up to 20 kg	18·0 μg	5·3 to 53·7 μg
	20–90 kg	10·0 μg	0·91 to 16·9 μg
	Sows	15·0 μg[3]	12·3 to 126·0 μg
Biotin	Up to 20 kg	May be a dietary requirement but insufficient data on which to base an estimate	
	20–90 kg	No dietary requirement established	
	Sows	No dietary requirement established	
Folic acid and *p*-aminobenzoic acid	Up to 90 kg	No dietary requirement established	
	Sows	May be a dietary requirement but insufficient data on which to base an estimate	
Inositol	Up to 90 kg	No dietary requirement established	
	Sows	No dietary requirement established	
Choline	Up to 90 kg	870 mg[4]	435 to 870 mg
	Sows	May be a dietary requirement but insufficient data on which to base an estimate	
Vitamin C	Up to 90 kg	No dietary requirement established[5]	
	Sows	No dietary requirement established[5]	

[1] Based on diets containing 0·27% tryptophan or less in the dry matter.
[2] Based on diets containing 0·21% tryptophan or less in the dry matter.
[3] Some evidence that a ten-fold increase in this level might be advantageous but further work is required.
[4] Based on a diet containing 0·8% methionine in the dry matter.
[5] The reservation is made that the possibility of a dietary requirement under certain stress conditions requires further study.

reduced by as much as 20 per cent. It has also been suggested that the requirement may, if anything, be reduced at low environmental temperatures.

Although there are no data available on which an estimate of the requirement for sows may be based, evidence from work with other animals suggests it is unlikely to differ greatly from that required by the growing pig.

Riboflavine

The estimated requirement for growing pigs up to 90 kg live weight of 2·5 mg riboflavine/kg dry matter in the diet is the mean of the range of 2·0 to 3·0 mg/kg dry matter indicated by the results of a majority of the experiments available.

Riboflavine requirements may be increased at low environmental temperatures, but the evidence is not considered to be sufficiently substantiated to permit requirements to be related to this variable. There is limited evidence that riboflavine requirements may be increased as the protein, and possibly also the fat, content of the diet are raised.

The requirement of 3·0 mg riboflavine/kg dry matter in the diet for sows is based on the results of one experiment only.

Nicotinic Acid

It is now established that almost all of the nicotinic acid present in maize, and in many other cereals, is in a bound form which is completely unavailable to the pig, and that tryptophan in excess of minimum requirements for protein synthesis can be converted by the pig into nicotinic acid. Estimates of requirements are given in terms of free (or available) nicotinic acid and are based on diets containing a maximum of 0·27 per cent tryptophan in the dry matter for pigs up to 20 kg live weight and of 0·21 per cent for pigs above this weight. If diets containing significantly higher levels of tryptophan are fed, the dietary nicotinic acid requirement will be reduced.

The estimated requirement for growing pigs up to 20 kg live weight of 20·0 mg nicotinic acid/kg dry matter in the diet is based on the results of two experiments in both of which relatively high fat basal diets were used. The requirement may be reduced if a lower energy diet containing less fat is fed.

The estimated requirement of 12·0 mg nicotinic acid/kg dry matter for growing pigs of 20 to 90 kg live weight is based on the results of three experiments considered to be the most reliable, suggesting a requirement lying within the range of 9 to 14 mg/kg dry matter. The requirement may be reduced by about 50 per cent at the higher live weights in the range, and there is some inconclusive evidence that it may also be slightly reduced in colder environments. No data are available on which an estimated requirement for breeding sows can be based.

Pantothenic Acid

The estimated requirement of 10·0 mg pantothenic acid/kg dry matter in the diet for growing pigs is the mean of the range of from 9·0 to 11·0 mg/kg suggested by a majority of the experimental results. There is evidence from one experiment that for optimum nitrogen retention the requirement may be about 20 per cent higher; the requirement may also be increased by some 70 per cent when a high-fat diet is fed.

It is possible that requirements may be reduced when a high-protein diet or one containing large amounts of vitamin C is fed, or when the animals are kept in a colder environmental temperature, for example at a temperature of 4°C instead of 29°C. The addition of vitamin B_{12} or an antibiotic to the diet may have a sparing effect on pantothenic acid requirements, but the evidence is conflicting.

The estimated requirement of 10·0 mg pantothenic acid/kg dry matter in the diet for breeding sows is the mean of the range of approximately 6·0 to 14·0 mg/kg dry matter suggested by the results of the three experiments reported.

Pyridoxine

Much of the available data suggests that for growing pigs the requirement is about 1·0 mg pyridoxine/kg dry matter in the diet. Evidence from a number of sources suggests, however, that for optimum nitrogen retention the requirement may be nearer 2·5 mg/kg dry matter and this figure has been tentatively given as the estimated requirement.

Recent evidence suggests that the level of dietary fat *per se* does not affect pyridoxine requirements.

The estimated requirement of 1·0 mg pyridoxine/kg dry matter in the diet for breeding sows is based on the results of one experiment only, which was of short duration and in which the animals used were not depleted of the vitamin prior to starting on the experiment.

Vitamin B_{12} (Cobalamin)

It is apparent that a dietary source of this vitamin is not always required; the needs presumably are fully met from other sources such as initial body reserves, presence in the environment of sources of the vitamin (for example, in moulds, soil, animal excreta) and bacterial synthesis in the intestinal tract.

A definite need for a dietary supply of the vitamin under some conditions is, however, well established.

For growing pigs up to 20 kg live weight, most of the available data suggests a requirement of between approximately 15 and 21 μg/kg dry matter, and the estimated requirement given of 18 μg vitamin B_{12}/kg dry matter in the diet, is the mean of this range.

The estimated requirement of 10·0 μg vitamin B_{12}/kg dry matter in the diet for growing pigs from 20 to 90 kg live weight is based on the range of approximately 5 to 17 μg/kg dry matter suggested by the experimental results available.

The addition of pantothenic acid or of an antibiotic to the diet may spare the requirements for vitamin B_{12} but the evidence for this is inconclusive. It is generally accepted that dietary requirements for vitamin B_{12} are reduced if the diet contains a good supply of compounds which can act as sources of methyl groups, such as choline and methionine.

The estimated requirement of 15μg vitamin B_{12}/kg dry matter in the diet for breeding sows is based on very limited data. It has been suggested that as much as a ten-fold increase in this level for optimum performance of the breeding sow is required, but confirmation of this evidence is necessary before it can be accepted with confidence.

Biotin

It is established that the pig has a definite requirement for this vitamin but normally there is no need to supply any in the diet of either the growing pig or the breeding sow, as the requirements are met from other sources, presumably bacterial synthesis in the intestinal tract.

A possible exception is the very young pig, but the data from the one experiment which suggests this dietary requirement for the young pig do not provide a sufficient basis for an estimate of the requirement.

Folic Acid and p-*aminobenzoic Acid*

The balance of evidence available suggests that there is normally no need to supply either folic acid or *p*-aminobenzoic acid (an essential group in the structure of folic acid) in the diet of the growing pig, as the needs for these vitamins are fully met from other sources, presumably bacterial synthesis in the intestinal tract.

For the breeding sow, there is evidence in one experiment that there may be a need for dietary folic acid but it provides no basis for an estimate of the requirement.

Inositol

Inositol should not properly be classified as a true B vitamin since as far as is known it does not appear to play any part in enzymic reactions, it can be synthesized in the body tissues, and is present in animal tissues in amounts much greater than usually associated with the true B vitamins.

For all classes of pig it appears that there is normally no dietary requirement for inositol; the needs are met from other sources, presumably tissue synthesis.

Choline

Unlike the true B vitamins, choline is not a metabolic catalyst, but forms an essential structural component of body tissues and can be synthesized in the body tissues from methionine. As the recommended dietary levels of methionine plus cystine are lower than those necessary to meet in full the requirements for choline synthesis, an estimated requirement of 870 mg choline/kg dry matter in the diet for growing pigs is tentatively given, based on the results of one experiment using pigs up to 8 weeks of age, in which the level of methionine in the dietary dry matter was 0·8 per cent. It is possible that the addition of vitamin B_{12} to the diet has a sparing effect on choline requirements.

Although there is some evidence of a dietary requirement for choline by breeding sows, it provides no basis for an estimate of requirement.

Vitamin C (Ascorbic Acid)

It appears that there is normally no need to supply any vitamin C in the diet of either the growing pig or the breeding sow; the requirements are fully met from other sources, presumably tissue synthesis. The reservation is, however, added that the possibility of a dietary need for the vitamin under certain stress conditions requires further investigation.

SECTION 5

REQUIREMENTS FOR WATER

Because of the number of variables involved, water requirements cannot be stated with precision. However, under normal conditions of feeding, environment and well-being, requirement will generally be related to feed intake. Excluding suckling sows and their offspring, the evidence suggests that the water requirement of pigs is met by the provision of 2 parts by weight of water for each 1 part of dry feed; for the lactating sow a wider ratio of 3:1 is indicated. Requirement is modified by dietary variants such as levels of protein, fibre and common salt, and by the condition of housing and state of health of the animal.

Under conditions of free access to water there are wide variations in individual consumption, particularly with sows. It is not possible to assess the significance of these variations.

Owing to lack of data no indication of the water requirements of boars can be given.

Estimated Requirements

Growing pigs	1 kg/day at 15 kg live weight increasing to 5 kg/day at 90 kg live weight
Sows:	
Non-pregnant	5 kg/day
Pregnant	5–8 kg/day
Lactating	15–20 kg/day

Note

When pigs are given their water mixed with the feed in a wet-feeding system, the following water: feed (dry matter) ratios by weight should meet the estimated requirements. Growing pigs 2:1; sows, non-pregnant 2:1; pregnant 2·5:1; lactating 3:1. For lactating sows, however, and also for early weaned pigs given dry feed, an unrestricted supply of water is preferred.

GLOSSARY

Ad lib. feeding. The feeding system under which the feed is available to appetite at all times, allowing pigs to eat at will. The feed is usually given dry. Synonymous with *Full feeding* and *Self feeding.*

Anabolism. The processes taking place in the body whereby substances are built up into more complex materials.

Apparent digestibility. The amount of a nutrient ingested minus the amount excreted in the faeces, whether a residue from the feed or of endogenous origin, expressed as a percentage of the intake.

$$\text{Apparent digestibility} = 100 \times \frac{\text{Amount in feed} - \text{amount in faeces}}{\text{Amount in feed}}$$

Apparently digestible energy. See *Digestible energy (DE).*

Availability. See *True digestibility.*

Basal metabolic rate. Synonymous with *Fasting heat metabolism* and *Fasting heat production.* It is the amount of heat produced by a resting animal per unit of time, measured sufficiently long after its last feed for the absorption of energy-yielding nutrients from its gut to be negligible. It is normally determined in the zone of thermal neutrality for the animal concerned.

Biological value (BV). The percentage of absorbed nitrogen from the feed protein that is retained by the animal.

$$\text{BV} = 100 \times \frac{\text{I} - (\text{F} - \text{MFN}) - (\text{U} - \text{EUN})}{\text{I} - (\text{F} - \text{MFN})}$$

$$= 100 \times \frac{\text{NR} + \text{MFN} + \text{EUN}}{\text{I} - (\text{F} - \text{MFN})}$$

Where I = dietary intake of nitrogen, F = total faecal nitrogen,
MFN = metabolic faecal nitrogen, U = total urinary nitrogen,
EUN = endogenous urinary nitrogen, and NR = nitrogen retention, all in g/day.

If correction for metabolic and endogenous losses is not made, the value is termed *Apparent biological value.*

Calorie (cal). The amount of heat required to raise the temperature of 1 g water from 14·5 to 15·5°C. By international agreement a calorie is now defined as being equal to 4·1852 International Joules.

Controlled feeding. The feeding system under which the feed given to an individual pig or group of pigs is under the daily control of the feeder of the pigs. Amounts given are usually restricted to less than the pigs' maximum voluntary intake and are most commonly given either once, or divided equally between 2 or 3, portions each day. The feed may be given either wet or dry. Synonymous with *Restricted feeding*.

Creep feed. Feed given to sucking pigs behind a barrier (or 'creep') which allows them access to the feed but excludes the sow.

Crude protein (*CP*). See *Protein*.

Diet. See *Feed*.

Digestible energy (*DE*). The gross energy (or heat of combustion) of a feed minus the gross energy of the corresponding faeces, expressed as Mcal or kcal/kg total feed. Synonymous with *Apparently digestible energy*.

Dressing percentage. The cold eviscerated carcass, usually including head, backbone, feet and kidneys, as a percentage of the life weight determined shortly before slaughter. Synonymous with *Killing-out percentage*.

(Note. *Dressing percentage* may sometimes be defined in other publications as the percentage of the eviscerated carcass put into bacon-curing tanks after trimming, which includes removal of the head, backbone, feet and kidneys. This definition has not been used herein.)

Early weaning. Weaning off the dam at any age before about 42 days of age. In experimental work it is usually at between 2 and 21 days. Under commercial conditions it is commonly at around 21–35 days of age.

Efficiency of feed conversion. See *Feed conversion efficiency*.

Efficiency of utilization of metabolizable energy (*ME*) *or digestible energy* (*DE*). The increase in energy retention, or of body energy saved at or below maintenance, resulting from an increase of 100 kcal in the intake of ME or DE. It is usually expressed as a percentage.

Endogenous loss or excretion. The loss of a nutrient originating from the tissues even when the nutrient is absent from the diet.

Energy retention. Metabolizable energy intake less body heat production.

Essential amino acids. Those indispensable amino acids which are not synthesized in the tissues of the animal, or by microorganisms in its alimentary tract, in amounts sufficient to meet the animal's requirements and which must therefore be present in the feed.

Eye muscle. The *longissimus dorsi* muscle exposed when a carcass is cut at right angles to the backbone at some defined point. Nowadays this is commonly at the level of the posterior edge of the head of the last rib.

Fasting heat metabolism (*or production*). See *Basal metabolic rate.*

Fat. Used to describe fatty tissues of the body in contrast to fat in the strictly chemical sense (see *Lipid*). Feeds such as tallow are not strictly fat or lipid as here defined, but because of common usage they are termed fats.

Fattening pig. See *Growing pig.*

Feed. That which is given to the animal to consume, whether a single feeding stuff or a mixture: used synonymously with *Diet* and should be assumed to be air-dry unless dry matter is specified.

Feed conversion efficiency (*FCE*). Conventionally taken to be the total weight of feed (usually air-dry) eaten over a period of time divided by the live weight gain over that period. It has little meaning unless the composition of the feed given is known since FCE varies with the dietary DE concentration, in general deteriorating as the latter is reduced, and improving as the DE concentration is increased. It should not be confused with efficiency of energy utilization which, for example, might be good at just above maintenance, when FCE would be very poor.

Full feeding. See *Ad lib. feeding.*

Gross energy (*GE*). The heat of combustion per unit weight of a feed, or of excreta or other products, determined with a bomb calorimeter.

Growing pig. Term used to describe an animal at any stage of life between weaning and slaughter. Used in preference to the term 'fattening pig'.

Heat increment. The increase in heat production which results when an increase occurs in the intake of metabolizable energy. Expressed as a per cent of the latter, it is defined as

$$\frac{\Delta \text{ heat production}}{\Delta \text{ ME intake}} \times 100 \text{ where } \Delta = \text{change in}$$

Katabolism. The processes taking place in the body whereby substances are broken down to less complex materials.

Killing-out percentage. See *Dressing percentage.*

Kilocalorie (*kcal*). One thousand calories.

Level of feeding. This refers to the weight of the daily intake of the complete diet and is not confined to energy intake. When the level of feeding is increased, daily intakes of all nutrients are thus increased. Synonymous with *Plane or scale of feeding* and *Ration.*

Lipid. Solvent-extractable 'fat'. Not synonymous with 'ether extract'.

Maintenance. At the maintenance level of feeding, the animal's requirement for a nutrient for the normal continuance of vital processes within the body, including the replacement of endogenous losses in faeces and urine and from the skin, is just met so that the net gain or loss of the nutrient from the tissues of the animal as a whole is zero.

Megacalorie (Mcal). One thousand kcal.

Metabolic faecal nitrogen (MFN). The nitrogen, other than the undigested feed nitrogen, which is excreted in the faeces. It arises from the unabsorbed constituents of the digestive juices, from cellular debris and mucus derived from the gastric and intestinal mucosa, from microorganisms and from nitrogenous compounds excreted into the gut.

Metabolizable energy (ME). The digestible energy (DE) of a unit of weight of feed less the heats of combustion of the corresponding urine and gaseous products of digestion. With pigs the gaseous products are usually considered to be insignificant and are ignored. At the maintenance level of feeding, ME, by definition, equals the daily heat production.

Metabolism. Term embracing the processes of *Anabolism* and of *Katabolism.*

Muscle. Used to describe muscle tissue of the body. Normally contains on average 19–23 per cent *Crude protein.*

Net energy (NE). The energy value of product formed, or of body substance saved at or below maintenance, per unit weight of feed consumed. It is ME less the *Heat increment.* NE value of a feed is not constant between pigs and ruminants, nor is it constant for maintenance and production. Thus it must be defined further, according to type of animal and the physiological processes for which the feed is being used:—

NE (MP): net energy determined at maintenance level with pigs.

NE (GP): net energy determined in young pigs laying down both muscle and fat.

NE (FP): net energy determined in mature pigs which are primarily laying down fat.

NE (MC): net energy determined at maintenance level with cattle.

NE (FC): net energy determined with mature cattle which are primarily laying down fat.

Nitrogen balance. See *Nitrogen retention.*

Nitrogen retention (*NR*). The net gain or loss of nitrogen (N) by the animal. NR = N in feed − (N in faeces + N in urine). Sometimes NR is divided by the live weight gain during the period of nitrogen retention measurement and the convention adopted of expressing this as NR% where:—

$$\text{NR\%} = \frac{\text{NR during } x \text{ days}}{\text{Live weight gain during } x \text{ days}} \times 100$$

(Note. Herein *Nitrogen retention* is considered as synonymous with *Nitrogen balance*. In other publications, *Nitrogen retention* may sometimes be defined as N in feed − (N in faeces − metabolic faecal N) − (N in urine − endogenous urinary N). If the correction for endogenous losses is not made, then the value is termed *Apparent nitrogen retention* or simply *Nitrogen balance*.)

Performance. As applied to growing pigs, is used to describe collectively their growth rate and feed conversion efficiency.

Plane or scale of feeding. See *Level of feeding*.

Pregnancy: gross weight gain during. Live weight of animal immediately before parturition less her weight at service.

Pregnancy: net weight gain during. Live weight of animal immediately after parturition less her weight at service.

Protein. Chemically determined protein, either crude or true, usually based on percentage nitrogen. Since proteins, on average, contain 16 per cent nitrogen, *Crude protein* is usually taken as being total nitrogen × 6·25. For milk, however, the factor generally adopted is 6·38. (See also *Muscle*.)

Ration. See *Level of feeding*.

Reproductive cycle. Period from conception of one litter to conception of the next.

Requirement. The *Requirement* for any given nutrient is the amount of that nutrient which must be supplied in the diet to meet the needs of the normal healthy animal given an otherwise completely adequate diet in an environment compatible with good health. It is the amount of a nutrient that must be supplied in the diet to meet the *Net requirement* of the animal.

Requirement (*Net*). The quantity of a nutrient that should be absorbed by a normal healthy animal given a completely adequate diet in an environment compatible with good health in order to meet its needs for maintenance, including the replacement of endogenous losses, and the amounts needed for a specified rate of growth, for a stated level of production and for reproduction.

Restricted feeding. See *Controlled feeding.*

Self feeding. See *Ad lib. feeding.*

Sucking pigs. Unweaned pigs receiving milk from their dam. They are also usually offered creep feed.

Total digestible nutrients (TDN). TDN per 100 lb of feed =
% apparently digestible crude protein
+% apparently digestible nitrogen-free extractives
+% apparently digestible crude fibre
+ 2·25 (% apparently digestible ether extract).

True digestibility. The amount of a nutrient ingested in a feed minus the amount in the residue of that feed which is excreted in the faeces, expressed as a percentage of intake. Synonymous with *Availability.*

True digestibility =

$$100 \times \frac{\text{Amount in feed} - (\text{amount in faeces} - \text{metabolic faecal loss})}{\text{Amount in feed}}$$

Vitamin Units

Vitamin A. 0·3 μg vitamin A alcohol = 0·344 μg vitamin A acetate = 1 i.u. vitamin A. It has been assumed that 1 μg vitamin A alcohol = 6 μg β-carotene and hence that 1 i.u. vitamin A alcohol = 1·8 μg β-carotene.

Vitamin D. 1 i.u. vitamin D_3 = 0·025 μg crystalline vitamin D_3. It has been assumed that vitamin D_2 is as effective as vitamin D_3.

Vitamin E. 1·0 i.u. vitamin E = 1 mg synthetic DL-α-tocopheryl acetate
1·1 i.u. vitamin E = 1 mg DL-α-tocopherol
1·49 i.u. vitamin E = 1 mg D-α-tocopherol.

Weights and Volumes

1 kilogram (kg) = 1000 grams (g) = 2·205 pounds (lb)
1 gram (g) = 1000 milligrams (mg)
1 milligram (mg) = 1000 micrograms (μg)
1 lb = 453·6 g

To convert:	*Multiply by:*
lb to kg	0·4536
kg to lb	2·205
amount/unit dry matter to amount/unit air-dry diet (13 per cent moisture)	0·87

TECHNICAL COMMITTEE ON THE NUTRIENT REQUIREMENTS OF FARM LIVESTOCK

Chairman	SIR DAVID CUTHBERTSON, C.B.E., M.D., D.SC., LL.D., F.R.C.P.E., F.R.S.E.	Formerly Director, Rowett Research Institute
	PROFESSOR SIR RONALD BASKETT, O.B.E., D.SC., F.R.I.C.	National Institute for Research in Dairying
	K. L. BLAXTER, Ph.D., D.SC., N.D.A., F.R.S.E.	Rowett Research Institute
	A. W. BOYNE, B.SC.	Rowett Research Institute
	R. BRAUDE, Ph.D., D.SC.	National Institute for Research in Dairying
	A. EDEN, M.A., Ph.D., F.R.I.C.	National Agricultural Advisory Service
	EMERITUS PROFESSOR S. K. KON, C.B.E., Ph.D., D.SC., F.R.I.C.	Formerly National Institute for Research in Dairying
	*F. H. MOULT, M.SC.	Agricultural Research Council
	PROFESSOR A. T. PHILLIPSON, M.A., Ph.D., D.V.SC., M.R.C.V.S., F.R.S.E.	School of Veterinary Medicine, Cambridge
	PROFESSOR ALEX. ROBERTSON, C.B.E., M.A., B.SC., Ph.D., M.R.C.V.S., F.R.I.C., F.R.S.H., F.R.S.E.	Royal (Dick) School of Veterinary Studies
	J. A. B. SMITH, C.B.E., Ph.D., D.SC., F.R.I.C., F.R.S.E.	Hannah Dairy Research Institute
	D. B. WALLACE, M.A.	School of Agriculture, Cambridge
Technical Secretaries	PROFESSOR I. A. M. LUCAS, B.SC., M.SC. (resigned September, 1961)	University College of North Wales
	J. L. CORBETT, B.SC., M.Agr. SC. (resigned December, 1962)	Formerly, Rowett Research Institute
	C. F. MILLS, M.SC., Ph.D., F.R.I.C., F.R.S.E.	Rowett Research Institute

* Now deceased.

WORKING PARTY ON NUTRIENT REQUIREMENTS OF PIGS

Chairman	PROFESSOR SIR RONALD BASKETT, O.B.E., D.SC., F.R.I.C.	National Institute for Research in Dairying
	A. W. BOYNE, B.SC.	Rowett Research Institute
	R. BRAUDE, PH.D., D.SC.	National Institute for Research in Dairying
	A. G. CHAMBERLAIN, B.SC., Ph.D.	University College of North Wales
	A. EDEN, M.A., Ph.D., F.R.I.C.	National Agricultural Advisory Service
	A. S. JONES, B.SC., Ph.D.	Rowett Research Institute
	G. A. LODGE, B.SC., Ph.D.	University of Nottingham
	PROFESSOR I. A. M. LUCAS, B.SC., M.SC.	University College of North Wales
	*F. H. MOULT, M.SC.	Agricultural Research Council
	PROFESSOR ALEX. ROBERTSON, C.B.E., M.A., B.SC., Ph.D., M.R.C.V.S., F.R.I.C., F.R.S.H., F.R.S.E.	Royal (Dick) School of Veterinary Studies
	K. L. ROBINSON, B.SC., B.Agr., A.R.I.C.	Queens University, Belfast
	D. B. WALLACE, M.A.	School of Agriculture, Cambridge
Technical Secretaries	W. E. COEY, M.A., M.Agr., Ph.D. (resigned March 1961)	Agricultural Research Institute, Hillsborough
	K. G. MITCHELL, B.SC., Ph.D.	National Institute for Research in Dairying

* Now deceased.

CHAPTER 30

Poultry Nutrition and Nutrient Requirements

F. W. HILL

Department of Nutrition, University of California, Davis, California

Adequate nutrition of any animal species requires a knowledge of the quantitative nutrient requirements at various stages of the life cycle and for various physiological functions, the feeding behaviour of the animal, and the quantitative distribution of the essential nutrients in the feeding materials which can be used. These kinds of information are critically necessary to achieve maximum biological and economic efficiency in the use of nutritional resources, considerations which will become even more important as problems of world food supply and the upgrading of human nutrition become increasingly urgent in the years ahead.

The purpose of this summary is to bring together in convenient form the information on nutrient requirements of poultry species and to discuss some of the considerations in applying this information to the nutrition of poultry in practice. The author has drawn liberally on the excellent summaries of the research literature pertaining to quantitative nutrient requirements which are consolidated in the recommendations of the Committee on Animal Nutrition of the National Academy of Sciences—National Research Council of the U.S.A.,[1] and the Technical Committee on Nutrient Requirements of Farm Livestock of the Agricultural Research Council, Great Britain[2] (see Chap. 32). These publications provide a key to the research literature for those who wish more detailed information than can be presented in the present summary. The quantitative aspects of nutrition have been given more detailed study and are probably better understood for poultry than for any other economically important animal species. This relative wealth of information for poultry is a part of the scientific advance in nutrition over the past 50 years, and has been derived in about equal measure from research directed to the solution of nutritional field problems and from basic research utilizing chickens and other experimental birds as laboratory animals. Some of the earliest experimental studies leading to the discovery of the vitamins were conducted with chickens—and the discovery of the most recent addition to the list of essential minerals, selenium, involved extensive basic studies with chickens and turkeys. Research with chickens and other birds has contributed significant information in the elucidation of essential nutrients and nutritional relationships, and in the process the understanding of practical nutrition of poultry has been enhanced. The development of the modern poultry industry is based to a considerable extent on the application of this basic knowledge.

FOOD INTAKE

The quantitative needs for the essential nutrients may be expressed in terms of concentration per unit weight of diet, or on a per animal per day basis. In either case, representative data on feed intake and its relation to productive function can be a useful starting point for discussion of nutritional requirements. Typical data are summarized for young and adult birds of various species in Table 1.

TABLE 1

Typical Rates of Feed Consumption for Poultry Species.

A.—CHICKENS: Growing Pullets, Egg Production Type

Age in weeks	Small strains		Large strains	
	Average weight	Cumulative feed	Average weight	Cumulative feed
	kg	kg	kg	kg
4	0·27	0·68	0·36	0·77
8	0·64	1·9	0·77	2·2
12	1·0	3·5	1·2	3·9
16	1·2	5·3	1·5	5·9
20	1·4	7·3	1·8	8·2

Based on diets containing approximately 2950 kcal metabolizable energy per kg.

B.—CHICKENS: Male and Female, Meat Production Type

Age in weeks	Males			Females		
	Average weight	Cumulative feed	Gain per kg feed consumed	Average weight	Cumulative feed	Gain per kg feed consumed
	kg	kg	kg	kg	kg	kg
2	0·18	0·23	0·78	0·18	0·23	0·78
4	0·55	0·91	0·60	0·50	0·86	0·58
6	1·1	2·0	0·55	0·91	1·8	0·51
8	1·6	3·5	0·46	1·3	3·0	0·43
10	2·2	5·3	0·42	1·8	4·6	0·39

Based on diets containing approximately 3000 kcal metabolizable energy per kg.

C.—CHICKENS: Layer Hens—Feed Consumption in relation to Body Size and Rate of Lay

Production rate[a]	Feed consumption[b] per 100 hens per day		
	1·8 kg hens	2·3 kg hens	2·7 kg hens
Per cent	kg	kg	kg
0	6·4	7·5	8·6
20	7·7	8·8	9·9
40	9·0	10·1	11·1
60	10·2	11·3	12·4
80	11·5	12·5	13·6

[a] Eggs produced per 100 hens per day.

[b] Based on a diet containing approximately 2860 kcal metabolizable energy per kilogram.

D.—TURKEYS: Males and Females, Large Type

Age in weeks	Males			Females		
	Average weight	Cumulative feed	Gain per kg feed consumed	Average weight	Cumulative feed	Gain per kg feed consumed
	kg	kg	kg	kg	kg	kg
4	0·77	0·96	0·80	0·64	0·82	0·78
8	2·3	4·0	0·58	2·0	3·6	0·56
12	4·6	11	0·42	3·5	9·0	0·39
16	7·0	19	0·37	5·0	15	0·33
20	9·3	28	0·33	6·2	22	0·28
24	11·4	39	0·29	7·2	31	0·23

Based on diets containing approximately 3000 kcal metabolizable energy per kg.

E.—DUCKS: Male and Female, White Pekin

Age in weeks	Males			Females		
	Average weight	Cumulative feed	Gain per kg feed consumed	Average weight	Cumulative feed	Gain per kg feed consumed
	kg	kg	kg	kg	kg	kg
2	0·5	0·77	0·65	0·5	0·77	0·65
4	1·6	3·9	0·41	1·6	3·9	0·41
6	2·6	7·3	0·36	2·5	7·3	0·34
8	3·3	11	0·30	3·1	11	0·28

Based on diets containing approximately 2900 kcal metabolizable energy per kg.

The feeding of poultry has two features which distinguish it from the nutrition of most other kinds of domestic animals. First, poultry are usually given free access to their diet, and feed intake is self-controlled. Secondly, poultry are often managed in very large populations, aggregating hundreds or thousands of individuals in a single unit. The performance of the nutrition programme is the result of two factors: the adequacy and unit value of the diet, and the level of voluntary feed intake. The task of the nutritionist is to develop the dietary composition and nutrient balance which will support maximum productive efficiency at the level of feed intake characteristic of the population under the given environmental conditions.

The data summarized in Table 1 are typical for modern commercial strains of poultry fed conventional diets in a temperate environment. As will be discussed in a later section, one of the primary factors influencing feed intake is the energy concentration of the diet; for this reason, the approximate dietary energy concentration is stated. Seasonal fluctuations also occur, and it can be expected that intake rates will be 5–7 per cent above the annual mean during the cold season, and similarly less than the annual mean during the warm season.

These data illustrate relationships between productive performance and feed intake which bear directly on the economic aspects of poultry production. In the growing birds, the daily feed intake in relation to body size markedly increases as the bird grows older and larger. The gross efficiency of growth (gain per unit feed intake) declines progressively as the bird grows larger, reflecting the progressively higher maintenance requirements of the larger birds. Within a species, the factors which increase rate of growth (genetic selection, improved diets, improved environment) will also improve the efficiency of growth; the converse is also generally true.

In the egg-producing chicken, feed requirement is related both to body size and rate of egg production. The economic maintenance requirement (zero egg production) is relatively high, and is directly related to metabolic body size; above this maintenance requirement, feed intake is linearly related to rate of egg production. Because the maintenance requirement is a relatively large fraction of the total need, the efficiency of egg production (eggs produced per unit feed consumed) increases markedly as the rate of egg production is increased. There is no intermediate point of maximum efficiency.

ENERGY

The largest nutrient requirement of poultry is for energy. Because this requirement is so dominant, the biological and economic efficiency of a feeding programme are largely determined by the dietary energy and its relation to the other essential nutrients.

The quantitative energy values of feeding materials can be determined and expressed in several ways. The energy value of the nutrients which are usable by the animal for all metabolic purposes is usually expressed as digestible energy (feed energy minus faecal losses) or metabolizable energy (feed energy minus faecal and urinary losses). Because of the technical difficulties of determining digestibility of diet components in birds, recent work in this field has been devoted mainly to the measurement of metabolizable energy values of feeding materials.

Studies of the digestive mechanisms in poultry (discussed elsewhere in this volume) and the metabolizable energy of well characterized reference materials (Table 2) show that chickens closely resemble other non-ruminants in their ability to utilize common nutrients.

TABLE 2
Metabolizable Energy of Some Reference Substances for Chickens

Substance	Metabolizable energy*	Utilization**
	kcal/g	Per cent
Glucose	3·64	97
Sucrose	3·80	95
Corn starch	4·08	97
Cellulose	0	0
Casein	4·50	78 (100)
Isolated soybean protein	3·83	67 (87)
Corn oil	8·80	93
Soybean oil	9·25	98

* All data on dry matter basis.
** Utilization expresses metabolizable energy as a percentage of gross combustible energy. Parenthetical figures are estimated values for proteins, with gross energy corrected for expected urinary loss.

Extensive data have been obtained in several laboratories on the metabolizable energy of feeding materials for chickens, and representative values are available for most feedstuffs. In addition to providing these quantitative data (Table 3), the research has shown that metabolizable energy (ME) values are highly reproducible, are additive in contributing to the ME of a diet, and are equal to the sum of the combustible energy of the digestible components of a material less the energy of the expected urinary wastes. ME values are commonly corrected to a condition of nitrogen equilibrium, which has the effect of assigning a value of about 4·3 kcal per gram to the digestible protein, instead of its total heat of combustion (5·7 kcal). This convention mathematically avoids the complicating effect of different rates of feed intake and growth or egg production, and different levels of dietary protein, in deriving ME values for dietary materials. ME values so determined have been found to be influenced very little by level of feed intake, age, sex, breed, rate of

it contains. Treatment with gum-splitting enzymes has been shown to be effective in improving the nutritional value of guar meal.[4]

The ME of milo varies among samples, and is inversely related to tannin content of the milo.[5]

Directly determined ME values are the preferred data for guiding the use of carbohydrate materials. Where considerable variation in the composition of a kind of material commonly occurs, differences in the crude fibre content can be used to estimate variation in ME value by assuming that fibre represents about 40 per cent of the indigestible components. Good correlation has been found between ME value and chemically estimated 'available carbohydrate' (starch and sugars) in carbohydrate-rich feeding materials.[6]

Fats. The utilization of fats and fatty acids can be expressed either in percent absorbability (digestibility) or in metabolizable energy; the two measures are equivalent, and agree well experimentally. ME can be computed from the product of percent absorbability and heat of combustion (about 9·4 kcal/g).

Natural fats differ widely in ME value for chickens. Beef tallow is approximately 70 per cent utilized by older chicks and adults; its utilization is lower in very young chicks. The fluid vegetable oils are highly absorbable. There is a rough general relationship between degree of saturation and absorbability, with the more highly saturated tallow and coconut fat showing lower absorbability. However, the distribution of fatty acids within the triglycerides also affects absorbability. Natural lard is highly absorbable despite its relatively high saturation, due to the predominant occurrence of the saturated fatty acids in the 2-position of the glycerides; rearrangement to a more random distribution reduces absorbability.[7]

The mixed fatty acids derived from natural fats when fed in free or monoester (methyl; ethyl) form are less well utilized than the fats from which they are derived (Table 4). The difference is relatively small in the case of a highly utilized vegetable oil, and is large for the more saturated tallow. The difference is due mainly to poorer utilization of the saturated fatty acids when fed in free form. Both stearic acid and palmitic acid are essentially unutilized when fed singly (Table 4) while oleic acid is well utilized. In mixtures, oleic acid enhances the utilization of saturated fatty acids.[8]

Generally, the utilization of mixtures of fats is equal to the weighted mean of their individual absorbabilities or ME values. Isolated exceptions have been reported, but no general method of improving the utilization of poorly absorbed fats, such as by mixing with a well utilized fat, is evident.

Interference with fat absorption can occur in several ways. High dietary calcium levels can increase faecal fat excretion. The inhibitor complex of soybeans includes a factor which markedly inhibits fat absorption, especially in young chicks, so raw or underheated soybean meal can exert such an effect. The fat of full-fat (unextracted) heated soybeans is also poorly utilized unless

TABLE 4

Utilization of Fats and Fatty Acids by Chickens

Substance	Metabolizable energy	Utilization
	kcal/g	Per cent
Beef tallow	6·30	71
Pork lard	8·76	93
Corn oil	8·80	94
Soybean oil	9·25	98
Tallow fatty acids	4·70	51
Lard fatty acids	6.30	68
Soybean oil fatty acids	8·15	88
Stearic acid	0	0
Palmitic acid	0–·68	0–7
Myristic acid	2·00	22
Lauric acid	5·84	66
Oleic acid	9·02	96

Utilization expresses metabolizable energy as percentage of gross energy; data agree well with absorbability measured in the same experiments.

the beans are subjected to mechanical compression by flaking, probably because disruption of cell structure is necessary to release the oil.[9]

Proteins. It is appropriate to consider the protein-rich feeding materials as energy sources also because (a) they usually contribute substantial amounts of carbohydrate and/or fat to the diet, (b) at least half of the absorbed amino acids are used for energy rather than for protein synthesis, and (c) the total metabolizable energy of the diet needs to be known. Furthermore, the ME of a protein source is a useful measure of the overall utilization of the protein.

ME data for common protein resources are included in Table 3. The ME of a diet or protein source is not influenced by amino acid deficiency, even of severe magnitude. Factors which can influence protein digestibility, and therefore ME value, include processing damage by overheating and the presence of inhibitory substances such as gossypol (cottonseed), and the inhibitor complex of unheated soybeans.

ENERGY IN NUTRITION AND DIET FORMULATION

For each kind of diet mixture, designed for a particular stage of life and function, there is a minimum dietary energy concentration below which productive performance is impaired because of inadequate feed intake. This minimum is difficult to define because it involves the bulk density of the diet and its physical nature, the nature of the indigestible components, and palatability as well as the ME per unit weight. The minimum is fairly high for diets designed for early rapid growth, and for high rate of egg production in a cool environment. A reasonable minimum for these purposes would be of the order of 2700 kcal/kg.

Equally as important as the concept of a necessary minimum energy concentration is the view that there is no physiological upper limit to the dietary energy level. To say it another way, there is no evidence that some minimum level of indigestible residue is needed, or that there is a limit to the amount of fat which can be incorporated in a diet which is otherwise satisfactory in texture and adequate to supply the nutritive essentials. This latter point is a departure from the view commonly held only a few years ago to the effect that poultry in particular and birds in general were limited in their ability to use fats. In practice, the physical problems of the preparation and handling of high fat mixtures are much more limiting than the physiological capacity of poultry.

When fed complete diets adequate in energy concentration, growing chickens and turkeys and production-bred laying hens regulate feed intake to obtain the amount of metabolizable energy needed. Over a wide range of ME concentration per unit weight of diet, most poultry will adjust feed intake to provide a fixed ME consumption. This regulation is illustrated by the data in Table 5, which show the productive performance and feed consumption of large groups of laying hens over a year-long period, fed diets differing widely in ME concentration. With the exception of the diet lowest in ME, which was inadequate during cold weather under the conditions used, all of the diets were consumed at levels which provided the same ME intake per day. These diets are representative of the range of ME concentration likely to be used in feeding practice.[10]

TABLE 5
Effect of Metabolizable Energy Concentration of Diet on Feed Consumption of Laying Hens

Dietary metabolizable energy	Egg production rate	Feed consumption		
		per dozen eggs	per hen per day	per hen per day
kcal/kg	Per cent	kg	g	kcal
2300	65	2·6	141	327
2660	67	2·4	131	351
2970	68	2·1	117	348
3060	70	2·0	114	351
3150	71	1·9	112	352

The level of ME intake in this example, 350 kcal/day, is typical for layers of moderate size in relatively uninsulated housing in northern United States. Under similar conditions in California the average consumption would approximate 300 kcal/day. Under intensified production conditions in temperature-controlled housing, lower consumption rates may be observed. In addition to the adjustment of ME intake in accordance with general environment, poultry also adjust seasonally in response to temperature changes; typically, ME consumption in winter will be 10–15 per cent higher than in summer, but the range of variation may be wider under extreme conditions.

The regulation of feed intake in accordance with energy need and the ME concentration of the diet is also evident in growing chickens and turkeys. Both in young growing birds and in adults, productive performance tends to be increased when diets high in ME concentration are used.

There are two notable exceptions to the general ability of poultry to regulate feed intake. One is the breeding hen of strains bred for the rapid growth rate necessary in meat production. Such hens when free-fed diets high in energy concentration tend to become obese. The problem can be met in part by using relatively low-energy diets, but the more successful procedure is to limit feed intake by controlled feeding. A second exception is the White Pekin duck used for meat production, which becomes quite fat during early growth.[11]

In addition to providing a quantitative basis for comparative evaluation of feeding materials and for establishing the total dietary energy level, ME data also provide a basis for relating the needs for other critical nutrients to feed intake. This is of value in providing for the protein and amino acid requirements in particular, as will be discussed in the next section, and also for such large needs as calcium for layers where the daily need varies according to conditions. Knowledge of the ME intake required under given conditions is necessary for this approach to the formulation of adequate diets.

PROTEIN AND AMINO ACID REQUIREMENTS

The dietary requirement for protein is made up of two components: the essential amino acids, and the total need for nitrogen in a suitable form. Both components require consideration, and they are interrelated in two ways:

(1) In practice the total protein level required in the diet depends on the amount needed to supply the essential amino acids. Though the essential amino-acid requirements aggregate less than half of the total amount of protein fed, the limitations of dietary proteins make relatively high feeding levels necessary. The estimated total protein requirements are therefore descriptive of successful feeding programmes, and do not necessarily define the minimum total needs in terms of ideal mixtures. It is fortunate that the protein requirements can be met by feeding a relatively high level of a poorly balanced protein, such as peanut meal, or a lower level of a more efficient

TABLE 6

Protein and Amino Acid Content of Selected Protein Resources and Grains

Source of protein	Crude protein	Metabolizable energy	Arginine	Lysine	Methionine and cystine	Tryptophan
	per cent	kcal/kg	per cent	per cent	per cent	per cent
Soybean meal	50	2500	3·6 (6·4)	3·2 (6·4)	1·57 (3·1)	0·64 (1·3)
Cottonseed meal	44	1980	4·3 (9·8)	1·8 (4·1)	1·60 (3·6)	0·53 (1·2)
Peanut meal	50	2640	6·2 (12·4)	1·8 (3·6)	1·23 (2·5)	0·51 (1·0)
Coconut meal	21	1500	2·4 (11·5)	0·7 (3·3)	0·33*	0·20 (1·0)
Corn gluten meal	42	3240	1·4 (3·3)	0·8 (1·9)	1·70 (4·0)	0·20 (0·5)
Fish meal	65	2900	3·8 (5·8)	6·0 (9·2)	3·00 (4·6)	0·85 (1·3)
Meat meal	50	1980	3·5 (7·0)	3·5 (7·0)	1·30 (2·6)	0·30 (0·6)
Corn	9	3370	0·5 (5·5)	0·2 (2·2)	0·36 (4·0)	0·09 (1·0)
Barley	9	3190	0·4 (4·4)	0·4 (4·4)	0·28 (3·1)	0·07 (1·5)
Milo	9	3300	0·3 (3·3)	0·2 (2·2)	0·22 (2·4)	0·08 (0·9)
Oats	12	2660	0·7 (5·8)	0·4 (3·3)	0·36 (3·0)	0·18 (1·6)
Wheat	13	3190	0·7 (5·4)	0·4 (3·1)	0·36 (2·8)	0·18 (1·4)
Wheat middlings	17	1890	0·9 (5·3)	0·7 (4·1)	0·40 (2·4)	0·20 (1·2)
Alfalfa meal	20	1580	1·0 (5·0)	1·0 (5·0)	0·68 (3·4)	0·24 (1·2)

( ) data indicate amino acid concentration as percentage of the feedstuff protein; other data are expressed as percentage of the feedstuff, air-dry basis.

* Value for methionine only; no cystine value available.

mixture such as soybean meal and fish meal, because this enables the use of biologically imperfect resources where they are economically feasible.

(2) The requirements for essential amino acids are in some instances related to the amount of protein in the diet. This has been found for the requirements of the chick for sulphur amino acids in diets based on soybean protein, and for lysine in a diet based on sesame protein, for example. The absolute amount of these amino acids required for growth increased as the dietary level of proteins was increased, even when protein levels greater than those usually adequate for maximum growth were used. The increased amino acid requirements were not proportional to the protein level, however, but declined as a percentage of the protein In such a situation, increasing the dietary level of a deficient protein is not an effective way to overcome an amino acid deficiency.

The requirements for protein and essential amino acids are complex and not fully understood. This discussion will emphasize present viewpoints in meeting the needs in practice, together with some distinctive aspects of the requirements of poultry.

Protein is usually the most expensive component of the diet. Accordingly, the usual practice is to provide the amount needed for the particular function without appreciable excess, and the requirements are treated in this context here. It has been customary to express protein and amino acid requirements

TABLE 7

Estimated Total Protein Requirements of Poultry Relative to Dietary Energy Concentration

Species and function	Interval	Grams diet protein per 1000 kcal metabolizable energy
		g
Growing chickens	1– 5 weeks	75
	5–10 weeks	65
	10–20 weeks	55
Laying chickens	Early production, >80 per cent	63
	Intermediate production, 70–80 per cent	54
	Late production, <70 per cent	48
Growing turkeys	1– 4 weeks	110
	4– 8 weeks	90
	8–12 weeks	75
	12–16 weeks	60
	16–20 weeks	55
	20+ weeks	47
Breeding turkeys		55
Growing ducks	1–2 weeks	65
	2–8 weeks	58

of poultry in terms of percentage of the diet; however, because of the wide range of dietary energy concentration possible in practice, it is preferable to express them in relation to ME or on a daily intake basis wherever possible. Estimated requirements for total protein are summarized in Table 7, stated in grams crude protein (N x 6·25) per 1000 kcal ME. For diets based on the commonly used grains and protein feedstuffs, these estimates provide reasonable formulation standards. The requirements are stated in terms of dietary crude protein; to state them in terms of digestible protein, it can be assumed that the proteins of common feedstuffs are 80–85 per cent digestible.

The total protein level must provide adequate amounts of the essential amino acids in order to support rapid efficient growth or a high rate of egg production. The estimated amino-acid requirements for growing chicks and turkey poults and for layer chickens are summarized in Table 9.[1, 2] The data are given in terms of percent of diet at defined dietary protein levels, from which they can be converted to equivalent levels in relation to ME or diet protein.

TABLE 8
Effect of Dietary Protein Relative to Energy Concentration on Fat Content of Growing Chickens

Dietary protein g per 1000 kcal metabolizable energy	Body fat (per cent of carcass dry matter)
g	Per cent
89	19
80	20
78	20
72	26
71	22
70	27
67	26
65	29
62	28
61	28
60	33
57	32
54	32
52	33
49	39
45	38

The consequences of a deficiency of dietary protein or an essential amino acid are generally similar. In growing birds, a deficiency results in retarded growth, reduction in gain per unit feed intake, and poor feathering; the vice of feather picking is more likely to occur. As protein concentration relative to ME is reduced, the deposition of fat in the tissues of growing chicks increases (Table 8).[12] This is directly related to ME intake, which is higher than needed for the reduced rate of growth, and the excess is deposited as fat because

TABLE 9

Estimated Amino-acid Requirements of Chickens and Turkeys

Amino acids	Growing chicks, 0–4 weeks	Laying hens	Growing turkeys, 0–6 weeks
	Per cent	Per cent	Per cent
Arginine	1·2	0·8	1·6
Lysine	1·0	0·5	1·5
Methionine + cystine	0·7	0·53	0·87
Tryptophan	0·2	0·15	0·26
Glycine	1·0	—	1·0
Histidine	0·4	?	?
Isoleucine	0·75	0·5	0·84
Leucine	1·4	0·7*–1·2	?
Phenylalanine + tyrosine	1·2	0·7*	?
Threonine	0·55	0·4	?
Valine	0·8	0·55*	?
Relative dietary protein level, g/1000 kcal ME	75	54	110
Dietary protein level, per cent	20	15	28

*Estimates based on composition of egg protein. Higher estimate for leucine based on limited work with highly purified diets.

protein synthesis is impaired. The feeding of excess protein (above 75 g per 1000 kcal ME, Table 8) reduces the fat content of the growing bird. In laying hens, protein deficiency causes a reduced rate of egg production and reduced egg size, and increases the likelihood of feather picking. Weight loss or impaired weight gains may be produced.

The amino-acid requirements for growth in chicks and turkey poults are based mainly on experiments in which the supplemental needs of diets based on deficient natural proteins were determined. These studies have been extended by experiments with purified diets based on amino acid mixtures, and it is possible to support rapid growth in short-term experiments with such diets. It is interesting, but of uncertain meaning, that amino-acid mixtures which support most rapid growth are quite different in amino-acid proportions from the pattern of requirements estimated from studies with diets based on proteins. For the present, we must depend mainly on the requirements estimated from classical studies as summarized in Table 9, and await the development of new concepts of optimum amino-acid balance which may come from continued basic studies.

Poultry have several unusual amino-acid requirements. Arginine is an absolute essential for growth of chicks and poults, and is also required by the hen; in contrast, rats have only a supplemental arginine requirement for growth, and no arginine requirement in the adult. The arginine requirement of the chick is variable; it is high when the diet is based on casein, due largely to the high lysine content characteristic of casein. Chicks and poults also require glycine, the simplest amino acid, and this appears to be due at least partly to the high demand for creatine synthesis during rapid growth. Except for these distinctive needs, the requirements of the chick resemble those of the rat, including the sparing relation of tyrosine to phenylalanine, and cystine to methionine.

In general, the amino acids most likely to be limiting in practice are methionine and lysine. The needs are most critical for growing birds. For the chick, the total sulphur amino-acid requirement (methionine and cystine) is 3·5 per cent of the diet protein, and the lysine requirement is 5·0 per cent of the protein. In the turkey poult, the respective requirements are 3·2 per cent and 5·4 per cent of diet protein. Two distinctive aspects of these requirements are of interest. The sulphur amino-acid requirement of the chick for maximum efficiency (gain per unit feed intake) is slightly higher than the requirement for maximum growth rate; this is entirely accounted for by a decrease in fat deposition at the higher methionine level.[13] Lysine deficiency in the Bronze turkey poult causes the development of white wing feathers, which can be used as a diagnostic criterion. Correction of the deficiency produces an immediate resumption of pigment deposition, and leaves the white area of the wing feather as a distinctive 'bar'.

Summarized in Table 6 are data for the protein content and distribution of critical amino acids in common feeding materials. They illustrate the widespread lysine deficiency characteristic of the grain proteins and most of the protein-rich feedstuffs of plant origin. The lysine-rich proteins are limited to soybean meal, and the animal proteins such as fish meals and meat meals. Of great potential importance is the recent discovery that a well-known genetic mutant trait of corn (maize), known as opaque-2, is associated with increased lysine content of corn protein.[14] The development of high-lysine corn, and extension of the principle to other plant materials, may greatly change the relative importance of lysine.

Methionine deficiency is characteristic of soybean meal, meat meals, and the grain proteins with the exception of corn (maize). The outstanding source of methionine is fish meal. Methionine is also available at moderate cost in synthetic form as either the DL-amino acid or its hydroxy analogue; they are interchangeable on a molar basis, and are biologically equivalent to L-methionine of natural proteins.

Because dietary protein is absorbed and utilized as amino acids, its value depends in part on the digestibility and absorbability (or 'availability') of its component amino acids. With the exception of lysine, it appears that the amino acids are 'available' in proportion to the overall digestibility of the feedstuff protein, which for the common feeding materials is of the order of 80–85 per cent. Lysine, on the other hand, can enter into chemical reactions with other substances, especially carbohydrate under the influence of heat and moisture, which produce an enzyme-resistant linkage involving the terminal epsilon-amino group of lysine. This renders the lysine unavailable to the animal, though the digestibility of the rest of the protein may be affected relatively little. The degree of lysine binding can be determined chemically by methods using dinitrofluorobenzene, and this is a useful tool for evaluating fish meals and meat meals.[15] The method is also applicable to the evaluation of cottonseed meals for poultry feeding, because the lysine of cottonseed is partly bound to gossypol during processing; the inactivation of the toxin requires the coupling with lysine, which unfortunately reduces the effective lysine value of the protein.

TABLE 10a

Comparative Amino-Acid Composition of Representative Diet Mixtures Formulated to contain 75 grams protein per 1000 kcal ME.

Proportions of diet components[1]	Composition of diet mixture[2]					
	ME	Protein	ARG	LYS	M + C	TRY
	kcal/kg	Per cent	Per cent	Per cent	Per cent	Per cent
(1) 63 Corn 32 Soybean meal	2910	22·0	1·47	1·15	*0·73*	0·26
(2) 53 Corn 5 Fat 37 Soybean meal	3110	23·3	1·60	1·29	*0·67*	0·29
(3) 54 Corn 5 Fat 31 Soybean meal 5 Fish meal	3140	23·6	1·58	1·40	0·83	0·30
(4) 63 Milo 32 Soybean meal	2880	21·7	1·34	1·15	*0·65*	0·25
(5) 62 Milo 33 Cottonseed meal	2700	20·1	1·62	*0·72*	*0·67*	0·23
(6) 63 Milo 26 Cottonseed meal 6 Fish meal	2770	21·0	1·54	*0·96*	0·75	0·24
(7) 62 Milo 18 Cottonseed meal 15 Soybean meal	2770	21·0	1·51	*0·92*	*0·67*	0·25
(8) 64 Milo 13 Cottonseed meal 12 Soybean meal 6 Fish meal	2840	21·4	1·41	1·10	0·73	0·25

[1]Components make up 95 per cent of diet by weight, and are assumed to provide all protein and ME.
[2]Values in italic are less than 95 per cent of estimated requirement.
ARG, arginine; LYS, lysine; M + C, methionine + cystine; TRY, tryptophan.

The approach and typical problems involved in the formulation of diets to meet the amino-acid requirements of growing chicks are illustrated in Tables 10a and 10b. The diet mixtures used as examples are based on either corn or

TABLE 10b
Estimated Requirements of the Young Chick for Critical Amino Acids at Various Dietary Protein Levels, assuming a constant relationship of 75 grams diet protein per 1000 kcal ME.

Amino acid	Estimated requirement at total protein level of:			
	18 per cent	20 per cent	22 per cent	24 per cent
Arginine	1·1	1·2	1·3	1·4
Lysine	0·9	1·0	1·1	1·2
Methionine + cystine	0·63	0·70	0·77	0·84
Tryptophan	0·18	0·20	0·22	0·24

milo as the grain component in combination with cottonseed meal, soybean meal and/or fish meal. The ME concentration of the mixture is unrestricted, but the protein level is formulated to a constant protein/energy balance of 75 grams crude protein per 1000 kcal ME, using the composition data in Table 6. The mixtures range from 20 per cent to nearly 24 per cent in protein content, but they would provide the same amounts of protein under conditions of constant energy intake, which is how they would be expected to be consumed. Their amino acid adequacy is evaluated by comparison to the estimated requirements in Table 10b, which are the same as the standards in Table 9. They have been adjusted to dietary protein levels above and below 20 per cent to provide each of the indicated amino acids at a constant percentage of the total dietary protein. Only the amino acids arginine, lysine, methionine and cystine, and tryptophan are considered, because these are of the greatest interest and likelihood of deficiency in diets used in practice.

The deficient amino acid levels in Table 10a are italicized, each of these providing less than 95 per cent of the estimated requirement. Mixture 1 based on corn and soybean meal is deficient in methionine. A similar mixture based on milo instead of corn is more deficient. This illustrates the basic resource combinations of much of the United States. The deficiency of the corn-soy mixture is made more severe by fat supplementation, Mixture 2, because it contains a higher proportion of protein from soybean meal, which is relatively more deficient in methionine than corn. Supplementing the corn-soy-fat mixture with 5 per cent fish meal overcomes the methionine deficiency; synthetic methionine would also be effective since the other amino acids are adequately suppled by Mixture 2.

A cottonseed meal-milo combination, Mixture 5, is deficient in both lysine and methionine. This is a common resource combination in the southwestern and Pacific coast regions of the United States. Supplementing with 6 per cen fish meal corrects the methionine deficiency, but is not sufficient to provide adequate lysine (Mixture 6); 9 per cent fish meal would correct both deficiencies. Replacing about half the cottonseed meal with soybean meal (Mixture 7) improves lysine but worsens the methionine deficiency. However, this combination is corrected for both lysine and methionine by 6 per cent fish meal (Mixture 8).

Although the requirements for the essential amino acids and for total protein cannot be considered entirely separately, it is probable that we supply more dietary protein than is strictly necessary when the common protein resources are used to meet the amino-acid requirements. To put it another way, any of the common dietary protein mixtures which adequately provides all of the essential amino acids, probably provides an excess of total protein. The use of supplementary amino acids to meet the needs for those which are ordinarily most limiting should allow some reduction in the amounts of protein needed. Such use of methionine and lysine in turkey rations, where protein requirements are relatively high, appears promising.

The protein requirement of the laying hen has been one of the most widely studied subjects in poultry nutrition. Defining this requirement is complicated because it is affected by the size of the hen, the rate of egg production, egg size, season, and environment, as well as by the type of diet. Expressed as a percentage of the diet, the protein requirement has variously been reported to range from as low as 11–12 per cent to as high as 18–20 per cent; all of these estimates were valid for the conditions involved. Because it also has been difficult to sustain high rates of egg production over reasonably long periods with diets based on amino acid mixtures, the protein and amino acid requirements for egg production have been difficult to define.

An approach which illustrates the problems and provides a means to formulate diets to meet the requirements under varying conditions follows. It is based on (a) a knowledge of the pattern of feed consumption to be expected, (b) a linear relation between egg production rate and daily protein requirement, and (c) a similarity between the amino acid requirement pattern and the amino acid composition of the egg.

Under California conditions, in which seasonal variations are fairly wide, it is reasonable to expect an annual average performance rate of 70 per cent egg production (255 eggs per year) with an average daily feed consumption of 310 kcal ME. During the first 60–90 days after the onset of egg production, the rate of egg laying rapidly rises to a 'peak' of perhaps 85 per cent for the flock, the birds are also increasing in body weight at a rate of about 2 grams per day, and the average rate of feed consumption is about 275 kcal per day. Following this, egg production will be held at a high rate (75–85 per cent) for about 90 days, egg size gradually increases to characteristic adult size, weight gains continue at about 1 gram per day, and the feed intake will be in the range 300–340 kcal per day depending on season. A gradual decline in egg production rate follows, and feed intake follows this decline with due variation according to season. It is particularly noteworthy that the early period of high egg production and continued weight gain is characterized by relatively low total feed intake. It is likely that the processes of egg production and weight gain are competitive during this period, so that maximum egg production can be attained only if the requirements for both are met.

The protein and amino acid requirements are estimated on a per day basis, and formulated into the expected daily feed intake. Inspection of the relation between protein intake and the peak rate of egg production supported by experimental diets, has led the author to estimate the daily protein needs summarized in Table 11a. Provision for weight gain has been made only in association with the higher production rates, because these are characteristic of the early part of the production period.

TABLE 11a

Estimated Daily Dietary Protein Requirement for Egg Production and Weight Gain

Percent production	Diet protein required for:	
	Egg formation (g)	Gain (g)
90	19	1
80	17	1
70	15	0·5
60	13	0
50	11	0
0	3·5	0

TABLE 11b

Assumptions

(1) One 58 g egg contains:
7 g protein
0·24 g methionine
0·17 g cystine
0·50 g lysine

(2) Daily maintenance of a 1700 g hen requires:
3 g protein
0·15 g methionine + cystine
0·05 g lysine

(3) Dietary protein and amino acids are digested and absorbed with 85 per cent efficiency.

(4) One g of weight gain requires 0·5 g of diet protein, 0·01 g sulphur amino acids and 0·02 g lysine.

TABLE 11c

Estimated Minimum Daily Requirements for Dietary Protein, Sulphur Amino Acids and Lysine

Egg production	Daily weight gain	Daily requirement for:		
		Protein	Sulphur amino acids	Lysine
Per cent	g	g	g	g
90	2	20	0·64	0·64
80	2	18	0·59	0·58
70	1	15·5	0·53	0·50
60	0	13	0·47	0·41
50	0	11	0·41	0·35
0	0	3·5	0·18	0·06

TABLE 11d

Estimated Diet Specifications to Sustain Various Production Rates at Different Energy Intakes

	Metabolizable energy intake, kcal/day		
	270	300	330
Crude protein required, per cent of diet			
90 per cent production	21	19	17
80 per cent production	19	17	16
70 per cent production	16	15	14
60 per cent production	14	13	11
50 per cent production	12	11	10
Sulphur amino acids required, per cent of diet			
90 per cent production	0·67	0·61	0·56
80 per cent production	0·62	0·56	0·51
70 per cent production	0·56	0·51	0·46
60 per cent production	0·50	0·45	0·41
50 per cent production	0·43	0·39	0·36
Lysine required, per cent of diet			
90 per cent production	0·67	0·61	0·56
80 per cent production	0·61	0·55	0·51
70 per cent production	0·53	0·48	0·44
60 per cent production	0·43	0·39	0·36
50 per cent production	0·37	0·33	0·30

The basic needs of the hen for dietary protein and amino acids may be considered as the sum of a maintenance (endogenous) need, the components necessary to synthesize 7 grams of protein per egg (this need being proportional to the rate of egg production), and the components necessary to support normal weight gains. The quantitative assumptions relative to protein, lysine and sulphur amino acid needs are summarized in Table 11b. The maintenance needs are estimated from studies with adult males which have indicated requirements for 280 mg nitrogen, 0·04-0·09 g sulphur amino acids and 0·03 g lysine per day per kg body weight.[16] Above the maintenance needs, the amounts of lysine and sulphur amino acids present in the amount of egg protein synthesized are assumed to be the amounts required, and the amounts supplied by the diet are considered to be utilized with 85% efficiency. These assumptions allow for very little loss of the critical amino acids, and therefore probably are estimates of minimum requirement. However, they lead to diet specifications which agree well with the requirements for lysine and sulphur amino acids shown in Table 9, and which are based on critical experiments.

The summed requirements for various rates of egg production are shown in Table 11c. These are translated into diet specifications in Table 11d for three different planes of feed intake; 270 kcal per day, for the early period of peak egg production; 330 kcal per day for cold weather feeding; and 300 kcal per day for the moderate environment, average condition. The most efficient feeding programmes will probably employ 2 or 3 different diets to meet changing conditions, commencing with a diet designed to meet the needs of the early peak period which has high demands, and following with one or more successive diets suitable to meet the lower needs of later stages of the production year. The same principle of formulation is useful in meeting other critical and variable needs, as will be referred to in a later section dealing with calcium requirements.

The choice of protein sources to be used in poultry rations is broad, and varies greatly among countries and regions. The cereals provide a substantial part of the diet protein (usually 20–50 per cent); their prevailing deficiencies of lysine and methionine place particular value on the supplementary protein sources which are rich in these amino acids. Amino acid composition and cost are the primary factors which determine the combination of choice. Some other considerations which apply to the use of common protein resources are:

(a) Meat meals: Limitation of use level is related to mineral content. The maximum level is that which supplies the supplemental phosphorus need. Meat meals are quite variable in feeding value, depending on the raw materials used and processing conditions.

(b) Peanut (groundnut) meals: Maximum use level depends on the nature of the supplemental proteins supplying the needed lysine and methionine. Another consideration is the possible presence of aflatoxins, a group of mould toxins, to which ducks and turkeys are particularly sensitive. Peanut meals are especially subject to mould damage, but toxic mould growth can also affect cottonseed and grains.

(c) Cottonseed meals: Both cyclopropenoid fatty acids (discussed in another section on lipids) and gossypol must be given attention. The former is avoided by using extracted meals containing less than 1 per cent residual fat. Gossypol is toxic to chicks and poults, producing retarded growth at dietary levels exceeding 0·02 per cent. Gossypol is deposited in the egg by hens, and causes an olive-green discoloration when the yolk pH rises on storage; very low dietary levels produce unmarketable eggs. Under well controlled processing conditions, the free gossypol content of prepress-solvent cottonseed meals can be held to less than 0·04 per cent; such meals are suitable for growing poultry, and can be used at 10 per cent of the diet for laying hens. The gossypol problem can be overcome by supplementing the diet with ferrous sulphate, using a level of the salt which provides 2 parts of iron for each 1 part of gossypol by weight. This apparently prevents the absorption of gossypol by forming an insoluble iron-gossypol complex.

VITAMINS

The quantitative requirements of poultry species for most of the vitamins have been studied extensively, with major attention to the needs for early growth and for reproduction. It is in these stages of the life cycle that the vitamin requirements are usually highest. Less information is available for the later stages of growth, and no data are available on the needs of breeding males or non-producing or molting hens. The estimated requirements for the various species and functions are summarized in Table 12, expressed in terms of concentration in the diet. An extensive bibliography of original research papers is contained in the reports of the U.S. National Research Council[1] and the Agricultural Research Council, Great Britain (Chapter 32),[2] and forms the basis of the estimated requirements established by these groups; the data in Table 12 largely follow their recommendations, and are intended to estimate the basic requirement for each nutrient. They do not contain any intentional added 'margin of safety' for use as feeding standards. They are intended to define dietary levels which will meet the requirements of a population for maximum average performance of the indicated function, and for prevention of all specific symptoms of deficiency.

The possible applications of the estimated requirements range over a much greater variety of conditions than the research on which they are based. For the most part, research on requirements has been conducted under laboratory conditions, using simplified or purified diets, and employing one or a very few breeds or strains of the species involved. Some of the effects of these variables may be of particular interest in some situations:

Environment. Laboratory environment is usually more nearly optimal than that of field production units. The effect of seasonal changes on feed intake has already been discussed; it is reasonable to expect that vitamin requirements, expressed in terms of diet concentration, would vary inversely with the level of feed intake as affected by physical environment. Under field

TABLE 12

Estimated Vitamin Requirements of Poultry Species

Vitamin	Species and age or function								
	Growing chickens 1–8 weeks	Growing chickens 8–18 weeks	Layer hens	Breeder hens	Growing turkeys 1–8 weeks	Growing turkeys 8–16 weeks	Breeder turkeys	Growing ducks	Growing pheasants
A (USP units)	1500	2000	4000	4000	1800	2200	2700	?	?
D (IC units)	200	200	500	500	1100	1100	2000	220	1200
K, mg	0·53	?	?	?	0·7	?	?	?	?
Thiamine, mg	1·8	?	?	0·8	2	?	?	?	?
Riboflavine, mg	3·6	1·8	2·2	3·8	3·6	?	3·8	4	3·5
Pantothenic acid, mg	10	10	2·2	10	11	?	16	11	?
Niacin, mg	27	11	?	?	70	?	?	55	60
Pyridoxine, mg	3	?	3	4·5	3	?	?	2·6	?
Folacin, mg	1·2	?	0·25	0·35	0·9	?	0·8	?	?
Biotin, mg	0·09	?	?	0·15	?	?	?	?	?
Vitamin B_{12}, mg	0·009	?	?	0·003	0·003	?	?	?	?
Choline, g	1·3	?	?	?	1·9	?	?	?	1·4

Requirements are stated per kg diet.

conditions, there may also be the added pressures of high population density and the presence of unfavourable microbial environment. Both of these conditions may increase the vitamin requirements. It has been observed that part of the growth-stimulating effect of dietary antibiotics may be accounted for by a sparing effect on the need for added vitamins, suggesting that unfavourable microbial environment can enhance the need for particular factors. On the other hand, where the field environment gives birds access to the litter of the poultry house under floor management conditions, coprophagy occurs and it is well established that substantial amounts of some of the vitamins (notably riboflavin, vitamin B_{12} and vitamin K) can be derived by birds in this way. Studies made under conditions which prevent coprophagy are given preference in estimating requirements.

Type of diet. Energy concentration directly influences the quantity of feed consumed; the indicated requirements are generally applicable to diets containing about 2800 kcal ME per kg. Purified or simplified diets are used extensively in requirement studies, and can differ from the diets of field practice in at least two important ways. The type of carbohydrate used may influence the dietary requirement for a vitamin, particularly when it modifies the intestinal microflora and the synthesis of the factor in the gut; in deriving the estimated requirements, preference has usually been given to work done with starch-based diets because this is the predominant carbohydrate in practical diets. Also, when purified diets are used the requirement is studied by adding the pure vitamin or a concentrated source of it, and all or most of the vitamin is supplied in this way. It is possible that a vitamin as it exists in natural feeding materials may be less available to the animal than the pure substance, so the total dietary level needed in practice could be greater than that indicated by experiment.

Disease. There is little information on the effects of acute and subacute infections on nutritional requirements, and the relation of nutrition to disease resistance. Because feed intake is reduced at the outset and during the course of a disease outbreak, it is often considered that increased dietary concentration of vitamins, minerals and proteins should be beneficial. Specific studies in experimental infections have shown that an increased level of vitamin A can be helpful in reducing mortality from coccidiosis[17]; a natural outbreak of this disease during a comprehensive study of vitamin D requirement showed that deficient chicks were less able to withstand the disease.[18] Studies with experimental fowl typhoid infection have shown beneficial effects from a supplementary mixture of vitamins; one of the beneficial components was vitamin C, which is not required normally.[19] The efficacy and required composition of the supplement varied between experiments. In contrast to these findings, it has been observed that the severity of coccidiosis infection can be reduced by *lowering* the concentration of dietary protein, because of a relationship between proteolytic enzymes and the infectious organism.[19] No specific recommendations can be made; it is one of the most important areas for further research.

Genetic differences. Differences between breeds and strains of chickens in their requirements for vitamins and other nutrients have been recognized. Current nutritional practice provides relatively generous levels of those vitamins which can be added at modest cost. Previously these could well have been the most limiting nutrients, and we can speculate that the present more generous (or unlimited) nutrition with respect to these factors has permitted progress through genetic selection which might otherwise not have been possible. The possibility must be recognized that substantial genetic changes in nutrient requirements may occur in poultry species which are subjected to intensive selection for production traits, and that this could occur over a relatively short time because of the rapid generation cycle and ease of multiplication of poultry. Present estimates of requirements, valid for the conditions under which they were determined, should not be considered as 'biological constants'.

Vitamin A

For poultry, beta-carotene has one half the activity of vitamin A alcohol per unit weight, at feeding levels approximating the dietary requirement (0·3 μg vitamin A = 1 International Unit = 0·6 μg beta-carotene). The older experiments on requirements were complicated by the instability of these compounds, especially vitamin A, in mixed diets. The estimated requirements have given greater weight to more recent studies conducted with synthetic vitamin A esters in mixtures protected against oxidation by antioxidants and gelatin coating; such mixtures are now widely used in feeding practice. The liver storage of vitamin A and its concentration in eggs are directly related to the dietary level. The requirement of the chick and its ability to withstand depletion are related to its stores at hatching time, which are dependent on the concentration in the egg. The requirement indicated for chicks assumes adequate but not excessive vitamin A nutrition of the hen.

Deficiency of vitamin A in chicks and poults produces retarded growth, weakness and ataxia. An opaque appearance of the eye may occur, due to abnormality of the third eyelid. Epithelial tissues of the mouth, esophagus and respiratory tract are affected. Urates accumulate in the ureters and kidneys. In adults, similar changes occur and egg production is markedly reduced.

Vitamin D

There is a large difference in efficacy between the D_2 (activated ergosterol) and D_3 (activated 7-dehydrocholesterol) forms of vitamin D. The requirements are stated in International Chick Units of vitamin D_3, the more effective form (1 I.C.U. = 0·025 μg vitamin D_2).

The requirement for vitamin D is related to dietary calcium and phosphorus. With optimal mineral levels and with the use of a phosphate of high availability, the D requirement is very low. In contrast, very high levels of D can improve the utilization of phosphate sources of relatively poor availability for calcification. The requirement estimates are based on adequate calcium and phosphorus levels from mixed plant and mineral sources.

Deficiency of vitamin D in growing birds produces impaired calcification (rickets) which is most evident in the legs, ribs and beak. The chicks are retarded in growth and walk with an ungainly gait and posture. In adults, egg production, shell calcification and hatchability are reduced, and the bones become fragile.

Vitamin E

Specific quantitative requirements for vitamin E (tocopherol) are not stated because the magnitude of the apparent requirement and the consequences of deficiency are interrelated with several other nutrients. The natural tocopherols which have vitamin E activity are antioxidants, and there is a fundamental question whether vitamin E is required as such or if its biological functions can be provided by other suitable antioxidants and the interrelated nutrients. Chickens have been carried through several generations on diets apparently devoid of vitamin E but containing synthetic antioxidants, without developing E deficiency symptoms.

A deficiency of vitamin E in chicks in the presence of unsaturated lipid (such as fish oil) and in the absence of other antioxidants produces a paralysis termed nutritional encephalomalacia, due to irreversible damage to the central nervous system. It can be prevented by vitamin E or any of several synthetic antioxidants; the latter exert their effects in the body tissues, rather than simply preventing fat peroxidation in the diet.

Deficiency of vitamin E and selenium in chicks produces an oedematous condition termed exudative diathesis. In turkeys, this deficiency causes a myopathy of the gizzard muscles. (See also Chapter 31.)

Deficiency of vitamin E and sulphur amino acids in chicks produces a nutritional muscular dystrophy, characterized by white striations in the breast muscle.

In adult chickens and turkeys, deficiency of vitamin E causes reduced hatchability, with little or no effect on egg production. Deficient turkey embryos exhibit abnormal eye development.

The most important natural sources of vitamin E for poultry are the cereal grains and alfalfa meal. Harvest damage and prolonged storage can greatly reduce their vitamin E content.

VITAMIN K

Deficiency of vitamin K causes delayed blood clotting. Deficient chicks succumb easily to haemorrhage. A deficiency has not been produced in adult poultry.

The vitamin K requirement can be greatly increased by certain drugs, particularly sulphaquinoxaline which has been widely used for coccidiosis prevention and treatment.

The major natural source of vitamin K is alfalfa meal, which contains the K_1 form. The synthetic compound 2-methyl-1,4-naphthoquinone in the form of its sodium bisulphite addition compound is widely used to supplement poultry diets.

RIBOFLAVINE

Riboflavine is one of the critical and sizeable requirements for growth and reproduction in all poultry species, and specific supplementation is usually required in diets for young birds and breeding hens. Deficiency in chicks causes retarded growth and 'curled-toe paralysis' in which they rest and walk on the hocks with their toes curling inward. In severe deficiency, hypertrophy of brachial and sciatic nerves is grossly evident. In adults, riboflavine deficiency causes poor hatchability, and in severe deficiency egg production is reduced. Riboflavine-deficient embryos show oedema and defective down development.

Milk products and fermentation products are rich sources of riboflavine. Synthetic riboflavine is commonly used in poultry diets.

NIACIN

Deficiency of niacin in chicks produces retarded growth, poor feathering and, in severe cases, inflammation of the tongue and mouth. Enlargement of the hock joint also occurs. Corn (maize), oats and soybean meal are among the poor sources of niacin, and diets based on these are marginal or deficient for chicks.

In turkeys and ducks, niacin deficiency causes a crippling leg disorder involving the hock joint, similar to perosis (see manganese). The requirement is high in these species, requiring liberal supplementation.

Synthetic nicotinic acid is commonly used to supplement poultry diets.

VITAMIN B_{12}

The presence of vitamin B_{12} (cobalamin) in animal tissues, and therefore in protein feedstuffs of animal origin, distinguishes them from the proteins of plant origin. Vitamin B_{12} accounts for some of the basis for the classical concept of the nutritional superiority of the animal proteins. Before the elucidation of vitamin B_{12}, it was known to be necessary to include some protein of animal origin in critical diets for poultry.

Deficiency of vitamin B_{12} produces growth retardation, anaemia, and lesions of the gizzard lining (gizzard erosion) in chicks. It is interrelated with choline and methionine in metabolism of 'single-carbon' units. The requirement is small, and is increased in diets high in fat or protein. In adults, vitamin B_{12} is required for normal hatchability.

In marked contrast to the other B-vitamins, vitamin B_{12} is tenaciously conserved. Reduction of hatchability requires a lengthy depletion of the reserves of the hen. Chicks from adequately fed hens have reserves which are also tenaciously held, and which will allow near-normal growth for an extended time with a deficient diet.

Sources of vitamin B_{12} for poultry diets include meat, fish and milk products, and supplements produced by fermentation. Under conditions of floor management, the requirement of breeder hens can largely or entirely be met by coprophagy, because microbial synthesis of the vitamin occurs in the excreta in the litter. (See also Chapter 7)

Choline

Choline is a dietary essential for young chicks and growing turkeys. Deficiency produces retarded growth and a leg abnormality (perosis) similar to that observed in manganese deficiency. Choline also serves as a methyl donor, a function shared by betaine which can spare part of the choline requirement. The requirement for choline is particularly high for the turkey.

The choline requirement of older growing poultry and adults is uncertain, but it appears that at least part of the need can be synthesized. Choline, vitamin B_{12} and folacin are interrelated in the metabolism of single carbon (methyl) units, and exhibit mutual sparing effects on requirements.

Synthetic choline chloride is commonly used to supplement critical poultry diets.

Pantothenic Acid

Deficiency of pantothenic acid in chicks, turkey poults and ducks causes retarded growth and the development of a characteristic dermatitis affecting the feet, the corners of the mouth, and the eyelids. Deficiency in adult hens causes lowered hatchability. Under field conditions in the U.S., a condition in chicks characterized by lack of vigour and high mortality in the first days following hatching has been found to be improved by dosing chicks with pantothenic acid, suggesting an unusual requirement due to genetic or environmental factors. Experimental pantothenic acid deficiency in breeding hens under laboratory conditions has shown similar effects.

Pantothenic acid is widely distributed in feeding materials, and milk products and fermentation products such as yeast are among the rich sources. Meat meals and corn (maize) are poor sources. Synthetic calcium pantothenate is commonly used to supplement poultry diets.

Folacin

Folacin deficiency in chicks produces retarded growth, poor feather development, anaemia and perosis (see manganese). In turkey poults, poor growth and feathering, a perosis-like hock disorder, and cervical paralysis are observed. Deficiency in adult chickens produces impaired hatchability.

Folacin deficiency may occur under field conditions when diets high in meat and fish meals and corn gluten are used. In general, the other vegetable proteins and alfalfa meal are rich sources. Synthetic folacin is available for use in poultry diets.

Thiamine

Deficiency of thiamine in young and adult chickens and turkeys causes inanition and the characteristic head retraction and convulsions of polyneuritis. Field deficiency of thiamine is unlikely, because the vitamin is normally supplied in liberal amounts by grains and grain products.

Vitamin B_6

Deficiency of vitamin B_6 (pyridoxine) has not been reported under field conditions, so this vitamin is not considered to be of significance in poultry feeding practice. Under laboratory conditions, a deficiency of vitamin B_6 produces retarded growth and characteristic convulsions which can be fatal. In adults, a deficiency produces inanition, weight loss, and failure of egg production and hatchability.

Biotin

Biotin deficiency in chicks and turkey poults produces a dermatitis affecting the feet, corners of the mouth, and the eyelids, and a perosis-like leg disorder (see Manganese). In adults, biotin deficiency causes reduced hatchability. Intestinal synthesis of biotin provides a significant amount of the vitamin to adults fed starch-based purified diets. Deficiency can be induced either by a low-biotin diet, or by feeding raw egg white which contains avidin, a protein which specifically binds biotin and makes it unavailable.

Biotin deficiency is unlikely under field conditions. Improvement of the growth of poults by supplementation of practical diets with biotin has been reported in Canada.

Vitamin Levels in Diet Formulation

The estimated requirements in Table 12 are intended to define dietary levels which will meet the requirements of a poultry population for maximum average performance of the indicated function (growth, etc.) and for prevention of all specific symptoms of deficiency. Since they do not contain any margin of operating safety such as to compensate for variations in feed composition, it is desirable to provide for somewhat higher levels in nutritional practice. The margin of excess to be used is a combined economic and nutritional question. It would be reasonable to provide 120–150 per cent of the estimated requirement for those nutrients of practical importance which can be added by synthetic or concentrated sources. Where stability and diet storage present problems, a somewhat higher excess of labile factors may be desirable.

Unidentified vitamins

It is now more than 15 years since the last of the known vitamins (B_{12}) was elucidated. During this time there have been many reports indicating the existence of additional uncharacterized nutrients, based on the growth-stimulating properties of various natural materials which are not explained on the basis of their known nutrient content. Fish meal, dried whey, fermentation products such as distillers' dried solubles and yeast, soybean meal, and alfalfa all have been found to produce such effects on growth in chicks and/or turkeys, and there could be as many as four distinct 'unidentified factors'. One difficulty in work on this problem is that the growth-stimulating effects are not large, and they are variable; however, the weight of evidence indicates a strong possibility that additional essential nutrients exist. Materials such as those cited are commonly included in critical diets (for early growth and reproduction) to provide for these possible needs.

MINERALS

Certain macroelements (calcium, phosphorus, sodium) present special problems in nutrition because it is desirable to meet the requirements without a large intentional excess. Excess calcium interferes with other nutrients; excess phosphate is costly; excess sodium may enhance water consumption

TABLE 13
Estimated Requirements[a] *of Poultry for Mineral Elements*

Mineral	Species and age or function							
	Growing chickens 1–8 weeks	Growing chickens 8–18 weeks	Layer chickens	Breeder chickens	Growing turkeys 1–8 weeks	Growing turkeys 8–16 weeks	Breeder turkeys	Growing pheasants
Calcium, per cent	1·0	1·0	3·0[b]	3·0[b]	1·2	1·2	2·25	1
Phosphorus, per cent	0·7[c]	0·6[c]	0·7[c]	0·7[c]	0·9[c]	0·8[c]	0·8[c]	0·6
'Available' phosphorus, per cent	0·45	0·4	0·5	0·5	0·6	0·5	0·5	0·5
Sodium, per cent	0·12	0·12	0·15	0·15	0·15	0·15	0·15	0·09
Potassium, per cent	0·2	0·16	?	?	0·4	?	?	?
Magnesium, per cent	0·05	?	?	?	?	?	?	?
Manganese, mg	55	?	?	33	55	?	33	?
Zinc, mg	35	?	?	?	70	?	?	?
Iron, mg	40	?	?	?	60	?	?	?
Copper, mg	4	?	?	?	6	?	?	?
Iodine, mg	0·4	0·4	0·3	0·3	?	?	?	0·3

[a] Requirements are stated in per cent of diet, or in mg per kg diet.
[b] See text for basis of estimating requirement according to conditions.
[c] See text for discussion of relative efficacy of forms of phosphorus.

and excretion, and create management problems. The microelements do not present the same problems, but it is desirable to avoid large excesses of them because physiological interactions are known to occur, and the interrelationships are not readily predictable. Where possible, the discussion which follows will indicate the desirable range of dietary concentration of each element. Estimated requirements are summarized in Table 13.

CALCIUM

Maximum growth rate and satisfactory bone calcification require an adequate level of calcium, a proper balance with phosphorus, and sufficient vitamin D. In practical diets containing 2700–3000 kcal ME per kg, the required calcium level for chicks is 0·9–1·0 per cent, and it is desirable not to exceed 1·2 per cent of the diet. For the turkey poult, the requirement is 1·2–1·5 per cent for early growth, and the diet level should not exceed 2 per cent. After 12 weeks of age, the turkey requirement is approximately 1 per cent of the diet. Excess calcium should be avoided because it can interfere with the utilization of manganese and zinc.

The calcium requirement of the adult hen depends on the rate of egg production and the level of feed intake. The daily need for calcium can be estimated as 2 grams per egg, and the efficiency of utilization of dietary calcium is estimated to be 50 per cent. To support an egg production level of 80 per cent would require 0·8 x 2 = 1·6 grams net calcium per hen, equal to twice this amount or 3·2 grams of diet calcium per day. If the feed intake were 106 grams per day, the diet should contain 3·0 per cent calcium.

In hot weather, both feed intake and the thickness of egg shells decline. A proportionate increase in diet calcium is desirable to compensate for the reduced feed intake. Further increase in calcium level will not improve the shell thickness because the decline under these conditions is due to the physiological effect of high temperature.

The calcium requirement of hens can also be met by providing them with free access to a source of calcium carbonate such as crushed oyster shells or calcite. A dolomite (mixed calcium and magnesium carbonates) should not be used because it would produce a magnesium excess.

PHOSPHORUS

The requirement for phosphorus (phosphate) is one of the most critical for all ages and types of poultry. It is needed both for calcification and for soft tissue metabolism and growth in the young; it serves the same functions in egg production and maintenance in adults.

There are wide differences in the availability (net value) of different phosphate sources. One of the reasons why phosphorus is a critical nutrient is that much of the phosphorus contained in plant materials is poorly utilized by birds. Measured by its comparative ability to promote bone calcification in growing chicks, phytin (mixed calcium–magnesium salt of phytic acid) is very poorly utilized.[20] On the other hand, recent evidence indicates that phytic acid is well utilized by chicks in the presence of liberal amounts of vitamin D.[21] The adult hen apparently can make better use of plant forms of phosphorus than the young chick. Under conditions of floor management, the dietary phosphorus requirement of hens is less than when they are in cages, suggesting that the plant phosphorus compounds are altered in the litter and coprophagy serves to provide a significant amount of phosphorus. The state of the phosphorus compounds in plant materials is not well understood, and more information on their nutritional value is needed. For practical formulation purposes the assumption is commonly made that the phosphorus of plant feedstuffs is 30 per cent utilized by poultry; 'available' phosphorus as used in Table 13 includes this amount of the plant phosphorus.

Inorganic phosphate sources such as dicalcium phosphate, steamed bone meal, defluorinated phosphates, and the feedstuffs of animal origin provide phosphorus in a highly available state. Unprocessed natural phosphates usually show lower availability for calcification in chicks. A by-product material from phosphate rock processing known as colloidal phosphate (or soft phosphate) has shown poor availability in most chick calcification tests. In recent work using high levels of vitamin D, higher availability for calcification comparable to dicalcium phosphate was found for this material; evidently the calcium level and vitamin D exert large effects on the utilization of this form of phosphorus. Questions of phosphorus availability and interaction will require more study.[22] Dicalcium phosphate, which is ordinarily a highly available form of phosphate, is poorly utilized by the turkey poult when fed as the anhydrous salt.

The estimated requirements for phosphorus are stated as total and available phosphorus, the latter term defined as the sum of the phosphorus from inorganic phosphates and animal products (assumed 100 per cent available) plus 30 per cent of the phosphorus from plant sources.

Sodium

The indicated sodium requirement levels can be met by the addition of sodium chloride, and this will adequately provide for the chloride requirement. Over the range from the required sodium level to approximately twice this amount in diets of chicks and hens, changes in sodium level will have little or no effect on water consumption. Higher sodium levels will tend to increase water intake and excretion. With free access to water, chicks can tolerate at least 2 per cent salt in the diet, and adults can tolerate much more. Turkey poults are more sensitive to salt and sodium excess.

Preliminary evidence has been reported indicating that limiting the chloride level in the diet of hens by providing part of the sodium as sodium bicarbonate enhances egg shell calcification, especially in warm weather. Anion-cation balance relationships may be more critical than has been recognized.

Manganese

Deficiency of manganese can be produced by low-manganese practical diets and accentuated by calcium excess. The primary symptom in chicks and poults is perosis, which is an enlargement and flattening of the hock joint followed by slipping of the Achilles tendon from the chondyles. The leg bones may bend and twist so that the lower leg is thrust to the side away from the body. The crippling is permanent and may affect one or both legs. Shortening of the long bones occurs, and there is some reduction in bone ash content. A similar disorder can be produced by deficiencies of zinc, choline, biotin, niacin or folacin. Prevention of a hock disorder in turkeys which resembles perosis also requires vitamin E.

In adults, a deficiency causes reduced egg production, shell thickness and hatchability. Deficient embryos exhibit chondrodystrophy, including a parrot-like beak and shortened limb bones.

Manganese sulphate and manganese carbonate are commonly used to supplement poultry diets.

Zinc

A deficiency of zinc in chicks and poults causes growth retardation, poor feather development, enlargement of the hock joint, and shortening of the limb bones. Skin changes (hyperkeratosis) occur. In adults, zinc deficiency causes reduced hatchability and limb abnormalities in the embryos.

The requirement of chicks for zinc is of the order of 15 mg per kg in a diet based on casein. In the presence of soybean protein, the requirement is increased by two-fold or more due to the binding of zinc by a complex of calcium, phytic acid and possibly other components of the protein. Even greater increase in requirement occurs in the turkey under these conditions.

Zinc availability can be improved in the presence of soybean protein by synthetic chelating agents such as ethylene diaminetetra-acetic acid.

The estimated requirements are sufficient to meet the needs in the presence of soybean protein. This high estimate was chosen because of the widespread use of soybean products. Sesame protein produces a similar effect, and other situations with a similar basis may be encountered under field conditions. The tolerance for excess zinc is quite high.

IRON

The requirement for iron is normally met easily by the amounts in common feeding materials.

Iron may be used to overcome the problem of gossypol toxicity in cottonseed meal feeding as discussed in another section.

SELENIUM

The newest addition to the list of required nutrients, selenium, has some unique nutritional properties (see also Chapter 31).

The apparent requirement of the chick for selenium for the prevention of exudative diathesis (related to vitamin E nutrition) is of the order of 0·1 mg per kg diet. The minimum toxic level is only about 10 times this amount. Under such circumstances, more information on function and need must be obtained before making an estimate of requirement which may be used in nutritional practice. It is of significant interest that a field deficiency of selenium affecting turkeys has been found in the United States, due to a local soil deficiency which produced selenium-low grain.

GRIT

The need for and function of grit is not a mineral 'requirement' in the ordinary sense, but can be appropriately considered with the minerals.

Insoluble grit (granite, pebbles) is readily consumed by young and adult poultry, and is retained in the gizzard over long periods. When birds have ready access to such grit, the consumption is usually far greater than the need. Grit is not required for the digestion and utilization of diets fed in ground meal form. Grit is needed for maximum digestibility when whole or cracked grains form part of the diet. It is common practice to provide grit periodically or continuously to growing and adult poultry, even when the diet is fed in meal form.

Calcite in grit form can be provided to hens to serve both as a grit and as a source of calcium. It may be used as a calcium source in the hen diet, or provided separately. It is inadvisable to use a calcite grit for growing birds; the proper calcium level should be provided in the diet, and the supplemental feeding of calcite could cause a serious calcium excess.

SPECIAL PROPERTIES OF LIPIDS

In addition to their function as a concentrated source of dietary energy, fats have several special effects in poultry nutrition.

Growth Stimulation

The growth rate of chicks is markedly improved when a well-utilized natural fat is added to a low-fat diet. Maximum growth rate is usually obtained with a total fat level of about 7 per cent. Part of the growth-stimulating effect is due to the essential fatty acid (linoleic acid) contribution of the fat, and it appears that the linoleic acid requirement is about 2 per cent of the diet (4 per cent of diet ME). Part of the growth stimulation is apparently due to the increased ME of the diet, because oleic acid is stimulatory in chicks which have good reserves of linoleic acid; in depleted chicks, linoleic acid is necessary to obtain improved growth. Fish oils are also effective growth stimulators under conditions in which linoleic acid is otherwise necessary, indicating a relationship between the polyunsaturated fatty acids of fish oils and linoleic acid.[23]

Increased egg size

Supplementing a low-fat diet for laying hens with an unsaturated fat during the early part of the egg production year produces a marked increase in egg size. The effect is mainly due to linoleic acid. Maximum effect requires about 2 per cent linoleic acid in the diet.[24]

Improved metabolic efficiency

Substitution of fat for carbohydrate on a ME basis in a chick diet produces significant improvement of the energetic efficiency of growth, measured as the recovery of dietary energy in the form of body tissue at a given energy intake. Similar findings have been reported in experiments with rats. The magnitude of improvement is such that the supplemental fat (at levels of 10–20 per cent of the diet) contributes an energetic effect 25 per cent greater than would be expected from its known ME content. This may be partly responsible for the growth-stimulating effect of fats in chick diets. The energetic improvement varies among fats. Corn oil, tallow and lard are effective; coconut oil has little or no effect.[25]

Tissue and egg composition

Because poultry, like other animal species, cannot synthesize linoleic acid, the amount of this fatty acid present in body and egg lipids depends on the dietary intake. When chicks are fed a low-fat diet, so that body lipids must largely be synthesized from carbohydrate, the body lipids contain approximately 40 per cent saturated fatty acids and 60 per cent oleic acid. Extracting this lipid and incorporating it in the diet of other chicks produces body lipid of similar composition, showing that presence or absence of dietary lipid does not affect this basic composition.[26] Adding a linoleate-rich fat to the diet causes the incorporation of linoleic acid in the body lipids primarily at the expense of oleic acid; the proportion of saturated fatty acids remains the same until quite high levels of unsaturated dietary fat are fed. A similar effect on egg fat occurs when a linoleate-rich fat is fed to hens; the linoleic acid content of egg fat can be raised to more than 40 per cent without greatly affecting the proportion of saturated fatty acids.[27] Typical practical diets produce eggs with about 15 per cent linoleic acid in the lipid. Some exceptions to this general pattern of fatty acid substitution also occur; for instance, when coconut oil is fed, myristic acid tends to be deposited in body fat at the expense of oleic acid.[28]

It has been known for many years from work with pigs that feeding cottonseed oil causes the deposition of a relatively hard (saturated) body fat, in spite of the highly unsaturated nature of the cottonseed oil. Recently it has been found that this is due to the presence of malvalic acid, a cyclopropenoid fatty acid, in cottonseed oil which causes the partial substitution of stearic acid for oleic acid in body lipids.[29] Cottonseed oil and other sources of cyclopropenoid fatty acids produce a similar effect on the body fat and egg fat of hens. In addition to the change in fatty acid composition of egg fat, feeding cottonseed oil or other cyclopropenoid sources produces a change in yolk structure such that iron diffuses from the yolk into the egg white where it reacts with conalbumin to produce a pink discoloration. An associated change is a rise in yolk pH. The discoloured eggs are considered damaged and unmarketable. Deposition of cyclopropenoid fatty acids in egg fat causes reduced hatchability.[30]

Changes in the amount and type of dietary fat, or even feeding cholesterol, produce little or no change in the cholesterol content of egg fat.

Pigmentation

The yellow colour of egg yolk and of the skin and shanks of yellow-skin varieties of poultry is produced by deposition of xanthophyll pigments derived directly from the diet. The intensity of yolk and skin colour is directly related to the dietary concentration of xanthophylls. Yellow corn (maize), corn gluten meal, and alfalfa meal are the major sources of xanthophylls among common feedstuffs; meals prepared from the petals of highly pigmented flowers are used in some countries. Pigmentation tends to be enhanced by the use of antioxidants in the diet, probably by preventing the oxidation of pigments. Supplementation of the diet with fat improves pigmentation by facilitating absorption. High dietary levels of vitamin A inhibit pigmentation. Regional consumer preferences relative to the degree of pigmentation vary widely.

ANTIBIOTICS

Small amounts of several antibiotics are capable of producing marked growth stimulation when added to the diet of the young chick. As little as 1–10 parts per million of such antibiotics as penicillin, bacitracin, tylosin and the tetracyclines are capable of giving maximum growth stimulation under many conditions. The concept is now generally accepted that antibiotics exert their favourable effect by overcoming the inhibitory effects of the micro-flora of the gut.[31] This is based on three kinds of evidence. First, the growth of chicks reared in clean quarters, uncontaminated by previous use for chickens, is not improved by antibiotic feeding; similar chicks in contaminated quarters commonly used for chickens generally show marked growth stimulation by antibiotic feeding, their maximum growth rate being equal to that of the chicks in the clean environment. This indicates a growth-retarding effect associated with environment. Second, the growth-stimulating effect does not occur when chicks are reared under germ-free conditions. Third, the establishment of a known controlled microflora in chicks has shown a marked inhibitory effect by certain organisms, such as *Clostridium welchii*, which can be prevented by feeding penicillin. The degree of response to small amounts of dietary antibiotics has been difficult to relate to specific changes in the intestinal flora, and it appears unlikely that a single kind of inhibitory effect is involved under all conditions.

Associated with the favourable effect of antibiotics on growth is a reduction in the weight and thickness of the intestine. The significance of this change is not known, but it suggests the possibility of improved gut function. It has recently been reported that dietary metabolizable energy was increased significantly by antibiotic supplementation under conditions in which the growth of chicks was improved 10–20 per cent.

Similar growth-stimulating effects produced by organic arsenical compounds and nitrofurans appear to be similar to the effects of antibiotics. Although the antibiotics and other growth stimulants are not nutrients, they are widely used as dietary components in countries which permit their use, and a part of their effect may be nutritional. There is experimental evidence that antibiotics tend to reduce the requirement for certain vitamins, possibly by preventing their uptake or destruction by the microflora.

SOME DIETETIC CONSIDERATIONS

Research with poultry has developed a substantial body of knowledge on their nutrition. Although many gaps exist in the tables of quantitative requirements, and the list of required nutrients may not be complete, enough is known about the most important needs to provide a sound basis for nutrition practice.

At the same time that research has developed the quantitative basis of poultry nutrition, industry has developed synthetic vitamins and amino acids, suitable phosphates and limestones, trace mineral salts, antioxidants, and the machinery and methods to process nutritional resources to fit the needs of poultry and other animals. This stands in marked contrast to the poultry nutrition of 35 years ago which depended largely on the by-products of other industries to meet the needs for poorly understood critical nutrients.

Because of these developments, poultry nutrition has become both simpler and more complex. The simplification is that the vitamin and trace minerals can be added as such, without having to choose the major components of the diet on the basis of their associated micronutrients. The protein and energy decisions which have the greatest bearing on least-cost formulation can largely be separated from micronutrient considerations. The complexities arise from the relatively large number of small components which are now included in a diet, and the control of dosage and uniformity.

There is a high degree of dietetic flexibility in the formulation of diets for poultry. The grain (carbohydrate) components can be chosen primarily on the basis of cost per unit metabolizable energy. The use of fats is optional because the basic requirement is usually met by the grains and other components, and can also be decided on the basis of cost per unit ME. The choice of proteins is more complicated, but can be decided primarily on the basis of amino acid needs.

The foundation of nutrition for poultry production is necessarily the cereals, proteins and forage resources of a region. Poultry are also efficient users of byproducts, and virtually any wholesome material with nutritional value can be utilized in poultry feeding according to its composition. Unique local resources can result in unusual but satisfactory diets for poultry, converting an otherwise wasted material into valuable human food. Only limited data on composition of feeding resources have been included in this discussion, but extensive summaries are available.[32]

The position of any component of our food production industries depends ultimately on its relative efficiency, taking into consideration all of the input requirements. The remarkable improvement in production efficiency which has been achieved by the poultry industry in the last 20 years has resulted from genetic improvement of stocks and improved disease control, as well as improved nutrition. Data illustrating the efficiency of food production from raw materials by meat chickens and laying hens are summarized in Tables 14a and 14b, based on rates of performance which are typical of well managed production units. These data illustrate the conversion of feedstuffs to human food on the bases of weight, energy, and protein. The recoveries of the most limiting essential amino acids, and many of the vitamins, are also relatively high. As it becomes increasingly necessary to examine critically the nutritional cost of converting plant materials to animal products to meet human requirements, a special function of the animal should not be overlooked. It frequently serves as a highly efficient 'filter', screening out from its tissues and products naturally occurring substances which would be harmful to humans if consumed directly. Two examples which come to mind are gossypol, the cottonseed toxin, and the aflatoxins. The process of conversion to animal products can serve as a highly effective nutritional barrier to some of the hazards of our environment.

TABLE 14a

Estimates of Nutritional Efficiency of Broiler Meat Production

Based on production of:		
1·6 kg live weight		
64·5 per cent water		
23 per cent protein		
8 per cent fat		
50 per cent edible meat yield (800 g)		
20 per cent protein		
7 per cent fat		
Requires:		
3·5 kg feeds, containing		
700 g protein		
10·5 mcal ME		
Efficiency:		
Chicken yield/feed input	Total	Edible
weight/weight	0·45	0·23
kcal/kcal	0·25	0·11
protein/protein	0·53	0·23

TABLE 14b

Estimates of Nutritional Efficiency of Egg Production

Based on production of:
250 eggs per layer-year
15 kg eggs
21·5 mcal energy
1·75 kg egg protein
Requires:
40 kg feeds, containing:
113 mcal metabolizable energy
6·5 kg protein
Efficiency:
0·38 kg egg/kg feed
0·19 kcal egg/kcal feed
0·27 g egg protein/g feed protein

REFERENCES

1. National Academy of Sciences—National Research Council, *Nutrient Requirements of Poultry.* Publication 1345 Washington, D.C. (1966).
2. Agricultural Research Council, London. *The Nutrient Requirements of Farm Livestock.* No. 1, *Poultry.* London (1963).
3. JENSEN, L. S., FREY, R. E., ALLRED, J. B. and MCGINNIS, J. *Poultry Science*, **36**, 919 (1957).
   LEONG, K. C., JENSEN, L. S. and MCGINNIS, J. *Poultry Science*, **41**, 36 (1962).
   RICKES, E. L., HAM, E. A., MASCATELLI, A. and OTT, W. H. *Arch. Biochem. Biophys.* **96**, 371 (1962).
4. KRATZER, F. H. and VOHRA, P. *Proc. 6th Intl. Congress of Nutrition*, p. 122 (1963).
   VOHRA, P. and KRATZER, F. H. *Poultry Science*, **43**, 502 (1964).
5. CHANG, S. I. and FULLER, H. L. *Poultry Science*, **43**, 30 (1964).
6. CARPENTER, K. J. and CLEGG, K. M. *J. Sci. Food and Agr.* **7**, 45 (1956).
   BOLTON, W. *Analyst*, **85**, 189 (1960).
7. RENNER, R. and HILL, F. W. *Poultry Science*, **39**, 849 (1960).
   RENNER, R. and HILL, F. W. *J. Nutrition*, **74**, 254 (1961).
8. YOUNG, R. J. and GARRETT, R. L. *J. Nutrition*, **81**, 321 (1963).
   RENNER, R. and HILL F. W. *J. Nutrition*, **74**, 259 (1961).
9. CAREW, L. B., Jr., HILL, F. W. and NESHEIM, M. C. *J. Am. Oil Chem. Soc.* **38**, 249 (1961).
10. HILL, F. W., *Poultry Science*, **35**, 59 (1956); also Chapter 1, pp. 3-17 in *Nutrition of Pigs and Poultry*, eds. Morgan, J. T. and Lewis, D., Butterworths, London (1962).
11. SCOTT, M. L., HILL, F. W., PARSON, E. H., Jr., BRUCKNER, J. H. and DOUGHERTY, E. *Poultry Science*, **38**, 497 (1959).
    SINGSEN, E. P., MATTERSON, L. D., TLUSTOHOWICZ, J. and POTTER, L. M. *Poultry Science*, **37**, 1243 (1958).
12. DONALDSON, W. E., COMBS, G. F. and ROMOSER, G. L. *Poultry Science*, **35**, 1100 (1956).
13. NELSON, T. S., YOUNG, R. J., BRADFIELD, R. B., ANDERSON, J. B., NORRIS, L. C., HILL, F. W. and SCOTT, M. L. *Poultry Science*, **39**, 309 (1960).
    CAREW, L. B., Jr. and HILL, F. W. *J. Nutrition*, **74**, 185 (1961).
14. NELSON, O. E., MERTZ, E. T. and BATES, L. S. *Science*, **150**, 1469 (1965).
15. CARPENTER, K. J. *Biochem. J.* **77**, 604, (1960).
16. LEVEILLE, G. and FISHER, H. *J. Nutrition*, **66**, 441 (1958).
    LEVEILLE, G., SHAPIRO, R. and FISHER, H. *J. Nutrition*, **72**, 8 (1960).
17. ERASMUS, J., SCOTT, M. L. and LEVINE, P. P. *Poultry Science*, **39**, 565 (1960).
    GERRIETS, E. *British Veterinary Journal*, **118**, 507 (1961).
18. MURPHY, R. R., HUNTER, J. E. and KNANDEL, H. C. *Pa Agr. Expt. Station Bulletin*, 334 (1936).
19. HILL, C. H., KELLY, J. W. and BARBER, C. W. *Proc. Soc. Exp. Biol. Med.* **88**, 535 (1955).
    HILL, C. H. and GARREN, H. W. *Annals N. Y. Acad. Sci.* **63**, 186 (1955).
    BRITTON, W. M., HILL C. H. and BARBER, C. W. *J. Nutrition*, **82**, 306, (1964).
20. GILLIS, M. B., NORRIS, L. C. and HEUSER, G. F. *Poultry Science*, **28**, 283 (1949).
    GILLIS, M. B., NORRIS, L. C. and HEUSER, G. F. *J. Nutrition*, **52**, 115 (1954).
21. TEMPERTON, H. and CASSIDY J. *British Poultry Science*, **5**, 75, 81, 87 (1964).
    WALDROUP, P. W., AMMERMAN, C. B. and HARMS, R. H. *Poultry Science*, **43**, 426 (1964): also *Poultry Science*, **44**, 880 (1965).
22. MOTZOK, I., ARTHUR D. and BRANION, H. D. *Poultry Science*, **44**, 1261 (1965).
23. RAND, N. T., SCOTT, H. M. and KUMMEROW, F. A. *Poultry Science*, **37**, 1075 (1958).
    DAM, R., LEACH, R. M., NELSON, T. S., NORRIS, L. C. and HILL, F. W. *J. Nutrition*, **68**, 615 (1959).
    EDWARDS, H. M., Jr. and MARION, J. E. *J. Nutrition*, **81**, 123 (1963).
24. SCHUTZE, J. V. and JENSEN, L. S. *Poultry Science*, **42**, 921 (1963).
25. CAREW, L. B., Jr., and HILL, F. W. *J. Nutrition*, **83**, 293 (1964).
    CAREW, L. B., Jr., HOPKINS, D. T. and NESHEIM, M. C. *J. Nutrition*, **83**, 300 (1964).
26. REISER, R. and DIECKERT, J. W. *J. Am. Oil Chem. Soc.* **31**, 625 (1954).
    REISER, R. *J. Nutrition*, **42**, 325 (1950).

27. Wheeler, P., Peterson, D. W. and Michaels, J. D. *J. Nutrition*, **69**, 253 (1959).
28. Feigenbaum, A. S. and Fisher, H. *Arch. Biochem. Biophys.* **79**, 302 (1959).
29. Evans, R. J., Bandemer, S. L., Anderson, M. and Davidson, J. A. *J. Nutrition*, **76**, 314 (1962).
30. Kemmerer, A. R., Heywang, B. W. and Vavich, M. G. *Poultry Science*, **40**, 1045 (1961). Kemmerer, A. R., Heywang, B. W., Vavich, M. G. and Phelps, R. A. *Poultry Science*, **42**, 893 (1963).
31. Coates, M. E., Dickinson, C. D., Harrison, G. F., Kon, S. K., Porter, J. W. G., Cummins, S. H. and Cuthbertson, W. F. J. *J. Sci. Food and Agr.* **3**, 43 (1952). Forbes, M. and Park, J. T. *J. Nutrition*, **67**, 69 (1959). Lev, M. and Forbes, M. *British J. Nutrition*, **13**, 78 (1959).
32. National Academy of Sciences—National Research Council, *Joint United States-Canadian Tables of Feed Composition.* Publication 1232 Washington, D.C. (1964).

CHAPTER 31

Antioxidants, Selenium and Sulphur Amino Acids in the Vitamin E Nutrition of Chicks

M. L. SCOTT

Department of Poultry Science and Graduate School of Nutrition, Cornell University, Ithaca, N.Y., U.S.A.

VITAMIN E, discovered over 40 years ago, has since been shown to be concerned in the normal functioning of most of the tissues of the animal body and in preventing many deficiency diseases in a variety of animal species. The signs of vitamin E deficiency, even in a single species, may differ with environmental or dietary conditions. Studies on the effects of feeding animals diets deficient in vitamin E have been reviewed by Mason (1944, 1949, 1954), Pappenheimer (1948), Blaxter and Brown (1952), Mackenzie (1953), Dam (1957) and Scott (1962 a,b). During the 1920's, vitamin E was recognized only as an unidentified fat-soluble factor required for reproduction in rats. After the classical experiments of Pappenheimer and Goettsch (1931, 1933 a,b) and Goettsch and Pappenheimer (1931) which showed that vitamin E is required also for prevention of encephalomalacia and nutritional muscular dystrophy, studies were made with a wide variety of experimental animals, which in time led to the recognition of vitamin E deficiency diseases as the cause of many different pathological conditions (Table 1).

Vitamin E was isolated as α-tocopherol by Evans, Emerson and Emerson (1936) about 13 years after its discovery in the same laboratory (Evans and Bishop, 1922; see also Evans and Burr, 1927). Its chemical structure was elucidated by Fernholz (1938) and its chemical synthesis was achieved in the same year by Karrer *et al.* (1938) and by Smith, Ungnade and Prichard (1938). Later it was found that other tocopherols with some vitamin E activity existed naturally or could be produced by chemical synthesis. None of them has been shown to equal D-α-tocopherol in biological activity. The chemistry and biochemistry of vitamin E have been reviewed recently by Vasington, Reichard and Nason (1960), Bro-Rasmussen and Hjarde (1961), Isler *et al.* (1962) and Kofler *et al.* (1962). In addition to a complete review of its chemistry, an excellent survey of the biological significance of the vitamin was made in a Symposium on Vitamin E and Metabolism in honour of Professor H. M. Evans at Zurich, Switzerland in 1962.

In view of the diversity of tissues affected by vitamin E deficiency it is logical that the vitamin has several different metabolic functions in the

TABLE 1
Pathology of Vitamin E Deficiency

Condition	Experimental animal	Tissue affected	Prevented by vitamin E	Prevented by selenium
Reproductive failure				
Embryonic degeneration	Female rat, hen, turkey, ewe	Vascular system of embryo	Yes	No
Sterility	Male rat, guinea pig, hamster, dog, cock	Male gonads	Yes	No
Muscle degeneration				
Muscular dystrophy	Rabbit, guinea pig, duck, chick, turkey poult	Skeletal muscle	Yes	No,or only partially
Stiff lamb disease	Lamb, kid	Skeletal muscle	Yes	Yes
White muscle disease	Calf, sheep, mouse, mink	Skeletal and heart muscle	Yes	Yes
Myopathy of gizzard	Turkey poult	Gizzard muscle	Yes	Yes
Necrotic liver degeneration	Rat, pig	Liver	Yes	Yes
Cerebellar degeneration				
Encephalomalacia	Chick	Cerebellum (Purkinje cells)	Yes	No
Exudative diathesis	Chick, turkey poult	Capillary walls	Yes	Yes
Erythrocyte destruction	Rat, chick, man (premature infant)	Blood, hemolysis of erythro-cytes	Yes	No
Blood protein loss	Chick, turkey poult	Blood, serum albumin severely reduced	Yes	Yes
Body lipid degeneration				
Steatitis (yellow fat disease)	Mink, pig, chick	Depot fat, oxida-tive rancidity *in vivo*	Yes	No
Depigmentation of maxillary incisors	Rat, hamster	Teeth	Yes	No
Kidney degeneration	Rat, monkey, mink	Tubular epithelium	Yes	Yes
Respiratory decline	Rat	Liver biochemistry	Yes	No

animal body, and that it affects the metabolism of other nutrients. Although no single primary metabolic role for vitamin E has yet been demonstrated, the studies have pointed to activity of the vitamin in several different metabolic systems: (1) as a biological antioxidant; (2) in normal tissue respiration, possibly to aid in some unknown way the functioning of the cytochrome *c* reductase system (Nason and Lehman, 1956; Vasington *et al.*, 1960) or to protect the lipid structures in the mitochondria from oxidative destruction (Zalkin and Tappel, 1960); (3) in normal phosphorylation reactions, especially of high energy phosphate compounds such as creatine phosphate and adenosine triphosphate (Carpenter, McCay and Caputto, 1958; Fitch and Dinning, 1959; Nesheim, Leonard and Scott, 1959; Calvert, Monroe and Scott, 1961); (4) in metabolism of nucleic acids (Dinning and Day, 1958); (5) in synthesis of ascorbic acid (Caputto, McCay and Carpenter, 1958; Kitabchi *et al.*, 1960); (6) in synthesis of ubiquinone (Edwin *et al.*, 1961; Green, *et al.*, 1961); and (7) in sulphur amino acid metabolism (Scott and Calvert, 1960, 1962; Desai, Calvert and Scott, 1964). The extent to which vitamin E is primarily and specifically involved in these functions remains to be determined.

VITAMIN E DEFICIENCY IN CHICKENS

Vitamin E is required for normal fertility in the cock (Adamstone and Card, 1934) and for normal reproductive performance in the hen (Adamstone, 1931). Deficiency of vitamin E in young chicks may result in body lipid degeneration (Dam and Granados, 1945) and erythrocyte hemolysis (Muytjens, 1956). Chicks suffering from acute vitamin E deficiency may show one or more of three well-defined deficiency diseases—encephalomalacia (Pappenheimer and Goettsch, 1931), exudative diathesis (Dam and Glavind, 1938), and muscular dystrophy (Dam, Prange and Søndergaard, 1952).

Although some similarities are observed in the tissues of chicks suffering from these diseases, it appears that no common metabolic defect can be held responsible for all three conditions, since various specific dietary changes unrelated to the vitamin E content of the diet can completely prevent one of the diseases without having any effect upon the other two. The most important of these are the prevention of encephalomalacia with synthetic antioxidants, the effectiveness of inorganic selenium in prevention of exudative diathesis, and the role of cystine in preventing muscular dystrophy.

BIOLOGICAL ANTIOXIDANT ACTION OF VITAMIN E IN PREVENTION OF ENCEPHALOMALACIA

Encephalomalacia (Fig. 1) is an ataxia resulting from haemorrhages and oedema within the molecular and/or granular layers of the cerebellum (Fig. 2) with pyknosis and eventual disappearance of the Purkinjé cells and separation of the molecular and granular layers of the cerebellar folia (Fig. 3) (Howes and Hutt, 1952; Scott and Stoewsand, 1961).

Dam and Granados (1945) showed that in certain of the body fats of vitamin E deficient chicks formation of lipoperoxides can occur *in vivo* and can be prevented by supplementing the diet with vitamin E. Michaelis and Wollman (1950) have shown that vitamin E can be converted readily to a free semiquinone radical. Since the current theory of protection by antioxidants against peroxidation is that it occurs through neutralization by the antioxidant of the free radical which is produced as one of the first changes in the auto-oxidation of fats, it appears that at least one important function of vitamin E is in the inhibition of auto-oxidation of fats, not only in the diet of animals but also within the animal body. A number of workers (Dam *et al.*, 1958; Century and Horwitt, 1959; Machlin, Gordon and Meisky, 1959; Machlin and Gordon, 1960) have provided evidence that in prevention of encephalomalacia in chicks, vitamin E functions as a biological antioxidant closely, if not specifically, related to linoleic acid metabolism. Sugai *et al.* (1960); Kokatnur, Kummerow and Scott (1960); and Nishida *et al.* (1960) have suggested that vitamin E protects chicks against encephalomalacia by preventing the breakdown of linoleic acid to 12-oxo-*cis*-9-octadecenoic (keto) acid, since α-tocopherol gave protection against encephalomalacia in the presence of 1·5 per cent methyl linoleate but not in the presence of 0·25 per cent of the keto acid.

From time to time during recent years many workers have presented evidence showing that antioxidants other than tocopherol are capable of preventing many signs of vitamin E deficiency in animals. The early work as described by Dam (1957), Singsen *et al.* (1955) and Bunnell *et al.* (1955) focused attention on this when they showed that the antioxidant, diphenyl-*p*-phenylenediamine (DPPD), prevented encephalomalacia when given by mouth or subcutaneously to chicks with values of vitamin E in blood too low to be detected by modern chemical procedures. Since then, a controversy has developed whether antioxidants prevent certain vitamin E deficiency diseases by completely replacing vitamin E in the metabolism of the animal, or whether they are effective only because they prevent destruction of traces of vitamin E in the diet or in the animal body. Recent studies by Machlin and Gordon (1960), Scott and Stoewsand (1961) and others provide evidence that antioxidants such as DPPD, 6-ethoxy-2:2:4-trimethyl-1:2-dihydroquinoline (ethoxyquin) and methylene blue are effective in preventing encephalomalacia in chicks with very low intakes of vitamin E over prolonged periods.

Further, Scott and Stoewsand (1961) found that DPPD was effective in preventing encephalomalacia even when it was given daily to chicks receiving a purified basal diet free from antioxidants and vitamin E, which was allowed to develop complete oxidative rancidity during the course of the experiment and on which, without DPPD in their diet, three-quarters of the chicks died of encephalomalacia.

The fact that low concentrations of such antioxidants as DPPD and ethoxyquin are capable of preventing encephalomalacia in chicks, but fail to prevent exudative diathesis or muscular dystrophy in the same chicks, strongly suggests that in preventing encephalomalacia vitamin E acts as an

FIG. 1. Encephalomalacia. Severe ataxia, tremors, retraction or twisting of the head, and clonus of the legs.

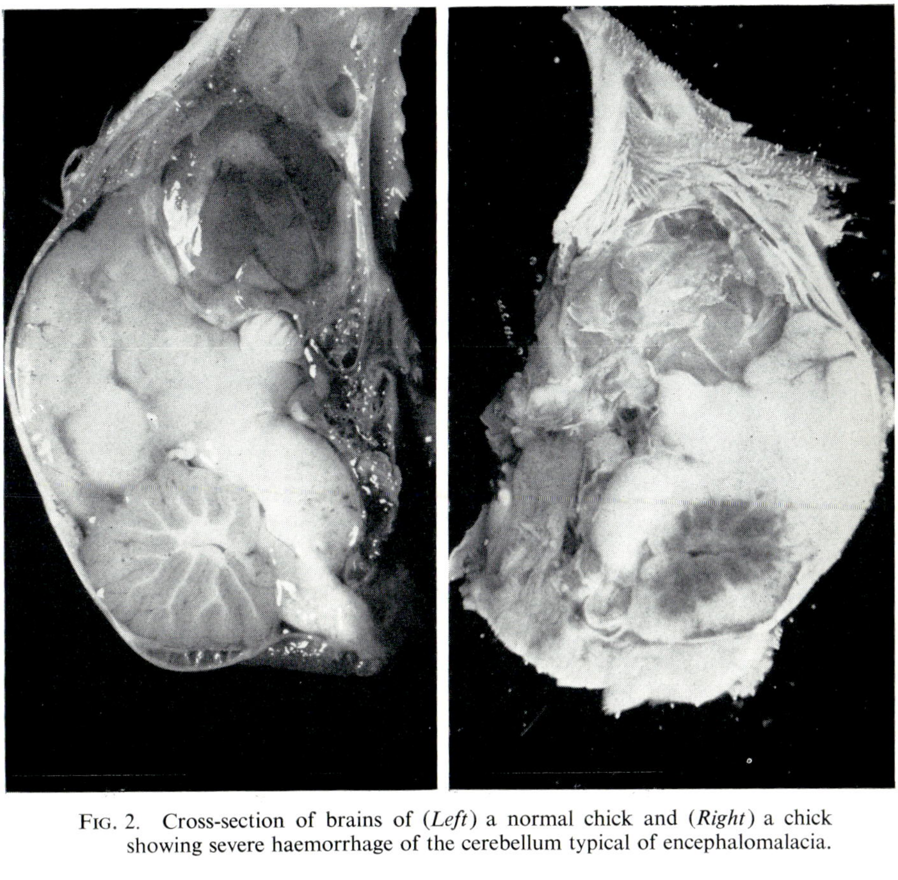

FIG. 2. Cross-section of brains of (*Left*) a normal chick and (*Right*) a chick showing severe haemorrhage of the cerebellum typical of encephalomalacia.

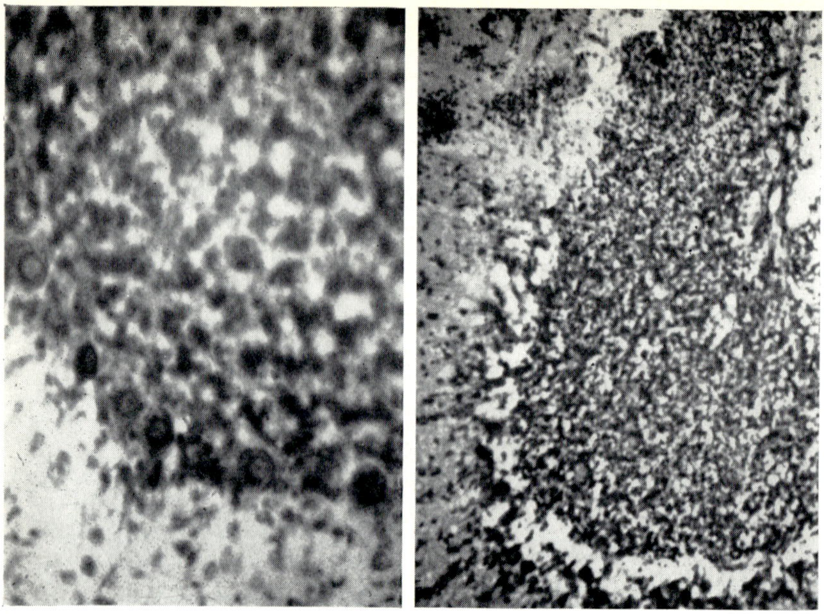

FIG. 3. Histological section of folia of cerebellum of (*Left*) normal chick showing intact Purkinje cells lying between the granular and molecular layers; and (*Right*) brain of chick suffering from encephalomalacia showing marked pyknosis and degeneration of Purkinje cells and oedema causing separation of molecular and granular layers of the cerebellum.

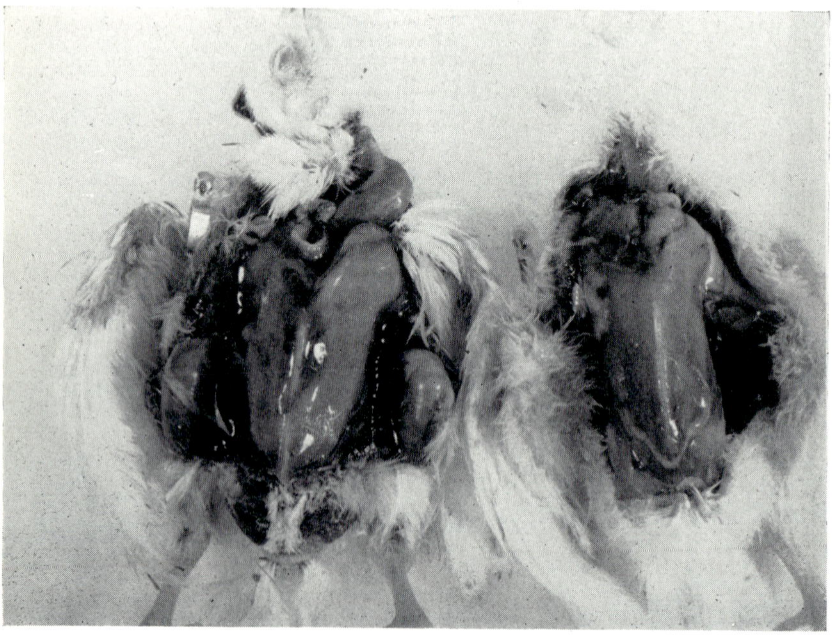

FIG. 4. Exudative diathesis. Severe oedema resulting in appearance of a greenish, viscous fluid accumulating largely under the skin covering the abdomen.

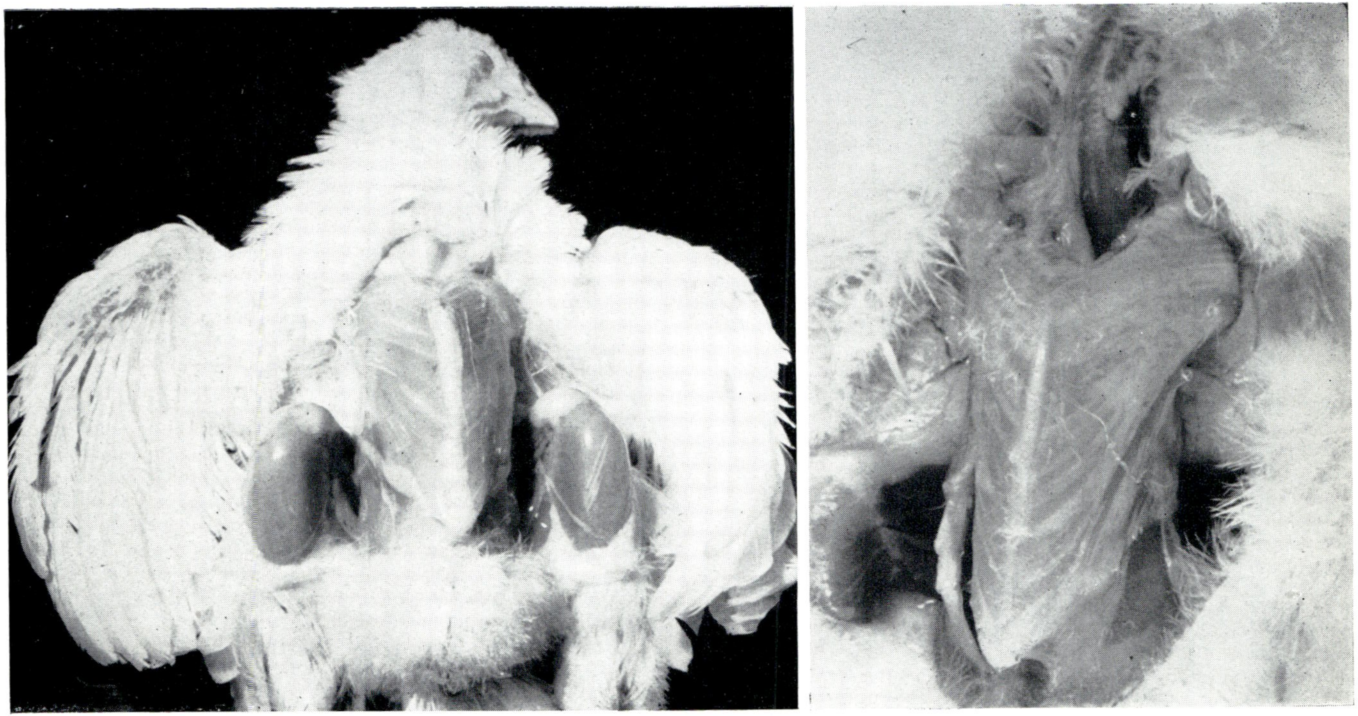

FIG. 5. Nutritional muscular dystrophy. Chick on left was given a severity score of 2; the one on right was scored 4.

antioxidant. This fact suggests that such action is nonspecific and can be performed by antioxidants which have chemical and biological properties so like those of the vitamin that they can enter into metabolic schemes where vitamin E acts as an antioxidant. Søndergaard, Scott and Dam (1962) found that phytyl-ubichromenol prevents encephalomalacia and suggested that the presence of this compound in brewer's and torula yeasts probably accounts for the effectiveness of yeasts in prevention of this vitamin E deficiency disease.

Machlin *et al.* (1959) reported prevention of exudative diathesis and muscular dystrophy by ethoxyquin, but the amount required was so high as to be almost toxic. Since vitamin E in extremely low concentrations easily prevents these diseases, its effect in preventing exudative diathesis and muscular dystrophy appears to be due to its participating in metabolic processes which are more specific than simple prevention of peroxidation within the animal body.

The discovery that a low concentration of ethoxyquin prevents encephalomalacia has aided studies of exudative diathesis and muscular dystrophy in chicks. Complete prevention of encephalomalacia has been regularly achieved in experiments on exudative diathesis and muscular dystrophy by supplementing the basal diets with 0·0125–0·025 per cent ethoxyquin.

SELENIUM IN PREVENTION OF EXUDATIVE DIATHESIS

Exudative diathesis in chicks, depicted in Fig. 4, is a severe oedema produced by a marked increase in capillary permeability. It was first recognized as a vitamin E deficiency disease in chicks by Dam and Glavind (1938, 1939, 1940). They obtained evidence of the abnormally high capillary permeability by injecting trypan blue intravenously into normal chicks and chicks suffering from exudative diathesis and showing rapid diffusion of the dye into the muscles of deficient chicks, but no diffusion from the capillaries of normal chicks. Goldstein and Scott (1956) presented evidence of a diffusion of blood proteins by showing that as exudative diathesis progresses, electrophoretic patterns of the blood show decreased albumin levels whereas paper electrophoresis of the exudate fluids indicates increases in tissue-fluid proteins. Dam and Glavind (1942) showed that although vitamin E is effective in preventing exudative diathesis, the disease can be prevented also by a number of other changes in the diet which are totally unrelated to its vitamin E content.

For some years several workers (Schwarz, 1951; Scott *et al.*, 1955) have recognized the existence in brewer's dried yeast and in other natural foodstuffs of an unidentified substance which reduces the amount of vitamin E required to prevent dietary necrosis of the liver in vitamin E deficient rats and exudative diathesis in vitamin E deficient chicks. Schwarz referred to the substance as Factor 3.

The effectiveness of dietary selenium, as inorganic sodium selenite, for prevention of exudative diathesis in chicks was reported simultaneously by Scott and associates (1957) and Stokstad *et al.* (1957) following the initial discovery by Schwarz and Foltz (1957) that inorganic selenium compounds prevent dietary liver necrosis in rats. Reviews on the nutritional significance of selenium have been presented by Schwarz (1961), Nesheim and Scott (1961), Hartley and Grant (1961), Schubert *et al.* (1961), Shrift (1961), and Scott (1962 a,c).

Although a number of inorganic and organic selenium compounds have been shown to be effective in preventing dietary necrosis of the liver, Schwarz and Foltz (1958) claimed that Factor 3 was more potent than any other selenium compound hitherto tested. Only 0·7 μg selenium per 100 g in the form of a Factor 3 concentrate from pig kidney powder was required to prevent necrosis of the liver in half the rats used, but with a large number of inorganic and organic compounds of selenium tested by Schwarz and Foltz about 2–3 μg selenium per 100 g was required to give the same degree of prevention of necrosis. They included selenic acid; selenium dioxide; sodium selenate and selenite; and potassium selenocyanate; selenium analogues of the sulphur amino acids (selenocystine, selenomethionine and seleno-cystathionine); as well as such organic compounds as dibenzyl diselenide; bis(2-nitro-4-methylphenyl)-triselenide; 2:4-dinitrobenzeneselenic acid and 6-selenopurine; all of which prevented necrosis of the liver in half the rats when they provided 2 to 2·5 μg selenium per 100 g of diet. As a result of their studies, Schwarz and Foltz suggested that the metabolically active form of selenium is an unknown specific organic selenium compound with about 2·5 to 3 times the potency of the inorganic and organic selenium compounds tested.

With a basal semi-purified diet severely deficient in both selenium and vitamin E in which torula yeast was the source of protein, Nesheim and Scott (1958) showed that selenium, 8 μg per 100 g as sodium selenite, not only completely prevented exudative diathesis, but had an effect on growth greater than that of vitamin E alone and equal to that obtained by giving both vitamin E and selenium. Inorganic selenite also was shown to be as effective as vitamin E in restoring the reduced values of blood albumin (Goldstein and Scott, 1956) in chicks given the basal vitamin E deficient diet. In the presence of adequate vitamin E, maximum growth was obtained by adding to the diet 4 μg selenium as selenite per 100 g diet; without selenium growth was poor and mortality between 5 and 8 weeks of age was high. The discovery that selenium is required for both growth and survival is strong evidence that it is indeed an essential nutrient.

Exudative diathesis has been studied also with purified diets containing no torula yeast (Nesheim and Scott, 1958). Purified diets deficient in vitamin E and selenium, containing either isolated soya bean protein or crystalline amino acids, produced a high incidence of exudative diathesis in chicks. In all instances addition of selenium as sodium selenite completely prevented it. It has not been possible to confirm the growth-promoting effect of selenium when added to such diets together with adequate amounts of vitamin E. However, neutron activation analysis (Leddicotte and Reynolds, 1951) of the basal diets showed that they contained at least 4 μg selenium per 100 g, the amount previously shown to give maximum growth when added in the presence of vitamin E to the torula yeast diet.

SELENIUM AND MUSCULAR DYSTROPHY

A myopathy in lambs generally referred to as 'stiff lamb disease' occurs sporadically in Western New York State. It was first recognized and described by workers at Cornell University (Metzger and Hagan, 1927; Willman, Asdell and Olafson, 1934). The disease was produced experimentally by feeding pregnant ewes on a diet of hay and cull beans obtained from certain areas of New York State. The hay found to be most effective in producing the disease year in, year out was a second-cut alfalfa-grass hay given with cull red kidney beans. In recent years, trefoil-grass hay has been used instead of alfalfa-grass hay with no change in the percentage of lambs affected, which usually ranged from 30 to 70; occasionally there were none. Vitamin E was known to be fairly effective in preventing the disease, but results were not always consistent. Further, it was shown that many feeding stuffs had profound effects on stiff lamb disease although they did not alter the vitamin E content of the diet of the ewes.

A similar nutritional myopathy of lambs and calves occurs in a number of specific areas in the Western United States, particularly in isolated areas of Oregon and Montana. Because of the white appearance of the muscles which have undergone the 'Zenker's degeneration' characteristic of the disease, it is commonly referred to as 'white muscle disease'.

After the reports on the effects of inorganic selenium compounds on dietary liver necrosis in rats and exudative diathesis in chicks, studies were initiated almost simultaneously by Proctor, Hogue and Warner (1958), of Cornell University and Muth and co-workers (1958) of Oregon State College to determine the effects of supplementing with selenium the basal diets for pregnant ewes which had previously been shown to produce the nutritional myopathy in lambs. Both research groups reported almost complete prevention of nutritional myopathy in newborn lambs by addition of 0·1 to 1·0 ppm selenium as sodium selenite to the diets of the ewes.

In New Zealand, where exudative diathesis has been seen in commercial flocks of chicks (Thompson and Smith, 1953) white muscle disease in lambs and calves has been a serious economic problem for years. On some farms 40 to 50 per cent of the lambs born to apparently healthy ewes died from congenital white muscle disease. Associated with the high incidence of white muscle disease was a high degree of infertility in the ewes; on six farms studied by Grant, Hartley and Drake (1960), 15 to 70 per cent of the ewes reared on those farms failed to produce young. In a series of studies by Drake, Grant and Hartley (1960) and Jolly (1960 a,b) giving selenium salts by mouth to ewes, unthrifty lambs and calves resulted in dramatic prevention of white muscle disease and marked improvement of growth of the lambs and calves. Giving 2·5 to 5 mg selenium monthly, as sodium selenate in aqueous solution by mouth, greatly improved fertility of the ewes, and improved rate of growth and reduced mortality of the lambs. This work shows that selenium must be included in the list of essential trace minerals.

OTHER STUDIES WITH SELENIUM

In addition to its effects on liver necrosis in rats, exudative diathesis in chicks and muscular dystrophy in lambs and calves, selenium has been shown to prevent 'multiple necrotic degeneration' (necrosis of heart, liver, muscle and kidney) in mice (DeWitt and Schwarz, 1958), exudative diathesis (Rahman *et al.*, 1959) and myopathy of the gizzard (Walter and Jensen, 1964) in turkey poults and dietary liver necrosis in pigs (Eggert *et al.* 1957).

Selenium did not prevent encephalomalacia in chicks. It had no effect in the gestation resorption test for vitamin E deficiency in rats (Harris, Ludwig and Schwarz, 1958) and failed to improve muscular dystrophy in vitamin E deficient rabbits (Hove, Fry and Schwarz, 1958; Draper, 1958).

SELENIUM AND MUSCULAR DYSTROPHY IN CHICKS

Nutritional muscular dystrophy in chicks, pictured in Fig. 5, is characterized by degeneration of the muscle fibres, especially of the breast (pectoral) muscles, but also occurring occasionally in the leg muscles. Histologically, these muscles show Zenker's degeneration with perivascular infiltration; marked accumulation of infiltrated eosinophils, lymphocytes, and histiocytes. These cells, together with the nuclei of the degenerated muscle fibres, present a picture of degeneration, the field containing large numbers of free nuclei.

The first studies of the effects of selenium on muscular dystrophy in chicks (Dam and Søndergaard, 1957; Nesheim and Scott, 1958) showed that addition of about 1 to 5 mg selenium per kg diet, much more than is required to prevent exudative diathesis, had some effect in reducing the incidence of muscular dystrophy in chicks receiving a vitamin E deficient diet low in methionine and cystine, but did not under any circumstances completely prevent the disease. In a field trial in north Scotland, Sharman, Blaxter and

Wilson (1959) prevented naturally occurring muscular dystrophy in calves by giving them 0·25 mg selenium as sodium selenite, daily. On the other hand, Maplesden and Loosli (1960) showed that vitamin E was required to prevent muscular dystrophy in calves given an experimental diet containing cod liver oil; 1 ppm selenium was not effective. The results presented in Table 2 (Scott, 1962a) show that selenium is completely effective in preventing muscular dystrophy in chicks when added in the presence of a low level of vitamin E which in itself has little or no effect upon the disease. These results demonstrate that *both* selenium and vitamin E are concerned in prevention of both exudative diathesis and nutritional muscular dystrophy. Further studies on the biological role of selenium in sparing the vitamin E requirement will be discussed in a later portion of this chapter.

TABLE 2
*Interaction of Vitamin E and Selenium in Prevention of Exudative Diathesis and Muscular Dystrophy in Chicks**

Dietary treatment	Incidence of	
	Exudative diathesis %	Muscular dystrophy %
Basal diet (deficient in vitamin E, selenium, and sulphur amino acids but containing either 0·02 per cent of DPPD or 0·0125 per cent of ethoxyquin)	100	80
+2·5 mg D-α-tocopheryl acetate (E)/kg	60	100
+5 mg E/kg	0	100
+10 mg E/kg	0	30
+20 mg E/kg	0	0
+0·1 mg Se (as sodium selenite)/kg	0	95
+0·1 mg Se/kg +2·5 mg E/kg	0	75
+0·1 mg Se/kg +5 mg E/kg	0	64
+0·1 mg Se/kg +10 mg E/kg	0	0
+1·0 mg Se/kg	0	47
+1·0 mg Se/kg +2·5 mg E/kg	0	5
+1·0 mg Se/kg +5 mg E/kg	0	0

*Average results from three experiments.

CYSTINE IN PREVENTION OF MUSCULAR DYSTROPHY

In their first work on muscular dystrophy in vitamin E deficient chicks, Dam, Prange and Søndergaard (1952) showed that the disease was not prevented by methylene blue or a variety of other antioxidants, but was prevented by giving vitamin E in small amounts or by adding 0·5 per cent cystine to the diet. The condition was accompanied by a high value for blood cholesterol and a low value for muscle creatine. Later, Machlin and Shalkop (1956) showed that muscular dystrophy in chicks given a casein diet deficient in vitamin E and low in cystine could be prevented by supplementing the diet with vitamin E or with either cystine or methionine.

In more recent studies, Scott and Calvert (1960) found that cystine appeared to be required specifically for prevention of muscular dystrophy in vitamin E deficient chicks. A basal diet lacking in vitamin E and low in both methionine and cystine could be supplemented with 0·2 per cent methionine without reducing the incidence of muscular dystrophy, although growth was improved, but when the same amount of sulphur was provided by 0·15 per cent cystine, muscular dystrophy was completely prevented.

Recent research in the author's laboratory has shown that methionine, at borderline dietary levels for prevention of nutritional muscular dystrophy, can be rendered ineffective by addition to the diet of choline, betaine, and creatine which supposedly act to prevent demethylation of methionine, a necessary step in the conversion of methionine to cystine. Borderline *ineffective* levels of methionine have been made effective by addition of guanidinoacetic acid, a methyl acceptor, thereby encouraging the conversion of methionine to cystine. Addition of D-α-tocopheryl acetate to the basal diet low in both methionine and cystine invariably brings about complete prevention of muscular dystrophy, indicating that vitamin E may be concerned in some unknown way in cystine metabolism.

ARGININE AND MUSCULAR DYSTROPHY

An interesting discovery was made by Nesheim, Calvert and Scott (1960) in the course of studies on interrelations between amino acids in vitamin E deficient, dystrophy-producing diets: when the arginine content of a casein diet was reduced to deficient levels muscular dystrophy did not occur even though the diet was grossly deficient in cystine and vitamin E. When that diet was supplemented with 1, 2 or 4 per cent arginine monohydrochloride, severe muscular dystrophy occurred in 90 per cent of the chicks by the time they were 5 weeks old. With 2 or 4 per cent arginine, incidence and severity of muscular dystrophy were no greater than with 1 per cent. Paired feeding with the arginine-deficient diet and with the dystrophy-producing diet containing arginine showed that rate of growth had no influence on the occurrence of muscular dystrophy, since the chicks receiving the arginine-containing diet developed dystrophy whether they were fed to appetite and so grew faster or had their intake restricted to that of the arginine-deficient chicks. When chicks had the basal arginine-deficient diet to 5 weeks of age and some then had a supplement of 1 per cent arginine, seven out of ten chicks given the supplement showed muscular dystrophy within the next 2 weeks, but none of those given the arginine-deficient diet alone developed signs of dystrophy in 7 weeks. When the basal diet containing arginine was supplemented with methionine, cystine or vitamin E, muscular dystrophy was prevented. These results demonstrate the existence of an important interrelation between arginine and the sulphur amino acids in preventing muscular dystrophy in chicks.

STUDIES ON METABOLIC FUNCTIONS OF VITAMIN E, SELENIUM AND SULPHUR AMINO ACIDS IN PREVENTION OF NUTRITIONAL MUSCULAR DYSTROPHY

Relative biopotencies of D-, DL- *and* L-α-*tocopherols.* The ratio of biological activity of D-α-tocopherol in comparison to DL-α-tocopherol has been established to be fairly constant at approximately 1·36:1 in studies of vitamin E deficiency diseases in a wide variety of animals ranging from erythrocyte hemolysis in rats (Friedman *et al.*, 1958), muscular dystrophy prevention in rabbits (Hove and Harris, 1947), tocopherol levels in human plasma (Week *et al.*, 1952) and plasma and liver tocopherol levels in chicks (Pudelkiewicz *et al.*, 1960). Its partial effectiveness in preventing this wide variety of vitamin E deficiency diseases indicates that L-α-tocopherol can perform all of the functions of D-α-tocopherol but that, because of some differentiating mechanism in the animal body which causes it to be less efficiently utilized, it is less active than the D- epimer.

With the successful isolation of the L- epimer of natural D-α-tocopherol by Robeson and Nelan (1962) and by Isler *et al.* (1962) it has been possible to conduct critical studies comparing the effectiveness of the pure L- epimer of α-tocopherol with natural D-α-tocopherol in a wide variety of vitamin E deficiency diseases in laboratory animals. The results of studies on the biological activity of the L- epimer by the resorption gestation bioassay in rats shows that 2*l*,4′-D,8′D-α-tocopheryl acetate is approximately 21–22 per cent as potent as synthetic 2*d*,4′D,8′D-α-tocopheryl acetate or as the equivalent D-α-tocopheryl acetate derived from natural sources (Ames *et al.*, 1963). Dam and Søndergaard (1964) studied the prevention of encephalomalacia in chicks reared on a diet with 30 per cent lard as the main source of linoleic acid, and found that the biological activities of the acetates of D-, DL- and L-α-tocopherols against encephalomalacia were 168 for the D- form, 100 for the DL- form, and 44 for the L- form of the vitamin (the activity of DL-α-tocopheryl acetate was arbitrarily set at 100). These results agree with those of Bruggemann and associates (1963) who reported that the L- isomer had approximately 50 per cent of the activity of DL-α-tocopheryl acetate, which in turn averaged a little more than half the activity of D-α-tocopheryl acetate for prevention of erythrocyte haemolysis in rats. These results, therefore, indicated that the acetate ester of the L- isomer is about 25 per cent as effective as D-α-tocopheryl acetate. The comparison of D- and DL-α-tocopheryl acetates in the above studies indicated a relative potency of 1·67 for the D-α-tocopheryl acetate as compared with the activity of the racemic mixture.

Using a purified basal diet containing casein, isolated soybean protein and torula yeast with sufficient antioxidant (ethoxyquin) to prevent encephalomalacia, and sufficient sodium selenite to prevent exudative diathesis, studies in our laboratory (Scott and Desai, 1964) showed that the relative antimuscular dystrophy activities of the D- and L- epimers of α-tocopheryl acetates determined in chick bioassay were 1·46 and 0·36, respectively, as compared with a reference standard DL-α-tocopheryl acetate (unofficial ANRC* standard) which was assigned an antimuscular dystrophy value of 1·0 unit/mg. Blood plasma tocopherol analyses showed that prevention of muscular dystrophy was correlated directly with total plasma tocopherol levels and that 900–1025 μg of tocopherol/100 ml of blood were present when muscular dystrophy was completely prevented regardless of the form of tocopherol included in the diet.

The results with muscular dystrophy in chicks, therefore, were in good agreement with those of the earlier studies of a variety of vitamin E deficiency diseases, indicating that D-α-tocopheryl acetate is superior to DL-α-tocopheryl acetate by a factor of 1·46, a value midway between the previously determined factors, which ranged from 1·22 to 1·67.

Several hypotheses can be advanced to explain the observed difference in the biological activity of the D- and L- epimers of α-tocopherol. These may be related to (1) differences in the degree of absorption of the epimers across the mucosa of the intestinal tract; (2) differences in rate of excretion; (3) differential rates of destruction, before or after absorption; (4) variations in the affinity of the epimers for a specific carrier required for their active transport to the proper loci or to differences in affinity by the tissues requiring tocopherol, or both; or (5) true differences in chemical activity influenced by the structural configurations.

Since the results of Scott and Desai (1964) showed that the prevention of muscular dystrophy in chicks was directly correlated with the total plasma tocopherol concentration with all forms of tocopherol administered, and since the plasma tocopherol levels must reflect the net difference between rate of absorption and rate of excretion of the tocopherols, or of differential destruction, it appears that their biological value for prevention of muscular dystrophy is largely dependent upon one of the first three possibilities listed above. It is also possible that differences in affinities for active sites may be responsible for the observed differences in retention of the D- and L- epimers and also for differences in their functioning at cellular and subcellular loci in the tissues throughout the body. Studies by Weber *et al.* (1964) indicated a synergistic relationship when D- and L-α-tocopherol were administered together as a racemic mixture as compared with the activity calculated for

DL-α-tocopherol from the biological activities of the epimers when fed singly. They attributed this synergistic relationship to an increased absorption of the racemic mixture as compared with the absorption of the separately administered epimers.

Studies by Desai, Parekh and Scott (1965), using randomly labelled ^{3}H-D-α- and ^{3}H-L-α-tocopheryl acetates showed that L-α-tocopherol is absorbed as well or better than the D- form of the vitamin. Although both D- and L-α-tocopherols were absorbed at a lower rate from duodenum, jejunum and ileum of chick suffering from nutritional muscular dystrophy, it was evident in both normal and deficient chicks that the L- epimer is absorbed at least as well as the D- epimer of α-tocopherol. These findings agree with those of Weber *et al.* (1964) that in rats L-α-tocopherol was absorbed better but at the same time disappeared from the blood and other tissues at a faster rate as compared with the D- form of the vitamin.

Studies conducted on the relative rate of absorption in duodenum, jejunum and ileum indicated no clear-cut preferential intestinal site of absorption, although it was slightly better in the jejunum than in either duodenum or ileum. This may have been due to the presence of bile in this portion of the intestinal tract, since Wiss *et al.* (1962) have indicated that although the specific site of absorption is not well established it appears to be linked with fat and oil absorption and is facilitated by bile.

The results of this study indicated that the differences in relative biopotency of D-, DL- and L-α-tocopherols are not due to decreased rate of absorption of the L- form of the vitamin. In view of the plasma tocopherol levels found by Scott and Desai (1964) it appears, therefore, that these differences in biopotency must be due to differences in retention whereby the D-epimer is retained much better than the L epimer in the blood and perhaps in other tissues of the body.

These results indicated the possibility of an active carrier of D-α-tocopherol in the blood and tissues which may have a greater affinity for the D epimer than for L-α-tocopherol.

Earlier studies in our laboratory (Calvert, Nesheim and Scott, 1962) had shown that the amount of D-α-tocopheryl acetate needed in the diet to prevent muscular dystrophy in chicks depends upon the selenium content of the diet. However, regardless of the level fed, selenium alone is ineffective in preventing dystrophy in chicks (Nesheim and Scott, 1958). These studies indicated that the biological role of dietary selenium in the prevention of nutritional muscular dystrophy in chicks might be due, at least in part, to its action in improving the retention of D-α-tocopherol in the lipoproteins of the blood.

A role of selenium in improving the metabolic utilization of D-*α-tocopherol.* The results presented above indicate that the variations observed in the biological activities of the D- and L- epimers of α-tocopherol, originally thought to be due to differences in absorption, now appear to be due to differences in *retention* brought about by differences in rate of destruction or excretion, or both.

Experiments were conducted in the author's laboratory (Scott, 1965; Desai and Scott, 1965) comparing the plasma tocopherol levels obtained from oral administration of D- and L-α-tocopheryl acetates in the presence of graded levels of dietary selenium. The plasma levels resulting from the administration of the tocopherols in the feed were compared with plasma values of chicks which had been injected with D- or L-α-tocopherol (mixed in chicken plasma) directly into the blood stream via heart puncture.

The results presented in Table 3 show that selenium improved the effectiveness of D-α-tocopherol much more than that of the L- epimer for prevention of dystrophy.

TABLE 3
Effect of Selenite on Tocopherol Requirement for Prevention of Nutritional Muscular Dystrophy

α-Tocopheryl acetate (mg/kg diet)	Muscular dystrophy score Selenium in diet (ppm)		
	None	0·1	1·0
0	4·0	3·5	2·2
2·5 D–	3·4	1·8	0·7
10·0 D–	1·4	0·0	0·0
10·0 L–	3·1	2·8	1·5
40·0 L–	0·4	0·0	0·0

The results presented in Table 4 show that in the absence of selenium, the retention of D- and L-α-tocopherols, as a percentage of the respective levels obtained by injection of these tocopherols, were approximately equal. As the dietary level of selenium was increased the plasma level of D-α-tocopherol increased markedly while that of L-α-tocopherol showed little or no increase.

TABLE 4
Influence of Selenite on Plasma Tocopherol Levels

α-Tocopheryl acetate administered (mg/kg diet)		Dietary selenium (ppm)		
		None	0·1	1·0
		(μg tocopherol/100 ml plasma)		
None		193	195	227
Oral	10·0 D–	591(65 per cent)*	795(87 per cent)	841(93 per cent)
	10·0 L–	370(60 per cent)	498(81 per cent)	341(56 per cent)
Injected †	10·0 D–		909	
	10·0 L–		614	

* $\frac{\text{Oral}}{\text{Injected}} \times 100.$

† Intra-cardiac in hen plasma.

*Animal Nutrition Research Council.

These results were in direct relation to the effects of selenium in enhancing the muscular dystrophy-preventing activity of D-α-tocopherol and its relative lack of effect in improving the anti-muscular dystrophy activity of L-α-tocopheryl acetate. These results, therefore, indicate further that a selenium compound in the blood may have a specific affinity for D-α-tocopherol, thereby aiding in its retention.

To obtain further information, a study was conducted to determine the effects of graded levels of dietary selenium on the retention of tritiated D- and L-α-tocopherols in chick serum. The results of this study (Table 5) show that the addition of 1 mg of selenium/chick in combination with ^{3}H-D- or ^{3}H-L-α-tocopheryl acetates improved the retention of tritium counts in serum to a greater extent when D-α-tocopherol was given than with L-α-tocopherol.

TABLE 5

Effect of Dietary Selenium on Retention of Tritiated D– *and* L–*α-tocopherols in Chick Serum**

Oral dose of selenium (mg/chick)	Beta counts/min/ml. of serum collected after 48 hr	
	^{3}H-D-α-tocopheryl acetate (5 mg/chick)	^{3}H-L-α-tocopheryl acetate† (5 mg/chick)
0	9146	2260
0·5	13224	2984
1·0	18396	3910

* Selenium and tocopherols administered orally in gelatin capsules in two doses at 24-hr intervals. Five chicks per treatment.

† Specific activity = 111 μc/mg.

At each level of selenium the retention of ^{3}H-L-α-tocopherol was approximately 25 per cent of that of ^{3}H-D-α-tocopherol and showed a growing increase in D-α-tocopherol compared to L-α-tocopherol retention as the level of selenium increased. This study with radioisotopes confirms the differences in relative biopotency of these forms of tocopherol described earlier and also provides further evidence for a specific 'carrier' of D-α-tocopherol in plasma.

A further study was conducted on the distribution of ^{75}Se and ^{3}H-tocopherols in various protein fractions of chick serum. The incorporation of ^{75}Se and ^{3}H-tocopherols was highest in the serum globulins and lowest in serum albumin. When expressed in terms of the ratio of ^{75}Se or ^{3}H per unit of protein, both selenium and tocopherols were shown to be concentrated in the γ-globulin of the serum. Dietary selenium markedly improved the amount of ^{3}H-D-α-tocopherol in the γ-globulin fraction but caused little if any improvement in the amount of ^{3}H-L-α-tocopherol present in this fraction. In the absence of selenium the amount of ^{3}H-D- and ^{3}H-L-α-tocopherol present in the γ-globulin fraction was approximately equivalent.

Recent studies in the author's laboratory (Thompson and Scott, 1968) have shown that when chicks obtained from selenium-depleted dams were given crystalline amino acid diets containing 0·0125% ethoxyquin, 5% stripped corn oil, graded levels of D-α-tocopheryl acetate and adequate amounts of all other recognized nutrients except selenium, all chicks receiving the basal diet with levels of vitamin E up to 200 mg per kg of diet grew poorly and died within the first 28 days of the experiment. Higher levels of vitamin E caused a graded decrease in mortality, but even 1000 mg D-α-tocopheryl acetate per kg of diet produced growth which was markedly inferior to that obtained with added selenium without added vitamin E. When the diet contained 100 mg D-α-tocopheryl acetate per kg, or more, selenium at 0·005 μg per gm of diet prevented symptoms and mortality and promoted good growth. However, 0·02 μg selenium per gm of diet was needed with diets supplemented with 10 mg D-α-tocopheryl acetate per kg.

Plasma and tissue uptake of oral doses of Dl-α-tocopherol 3,4-^{14}C, or the corresponding acetate, were very low in chicks receiving the basal, selenium-low diet. Plasma and tissue levels of Dl-α-tocopherol-3,4-^{14}C were up to 100 times higher in chicks receiving similar diets supplemented with selenium.

These results further demonstrate that selenium is an essential trace element for the chicken.

Thus one biological role of selenium appears to lie in a selenium-containing compound which may act as either a carrier or a protector of vitamin E and which may function in absorption, retention, prevention of destruction, and perhaps transfer across cell membranes of D-α-tocopherol, thereby enhancing its biological activity in the blood and perhaps in cells throughout the body. In view of this, the 'vitamin' role of D-α-tocopherol may not reside in a specific metabolic function, since L-α-tocopherol at a dietary level which provides an adequate blood level of tocopherol is as effective as D-α-tocopherol at the same blood level for prevention of muscular dystrophy in the chick. The vitamin role of D-α-tocopherol, therefore, appears to reside in its specificity of retention and transport, through mechanisms which are concerned in some unknown way with selenium.

STUDIES DIRECTED TOWARD DETERMINATION OF THE CAUSE OF NUTRITIONAL MUSCULAR DYSTROPHY IN CHICKS

Linoleic acid. Since vitamin E is known to act both *in vivo* and *in vitro* as an antioxidant, a natural hypothesis to explain the function of vitamin E is that it is required to prevent peroxidation of the unsaturated fatty acids of the lipoproteins and phospholipids in the cell walls of the tissues. This antioxidant effect of vitamin E appears to be its main function in the prevention of encephalomalacia. Although selenium is of primary importance in the prevention of exudative diathesis (Nesheim and Scott, 1958), the antioxidant properties of vitamin E may have some bearing on the prevention of this disease. However, the effect of oxidation of fatty acids upon production of muscular dystrophy in the chick has not been investigated extensively, primarily because the work of Dam *et al.* (1952) and Machlin and Shalkop (1956) did not demonstrate an effect of dietary fat on this disease.

Experiments were conducted by Calvert, Desai and Scott (1964) which showed that in the absence of dietary linoleic acid, muscular dystrophy does not occur in chicks even when the diet is low in both vitamin E and sulphur amino acids. The addition of graded levels of linoleic acid to the diet resulted in proportionate increases in the incidence and severity of muscular dystrophy. The vitamin E requirement for prevention of muscular dystrophy was increased with increasing levels of linoleic acid up to 0·5 per cent linoleic acid. No further increase was observed, however, in the vitamin E requirement for prevention of muscular dystrophy when the level of linoleic acid was increased from 0·5 to 2·5 per cent of the diet. Vitamin E at a level which completely prevented muscular dystrophy had no effect on the level of linoleic acid in muscle lipids, which increased in both dystrophic and non-dystrophic chicks in proportion to the amount of linoleic acid in the diet.

At the same time, Desai, Calvert, Scott and Tappel (1964) showed that muscle tissues of chicks with nutritional muscular dystrophy showed significant increases in peroxidation as evidenced by thiobarbituric acid (TBA) index of the unincubated muscle tissues and increased activities of the lysosomal enzymes, acid phosphatase, β-glucuronidase, cathepsin, β-galactosidase, and ribonuclease. The increases in peroxidation and lysosomal enzymes were in direct proportion to the degree of muscular dystrophy, as measured in chicks, at 4 weeks of age. Addition of either DL-methionine or D-α-tocopoeryl acetate to the diet prevented the dystrophy, prevented the peroxidation, and prevented the increases in lysosomal enzyme activities. These results support the hypothesis, therefore, that nutritional muscular dystrophy in the chick is initiated by peroxidative tissue damage, and that possibly the increased tissue levels of lysosomal enzymes may hydrolyze *in situ* the components of the muscle tissues. This is subject to question, however, since it does not explain the biochemical mechanism responsible for the protection of the muscle tissues from degeneration, especially the role of sulphur amino acids which prevent dystrophy in the absence of vitamin E, and of selenium which markedly spares the vitamin E requirement.

When the basal, vitamin E-deficient diet is supplemented with methionine until the chicks are 3 weeks of age, dystrophy is prevented as long as the methionine is present. Sudden removal of the supplemental methionine causes immediate precipitation (within 48 hr) of severe muscular dystrophy. This technique was used by Desai, Calvert and Scott (1964) in the conduct of a time-sequence study to determine whether increases in peroxidizability of the tissues and in lysosomal enzymes *precede* the onset of muscular dystrophy *or occur concurrently* with its initiation. At the same time, studies were made of the peroxidizability and lysosomal enzyme content of the tissues during the period of regression of dystrophy following supplementation of the diet of dystrophic chicks with methionine or vitamin E.

When methionine was removed from the vitamin E-deficient diet, symptoms of muscular dystrophy occurred concurrently with the appearance of peroxidation and lysosomal enzymes. Upon returning methionine or upon adding vitamin E to the diet of dystrophic chicks, recovery from muscular dystrophy began to occur in a short period of time. The rate of recovery was faster following resupplementation with methionine than with vitamin E.

These results showed no significant induction of lysosomal enzymes before the visible symptoms of the dystrophy appeared, indicating that the lysosomal enzymes are not the primary cause of the dystrophy, but that their high activity during dystrophy apparently is involved mainly in the hydrolytic breakdown and removal of the degradative products of the tissues. Such a process, in turn, may facilitate a more rapid regeneration of the muscle tissues in the presence of added methionine.

The finding that methionine is completely effective, not only in prevention but also in reversion of nutritional muscular dystrophy under conditions of high peroxidizability of the muscle tissues indicates that the sulphur amino acids play a primary role in prevention of nutritional muscular dystrophy in the chick.

Studies on sulphydryl-containing enzymes and glutathione in dystrophic chicks. In an attempt to determine a possible role of sulphur in the prevention of dystrophy, experiments were undertaken by Scott, Søndergaard and Dam (1962) to investigate whether or not additions of cystine, vitamin E, selenium, ascorbic acid or $CuSO_4$ to a vitamin E-deficient, dystrophy-producing chick diet would influence the sulphydryl-containing enzyme, succinic dehydrogenase, in the pectoral muscle, and if any influence upon this enzyme could be correlated with incidence of muscular dystrophy in the chicks. Various treatments produced marked effects on muscular dystrophy, but this in no way correlated with the activity of the muscle succinic dehydrogenase. The enzyme content of the muscles of dystrophic chicks was approximately the same as in chicks in which dystrophy was prevented by vitamin E supplementation. Addition of $CuSO_4$ increased severity of muscular dystrophy and increased succinic dehydrogenase in the muscles. On the other hand, addition of cystine prevented muscular dystrophy and slightly increased the succinic dehydrogenase. Selenium decreased muscular dystrophy and somewhat decreased the succinic dehydrogenase. Ascorbic acid was without effect upon either dystrophy or succinic dehydrogenase content of the muscle.

Studies of the glutathione content of muscle and liver tissues of dystrophic and non-dystrophic chicks conducted by Calvert and Scott (1962) showed that the level of reduced glutathione in muscle was markedly increased in dystrophy and returned to normal lower levels when dystrophy was prevented by supplementation of the diet with vitamin E.

These results show, therefore, that the primary cause of nutritional muscular dystrophy is not due to insufficient utilization of sulphur for sulphydryl enzymes as represented by succinic dehydrogenase or for formation of glutathione. Indeed, the results indicated that in the absence of dietary vitamin E the animal body may compensate for a lack of normal metabolic antioxidant or reducing substance by synthesis of extra amounts of reduced glutathione.

Yang, Riegl and Oleson (1958) found that selenite, selenate, a synthetic antioxidant and cystine, all protected rats against dietary liver necrosis and also caused significantly elevated liver levels of coenzyme A; but no increase in coenzyme A occurred with dietary vitamin E which also prevented liver necrosis. Nevertheless, these authors postulated that selenite and antioxidants may function by making more amino acid sulphur available to the liver cell.

Thus, the recent studies with chicks on the interrelationships between antioxidants, selenium and sulphur amino acids have succeeded in bringing some order out of the previously apparently hopeless confusion regarding vitamin E nutrition. It is now apparent that selenium and vitamin E are mutually interdependent in the performance of several 'antioxidant' roles, one of which is intimately concerned with sulphur metabolism in the prevention of nutritional muscular dystrophy.

REFERENCES

ADAMSTONE, F. B. (1931) *J. Morphol.* **52**, 47.

ADAMSTONE, F. B. and CARD, L. E. (1934) *J. Morphol.* **56**, 339.

AMES, S. R., LUDWIG, M. I., NELAN, D. R. and ROBESON, C. D. (1963) *Biochemistry* **2**, 188.

BLAXTER, K. L. and BROWN, F. (1952) *Nutr. Abstr. Rev.* **22**, 1.

BRO-RASMUSSEN, F. and HJARDE, W. (1961) *Ann. Rev. Biochem.* **30**, 458.

BRUGGEMANN, J., NIESAR, K. H. and ZENTZ, Ch. (1963) *Intern. Z. Vitaminforsch.* **33**(2),180.

BUNNELL, R. H., MATTERSON, L. D., SINGSEN, E. P., POTTER, L. M., KOZEFF, A. and JUNGHERR, E. L. (1955) *Poultry Sci.* **34**, 1068.

CALVERT, C. C., DESAI, I. D. and SCOTT, M. L. (1964) *J. Nutr.* **83**, 307.

CALVERT, C. C., MONROE, R. A. and SCOTT, M. L. (1961) *J. Nutr.* **73**, 355.

CALVERT, C. C., NESHEIM, M. C. and SCOTT, M. L. (1962) *Proc. Soc. Exp. Biol. Med.* **109**, 16.

CALVERT, C. C. and SCOTT, M. L. (1962) *Poultry Sci.* **41**, 1633.

CAPUTTO, R., MCCAY, P. B. and CARPENTER, M. P. (1958) *J. Biol. Chem.* **233**, 1025.

CARPENTER, M. P., MCCAY, P. B. and CAPUTTO, R. (1958) *Proc. Soc. Exp. Biol. Med.* **97**, 205.

CENTURY, B. and HORWITT, M. K. (1959) *Proc. Soc. Exp. Biol. Med.* **102**, 375.

DAM, H. (1957) *Pharmacol. Rev.* **9**, 1.

DAM, H. and GLAVIND, J. (1938) *Nature* **142**, 1077.

DAM, H. and GLAVIND, J. (1939) *Nature* **143**, 810.
DAM, H. and GLAVIND, J. (1940) *Naturwissenschaften* **28**, 207.
DAM, H. and GLAVIND, J. (1942) *Science* **96**, 235.
DAM, H. and GRANADOS, H. (1945) *Acta Physiol. Scand.* **10**, 162.
DAM, H., NIELSEN, G. K., PRANGE, I. and SONDERGAARD, E. (1958) *Nature* **182**, 802.
DAM, H., PRANGE, I. and SONDERGAARD, E. (1952) *Acta Pathol. Microbiol. Scand.* **31**, 172.
DAM, H. and SONDERGAARD, E. (1957) *Experientia* **13**, 494.
DAM, H. and SONDERGAARD, E. (1964) *Z. Ernahrungswiss.* **5**, 73.
DESAI, I. D., CALVERT, C. C. and SCOTT, M. L. (1964) *Arch. Biochem. Biophys.* **108**, 60.
DESAI, I. D., CALVERT, C. C., SCOTT, M. L. and TAPPEL, A. L. (1964) *Proc. Soc. Exp. Biol. Med.* **115**, 462.
DESAI, I. D., PAREKH, C. K. and SCOTT, M. L. (1965) *Biochim. Biophys. Acta* **100**, 280.
DESAI, I. D. and SCOTT, M. L. (1965) *Arch. Biochem. Biophys.* **110**, 309.
DEWITT, W. B. and SCHWARZ, K. (1958) *Experientia* **14**, 28.
DINNING, J. S. and DAY, P. L. (1958) *J. Biol. Chem.* **233**, 240.
DRAKE, C., GRANT, A. B. and HARTLEY, W. J. (1960) *New Z. Vet. J.* **8**, 4.
EDWIN, E. E., BUNYAN, J., DIPLOCK, A. T. and GREEN, J. (1961) *Nature*, **189**, 747.
EGGERT, R. G., PATTERSON, E., AKERS, W. T. and STOKSTAD, E. L. R. (1957) *J. Animal Sci.* **16**, 1037.
EVANS, H. M. and BISHOP, K. S. (1922) *Am. J. Physiol.* **63**, 396.
EVANS, H. M. and BURR, G. O. (1927) *Mem. Univ. Calif.* No. 8.
EVANS, H. M., EMERSON, O. H. and EMERSON, G. A. (1936) *J. Biol. Chem.* **113**, 319.
FERNHOLZ, E. (1938) *J. Am. Chem. Soc.* **60**, 700.
FITCH, C. D. and DINNING, J. S. (1959) *Proc. Soc. Exp. Biol. Med.* **100**, 201.
FRIEDMAN, L., WEISS, W., WHERRY, F. and KLINE, O. L. (1958) *J. Nutr.* **65**, 143.
GOETTSCH, M. and PAPPENHEIMER, A. M. (1931) *J. Exp. Med.* **54**, 145.
GOLDSTEIN, J. and SCOTT, M. L. (1956) *J. Nutr.* **60**, 349.
GRANT, A. B., HARTLEY, W. J. and DRAKE, C. (1960) *New Z. Vet. J.* **8**, 1.
GREEN, J., EDWIN, E. E., DIPLOCK, A. T. and BUNYAN, J. (1961) *Nature*, **189**, 78.
HARRIS, P. L., LUDWIG, M. I. and SCHWARZ, K. (1958) *Proc. Soc. Exp. Biol. Med.* **97**, 686.
HARTLEY, W. J. and GRANT, A. B. (1961) *Federation Proc.* **20**, 679.
HOVE, E. L., FRY, G. S. and SCHWARZ, K. (1958) *Proc. Soc. Exp. Biol. Med.* **98**, 27.
HOVE, E. L. and HARRIS, P. L. (1947) *J. Nutr.* **33**, 95.
HOWES, C. E. and HUTT, F. B. (1952) *Poultry Sci.* **31**, 360.
ISLER, O., SCHUDEL, P., MAYER, H., WURSCH, J. and RUEGG, R. (1962) *Vitamins Hormones* **20**, 389.
JOLLY, R. D. (1960a) *New Z. Vet. J.* **8**, 11.
JOLLY, R. D. (1960b) *New Z. Vet. J.* **8**, 13.
KARRER, P., FRITZSCHE, H., RINGIER, B. H. and SALOMON, N. H. (1938) *Nature*, **141**, 1057.
KITABCHI, A. E., MCCAY, P. B., CARPENTER, M. P., TRUCCO, R. E. and CAPUTTO, R. (1960) *J. Biol. Chem.* **235**, 1591.
KOFLER, M., SOMMER, P. F., BOLLIGER, H. R., SCHMIDLI, B. and VECCHI, M. (1962) *Vitamins Hormones* **20**, 407.
KOKATNUR, S. O., KUMMEROW, F. A. and SCOTT, H. M. (1960) *Proc. Soc. Exp. Biol. Med.* **104**, 170.
LEDDICOTTE, G. W. and REYNOLDS, S. A. (1951) *Nucleonics* **8**, 62.
MACHLIN, L. J. and GORDON, R. S. (1960) *Proc. Soc. Exp. Biol. Med.* **103**, 659.
MACHLIN, L. J., GORDON, R. S. and MEISKY, K. H. (1959) *J. Nutr.* **67**, 333.
MACHLIN, L. J. and SHALKOP, W. T. (1956) *J. Nutr.* **60**, 87.
MACKENZIE, C. G. (1953) *Symposium on Nutrition*, p.136. Ed. Herriott, R. M. Johns Hopkins Press, Baltimore.
MAPLESDEN, D. C. and LOOSLI, J. K. (1960) *J. Dairy Sci.* **43**, 645.
MASON, K. E. (1944) *Vitamins Hormones* **2**, 107.
MASON, K. E. (1949) *Ann. N. Y. Acad. Sci.* **52**, 63.
MASON, K. E. (1954) *The Vitamins*, Vol. 3, p. 514. Ed. Sebrell, W. H. and Harris, R. S., Academic, New York.
METZGER, H. J. and HAGAN, W. A. (1927) *Cornell Vet.* **17**, 35.

MICHAELIS, L. and WOLLMAN, S. H. (1950) *Biochim. Biophys. Acta* **4**, 1956.
MUTH, O. H., OLDFIELD, J. E., REMMERT, L. F. and SCHUBERT, J. R. (1958) *Science*, **128**, 1090.
MUYTJENS, E. E. (1956) *Biochim. Biophys. Acta* **20**, 553.
NASON, A. and LEHMAN, I. R. (1956) *J. Biol. Chem.* **222**, 511.
NESHEIM, M. C., CALVERT, C. C. and SCOTT, M. L. (1960) *Proc. Soc. Exp. Biol. Med.* **104**, 783.
NESHEIM, M. C., LEONARD, S. L. and SCOTT, M. L. (1959) *J. Nutr.* **68**, 359.
NESHEIM, M. C. and SCOTT, M. L. (1958) *J. Nutr.* **65**, 601.
NESHEIM, M. C. and SCOTT, M. L. (1961) *Federation Proc.* **20**, 674.
NISHIDA, T., TSUCHIYAMA, H., INONE, M. and KUMMEROW, F. A. (1960) *Proc. Soc. Exp. Biol. Med.* **105**, 308.
PAPPENHEIMER, A. M. (1948) *Amer. Lecture Ser.* No. 17. Thomas, Springfield, Ill.
PAPPENHEIMER, A. M. and GOETTSCH, M. (1931) *J. Exp. Med.* **53**, 11.
PAPPENHEIMER, A. M. and GOETTSCH, M. (1933a) *J. Exp. Med.* **57**, 365.
PAPPENHEIMER, A. M. and GOETTSCH, M. (1933b) *J. Exp. Med.* **59**, 35.
PROCTOR, J. F., HOGUE, D. E. and WARNER, R. G. (1958) *J. Animal Sci.* **17**, 1183.
PUDELKIEWICZ, W. J., MATTERSON, L. D., POTTER, L. M., WEBSTER, L. and SINGSEN, E. P. (1960) *J. Nutr.* **71**, 115.
RAHMAN, M. M., DAVIES, R. E., REID, B. L. and COUCH, J. R. (1959) *Poultry Sci.* **38**, 1239.
ROBESON, C. D. and NELAN, D. R. (1962) *J. Am. Chem. Soc.* **84**, 3196.
SCHUBERT, J. R., MUTH, O. H., OLDFIELD, J. E. and REMMERT, L. F. (1961) *Federation Proc.* **20**, 689.
SCHWARZ, K. (1951) *Proc. Soc. Exp. Biol. Med.* **78**, 852.
SCHWARZ, K. (1961) *Federation Proc.* **20**(1), 665.
SCHWARZ, K. and FOLTZ, C. M. (1957) *J. Am. Chem. Soc.* **79**, 3292.
SCHWARZ, K. and FOLTZ, C. M. (1958) *J. Biol. Chem.* **233**, 245.
SCOTT, M. L. (1962a) *Nutr. Abstr. Rev.* **32**, 1.
SCOTT, M. L. (1962b) *Vitamins Hormones* **20**, 621.
SCOTT, M. L. (1962c) *Mineral Metabolism*, 2B, Chap. 37. Ed. Comar, C. L. and Bronner, F. Academic Press, New York.
SCOTT, M. L., BIERI, J. G., BRIGGS, G. M. and SCHWARZ, K. (1957) *Poultry Sci.* **36**, 1155.
SCOTT, M. L. and CALVERT, C. C. (1960) *Abstr. 5th Intern. Congr. on Nutr.*, p.77.
SCOTT, M. L. and CALVERT, C. C. (1962) *J. Nutr.* **77**, 105.
SCOTT, M. L. and DESAI, I. D. (1964) *J. Nutr.* **83**, 39.
SCOTT, M. L., HILL, F. W., NORRIS, L. C., DOBSON, D. C. and NELSON, T. S. (1955) *J. Nutr.* **56**, 387.
SCOTT, M. L., SØNDERGAARD, E. and DAM, H. (1962) *Federation Proc.* **21**, 473.
SCOTT, M. L. and STOEWSAND, G. S. (1961) *Poultry Sci.* **40**, 1517.
SHARMAN, G. A. M., BLAXTER, K. L. and WILSON, R. S. (1959) *Vet. Rec.* **71**, 536.
SHRIFT, A. (1961) *Federation Proc.* **20**, 695.
SINGSEN, E. P., BUNNELL, R. H., MATTERSON, L. D., KOZEFF, A. and JUNGHERR, E. L. (1955) *Poultry Sci.* **34**, 262.
SMITH, L. I., UNGNADE, H. E. and PRICHARD, W. W. (1938) *Science* **88**, 37.
SØNDERGAARD, E., SCOTT, M. L. and DAM, H. (1962) *J. Nutr.* **78**, 15.
STOKSTAD, E. L. R., PATTERSON, E. L. and MILSTREY, R. (1957) *Poultry Sci.* **36**, 1160.
SUGAI, M., INONE, M., TSUCHIYAMA, H. and KUMMEROW, F. A. (1960) *Federation Proc.* **19**, 421.
SYMPOSIUM ON VITAMIN E AND METABOLISM IN HONOUR OF PROFESSOR H. M. EVANS. (1962) *Vitamins Hormones* **20**, 371.
THOMPSON, J. J. and SMITH, N. C. (1953) *Australian Vet. J.* **29**, 89.
THOMPSON, J. N. and SCOTT, M. L. (1968) *Federation Proc.* **27**, 1145 (*Abstr.*).
VASINGTON, F. D., REICHARD, S. M. and NASON, A. (1960) *Vitamins Hormones* **18**, 43.
WALTER, E. D. and JENSEN, L. S. (1964) *Poultry Sci.* **43**, 919.
WEBER, F., GLOOR, U., WURSCH, J. and WISS, O. (1964) *Biochem. Biophys. Res. Commun.* **14**, 186.
WEEK, E. F., SEVIGNE, F. J. and ELLIS, M. E. (1952) *J. Nutr.* **46**, 353.
WILLMAN, J. P., ASDELL, S. A. and OLAFSON, P. (1934) *Cornell Univ. Agr. Exp. Sta. Bull.* **603**, 20.

WISS, O., BUNNELL, R. H. and GLOOR, U. (1962) *Vitamins Hormones* **20**, 441.
YANG, C. S., RIEGL, MARCIA and OLSON, R. E. (1958) *Federation Proc.* **17**, 498.
ZALKIN, H. and TAPPEL, A. L. (1960) *Arch. Biochem. Biophys.* **88**, 113.

CHAPTER 32

The Nutrient Requirements of Farm Livestock

*1. Poultry**

FOREWORD

By SIR DAVID CUTHBERTSON, C.B.E.

Department of Pathological Biochemistry, Royal Infirmary, Glasgow, C.4

Following representations that there was a need for a reappraisal of the accepted standards for the feeding of farm livestock, the Agricultural Research Council set up in 1960 a Technical Committee charged with the task of reviewing periodically the literature on the nutrient requirements of various classes of stock, and of making recommendations or alterations that seem necessary in published feeding standards. To assist it in discharging its duties, the Committee established Working Parties to consider the requirements of ruminants, of poultry, and of pigs, as well as one to consider the nomenclature to be used in the Committee's reports.

The findings of these Working Parties form the basis of the present series of Summaries of Recommendations of which this is one. It is hoped that these Summaries will be of direct assistance to those engaged in the laying down of ration scales and the compounding of feeding stuffs on or off the farm. In the course of their studies the Working Parties have amassed much information of interest which could not be included in the Summaries; where appropriate this information is to be published in the form of detailed Technical Reviews.

It is the Committee's intention to keep under review future developments in the study of animal nutrient requirements, and to publish from time to time revised Summaries of Recommendations and, when the material warrants, further Technical Reviews. It would therefore be appreciated if persons possessing unpublished data pertinent to any aspect of the Committee's task would make them available to the Technical Secretary.

INTRODUCTION

By DR. S. K. KON, *Chairman of Working Party on Nutrient Requirements of Poultry*

By 'poultry' we mean all those tame birds reared for their flesh, eggs or feathers, and by 'fowl' the barndoor fowl (*Gallus gallus domesticus*). Some information is given about the pheasant.

**This chapter is reprinted from A.R.C. publication No. 1, H.M.S.O. 1963*

We define the term 'requirement' as the quantity of a nutrient or of energy that should be supplied in the diet to meet the net requirement of the bird. The net requirement of a normal, healthy bird is the quantity of a nutrient or of energy that it should absorb, given a completely adequate diet in an environment compatible with good health, in order to meet its needs. Depending upon the physiological state and function of the bird, its needs may be for maintenance including the replacement of endogenous losses, for a specified rate of growth, for a stated level of production or for reproduction. Our recommendations do not make any allowance for variations in availability of nutrients or for their possible losses during manufacture and storage of feeding-stuffs.

Our method of approach has been to review the published literature on each nutrient, taking note particularly of the composition of the diet against the background of other essential and non-essential nutrients in it and, when possible, of breed of bird and management. Special attention has been paid to investigations in the United Kingdom which reflected conditions prevailing there.

It has been customary to express nutrient requirements of poultry in terms of amount of nutrient per unit weight of air-dry diet and we have mostly used this method in this Summary of Recommendations. Departures from it are stated.

It is established that the requirements for certain nutrients are influenced by the concentration of energy in the diet; for this reason, except when otherwise stated, all our estimates relate to diets of energy content around 2·8 Mcal metabolizable energy (ME)/kg, which we consider representative of poultry diets normally used in the U.K. Experiments known to have been complicated by ill-health and experiments with diets obviously unbalanced in nutrients other than that under consideration have been disregarded. For most nutrients a range of values has been found and the median taken as the best estimate of requirement, unless the conditions of some experiment seemed to justify greater emphasis on a particular value. (Median: if a set of values is arranged in order of magnitude the median is either the central value if the set is odd in number or the arithmetic mean of the two central values if the set is even in number.)

In arriving at our estimates of requirements we have been restricted by the limitations of the methods of experimental approach. Thus, in attempting to define requirements for growth we had to base our recommendations mostly on experiments dealing with the rate of live weight gain without any separate assessment of the composition of this gain in terms of proportions of skeleton, muscle, fat or feather. In dealing with the requirements of the laying fowl we had to use results of experiments assessed by different criteria: for example, requirements for egg production have been assessed by nitrogen balance as well as by the more customary methods of number of eggs laid or total egg mass produced over various periods of time.

The requirements of the bird are known to vary with age and with type of physiological activity. For most nutrients we have found enough information to allow us to subdivide the estimates of requirements into those of young birds, layers and breeders. We have taken no account of the nutrient content of the egg in assessing requirements of laying hens.

The rate of live weight increase for growing birds, the rate of egg production for laying birds or the rate of egg production and hatchability for breeding stock are various ways of expressing the yield of the bird. Other concepts may also be involved as, for example, that of nitrogen retention.

When several nutrients might interact in their influence upon yield it is theoretically possible to set up factorial experiments to investigate this interaction, and from the results to derive equations stating the different combinations of nutrients associated with different levels of yield. Such equations would indicate the requirements for specified yields. We found no reports of such comprehensive experiments, and in the published work the relationship between levels of specific nutrients in the diet and the resulting yield has usually been limited to an assessment under some restricted set of conditions of the lowest level of the nutrient under test that gave maximal yield.

In using such estimates as a basis for our stated requirements for any one nutrient we have assumed that all other nutrients are supplied in adequate amounts.

It has already been made clear that such estimates as we give refer only to the requirements of the bird for specific nutrients, and we have left untouched all the economic considerations that would arise in practice. Outside our terms of reference are such matters as the most profitable level of yield and, when more than one foodstuff may supply a specific nutrient, the substitution rates between them. These and other similar questions are highly relevant to the feed compounder and poultry farmer, and their resolution requires close co-operation at all stages between biologists, economists and statisticians, but the amount of work published on them is meagre, and appears to have been restricted to broiler and turkey meat production under American conditions.

The metric system of measurement has been adopted in this Summary of Recommendations. For weight, microgram (μg), milligram (mg), gram (g) and kilogram (kg) have been used; for energy, megacalorie and kilocalorie (Mcal = 1000 kcal = 1 million cal), and for biological units, the international unit (i.u.).

All the references consulted are presented at the end of each main division of this Summary of Recommendations, with sub-divisions within which the references are arranged alphabetically; they are numbered consecutively through each main division. These numbers are used in the text and tables to identify the references on which statements or values are based.

WATER REQUIREMENTS

The need for water varies widely and no values for requirement can be stated. The amount needed increases with increasing environmental temperature and content of indigestible organic matter in the diet and, with adults, is greater in laying birds than in non-layers. Water should be freely available at all times; restriction is undesirable and, with young birds, can have disastrous consequences.

References

1. Arscott, G. H., McCluskey, W. H. and Parker, J. E. (1958) *Poult. Sci.* **37**, 117.
2. Arscott, G. H. and Rose, R. J. (1960) *Poult. Sci.* **39**, 93.
3. Barott, H. G. and Pringle, E. M. (1949) *J. Nutr.* **37**, 153.
4. Fisher, H., Griminger, P., Weiss, H. S. and Hudson, C. B. (1961) *Poult. Sci.* **40**, 813.
5. Fuerst, W. F. Jr. and Kare, M. R. (1962) *Poult. Sci.* **41**, 71.
6. Gates, J. D. and Kare, M. R. (1961) *Poult. Sci.* **40**, 1407.
7. Kare, M. R. and Pick, H. L. Jr. (1960) *Poult. Sci.* **39**, 697.
8. Lepkovsky, S., Chari-Bitron, A., Lyman, R. L. and Dimick, M. K. (1960) *Poult. Sci.* **39**, 390.
9. Lepkovsky, S. and Furuta, F. (1960) *Poult. Sci.* **39**, 394.
10. McCluskey, W. H. and Johnson, L. E. (1958) *Poult. Sci.* **37**, 729.
11. Maxwell, B. F. and Lyle, J. B. (1957) *Poult. Sci.* **36**, 921.
12. Medway, W. and Kare, M. R. (1959) *Poult. Sci.* **38**, 631.
13. Milligan, J. L., Eaton, R. C. and Wilcke, H. L. (1960) *Poult. Sci.* **39**, 1276.
14. Patrick, H. and Ferrise, A. (1962) *Poult. Sci.* **41**, 1363.
15. Ranby, P. D. (1960) *Qd Agric. J.* **86**, 363.
16. Ross, E. (1960) *Poult. Sci.* **39**, 999.
17. Siegel, P. B., Siegel, H. S., Kramer, C. Y. and Howes, C. E. (1961) *Poult. Sci.* **40**, 201.
18. Sturkie, P. D. (1958) *Poultry Comment* (E. I. Dupont de Nemours and Co., Industrial and Biochemicals Dept., Wilmington, Delaware) **15**, no. 2, p. 3.
19. Tyler, C. (1958) *J. Agric. Sci.* **51**, 237.
20. Wilson, W. O., McNally, E. H. and Ota, H. (1957) *Poult. Sci.* **36**, 1254.

ENERGY REQUIREMENTS

In practice birds are fed *ad libitum* and, therefore, the necessity for setting a requirement figure for energy is probably largely academic. On *ad libitum* feeding, birds will normally tend to adjust intake voluntarily to their individual requirements for energy which can vary with factors such as environment, breed, live weight, balance of nutrients in the diet and, for layers, rate of egg production. We express these requirements in terms of metabolizable energy (ME), that is the gross energy (total heat of combustion) of the food minus the gross energies of faeces, urine and any gaseous products of digestion. We have information only for the fowl.

Growth

Broilers. The U.K. data we have found enable us to provide no more than a guide to the voluntary energy intakes of fast-growing birds. The most comprehensive[6] were derived from experiments with male broiler-type chickens fed on cereal-based diets with or without added lard or starch to alter the ME content. Weekly records were obtained for sixteen groups of a commercial type of fast-growing broiler chicken, forty-four groups of Rhode Island Red × Light Sussex and fifteen groups of Brown Leghorn × Light Sussex, with twelve cockerels in each group. Apart from the small group size, the conditions were essentially those encountered in practice. The diets, which ranged in fat content from 2 to 11 per cent and in concentration of ME from 2·6 to 3·2 Mcal/kg, were divided into categories according to fat and ME content. From separate statistical analysis of weekly weights, weight gains and ME intakes it was clear that differences between diets within each category were small compared with differences between experiments, which were in themselves quite as large as differences between categories. We present in two graphs in Fig. 1 the overall mean weights and cumulative ME intakes. The hatched area in each graph shows the extreme values obtained for the separate groups of twelve birds. Because of the range in environmental conditions and genetic material obtaining in the broiler industry, it is again emphasized that these figures can be no more than a guide to the growth rates and energy intakes of growing birds.

In another series of experiments in the U.K.[16, 17, 18, 21] both male and female White Rock × Light Sussex and Indian Game (White Cornish) × White Rock chicks were given practical broiler diets ranging in ME content from 2·9 to 3·4 Mcal/kg. In two experiments[16, 18] the values for live weight and age fell within the limits given in the upper graph of Fig. 1, but in a third experiment[21] in which very rapid growth was obtained, the cockerels were heavier at 9 weeks than would be expected from the graph. In all experiments the live weights obtained for different intakes of ME fell well within the limits given in the lower graph of Fig. 1.

We found only one other recent report on growth rate and food consumption,[5] and the growth rates given in it were similar to those above, as were the estimated ME intakes for the first 5 weeks. After this age, however, the intakes of food and ME were high compared with those referred to above, perhaps because spillage was not measured.

Layers. The values in the literature for kcal ME/g live weight gain fall into two groups with ranges of 2·9–3·6 (mean 3·2)[1, 3, 15] and 8·7–9·5 (mean 9·0).[2, 12, 22] The information given suggests that the former refers to the maturing pullet and the latter to the laying hen putting on fat.

For young birds destined as future egg producers there is insufficient evidence to provide curves of growth and food consumption similar to those produced for broilers.

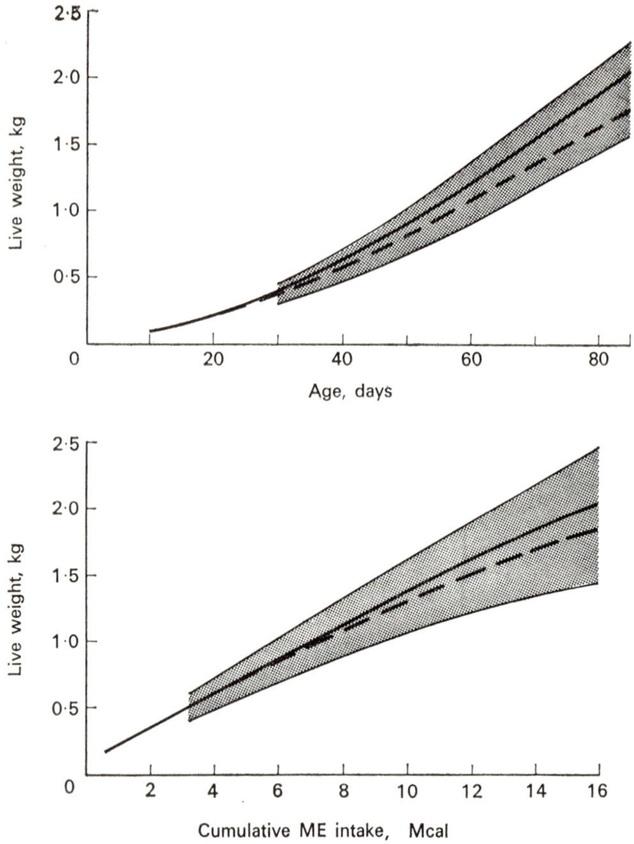

FIG. 1. Overall mean weights and cumulative ME intakes of chickens. Upper graph, live weight gain plotted against age in days; lower graph, cumulative ME intake plotted against live weight. ——, mean values for fast-growing chickens; — — —, mean values for RIR × LS and BL × LS; ▦, range of group means recorded.

EGG PRODUCTION

The requirement for ME by the adult is made up of that for maintenance and that for production. When pullets come into lay they have not reached mature weight and ME is needed for growth. Once mature weight has been reached the excess of ME ingested above that needed for maintenance and egg production is laid down as fat.

The ME required for maintenance will depend upon the microclimate surrounding the laying hen, hence will generally be higher (by about 5 per cent) in winter and lower (by about 5 per cent) in summer than in spring and autumn.[13] The values set out below relate to spring and autumn.

kcal ME required per day for maintenance and stated rates of egg production

Live weight (g)	For maintenance only	References	For maintenance + production* at rate of production of			
			20 per cent†	40 per cent†	60 per cent†	80 per cent†
1500	185	2, 3, 4	210	235	260	285
2000	250		275	300	325	350
2500	310		335	360	385	410
3000	370		395	420	445	470

*Energy required per 'standard' (2 oz) egg: 122 ± 10 kcal ME[1, 2, 3, 4, 9, 15, 22]

$$\dagger \text{ per cent production} = \frac{\text{No. of eggs laid}}{\text{No. of days in observation period}} \times 100.$$

Sources of Energy

Carbohydrate. In conventional poultry diets carbohydrate is the main source of energy. There is no clear evidence of nutritional distinction in this respect between starches, simpler polysaccharides and sugars.

Fat. There is no proof that fat is needed by poultry otherwise than as a source of the essential fatty acids, linoleic and arachidonic. Deficiency of essential fatty acids has been produced in fowls with highly purified synthetic diets,[60, 61] but information is lacking on which to base a statement of requirement.

Protein. Protein in excess of that needed to cover the amino acid requirements (see below) is used as a source of energy.

References

Energy Requirements

1. Bird, S. and Sinclair, J. W. (1938–9) *Sci. Agric.* **19**, 542.
2. Bolton, W. (1959) *J. Agric. Sci.* **52**, 364.
3. Brody, S., Funk, E. M. and Kempster, H. L. (1938) *Res. Bull. Mo. Agric. Exp. Sta.* No. 278.
4. Byerly, T. C. (1941) *Bull. Md Agric. Exp. Sta.* No. A1.
5. Coles, R. (1953) *J. Sci. Fd Agric.* **4**, 532.
6. Davidson, J. and Boyne, A. W. (1962) Rowett Research Institute; unpublished.
7. Ericksson, S. (1954) *K. LantbrHogsk. Ann.* **21**, 385.
8. Gerhartz, H. (1914) *Pflug. Arch. ges. Physiol.* **156**, 1.
9. Halnan, E. T. (1939) *World's Poult. Congr. VII. Cleveland*, p. 145.
10. Hari, P. (1917) *Biochem. Z.* **78**, 313.
11. Hari, P. and Kriwuscha, A. (1918) *Biochem. Z.* **88**, 345.
12. Hendricks, W. A., Jull, M. A. and Titus, H. W. (1932) *Poult. Sci.* **11**, 74.
13. Hill, F. W. (1956) *Poult. Sci.* **35**, 59.
14. Hill, F. W., Anderson, D. L. and Dansky, L. M. (1956) *Poult. Sci.* **35**, 54.
15. Joshi, B. C., Shaffner, C. S. and Jull, M. A. (1948) *Poult. Sci.* **27**, 670.

16. LEWIS, D. and MORGAN, J. T. (1963) *Brit. Poult. Sci.* **4**, 3.
17. LEWIS, D., MORGAN, J. T. and PAYNE, C. G. (1962) *Brit. Poult. Sci.* **3**, 141.
18. LEWIS, D. and PAYNE, C. G. (1963) *Brit. Poult. Sci.* **4**, 13.
19. MITCHELL, H. H. and HAINES, W. T. (1927) *J. Agric. Res.* **34**, 549.
20. MITCHELL, H. H. and HAINES, W. T. (1927) *J. Agric. Res.* **34**, 927.
21. PAYNE, C. G. and LEWIS, D. (1963) *Brit. Poult. Sci.* **4**, 179.
22. TITUS, H. W. (1928) *Poult. Sci.* **8**, 80.

Sources of Energy

Carbohydrates

23. ARSCOTT, G. H., MCCLUSKEY, W. H. and PARKER, J. E. (1958) *Poult. Sci.* **37**, 117.
24. BLAKELY, R. M. and MACGREGOR, H. I. (1953) *Poult. Sci.* **32**, 930.
25. BRIGGS, G. M. and SPIVEY, M. R. (1954) *Fed. Proc.* **13**, 452.
26. COUCH, J. R., BARKI, V. H., SUNDE, M. L., CRAVENS, W. W. and ELVEHJEM, C. A. (1949) *J. Nutr.* **38**, 105.
27. DA ROCHA, G. L. and RAIMO, H. F. (1952) *Bol. Industr. Anim.* **13**, 119.
28. DUNCAN, D. C., GREENAWAY, B. and FORGE, R. (1953) *Agric. Gaz. N.S.W.* **64**, 6.
29. EDWARDS, H. M., MARIEN, J. E. and DRIGGERS, J. C. (1961) *Poult. Sci.* **40**, 811.
30. FOX, M. R. S. and BRIGGS, G. M. (1959) *Poult. Sci.* **38**, 964.
31. FRY, R. E., JENSEN, L. S. and MCGINNIS, J. (1958) *Poult. Sci.* **37**, 624.
32. HARMS, R. H., RASKOPF, B. D., HOCHVEICH, H. J., KIDD, I. H. and DOUGLAS, C. R. (1958) *Poult. Sci.* **37**, 359.
33. KARE, M. R. (1958) *Poult. Sci.* **37**, 1216.
34. KNABE, B. (1958) *Wiss. Abh. dtsch. Akad. Landwirtschaftswissensch., Berlin*, No. 37, p. **116**.
35. KRATZER, F. H., MARSHALL, B. and WILLIAMS, D. E. (1949) *Calif. Agric.* **3**, No. 12, p. 10.
36. MCCLYMONT, G. L. and DUNCAN, D. C. (1952) *Aust. Vet. J.* **28**, 229.
37. MCGINNIS, J., MACGREGOR, H. I. and CARVER, J. S. (1948) *Poult. Sci.* **27**, 459.
38. MEHRING, A. L. Jr., JOHNSON, D. Jr., NESBITT, L. L. and TITUS, H. W. (1962) *Poult. Sci.* **41**, 249.
39. MILLER, O. N. (1953) *J. Nutr.* **50**, 13.
40. MONSON, W. J., DIETRICH, L. S. and ELVEHJEM, C. A. (1950) *Proc. Soc. Exp. Biol., N.Y.*, **75**, 256.
41. PALAFOX, A. L. and ROSENBERG, M. M. (1954) *Poult. Sci.* **33**, 127.
42. ROSENBERG, M. M. (1954) *Poult. Sci.* **33**, 382.
43. ROSENBERG, M. M. and SEU, J. (1952) *World's Poult. Sci. J.* **8**, 93.
44. SCOTT, H. M., MATTERSON, L. D. and SINGSEN, E. P. (1947) *Poult. Sci.* **26**, 554.
45. SMITH, R. E. and MACINTYRE, T. M. (1959) *Poult. Sci.* **38**, 1248.
46. SMITH, W. G. and STEPHENSON, E. L. (1960) *Poult. Sci.* **39**, 1296.
47. SOEDARMO, D., KARE, M. R. and WASSERMAN, R. H. (1961) *Poult. Sci.* **40**, 123.
48. SZEPSENWOL, J. (1953) *Fed. Proc.* **12**, 141.
49. THOMPSON, C. R., TAYLOR, K. W. and KOHLER, G. O. (1961) *Poult. Sci.* **40**, 128.

Fats

50. BIERI, J. E., BRIGGS, G. M., FOX, M. R. S., POLLARD, C. J. and ORTIZ, L. O. (1956) *Proc. Soc. Exp. Biol., N.Y.* **93**, 237.
51. CARVER, D. S. and JOHNSON, E. L. (1954) *Poult. Sci.* **33**, 543.

PROTEIN AND AMINO ACID REQUIREMENTS

It has been customary in the past to state the total requirement for nitrogen in terms of undefined protein, e.g. 18 per cent for young chicks, 15 per cent for layers and 24 per cent for turkey poults. It is clear, however, that a quantitative requirement for protein cannot be stated without reference to its amino acid content. It is probable that the quantities of normal ingredients required to provide the essential amino acids in sufficient amounts will at the same time provide the nitrogen necessary for the synthesis of the non-essential amino acids. We only state, therefore, requirements for essential amino acids.

The requirement for amino acids is also a function of the energy content of the diet and of the class of bird to be fed.

We state the requirements (Tables 1 and 3) for the L form of all amino acids except methionine and phenylalanine, of which the D and L forms are of equal biological value.[3] The D forms of leucine, threonine and tryptophan are of no value on their own but have up to 40 per cent of the value of the L form when both are present in the diet.[3] On the other hand, D-lysine antagonizes the action of the L form and acts as a growth depressant, whereas D-histidine is without effect on growth.[3] For the D form of isoleucine and of valine the evidence is contradictory.[3, 6, 8]

Ideally, many factors should be taken into account in determining amino acid requirements. For example, light breeds are more efficient in egg production than heavy breeds; the energy necessary for maintenance is greater in a cold than in a hot environment, but that for production remains unaffected. Theoretically, therefore, the requirements should be qualified by all such factors, but we are unable to suggest appropriate corrections in view of the wide range of most estimates on which we had to base our values.

Apart from its energy content, the amino acid pattern of the diet is perhaps the most important factor that influences the performance associated with different levels of a particular amino acid. After absorption from the alimentary canal, amino acids are transported to the sites of anabolism and formed into the tissue proteins required. It is generally assumed that mixtures of free amino acids are completely absorbed, and at first sight values based on experiments with such mixtures should be a particularly reliable guide to requirements as defined in this Summary of Recommendations. However, findings with rats[4, 9] have indicated that if all the amino acids are given together in the free form they may be too rapidly absorbed and arrive at the sites of anabolism at rates greater than can be dealt with efficiently by the anabolic processes. Earlier work with chicks suggests that the same may obtain for poultry.[23] On the other hand, amino acids as present in food proteins may not be fully available for tissue synthesis, and experiments with conventional diets are beset with such problems of availability of amino acids. We have therefore relied to the same extent on studies in which free amino acids were used as on those with complete proteins as we find no reason to favour either type of experiment.

It is generally considered that for any particular type and intensity of physiological function a specific amount of each amino acid is required. It follows that the requirement expressed as a percentage of the diet will vary with the amount of diet eaten, which in turn will be related to the concentration of energy in it. In most experiments the amino acid pattern has been kept constant apart from the amino acid under test, and amino acid and energy levels of the diets have been estimated from tables and not by direct analysis. The energy level has been altered intentionally in only a few studies with lysine and methionine[1, 2, 10] for which the relationship between the requirement and the energy level of the diet has been experimentally established (Table 2), but there is no direct evidence for the other essential amino acids. For them, by analogy, those requirements which we have stated for diets of 2·8 Mcal ME/kg could possibly be modified in direct proportion to the energy concentration. However, in view of the uncertainty, we have merely indicated in the tables those values based on experiments with purified diets of calorie concentration much higher than 2·8 Mcal ME/kg.

Notes on Individual Amino Acids

Arginine. The chick is unable to synthesize arginine at a rate sufficient for normal growth.[16, 23] Its requirement appears to be much greater when the diet is purified and the protein is supplied as casein or a casein-gelatin mixture than when the diet is a practical one based on maize and soya-bean meal.[25, 26, 37] Requirement values of up to 2 per cent of the diet have been found with purified diets, whereas concentrations of arginine less than half that have proved adequate in practical diets. In a recent study in the U.K.[29] an arginine content of 0·8 per cent of the diet was found to be sufficient for young chicks fed on practical diets containing 21 per cent crude protein. The discrepancy has not yet been fully explained, although growth-factor deficiency,[25, 26, 27, 39] low concentrations of creatine,[17, 19, 33, 37, 38] amino acid balance[13, 18] and amino acid availability[30, 31] have all been implicated.

Glycine. The requirement of the chick appears to be greater when the rate of feather development is rapid[44, 46] and when the dietary concentration of arginine,[53] methionine,[42] or vitamins of the B complex[43] is low. There is also evidence that in early life the free amino acid is more effective than that combined in protein, because of a slow rate of liberation of the latter,[50] although this evidence is contrary to earlier findings.[41]

Many studies on the glycine requirement of the chick have been carried out with purified diets in which the main source of protein has been casein. The requirement based on such studies has varied from 0·2 per cent to between 1·5 and 1·9 per cent. When practical-type diets have been used, a value of about 1 per cent has proved adequate.[42, 52] Because of the wide divergence in glycine levels found satisfactory in different studies, and because it is known that the chick can synthesize glycine to a limited extent,[51] it is difficult to state a definite figure for requirement. However, on the basis that the value of 1 per cent found with practical diets is about the same as the median of those obtained in various studies with purified diets, it is given as a guide to requirement.

It is usually assumed that adult fowls can synthesize glycine at a rate commensurate with their needs.

The requirement of 0·9 per cent suggested for turkey poults is based on results of one study[48] in which the growth rate on a purified basal diet increased slightly but progressively with the increase of the glycine content from 0·12 to 0·9 per cent. Even with the lowest level growth was as good as on a practical, unspecified diet, and the growth rate decreased when more than 0·9 per cent of glycine was included in the diet.

Histidine. For laying hens 0·2 per cent in a purified diet appeared to be adequate,[56] but since a lower level was not tested this value cannot be accepted as the requirement.

Isoleucine. We have assessed the chick's requirement from three studies with purified diets. The study[60] with the diet containing least ME (3·4 Mcal/kg) yielded a requirement of 0·5 per cent and this value was not at variance with those obtained in the other two studies,[59, 62] when correction was made by direct proportion for the differences in dietary energy concentration. However, all diets derived their nitrogen from a mixture of pure amino acids, and work with rats,[65] probably also applicable to chicks, has shown that there can be greater loss of energy in the urine with such diets than with those containing whole protein. Extrapolation of the data to calculate a requirement corresponding to 2·8 Mcal ME/kg in practical diets did not thus seem justified and the value given refers to a highly purified diet having a predicted ME content of around 3·5 Mcal/kg, a prediction that may err on the high side.

Lysine. After 4 weeks of age the requirement of the chick is said to decrease by 0·05 per cent per week to 0·8 per cent at 9 weeks of age.[80] It has been suggested that the adult rooster does not require lysine.[79] We have adopted the tentative suggestion[77] that the requirement of the laying hen is 0·5 per cent.

Methionine and cystine. Methionine, the essential sulphur amino acid for poultry, can be replaced in part by cystine, hence both are considered together. The many studies on the requirement of the chick for sulphur amino acids were divided into those with purified diets and those with practical diets. The median for each group was approximately the same at about 0·7 per cent. The range for practical diets was from 0·6 to 0·8 per cent and for purified diets from 0·5 to 0·9 per cent. The available evidence would suggest that at least 40 per cent of the quantity of sulphur amino acids required by the chick should be in the form of methionine. There is no such information for the laying hen or the turkey poult, but by analogy a figure of about 40 per cent is probably reasonable.

Threonine. Our information is only about chicks given purified diets. As these diets differed in nitrogen content and as the requirement appeared to be proportional to the dietary nitrogen concentration, all figures were adjusted to a content of 3·2 per cent nitrogen (equivalent to 20 per cent crude protein) before the median value was taken. The range given refers to adjusted values.

References

General

1. Baldini, J. T., Marvel, J. P. and Rosenberg, H. R. (1957) *Poult. Sci.* **36**, 1031.
2. Baldini, J. T. and Rosenberg, H. R. (1955) *Poult. Sci.* **34**, 1301.
3. Fell, R. V., Wilkinson, W. S. and Watts, A. B. (1959) *Poult. Sci.* **38**, 1203.
4. Geiger, E., El Rawi, I. and Geiger, L. E. (1952) *Fed. Proc.* **11**, 444.
5. Grau, C. R. and Almquist, H. J. (1943) *J. Nutr.* **26**, 631.
6. Grau, C. R. and Peterson, D. W. (1946) *J. Nutr.* **32**, 181.
7. Kratzer, F. H., Williams, D. E. and Marshall, B. (1951) *J. Nutr.* **43**, 223.
8. Leveille, G. A. and Fisher, H. (1960) *J. Nutr.* **70**, 135.
9. Maddy, K. H. and Swift, R. W. (1952) *J. Nutr.* **47**, 243.
10. Schwartz, H. G., Taylor, M. W. and Fisher, H. (1958) *J. Nutr.* **65**, 25.
11. Wilkening, M. C. and Schweigert, B. S. (1947) *J. Biol. Chem.* **171**, 209.

Arginine

12. Almquist, H. J. and Merritt, J. B. (1950) *Proc. Soc. Exp. Biol., N.Y.* **73**, 136.
13. Anderson, J. O. and Dobson, D. C. (1959) *Poult. Sci.* **38**, 1140.
14. Arnold, A., Kline, O. L., Elvehjem, C. A. and Hart, E. B. (1936) *J. Biol. Chem.* **116**, 699.
15. Fisher, H., Grimminger, P. and Lutz, H. (1960) *Poult. Sci.* **39**, 173.
16. Fisher, H., Leveille, G. A. and Griminger, P. (1959) *Poult. Sci.* **38**, 231.
17. Fisher, H., Salander, R. C. and Taylor, M. W. (1956) *J. Nutr.* **59**, 491.
18. Fisher, H., Shapiro, R. and Griminger, P. (1960) *J. Nutr.* **72**, 16.
19. Griminger, P., Fisher, H. and Scott, H. M. (1955) *Poult. Sci.* **34**, 1247.
20. Johnson, D., Jr. and Fisher, H. (1956) *J. Nutr.* **60**, 275.
21. Johnson, D., Jr. and Fisher, H. (1958) *Brit. J. Nutr.* **12**, 276.
22. Klain, G. J., Scott, H. M. and Johnson, B. C. (1959) *Poult. Sci.* **38**, 488.
23. Klose, A. A., Stokstad, E. L. R. and Almquist, H. J. (1938) *J. Biol. Chem.* **123**, 691.
24. Kratzer, F. H., Bird, F. H., Asmundson, V. S. and Lepkovsky, S. (1947) *J. Nutr* **34**, 167.
25. Krautmann, B. A., Hauge, S. M., Mertz, E. T. and Carrick, C. W. (1956) *Poult Sci.* **35**, 1153.
26. Krautmann, B. A., Hauge, S. M., Mertz, E. T. and Carrick, C. W. (1957) *Poult Sci.* **36**, 935.
27. Krautmann, B. A., Hauge, S. M., Mertz, E. T. and Carrick, C. W. (1958) *Poult Sci.* **37**, 530.
28. Leveille, G. A. and Fisher, H. (1959) *J. Nutr.* **69**, 289
29. Lewis, D., Smith, G. H. and Payne, C. G. (1963) *Brit. J. Nutr.* **17** (in press).
30. O'Dell, B. L., Laedal, O. A., Jeffay, A. M. and Savage, J. E. (1958) *Poult. Sci.* **37**, 817.
31. Ousterhout, L. E., Grau, C. R. and Lundholm, B. D. (1959) *J. Nutr.* **69**, 65.
32. Owings, W. J. and Balloun, S. L. (1959) *Poult. Sci.* **38**, 1285.
33. Savage, J. E. and O'Dell, B. L. (1960) *J. Nutr.* **70**, 129.
34. Scott, H. M. and Forbes, R. M. (1958) *Poult. Sci.* **37**, 1347.
35. Smith, G. H. and Lewis, D. (1963) *Brit. J. Nutr.* **17** (in press).
36. Snyder, J. M., Morrison, W. D. and Scott, H. M. (1954) *Poult. Sci.* **33**, 1082.

37. SNYDER, J. M., MORRISON, W. D. and SCOTT, H. M. (1956) *Poult. Sci.* **35**, 852.
38. WIETLAKE, A. W., HOGAN, A. G., O'DELL, B. L. and KEMPSTER, H. L. (1954) *J. Nutr.* **52**, 311.
39. YOUNG, R. J., GILLIS, M. B. and NORRIS, L. C. (1953) *J. Nutr.* **50**, 291.

Glycine

40. ALMQUIST, H. J. and MECCHI, E. (1940) *J. Biol. Chem.* **135**, 355.
41. ALMQUIST, H. J. and MECCHI, E. (1942) *Proc. Soc. Exp. Biol., N.Y.* **49**, 541.
42. DOUGLAS, C. R., HOCHREICH, H. J. and HARMS, R. H. (1958) *Poult. Sci.* **37**, 620.
43. FISHER, H., SCOTT, H. M. and JOHNSON, B. C. (1955) *J. Nutr.* **55**, 415.
44. GLISTA, W. A. (1951). The amino acid requirements of the chick: method and application to some of the amino acids. Doctoral Thesis, University of Illinois.
45. GREENE, D. E., SCOTT, H. M. and JOHNSON, B. C. (1960) *Poult. Sci.* **39**, 512.
46. HEGSTED, D. M., BRIGGS, G. M., ELVEHJEM, C. A. and HART, E. B. (1941) *J. Biol. Chem.* **140**, 191.
47. HOGAN, A. G., CRAGHEAD, R. W., SAVAGE, J. E., COLE, J. J. and O'DELL, B. L. (1957) *J. Nutr.* **62**, 97.
48. KRATZER, F. H. and WILLIAMS, D. (1948) *J. Nutr.* **35**, 315.
49. MENGE, H., MACHLIN, L. J. and LILLIE, R. J. (1956) *Poult. Sci.* **35**, 1159.
50. MONSON, W. J., HARPER, A. E., BENTON, D. A., WINJE, M. and ELVEHJEM, C. A. (1955) *Poult. Sci.* **34**, 186.
51. PATTON, A. R. and PALMER, L. S. (1936) *J. Nutr.* **11**, 129.
52. RIEDEL, B. B. (1954) *Poult. Sci.* **33**, 742.
53. SAVAGE, J. E., CRAGHEAD, R. W., O'DELL, B. L. and HOGAN, A. G. (1955) *Poult. Sci.* **34**, 1219.
54. WIETLAKE, A. W., HOGAN, A. G., O'DELL, B. L. and KEMPSTER, H. L. (1954) *J. Nutr.* **52**, 311.
55. WIXOM, R. L., PIPKIN, G. E. and DAY, P. L. (1955) *J. Nutr.* **56**, 409.

Histidine

56. JOHNSON, D., Jr. and FISHER, H. (1958) *Brit. J. Nutr.* **12**, 276.
57. KLAIN, G. J., SCOTT, H. M. and JOHNSON, B. C. (1960) *Poult. Sci.* **39**, 39.
58. ROSENBERG, H. R., BALDINI, J. T. and TOLLEFSON, C. I. (1957) *Poult. Sci.* **36**, 1381.

Isoleucine

59. GLISTA, W. A. (1951). The amino acid requirements of the chick: method and application to some of the amino acids. Doctoral Thesis, University of Illinois.
60. GRAU, C. R. and PETERSON, D. W. (1946) *J. Nutr.* **32**, 181.
61. JOHNSON, D., Jr. and FISHER, H. (1958) *Brit. J. Nutr.* **12**, 276.
62. KLAIN, G. J., SCOTT, H. M. and JOHNSON, B. C. (1960) *Poult. Sci.* **39**, 39.
63. KRATZER, F. H., WILLIAMS, D. E. and MARSHALL, B. (1952) *J. Nutr.* **47**, 631.
64. LEVEILLE, G. A. and FISHER, H. (1960) *J. Nutr.* **70**, 135.
65. MADDY, K. H. and SWIFT, R. W. (1952) *J. Nutr.* **47**, 243.
66. MILLER, E. C., SUNDE, M. L., BIRD, H. R. and ELVEHJEM, C. A. (1954) *Poult. Sci.* **33**, 1201.

Leucine

67. ANDERSON, J. O. and DRAPER, C. I. (1956) *Poult. Sci.* **35**, 562.
68. CRAVENS, W. W. (1948) *Poult. Sci.* **27**, 562.
69. GLISTA, W. A. (1951). The amino acid requirements of the chick: method and application to some of the amino acids. Doctoral Thesis, University of Illinois.
70. GRAU, C. R. and PETERSON, D. W. (1946) *J. Nutr.* **32**, 181.
71. JOHNSON, D., Jr. and FISHER, H. (1958) *Brit. J. Nutr.* **12**, 276.
72. KLAIN, G. J., SCOTT, H. M. and JOHNSON, B. C. (1958) *Poult. Sci.* **37**, 976.
73. KLAIN, G. J., SCOTT, H. M. and JOHNSON, B. C. (1960) *Poult. Sci.* **39**, 39.

Lysine

74. Balloun, S. L. and Phillips, R. E. (1957) *Poult. Sci.* **36**, 884.
75. Edwards, H. M., Jr., Norris, L. C. and Heuser, G. F. (1956) *Poult. Sci.* **35**, 385.
76. Grimminger, P. and Scott, H. M. (1959) *J. Nutr.* **68**, 429.
77. Johnson, D., Jr. and Fisher, H. (1958) *Brit. J. Nutr.* **12**, 276.
78. Kratzer, F. H., Davis, P. N. and Marshall, B. J. (1956) *Poult. Sci.* **35**, 197.
79. Leveille, G. A. and Fisher, H. (1959) *J. Nutr.* **69**, 289.
80. Schwartz, H. G., Taylor, M. W. and Fisher, H. (1958) *J. Nutr.* **65**, 25.

Methionine and Cystine

81. Adams, R. L., Andrews, F. N., Rogler, J. C. and Carrick, C. W. (1962) *Poult. Sci.* **41**, 1801.
82. Almquist, H. J. (1952) *Poult. Sci.* **31**, 966.
83. Baldini, J. T., Marvel, J. P. and Rosenberg, H. R. (1957) *Poult. Sci.* **36**, 1031.
84. Baldini, J. T. and Rosenberg, H. R. (1955) *Poult. Sci.* **34**, 1301.
85. Davidson, J. and Mathieson, J. (1961). Unpublished.
86. Demers, J. M. and Bernard, R. (1950) *Canad. J. Res.* E, **28**, 202.
87. Evans, R. J., Bandemer, S. L. and Bauer, D. H. (1956) *Poult. Sci.* **35**, 174.
88. Evans, R. J. and McGinnis, J. (1948) *J. Nutr.* **35**, 477.
89. Fisher, H., Griminger, P. and Leveille, G. A. (1959) *J. Nutr.* **69**, 117.
90. Fisher, H., Scott, H. M. and Johnson, B. C. (1955) *J. Nutr.* **55**, 415.
91. Ford, J. E. and Porter, J. W. G. (1961) *Int. Congr. Biochem. V. Moscow*, Abstracts of Communications, p. 242.
92. Glista, W. A. (1951). The amino acid requirements of the chick: method and application to some of the amino acids. Doctoral Thesis, University of Illinois.
93. Glista, W. A., Mitchell, H. H. and Scott, H. M. (1951) *Poult. Sci.* **30**, 915.
94. Grau, C. R. and Kamei, M. (1950) *J. Nutr.* **41**, 89.
95. Grimminger, P., Forbes, R. M. and Scott, H. M. (1954) *Poult. Sci.* **33**, 1057.
96. Ingram, G. R., Cravens, W. W., Elvehjem, C. A. and Halpin, J. G. (1951) *Poult. Sci.* **30**, 431.
97. Johnson, D., Jr. and Fisher, H. (1958) *Brit. J. Nutr.* **12**, 276.
98. Klain, G. J., Scott, H. M. and Johnson, B. C. (1960) *Poult. Sci.* **39**, 39.
99. Kratzer, F. H., Williams, D. E. and Marshall, B. (1949) *J. Nutr.* **37**, 377.
100. Leong, K. C. and McGinnis, J. (1952) *Poult. Sci.* **31**, 692.
101. Leveille, G. A., Shapiro, R. and Fisher, H. (1960) *J. Nutr.* **72**, 8.
102. McGinnis, J. and Evans, R. J. (1947) *J. Nutr.* **34**, 725.
103. Milligan, J. L., Machlin, L. J., Bird, H. R. and Heywang, B. W. (1951) *Poult. Sci.* **30**, 578.
104. Nelson, T. S., Young, R. J., Bradfield, R. B., Anderson, J. B., Norris, L. C., Hill, F. W. and Scott, M. L. (1960) *Poult. Sci.* **39**, 308.
105. Pepper, W. F. and Slinger, S. J. (1955) *Poult. Sci.* **34**, 957.
106. Scott, M. L., Parsons, E. H., Jr. and Dougherty, E. III. (1957) *Poult. Sci.* **36**, 1181.
107. Titus, H. W. (1955) *The Scientific Feeding of Chickens*, 3rd ed. Danville, Ill.: Interstate.
108. West, J. W., Carrick, C. W., Hauge, S. M. and Mertz, E. T. (1951) *Poult. Sci.* **30**, 880.
109. Williams, H. H., Curtin, L. V., Abraham, J., Loosli, J. K. and Maynard, L. A. (1954) *J. Biol. Chem.* **208**, 277.

Phenylalanine and Tyrosine

110. Almquist, H. J. and Grau, C. R. (1944) *J. Nutr.* **28**, 325.
111. Fisher, H. (1956) *Arch. Biochem. Biophys.* **60**, 108.
112. Fisher, H. and Johnson, D., Jr. (1957) *Poult. Sci.* **36**, 444.
113. Fisher, H., Johnson, D., Jr. and Leveille, G. A. (1957) *J. Nutr.* **62**, 349.
114. Grau, C. R. (1947) *J. Biol. Chem.* **170**, 661.

115. JOHNSON, D., Jr. and FISHER, H. (1958) *Brit. J. Nutr.* **12**, 276.
116. KLAIN, G. J., SCOTT, H. M. and JOHNSON, B. C. (1958) *Poult. Sci.* **37**, 976.
117. KLAIN, G. J., SCOTT, H. M. and JOHNSON, B. C. (1960) *Poult. Sci.* **39**, 39.
118. KRAUTMANN, B. A., HAUGE, S. M., MERTZ, E. T. and CARRICK, C. W. (1958) *Poult. Sci.* **37**, 535.

Threonine

119. ADKINS, J. S., MILLER, E. C., BIRD, H. R., ELVEHJEM, C. A. and SUNDE, M. L. (1957) *Poult. Sci.* **36**, 1101.
120. ADKINS, J. S., MILLER, E. C., BIRD, H. R., ELVEHJEM, C. A. and SUNDE, M. L. (1958) *Poult. Sci.* **37**, 1362.
121. ALMQUIST, H. J. (1947) *J. Nutr.* **34**, 543.
122. ALMQUIST, H. J. (1952) *Poult. Sci.* **31**, 966.
123. ALMQUIST, H. J. and GRAU, C. R. (1944) *J. Nutr.* **28**, 325.
124. GLISTA, W. A. (1951). The amino acid requirements of the chick: method and application to some of the amino acids. Doctoral Thesis, University of Illinois.
125. GRAU, C. R. (1947) *J. Biol. Chem.* **170**, 661.
126. GRAU, C. R. (1949) *J. Nutr.* **37**, 105.
127. JOHNSON, D., Jr. and FISHER, H. (1958) *Brit. J. Nutr.* **12**, 276.
128. KLAIN, G. J., SCOTT, H. M. and JOHNSON, B. C. (1960) *Poult. Sci.* **39**, 39.
129. KRAUTMANN, B. A., HAUGE, S. M., MERTZ, E. T. and CARRICK, C. W. (1958) *Poult. Sci.* **37**, 535.
130. LEVEILLE, G. A. and FISHER, H. (1960) *J. Nutr.* **70**, 135.
131. LEWIS, E. W., Jr., ELVEHJEM, C. A. and HART, E. B. (1951) *J. Nutr.* **43**, 113.
132. WILKENING, M. C., SCHWEIGERT, B. S., PEARSON, P. B. and SHERWOOD, R. M. (1947) *J. Nutr.* **34**, 701.

Tryptophan

133. BENTON, D. A., HARPER, A. E. and ELVEHJEM, C. A. (1954) *Poult. Sci.* **33**, 1279.
134. BIRD, F. H. (1950) *Poult. Sci.* **29**, 737.
135. CHILDS, G. R., CARRICK, C. W. and HAUGE, S. M. (1952) *Poult. Sci.* **31**, 551.
136. FISHER, H., SCOTT, H. M. and JOHNSON, B. C. (1954) *Poult Sci.* **33**, 1054.
137. FISHER, H., SCOTT, H. M. and JOHNSON, B. C. (1955) *Brit. J. Nutr.* **9**, 340.
138. GRIMINGER, P., SCOTT, H. M. and FORBES, R. M. (1956) *J. Nutr.* **59**, 67.
139. INGRAM, G. R., CRAVENS, W. W., ELVEHJEM, C. A. and HALPIN, J. G. (1951) *Poult. Sci.* **30**, 426.
140. INGRAM, G. R. and LITTLE, P. L. (1958) *Poult. Sci.* **37**, 1214.
141. JOHNSON, D., Jr. and FISHER, H. (1958) *Brit. J. Nutr.* **12**, 276.
142. KLAIN, G. J., SCOTT, H. M. and JOHNSON, B. C. (1958) *Poult. Sci.* **37**, 976.
143. KRATZER, F. H., WILLIAMS, D. E. and MARSHALL, B. (1951) *J. Nutr.* **43**, 223.
144. PATTERSON, E. B., HUNT, J. R., VOHRA, P., BLAYLOCK, L. G. and MCGINNIS, J. (1954) *Poult. Sci.* **33**, 1075.
145. PATTERSON, E. B., HUNT, J. R., VOHRA, P., BLAYLOCK, L. G. and MCGINNIS, J. (1956) *Poult. Sci.* **35**, 499.
146. WEST, J. W., CARRICK, C. W., HAUGE, S. M. and MERTZ, E. T. (1952) *Poult. Sci.* **31**, 479.
147. WILKENING, M. C. and SCHWEIGERT, B. S. (1947) *J. Biol. Chem.* **171**, 209.

Valine

148. JOHNSON, D., Jr. and FISHER, H. (1958) *Brit. J. Nutr.* **12**, 276.
149. KLAIN, G. J., SCOTT, H. M. and JOHNSON, B. C. (1960) *Poult. Sci.* **39**, 39.
150. LEVEILLE, G. A. and FISHER, H. (1960) *J. Nutr.* **70**, 135.

VITAMIN REQUIREMENTS

There is little direct information on the interrelationship between vitamin requirements and dietary energy but, since birds eat less as the energy content of the diet increases, it may be advisable to alter the requirements stated in Tables 4—7 in direct proportion to the ME content of the diet if it differs from 2·8 Mcal/kg. The specific problem of thiamine is discussed on p. 1221.

The requirements of young birds are obviously affected by the carry-over of vitamins to the egg and thence to the hatched chick. We have usually based our estimates on the assumption that there was a normal carry-over and have disregarded experiments in which hens were known to have received excessively high levels of the vitamin in question.

NOTES ON INDIVIDUAL VITAMINS

Vitamin A. Data have been taken from experiments in which the source of vitamin A was either the preformed vitamin or its carotenoid precursors. The requirements stated may be somewhat high, as much of the work was done before stabilized preparations were developed and, since in most instances the vitamin A supplement was given in the diet, some losses may have occurred during mixing or storage. It is known that antioxidants have a sparing effect on vitamin A.

In general, β-carotene is more stable than the preformed vitamin. By definition, the activity of β-carotene, on a weight basis, is half that of vitamin A, but its biological value in terms of vitamin A may be reduced by a variety of factors affecting the utilization of carotene in the diet. We suggest therefore that, in practice, β-carotene may have weight for weight about one-sixth of the value of vitamin A, although opinion on this matter is still divided.

Vitamin D. Vitamins D_2 (ergocalciferol) and D_3 (cholecalciferol) are the two forms of special significance in animal nutrition. There are other antirachitic substances but their potencies are low compared with those of ergocalciferol and cholecalciferol. Whereas the last two forms are approximately equal in antirachitic activity for mammals, for the fowl and other avianspecies cholecalciferol is distinctly more effective. The ratio of activity for chicks of vitamin D_3 to vitamin D_2 is approximately 35 : 1.[51, 56]

The requirement for vitamin D is intimately bound up with the calcium and phosphorus levels in the diet. The values we give for this vitamin refer to diets supplying the estimated requirements for calcium and phosphorus stated in Tables 8 and 9. In the presence of sunlight the vitamin can be formed in the bird and the dietary requirement is, therefore, less. There is evidence that, for turkey poults, irradiated animal sterols may be more effective in terms of international units than vitamin D_3 in fish oils,[17, 39, 68] particularly when the diet is low in inorganic phosphorus.

Vitamin E (the tocopherols). So far several natural tocopherols have been described which vary considerably in their potency for mammals. Of them, α-tocopherol is the most potent and is also the most widespread and commonly used. It is the only tocopherol so far studied extensively in poultry nutrition.

Experiments to determine requirements have produced widely different results, varying from almost zero to 92 mg DL-α-tocopheryl acetate/kg diet, depending on the criterion of deficiency, the nature of the diet and other environmental factors. It has therefore not been possible to give reliable estimates.

As one of the biochemical functions of vitamin E is as an antioxidant, the need for it is increased by high levels of unsaturated fats in the diet. Conversely, the presence of other antioxidants will decrease the requirement for this vitamin. There is evidence that some effects of vitamin E deficiency may be prevented by selenium; different types of organic and inorganic selenium compounds have been found to vary widely in their biological activity, but the most active are the simple inorganic forms such as selenite and selenate.[79]

Vitamin K. Of the two naturally occurring forms of the vitamin, K_1 (2-methyl-3-phytyl-1,4-napthoquinone) and K_2 (2-methyl-3-difarnesyl-1,4-naphthoquinone), the second can be synthesized by intestinal microorganisms, but whether it is then available to poultry without ingestion of the droppings is questionable. Experiments to determine the requirements have used the synthetic vitamin K_3 (menaphthone, 2-methyl-1,4-napthoquinone) or its water-soluble form, menaphthone sodium bisulphite, as the source of this vitamin. Measurement of prothrombin time or blood clotting time has been used as an index of the vitamin deficiency with purified diets, and requirements have been expressed in terms of menaphthone and not natural vitamin K. There is evidence that the requirement may be less with diets containing more than 5 per cent fat.[91]

Thiamine. This vitamin plays a specific part in the metabolism of carbohydrate, hence the requirement is largely determined by the carbohydrate content of the diet. The practical implication is that with diets of energy content higher than 2·8 Mcal ME/kg the stated requirement should be increased only if the additional energy is supplied as carbohydrate.

Riboflavin. The slightly lower value of our estimate for turkey poults compared with that for young chicks probably reflects the tendency to include generous amounts of this vitamin in the diets of breeding turkeys, thus increasing the reserves in the eggs, and does not necessarily imply a lower net requirement (see p. 1220).

Nicotinic acid. The values given in the tables are for diets containing about 0·15 per cent tryptophan; with increasing dietary levels of this amino acid less nicotinic acid is needed, probably because of its synthesis from tryptophan.

TABLE 4
Fat-soluble Vitamin Requirements of Fowls

Vitamin	Chick (first few weeks of life)				Layer				Breeder			
	Requirement	Values on which requirement based			Requirement	Values on which requirement based			Requirement	Values on which requirement based		
		Range	No.	References		Range	No.	References		Range	No.	References
Vitamin A (i.u./kg diet)	1500	1300—4200	11	5, 6, 7, 13, 19, 20, 24, 25, 29, 34	4300	3200—8400	7	1, 13, 15, 17, 21, 23, 25	6600	4000—10 700	11	1, 4, 5, 15, 19, 23, 25, 27
Vitamin D_3 (i.u./kg diet)	300	200—600	6	40, 41, 48, 49, 55, 57	600	500—780	4	36, 42, 44, 55	600	500—780	4	36, 42, 44, 55
Vitamin E (see text)												
Vitamin K* (mg/kg diet)	0·99	0·59—1·39	2	88, 91	?				?			

* As menaphthone

TABLE 5
Fat-soluble Vitamin Requirements of Turkeys, Ducks, Geese and Pheasants

Vitamin	Turkey poult				Breeding turkey			Duckling				Gosling			Pheasant chick		
	Requirement	Values on which requirement based			Requirement	Values on which requirement based		Requirement	Values on which requirement based			Requirement	Value on which requirement based		Requirement	Value on which requirement based	
		Range	No.	References		No.	References		Range	No.	References		No.	Reference		No.	Reference
Vitamin A (i.u/kg diet)	1800		1	26	2700	1	26	?				?					
Vitamin D_3 (i.u./kg diet)	1100	670–2000	14	37, 39, 46, 47, 50, 52, 58, 60, 61	2000	1	61	300	280–350	3	38, 45, 53	300	1	45	1100	1	59
Vitamin E	(see text)																
Vitamin K	?				?			?				?					

TABLE 6

Vitamin B Complex Requirements of Fowls (mg/kg diet)

Vitamin	Chick (first few weeks of life)				Layer				Breeder			
	Requirement	Values on which requirement based			Requirement	Values on which requirement based			Requirement	Values on which requirement based		
		Range	No.	References		Range	No.	References		Range	No.	References
Thiamine	1·0	0·7–1·4	5	94, 95, 96, 99, 100	?				?			
Riboflavin	4·0	3·0–5·0	5	101, 103, 104, 108, 112	2·5	2·5–3·6	2	105, 117	4·0	3·6–5·0	3	109, 113, 118
Nicotinic acid	28	25–33	4	125, 126, 127, 132	?				?			
Pantothenic acid	10	6·0–10·0	3	137, 138, 143	1·5		1	139	6·5	6·5–8·0	3	136, 139, 140
Vitamin B_6	3·5	2·0–4·0	5	146, 149, 151, 153, 154	2·0	2·0–2·3	2	147, 150	4·0	2·0–5·5	3	147, 149, 150
Biotin	$>0·07$ $\not>0·10$		1	157	?				0·15	0·04–0·2	2	155, 156
Folic acid	1·5	0·6–2·0	4	160, 163, 166, 167	0·03	0·012–0·42	6	158, 159, 164, 169, 171, 172	0·5	0·3–1·0	4	158, 159, 164, 171
Vitamin B_{12}	0·02*	0·015–0·027	2	181, 186	?				0·002	0·001–0·006	8	174, 176, 178, 180, 182, 183, 187, 188
Choline	1300	1000–1500	7	190, 196, 197, 198, 199, 201, 205	?				1100		1	189

* Negligible reserves at hatching assumed.

TABLE 7

Vitamin B Complex Requirements of Turkeys, Ducks, Geese and Pheasants

(mg/kg diet)

Vitamin	Turkey poult				Breeding turkey				Duckling			Gosling			Pheasant chick			
	Require-ment	Values on which requirement based			Re-quire-ment	Values on which requirement based			Re-quire-ment	Values on which requirement based		Re-quire-ment	Value on which requirement based		Re-quire-ment	Values on which requirement based		
		Range	No.	Refer-ences		Range	No.	Refer-ences		No.	Refer-ences		No.	Refer-ence		Range	No.	Refer-ences
Thiamine	1·0		1	97	?				?			?			?			
Riboflavin	3·5	2·7–4·0	3	106, 116, 121	3·5	3·0–3·5	1	106	4·0	1	111	?			3·5		1	120
Nicotinic acid	>20 ≯ 50		1	130	?				25·0	1	128	65	1	122	60	50–70	2	133, 135
Pantothenic acid	10·5		1	145	16·0		1	144	11·0	1	141	?			?			
Vitamin B_6	3·0		1	154	?				2·5	1	152	?			?			
Biotin	?				?				?			?			?			
Folic acid	1·5	0·8–2·0	4	161, 168, 169, 170	0·55	0·4–0·7	2	162, 169	?			?			?			
Vitamin B_{12}	0·004	0·002–0·01	2	177, 184	?				?			?			?			
Choline	1500	900–2500	6	191, 192, 193, 194, 202, 203	1700		1	204	?			?			1400		1	200

It has been shown that nicotinic acid in cereals is not available to several mammalian species. There is some evidence that the same may be true for the chick[131] and the duckling.[129]

Folic acid. The requirement of the chick may be increased with diets rich in protein and fat.[166]

Vitamin B_{12}. Our estimate of the chick's requirement is based on data obtained from experiments in which hens were depleted of this vitamin, and hence the figure given is many times higher than would be necessary under practical conditions, when little or none may be needed in the diet. There is evidence that the requirement may be increased with diets rich in protein[188] or fat[185] and that there is a mutually sparing effect between vitamin B_{12} and methionine.[173]

References

Vitamin A

1. Almquist, H. J. and Mecchi, E. (1939) *Poult. Sci.* **18**, 129.
2. Ascarelli, I. (1957) *Poult. Sci.* **36**, 549.
3. Asmundson, V. S. and Kratzer, F. H. (1939) *Circ. Calif. Agric. Ext. Serv.* No. 110.
4. Baelum, J. (1948) *World's Poult. Congr. VIII. Copenhagen* **1**, 81.
5. Bearse, G. E. and Miller, M. W. (1937) *Poult. Sci.* **16**, 39.
6. Biely, J. and Chalmers, W. (1936) *Canad. J. Res.* D, **14**, 21.
7. Bolin, D. W., Lampman, C. E. and Berg, L. R. (1943) *Poult. Sci.* **22**, 348.
8. Brown, P. S., Blum, W. P. and Stern, M. H. (1959) *Nature, Lond.* **184**, 1377.
9. Davies, A. W. and Worden, A. N. (1954) *J. Sci. Fd Agric.* **5**, 107.
10. Gledhill, R. H. and Smith, S. B. (1955) *Poult. Sci.* **34**, 942.
11. Gordon, R. S. and Machlin, L. J. (1959) *Poult. Sci.* **38**, 1463.
12. Halverson, A. W. and Hendrick, C. M. (1955) *Poult. Sci.* **34**, 355.
13. Hill, F. W., Scott, M. L., Norris, L. C. and Heuser, G. F. (1961) *Poult. Sci.* **40**, 1245.
14. Hinshaw, W. R. and Lloyd, W. E. (1934) *Hilgardia* **8**, 281.
15. McClary, C. F., Miller, V. L. and Bearse, G. E. (1942) *Bull. Wash. Agric. Exp. Sta.* December 1942, p. 56.
16. Marsden, S. J. and Martin, J. H. (1955) *Turkey Management.* 6th ed. Danville, Ill.: Interstate.
17. Matterson, L. D., Bunnell, R. H., Stinson, L. D., Singsen, E. P. and Potter, L. M. (1955) *Poult. Sci.* **34**, 1080.
18. Paredes, J. R. and Garcia, T. P. (1959) *Poult. Sci.* **38**, 3.
19. Record, P. R., Bethke, R. M., Wilder, O. H. M. and Chamberlin, V. D. (1937) *Bull. Ohio Agric. Exp. Sta.* No. 579, p. 91.
20. Ringrose, R. C. and Norris, L. C. (1936) *Poult. Sci.* **15**, 390.
21. Russell, W. C., Platt, C. S., Taylor, M. W. and Chichester, D. F. (1936) *Circ. N. J. Agric. Exp. Sta.* No. 369.
22. Scott, H. M. (1937) *Bull. Kans. Agric. Exp. Sta.* No. 276.
23. Sherwood, R. M. and Fraps, G. S. (1935) *Bull. Tex. Agric. Exp. Sta.* No. 514.
24. Sherwood, R. M. and Fraps, G. S. (1936) *Bull. Tex. Agric. Exp. Sta.* No. 528.
25. Sherwood, R. M. and Fraps, G. S. (1940) *Bull. Tex. Agric. Exp. Sta.* No. 583.
26. Stoewsand, G. S. and Scott, M. L. (1961) *Poult. Sci.* **40**, 1255.
27. Taylor, M. W., Stern, J. R., Russell, W. C. and Jungherr, E. (1947) *Poult. Sci.* **26**, 243.
28. Temperton, H. and Dudley, F. J. (1948) *Harper Adams Util. Poult. J.* **33 N.S.**, 4.

29. TEPPER, A. E. and DURGIN, R. C. (1938) *Sta. Bull. N.H. Agric. Exp. Sta.* No. 310.
30. THORNTON, P. A. and WHITTET, W. A. (1962) *Poult. Sci.* **41**, 32.
31. VAN REEN, R., TAYLOR, M. W. and RUSSELL, W. C. (1951) *J. Nutr.* **43**, 235.
32. WHARTON, F. D., MATTERSON, L. D., SCOTT, H. M. and BLISS, C. I. (1949) *J. Nutr.* **39**, 543.
33. WILGUS, H. S., Jr. (1940) *Colo. Fm Bull.* **2**, 3.
34. WILLIAMS, J. K., LAMPMAN, C. E. and BOLIN, D. W. (1939) *Poult. Sci.* **18**, 268.

Vitamin D

35. BERG, L. R., BEARSE, G. E. and MILLER, V. L. (1951) *Poult. Sci.* **30**, 799.
36. BETHKE, R. M., RECORD, P. R., WILDER, O. H. M. and KENNARD, D. C. (1937) *Poult. Sci.* **16**, 438.
37. BIRD, H. R. (1944) *J. Nutr.* **27**, 377.
38. BLACK, D. J. G. and COATES, M. E. (1948) *World's Poult. Congr. VIII. Copenhagen*, **1**, 96.
39. BOUCHER, R. V. (1944) *J. Nutr.* **27**, 403.
40. BUCKNER, G. D., INSKO, W. M., Jr., HENRY, A. H. and WACHS, E. F. (1952) *Poult. Sci.* **31**, 628.
41. CARVER, J. S., EVANS, R. J. and MCGINNIS, J. (1946) *Poult. Sci.* **25**, 294.
42. CORRELL, J. T. and WISE, E. C. (1938) *J. Biol. Chem.* **126**, 573.
43. COUCH, J. R., JAMES, L. E. and SHERWOOD, R. M. (1947) *Poult. Sci.* **26**, 30.
44. EVANS, R. J., CARVER, J. S. and WILHELM, L. A. (1944) *Poult. Sci.* **23**, 234.
45. FRITZ, J. C., ARCHER, W. and BARKER, D. (1941) *Poult. Sci.* **20**, 151.
46. FRITZ, J. C., HOOPER, J. H. and MOORE, H. P. (1945) *Poult. Sci.* **24**, 324.
47. HAMMOND, J. C. (1941) *Poult. Sci.* **20**, 204.
48. HEIMAN, V. and TIGHE, L. W. (1942) *Poult. Sci.* **21**, 511.
49. JONES, J. I. M. (1945) *Biochem. J.* **39**, 324.
50. JUKES, T. H. and SANFORD, T. D. (1939) *J. Nutr.* **18**, 71.
51. MCCHESNEY, E. W. (1943) *J. Nutr.* **26**, 81.
52. MCGINNIS, J. and EVANS, R. J. (1946) *Poult. Sci.* **25**, 521.
53. MOTZOK, I. and BRANION, H. D. (1948) *Poult. Sci.* **27**, 482.
54. MOTZOK, I., GRAHAM, W. D., BRANION, H. D. and SLINGER, S. J. (1946) *Poult. Sci.* **25**, 298.
55. MURPHY, R. R., HUNTER, J. E. and KNANDEL, H. C. (1936) *Bull. Pa Agric. Exp. Sta.* No. 334.
56. REMP, D. G. and MARSHALL, I. H. (1938) *J. Nutr.* **15**, 525.
57. ROSS, E. and YACOWITZ, H. (1954) *Poult. Sci.* **33**, 262.
58. SANFORD, T. D. and JUKES, T. H. (1944) *Poult. Sci.* **23**, 221.
59. SCOTT, M. L., HOLM, E. R. and REYNOLDS, R. E. (1958) *Poult. Sci.* **37**, 1419.
60. SINGSEN, E. P., MATTERSON, L. D. and SCOTT, H. M. (1947) *J. Nutr.* **33**, 13.
61. STADELMAN, W. J., BOUCHER, R. V. and CALLENBACH, E. W. (1950) *Poult. Sci.* **29**, 146.
62. WILHELM, L. A., ROBERTSON, E. I. and RHIAN, M. (1941) *Poult. Sci.* **20**, 565.

Vitamin E (the tocopherols)

63. ATKINSON, R. L., FERGUSON, T. M., QUISENBERRY, J. H. and COUCH, J. R. (1955) *J. Nutr.* **55**, 387.
64. BIERI, J. G., BRIGGS, G. M., POLLARD, C. J. and FOX, M. R. S. (1960) *J. Nutr.* **70**, 47.
65. BIRD, H. R. (1943) *Science* **97**, 98.
66. BIRD, H. R. and CULTON, T. G. (1940) *Proc. Soc. Exp. Biol., N.Y.*, **44**, 543.
67. BIRD, H. R. and CULTON, T. G. (1940) *J. Nutr.* **19**, Suppl. p. 15.
68. BUNYAN, J., DIPLOCK, A. T., EDWIN, E. E. and GREEN, J. (1962) *Brit. J. Nutr.* **16**, 519.
69. COATES, M. E. (1949) *Brit. J. Nutr.* **3**, 261.
70. CREGER, C. R., MITCHELL, R. H., ATKINSON, R. L., FERGUSON, T. M., REID, B. L. and COUCH, J. R. (1960) *Poult. Sci.* **39**, 59.

71. DAM, H., KRUSE, I., PRANGE, I. and SONDERGAARD, E. (1948) *Biochim. Biophys. Acta* **2**, 501.
72. DIPLOCK, A. T., BUNYAN, J., EDWIN, E. E. and GREEN, J. (1962) *Brit. J. Nutr.* **16**, 109.
73. GERRIETS, E. (1954) *Arch. Geflugelz. Kleintierk.* **3**, 54.
74. GRAU, C. R., KRATZER, F. H. and NEWLON, W. E. (1949) *Circ. Calif. Agric. Ext. Serv.* No. 159.
75. GRIFFITHS, T. W. (1961) *Brit. J. Nutr.* **15**, 271.
76. JENSEN, L. S. and MCGINNIS, J. (1957) *Poult. Sci.* **36**, 1344.
77. JENSEN, L. S., SCOTT, M. L., HEUSER, G. F., NORRIS, L. C. and NELSON, T. S. (1956) *Poult. Sci.* **35**, 810.
78. PAPPENHEIMER, A. M. (1940) *Proc. Soc. Exp. Biol., N.Y.*, **45**, 457.
79. SCOTT, M. L. (1962) *Nutr. Abstr. Rev.* **32**, 1.
80. SCOTT, M. L., HILL, F. W., NORRIS, L. C., DOBSON, D. C. and NELSON, T. S. (1955) *J. Nutr.* **56**, 387.
81. SCOTT, M. L. and NELSON, T. S. (1956) *Fed. Proc.* **15**, 572.
82. SINGSEN, E. P., BUNNELL, R. H., MATTERSON, L. D., KOZEFF, A. and JUNGHERR, E. L. (1955) *Poult. Sci.* **34**, 262.
83. SINGSEN, E. P., MATTERSON, L. D., KOZEFF, A., BUNNELL, R. H. and JUNGHERR, E. L. (1954) *Poult. Sci.* **33**, 192.

Vitamin K

84. FERNHOLZ, E. and ANSBACHER, S. (1939) *Science* **90**, 215.
85. FIESER, L. F. (1939) *J. Amer. Chem. Soc.* **61**, 2559.
86. FROST, D. V., PERDUE, H. S. and SPRUTH, H. C. (1956) *J. Nutr.* **59**, 181.
87. NELSON, T. S. and NORRIS, L. C. (1959) *Poult. Sci.* **38**, 1094.
88. NELSON, T. S. and NORRIS, L. C. (1960) *J. Nutr.* **72**, 137.
89. PERDUE, H. S., SPRUTH, H. C. and FROST, D. V. (1957) *Poult. Sci.* **36**, 633.
90. QUICK, A. J. and STEFANINI, M. (1948) *J. Biol. Chem.* **175**, 945.
91. SHELTON, D. C., ANDERSON, G. C., CLARK, T. B. and WEAKLEY, C. E. Jr. (1956) *Poult. Sci.* **35**, 1171.
92. STAMLER, F. W., TIDRICK, R. T. and WARNER, E. D. (1943) *J. Nutr.* **26**, 95.
93. THAYER, S. A., MCKEE, R. W., BINKLEY, S. B. and DOISY, E. A. (1940) *Proc. Soc. Exp. Biol., N.Y.*, **44**, 585.

Thiamine

94. ABBOTT, O. J., BIRD, H. R. and CRAVENS, W. W. (1954) *Poult. Sci.* **33**, 1245.
95. ARNOLD, A. and ELVEHJEM, C. A. (1938) *J. Nutr.* **15**, 403.
96. JUKES, T. H. and HEITMAN, H. Jr. (1940) *J. Nutr.* **19**, 21.
97. ROBENALT, R. C. (1960) *Poult. Sci.* **39**, 354.
98. THORNTON, P. A. (1960) *Poult. Sci.* **39**, 440.
99. THORNTON, P. A. and SHUTZE, J. V. (1960) *Poult. Sci.* **39**, 192.
100. THORNTON, P. A., SHUTZE, J. V. and MORENG, R. E. (1957) *Poult. Sci.* **36**, 1164.

Riboflavin

101. BETHKE, R. M. and RECORD, P. R. (1942) *Poult. Sci.* **21**, 147.
102. BIRD, H. R. and MARVEL, J. A. (1943) *Poult. Sci.* **22**, 403.
103. BOLTON, W. (1944) *J. Agric. Sci.* **34**, 198.
104. BOLTON, W. (1947) *J. Agric. Sci.* **37**, 316.
105. BOLTON, W. (1959) *J. Agric. Sci.* **52**, 364.
106. BOUCHER, R. V., PATRICK, H. and KNANDEL, H. C. (1942) *Poult. Sci.* **21**, 466.
107. BROWN, W. O. (1951) *J. Sci. Fd Agric.* **2**, 43.
108. COATES, M. E., KON, S. K. and SHEPHEARD, E. E. (1950) *Brit. J. Nutr.* **4**, 203.
109. COLES, R. and CUMBER, F. (1955) *J. Agric. Sci.* **46**, 191.
110. DAVIS, H. J., NORRIS, L. C. and HEUSER, G. F. (1938) *Poult. Sci.* **17**, 87.
111. HEGSTED, D. M. and PERRY, R. L. (1948) *J. Nutr.* **35**, 411.

112. HEUSER, G. F., WILGUS, H. S. and NORRIS, L. C. (1938) *Poult. Sci.* **17**, 105.
113. HILL, F. W., NORRIS, L. C. and SCOTT, M. L. (1954) *Poult. Sci.* **33**, 1059.
114. JUKES, T. H. (1937) *J. Nutr.* **14**, 223.
115. JUKES, T. H. (1938) *Poult. Sci.* **17**, 227
116. PATRICK, H., DARROW, M. I. and MORGAN, C. L. (1944) *Poult. Sci.* **23**, 146.
117. PETERSEN, C. F., LAMPMAN, C. E. and STAMBERG, O. E. (1947) *Poult. Sci.* **26**, 180.
118. PETERSEN, C. F., LAMPMAN, C. E. and STAMBERG, O. E. (1947) *Poult. Sci.* **26**, 187.
119. SCHUMACHER, A. E. and HEUSER, G. F. (1939) *Poult. Sci.* **18**, 369.
120. SCOTT, M. L., HOLM, E. R. and REYNOLDS, R. E. (1959) *Poult. Sci.* **38**, 1344.
121. SLINGER, S. J., PEPPER, W. F. and EVANS, E. V. (1955) *Poult. Sci.* **34**, 1222.

Nicotinic Acid

122. BATTIG, M. J., HILL, E. G., CANFIELD, T. H. and SLOAN, H. J. (1953) *Poult. Sci.* **32**. 550.
123. BRIGGS, G. M., Jr., LUCKEY, T. D., TEPLY, L. J., ELVEHJEM, C. A. and HART, E. B. (1943) *J. Biol. Chem.* **148**, 517.
124. BRIGGS, G. M., Jr., MILLS, R. C., ELVEHJEM, C. A. and HART, E. B. (1942) *Proc. Soc. Exp. Biol., N.Y.*, **51**, 59.
125. CHILDS, G. R., CARRICK, C. W. and HAUGE, S. M. (1952) *Poult. Sci.* **31**, 551.
126. COATES, M. E., KON, S. K. and SHEPHEARD, E. E. (1950) *Brit. J. Nutr.* **4**, 203.
127. FISHER, H., SCOTT, H. M. and JOHNSON, B. C. (1954) *Poult. Sci.* **33**, 1054.
128. HEGSTED, D. M. (1946) *J. Nutr.* **32**, 467.
129. HEUSER, G. F. and SCOTT, M. L. (1953) *Poult. Sci.* **32**, 137.
130. JUKES, T. H., STOKSTAD, E. L. R. and BELT, M. (1947) *J. Nutr.* **33**, 1.
131. MCDANIEL, E. G. and HUNDLEY, J. M. (1958) *Fed. Proc.* **17**, 484.
132. PATTERSON, E. B., HUNT, J. R., VOHRA, P., BLAYLOCK, L. G. and MCGINNIS, J. (1956) *Poult. Sci.* **35**, 499.
133. SCOTT, M. L., HOLM, E. R. and REYNOLDS, R. E. (1959) *Poult. Sci.* **38**, 1344.
134. SUNDE, M. L. (1955) *Poult. Sci.* **34**, 304.
135. SUNDE, M. L. and BIRD, H. R. (1957) *Poult. Sci.* **36**, 34.

Pantothenic Acid

136. BALLOUN, S. L. and PHILLIPS, R. E. (1957) *Poult. Sci.* **36**, 929.
137. BAUERNFEIND, J. C., NORRIS, L. C. and HEUSER, G. F. (1942) *Poult. Sci.* **21**, 142.
138. COATES, M. E., KON, S. K. and SHEPHEARD, E. E. (1950) *Brit. J. Nutr.* **4**, 203.
139. GILLIS, M. B., HEUSER, G. F. and NORRIS, L. C. (1947) *Poult. Sci.* **26**, 540.
140. GILLIS, M. B., HEUSER, G. F. and NORRIS, L. C. (1948) *J. Nutr.* **35**, 351.
141. HEGSTED, D. M. and PERRY, R. L. (1948) *J. Nutr.* **35**, 411.
142. JUKES, T. H. (1939) *J. Biol. Chem.* **129**, 225.
143. JUKES, T. H. and MCELROY, L. W. (1943) *Poult. Sci.* **22**, 438.
144. KRATZER, F. H., DAVIS, P. N., MARSHALL, B. J. and WILLIAMS, D. E. (1955) *Poult. Sci.* **34**, 68.
145. KRATZER, F. H. and WILLIAMS, D. (1948) *Poult. Sci.* **27**, 518.

Vitamin B_6

146. BRIGGS, G. M., Jr., MILLS, R. C., HEGSTED, D. M., ELVEHJEM, C. A. and HART, E. B. (1942) *Poult. Sci.* **21**, 379.
147. CRAVENS, W. W., SEBESTA, E. E., HALPIN, J. G. & HART, E. B. (1946). *Poult. Sci.* **25**, 80.
148. DILLON, J. F. (1962) *World's Poult. Congr. XII. Sydney*, Sect. Papers, p. 157.
149. FULLER, H. L. and FIELD, R. C. (1958) *Poult. Sci.* **37**, 1206.
150. FULLER, H. L., FIELD, R. C., RONCALLI-AMICI, R., DUNAHOO, W. S. and EDWARDS, H. M., Jr. (1961) *Poult. Sci.* **40**, 249.
151. FULLER, H. L. and KIFER, P. E. (1959) *Poult. Sci.* **38**, 255.
152. HEGSTED, D. M. and RAO, M. N. (1945) *J. Nutr.* **30**, 367.
153. HOGAN, A. G., RICHARDSON, L. R., PATRICK, H., O'DELL, B. L. and KEMPSTER, H. L. (1941) *Poult. Sci.* **20**, 180.

154. Kratzer, F. H., Bird, F. H., Asmundson, V. S. and Lepkovsky, S. (1947) *Poult. Sci.* **26**, 453.

Biotin

155. Couch, J. R., Cravens, W. W., Elvehjem, C. A. and Halpin, J. G. (1948) *Anat. Rec.* **100**, 29.
156. Cravens, W. W., Sebesta, E. E., Halpin, J. G. and Hart, E. B. (1942) *Proc. Soc. Exp. Biol., N.Y.*, **50**, 101.
157. Hegsted, D. M., Mills, R. C., Briggs, G. M., Elvehjem, C. A. and Hart, E. B. (1942) *J. Nutr.* **23**, 175.

Folic Acid

158. Couch, J. R. and German, H. L. (1950) *Poult. Sci.* **29**, 539.
159. Cravens, W. W. and Halpin, J. G. (1949) *J. Nutr.* **37**, 127.
160. Hutchings, B. L., Oleson, J. J. and Stokstad, E. L. R. (1946) *J. Biol. Chem.* **163**, 447.
161. Jukes, T. H., Stokstad, E. L. R. and Belt, M. (1947) *J. Nutr.* **33**, 1.
162. Kratzer, F. H., Davis, P. N. and Abbott, U. K. (1956) *Poult. Sci.* **35**, 711.
163. Lillie, R. J. and Briggs, G. M. (1947) *Poult. Sci.* **26**, 295.
164. Lillie, R. J., Combs, G. F. and Briggs, G. M. (1950) *Poult. Sci.* **29**, 115.
165. Luckey, T. D., Moore, P. R., Elvehjem, C. A. and Hart, E. B. (1946) *Science*, **103**, 682.
166. March, B. and Biely, J. (1956) *Poult. Sci.* **35**, 550.
167. Robertson, E. I., Daniel, L. J., Farmer, F. A., Norris, L. C. and Heuser, G. F. (1946) *Proc. Soc. Exp. Biol., N.Y.*, **62**, 97.
168. Russell, W. C., Taylor, M. W. and Derby, J. V. Jr. (1947) *J. Nutr.* **34**, 621.
169. Schweigert, B. S., German, H. L., Pearson, P. B. and Sherwood, R. M. (1948) *J. Nutr.* **35**, 89.
170. Scott, M. L., Heuser, G. F. and Norris, L. C. (1948) *Poult. Sci.* **27**, 770.
171. Sunde, M. L., Cravens, W. W., Bruins, H. W., Elvehjem, C. A. and Halpin, J. G. (1950) *Poult. Sci.* **29**, 220.
172. Taylor, L. W. (1947) *Poult. Sci.* **26**, 372.

Vitamin B_{12}

173. Briggs, G. M. (1952) *Trans. Amer. Ass. Cereal Chem.* **10**, 31.
174. Chin, D., Anderson, J. B., Miller, R. F., Norris, L. C. and Heuser, G. F. (1958) *Poult. Sci.* **37**, 335.
175. Davis, R. L. and Briggs, G. M. (1951) *Poult. Sci.* **30**, 628.
176. Johnson, E. L. (1954) *Poult. Sci.* **33**, 100.
177. Johnson, E. L. (1955) *Poult. Sci.* **34**, 1013.
178. Mariakulandai, A. and McGinnis, J. (1953) *Poult. Sci.* **32**, 3.
179. Miller, R. F., Norris, L. C. and Heuser, G. F. (1956) *Poult. Sci.* **35**, 342.
180. Milligan, J. L. and Combs, G. F. (1950) *Poult. Sci.* **29**, 772.
181. Ott, W. H. (1951) *Poult. Sci.* **30**, 86.
182. Petersen, C. F., Wiese, A. C., Dahlstron, R. V. and Lampman, C. E. (1952) *Poult. Sci.* **31**, 129.
183. Petersen, C. F., Wiese, A. C., Milne, G. E. and Lampman, C. E. (1953) *Poult. Sci.* **32**, 535.
184. Sherwood, D. H. and Briggs, G. M. (1951) *Poult. Sci.* **30**, 902.
185. Spivey, M. R., Briggs, G. M. and Ortiz, L. O. (1954) *Proc. Soc. Exp. Biol., N.Y.*, **85**, 451.
186. Stokstad, E. L. R., Jukes, T. H., Pierce, J., Page, A. C., Jr. and Franklin, A. L. (1949), *J. Biol. Chem.* **180**, 647.
187. Wiese, A. C., Petersen, C. F., Dahlstrom, R. V. and Lampman, C. E. (1952) *Poult. Sci.* **31**, 851.
188. Yacowitz, H., Miller, R. F., Norris, L. C. and Heuser, G. F. (1952) *Poult. Sci.* **31**, 89.

Choline

189. BALLOUN, S. L. (1956) *Poult. Sci.* **35**, 737.
190. DEEB, S. S. and THORNTON, P. A. (1959) *Poult. Sci.* **38**, 1198.
191. EVANS, R. J. (1943) *Poult. Sci.* **22**, 266.
192. GARTLEY, K. M., SLINGER, S. J., MORPHET, A. M. and CAMERON, C. R. (1951) *Poult. Sci.* **30**, 913.
193. GOGUS, K. and GRIMINGER, P. (1957) *Poult. Sci.* **36**, 1121.
194. JUKES, T. H. (1940) *J. Nutr.* **20**, 445.
195. JUKES, T. H. (1947) *Ann. Rev. Biochem.* **16**, 193.
196. JUKES, T. H. and STOKSTAD, E. L. R. (1951) *J. Nutr.* **43**, 459.
197. MCKITTRICK, D. S. (1947) *Arch. Biochem.* **15**, 133.
198. MARCH, B. and BIELY, J. (1956). *Poult. Sci.* **35**, 545.
199. RECORD, P. R. and BETHKE, R. M. (1942) *Poult. Sci.* **21**, 271.
200. SCOTT, M. L., HOLM, E. R. and REYNOLDS, R. E. (1959) *Poult. Sci.* **38**, 1344.
201. SHAEFER, A. E., SALMON, W. D. and STRENGTH, D. R. (1949) *Proc. Soc. Exp. Biol., N.Y.*, **71**, 202.
202. SHERWOOD, D. H. and SLOAN, H. J. (1954) *Poult. Sci.* **33**, 1015.
203. SINGSEN, E. P., MATTERSON, L. D., KOZEFF, A. and DECKER, L. (1950) *Poult. Sci.* **29**, 468.
204. SINGSEN, E. P., MATTERSON, L. D., KOZEFF, A. and DECKER, L. (1950) *Poult. Sci.* **29**, 780.
205. WEST, J. W., CARRICK, C. W., HAUGE, S. M. and MERTZ, E. T. (1951) *Poult. Sci.* **30**, 880.

INORGANIC ELEMENT REQUIREMENTS

The requirements of poultry for inorganic elements, with the exception of those for phosphorus and calcium, have been studied in only a few critical experiments. We have considered these requirements from two aspects. We have endeavoured to assess the requirements for those elements that are known to be essential for life and present them in Tables 8—10. Secondly, because some elements are known to be toxic in excess, we have estimated their minimal toxic levels and show them in Table 11.

It has been shown that the percentage of calcium and of potassium needed by the chick in the diet increases with the energy content.[7, 175] It is possible that there may be similar relationships for the other essential inorganic elements, although there is no information on this point. With laying birds the situation is somewhat different, since most of the calcium required is to satisfy the demand for egg-shell production, and hence the percentage needed in the diet increases with the rate of lay and decreases with increase in food intake. With some of the other elements, the amounts in the egg represent a significant proportion of the intake and the situation may be similar to that with calcium although, as yet, there is no experimental evidence to support this contention.

NOTES ON INDIVIDUAL INORGANIC ELEMENTS

Calcium. As the requirement of the laying hen is related to the rate of egg-shell formation, we have derived the following equation from Common's[5] data which cover a range of egg production from 25 to 87 per cent, and we recommend it for calculating requirement:

Daily requirement (g) = 0·049 × percentage production + 0·1.
(See also references 12, 15, 32.)

Phosphorus. One of the difficulties in suggesting reliable estimates of phosphorus requirements is the problem of the availability of phytate phosphorus. Most workers have assumed that phytate phosphorus is only slightly hydrolysed in the intestine, but there is evidence that poultry can utilize a considerable proportion of it.[41] Almost all the published work has demonstrated the necessity for adding inorganic phosphorus to diets of plant origin. Several other factors are known to influence requirements, namely the calcium, vitamin D_3, fat and energy levels of the diet and the source in it of inorganic phosphorus. All these factors have varied markedly in the experiments on which our assessments are based.

Bromine. There is some evidence that bromine is an essential nutrient but because of the paucity of published data[85, 86] we make no recommendations.

Chlorine. This element is considered under *Sodium salts.*

Cobalt. This element appears to be necessary only as a constituent of vitamin B_{12}.

Fluorine. It is not yet known whether fluorine is essential for poultry and it is not a recognized constituent of mineral mixtures. Possible toxic effects must be considered when rock phosphate is used in them instead of steamed bone flour.

Manganese. Requirements of the fowl for this element are influenced by breed differences (and even strains within a breed), and by the calcium, phosphorus, choline and nicotinic acid levels of the diet. In addition, the basal diets in the various experiments surveyed have been so different in composition that it has been found impossible to make valid comparisons. Consequently, the estimates put forward must be considered only very tentative.

Molybdenum. The essential nature of traces of the element for poultry is reasonably established[166, 170] but information is insufficient for a statement of requirement. Larger quantities are toxic (Table 11). Inorganic sulphate is effective in counteracting the toxicity.[163] Molybdenum as present in soya beans is probably largely unavailable to the chick.[166]

Selenium. At low levels selenium can replace vitamin E in some of its functions[185] but as it is toxic at higher levels it is safer to use the vitamin to prevent a deficiency that can be overcome by either.

Sodium. This element is considered under *Sodium Salts.*

Sodium salts. The maximal safe level for common salt in the food of chicks is about 2 per cent.[187, 201, 208] Adult birds are able to withstand 20 per cent

TABLE 8

Calcium Requirements of Young Fowls, Turkeys and Pheasants
(g/100g diet)

Kind	Energy content of diet (Mcal ME/kg)	Requirement	Values on which requirement based		
			Range	No.	References
Chick	2·5	0·9	0·4–3·5	17	1, 3, 4, 7, 9, 14, 16, 17, 18, 22, 23, 26, 27, 31, 34, 35, 37
	3·2	1·0		1	7
	3·9	1·5		1	7
Turkey poult	2·5	1·5	0·8–2·5	15	2, 8, 9, 10, 11, 13, 19, 20, 21, 24, 28, 29, 30, 33, 36
Pheasant					
0–5 weeks	2·85	1·1		1	25
5–14 weeks	2·95	0·5		1	25

TABLE 9

*Phosphorus Requirements of Fowls, Turkeys, Geese and Pheasants expressed as Inorganic Phosphorus**
(g/100g diet)

Kind	Requirement	Values on which requirement based		
		Range	No.	References
Fowl				
0–4 weeks	0·43	0·2–0·6	10	42, 48, 51, 52, 53, 54, 57, 63, 65, 68
4–10 weeks	0·27		1	63
10–18 weeks	0·19		1	64
Layer	0·3 g per bird per day		1	56
Turkey				
0–8 weeks	0·6	0·5–0·8	8	9, 39, 50, 58, 60, 75, 76, 84
8–16 weeks	0·42	0·33–0·50	2	46, 77
16–24 weeks	0·20		1	46
Breeder	?			
Gosling	0·20		1	38
Pheasant				
0–5 weeks	0·25	0·2–0·3	2	67, 79
5–14 weeks	0·10		1	67

* These quantities, when supplemented by the phosphorus contributed by the plant materials in the diet, will satisfy the requirements, although of the plant phosphorus only up to 40 per cent may be available.

in the diet if adequate drinking water is available.[188] When given in drinking water, 0·9 per cent is highly toxic to young chicks.[188, 206]

Poults are less tolerant than chicks, and 1 per cent of common salt or sodium bicarbonate is the most allowable in the diet. Sodium citrate, iodide, sulphate and carbonate proved highly toxic at 0·75 per cent.[205]

Zinc. It is well established that soya bean protein contains some compound that renders zinc unavailable and increases the apparent requirement.[224] Consequently, all experiments in which the diet contained this protein or its derivatives have been disregarded. Excess calcium in the diet may also depress the absorption of zinc, thus apparently increasing the requirement.[214]

TABLE 10

Other Inorganic Element Requirements of Chicks

Element	Requirement	Values on which requirement based		
		Range	No.	References
Chlorine (g/100g diet)	0·06		1	87
Copper (mg/kg diet)	3·2		2	93, 94
Fluorine	?			
Iodine (mg/kg diet)	0·35	0·3–0·4	3	100, 101, 102
Iron (mg/kg diet)	40		2	106, 107
Magnesium (mg/kg diet)	250	200–400	4	109, 110, 112, 113
Manganese (mg/kg diet)	35	35–60	3	131, 138, 154
Potassium (g/100g diet) at energy level (Mcal ME/kg) of diet of:				
2·80	0·20			
3·35	0·24		1	175
3·72	0·27			
4·25	0·34			
Sodium (g/100g diet)	0·11		1	186
Zinc (mg/kg diet)	15	10–36	5	212, 213, 214, 223, 224

TABLE 11

Harmful Levels of Certain Inorganic Elements for Fowls

(mg/kg diet)

Element	Chick	Layer	Values on which estimate based	
			No.	References
Copper	325		1	95
Fluorine	<700		2	98,99
Iodine	180		1	104
Molybdenum	<300		4	158, 159, 162, 164
Selenium	<10	8	1/1	180/181 (layer)
Zinc	>1000		2	212, 216

REFERENCES

Calcium

1. BETHKE, R. M., KENNARD, D. C. and KICK, C. H. (1929–30), *Poult. Sci.* **9**, 45.
2. BLAYLOCK, L. G., NEAGLE, L. H. and LE FEVRE, C. F. (1961) *Poult. Sci.* **40**, 1381.
3. BRANION, H. D. (1938) *Sci. Agric.* **18**, 217.
4. CARVER, J. S., EVANS, R. J. and MCGINNIS, J. (1946) *Poult. Sci.* **25**, 294.
5. COMMON, R. H. (1943) *J. Agric. Sci.* **33**, 213.
6. COUCH, J. R., FRAPS, G. S. and SHERWOOD, R. M. (1937) *Poult. Sci.* **16**, 106.
7. EDWARDS, H. M., Jr., DUNAHOO, W. S., CARMON, J. L. and FULLER, H. L. (1960) *Poult. Sci.* **39**, 1389.
8. EVANS, R. J. and BRANT, A. W. (1945) *Poult. Sci.* **24**, 404.
9. FORMICA, S. D., SMIDT, M. J., BACHARACH, M. M., DAVIN, W. F. and FRITZ, J. C. (1961) *Poult. Sci.* **40**, 1402.
10. FRITZ, J. C., HALPIN, J. L. and HOOPER, J. H. (1947) *Poult. Sci.* **26**, 78.
11. FRITZ, J. C., HOOPER, J. H. and MOORE, H. P. (1945) *Poult. Sci.* **24**, 324.
12. GRIMMINGER, P. (1961) *Poultry Comment* (E. I. Dupont de Nemours & Co., Industrial and Biochemicals Dept., Wilmington, Delaware) **18**, No. 3, p. 1.
13. HAMMOND, J. C., MCCLURE, H. E. and KELLOGG, W. L. (1944) *Poult. Sci.* **23**, 239.
14. HELLER, V. G. and PENQUITE, R. (1942) *Poult. Sci.* **21**, 247.
15. HURWITZ, S. and GRIMMINGER, P. (1960) *J. Agric. Sci.* **54**, 373.
16. LILLIE, R. J., TWINING, P. F., O'BARR, J. S. and DENTON, C. A. (1961) *Poult. Sci.* **40**, 1423.
17. MCGINNIS, J., NORRIS, L. C. and HEUSER, G. F. (1944) *Poult. Sci.* **23**, 157.
18. MITCHELL, H. H. and MCCLURE, F. J. (1937) *Bull. Nat. Res. Coun., Wash.*, No. 99.
19. MOTZOK, I. and SLINGER, S. J. (1948) *Poult. Sci.* **27**, 486.
20. MUSSEHL, F. E. and ACKERSON, C. W. (1935) *Poult. Sci.* **14**, 147.
21. NELSON, F. E., JENSEN, L. S. and MCGINNIS, J. (1961) *Poult. Sci.* **40**, 407.
22. NOWOTARSKI, J. S. and BIRD, H. R. (1943) *Poult. Sci.* **22, 72.**
23. ONTARIO AGRICULTURAL COLLEGE (1958) *Rep. Ont. Agric. Coll.* p. 28.
24. PENSACK, J. M. and WHITE-STEVENS, R. H. (1961) *Poult. Sci.* **40**, 1443.
25. SCOTT, M. L., HOLM, E. R. and REYNOLDS, R. E. (1958) *Poult. Sci.* **37**, 1419.
26. SIMCO, T. F. and STEPHENSON, E. L. (1960) *Poult. Sci.* **39**, 1295.
27. SIMCO, T. F. and STEPHENSON, E. L. (1961) *Poult. Sci.* **40**, 1188.
28. SINGSEN, E. P., MATTERSON, L. D. and SCOTT, H. M. (1947) *J. Nutr.* **33**, 13.
29. SLINGER, S. J., PEPPER, W. F., MOTZOK, I. and SIBBALD, I. R. (1961) *Poult. Sci.* **40**, 1458.
30. STEPHENSON, E. L., RINEHART, K. E. and BRAGG, D. B. (1962) *Poult. Sci.* **41**, 1686.
31. TULLY, W. C., HAUGE, S. M., CARRICK, C. W. and ROBERTS, R. E. (1930–1) *Poult. Sci.* **10**, 310.
32. TYLER, C. and WILLCOX, J. S. (1942) *J. Agric. Sci.* **32**, 62.
33. WAIBEL, P. E., JOHNSON, E. L. and PILKEY, A. M. (1961) *Poult. Sci.* **40**, 256.
34. WALDROUP, P. W., AMMERMAN, C. B. and HARMS, R. H. (1962) *Poult. Sci.* **41**, 1433.
35. WHITE-STEVENS, R. H., PENSACK, J. M. and STOKSTAD, E. L. R. (1960) *Poult. Sci.* **39**, 1305.
36. WILCOX, R. A., CARLSON, C. W., KOHLMEYER, W. and GASTLER, G. F. (1953) *Poult. Sci.* **32**, 1030.
37. WILGUS, H. S., Jr. (1930–1) *Poult. Sci.* **10**, 107.

Phosphorus

38. AITKEN, J. R., LINDBLAD, G. S. and HUNSAKER, W. G. (1958) *Poult. Sci.* **37**, 1180.
39. ALMQUIST, H. J. (1954) *Poult. Sci.* **33**, 936.
40. AMMERMAN, C. B., DOUGLAS, C. R., DAVIS, G. K. and HARMS, R. H. (1961) *Poult. Sci.* **40**, 548.
41. ASHTON, W. M., EVANS, C. and WILLIAMS, P. C. (1960) *J. Sci. Fd. Agric.* **11**, 722.
42. BARUAH J. N. DAVIES. R. E., REID, B. L. and COUCH, J. R. (1960) *Poult. Sci.* **39**, 843.

43. Crowley, T. A., Kemmerer, A. R. and Kurnick, A. A. (1960) *Poult. Sci.* **39**, 1240.
44. Crowley, T. A., Pasvogel, M. W., Kemmerer, A. R., Vavich, M. G. and Kurnick, A. A. (1961) *J. Nutr.* **73**, 33.
45. Crowley, T. A., Pasvogel, M. W., Kemmerer, A. R., Vavich, M. G. and Kurnick, A. A. (1961) *Poult. Sci.* **40**, 74.
46. Day, E. J. and Dilworth, B. C. (1962) *Poult. Sci.* **41**, 1324.
47. Evans, R. J. and Brant, A. W. (1945) *Poult. Sci.* **24**, 404.
48. Fisher, H., Singsen, E. P. and Matterson, L. D. (1953) *Poult. Sci.* **32**, 749.
49. Formica, S. D., Smidt, M. J., Bacharach, M. M., Davin, W. F. and Fritz, J. C. (1962) *Poult. Sci.* **41**, 771.
50. Fritz, J. C., Halpin, J. L. and Hooper, J. H. (1947). *Poult. Sci.* **26**, 78.
51. Gardiner, E. E. (1962). *Poult. Sci.* **41**, 1156.
52. Gillis, M. B. (1948). *J. Nutr.* **36**, 351.
53. Gillis, M. B., Norris, L. C. and Heuser, G. F. (1949). *Poult. Sci.* **32**, 283.
54. Grau, C. R. and Zweigart, P. A. (1953). *Poult. Sci.* **32**, 500.
55. Hammond, J. C., McClure, H. E. and Kellogg, W. L. (1944) *Poult. Sci.* **23**, 239.
56. Hurwitz, S. and Griminger, P. (1962) *J. Sci. Fd. Agric.* **13**, 185.
57. Lindblad, G. S., Slinger, S. J. and Motzok, I. (1954) *Poult. Sci.* **33**, 482.
58. Matterson, L. D., Scott, H. M. and Singsen, E. P. (1946) *J. Nutr.* **31**, 599.
59. Miller, M. W. and Joukovsky, V. V. (1953) *Poult. Sci.* **32**, 78.
60. Motzok, I. and Slinger, S. J. (1948) *Poult. Sci.* **27**, 486.
61. Nelson, F. E., Jensen, L. S. and McGinnis, J. (1961) *Poult. Sci.* **40**, 407.
62. Nelson, T. S. and Peeler, H. T. (1961) *Poult. Sci.* **40**, 1321.
63. O'Rourke, W. F., Phillips, P. H. and Cravens, W. W. (1952) *Poult. Sci.* **31**, 962.
64. O'Rourke, W. F., Phillips, P. H. and Cravens, W. W. (1955) *Poult. Sci.* **34**, 47.
65. Peeler, H. T., Nelson, T. S. and Storer, N. (1960) *Poult. Sci.* **39**, 1282.
66. Pepper, W. F., Slinger, S. J. and Motzok, I. (1952) *Poult. Sci.* **31**, 1054.
67. Scott, M. L., Holm, E. R. and Reynolds, R. E. (1958) *Poult. Sci.* **37**, 1419.
68. Sieburth, J. F., McGinnis, J., Wahl, T. and McLaren, B. A. (1952) *Poult. Sci.* **31**, 813.
69. Simco, T. F. and Stephenson, E. L. (1960) *Poult. Sci.* **39**, 1295.
70. Singsen, E. P. (1948) *Bull. Storrs Agric. Exp. Sta.* No. 260.
71. Singsen, E. P., Matterson, L. D. and Kozeff, A. (1950) *Poult. Sci.* **29**, 635.
72. Singsen, E. P., Matterson, L. D. and Scott, H. M. (1947) *J. Nutr.* **33**, 13.
73. Singsen, E. P. and Mitchell, H. H. (1945) *Poult. Sci.* **24**, 479.
74. Singsen, E. P., Spandorf, A. H., Matterson, L. D., Serafin, J. A. and Tlustohowicz, J. J. (1962) *Poult. Sci.* **41**, 1401.
75. Slinger, S. J., Pepper, W. F. and Motzok, I. (1954) *J. Nutr.* **52**, 395.
76. Stadelman, W. J., Boucher, R. V. and Callenbach, E. W. (1950) *Poult. Sci.* **29**, 146.
77. Sullivan, T. W. (1960) *Poult. Sci.* **39**, 1321.
78. Sullivan, T. W. and Kingan, J. R. (1962) *Poult. Sci.* **41**, 1687.
79. Sunde, M. L. and Bird, H. R. (1956) *Poult. Sci.* **35**, 424.
80. Waibel, P. E., Johnson, E. L. and Pilkey, A. M. (1961) *Poult. Sci.* **40**, 256.
81. Waldroup, P. W., Ammerman, C. B. and Harms, R. H. (1962) *Poult. Sci.* **41**, 1433.
82. Walter, E. D. and Aitken, J. R. (1962) *Poult. Sci.* **41**, 386.
83. Wilcox, R. A., Carlson, C. W. and Kohlmeyer, W. (1961) *Poult. Sci.* **40**, 1533.
84. Wilcox, R. A., Carlson, C. W., Kohlmeyer, W. and Gastler, G. F. (1953) *Poult. Sci.* **32**, 1030.

Bromine

85. Bosshardt, D. K., Huff, J. W. and Barnes, R. H. (1956) *Proc. Soc. Exp. Biol., N.Y.* **92**, 219.
86. Norris, L. C., Leach, R. M., Jr. and Zeigler, T. R. (1958) *Proc. 13th Distillers Feed Conference*, p. 63.

Chlorine

87. Burns, C. H., Cravens, W. W. and Phillips, P. H. (1953) *J. Nutr.* **50**, 317.

Cobalt

88. Burns, M. J. and Salmon, W. D. (1956) *J. Agric. Fd. Chem.* **4**, 257.
89. Davis, R. L., Briggs, G. M. and Sloan, H. J. (1953) *Proc. Soc. Exp. Biol., N.Y.* **82**, 175.
90. Malaishkaite, B. (1956). Quoted in *Chem. Abstr.* (1959) **53**, 1561.
91. Popescu, F., Nedelniuc, V., Dimovici, S. and Tacu, A. (1958) *Lucr. sti. Inst. Cercet. zooteh.* **16**, 545.

Copper

92. Anderson, R. L. and Tove, S. B. (1958) *Nature, Lond.* **182**, 315.
93. Hill, C. H. and Matrone, G. (1958) *Poult. Sci.* **37**, 1211.
94. Hill, C. H. and Matrone, G. (1961) *J. Nutr.* **73**, 425.
95. Mayo, R. H., Hauge, S. M., Parker, H. E., Andrews, F. N. and Carrick, C. W. (1956) *Poult. Sci.* **35**, 1156.
96. Mehring, A. L. Jr., Brumbaugh, J. H., Sutherland, A. J. and Titus, H. W. (1960) *Poult. Sci.* **39**, 713.
97. Norris, L. C., Leach, R. M., Jr. and Zeigler, T. R. (1958) *Proc. 13th Distillers Feed Conference*, p. 63.

Fluorine

98. Kick, C. H., Bethke, R. M. and Record, P. R. (1933) *Poult. Sci.* **12**, 382.
99. Motzok, I. and Branion, H. D. (1958) *Poult. Sci.* **37**, 1469.

Iodine

100. Creek, R. D., Parker, H. E., Hauge, S. M., Andrews, F. N. and Carrick, C. W. (1954) *Poult. Sci.* **33**, 1052.
101. Creek, R. D., Parker, H. E., Hauge, S. M., Andrews, F. N. and Carrick, C. W. (1957) *Poult. Sci.* **36**, 1360.
102. Godfrey, P. R., Carrick, C. W. and Quackenbush, F. W. (1953) *Poult. Sci.* **32**, 394.
103. Scott, M. L., Van Tienhoven, A., Holm, E. R. and Reynolds, R. E. (1960) *J. Nutr.* **71**, 282.
104. Wilgus, H. S., Gassner, F. X., Patton, A. R. and Harshfield, G. S. (1948) *Poult. Sci.* **27**, 686.
105. Wilgus, H. S., Harshfield, G. S., Patton, A. R., Harris, L. P. and Gassner, F. X. (1941) *Poult. Sci.* **20**, 477.

Iron

106. Hill, C. H. and Matrone, G. (1958) *Poult. Sci.* **37**, 1211.
107. Hill, C. H. and Matrone, G. (1961) *J. Nutr.* **73**, 425.
108. Schaible, P. J., Davidson, J. A. and Bandemer, S. L. (1944) *Poult. Sci.* **23**, 441.

Magnesium

109. Almquist, H. J. (1942) *Proc. Soc. Exp. Biol., N.Y.* **49**, 544.
110. Gardiner, E. E., Rogler, J. C. and Parker, H. E. (1960) *Poult. Sci.* **39**, 1111.
111. Keene, O. D. and Combs, G. F. (1962) *Poult. Sci.* **41**, 1654.
112. McWard, G. W. and Scott, H. M. (1961) *Poult. Sci.* **40**, 1174.
113. Nugara, D. and Edwards, H. M. Jr. (1961) *Poult. Sci.* **40**, 1438.
114. Scott, H. M., Moeller, M. W. and Hinners, S. W. (1956) *Poult. Sci.* **35**, 1169.
115. Sullivan, T. W. (1962) *Poult. Sci.* **41**, 1686.
116. Van Reen, R. and Pearson, P. B. (1953) *J. Nutr.* **51**, 191.

Manganese

117. BOLTON, W. (1957) *Poult. Sci.* **36,** 732.
118. BRIGGS, G. M. and LILLIE, R. J. (1946) *Proc. Soc. Exp. Biol., N.Y.*, **61**, 430.
119. CASKEY, C. D. and NORRIS, L. C. (1938) *Poult. Sci.* **17**, 433.
120. CASKEY, C. D. and NORRIS, L. C. (1939) *Proc. Soc. Exp. Biol., N.Y.*, **40**, 590.
121. CASKEY, C. D. and NORRIS, L. C. (1940) *Proc. Soc. Exp. Biol., N.Y.*, **44**, 332.
122. CASKEY, C. D., NORRIS, L. C. and HEUSER, G. F. (1944) *Poult. Sci.* **23**, 516.
123. CHRISTIANSEN, J. B., HALPIN, J. G. and HART, E. B. (1939) *Science*, **90**, 356.
124. CHUBB, L. G. (1954) *World's Poult. Congr. X. Edinburgh*, p. 164.
125. COMBS, G. F., NORRIS, L. C. and HEUSER, G. F. (1942) *J. Nutr.* **23**, 131.
126. COOPER, D. M., CHUBB, L. G. and ROWELL, J. G. (1963) *Brit. Poult. Sci.* **4**, 83.
127. COUCH, J. R., JAMES, L. E. and SHERWOOD, R. M. (1947) *Poult. Sci.* **26**, 30.
128. EVANS, R. J., RHIAN, M. & DRAPER, C. I. (1943) *Poult. Sci.* **22**, 88.
129. EVANS, R. J. ROBERTSON, E. I., RHIAN, M. and WILHELM, L. A. (1942) *Poult. Sci.* **21**, 422.
130. GALLUP, W. D. and NORRIS, L. C. (1938) *Science*, **87**, 18.
131. GALLUP, W. D. and NORRIS, L. C. (1939) *Poult. Sci.* **18**, 76.
132. GALLUP, W. D. and NORRIS, L. C. (1939) *Poult. Sci.* **18**, 83.
133. GALLUP, W. D. and NORRIS, L. C. (1939) *Poult. Sci.* **18**, 99.
134. GOLDING, W. V., SCHAIBLE, P. J. and DAVIDSON, J. A. (1940) *Poult. Sci.* **19**, 263.
135. GUTOWSKA, M. S. and PARKHURST, R. T. (1942) *Poult. Sci.* **21**, 277.
136. GUTOWSKA, M. S., PARROTT, E. M. and SLESINSKI, F. A. (1941) *Poult. Sci.* **20**, 379.
137. HOGAN, A. G., RICHARDSON, L. R., PATRICK, H. and KEMPSTER, H. L. (1941) *J. Nutr.* **21**, 327.
138. INSKO, W. M., Jr., LYONS, M. and MARTIN, J. H. (1938) *Poult. Sci.* **17**, 264.
139. JUKES, T. H. (1939) *Poult. Sci.* **18**, 405.
140. LYONS, M. (1939) *Bull. Ark. Agric. Exp. Sta.* No. 374.
141. LYONS, M., INSKO, W. M., Jr. and MARTIN, J. H. (1938) *Poult. Sci.* **17**, 12.
142. MUSSEHL, F. E. and ACKERSON, C. W. (1939) *Poult. Sci.* **18**, 408.
143. NIELSEN, H. M. and MADSEN, D. E. (1940) *Poult. Sci.* **19**, 198.
144. NORRIS, L. C. and CASKEY, C. D. (1939) *J. Nutr.* **17**, Suppl. p. 16.
145. PENQUITE, R., HELLER, V. G. and THOMPSON, R. B. (1940) *Bull. Okla. Agric. Exp. Sta.* No. 243.
146. PEPPER, W. F., SLINGER, S. J. and MOTZOK, I. (1952) *Poult. Sci.* **31**, 1054.
147. PEPPER, W. F., SLINGER, S. J. and MOTZOK, I. (1953) *Poult. Sci.* **32**, 656.
148. RINGROSE, A. T., MARTIN, J. H. and INSKO, W. M., Jr. (1939) *Poult. Sci.* **18**, 409.
149. SCHAIBLE, P. J. and BANDEMER, S. L. (1942) *Poult. Sci.* **21**, 8.
150. SCHAIBLE, P. J., BANDEMER, S. L. and DAVIDSON, J. A. (1938) *Tech. Bull. Mich. Agric. Exp. Sta.* No. 159.
151. SCOTT, M. L., HOLM, E. R. and REYNOLDS, R. E. (1959) *Poult. Sci.* **38**, 1344.
152. VAN DER HOORN, R., BRANION, H. D. and GRAHAM, W. R., Jr. (1938) *Poult. Sci.* **17**, 185.
153. WIESE, A. C., BENHAM, G. H., ELVEHJEM, C. A. and HART, E. B. (1941) *Poult. Sci.* **20**, 255.
154. WILGUS, H. S., Jr., NORRIS, L. C. and HEUSER, G. F. (1937) *J. Nutr.* **14**, 155.
155. WILGUS, H. S., Jr., NORRIS, L. C. and HEUSER, G. F. (1939) *World's Poult. Congr. VII. Cleveland*, p. 171.

Molybdenum

156. COUCH, J. R., KURNICK, A. A., REID, B. L. and SVACHA, R. L. (1957) *Fed. Proc.* **16**, 383.
157. DAVIES, R. E., BURROUGHS, R. N., REID, B. L. and COUCH, J. R. (1958) *Poult. Sci.* **37,** 1196.
158. DAVIES, R. E., KURNICK, A. A., REID, B. L. and COUCH, J. R. (1957) *Poult. Sci.* **36**, 1111.
159. HIGGINS, E. S., RICHERT, D. A. and WESTERFELD, W. W. (1956) *Fed. Proc.* **15**, 274.
160. KRATZER, F. H. (1952) *Proc. Soc. Exp. Biol., N.Y.*, **80**, 483.
161. KURNICK, A. A., REID, B. L., BURROUGHS, R. N., STELZNER, H. D. and COUCH, J. R. (1957). *Proc. Soc. Exp. Biol., N.Y.*, **95**, 353.

162. MILLER, E. C. and DENTON, C. A. (1959) *Poult. Sci.* **38**, 910.
163. MILLER, E. C., PEARSON, P. B. and DENTON, C. A. (1958) *Fed. Proc.* **17**, 485.
164. MOTZOK, I., ARTHUR, D. and BRANION, H. D. (1957) *Poult. Sci.* **36**, 1144.
165. REID, B. L., KURNICK, A. A., BURROUGHS, R. N., SVACHA, R. L. and COUCH, J. R. (1957) *Proc. Soc. Exp. Biol., N.Y.*, **94**, 737.
166. REID, B. L., KURNICK, A. A., SVACHA, R. L. and COUCH, J. R. (1956) *Poult. Sci.* **35**, 1167.
167. REID, B. L., KURNICK, A. A., SVACHA, R. L. and COUCH, J. R. (1956) *Proc. Soc. Exp. Biol., N.Y.*, **93**, 245.
168. RICHERT, D. A. and WESTERFELD, W. W. (1953) *J. Biol. Chem.* **203**, 915.
169. TEEKELL, R. A. and WATTS, A. B. (1959) *Poult. Sci.* **38**, 791.
170. TEEKELL, R. A. and WATTS, A. B. (1959) *Poult. Sci.* **38**, 1127.

Potassium

171. BEN DOR, B.-A. (1941) *Proc. Soc. Exp. Biol., N.Y.*, **46**, 341.
172. BURNS, C. H., CRAVENS, W. W. and PHILLIPS, P. H. (1953) *J. Nutr.* **50**, 317.
173. GILLIS, M. B. (1948) *J. Nutr.* **36**, 351.
174. GILLIS, M. B. (1950) *J. Nutr.* **42**. 45.
175. LEACH, R. M., Jr., DAM, R., ZEIGLER, T. R. and NORRIS, L. C. (1959) *J. Nutr.* **68**, 89.
176. LEACH, R. M., Jr. and NORRIS, L. C. (1958) *Poult. Sci.* **37**, 1220.
177. MCWARD, G. W. and Scott, H. M. (1960) *Poult. Sci.* **39**, 1274.
178. SUPPLEE, W. C., BLAMBERG, D. L. and COMBS, G. F. (1958) *Fed. Proc.* **17**, 495.
179. SUPPLEE, W. C. and COMBS, G. F. (1959) *Poult. Sci.* **38**, 833.

Selenium

180. CARLSON, C. W. and LEITIS, E. (1957) *Poult. Sci.* **36**, 1108.
181. GRUENWALD, P. (1958) *Amer. J. Path.* **34**, 77.
182. KOHLMEYER, W. and MOXON, A. L. (1948) *Poult. Sci.* **27**, 670.
183. NESHEIM, M. C. and SCOTT, M. L. (1958) *J. Nutr.* **65**, 601.
184. POLEY, W. E. and MOXON, A. L. (1938) *Poult. Sci.* **17**, 72.
185. SCOTT, M. L. (1962) *Nutr. Abstr. Rev.* **32**, 1

Sodium

186. MCWARD, G. W. and SCOTT, H. M. (1961) *Poult. Sci.* **40**, 1026.

Sodium salts

187. BARLOW, J. S., SLINGER, S. J. and ZIMMER, R. P. (1948) *Poult. Sci.* **27**, 542.
188. BLAXLAND, J. D. (1946) *Vet. J.* **102**, 157.
189. COLLIER, P. (1892) *Bull. N.Y. St. Agric. Exp. Sta.* No. 39, p. 22.
190. DAM, H. (1944) *J. Nutr.* **27**, 193.
191. EDWARDS, J. T. (1918) *J. Comp. Path.* **31**, 40.
192. FIELD, H. I. and EVANS, E. T. R. (1946) *Vet. Rec.* **58**, 253.
193. HALPIN, J. G., HOLMES, C. E. and HART, E. B. (1934) *Poult. Sci.* **13**, 308.
194. HALPIN, J. G., HOLMES, C. E. and HART, E. B. (1936) *Poult. Sci.* **15**, 99.
195. HARPER, J. A. and ARSCOTT, G. H. (1962) *Poult. Sci.* **41**, 497.
196. KARE, M. R. and BIELY, J. (1948) *Poult. Sci.* **27**, 751.
197. KRAKOWER, C. A. and HEINO, H. E. (1947) *Arch. Path.* **44**, 143.
198. KRISTA, L. M., CARLSON, C. W. and OLSON, O. E. (1961) *Poult. Sci.* **40**, 938.
199. MATTERSON, L. D., SCOTT, H. M. and JUNGHERR, E. (1946) *Poult. Sci.* **25**, 539.
200. MITCHELL, H. H., CARD, L. E. and CARMAN, G. G. (1926) *Bull. Ill. Agric. Exp. Sta.* No. 279, p. 135.
201. PAVER, H., ROBERTSON, A. and WILSON, J. E. (1953) *J. Comp. Path.* **63**, 31.
202. PETERSON, E. H. (1945) *N. Amer. Vet.* **26**, 37.
203. QUIGLEY, G. D. and WAITE, R. H. (1932) *Bull. Md. Agric. Exp. Sta.* No. 340, p. 343.

204. Roberts, R. E. (1957) *Poult. Sci.* **36**, 672.
205. Scrivner, L. H. (1946) *J. Amer. Vet. Med. Ass.* **108**, 27.
206. Selye, H. (1943) *J. Amer. Vet. Med. Ass.* **103**, 140.
207. Sibbald, I. R., Pepper, W. F. and Slinger, S. J. (1962) *Poult. Sci.* **41**, 541.
208. Sjollema, B. (1935) *Tierernahrung* **7**, 184.
209. Slinger, S. J., Pepper, W. F. and Motzok, I. (1950) *Poult. Sci.* **29**, 780.
210. Suffran, F. (1909) *Rev. gen. Med. vet.* **13**, 698. Abstract in *J. Comp. Path.* (1910) **23**, 71.

Zinc

211. Day, E. J. and Hill, J. E. (1959) *Poult. Sci.* **38**, 646.
212. Mehring, A. L., Jr., Brumbaugh, J. H. and Titus, H. W. (1956) *Poult. Sci.* **35**, 956.
213. Moeller, M. W. and Scott, H. M. (1958) *Poult. Sci.* **37**, 1227.
214. Morrison, A. B. and Sarett, H. P. (1958) *J. Nutr.* **65**, 267.
215. Pensack, J. M., Henson, J. N. and Bogdonoff, P. D. (1958) *Poult. Sci.* **37**, 1232.
216. Roberson, R. H. and Schaible, P. J. (1960) *Poult. Sci.* **39**, 893.
217. Scott, M. L., Holm, E. R. and Reynolds, R. E. (1959) *Poult. Sci.* **38**, 1344.
218. Slinger, S. J., Summers, J. D. and Pepper, W. F. (1956) *Poult. Sci.* **35**, 1172.
219. Supplee, W. C., Combs, G. F. and Blamberg, D. L. (1958) *Poult. Sci.* **37**, 63.
220. Supplee, W. C., Creek, R. D., Combs, G. F. and Blamberg, D. L. (1961) *Poult. Sci.* **40**, 171.
221. Turk, D. E., Sunde, M. L. and Hoekstra, W. G. (1959) *Poult. Sci.* **38**, 1256.
222. Whiteside, C. H., Hulett, B. J., Ferguson, T. M., Davies, R. E. and Couch, J. R. (1960) *Poult. Sci.* **39**, 1305.
223. Zeigler, T. R., Leach, R. M., Jr. and Norris, L. C. (1958) *Fed. Proc.* **17**, 498.
224. Zeigler, T. R., Leach, R. M., Jr., Norris, L. C. and Scott, M. L. (1961) *Poult. Sci.* **40**, 1584.

TECHNICAL COMMITTEE ON THE NUTRIENT REQUIREMENTS OF FARM LIVESTOCK

Chairman	D. P. CUTHBERTSON, C.B.E., M.D., D.SC., LL.D., F.R.S.E.	Rowett Research Institute
	PROFESSOR R. G. BASKETT, O.B.E., M.SC., D.SC., F.R.I.C.	National Institute for Research in Dairying
	A. W. BOYNE, B.SC.	Rowett Research Institute
	R. BRAUDE, PH.D., D.SC.	National Institute for Research in Dairying
	A. EDEN, M.A., PH.D., F.R.I.C.	National Agricultural Advisory Service
	S. K. KON, PH.D., D.SC., F.R.I.C.	National Institute for Research in Dairying
	F. H. MOULT, M.SC.	Agricultural Research Council
	PROFESSOR ALEX. ROBERTSON, C.B.E., M.A., B.SC., PH.D., M.R.C.V.S., F.R.I.C., F.R.S.E.	Royal (Dick) School of Veterinary Studies
	J. A. B. SMITH, C.B.E., PH.D., D.SC., F.R.I.C , F.R.S.E.	Hannah Dairy Research Institute
	D. B. WALLACE, M.A.	School of Agriculture, Cambridge
Technical Secretaries	PROFESSOR I. A. M. LUCAS, B.SC., M.S. (resigned September, 1961)	Rowett Research Institute
	J. L. CORBETT, B.SC., M.AGR.SC. (resigned December, 1962)	Rowett Research Institute
	C. F. MILLS, M.SC., PH.D., F.R.I.C.	Rowett Research Institute

WORKING PARTY ON NUTRIENT REQUIREMENTS OF POULTRY

Chairman	S. K. Kon, PH.D., D.SC., F.R.I.C.	National Institute for Research in Dairying
	W. Bolton, D.SC., F.R.I.C., F.C.S., F.R.S.E.	Poultry Research Centre, Edinburgh
	A. W. Boyne, B.SC.	Rowett Research Institute
	K. J. Carpenter, M.A., B.SC., PH.D. (resigned June, 1961)	School of Agriculture, Cambridge
	Marie E. Coates, PH.D., F.P.S.	National Institute for Research in Dairying
	J. Davidson, B.SC., PH.D., F.R.I.C.	Rowett Research Institute
	A. Eden, M.A., PH.D., F.R.I.C., DIP. AGRIC.	National Agricultural Advisory Service
	F. H. Moult, M.SC.	Agricultural Research Council
	Professor Alex. Robertson, C.B.E., M.A., B.SC., PH.D., M.R.C.V.S., F.R.I.C., F.R.S.E.	Royal (Dick) School of Veterinary Studies
	H. Temperton, D.SC., N.D.A., N.D.P., J.P.	National Institute of Poultry Husbandry
	S. R. Wragg, M.A.	University of Bristol
Technical Secretary	L. G. Chubb, M.SC., PH.D., A.R.I.C.	Houghton Poultry Research Station

CHAPTER 33

The Metabolic Diseases of Ruminants

A. T. PHILLIPSON

Department of Veterinary Clinical Studies, University of Cambridge

THE expression 'The metabolic diseases of ruminants' is a loose one that is used generally to classify conditions in which metabolic disturbances occur which are characterized by well-recognized signs and are common all over the world. The expression could, by its implication, include nutritional deficiencies, congenital diseases, photosensitizations and many other conditions the origin of which is not the result of specific infective or toxic agents.

It is the custom to include milk fever, lactation tetany, ketosis of cattle, pregnancy toxaemia of ewes, bloat and acute indigestion under the general title of metabolic diseases and this custom is followed in this chapter. There is no logic in this custom, but as these conditions are common sources of loss to farmers, and because there is an element of uncertainty in the aetiology of all of them, they can conveniently be considered under one heading even though the impression it gives of some underlying uniformity is misleading. In practice, these conditions fall into three groups; namely, those which are a form of acute indigestion, those which are characterized by a disturbance of the major elements of the blood and those which are characterized by inappetance and the accumulation of ketone bodies in the blood.

ACUTE INDIGESTION DUE TO OVEREATING OF CEREALS

Acute and often fatal indigestion in ruminants due to eating large quantities of cereal grains or foodstuffs rich in soluble sugars, particularly when unaccustomed to such foodstuffs, has been recognized for a long time. In the past, the syndrome has been described as one that may occur when cattle gain entry to a barn and gorge themselves on grains or meals mixtures as a welcome change from their usual diet; more recently, the condition has also been described in Australia where wheat has been used during droughts to feed sheep and, in the United Kingdom, a condition known as 'barley poisoning' has occurred when high-energy diets with bruised barley as the principal ingredient have been used for the production of baby beef.

(*a*) The Experimental Production of the Overeating Syndrome

The effects of feeding sheep on wheat or maize in large quantities or of introducing glucose or starch into the rumen in large quantities *via* a rumen cannula have been studied by Hungate, Dougherty, Bryant and Cello (1952), Turner and Hodgetts (1948–52) and by Broberg (1960). These procedures lead to a massive production of lactic acid in the rumen as a result of fermentation of the soluble carbohydrates and a rapid increase in the hydrogen ion concentration in the rumen, so that the contents are unduly acidic. According to Dunlop and Hammond (1965) the concentration of lactic acid in the rumen may reach 240 m-mole/l. and the pH of the rumen fluid fall to between 4·0 and 4·5. Both the L and D forms of lactic acid were found. There seems to be little doubt that the excessive production of acidic products of fermentation is the primary cause of the overeating syndrome but the other effects that such a fermentation may have on the animal need some discussion.

The production of appreciable quantities of lactic acid in the rumen when sheep are introduced to diets rich in starch under normal circumstances, is rather capricious, and in a group of animals high concentrations of lactic acid may appear in some animals and comparatively insignificant concentrations in others (Phillipson, 1952). Animals which at first react strongly, do not necessarily produce later any appreciable concentrations of lactic acid, while lactate introduced into the rumen may disappear rapidly from the contents, although at other times it may persist for longer periods (Phillipson and McAnally, 1942; Phillipson, 1952). The occurrence of lactic acid is less predictable than is the appearance of the three principal acids formed during fermentation, namely, acetic, propionic and butyric acids. Further fermentation of lactic acid produces propionic and acetic acids and the accumulation of lactic acid in the rumen is the balance between production, further fermentation and absorption. It is probable that the relative speeds of production and fermentation are the two most important factors concerned, and the differences exhibited between animals is likely to be due to the activities and numbers of bacteria present that produce lactic acid or ferment it. Baldwin, Wood and Emery (1962–3) examined the two principal routes by which propionic acid is formed, the one *via* succinate and the other *via* acrylate. Both kinds of fermentation have been found in the rumen, and by differential labelling of substrate these workers conclude that in the fermentation of glucose the succinate pathway is predominant, while in the fermentation of lactate the acrylic pathway is dominant. Organisms exhibiting both enzyme systems can ferment glucose to propionic acid—some acetic acid being formed —but the emphasis on the two main stages of the fermentation can clearly alter according to the dominant species of bacteria present.

(*b*) Absorption of Lactic Acid

An increase in the concentration of lactate in the peripheral blood has been shown to occur in animals in which large quantities of lactate are formed in the rumen (Hueter, Shaw and Doetsch, 1956; Broberg, 1960), but this does not necessarily mean that it is directly absorbed from the rumen. Tsuda (1956) observed a fall in concentration of lactate in solutions introduced in a pouch of the rumen, but he made no quantitative estimate of the apparent loss, while Shoji, Miyazaki and Umezu (1964) observed in goats with a normally filled rumen that a higher concentration of lactate was present in the venous blood leaving the rumen than was present in arterial blood when the blood was drawn. As they point out, the metabolism of propionate by the rumen epithelium produces lactate in considerable quantity *in vitro* (Pennington and Sutherland, 1956) and an increased quantity of lactate in venous blood from the rumen may reflect propionate metabolism rather than lactate absorption.

The blood flow from the rumen of anaesthetized sheep is sensitive to the pH of buffered solutions placed within it when acetate, propionate or butyrate are included; thus blood flow is more rapid when acidic solutions are introduced than when they are included at a neutral pH, and the speed with which the fatty acids as such are absorbed is reflected in a slight decrease of the pH of the venous blood to the region of 7. When lactic acid buffers are introduced at as low a pH as 3, no change in blood flow can be detected and no evidence of the absorption of lactic acid can be detected from any change in blood pH (Dobson and Phillipson, 1956). This test suggests that if lactic acid is absorbed from the rumen, it is only absorbed very slowly. Recently Williams and Mackenzie (1965) have carried out direct experiments on the loss of lactic acid from the empty washed rumen of anaesthetized sheep under a variety of conditions, and they compared losses of lactic acid to the loss of the short-chain fatty acids. They found that the rate at which it disappeared was about one-sixth to one-tenth of that of the short-chain acids and was unaffected by the pH of the solutions introduced. Thus lactic acid was lost from 0·06 M solution at a rate of 0·1 to 0·3/m-mole/kg/hr. in comparison to losses of 1·0 to 1·9 from equimolar solutions of short-chain fatty acid/kg/hr.

The increase in acidity of the rumen contents when lactic acid accumulates as a result of fermentation may cause the pH to fall to between 4 and 5 in extreme cases. Under these circumstances, the rates at which acetic, propionic and butyric acids are absorbed, are known to be much increased even though the rate of lactic acid itself is unaffected. Reduction of the pH of the rumen contents to these levels has been shown to be accompanied by loss of movement of the rumen (Broberg, 1960) and the introduction of solutions of acetate, propionate or butyrate buffered to pH 4, or thereabouts, into the empty rumen of conscious sheep inhibits the normal contractions of the reticulum and rumen (Ash, 1959). It is not clear whether this is a central or peripheral effect for the intravenous injection of the sodium salts of these acids also inhibits gastric movements (Le Bars *et al.*, 1954). In any event excessive acidification of the rumen by decreasing or inhibiting rumen contractions is likely to decrease the speed with which food passes to the abomasum and intestine.

(*c*) The Systemic Effects of Overeating

Starches and sugars are fermented in the rumen more rapidly than cellulose, consequently the greater the proportion of these carbohydrates to cellulose in the diet the greater is the speed of fermentation and the amount of organic matter digested in the rumen. The opinion that is sometimes asserted by enthusiasts for the barley feeding system for young cattle, that it encourages intestinal digestion rather than fermentation in the rumen, has no foundation to support it; indeed the reverse is true as it encourages fermentation in the rumen and decreases the quantity of organic matter that leaves the rumen for the lower parts of the alimentary tract. In addition the concentration of short-chain fatty acids that appear in the rumen contents is greater than is found in animals fed in a traditional way.

The effect of feeding cereal grains without additional roughage other than that present in the husk, can be deduced from a study of the urine. Determinations of the pH of the urine of cattle fed on bruised barley supplemented with protein rich meals when no roughage was allowed, revealed that it was usually between 5 and 6; out of 47 determinations, only 10 samples were higher than pH 6·0. In contrast the pH of urine of cows fed in a traditional manner was between 7·0 and 8·0. This is an indication that the acid-base equilibrium of cattle fed in this way is under greater pressure than in traditionally fed animals. Experimentally, Dunlop and Hammond (1965) have shown that the pH of the blood of steers after engorgement of barley may decrease to a value of 7·1 which is accompanied by a threefold decrease in the alkali reserve of the plasma.

Examination of the rumen contents of young cattle that died suddenly when eating cereal-rich rations have shown abnormally large concentrations of short-chain fatty acids in the rumen; thus in five such animals the concention of the steam volatile fatty acids were from 252 to 359 m-equiv./1. of rumen liquor—values which are two to three times the usual concentration—but lactic acid amounted to only 0·6 to 14·4 m-equiv./1. As these figures were obtained from animals that died, they may well be greater than they were in life; in addition, some lactic acid may have been destroyed. They serve to indicate the trend that may be expected when starch-rich rations are given without roughage.

The osmotic relations between the plasma and the rumen content are disturbed by concentrations in the rumen that deviate far from 0·18 M. Greater concentrations of ions in the rumen will have the effect of causing water to pass from the blood to the rumen and so inducing a haemo-concentration in the animal. Broberg (1960) records increases in the osmotic pressure of the rumen contents of animals suffering from acute overeating, while Turner and Hodgetts (1953) found haemo-concentration to occur in wheat-fed sheep and state that the carcasses of sheep dying from overeating wheat showed signs of dehydration. In addition anuria has been noted as part of the overeating syndrome.

In practice, sheep and cattle that are introduced slowly to high-starch and low-roughage diets do not become ill. They may be temporarily fickle in appetite but as they become accustomed to the regimen they tolerate it without signs of illness. If they are abruptly changed to this kind of ration then the acute overeating syndrome may appear. It is characteristic that they do not feed, they lack movement of the rumen, some degree of tympanitis may be seen, anuria and some diarrhoea may occur. An increased pulse rate and signs of cardiac irregularity have also been noted. Death is common in acute cases.

Pharmacological examination of the rumen liquor of sheep in which the overeating syndrome has been induced experimentally has suggested that some toxic factor is present which depresses blood pressure and Dain, Neal and Dougherty (1955) were able to demonstrate the presence of histamine and tyramine in the rumen contents of experimentally overfed sheep. They noted a correlation between the quantities of histamine present and the condition of the animal. Concentrations of about 70 μg of histamine/ml were found in sheep that died when the pH of the rumen was 4·5 or less. They found no histamine in the rumen before the induction of the overeating syndrome. Scarisbrick (1954) describes a case of acute overeating of mangolds in a sheep and found concentrations of lactic acid as high as 90mM. He observed the same signs of sickness and calls attention to the fact that bacterial decarboxylation of histidine to histamine can occur and that the optimal pH for this reaction is between 4 and 5. According to a recent report from the Rowett Institute, however, 80-90 per cent of histamine in the rumen of normal sheep is destroyed. Laminitis or tenderness of the feet may accompany the overeating syndrome and this has been attributed to the intestinal production of histamine by Akerblom (1934).

It has been postulated by Broberg (1960) that lack of thiamine may also be concerned and that this is indicated by the lactacidaemia that appears when large quantities of lactic acid are formed (Dunlop and Hammond, 1965). Broberg found that the capacity of sheep to metabolize lactate when this is given slowly by an intravenous drip was improved if additional thiamine was given. In this respect Phillipson and Reid (1957) found that when sheep were given a low roughage and high flaked maize ration the concentration of thiamine in the rumen was halved for the next week when sheep were introduced to this diet before it returned to the usual value.

An animal that survives the initial acute acidosis remains in a precarious condition and refuses to eat. This has been attributed to several other possible causes such as profound modification of the rumen flora; to damage to the rumen epithelium; to the systemic effects of toxic or pyrogenic substances produced in the rumen. It is recorded that a sheep that voluntarily overate and showed all the signs of acute acidosis could never be persuaded to overeat again (Scarisbrick, 1954). If the results of overeating were less severe this might be a useful prophylactic procedure. It is possible that mild bouts of overeating do occur in young cattle and pass unnoticed when they are introduced gradually on to high concentrate diets and the animals learn by experience to regulate their appetite to their metabolism.

The logical treatment of this condition is to give large quantities of sodium bicarbonate by mouth in isotonic solution. Relief of dehydration should be instigated at once by intravenous infusions of suitable isotonic solutions. An old custom in Scandinavia is to give bakers' yeast which is compatible with Broberg's view that the metabolism of lactic acid should be encouraged by the administration of thiamine; there are no systematic observations on the use of thiamine as a therapeutic agent in this condition but it may be expected to be more effective than yeast if this is really the crux of the matter. Antihistamines have proved ineffective in treating animals suffering from barley poisoning.

BLOAT

The condition known as 'bloat' occurs in all countries; its official name is 'tympanitis of the rumen' which means distension of the rumen with gas. It is more common in cattle than in sheep. The fact that carbon dioxide and methane are evolved in the rumen during the normal fermentation of foodstuffs means that any condition in which the escape of gas is hindered or prevented may lead to tympanitis; for example, it may complicate conditions in which the movements of the rumen are disturbed, it occurs in animals that choke themselves by trying to swallow large pieces of turnip or other chunky foods and it is characteristic of the animal that falls into a ditch and, being unable to regain its feet, lies on its back. The expression 'bloat', however, is used to distinguish the acute tympanitis which occurs as a result of grazing young summer pastures that contain legumes or which may occur in animals fed in yards on starch-rich foodstuffs with only small quantities of roughage, from tympanitis due to other causes.

The present discussion is confined to the acute form of tympanitis known as 'bloat'.

Bloat is characterized by distension of the rumen by a foaming mass of contents and this is the reason why animals so affected cannot relieve themselves by belching as only a small part of the gas present is free in the upper part of the dorsal sac of the rumen. Most of the recent work on the subject has been directed towards seeking an explanation for the sudden occurrence of excessive foaming in the rumen.

(*a*) The Occurrence of Bloat

It is characteristic of bloat that it occurs suddenly and in an unpredictable manner. For this reason it is impossible to make any clear statement concerning the weather, the pasture, the soil or manurial treatment in relation to the onset of bloat. All that can be said is that pasture in the rapid phase of growth in the early summer may be liable to cause bloat and that bloat is usually associated with pastures containing white or other species of clover. As bloat has been observed on purely grass leys the presence of clover is not an essential factor in rendering a pasture 'bloat prone'. In the United States and elsewhere bloat is usually associated with cattle grazing lucerne in the preflowering phase or when they are maintained in 'feed lots' on diets containing a lot of maize and little roughage.

It is well recognized that some animals are more susceptible to bloat than others. Johns (1954), however, showed that the susceptibility of individuals to bloat was not related to the rate at which they ate or the quantities of food they consumed. The distinction between susceptible and non-susceptible animals is a matter of degree and if the conditions are favourable enough then most animals in a herd will develop some degree of bloat. Johns (1963) also points out that resistance and non-resistance to bloat is temporary and that when identical twin pairs are observed closely, their resistance to bloat is different at different times even though both animals of a twin pair vary in harmony with each other. It is difficult to imagine that the physiological responses of animals vary very much from period to period and it is more probable that variation in the nature of the mixed population of bacteria and protozoa within the rumen occurs. There is evidence that even on one feeding regime some animals harbour organisms which are not present or hard to find in others. Similarly, the production and fermentation of lactate is somewhat unpredictable as we have seen, and this is an indication that the bacterial population is not always a stable entity.

(*b*) Plant Factors

It was shown many years ago by Mead and Goss (1936) that cattle reared and maintained on diets devoid of roughage were liable to bloat. This observation led to the use of hays and straws as a means of preventing bloat in cattle grazing on young succulent lucerne (alfalfa). Cole, Mead and Regan (1943) obtained good protection against alfalfa bloat when dairy cattle were given 14 lb of Sudan Grass hay overnight or in the morning before they were turned on to alfalfa. This treatment resulted in the cattle eating about 25 per cent less alfalfa than they would otherwise have done but it also prevented bloating. Smaller quantities of barley straw were less effective and lucerne (alfalfa) hay less effective still. They attributed the effect of Sudan Grass hay to its rough texture inducing a reflex belching but it is not possible to distinguish the effect of this from the effects of the lowered quantities of alfalfa eaten.

Evans and Evans (1949) discussed the possible role of the cyanogenic glucosides that are present in white clover as they were impressed by the inhibition of rumen movements that occurred during cyanide poisoning. Bloat may also occur when red clovers, which contain no cyanogenic glucosides, are eaten, so that the presence of these substances in white clovers may have no relation to bloat. In addition, the rumen movements of animals in the early stages of bloat are more obvious and seem to be stronger than is usual and inhibition only occurs later when the pressure within the rumen becomes severe. Ruminants seem to be able to detoxify comparatively large quantities of cyanide and the likelihood of cyanide poisoning, as such, occurring when white clovers are eaten seems to be rather remote. Even so, cyanogenic glucosides are not desirable features of green fodders.

Considerable attention has been given to the saponins present in clovers and other legumes. There are conflicting reports whether the bloating properties of lucerne and other legumes are related to their saponin content (Cole, Dougherty, Huffman, Hungate, Kleiber and Maclay, 1956). A study of the foaming properties of saponins isolated from lucerne, red clover and other legumes was carried out by Mangan (1958, 1959), who found that the strength of the foam, induced by passing gas in fine bubbles at a standard pressure and rate of flow for a standard time, was markedly influenced by the pH of the solution; above pH 6 the strength of the foam was negligible and it was greatest at pH 4·5 to 5·0. The saponins from lucerne and red clover had very similar properties and the strength of these foams at their pH optima was much greater than foams prepared from white clover or subterranean clover saponins. It was found that the presence of calcium was essential in order to produce a strong foam and its removal reduced foam strength practically to zero. It has been noted by Reid (quoted by Mangan, 1959) that bloating or foaming of the rumen contents of cattle fed on red clover occurs only when the pH of the rumen contents drops to pH 6·3 or lower and that the lower limit, as fermentation proceeds, is about pH 5·7. This range is one in which foam strength of saponin solutions is comparatively small. Mangan (1959) also examined the foaming properties of the cytoplasmic protein of red clover, of the salivary mucoproteins and of the rumen liquor of cattle fed on red clover. Solutions of the cytoplasmic proteins foamed satisfactorily over a pH range of 4·5 to 7·0 but the strength of the foams formed had a marked pH optima at pH 5·4; foam strength was negligible above pH 6·0 and below pH 5·0. The strength of the foam increased with the concentration of nitrogen and sodium and the optimal pH also shifted to the region of pH 6·0. At low cation concentrations there is little to choose between sodium and calcium in promoting foam strength, although at 0·1 M concentrations calcium is much more effective than sodium.

Foam stability of the salivary mucoproteins was high and in the range of pH 6 to 8 but foam strength was optimal at pH 8·0. Rumen liquor had a pH optimum that was the same as that for solutions of the cytoplasmic proteins but at pH 6·0 or over its strength was negligible. Complete removal of particulate matter by high speed centrifugation increased foam stability considerably but the volume of foam produced was unchanged. These experiments led Mangan (1959) to conclude that the cytoplasmic proteins of red clover were probably the most important plant component concerned with foam formation in the rumen. This research was combined with the work of Mangan, Johns and Bailey (1959) who studied the concentration of soluble protein and other plant constituents in the rumen liquor, and of the products of fermentation in cows fed on red clover with and without orally administered penicillin. Soluble protein was estimated on cell-free rumen liquor obtained by high-speed centrifugation. In cattle treated with penicillin, the concentration of soluble protein, of amino-nitrogen and of reducing substances all increased markedly during the two hours while the cattle were feeding, so that there can be no doubt that the passage of plant cell constituents into solutions in the rumen fluid occurs on a large scale within a short time; changes in the fermentation products were comparatively small. The reverse situation occurred in the untreated animals, but even so, soluble protein doubled in concentration although the changes occurred at lower concentrations than in the treated animals. The greatest values for untreated animals were 12 mg N/100 ml, and for treated animals 102·6 mg N/100 ml. Mangan's work showed that solutions of 10 mg protein N/100 ml would produce foams of considerable strength and the foam formation in the untreated animals was much greater than in the treated animals. There was some indication that the foam was stronger when crude rumen liquor was used; however, the same tests applied to the cell-free rumen liquor showed that after the removal of particulate matter the treated animal produced as much foam and a much stronger foam than the untreated animal in accordance with the higher protein concentration. The particulate matter of rumen liquor of the treated animals therefore seemed to have anti-foaming properties which were in greater concentration in the treated than in the untreated animals.

Mangan (1959) mentioned that the chloroplasts of plant juice, which have a layer of lipid surrounding them, materially weaken the strength of foams but if they are defatted they then have a foam stabilising effect. The implications of these observations are not as yet clear. Both Fraser (1961) and Wright (1961) have examined this proposition; Fraser was unable to find any difference in the lipid fractions of the rumen liquor of bloating and non-bloating cows, but points out that this does not preclude the possibility of chloroplasts being taken out of circulation by ingestion by rumen protozoa, as was observed by Oxford (1958). He does, however, point out that the lipid contents of rye grass on which bloat seldom occurs is greater than it is in clovers. Wright examined the hydrolysis of various lipids by rumen bacteria and noted that penicillin and terramycin considerably reduced lipolytic activity in the rumen but as the chloroplastic fat is not a simple glycerol ester, as were those that were tested, the enzymes needed for its hydrolysis may not be the same.

Other investigations have suggested that mucins from saliva or plants decreased the stability of the foam of rumen contents and that if saliva plays a role, then it is in reducing the stability of foam formation in the rumen (Van Horn and Bartley, 1961; Bartley and Yadova, 1962). Thus the addition of saliva or linseed meal mucin increased the rate of gas released from frothing rumen contents and from foams made from alfalfa saponin *in vitro*. The addition of mucins to cattle pastured on green alfalfa reduced foaming in the rumen. Various bacteria capable of destroying salivary mucins were studied by Fina, Hay, Bartley and Mishra (1961) and when the rumen of cattle were inoculated with these organisms the extent of frothing was somewhat increased. This work puts a rather different complexion on the possible role of saliva which will be discussed later.

Gutierrez, Davis and Lindahl (1959) and Gutierrez and Davis (1962), however, put forward further evidence concerning the possible role of saponins for they were able to isolate and study an organism similar to *B. butyrivibris* found by Bryant, Barrentine, Sykes, Robinson, Shawver and Williams (1960) to be present in unusually large numbers of cattle fed on Ladino clover which produced a 'slime' from saponins. They hesitate to claim an outstanding role for this organism but consider it may be one of a complicated balance of factors. Bailey (1959), however, found no evidence that the soluble polysaccharides of rumen liquor which may be of plant, bacterial or protozoa origin were related to the foaming properties of the rumen liquor.

A recent observation of Harris and Sebba (1965) introduces a further factor for consideration, namely, the presence of metallic ions in very small traces. These workers studied the stability of foam produced from lucerne extracts and observed that nickel and zinc were concentrated in the foams. As the foams aged, they became weaker and the nickel content was reduced. The addition of traces of nickel sulphate to weak foams restored their stability but zinc had only a slight effect. Alkalie metals were ineffective and so was manganese. The foams formed were primarily proteinaceous in nature. According to Mitchell (1944), nickel is present in Scottish pastures to the extent of 0·8 to 3·7 μg/g of dry matter which is ten times more than may be expected for cobalt. The concentration of nickel in the foams from lucerne extracts studied were about 0·5 to 5·0 μg/ml of extract and presumably the concentration per gramme of dry matter would be about twenty times greater, which supports the contention that this element is concentrated in the foam.

It can be seen from this summary that very few parts of the plant have not been suspected as an aetiological factor in foam formation at one time or another. All that remains to be said is that the soluble carbohydrates of green fodders, by being easily and rapidly fermented substrates, may be largely responsible for gas formation in the first place and so should also be included, for without gas formation, there can be no foam. Rapid fermentation also leads to a rapid accumulation of short-chain fatty acids and their absorption is associated with the passage of carbon dioxide from the blood into the rumen. A general suppression of bacterial metabolism, therefore, is the prime reason why antibiotics can prevent bloat.

(*c*) Animal Factors

When the rumen of an anaesthetized sheep is inflated after the oesophagus is ligated to prevent the escape of gas, it is found that the quantities of gas needed to increase intra-ruminal pressure to 10 mm Hg is much greater than the quantities needed to raise the pressure from 10 to 20 mm Hg, and so on. This is illustrated by the following measurements:

Volumes of carbon dioxide in litres introduced into the rumen.

	Sheep Kay and Phillipson (1959a)	Five-month-old calf Johns (1963)
0 to 10 mm Hg	5·0	—
2·5 to 10 mm Hg	—	20·8
10 to 20 mm Hg	3·6	4·8
20 to 30 mm Hg	2·6	2·7
30 to 40 mm Hg	1·1	2·2

Lindahl, Davis, Jacobson and Shaw (1957) classify the severity of bloat according to the pressure induced in the rumen and consider that cases are mild when intra-rumenal pressures are less than 15 mm Hg, moderate when the pressure rises to 35 mm Hg, and severe when pressures are greater than 45 mm Hg. The experiments given above show that the quantity of gas needed to turn a moderate case into a severe one is relatively small compared to the quantities needed to induce a mild case.

(*d*) The Influence of Saliva

The introduction of gas into the rumen of calves and sheep when the oesophagus is ligated in the neck to prevent its escape first stimulates salivary secretion but as pressure increases to the region of 15 to 20 mm Hg secretion of saliva is inhibited (Phillipson and Reid, 1958; Kay and Phillipson, 1959b). There is considerable variation from animal to animal in the pressure needed to produce the secondary inhibition. Further analysis showed that stimulation was evoked by distension of the thoracic oesophagus and inhibition by distension of the rumen. The effects were pronounced on the secretions from the parotid, buccal, palatine and inferior molar glands but insignificant upon the sub-mandibular and sub-lingual glands. The stimulative effect induced a five-fold or even greater increase in salivation. Kleiber (1956) has calculated that a litre of saliva containing about 112 m-moles of bicarbonate buffer and secreted at a pH of 8·0 would liberate about 1·9 litres of carbon dioxide when it passed to the rumen if the pH of the contents was 5·7 and the gas phase consisted of 70 per cent carbon dioxide. This volume would substantially assist in turning a mild case of bloat into a moderate case or a moderate case into a severe one. There is no information to show whether salivary stimulation does occur in the early stages of bloat, but should the oesophagus become filled with frothing digesta in unsuccessful attempts to belch, stimulation may occur. At a maximal rate of salivation a cow may secrete 5 litres or more of saliva per hour.

(*e*) The Absorption of Short-chain Fatty Acids

Carbon dioxide passes from the blood to the rumen during the absorption of short-chain fatty acids (Masson and Phillipson, 1951) and the two events are related in a linear manner. As about one mole of carbon dioxide appears in the rumen for every two moles of fatty acid absorbed (Ash and Dobson, 1963), it is possible to make an assessment of the possible accumulation of gas from this source. Using the data of Kleiber (1956) and again assuming that the pH of the rumen contents is 5·7, it can be calculated that about 1·3 to 2·0 litres of carbon dioxide would be liberated per hour if the adult cow is absorbing 3·3 to 5·0 moles of short-chain fatty acid within this period, which is an estimate based on the data summarized by Annison (1965). Absorption of fatty acids is likely to be considerable for animals eating the diets that may lead to bloating and these quantities of gas would make a material difference to the pressure within the rumen. It should be recalled that the New Zealand work showed that cows eating red clover did not bloat unless the pH fell to the region of 6·0 or lower. This reduction in pH would materially increase the speed of absorption of short-chain fatty acids and increase the liberation of carbon dioxide from bicarbonate from entering the rumen.

Saliva may form an appreciable source of carbon dioxide and may aggravate the condition if secreted in sufficient quantity; on the other hand, the salivary mucoproteins tend to decrease foam stability. Absorption of fatty acids certainly will increase the gas content of the rumen but neither saliva nor absorption could in themselves create bloat unless the factors leading to foaming were present for without foam there would be no impediment to the eructation of gas.

(*b*) Eructation

The removal of gas from the rumen by belching is part of the co-ordinated movements of the reticulum and rumen which have been described many times. The movements of the rumen consist essentially of a backward moving wave of contraction that follows immediately the biphasic contraction of the reticulum. This may or may not be followed by a forward moving wave of contraction. Belching usually occurs during the forward moving contraction although it can be observed occasionally during the biphasic contraction of the reticulum.

The forward moving contraction usually occurs after every other backward moving contraction of the rumen when the animal is ruminating and resting so that the incidences of the first to second rumen movement are in the ratio of 2 : 1. This ratio is decreased when the animal eats and becomes approximately 3 : 2 (Phillipson and Reid, 1960).

The passage of gas into the oesophagus has been studied by Dougherty (1960) in detail. He describes three areas in the oesophagus when obstructions to the passage of gas may occur. The first is about two inches anterior to the cardia where the oesophagus actively contracts and causes whatever is in the oesophagus to pass orally or caudally, depending upon its position. The second is at the level of the first rib and may merely reflect the narrow entrance to the chest and the third is immediately below the pharynx. None of these structures present an obstruction to the passage of gas during belching but if the oesophagus is filled with liquid then the upper area of the oesophagus below the pharynx does not allow liquid to pass to the mouth. It is because of this, that it is suggested that if bloating animals attempt to belch when there is little or no free gas in the rumen but a mass of foaming contents, this foam may pass into the oesophagus but cannot be expelled. Presumably relaxation of this part of the oesophagus is only part of the ruminating or swallowing mechanism as far as liquids are concerned.

In the early stages of bloat, rumen movements and belching are not inhibited, indeed the movements of the rumen as far as these can be judged by the movements of the left lumbar triangle are more pronounced than usual. The frequency has been described as being increased but this may be due to the rumen becoming distended by foaming digesta so that the pressure changes of the ventral sac are transmitted to the dorsal sac which does not occur under normal circumstances. It is only in the later stages of bloat when the rumen is severely distended that movements cease.

(*g*) The Cause of Death

Distension of the rumen impedes the blood flow through the wall of the organ; in addition it may be expected to impede the venous return from the inferior vena cava. The tidal volume of the animal is reduced as pressure in the abdomen increases and anoxia develops. Distension of the rumen in anaesthestized animals leads to respiratory failure and the pressure in the rumen at which respiratory failure occurs is somewhat variable. Arterial blood pressure tends to rise as pressure in the rumen increases; with respiratory failure, the arterial pressure falls and an anoxic crisis develops which is characteristic of an adrenaline discharge into the blood causing a dramatic rise in blood pressure which is accompanied by the return of respirations, which persist for only a short while. Blood pressure then falls again until another anoxic crisis sets in. These observations suggest that death in bloating animals is due to respiratory failure leading to anoxia.

(*h*) The Prevention of Bloat

Two principal ways of preventing bloat have been explored; the first is to administer an antibiotic *via* the mouth to the rumen before turning animals on to a bloat-prone pasture. Comparatively small doses of penicillin have been effective in preventing foaming in the rumen; its presence allows a greater accumulation of soluble sugars and plant proteins in the rumen liquor than would otherwise take place. The method is hardly practical as no farmer can contemplate the regular oral dosing of all his cows and because continued dosing would lead to development of a resistant population in the rumen and so defeat its purpose.

The administration of vegetable oils by mouth to mildly bloating cows is effective in liberating gas from foam, and this has led to the introduction of spraying pastures with oil emulsions. This seems to be effective in preventing bloat but it has to be done once every three days and so is a considerable undertaking. It is applicable where strip grazing is practised.

Various foam 'breaking' preparations have been recommended for treatment of bloat. Turpentine has been the traditional medicament and even paraffin oil is sometimes used.

Restriction of the grazing time or of the grazing area will not necessarily be effective unless the restriction is severe. When bloat occurs, certain pastures are found to be safe and some unsafe, but there is no known way of predicting the safe ones except by experience. A change of pasture, however, is often the best course to adopt. Cutting a bloat-forming pasture and feeding the herbage, indoors, but in safe quantities over the dangerous period is another way of meeting the situation; the safety of this procedure is increased if the cut grass is allowed to wilt for 12 hours before it is eaten. The losses due to bloat are small in the United Kingdom but the nuisance it causes by upsetting the plans of a farmer for his dairy herd, may be considerable.

DISORDERS ASSOCIATED WITH DISTURBANCE TO THE CALCIUM AND MAGNESIUM CONTENTS OF THE BLOOD

Ruminants, in particular, are susceptible to the conditions which are characterized by 'hypocalcaemia' or by 'hypomagnesaemia' which mean a depression in the concentration of calcium or of magnesium in the blood. These conditions may arise from several causes and can occur in other animals especially the 'hypocalcaemias'. Here the common form of hypocalcaemia, namely, 'milk fever' in dairy cows, will be considered as it is the syndrome that is best known to farmers. Hypomagnesaemia may be the result simply of an inadequate quantity of magnesium in the diet and under these circumstances can occur in any species of animal of any age and in any state or it may develop within a few days. The latter acute condition is seen principally in lactating dairy cattle when they are grazing early summer pastures but it may also occur in non-lactating animals at that time of year and sometimes in the autumn. The principal difference between the acute and delayed hypomagnesaemic states is that inadequacy in the diet results in a slow progression to the state when the outward signs of hypomagnesaemia appear without any further change in the animals' dietary or environmental circumstances, while the hypomagnesaemia associated with cattle in the early summer occurs when lactating cows are turned out on to rapidly growing pastures in the early summer and the clinical tetany as such may then develop acutely within a few days.

(*a*) Milk Fever in Cattle

This condition occurs immediately after parturition; the animal usually becomes semi-comatose and dies if treatment is not given. The concentration of calcium and usually of inorganic phosphorus in the blood plasma is

(*b*) The Absorption of Calcium

In the ruminant most of the calcium in the first three parts of the stomach is bound to large organic molecules so that it will not diffuse through colloidion membranes. The concentrations of free ionized calcium are so small that it is inconceivable that absorption of calcium from the rumen reticulum and omasum can occur. In anaesthetized sheep, absorption of ionized calcium from the rumen when present in physiological concentrations is negligible (Phillipson and Storry 1965). It is only when the material enters the abomasum that calcium passes to an ionizable state owing to the acidity within the abomasum being close to pH 3. In the duodenum the acidity of the contents increases to the region of pH 5 but it does not reach neutrality until well into the jejunum. Calcium is bound to a nitrogen-rich material and so disintegration of this compound may occur as the material passes caudally, so releasing calcium.

Absorption studies with temporarily isolated loops of jejunum of sheep show that calcium in the ionic state is absorbed roughly in relation to concentration introduced (Scott, 1965; Phillipson and Storry, 1965). Absorption studies on ileal loops, however, have shown both positive and negative results. The method used in these experiments measured only the total loss or gain from the solutions introduced and the results were influenced by the quantity of calcium secreted during the experimental period. Scott (1965) found the secretions from unstimulated loops of jejunum and ileum to contain 0·5 to 2·1 and 1·9 to 2·4 m-mole of calcium per litre. The volume of juice secreted, from loops that may be assumed to be from 30 to 90 cm long, was very variable and never exceeded 30 ml/hour. From a study of the electrical potential difference between the lumen of the intestine and the blood, Scott (1965) calculated from the data given by Storry (1961) for the concentrations of ultra filterable calcium present in the intestinal contents of sheep, that the electrical potential gradient was usually unfavourable for the absorption of ionized calcium from the concentrations normally present. This conclusion is in keeping with the experimental evidence that indicates an 'active' uptake of calcium from the intestine.

The quantity of food flowing to the intestine is an important factor that will influence the calcium available to the animal. Normally in sheep about 8 to 10 litres of contents flow to the intestine per 24 hours (Harris and Phillipson, 1962) while about 2–4 litres enter the caecum (Hogan and Phillipson, 1960; Goodall and Kay, 1963). The continuance of this flow appears to depend upon the regular cycle of contractions in the reticulo-rumen sac, for the quantity of material in the duodenum or in the abomasum influences the contractions of the reticulum (Phillipson and Ash, 1965) and the outflow from the omasum (Ash, 1962). Similarly, the quantity of food eaten is related to the output by the omasum (Ash, 1962). Reduction in food eaten or stasis of the stomach can be expected to cause reduction or cessation of the flow of food to the intestine and consequently of the calcium available for absorption, as absorption from the rumen itself can be discounted as negligible. Moodie and Robertson (1962) have shown that inducing alimentary stasis by repeated injections of hyoscine hydrobromide in cows causes a reduction in rumen movements and depression of the concentrations of calcium and inorganic phosphate in the blood. The effect was much more marked in old cows than in young ones. This work, together with earlier observations (Moodie and Robertson, 1961) which recorded a reduction in food eaten as calving approached and which increased with the age of the cow, gives powerful support to the concept that dietary sources of calcium and the appetite of the animal are concerned with the onset of milk fever. It is also of interest that a comparison of two milk fever-prone cows with two cows of the same age which had no history of milk fever showed that the former retained less calcium during the 45 days before calving than the latter. This indicates either that the intestinal capacity of the milk fever-prone cows to absorb calcium was less than the cows who were not susceptible or else that the endogenous excretion of calcium was greater. In the sheep, the endogenous excretion of calcium into the alimentary tract has been estimated to equal the calcium in the extra-cellular fluids of the body (Storry, 1961) so that an impairment of absorption would demand mobilization of calcium from bone if equilibrium in the blood is to be maintained, for impaired absorption will affect equally calcium from dietary and endogenous sources. As Payne (1964) has found, hypocalcaemia following the infusion of EDTA reduces rumen motility so that once hypocalcaemia sets in its extent may be increased for alimentary reasons.

Less is known of the control exerted by the parathormone and the calcitonin factors on the mobilization of skeletal calcium than is known on the availability of calcium from the food, but there seems little doubt that interplay of hormonal and alimentary effects are concerned with the occurrence of milk fever. Outstanding areas that need further definition are the effects of the hormonal factors on the absorption of calcium.

(*c*) Hypomagnesaemia

The same considerations have to be given the question of hypomagnesaemia as have been given to hypocalcaemia. The difference is that there is no secure information to show that there is an hormonal factor produced by the endocrine system which has a direct control over the magnesium balance of the body. For this reason, dietary influences have received most attention.

Hypomagnesaemia which is the result of a deficiency of magnesium in the diet has been investigated extensively in rats and if it is sufficiently severe, ends in death which is preceded by increased sensitivity leading to tetanic plasma protein and the packed cell volume (Scott and Dobson, 1965). Observations on sheep grazing heavily fertilized pasture showed that there was a correlation between a depressed concentration of magnesium in the plasma and a high concentration of potassium in the saliva, which supports the concept that a part of the hypomagnesaemia observed when cattle or sheep graze early summer highly fertilized pasture may be due to enhanced aldosterone secretion. Even so, aldosterone does not seem able to produce more than a mild degree of hypomagnesaemia (Dobson, Scott and McDonald, 1966).

There is some evidence that excessive thyroid activity may lower the concentration of magnesium in the blood of cattle, thus Swan and Jamieson (1956) observed such a depression when iodinated casein was added to the diet of dairy cattle.

A dietary factor may also be concerned as cattle maintained on winter feeds usually consume from 16 to 33 g of magnesium daily whereas, when they go out to pasture, the intake may fall to the region of 10 to 16 g daily. This in itself does not seem sufficient to induce hypomagnesaemia within a few days but it may predispose to lactation tetany, for there is little doubt from balance studies that the concentration of magnesium in the plasma reflects the quantity of magnesium available to the animal. The availability of magnesium in grass has been called in question in this respect and many investigations examined the possible reasons for a decreased availability of magnesium in grass. None, however, have been based on sufficiently secure evidence to give confidence in their reality. These are fully discussed by Care (1964). The quantity of magnesium secreted into the alimentary tract is sufficient to account for all the extracellular magnesium of the body (Storry, 1961) so it is clear that any impediment to reabsorption as well as to the absorption of dietary magnesium would be sufficient to precipitate a severe hypomagnesaemia.

There seems no doubt that adequate treatment of the pasture with magnesium sulphate so that the magnesium content of the herbage is greater than 0·2 g per 100 g dry matter will prevent hypomagnesaemia. Even so, the view that the rapid development of hypomagnesaemia in cattle grazing early summer pasture should be considered to be an uncomplicated deficiency syndrome is not convincing; endocrine factors cannot be entirely excluded and the possibility of impeded absorption of magnesium is still an open question. It may also be that tetanies associated with hypomagnesaemia are still confused with tetanies due to other causes.

KETOSIS

This expression means in biochemical terms that the concentrations of β-hydroxybutyrate, acetoacetate and acetone in the blood are greater than normal. In practice, it is applied to a condition seen in dairy cattle usually during the early stages of lactation and to a condition found in pregnant ewes during the last month of gestation especially when the ewes are carrying twins or triplets.

Some degree of ketosis may be expected in any animal which is inadequately fed or which is fasting and it indicates that the animal is using its reserves of fat to supply the necessary energy for its maintenance. Ketosis as such is not a well defined entity and may complicate many disease syndromes. The severe ketosis, however, that occurs in dairy cattle or in pregnant ewes is distinct from ketosis from other causes and, consequently, the two conditions are recognized as clinical entities, even though there is still uncertainty whether they may be explained entirely in terms of the animals' dietary needs in relation to the demands made upon it either for lactation or for the nourishment of the foetus. In the absence of any positive evidence to show that a toxic agent is causing a metabolic block or that a pathological condition of an endocrine gland or of any other organ is the primary cause, there is no alternative but to consider the metabolism of the animal in relation to the formation of ketone bodies and the possible ways in which this may be exaggerated due to lack of suitable substrates, to the demands made by the secretion of milk or by the foetus and to the influence of the environment on the animal itself. In doing so, it is as well to recognize that no two animals are alike with the exception of identical twins, and that different animals may be susceptible in different ways to the stresses that are put upon them even though the outward clinical signs follow a similar pattern. In all the conditions considered here, the question may be asked why it is that in a group of animals only a few develop illness even under closely controlled conditions. There is no answer to this question except by making the assumption that genetic differences ought to be demonstratable in the metabolic terms concerned. The fact that genetic metabolic differences between strains of mice are recognized and that adaptation to a given set of circumstances may occur at different metabolic levels may be taken as a model of what is possible. If in an experimental group of sheep an identical response was obtained to a given experimental procedure there would be no need to provide a statistical analysis of the results. This never occurs; for example, if the quantities of calcium, magnesium and phosphorus, excreted in the urine of sheep maintained on the same food which is eaten in the same amounts, is determined, it is found that different animals regularly excrete different amounts of each element daily in the urine. Under the circumstances, a clinical syndrome which is indicative of metabolic disturbance need not be the result of exactly the same cause. The illusive nature of the evidence concerning the aetiology of the two ketotic conditions under discussion is understandable when thought of in these terms.

observations, although a first step in the analysis of appetite, do not explain the many vagaries which deficiencies or illnesses of one sort or another impose on an animal's appetite, and this remains one of the most illusive physiological properties of the animal and deserves more attention than it has yet received.

SUMMARY

Acute indigestion may occur in cattle from the excessive ingestion of cereals when they are unaccustomed to such a diet and when grazing young summer pastures which contain considerable quantities of clovers. The first condition is characterized by excessive fermentation within the rumen which may lead to the accumulation of lactic acid, excessive concentrations of lactate in the blood, dehydration and general signs of acidosis. Animals that are accustomed to a high cereal diet show evidence of a greater production of short chain fatty acids in the rumen than is usual, although they adapt to this situation if the transition is gradual.

Indigestion occurring on young summer pastures is characterized by distension of the rumen due to the gasses produced during fermentation being trapped as a foam so that the animals cannot relieve themselves by belching. The constituents of young grasses and legumes are easily fermented and the condition develops rapidly; respiration is impeded and death occurs if the condition becomes acute. The evidence suggests that the cytoplasmic proteins of the young herbage are probably responsible for foam formation but there is also evidence of an antifoaming agent in the particulate parts of the plants which, under normal circumstances, prevents excessive foaming. Variation in the bacterial population of the rumen has been explored as a possible reason for the somewhat unpredictable nature of outbreaks of 'bloat', as this condition is called, but supporting evidence is circumstantial.

The conditions characterized by low concentrations of calcium or magnesium in the blood of cattle and sheep are discussed in relation to the dietary needs of the animal and to the endocrine factors that influence the availability of those elements and their concentrations in the plasma. The two factors are discussed in relation to the conditions under which the two principal conditions occur, namely, milk fever of dairy cows and the lactation tetany or grass staggers of cattle on early summer grazings.

Ketotic conditions are common in both cattle and sheep. Two principal biochemical situations can explain the genesis of ketosis and it should be borne in mind that ketosis need not be the result of a single cause. Ketosis of dairy cows has not been produced experimentally and so its possible aetiology is still uncertain. The field and experimental conditions that lead to ketosis in pregnant ewes are associated with sudden fasts although it has also been seen in ewes to which ample, if unsuitable, food supplies are available. Acute ketosis in both cattle and sheep occurs when the animals are usually either in the early stages of lactation or during the later stages of a multiple pregnancy.

REFERENCES

AKERBLOM, E. (1934) *Skand. Arch. Physiol.* **68** (Suppl), 168.
ALLCROFT, W. M. (1947) *Vet. J.* **103**, 75–100.
ANNISON, E. F. (1965) In *Physiology of Digestion in the Ruminant*, p. 185, R. W. Dougherty, R. S. Allen, W. Burroughs, N. L. Jacobson and A. D. McGilliard (Ed.), Butterworths, Washington.
ASH, R. W. (1959) *J. Physiol.* **147**, 58–73.
ASH, R. W. (1962) *J. Physiol.* **164**, 25–26P.
ASH, R. W. and DOBSON, A. (1963) *J. Physiol.* **169**, 39–61.
BAILEY, R. W. (1959) *N.Z.Jl. agric. Res.* **2**, 355–64.
BACH, S. J. and HIBBIT, K. G. (1959) *Biochem. J.* **72**, 87–92.
BALDWIN, R. L., WOOD, W. A. and EMERY, R. S. (1962) *J. Bact.* **83**, 907–13.
BALDWIN, R. L., WOOD, W. A. and EMERY, R. S. (1963) *J. Bact.* **85**, 1346–9.
BARTLEY, E. E. and YADOVA, I. S. (1961) *J. Anim. Sci.* **20**, 648–53.
BENZIE, D., BOYNE, A. W., DALGARNO, A. C., DUCKWORTH, J., HILL, R. and WALKER, D. M. (1955) *J. Agric. Sci.* **46**, 425–40.
BENZIE, D., BOYNE, A. W., DALGARNO, A. C., DUCKWORTH, J., HILL, R. and WALKER, D. M. (1956) *J. Agric. Sci.* **48**, 175–86.
BLAXTER, K. L., ROOK, J. A. F. and MACDONALD, A. M. (1954) *J. Comp. Path.* **64**, 157–75.
BROBERG, G. (1960) Thesis. Ab. lovisa nya tryckeri—Lovisa.
BRYANT, M. P., BARRENTINE, B. F., SYKES, J. F., ROBINSON, I. M., SHAWVER, C. V. and WILLIAMS, L. W. (1960) *J. Dairy Sci.* **43**, 1435–44.
CARE, A. D. (1964) *Vet. A.* **5**, 222–7.
CARE, A. D. (1965) *Nature, Lond.* **205**, 1289–91.
COGHLAN, J. P., DENTON, D. A., GODING, J. R. and WRIGHT, R. D. (1960) *Post-grad. med. J.* **36**, 76–102.
COLE, H. H., DOUGHERTY, R. W., HUFFMAN, C. F., HUNGATE, R. E., KLEIBER, M. and MACLAY, W. D. (1956) *Nat. Acad. Sci., Natn. Res. Counc.* Publ. 388. Washington 25, D.C.
COLE, H. H., MEAD, S. W. and REGAN, W. M. (1943) *J. Anim. Sci.* **2**, 285–94.
COPP, D. H., CAMERON, E. C., CHENY, B. A., DAVIDSON, A. G. F. and HENZE, K. G. (1962) *Endocrinology*, **70**, 638–49.
DAIN, J. A., NEAL, A. L. and DOUGHERTY, R. W. (1955) *J. Anim. Sci.* **14**, 930–5.
DOBSON, A. (1963) *Res. vet. Sci.* **4**, 238–46.
DOBSON, A. and MCDONALD, I. (1963) *Res. vet. Sci.* **4**, 247–57.
DOBSON, A. and PHILLIPSON, A. T. (1956) *J. Physiol.* **153**, 76–77P.
DOBSON, A., SCOTT, D. and MCDONALD, I. (1966) *Res. vet. Sci.* **7**, 94–100.
DOUGHERTY, R. W. (1960) In *Digestive Physiology and Nutrition of the Ruminant*, p. 79, D. Lewis (Ed.) Butterworths, London.
DUCKWORTH, J. and GODDEN, W. (1941) *Biochem. J.* **35**, 816–23.
DUCKWORTH, J., GODDEN, W. and WARNOCK, G. McM. (1940) *Biochem. J.* **34**, 97–108.
DUNLOP, R. H. and HAMMOND, P. B. (1965) *Ann.N.Y. Acad. Sci.* **119**, 1109.
ENDERS, F., DISHINGTON, I. W. and HELGEBOSTAD, A. (1962) *Acta vet. Scand.* **3** (Suppl), 5–52.
EVANS, W. C. and EVANS, E. T. R. (1949) *Nature, Lond.* **163**, 373–75.
FINA, L. R., HAY, C. A., BARTLEY, E. E. and MISHRA, B. (1961) *J. Anim. Sci.* **20**, 654–8.
FRASER, J. G. (1961) *N.Z.Jl. Agric. Res.* **4**, 224–31.
GOODALL, E. D. and KAY, R. N. B. (1963) *J. Physiol.* **176**, 12–23.
GUTIERREZ, J. and DAVIS, R. E. (1962) *J. Anim. Sci.* **21**, 819–23.
GUTIERREZ, J., DAVIS, R. E. and LINDAHL, I. L. (1959) *Appl. Microbiol.* **7**, 304–8.
HARRIS, L. E. and PHILLIPSON, A. T. (1962) *Anim. Prod.* **4**, 97–116.

HARRIS, P. J. and SEBBA, F. (1965) *Nature, Lond.* **208**, 869–71.
HOGAN, J. P. and PHILLIPSON, A. T. (1960) *Br. J. Nutr.* **14**, 147–55.
HUETER, F. G., SHAW, J. C. and DOETSCH, R. N. (1956) *J. Dairy Sci.* **39**, 1430–7.
HUNGATE, R. E., DOUGHERTY, R. W., BRYANT, M. P. and CELLO, R. M. (1952) *Cornell Vet.* **62**, 423–9.
INGLIS, J. S. S. (1960) *Proc. 8th int. Grassland Congr. Reading*, pp. 558–64.
JOHNS, A. T. (1954) *N.Z.Jl. Sci. Technol. A.* **36**, 289–320.
JOHNS, A. T. (1963) In *Animal Health and Production*, p. 384, A. N. Worden, K. C. Sellars and D. E. Tribe (Ed.) Longman Green & Co. Ltd., London.
KAY, R. N. B. and PHILLIPSON, A. T. (1959a) *Proc. R. Soc. Med.* **52**, 374–76.
KAY, R. N. B. and PHILLIPSON, A. T. (1959b) *J. Physiol.* **148**, 507–23.
KLEIBER, M. (1956) *Nat. Acad. Sci. Natn. Res. Counc.* Publ. 388, 10–18.
KREBS, H. (1960) *Proc. R. Soc. Med.* **53**, No. 2, 71–80.
LE BARS, H., LEBRUMENT, J., NITESCU, R. and SIMMONET, H. (1954) *Bull. Acad. vet. Fr.* **27**, 53–67.
LINDAHL, I. L., DAVIS, R. E., JACOBSON, D. R. and SHAW, J. C. (1957) *J. Anim. Sci.* **16**, 165–78.
MCCLYMONT, G. L. and SETCHELL, B. P. (1955) *Aust. vet. J.* **31**, 53–68.
MANGAN, J. L. (1958) *N.Z.Jl. Agric. Res.* **1**, 140–7.
MANGAN, J. L. (1959) *N.Z.Jl. Agric. Res.* **2**, 47–61.
MANGAN, J. L., JOHNS, A. T. and BAILEY, R. W. (1959) *N.Z.Jl. Agric. Res.* **2**, 342–54.
MASSON, M. J. and PHILLIPSON, A. T. (1951) *J. Physiol.* **113**, 189–206.
MEAD, S. W. and GOSS, H. (1936) *J. Dairy Sci.* **19**, 465 (abstr.).
MITCHELL, R. L. (1944) *Proc. Nutr. Soc.* **1**, 183–9.
MOODIE, E. W., MARR, A. and ROBERTSON, A. (1955) *J. comp. Path.* **65**, 20–36.
MOODIE, E. W. and ROBERTSON, A. (1961) *Res. vet. Sci.* **2**, 217–26.
MOODIE, E. W. and ROBERTSON, A. (1962) *Res. vet. Sci.* **3**, 470–84.
OXFORD, A. E. (1958) *N.Z.Jl. Agric. Res.* **1**, 809–24.
PAYNE, J. M. (1964) *Vet. Rec.* **76**, 1275–9.
PAYNE, J. M. (1964) *Vet. Rec.* **76**, 77–81.
PAYNE, J. M. and SANSOM, B. F. (1966) *J. Physiol.* **184**, 433–43.
PENNINGTON, R. J. and SUTHERLAND, T. M. (1956) *Biochem. J.* **63**, 618–28.
PHILLIPSON, A. T. (1952) *Br. J. Nutr.* **6**, 190–8.
PHILLIPSON, A. T. and ASH, R. W. (1965) In *Physiology of Digestion in the Ruminant*, p. 97, R. W. Dougherty, R. S. Allen, W. Burroughs, N. L. Johnson and A. D. McGilliard (Ed.) Butterworths, Washington.
PHILLIPSON, A. T. and MCANALLY, R. A. (1942) *J. exp. Biol.* **19**, 199–214.
PHILLIPSON, A. T. and REID, R. S. (1957) *Br. J. Nutr.* **11**, 27–41.
PHILLIPSON, A. T. and REID, C. S. W. (1958) *Nature, Lond.* **181**, 1722–3.
PHILLIPSON, A. T. and REID, C. S. W. (1960) *Proc. Nutr. Soc.* **19**, XXVII.
PHILLIPSON, A. T. and STORRY, J. E. (1965) *J. Physiol.* **181**, 130–50.
REID, R. L. (1960a) *Aust. J. Agric. Res.* **11**, 346–63.
REID, R. L. (1960b) *Aust. J. Agric. Res.* **11**, 364–82.
REID, R. L. (1960c) *Aust. J. Agric. Res.* **11**, 530–38.
ROWETT RESEARCH INSTITUTE (1966) *Annual Report* **22**, p.31.
SCARISBRICK, R. (1954) *Vet. Rec.* **66**, 131–2.
SCOTT, D. (1965) *Q.Jl. exp. Physiol.* **50**, 312–29.
SCOTT, D. and DOBSON, A. (1965) *Q. Jl. exp. Physiol.* **50**, 42–56.
SHOJI, Y., MIYAZAKI, K. and UMEZU, M. (1964) *Tohoku J. Agric. Res.* **15**, 91.
STEVENSON, D. E. and WILSON, A. A. (1963) *Metabolic Disorders of Domestic Animals*, p. 198, Blackwell, Oxford.
STEWART, J. (1954) *Scott. Agric.* **34**, 68–73.
STEWART, A. B. and HOLMES, W. (1953) *J. Sci. Fd. Agric.* **4**, 401–8.
STORRY, J. E. (1961) *J. Agric. Sci.* **57**, 97–102.
SWAN, J. B. and JAMIESON, N. D. (1956) *N.Z.Jl. Sci. Technol. A.* **38**, 363–82.
TSUDA, T. (1956) *Tohoku J. Agric. Res.* **7**, 241–56.

TURNER, A. W. and HODGETTS, V. E. (1948–52) Toxicity of Large Rations of Wheat. Annual Reports of Commonwealth Scientific & Industrial Research Organization: 1st Report (1948–9) pp. 7; 3rd Report (1950–51) pp. 44–45; 4th Report (1951–2) pp. 48–49.
VAN HORN, H. H. and BARTLEY, E. E. (1961) *J. Anim. Sci.* **20**, 85–87.
WALSHE, M. J. and CONWAY, A. (1960) *Proc. 8th int. Grassland Congr. Reading*, pp. 548–53.
WATCHORN, E. and MCCANCE, R. A. (1937) *Biochem. J.* **31**, 1379–90.
WILLIAMS, V. J. and MACKENZIE, D. D. S. (1965) *Aust. J. Biol. Sci.* **18**, 917–34.
WORDEN, A. N., SELLARS, K. C. and LEAHY, J. S. (1963) In *Animal Health, Production and Pasture*, p. 409, Longmans, London.
WRIGHT, D. E. (1961) *N.Z.Jl. Agric. Res.* **4**, 216–23.

CHAPTER 34

The Economics of Animal Nutrition

S. R. WRAGG

Department of Economics (Agricultural Economics), the University, Bristol

FOREWORD

To write about the economics of animal nutrition in the sense of describing the systems and levels of feeding that are most efficient and profitable is quite impossible except in the very limited context of a single set of price relationships. To do so would not be particularly helpful because prices are never completely stable for longer than a point in time. Short-run oscillations superimpose themselves on long-term changes while underlying both of these there may be even longer-term changes taking place.

For any given set of production conditions and price relationships there are some policies more profitable than others, but there will always be one particular policy which leads to an optimum result with respect to the objectives laid down. When a significant change in relative prices occurs, adjustments to the existing policy become necessary in order to re-direct production and resource use to a new optimum position—which may be above or below the previous optimum. The principles and methods of production economics and the relevant facts about the physical production relationships together provide a system of guidance in relation to these problems of adjustment. This chapter is devoted mainly to explaining the system and its application.

The economics of animal nutrition falls naturally into that sector of economic studies which deals with production and resource use, usually referred to quite simply as 'production economics'. Production economics is concerned with the productivity and organization of relatively scarce resources; resources to which can be assigned costs of production or procurement and disposal values; resources for which there may be several different possible uses; resources which can be substituted one for another in the production process.

If, for example, the nutrient requirements of farm animals were freely available in unlimited quantities, like oxygen in the atmosphere, then there would be very little in the subject of animal nutrition of professional interest to economists. But the various materials which provide the nutrient requirements of farm livestock are not what the economist calls 'free' goods, available in unlimited quantities. They are themselves the result of production processes which involve the use of scarce resources—land, labour, capital and management. Like other 'economic' goods they have their production or procurement costs and their disposal values. They also have many different possible uses, both between animals of the same species and between animals of different species. The possible advantage of transferring some of the resources used in producing animal feeding stuffs to other sectors of the farming industry, or to some production process outside the agricultural industry must also be considered. Finally, nutrient requirements for a given quantity of product may be supplied through a number of different feeding stuffs, used either singly or in various combinations, so it can be said that there is substitutability between nutrient sources. Thus, decision making with regard to resource use in feeding farm animals may involve the choice of animal, the choice of feeding stuffs and the choice of quantities to be fed.

Rational decisions about the use of scarce resources cannot be made without reference to some goal which the decision maker is seeking to attain. In the very broadest sense possible, producers are said to aim at maximizing their satisfaction. But, however apposite and true this description of what producers actually strive to do may be, it is not a concept which lends itself very easily to quantification. In practice, therefore, any element in the objective function which cannot be quantified by some means or another has to be excluded, even though to do so may involve some loss of accuracy and realism in the formulation of the model. However, such a course does not mean that the non-quantifiable elements are lost sight of altogether or that they cannot be considered at some stage in the practical interpretation of the solution.

As far as farming is concerned, in countries where production is both on a commercial scale and in private hands, the goal of management (the decision maker) is generally assumed to be the long-term maximum profit. The goal of subsistence farmers, on the other hand, is more likely to be the maximization of physical production. But whatever values or quantities producers seek to maximize, the same maximizing principles can still be applied in solving problems as to how best they can organize their resources to achieve the desired ends.

If profit maximization is the producers' goal then the criteria, or choice indicators, required for guiding the decision-making processes must be stated in economic terms, because the problem is itself an economic one. Production coefficients expressed in physical terms do not, by themselves, provide the answers, although they are a very essential step to their provision. It is a combination of physical input/output coefficients and the appropriate cost and price information that provides the economic choice indicators used by decision makers trying to maximize their total profits.

Investigations into the economics of livestock production, based on enterprise and whole farm cost studies, have been pursued in several countries over nearly half a century. Many of them have provided, and still continue to provide, useful information, particularly in the form of simple input/output coefficients which have at times proved themselves to be more reliable estimators for farm planning purposes than coefficients derived from technical sources. But the vast majority of these studies have not, however, been able to provide much more than this, because of their limitations in concept, design and statistical treatment.[1,2] Until quite recently the analysis of farm costings data followed the simple classic routine in which profit is related first to one single variable and then to another until every likely relationship had been tested in the hope of identifying the 'factors in production affecting profits'. The expectation, or hope, is that there will emerge from the analysis sufficient evidence to enable broad generalizations to be made about the conditions or methods of production found to be associated with higher than average profits or lower than average costs. All too often this procedure has yielded disappointing and, at times, contradictory results. But even where the analysis has apparently succeeded and the indicated relationships look plausible, the results do not provide a really effective basis for decision making. In the first place, being themselves conditioned by a single set of cost and price factors they cannot provide a framework for guiding production and resource use adjustments to meet a new set of costs and prices. Secondly, the concepts from which they are derived have little in common with profit maximizing and cost minimizing principles.

Nevertheless, the view is widely held that in spite of their limitations and defects, the traditional type of input/output studies are not without their uses. Apart from providing simple bench marks in input/output relationships,[3] which are frequently not available from any other source, they do on occasions lend themselves to more advanced statistical treatment than is normally used.[4] As a means of keeping in touch with what really happens at the farm level they are irreplaceable.

In recent years the conventional type of input/output study has begun to give way to a more sophisticated approach in which the concepts and theories of production economics play a more positive role while at the same time more advanced and rigorous mathematical techniques are applied to the testing of hypotheses and the quantification of problem solutions. Part at least of the theoretical framework will be familiar to agricultural scientists because it is concerned with the physical relationship between input and output—the physical production function. This brief description of the theoretical framework commences with the single variable model.

It can be stated, with respect to a technical unit, such as an acre, a farm, an animal, a herd or flock, or indeed the whole agricultural industry of a country, that if increased amounts of one factor of production are used, while all other factors and conditions are held constant, then its physical productivity must take one or other of three forms, namely (1) increasing returns, (2) constant returns or (3) diminishing returns. These three conditions are represented graphically in Figs. 1, 2 and 3. Inputs of the variable factor are measured on the X-axis and production is measured on the Y-axis. Increasing returns to food in animal production, even granting their presence, would, it is assumed, very soon give way to either constant or diminishing returns. On the basis of this assumption, further discussion is limited to these two latter situations.

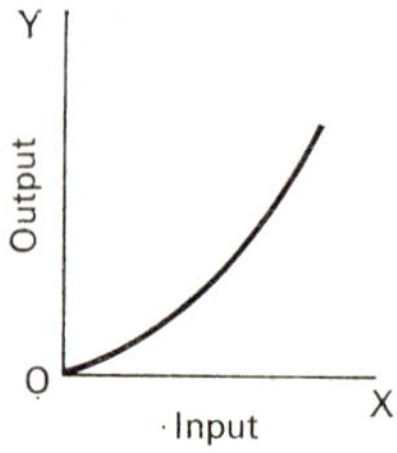

FIG. 1

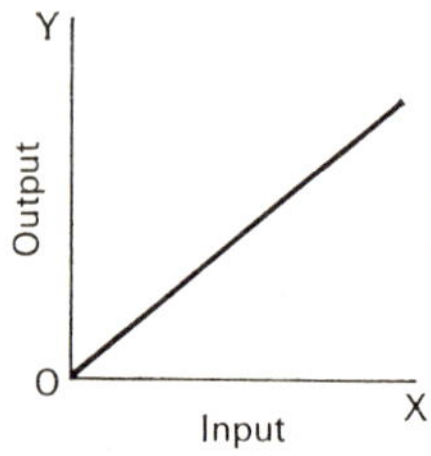

FIG. 2

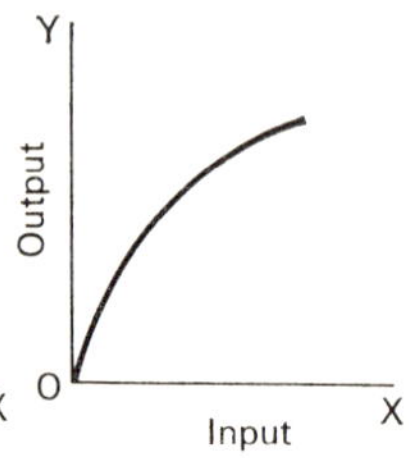

FIG. 3

Moving from left to right along the X-axis in Figs. 2 and 3, the productivity of each additional unit of input is measured by the ratio $\triangle Y/\triangle X$, or the marginal product of X. In Fig. 2 it will be seen that the value of the marginal product $\triangle Y/\triangle X$ is constant for all levels of X. It follows from this that the marginal product will be equal to the average product Y/X. Economists frequently draw attention to the fact that most recommendations made to farmers by animal husbandry scientists in connection with feeding farm livestock are framed in the context of a linear production function which implies a constant return to increasing feed inputs. Such a situation, if it were true, would greatly simplify the decision-making process; for it follows that if it paid to use the first unit of input it would pay to use all possible units because they all yield the same amount of physical product. But the question then arises as to what is meant by the word 'all'. There must be some limiting factor operating, otherwise the absurd conclusion is reached that it would, for example, be both possible and profitable to produce a whole country's milk supply from one single cow. The finite aspect of the function is provided by the animal's limited capacity, or willingness, to consume additional feed inputs. The linear hypothesis can, therefore, be restated as follows. Assuming constant returns to increasing units of input and an upper limit on the number of units which can or will be consumed by a single animal, then if it pays to produce at all it will pay to produce the maximum quantity possible, this being the point on the physical production function at which profit is also maximized.

Conventional feeding standards for dairy cows, which prescribe a constant rate of feeding, such as 2·5 lb of S.E. or 3·2 lb of T.D.N. per gallon of milk, are based on this linear model. True these conventions do not imply that all cows of similar weight and similar willingness and capacity to consume feed will produce the same quantities of milk when fed to appetite. But the underlying assumption is, that up to the limit of their individual capacities to produce milk, the feed input/milk output relationship is constant.

The alternative hypothesis to a function implying constant returns is that shown in Fig. 3 where, after an initial period of approximately constant returns, the marginal product $\triangle Y/\triangle X$ begins to decline as additional units of input are applied. If the marginal product is decreasing it follows that the average product Y/X is also decreasing, because each additional unit of input yields a progressively smaller product, so proportionately less is being added to the numerator Y than to the denominator X. The size of the total product Y will, of course, continue to increase so long as there is a positive response in production to increasing inputs of X, however slight.

Under a diminishing returns assumption, production and resource use decisions are by no means as simple as they are under the constant returns assumption. Information about the level of average resource productivity, which is comparatively easy to obtain, provides no quantitative guide as to what happens to production when one unit more, or one unit less, of the input factor is employed; but it is necessary to have this latter type of information in order to determine what quantity of input and related level of output will result in maximum profits. The physical production function by itself does not provide all the information required for a profit maximizing solution. The other essential ingredient is information about input factor and product prices. It is the physical input/output relationship combined with the factor/product price ratio that enable the conditions of an economic optimum to be specified. The important consideration is not the size of the marginal product in physical terms but its value in relation to the cost of a unit of the input factor.

An arithmetical example, in which hypothetical data are used to illustrate diminishing marginal and average productivity per feed unit in milk production, is given in Table 1.[5] The figures show very clearly the distinction between the average and the marginal product.

In Table 1 total profit is maximized when the marginal product of the feed input $\triangle Y/\triangle X$ is equal to the feed/milk price ratio Px/Py. If the price per unit of feed is 10 pence and the price per unit of milk is 5 pence then $Px/Py = 10/5 = 2$. It will be seen that the marginal product $\triangle Y/\triangle X$ is equal to 2 at 14 units of feed over maintenance which produces a total of 47·4 units of milk. These quantities represent the economic optimum for total feed use and total production at the specified factor/product price ratio. If, however, the value of Px falls and/or Py rises then the price ratio Px/Py will fall so that it becomes economically worthwhile to add further inputs of feed until $\triangle Y/\triangle X$ is again equal to Px/Py. Conversely, if Px rises and/or Py falls the price ratio Px/Py will rise, so in order to maximize profits under this new price situation, feed inputs must be reduced until a new point of equilibrium is reached between the marginal productivity of feed and the new price ratio Px/Py.

TABLE 1

Hypothetical Data illustrating Total, Average and Marginal Productivity

Feed over maintenance X	Total milk yield Y	Yield increase $\triangle Y$	Feed increase $\triangle X$	Average product $\frac{Y}{X}$	Marginal product $\frac{\triangle Y}{\triangle X}$
units	units	units	units		
0	0·0	—	—		
2	8·0	8·0	2	4·0	4·0
4	16·0	8·0	2	4·0	4·0
6	24·0	8·0	2	4·0	4·0
8	31·4	7·4	2	3·9	3·7
10	38·0	6·6	2	3·8	3·3
12	43·4	5·4	2	3·6	2·7
14	47·4	4·0	2	3·4	2·0
16	49·2	1·8	2	3·1	0·9
18	49·6	0·4	2	2·8	0·2
20	49·6	—	2	2·5	—

The figures in Table 1 also illustrate how misleading the use of the average product can be for decision making when the production function is curvilinear. Assuming, as before, an input factor/product price ratio of 2 : 1, the average product would still show a margin over feed cost at the 20 unit level. But according to the marginal productivity figures, each unit of feed in excess of the fourteenth would cost more than the value of the resulting product. Clearly, it is impossible, except by chance, to predict the level of input which results in the economic optimum from information about the average product.

The assumption of a production function related to a single feed input variable must now be changed in order to consider the more probable situation in which there are two or more feed variables. Under this new assumption the problems facing the decision maker are considerably more complex. In addition to wanting to know at what level of input and output profit will be maximized it will now be necessary to know also the combination of the different feed inputs which will result in minimizing feed costs for a given level of output.

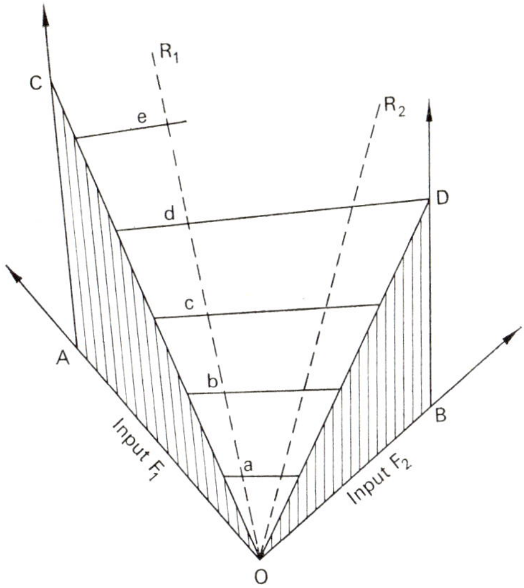

FIG. 4

A production function containing two variable inputs can be represented by a three-dimensional perspective drawing such as the one shown in Fig. 4. It will be seen that the slope of the function is linear. Inputs of two feed variables F_1 and F_2 are measured on the axis *OA* and *OB*, which lie on the horizontal plane, while output is measured on the vertical axis *AC* and *BD*. The production surface is represented by the area *CDO* which is positioned between the horizontal and the vertical planes. Each point on the horizontal feed plane represents a combination of the feeds F_1 and F_2, the product of which is located on the production surface. It is measured by the vertical distance between the selected point on the feed plane and the corresponding point, immediately above, on the production surface. The line *OC* represents the production response to varying quantities of F_1 when no F_2 is fed and *OD* represents the production response to varying quantities of F_2 when no F_1 is fed.

The lines drawn across the production surface are isoquants or isoproduct contours joining up all the points located at the same vertical distance from the horizontal plane, thereby denoting the same level of output. When these lines are projected onto the horizontal feed plane they represent all the various combinations of the two feeds F_1 and F_2 which could be used to produce the fixed amount of product. The isoquants in Fig. 4, drawn to represent successive equal increments of production, will be seen to be straight lines, running parallel to and equi-distant from one another, conditions which are characteristic of a linear production surface with constant rates of marginal substitution between the inputs.

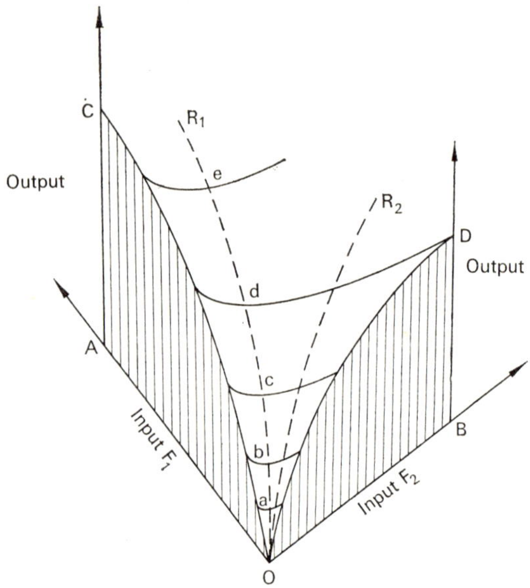

FIG. 5

The non-linear hypothesis is demonstrated in Fig. 5. Here the production surface rises quite steeply to begin with but later tends to flatten out. The isoquants, as in Fig. 4, represent successive equal increments of production, but it will be seen that the distances between them increase at higher rates of

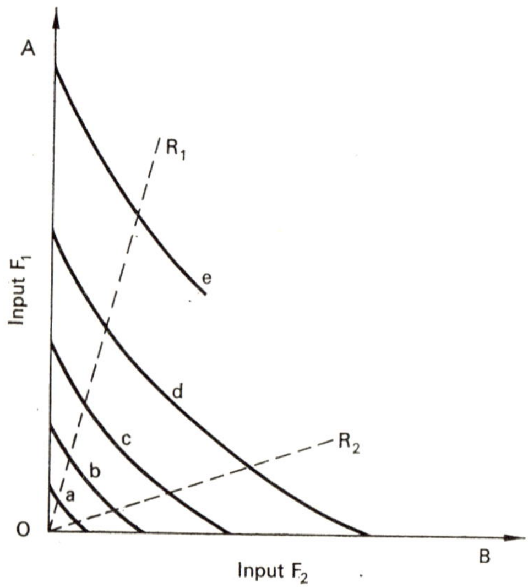

FIG. 6

input and also that they are curvi-linear. These conditions denote both diminishing returns to the use of additional inputs and diminishing marginal rates of substitution between the inputs.

Further exposition of the non-linear model is made somewhat easier if the three-dimensional representation is reduced to two-dimensional form. In Fig. 6 feed inputs F_1 and F_2 are measured on the axis *OA* and *OB* respectively and *a*, *b*, *c*, *d*, *e* are isoquants which are non-linear. Taking the isoquant *d* as an example, it will be noted that the rate at which F_1 substitutes for F_2, diminishes as the ratio of F_2 to F_1 in the ration declines; that is, as increasing quantities of F_1 are substituted for F_2. The particular combination which the profit maximizing decision maker will choose is the one that minimizes cost for any given level of production. This objective is achieved when the substitution ratio of the two feeds is equal to the inverse ratio of their prices: that is, when $\triangle F_2/\triangle F_1 = Pf_1/Pf_2$. The other profit maximizing condition relates to the optimum quantity of the ration which can be fed. This is defined by the equation $\triangle Y/\triangle R = P_r/P_y$, where Y is the product, R is the ration and P_y and P_r are their respective prices per unit.

Lines, such as R_1 and R_2, can be drawn across the isoquants in such a manner that they pass through points on successive isoquants at which their slopes are all equal; that is, where the marginal rates of substitution between the two feed inputs are equal. Such lines are called isoclines. Given the price ratio between two feed inputs, the least cost combination for any given level of production is specified by that particular isocline which has a slope equal to the price ratio. For the corresponding feed price ratios, isoclines delineate the most profitable combination of feed inputs to be used in feeding for progressively higher levels of production. For this reason they are sometimes referred to as 'expansion paths'.

A hypothetical set of data illustrating feed combinations and marginal rates of substitution for two milk isoquants is given in Table 2.[5] Diminishing rates of substitution between F_1 and F_2 are indicated by the declining values of $\triangle F_2/\triangle F_1$. Given that the prices of feeds F_1 and F_2 are 15 pence and 5 pence per unit respectively, then the price ratio Pf_1/Pf_2 is equal to $15/5 = 3$. For the 500 unit isoquant the marginal rate of substitution $\triangle F_2/\triangle F_1$ is equal to 3 when the feed combination is 150 units of F_1 and 400 units of F_2. A similar marginal rate of substitution on the 700 unit isoquant occurs where the feed combination is 200 units of F_1 and 775 units of F_2. For the given price ratio $Pf_1/Pf_2 = 3$, these two rations specify the minimum feed cost for producing 500 and 700 units of milk. An isocline having a slope equal to the price ratio of the two feeds would pass through the points on the two isoquants where these rations are located. A change in the price ratio Pf_1/Pf_2 would change the mix of the least-cost ration which could be identified by an isocline connecting points on the isoquants where $\triangle F_2/\triangle F_1$ coincided with the new value of Pf_1/Pf_2.

TABLE 2

Hypothetical Data representing Feed Combinations and Marginal Rates of Feed Substitution for 500 and 700 unit isoquants

Feed combinations to produce 500 units of milk			Feed combinations to produce 700 units of milk		
Feed F_1	Feed F_2	$\frac{\triangle F_2}{\triangle F_1}$	Feed F_1	Feed F_2	$\frac{\triangle F_2}{\triangle F_1}$
units	units		units	units	
50	718	—	50	1292	—
100	550	3·37	100	1100	3·85
150	400	3·00	150	930	3·40
200	255	2·65	200	775	3·00
250	130	2·33	250	635	2·60
300	30	1·95	300	510	2·25
—	—	—	350	410	1·95

Marginal productivities, isoquants, isoclines and other relevant parameters not mentioned in the foregoing discussion can be obtained as partial derivatives from the main production function.

Assuming that the empirical data needed for estimating the coefficients of a production function are available, a decision has to be made regarding the form of the function which will best represent the data. This implies that not only should the fitted function explain a large proportion of the variance present in the dependent variable and provide statistically significant regression coefficients, but also that its characteristics must be consistent with generally accepted fundamental principles in animal physiology and other relevant sciences. Unless there are very sound *a priori* grounds which point unerringly to one particular algebraic form of the function, when actual attempts are made to estimate the feed input/livestock output relationship from a set of data, a number of different equations may need to be fitted and the results appraised. The more conventional forms used are linear, power, quadratic and square root functions, but there are several other polynomial forms which may also be tried.

For two feed variables, X_1 and X_2, a linear function implies a constant rate of response to changes in the quantities of X_1 and X_2 used, taken either individually or in combination. The upper limit of the input/output relationship is not specified by the function, but as it cannot be infinite it is assumed to be determined either by the ability or willingness of the animal in question to increase its feed intake, or by its inherent productive capacity. This function also has linear isoquants, thereby implying that the marginal rate of substitution between X_1 and X_2 is constant at all levels of output and for all proportions of X_1 and X_2. Bearing in mind that the equation $dX_1/dX_2 = Px_2/Px_1$ specifies the conditions for the least-cost ration for a given output, it follows that if the ratio dX_1/dX_2 is greater than Px_2/Px_1 then the margin over feed cost would be maximized by feeding all X_1 and no X_2. Alternatively, if the ratio dX_1/dX_2 is less than Px_2/Px_1 then the margin over feed cost would be maximized by feeding all X_2 and no X_1. A constant rate of substitution between X_1 and X_2 also means in effect that there are no isoclines because at every point on the production surface the ratio dX_1/dX_2 has the same value.

For two feed variables, X_1 and X_2 a power function, commonly referred to as a Cobb–Douglas function after two early pioneers of production function analysis, has the form

$$Y = aX_1^{b1} \quad X_2^{b2}. \tag{1}$$

It is estimated from the logarithms of the variables. The regression coefficients b_1 and b_2 are the production elasticities of X_1 and X_2 respectively. The function allows for either increasing, constant or diminishing marginal productivity depending on the sum of the elasticities. If $b_1 + b_2 > 1$ this indicates that a one per cent increase in both inputs will result in a greater than one per cent increase in output. If $b_1 + b_2 < 1$ then a one per cent increase in both inputs will produce a less than one per cent increase in output. The isoquants of the Cobb–Douglas function allow diminishing rates of marginal substitution between X_1 and X_2; but one very serious defect, from the point of view of its application to livestock production functions, is that if the value of either X_1 or X_2 happens to be zero then the value of Y, the product of the function, must also be zero. This condition clearly violates both accepted scientific principles and common sense. In practice, however, the problem is met by replacing zero observations with values suitably small but significantly larger than zero. Because of its constant elasticities the Cobb–Douglas function cannot adequately represent the whole of the production surface which might include conditions of increasing, constant and diminishing returns. Moreover, the isoclines, which are linear from the origin, imply that the optimum ration will remain constant for all levels of production.

The Cobb–Douglas function has been extensively used by agricultural economists, partly because it is relatively easy to compute, partly because it is sparing in its use of degrees of freedom and partly because of some simplicity in its interpretation. In the main, it has been used to estimate functions for farms or single enterprises as a whole. The purpose behind such exercises has usually been to examine the marginal productivities of the various input factors employed in order to be able to pass judgment on the efficiency of resource allocation. It should also be stressed that the marginal resource productivities are measured at their geometric means, hence there is an increasing tendency for the coefficients to overestimate the marginal productivity of the inputs the farther out they reach beyond their mean values.

The quadratic function for two feed variables, X_1 and X_2 may take the form

$$Y = a + b_1 X_1 + b_2X_2 - b_3X^2{}_1 - b_4X^2{}_2 + b_5X_1X_2. \quad (2)$$

The presence of the squared terms with negative coefficients allows diminishing returns to operate for either one or both input variables. The interaction term could also have a negative coefficient. The isoquants are curved denoting diminishing rates of substitution between X_1 and X_2, but, unlike the power function, some production could be forthcoming even when one or other of the variables is set at zero. The isoclines are linear and converge from the input axis to a position in the input plane which would coincide with a point indicating maximum output on the production surface. As production increases along the expansion paths denoting least-cost rations for any given price ratio of the inputs, the optimum proportions of X_1 and X_2 in the ration change.

The square root function for two feed variables X_1 and X_2 can take the form

$$Y = a - b_1X_1 - b_2X_2 + b_3X_1{}^{0.5} + b_4X_2{}^{0.5} + b_5X_1{}^{0.5}X_2{}^{0.5} \quad (3)$$

This function allows the marginal product of any given combination of X_1 and X_2 to decline at a decreasing rate. The isoquants are curved and so allow for diminishing marginal rates of substitution between X_1 and X_2. The isoclines are also curved, passing through the origin and converging to a point in the input plane coinciding with a maximum output on the production surface. They indicate that the least cost expansion path, corresponding to a given price ratio of the inputs, involves changes in the ration as production is expanded towards the maximum.

A large number of production functions involving livestock have been estimated since the techniques were first developed during the nineteen-twenties. Most of the early examples are of United States origin calculated from survey data obtained from comparatively large areas of fairly homogeneous farming and environmental conditions. As will be seen in the example illustrated in Tables 3 and 4, the input and output categories used tend to be highly aggregated.

In 1946 Heady published the results of a pioneer study[6] in which he had estimated numerous production functions using input/output data from a

TABLE 3

Production Elasticities (Regression Coefficients) for Various Input Categories used on Certain Types of Livestock Farms in Iowa in 1939

Type group	Land	Labour	Equipment	Livestock and feed	Sundry inputs	Sum of elasticities
Northeast dairy	0·2081	−0·0449*	0·0871	0·6136	0·0223	0·8862
Western meat	0·1037	0·0107	0·0888*	0·5511	0·0310	0·7853
Eastern meat	0·1918	0·2364*	0·0952*	0·4263	0·0245	0·9742
Southern pasture	0·1624	−0·0508*	0·1325*	0·4944	0·0438	0·7823

*Not significant at the 5 per cent level.

random sample of 738 Iowa farms. Cobb–Douglas functions were estimated for farm sub-groups classified by type and location. These included a number of mainly livestock producing types the results for which are summarized in Tables 3 and 4.

The data above relate to whole farm functions. Many of the variables used are highly aggregated.† Both inputs and outputs were measured in dollar values. The sum of the elasticities, being in all cases less than one, indicated that there were diminishing returns to scale in all sub-groups; i.e. a simultaneous one per cent increase in all inputs resulted in a less than one per cent increase in output. Marginal productivities for each input category were derived from the main regression equation and these are set out in Table 4.

From the figures in this table the conclusion seems to be that for the first three type groups, the marginal productivity of all the inputs are below their cost. But in all instances the marginal productivity of a dollar spent on livestock and feed is much higher than the marginal productivity of any other input category. Indeed, on the Southern Pasture farms, an additional dollar spent on livestock and feed is calculated (at the geometric mean) to yield a marginal product of 2·64 dollars.

TABLE 4

Average Marginal Productivities for Various Input Categories used on Certain Types of Livestock farms in Iowa in 1939 (measured in dollar output per dollar input)

Type group	Land	Labour	Equipment	Livestock and feed	Sundry inputs
Northeast dairy	0·0331	0·0973	0·1484	0·6588	0·3783
Western meat	0·0382	0·0302	0·2410	0·7130	0·4037
Eastern meat	0·0398	0·0685	0·2417	0·5012	0·3407
Southern pasture	0·0187	−0·1091	0·3133	2·6419	0·4025

It may be argued that even if the methods and results of this type of analysis are accepted as being scientifically correct they do nothing to explain why certain results occur: nor do they provide detailed guidance as to how profits can be raised, on the Southern Pasture farms, for example, beyond the general recommendation to increase expenditure on livestock and feed—and, of course, to reduce expenditure on all other inputs because their marginal productivities are below their cost of acquisition. In practice, institutional impediments and the lumpy nature of many types of inputs would make substantial short-run downward adjustments in expenditure on most inputs very difficult to achieve. Notwithstanding these reservations this kind of information can be most useful for policy making at national and regional

† Livestock and feed were too highly correlated for inclusion as separate independent variables.

level. It could also be of considerable value to research scientists in developing new hypotheses and in carrying out a reappraisal of the old ones.

A somewhat similar analysis to the one just described was carried out by Godsell and Wragg[7] in 1956 using data relating to two groups of specialist dairy farms in the West of England. The choice of variables was dictated by the purpose of the investigation which was to test the hypothesis that, assuming fixed land resources, cow stocking density was the most important determinant of farm profit. The two farm groups represented low-yield and high-yield herds. Output was measured as margin over purchased feed cost. The three variables were cows, other cattle and other inputs which included all costs other than rent and purchased feed. The marginal productivities for the three input categories, measured in £'s per £ of input, were as follows.

TABLE 5
Marginal Productivities in Milk Production

Resource	Marginal productivities per £1 input	
	Low yielding herds	High yielding herds
	£	£
Cows	3·71	3·41
Other Cattle (per cow equivalent)	0·01	0·66
Other inputs	0·68	0·08

The results, although not satisfactory in all respects, fully confirmed the original hypothesis. They are also very much in line with the now widely accepted conclusion that, under average farm conditions in this country, the quantity of utilized nutrients per acre of crops and grass land devoted to ruminants is mainly determined by the number and class of ruminants present. While these results could be most useful as a basis for making general recommendations to farmers they can be criticized on the grounds that they neither prove the existence of any underlying relationships nor, if such relationships exist, do they in any way contribute to their explanation. Moreover, in the variable described as 'other inputs' there is a very large amount of aggregation of inputs, a point of criticism frequently made in connection with production function estimates.[8]

Rather less aggregation of variables was carried out by workers in Sweden in estimating a Cobb–Douglas function from farm survey data containing seventeen independent variables.[9] These, however, only included three feed variables which were in fact aggregates of feeding stuffs classified under three sub-headings, viz. purchased feed, home-produced feed grain and roughages; the latter included hay, silage, roots, potatoes and pasture. It is interesting to note that, as in the United Kingdom study referred to earlier, the resource yielding by far the highest marginal productivity was investment in livestock, viz. 7·9 Swedish Crowns per Swedish Crown input.

There are many examples of farm survey data having been used by economists for estimating more specific types of production functions, in which the relationships studied are confined to livestock production as a function of nutrition. The most popular subject for this kind of investigation has been milk production, the reason for which is quite simple. Farm records relating to dairy cow feeding and milk production provide abundant evidence concerning the existence of wide variations between recommended feeding standards and the input/output ratios commonly achieved at the farm level. Naturally, economists have been anxious to know how to interpret this situation. Are the quantities of food that are fed in excess of the recommended requirements technically unnecessary and therefore wholly uneconomic, or alternatively, if all or a substantial part of the excess food were to be withdrawn, would output decline, and if so, by how much? At the time when economists first started to make these enquiries nutrition scientists themselves had only just begun to ask themselves the same questions.

Few of these studies have included as a major objective the estimation of input/output coefficients for direct use in making specific recommendations to farmers concerning optimal quantities and combinations of foods for dairy cows. They have been directed more towards testing the basic hypothesis that the input/output function for milk and the rates of substitution between foods fed to dairy cows are non-linear.

A study by Orton,[10] using both inter-farm and time series data taken from the National Milk Costs Investigation relating to England and Wales, and also time series data taken from Danish farm account records, produced conflicting results. Analysis of the inter-farm data for England and Wales, 1951–52, showed that the average milk response to an additional 1 lb (0·45 kg) of S.E. fed per cow was only 0·07 gallons (0·32 litres) as compared with an expected 0·4 gallons (1·82 litres) according to the orthodox feeding standard. However, analysis of the time series data for England and Wales indicated a constant marginal return to increased inputs of S.E. which in magnitude corresponded quite closely to the orthodox feeding standard. But this result was, in turn, contradicted by the analysis of the Danish time series which indicated that the marginal response to additional fodder units was considerably below the accepted Scandinavian standard of 2·5 kg of milk per F.U.

Using food consumption and production records derived from bull progeny testing records, Gardiner[11] estimated production functions for five groups of heifers, each consisting of 40–50 animals. Both linear and quadratic equations were used to estimate relationships between milk yield and food consumed measured in lb of S.E. This study also produced conflicting results concerning the nature of the production function. For some groups the linear form gave the best fit while for others the quadratic gave a significantly better fit. For four out of the five groups the squared term in the quadratic equation had a positive sign thus indicating increasing marginal returns to additional feed inputs, a phenomenon which is, in the long run, much less plausible than diminishing returns. Gardiner himself suggests that the levels of feeding employed in these tests may not have been sufficiently high for diminishing returns to set in. But one very curious feature of the results is that while the function itself is predicting increasing marginal productivity, within the range of inputs observed, the average response and the marginal response to additional S.E. inputs both appear to have been significantly below what could have been expected on the basis of orthodox feeding standards.

Using a sample of milk costings data relating to West of England farms included in the National Milk Costs Investigation for 1949, Jawetz[12] estimated a number of functions for Shorthorn and Friesian herds using average daily inputs over maintenance and pregnancy requirements of S.E. and P.E. per cow as independent variables. The measurements related to the daily average over a period of 91 days from the beginning of January to the end of March. Within the range of the observations studied the isoquants are virtually linear, but their slopes decline as both inputs and output increase, thereby indicating diminishing marginal rates of substitution of S.E. for P.E. These two conditions preclude the existence of isoclines. The regression equations for the two breeds are as follows, with X_1 representing average daily yield per cow in milk measured in gallons, and X_2 and X_3 representing average daily inputs of S.E. and P.E. over maintenance and pregnancy requirements per cow in milk measured in lb.

Shorthorn cows: $$X_1 = 0{\cdot}504602 + 0{\cdot}176979X_2 + 0{\cdot}964188X_3 - 0{\cdot}06828X_2X_3 \quad (4)$$

Friesian cows: $$X_1 = -0{\cdot}027 + 0{\cdot}2354X_2 + 1{\cdot}1895X_2 - 0{\cdot}063834X_2X_3 \quad (5)$$

The standard errors of the estimate; $S_{1.23}$, are given as 0·08103 for the Shorthorn cow equation and 0·1695 for the Friesian cow equation, but the standard errors of the regression coefficients are not given.

The objectives of this particular study go far beyond the mere testing of a diminishing returns hypothesis. Jawetz used the results of this analysis as a basis for promulgating what he himself describes as 'virtually a complete, although roughly calculated, new set of standards for feeding dairy cows for production'.

The estimation of production functions from farm survey data has not by any means been confined to dairy cows and milk production. Godsell and Wragg[13] fitted three types of functions; linear, Cobb–Douglas, and quadratic to survey data relating to beef production from grass and forage crops with supplementary concentrate feeding on 25 farms in 1960/61. The best fit was obtained with the Cobb-Douglas function. The coefficients of the regression equation, in which $Y =$ output measured in cwt liveweight, $X_1 =$ acres of grass and forage and $X_2 =$ concentrates fed measured in terms of cost, were as follows.

$$Y = 1{\cdot}45\, X_1^{0.819}\, X_2^{0.261} \qquad R^2 = 0{\cdot}79 \qquad (6)$$

The sum of the elasticities is slightly in excess of unity thus indicating that returns to scale are at least constant. The marginal physical productivities of grass and forage acres and of concentrates, measured at their geometric means, are shown in Table 6. The marginal productivities expressed in monetary value terms were based on 1960/61 prices.

TABLE 6

Marginal Productivities of Grazing and Forage Acres and of Concentrates used in Beef Production

Resource	Marginal productivity	
	Liveweight output	Value
	cwt	£
Grazing and forage per acre (0·404-ha)	2·61 (132·6 kg)	21·8
Concentrates per £1	0·13 (6·6 kg)	1·1

Such estimates as these could be used in farm management for both diagnostic and planning purposes but, as with any other input/output coefficients, their use in circumstances outside the range of observations recorded in the original data would be extremely unwise. The technical coefficients are, of course, considerably more stable than the value coefficients. Once a set of technical coefficients has been satisfactorily established, the corresponding value coefficients can be very easily calculated for a whole range of prices; past, present or future. If, however, production techniques themselves change markedly, then quite clearly this would call for new estimates to be made of the technical coefficients. It must also be added that because in beef production, more than in any other branch of livestock, there are so many possible variations on a theme, in terms of the system of production and management, a great many more production functions need to be estimated in order to have a measure of the whole range of resource use and production opportunities available to producers. But it does not by any means follow that all the different systems of beef production will produce input/output coefficients that are significantly different from one another.

A simple production function such as the one illustrated above is likely to be criticised because there has been a considerable aggregation of the variables;[8] particularly is this true of X_1 which relates in a comprehensive manner to grazing and forage acres. It would clearly be impossible to base very detailed recommendations to farmers on a simple result of this kind. On the other hand, there is not very much evidence as yet to support the belief that the disaggregation of variables will provide functions with greater explanatory powers or capable of providing more practical guide lines.

proportionate increase in labour, maize and protein supplement, with housing space held constant at the mean for the 140 farms. The results of this calculation are shown in Table 8 below.

TABLE 8

Predicted Effect of Increase in Hen Numbers with Proportionate Increase in Labour Hours and Quantities of Maize and Protein Supplement but with floor space constant at 980 ft² (91 m²)

No. of hens	Space per hen	Marginal Production of hens‡	Total egg Production **	Increase in total egg Production**
No.	ft²	eggs	eggs	eggs
288*	3·40* (0·316)†	81·7*	61,170*	—
308	3·18 (0·295)	80·7	64,568	3,398
328	2·98 (0·277)	79·9	67,920	3,352
348	2·81 (0·261)	78·8	71,239	3,319

*Geometric mean values.
†Square metres.
‡With all other variables held constant at their geometric mean values.
**With proportionate increases in labour, maize and protein.

Economists have been far from satisfied with the results obtained from their attempts to estimate input/output relationships from farm survey data, especially where it was intended that they should be used as a basis for making recommendations about production and resource use to individual farmers. There are several reasons for this disquiet, the most important being the hybrid nature of inter-farm production functions, the problem presented by management as a variable, the aggregation of variables and, with special reference to ruminants, the difficulties involved in the specification and measurement of certain inputs, notably grazing and grass products.

Every farm, it is said, has its own production function which is to some degree, however slight, unique to that particular farm; and what a farm survey does in fact do, is record information about the individual sample farms which relates to the particular point on their own unique production functions at which they were operating during the period to which the survey relates. A function is fitted to these observations and may then be used as a basis for predicting what would happen to production on any one farm of a similar type if a specified quantity of inputs were used. To do so involves the assumption that as inputs on a given farm are raised or lowered, an acceptably close approximation to that farm's own response function can be obtained from a line drawn somewhere across a whole series of individual farm functions. Among agricultural economists, opinions are very much divided on this issue. Variations in management present problems in estimating input/output relationships for two reasons. In the first place a unit of input in the presence of good management will be more productive than it would be in the presence of inferior management. This might not be too great an embarrassment if it could be assumed that variations in the level of inputs were independent of variations in the level of management, but one cannot be sure about this. Indeed, most investigators would tend to the opinion that inputs and management are positively correlated which, if true, would mean that production coefficients from inter-farm functions will tend to be over-estimated to an extent depending upon the degree of positive association between the level of inputs and management. As far as the United Kingdom is concerned, however, it is by no means certain that the relationship is particularly strong.

Again, aggregation of variables presents difficulties when attempting to base recommendations on the input/output coefficients, as the very general and even vague terms in which the inputs and outputs are often described is inevitably carried through into the advice offered. The difficulties frequently experienced in the measurement and specification of variables during the survey itself are part of the same problem. Grazing and grass products present special difficulties for production function estimates. Both hay and silage production and disposal are difficult to measure accurately and their qualities vary enormously. Grazing usually has to be allocated between two or more different groups or species of animals on the basis of cow equivalent grazing days, while a good deal of additional guesswork goes into apportioning grassland when it is used for both grazing and conservation in the same season.

One consequence of dissatisfaction with farm survey data is that economists have begun to look to the results of technical experiments as a possible source of data from which to calculate input/output and other similar relationships, because data from controlled experiments are considered to be virtually free from the kind of biases and other limitations which afflict farm survey data. This does not mean that economists have found the results of many past experiments much better suited to estimating input/output relationships than survey data, primarily because this was not part of their original purpose. What is in fact being suggested by economists and others, is that new experiments should be carried out which include among their objectives the estimation of the physical production function. There are already a few investigations, some completed, some in hand, which fall into this category, but they are almost exclusively of United States origin.[16]

The first co-ordinated attempt by scientists and economists to use controlled experiments as a means of estimating a livestock production function was the now well-known series of experiments in dairy cow feeding by Jensen and Woodward *et al.* in the United States, the results of which were published in 1942.[17] The investigations were carried out at ten centres over a period of three years and involved a total of 346 cows and 469 annual lactations. Their purpose was to measure the extent to which the production of cows of known milking ability varied with changes in the level of feeding above and below standard feed requirements. A Spillman type exponential function was chosen make their predictions concerning optimum levels of feeding and least-cost rations, with an acceptably high degree of confidence under widely differing conditions.

Blaxter's criticisms concerning the practice of estimating input/output relationships from feeding experiments by the application of econometric forms of analysis are widely shared by other animal nutrition scientists, particularly in Europe. Nevertheless, reporting to an O.E.C.D. seminar in 1964,[20] Heady was able to refer to eight projects, five in North America and three in Europe, then currently under way, in which co-ordinated attempts were being made by animal scientists and agricultural economists to estimate production functions for milk. Feeding experiments, planned specifically for this purpose, are being carried out, and when the results become available, econometric analysis will certainly be used to estimate production functions and their various derivatives. Indeed, suggestions have already been made for pooling national data to form a large inter-continental set of observations. This could lead to the 'collectivization' of variables on an international scale.[20]

Godsell and Preston fitted equations to cattle feeding data from experiments conducted at the Rowett Research Institute in 1961/62.[21] Inputs were measured by a single feed variable. The ration consisted of a mixture containing 85 per cent of barley and 15 per cent of protein feed (fish meal and soya bean meal) to which were added mineral and vitamin supplements. A quadratic function provided the best fit, giving highly significant regression coefficients and a value of $R^2 = 0{\cdot}981$.

It has since been possible to test the value of this equation in terms of its ability to predict results in general farming practice. Data relating to foods fed and liveweight gains per beast were collected from a sample of 17 barley beef units on commercial farms during 1964.[22] The average quantities of concentrates fed per beast in each of these units was entered in the regression equation and estimated values for Y, the liveweight gain, were obtained. Two of the 17 units were rejected because on inspection their systems of management were markedly different from that on which the equation was based. The equation predicted the results on six out of fifteen farms within $\pm$ 5 per cent of actual and on fourteen out of the fifteen farms within $\pm$ 10 per cent of actual. There was a general tendency for the equation to underestimate commercial performance. Revised equations based on more recent experimental work on intensively fed animals will shortly be available. Preliminary tests suggest that these will tend to over-estimate the results achieved in commercial practice.

Using the Godsell–Preston quadratic function, the marginal productivities for intensively fed beef cattle at different levels of total feed intake were computed in order to calculate the optimum liveweight gain for different feed and meat prices. The results are given in Table 10 below.

TABLE 10

Optimum Liveweight Gain per Beast at Three Price Levels per 100 kg of feed and per live 100 kg of beef

Price per 100 kg of concentrate	Price per live 100 kg		
	£17·7	£18·7	£19·7
	100 kg	100 kg	100 kg
£2·36	2·88	2·99	3·09
£2·46	2·77	2·21	3·00
£2·56	2·68	2·80	2·93

The average weight at slaughter of the cattle in this experiment was approximately 410 kg. If an initial calf weight of 45 kg is added to the figures in the preceding table, it will be seen that, at these optimum weight gains, all of the totals would be less than 360 kg liveweight which clearly cannot be regarded as representing a finished beast fit for slaughter. Using the regression coefficients it was possible to estimate for each situation described in Table 10 the quantity of additional food required to raise the total liveweight per animal from its optimum liveweight gain plus 45 kg initial calf weight to a finished liveweight of 410 kg. Table 11 shows the differences between the value of added liveweight and the cost of the additional feed required. In all cases these differences are, of course, negative.

TABLE 11

Cost of Additional Feed required to Raise Cattle from Optimum Liveweight, as shown in Table 10, to Actual Finished Liveweight of 410 kg, less the value of the Additional Liveweight Gain

Price per 100 kg of concentrates	Price per live 100 kg		
	£17·7	£18·7	£19·7
	£	£	£
£2·36	−6·2	−5·1	−4·2
£2·46	−7·5	−6·0	−4·9
£2·56	−7·9	−6·8	−5·7

Within the range of input and output prices used, an econometric analysis of these experimental results would indicate that intensively fed cattle should be sent for slaughter as soon as possible after they reach 360 kg, given a feed price of £2·36 per 100 kg and a beef price of £19·7 per live 100 kg, or as soon after 320 kg as possible, given a feed price of £2·56 per 100 kg and a beef price of £18·7 per live 100 kg. This kind of information would not be revealed in the conventional presentation of experimental results, nor is it likely that it would come to light through a conventional application of cost accounting to the physical data obtained from the experiment.

Technical research in poultry nutrition is relatively cheap in its demands on research resources. This places fewer hindrances in the way of adopting the most efficient experimental design and scale. A further stimulus should be drawn from the fact that the cost of feeding poultry, whether they be kept for meat or egg production, accounts for between 70 per cent and 80 per cent of total costs. As competition in the industry becomes keener, so the drive for cost reduction becomes more insistent and this, not unnaturally, is directed at feed costs.

As there is no practical alternative to *ad lib.* feeding in commercial poultry production, the approach to feed input/output relationships must be made through variations which it is possible to introduce both in the nutrient composition of the ration and in the mix of possible ingredients used for providing a particular nutrient specification. Linear programming, about which more will be said in a later section, can be used, and indeed is used by many feed compounders, to adjust ration compositions to periodic changes in the price of possible ingredients. But this is usually done in relation to a given standard set of nutrient requirements and a number of associated assumptions concerning the additivity of nutrients in feedstuffs, linear response functions and linear rates of substitution between possible ingredients, none of which may be valid. As alternatives, either the optimum level of feeding and least-cost rations can be estimated direct from the appropriate production function and its derivatives or, having estimated the production function, it can be used as a basis for obtaining technical coefficients for use with linear programming models.

In spite of what has been said above concerning the relative cheapness and simplicity of experimental work in poultry nutrition very few projects of the type referred to above have been undertaken.[23] As at the end of 1963, published results appear to be confined to four investigations relating to broilers and one relating to turkeys.[24] Nothing has been published so far concerning egg production; the first experimental programme designed specifically for estimating a production function for eggs was put in hand in 1965. Perhaps not surprisingly, all the projects referred to above are of North American origin.[23]

In problem formulation and the design of these experiments, the investigators had as their objective the prediction of liveweight gains in response to different mixes and levels of feeding. The two major variables were maize and soya bean oil meal. Information about the basic relationships was used to specify least-cost rations, least-time rations (i.e. maximum rates of gain, knowledge of which might be used to counter an anticipated reduction in product price) and optimum marketing weights.

In one such experiment[24] 600 chicks were used, randomly assigned to 30 pens, each pen being regarded as an experimental unit and each weekly weighing of the birds in a pen being one observation. There were six basic treatments representing six levels of protein ranging from 16 per cent to 24 per cent. In one part of the experiment two pens were put on each treatment. The other eighteen pens were similarly assigned to the six treatments, but as the birds in each of these pens reached the approximate weight of 0·6 kg per bird they were transferred on to a lower protein ration.

The analysis of the results followed the usual procedure for estimating production functions and the derived relationships and these were then used for predicting economic optima under various price conditions. The figures in Table 12 are given to illustrate the kind of information which this type of analysis provides. Their relevance to production and resource use planning will be self-evident.

Experiments of a very similar kind have been carried out in the United States in pig feeding with a view to estimating production functions, and their derivatives for pork pigs.[25] As in the poultry experiments the basic feed stuffs were maize and soya bean meal, combined in varying quantities so as to provide six protein levels. These eventually produced a large amount of information concerning input/output relationships including least-cost rations for three pig weight ranges at various soya bean meal/maize price ratios, optimum marketing weights and various time/weight gain/feed relationships, all of which would be most useful to producers in aiding the decision-making process.

TABLE 12

Optimum Broiler Weights and Feed required at Varying Protein Levels for Varying Feed/Broiler Price Ratios

Feed broiler price ratio	Per cent protein									
	16	18	20	22	24	16	18	20	22	24
	Broiler weight (kg)					Feed required (kg)				
0·250	1·21	1·25	1·27	1·26	1·24	3·35	3·43	3·44	3·39	3·28
0·227	1·30	1·35	1·36	1·35	1·35	3·75	3·83	3·83	3·76	3·58
0·208	1·37	1·42	1·43	1·42	1·38	4·08	4·16	4·15	4·06	3·91
0·192	1·43	1·47	1·48	1·47	1·43	4·37	4·44	4·42	4·31	4·14
0·179	1·48	1·52	1·52	1·51	1·47	4·61	4·67	4·65	4·53	4·35
0·167	1·52	1·55	1·56	1·54	1·49	4·83	4·85	4·85	4·72	4·53
0·156	1·55	1·58	1·59	1·56	1·52	5·01	5·06	5·43	4·89	4·68

The work carried out in the United States in connection with the estimation of production functions for all classes of farm livestock has been greatly facilitated by the extreme simplicity of the diets used, both in the actual feeding experiments undertaken and in ordinary commercial farming practice.

The most complex, confused and controversial sector of animal nutrition economics is that which includes grassland and ruminants. There are many reasons why this is so, but primarily it is due to inadequate specification of the problems in economic terms, the use of under-developed models in proposing solutions and the difficulties involved in obtaining the right kind and range of quantitative data for testing hypotheses.

Commencing with the proposition that fresh and conserved grass provide the cheapest possible source of energy and protein for ruminants, the whole strategy of production and resource use is placed in a straight jacket by the use of an over-simplified cost-minimizing model which leads inevitably to the restrictive conclusion that the proper role of grazing and grass conservation is to provide a substitute for more expensive sources of nutrients. This could only be true if the supply of land was in no way a limiting factor, but most farmers have to operate in a situation in which land is the least elastic of all their resources. Having regard, therefore, to possible scale effects in relation to other relatively fixed inputs, the total supply of nutrients may be no less important to profit maximizing strategies than their unit cost of production or purchase. Economists themselves are not without some responsibility for the ambiguity and confusion which arises in connection with this cost minimizing approach, for they have, from time to time, produced figures purporting to show the comparative cost of providing starch equivalent from various sources, including the purchase of a standard dairy compound cake. One such set of calculations is illustrated in Table 13.[26]

TABLE 13
Estimated Average Costs of Production of S.E. from Various Crops, 1960/61

	Cost per 100 kg S.E.
	£
Grazing	1·64
Kale (grazed)	1·70
Hay	1·79
Grass silage	2·05
Kale (carted)	2·56
Arable silage	3·19
Mangolds	3·39
Turnips and swedes	3·88
Barley	1·64
Oats	2·13
Standard dairy cake (63% S.E.)	4·78

These figures do little more than enable one to conclude, with a considerable degree of confidence, that any of the home-grown foods specified in the table would provide S.E. at a lower cost per unit than would a purchased compound dairy cake. The only other kind of conclusion one might logically draw concerns the difference in cost per unit of S.E. one could expect as a result of growing one crop rather than another. But few people ever use this or similar evidence to make out an economic case for transferring land from arable to grass. It is, however, quite frequently and most incorrectly used to support the argument that more resources should be devoted to the existing grassland acreage in order to produce more low cost S.E. for use as a substitute for expensive sources of S.E., notably purchased dairy cake.

The first error in this line of approach arises from the fact that it fails to distinguish between average costs and marginal costs. The figures in Table 13 are quite clearly labelled 'average' costs; and while it would be correct to assume that the individual farmer could purchase (in 1960/61) as much dairy cake as he could possibly require at a constant cost of £4·78 per 100 kg of S.E., the same as the average cost shown in Table 13, it might not be correct to assume, that by increasing expenditure on a fixed grassland acreage, additional units of S.E. could be produced at or below the average cost per unit. On the contrary; as in the United Kingdom outputs of 1500–1700 kg of S.E. per hectare are not at all uncommon from grassland which has received little or nothing in the way of investment for the purpose of raising grass output, and as outputs of S.E. in excess of 2500 kg per hectare are comparatively uncommon, even on farms where investment in grass production is relatively high, it is probable that the marginal cost of S.E. from grass is higher than its average cost. This should not, of course, be used as an argument against increasing expenditure on grassland in order to raise its productivity. If capital is not a limiting consideration it will pay a farmer to continue investing in grass production until such time as the marginal cost per unit of S.E. from grass exceeds the marginal cost of producing S.E. from other sectors of the farm cropping programme, or until it exceeds the marginal cost per unit of S.E. from off farm purchases of concentrates or fodder. If, on the other hand, capital is a limiting factor, then it should be so distributed between grassland, other fodder crops, cereals and purchase concentrates as to equate their marginal costs per unit of S.E.

Although the above model is much more relevant to the real farm situation than the earlier model based on a simple comparison of average costs it still makes use of other assumptions which are of doubtful validity. It is implicit in both models that all units of S.E., from whatever source they are available, are perfect substitutes one for the other and their marginal rates of substitution are linear however far in one direction the process of substitution is extended. This is certainly not true at the farm level, and it is probably not true in principle either. Furthermore, both models conveniently ignore the most important part of the strategy involved in increasing S.E. production from grassland. It is the stocking rate rather than the level of technical investment in grassland that determines S.E. production. If the stocking rate is zero the output of S.E. will also be zero, no matter how high the level of investment in grass production. On the other hand, there will generally be a significant production of S.E. if stock are present in sufficient numbers even though investment in grass production is zero. In this country at least, it is most probable, in the average farm situation, that the marginal productivity of additional investment in ruminant livestock, with grassland inputs held constant, is greater than the marginal productivity of investment in grassland with stocking rate held constant. This appears to be borne out by the results of investigations based on farm surveys and grassland recording schemes.[27,28]

Assuming a fixed area of land, the production function for grassland and ruminants could be described in the general form

$$Y = f(A, I, C)$$

Where Y is output measured in £'s, units of S.E. or some other convenient term, A is the number of animal units, or the stocking rate, I is the level of investment in grass production and conservation and C is the quantity of concentrate used. This equation could, of course, be expanded to include as separate input categories, such grassland investments as fertilizers, weed control measures, irrigation, fencing and conservation equipment. Separate functions could be estimated for each class of ruminant and also for mixed stocking regimes. Although it may perhaps not accord with the kind of models which would be proposed by grassland specialists, the feeding of concentrates cannot be excluded from the model if it is to be efficient in helping to define maximum profit situations. In the world of grassland experts one is apt to be reminded of the almost infinite number of variables and interactions that are present in the grassland, ruminant complex. But without denying that this is so, one can ask that these be reduced to feasible proportions so that some broad relationships can be investigated which would fit reasonably well into a profit maximizing model. Dillon and Burley[29] have expressed similar views about the need to restrict consideration of the relationships to the three or four basic variables. They suggest, moreover, that a second order rotatable design would not only prove to be very efficient for estimating the production function but would also make this possible at comparatively small cost in the number of treatments required.

As we have seen earlier, a great deal of the economists' efforts in studies of animal production have been devoted to attempts to measure the effects on production of changes in the amount and composition of animal diets. Animal nutritionists, on the other hand, as the titles of many of their publications indicate, have a rather more definitive approach in which the basic problem of animal nutrition is expressed in terms of the animal's nutrient requirements. Although economists may feel that there are numerous occasions on which this concept fails to provide all the relevant information required for the study of profit maximization, it does at least underline the need and the scope for the application of cost minimizing principles.

If, in relation to a given production situation, a set of nutrient requirements can be specified, then having regard to the many possible ingredients that there are available to choose from, it is clear that the requirements could be met in a wide variety of ways, each of which will have a different cost per unit from all the other possibilities. It does not require a profound knowledge of economic principles to reach the conclusion that if, as presupposed, all the possible mixtures fulfilling the specification are equally effective in use, then the interests of producers and consumers alike will best be served by selecting the one that is cheapest.

An approximation to a least-cost ration can be reached using trial and error methods; but this is very time consuming and likely to encourage over-simplification and orthodoxy. These troubles, however, are now a thing of the past. The orderly and precise procedure of linear programming and the ability of modern high speed electronic computers to provide accurate and rapid solutions to large and complex problems have made the whole operation of deriving least-cost ration solutions both simple and commonplace. But the ease with which the computational side of the process can now be handled has tended to concentrate attention once more on the alleged gaps in knowledge of both animal nutrient requirements and the ability of different foodstuffs to supply them. Human errors excepted, the solution provided by the computer will be the least-cost ration. No other ration composed from the same list of ingredients and meeting the same set of constraints could be cheaper. However, it is not uncommon for least-cost ration solutions to be rejected as being technically unfeasible or even completely nonsensical. Such situations can occur as a result of using unreliable technical data or of failing to make a complete and correct preliminary diagnosis of the problem; for it is on these that the credibility of the solution will to a large extent depend.

It is not proposed to enter here into a detailed explanation of either the principles or the processes of linear programming as there is already a large amount of literature on the subject, ranging from simplified presentations, using simple geometry and arithmetic, to advanced treatments the comprehension of which requires some knowledge of matrix algebra.[30-32] What follows is a brief description of some of the more important considerations involved in using linear programming for solving least-cost feed mix problems.

The problem can be expressed mathematically in the form of a cost function

$$Z = \sum_{j=1}^{n} X_j P_j \tag{8}$$

where Z is the cost of the mixture, X_j and P_j are the quantity and price per unit of each ingredient included and n is the number of ingredients. The objective is to minimize the value of Z subject to k linear constraints.

In order to employ the logic of linear programming and make a solution to a least-cost feed mix problem mathematically possible, a number of technical assumptions have to be made. Two of these, divisibility of ingredients and a finite number of possible ingredients and nutritional constraints are referred to here but need no further comment. Those of more relevance to the subject of animal nutrition as such are as follows:

(i) The values for each type of nutrient contained in the several components of the mixture are assumed to be additive, so that the total value of a nutrient contained in a mixture is equal to the sum of the nutrient values of the individual ingredients.

(ii) It is assumed that the nutritive coefficients of all possible ingredients are constants and remain so irrespective of scale and of the level at which other ingredients are included in the mixture.

The programming matrix is built up from three sets of quantitative data as follows:

(1) Nutritional constraints: these are expressed as maximum or minimum values or equalities, to which the ration must conform. They relate to the quantities of the various nutrients which a unit of the mixture must contain and to the weight of the unit itself. It may be further stipulated, for example, that the mixture must not contain more than a specified proportion of fibre or not more than a specified proportion of a particular ingredient. Any important requirement relating to the ration which is quantifiable and which conforms to the basic assumptions of linearity and independence can and should be specified as a constraint if its value is thought to be in any way critical. In this way it is possible to avoid wasting time on producing solutions which will be later rejected because they fail to conform to some requirement or condition which was overlooked at the planning stage.

(2) The nutritive coefficients of all the possible ingredients.

(3) The price per unit of all the possible ingredients.

In addition to providing the solution to the least-cost ration, the output from the computer may also give the range of ingredient prices within which the final solution remains valid and also the marginal cost of each ingredient excluded: this latter represents the additional cost incurred if one unit of the particular excluded ingredient is forced into the mixture. From the print-out of the step-by-step solving process (iterations) it is possible to measure the difference in cost between the final solution and close approximations.

The view is sometimes expressed that least-cost rations should be tested empirically before the details are released for general publication; this could presumably be done by comparing them with proven mixtures having the same nutrient specification. There is very little evidence, in the form of published material, of such tests having been carried out. One example, relating to broiler rations, is quoted by Taylor[33] in which preliminary figures showed that the computed least-cost ration was exactly equal to the control ration in terms of conversion, thus achieving a substantial saving in cost per lb liveweight gain by virtue of its lower cost per unit.

Least-cost rations for fattening pigs within various weight ranges and for three sets of assumptions concerning the level of feed prices were calculated by Godsell,[34] many of which proved to be highly unconventional as the examples illustrated below indicate. This has resulted in some exploratory experiments being commenced at several research centres in England and Wales as a preliminary to the possible full-scale testing of least-cost pig rations.

If the activities of feed compounders are excluded it would almost certainly be true to say that most linear programming studies have been made with a view to investigating maximum profit situations rather than least-cost situations. As far as principles and methods are concerned, these remain essentially the same for both problems except that the equation

$$Z = \sum_{j=1}^{n} X_j P_j \qquad (8)$$

now represents a profit function where X_j measures the level at which each activity is included in the solution and P_j is the gross profit per unit of each activity. The objective this time is to *maximize* the value of Z subject to k linear constraints. The latter includes the available supply of the given fixed resources. For each of the various activities, P_j, the gross profit, is the difference between the gross output per unit of the activity and the variable costs incurred per unit of the activity; such as purchased feed, seed, fertilizers, fuel, etc. Studies of this kind are conceived and pursued in a much more general and aggregative context than is usually implied in least-cost feed mix problems. Whereas the latter are very closely identifiable with the fundamental principles of animal nutrition in its most detailed aspects, the former are approached in a much less rigorous, but perhaps slightly more practical manner. The technical coefficients and constraints are more home-spun, being expressed somewhat loosely in terms of the actual feeding stuffs themselves; and, in the case of the really 'difficult to measure' feed resources, grass for example, inputs may even be measured in terms of acres. In the United Kingdom, input/output coefficients used in connection with profit maximizing exercises are likely to have originated mainly from farm survey data. In the United States, where the bulk of linear programming in agriculture has been carried out, both farm survey data and formal production function estimates, based on experimental data, are used as sources of information. Other countries most frequently use the normative input/output standards provided by agronomists and animal husbandry scientists; partly, no doubt, for the reason that they do not have the same ample supply of alternative data.

Typical of profit maximizing studies involving livestock is one carried out by Heady and Gilson at Iowa State College[35] in which linear programming

TABLE 14
Least Cost Pig Mixtures

Ingredients	30–120 lb liveweight (14–54 kg)						120–200 lb liveweight (54–91 kg)					
	1·3 Mcal/lb (2·86 Mcal/kg)			1·4 Mcal/lb (3·08 Mcal/kg)			1·3 Mcal/lb (2·86 Mcal/kg)			1·4 Mcal/lb (3·08 Mcal/kg)		
	Price assumption			Price assumption			Price assumption			Price assumption		
	1	2	3	1	2	3	1	2	3	1	2	3
	%	%	%	%	%	%	%	%	%	%	%	%
Barley meal			77·8			77·8		11·6	13·8		79·4	79·4
Ground oats		78·6			48·2			70·0	69·3			
Wheat					30·2						1·2	1·2
Bran	2·6							1·8				
Weatings	70·0			3·7			83·3			2·8		
Fine weatings				69·6						80·8		
White fish meal		2·4	2·4			2·4						
Meat and bone meal		2·9	11·6	1·5	6·4	11·6		5·5	6·0		14·1	14·0
Soya bean meal	24·7	14·2	8·2	23·4	13·7	8·2	14·9	7·5	7·1	14·8	1·2	1·3
Bean meal								1·5	1·9		4·1	4·1
Salt	0·3	0·4		0·3	0·2		0·3	0·6	0·6	0·3		
Calcium carbonate	2·4	1·5		1·5	1·3		1·5	1·5	1·3	1·3		
Cost per cwt (£)	1·362	1·292	1·233	1·412	1·300	1·233	1·254	1·187	1·129	1·308	1·204	1·133
Cost per 100 kg (£)	2·861	2·543	2·427	2·780	2·560	2·427	2·468	2·336	2·220	2·575	2·366	2·230

Notes:
(1) Price assumptions 1, 2 and 3 relate to prices of feeding stuffs in August 1963.
(2) It will be observed that the least-cost ration supplying 1·3 Mcal/lb (2·68 Mcal/kg) for the weight range 30–120 lb (14–54 kg) liveweight under price assumption 3 actually provides enough energy to meet the next requirement of 1·4 Mcal/lb (3·08 Mcal/kg).

was used to determine the maximum profit combination of livestock enterprises and pattern of feed allocation for a 'typical' farm with 'typical feed supplies' in Northeast Iowa. A total of 43 different activities were considered. These included three different levels of productivity for dairy cows, pigs and poultry and five different levels of grain feeding for dairy cows. The latter were included as a means of determining whether or not the most profitable farm organization in the region was one with most of the grain going to pigs and the forage going mainly to the dairy cows. The following extract from the report on this project gives a fair indication of the type of input/output coefficients in common use for linear programming by workers in the United States.

P_3 = Above-average dairy cow fed 7·11 tons of hay equivalent and 1940 pounds of grain. Milk production 7650 pounds.

P_7 = Average dairy cow fed 6·73 tons of hay equivalent and 963 pounds of grain. Milk production 5700 pounds.

P_{19} = Above-average fall hogs fed 345 pounds of grains and supplement per 100 pounds pork produced; 6·7 pigs saved per litter. Pigs raised on pasture.

The capital letter P with a subscript is used to denote an activity. Those briefly described above are just three out of the total of 43 activities considered.

Doubts have been raised as to whether linear programming of least-cost rations can provide anything like a complete answer to the problem of choice facing the producer. It is not the technique itself that is open to question, but rather the conceptual framework in which it is usually applied. Most linear programming of least-cost rations commences with the basic assumption of a single set of nutrient requirements and constraints related to one particular level of output, which may be either explicit or merely implied. There is in fact nothing left for the econometrician to solve except the mix of ingredients which fulfils the stipulated requirements at minimum cost. However, it is possible that the same level of animal performance could be obtained using any one of several rations, for each of which it would be possible to compute a least-cost mix; if, therefore, the cost of the ration were the only criterion, clearly the most profitable choice would be the cheapest of all the possible least-cost rations. However, the objective of the profit maximizer is not to minimize costs as such but to maximize the difference between costs and output value, a situation that can only be reached through a careful consideration of both these terms in the profit equation simultaneously, because they are not independent of one another.

It has been suggested by Dent,[36] with particular reference to pig feeding, that this complicated problem of choice could best be solved by means of a two stage analysis. In the first stage, multiple regression analysis would be applied to the results of specially designed feeding experiments in order to derive production functions for pigs within various weight ranges. From the regression equations a series of gain isoquants could then be calculated. In stage two, linear programming would be used to calculate least-cost rations for a number of points on each isoquant. This combination of regression analysis and linear programming would thus provide information concerning both the output function and the corresponding cost function. A very close approximation to the maximum profit situation could then be reached by selecting the isoquant and relevant least-cost ration which resulted in the highest margin between the market value of the product and its minimum food cost.

The standard simplex method used in linear programming has been subjected to a great deal of modification and development during recent years with the result that quite complex planning situations can now be dealt with, provided there is sufficient basic information available to feed into the programming matrix. The new models include quadratic programming which can take into account the risk element arising out of variability in yields; parametric programming which is capable of handling variable resources or prices, and dynamic programming which can handle the problem of optimizing some objective over time.[32] At the present rate of development of new programming models the real handicap to progress in providing practical solutions is, as always seems to be the case, a chronic shortage of sufficiently reliable and refined input/output coefficients.

Recent developments in the theory of ruminant nutrition and in the evolution of a new system of measuring nutrient requirements and feeding values have a very special significance for agricultural economics. The new theory of energy metabolism in ruminants, as developed by Blaxter[37,38] lends credence to the existence of both diminishing returns in animal production and diminishing rates of marginal substitution between feeding stuffs. In almost every way it conforms very closely indeed to ideas about ruminant production and nutrition in which economists have long believed and have spent many years trying to prove, with the aid of whatever input/output data have come to hand, combined with econometric forms of analysis. These new discoveries will also make cheering news for those economists who find the thought of complete linearity too monotonous even to contemplate. But however much the practice is frowned on by scientists, economists are unlikely to relinquish at once their own efforts in this field in favour of an exclusively physiological approach. Indeed, they may regard current progress in the elucidation of the physiological processes as a means of making the direct econometric approach more effective through a better understanding of the underlying principles of the input/output relationships which this new knowledge provides. This will certainly be true in the econometric interpretation of farm survey data. On the other hand, it is probable that many efforts will be made to use the actual data on metabolizable energy (M.E.) requirements and feed values published by the Agricultural Research Council[38] in some form of econometric model in order to provide the kind of quantities and relationships needed for making recommendations about the optimum use of resource. Blaxter himself has already made estimates of the marginal productivities of M.E. under different milk yield assumptions and used them to predict 'the economic allowance of concentrates' for various milk/concentrate price ratios.[38] This kind of approach has a sound technical basis and could be expected to receive the general approval of scientists on this account.

The new theory poses some special problems for linear programmers in the computation of least-cost rations because it violates a number of the basic assumptions made in linear programming, namely linearity of input/output ratios and return to scale and additivity of inputs. The first two problems may be overcome in a variety of ways. But the non-additivity of nutrients is something which may prove more difficult to handle.

REFERENCES*

1. SCHULTZ, T. W., Theory of the firm and farm management research. *J.F.E.*, Vol. 21, No. 3, 1939.
2. COZENS, L. E., Production economics, averages and standards in research and extension. *A.J.A.E.*, Vol. 9, No. 2, 1965.
3. *The Farm as a Business*. H.M.S.O. (London), 1963.
4. RASMUSSEN, K. and SANDILANDS, M. M., Production function analyses of British and Irish farm accounts. Nottingham University, 1962.
5. WRAGG, S. R., Review of basic econometric concepts and their implication and application in inter-disciplinary research. *O.E.C.D.; Doc. F. and A.*, No. 71, 1965.
6. HEADY, E. O., Production functions from a random sample of farms. *J.F.E.*, Vol. 28, No. 4, 1946.
7. GODSELL, T. E. and WRAGG, S. R., Production functions for dairy farming and their application. *The Farm Economist*, Vol. 8, No. 5, 1956.
8. BLAXTER, K. L., Technical experimentation as a source of input/output data. *O.E.C.D.; Doc. F. and A.*, No. 50, 1961.
9. HJELM, L., *Input/Output Relationships in Milk Production*. Agricultural Economics Research Institute, Uppsala, 1953.
10. ORTON, F. J., The economy of feed input in milk production. *The Farm Economist*, Vol. 9, Nos. 1 and 2, 1958.
11. GARDINER, T. W., Levels of feeding and milk yields per cow. *J.A.E.*, Vol. 13, No. 4, 1960.
12. JAWETZ, M. B., *The Joint Production function of Starch and Protein Equivalent in Feeding Dairy Cows*. Univ. Coll. Wales, 1961.
13. GODSELL, T. E. and WRAGG, S. R., Unpublished, Bristol University.
14. WRAGG, S. R., Unpublished, Bristol University.
15. HEADY, E. O. and BROWN, D. W., Unpublished, Iowa State College.
16. WRAGG, S. R., The integration of technical experimentation and production economics in agriculture. *O.E.C.D.; Doc. F. and A.*, No. 50, 1962.
17. JENSEN, E., WOODWARD, T. E., *et al.* Input/output relationships in milk production. *U.S.D.A. Tech. Bull.*, No. 815, 1942.
18. HEADY, E. O., JACOBSON, N. L., *et al.* Milk production functions incorporating variables for cow characteristics and environment. *J.F.E.*, Vol. 46, No. 1, 1964.
19. BLAXTER, K. L., Economics and animal husbandry. *J.A.E.*, Vol. 14, No. 3, 1961.
20. Co-operative research to improve input/output data in cow milk production. Recommendations and conclusions, pp. 19–22. *O.E.C.D.; Doc. F. and A.*, No. 71, 1965.
21. GODSELL, T. E. and PRESTON, T. R., Input/output relationships in beef production. *O.E.C.D.; Doc. F. and A.*, No. 65, 1964.

22. Broughton, R. C., Beef costings survey; summary of individual farm results. Mimeo., Bristol University, 1965.
23. Co-operation between research in agricultural natural sciences and agricultural economics. Consultant's report, pp. 76–107. *O.E.C.D.; Doc. F. and A.*, No. 65, 1964.
24. Heady, E. O. and Balloon, S., Input/output relationships in poultry production. *O.E.C.D.; Doc. F. and A.*, No. 65, 1964.
25. Heady, E. O., Woodworth, R., Catron, D. V. and Ashton, G. C., New procedures in estimating feed substitution rates and in determining economic efficiency in pork production. *Iowa State Coll: Res: Bull.*, No. 409, 1954.
26. *Costs and Efficiency in Milk Production.* H.M.S.O. (London), 1964.
27. Progress in milk production. *Agric. Dev. Dept., I.C.I., Bull.*, No. 5, 1954.
28. Baker, R. D. and Baker, H. K., The output of swards on commercial farms in relation to fertilizer and other management factors. *J. Brit. Grassland Soc.*, Vol. 30, No. 3, 1965.
29. Dillon, J. L. and Burley, H. T., A note on the economics of grazing and its experimental investigation. *A.J.A.E.*, Vol. 5, No. 2, 1961.
30. Heady, E. O., Simplified presentation and logical aspects of linear programming technique. *J.F.E.*, Vol. 36, No. 5, 1954.
31. Gilson, J. C., Yeh, M. H. and Hodgson, G. C., The use of linear programming to determine least cost poultry rations. *Tech. Bull.*, No. 7, University of Manitoba, 1963.
32. Heady, E. O. and Candler, W., *Linear Programming Methods.* Iowa State College Press, 1958.
33. Taylor, N. W., The use of linear programming in least-cost feed compounding. *Agric. Econ. Res. Unit, Bull.*, No. 20, 1965, Lincoln College, Canterbury.
34. Godsell, T. E., P.I.D.A. Conference, Brighton, mimeo., 1966.
35. Heady, E. O. and Gilson, J. C., Optimum combinations of livestock enterprises and management practices. *Iowa Agric. Exp. Sta. Res. Bull.*, 437, 1956.
36. Dent, J. B., Optimal rations for livestock with special reference to bacon pigs. *J.F.E.*, Vol. 16, No. 1, 1964.
37. Blaxter, K. L., *The Energy Metabolism of Ruminants.* Hutchinson, 1962.
38. *The Nutrient Requirements of Farm Livestock, No. 2. Ruminants.* Agricultural Research Council, 1965.

* Abbreviations used: *J.F.E.*, American Journal of Farm Economics; *J.A.E.*, *Journal of Agricultural Economics; A.J.A.E., Australian Journal of Agricultural Economics; O.E.C.D.; Doc. F. and A., Organization for Economic Co-operation and Development; Documentation in Food and Agriculture.*

CHAPTER 35

Animal Production and World Needs

SIR DAVID CUTHBERTSON, C.B.E., F.R.S.E.
Department of Pathological Biochemistry, Royal Infirmary, Glasgow C.4.

INTRODUCTION

HUNGER is a real experience or a constant threat for much of mankind. In the less developed countries of the world there are regions and sections of communities where there is continuous undernourishment. The situation is being aggravated by the geometric increase in world population (1·9 per cent per annum) which will soon reach explosive proportions. It looks as if the Malthusian prognostication of wars, pestilence and famine will supervene unless restraint in population growth is practised for currently the increase in the agricultural production of food in the developing countries is not catching up sufficiently quickly or indeed is frequently lagging behind population increase. The world agricultural situation as related to knowledge of science and its application was briefly reviewed by the author in 1962 at the World Food Programme commemorating the centenary of the U.S. Department of Agriculture, and in respect of animal management at the United Nations Conference on the Application of Science and Technology for the Benefit of the Less Developed Areas (Cuthbertson, 1962*a*, *b*).

India and China are the two major areas where there will be maximum increase in population in the years ahead. In India there will be an expected rise from 458 millions in 1963 (FAO, 1964) to 600 millions in 1981: China's increase will be from 686 millions in 1960 (FAO, 1964) to 860 millions. Growth rate in Asia is 2·1 per cent per annum, in Africa 1·7, in Latin America 2·7 and the world average is, as noted above, 1·9 per cent.

Japan is an example of what can be done, given education, research, advisory services, appropriate price and credit policies and a generally accepted national objective and Japan's population growth is slowing. Israel is another shining example of what an educated people can accomplish in a congested and relatively difficult terrain.

The basic problem is that more food is needed to feed more people better diets; that land resources to this end are very limited; that input sources to raise yields are also limited and depend to a considerable extent on industrialization, education, social and cultural factors. It is man who has created the problem: can he, through his science, now solve it? Malthus thought man could and would check the need for wars, pestilence and famine, and correct the balance. But we note that by entirely understandable steps the scientfiic industrial revolution in the past has multiplied by a factor of 10 or more the material wealth of about a quarter of the world's population and in doing so has increased the production of food—in some countries greatly. The other three-quarters are setting out on the same road and are in a hurry, but still the growth of their agriculture is, at most, linear compared to the exponential increase in their population. By better diets in this context is meant diets with increased animal protein content, e.g. milk, meat and eggs. These also supply other nutrients commonly lacking in their absence.

The existing level of scientific technology is, in the main, sufficient to allow the solution of the world's food problems over the next decade, if it can only be applied. There's the rub! Further scientific and technological advances must continually be sought and not awaited as windfalls.

BASIC LIMITATIONS

Lack of indigenous drive and lack of acceptance by governments that they have duties other than the preservation of law and order are real obstacles to progress. Most of the highly educable portion must become interested in the natural sciences and in chemistry. If favourable conditions for attracting and retaining scientists, technologists and, above all, for producing agricultural general practitioners of the necessary calibre can be achieved there would be solid reasons for believing that the problem of hunger and malnutrition in our time may, in part, be solved by the application of existing knowledge.

THE FOOD SITUATION

By and large food deficits are greatest in the countries that have currently the least means of obtaining adequate supplies. This seems so obvious but it is the circling of the vicious circle. Total calories, total proteins and particularly animal proteins are, on the whole, very low for populations of Western Asia, Africa, the Far East, mainland China and large parts of Latin America.

Protein Deficiency

As the major organic substance of the body is protein and food protein is its main source, lack of adequate protein in the diet will lead to malnutrition and affect children in particular. Protein nutrition in young children is, indeed, the major nutritional problem of the world and protein malnutrition in both its direct and indirect effects, is a major cause of ill health. Frank protein deficiency is common in the less developed countries and latent or subclinical protein deficiency is probably even more prevalent.

Protein and Energy Deficiency

Although protein deficiency may predominate more often calorie deficiency contributes important effects and the simultaneous insufficiency of other nutrients complicates the picture in varying degree (FAO, 1965).

A deficiency of protein and energy may occur at any age but its incidence is greatest in the weanling and immediate post-weaning periods. As many reliable investigators have pointed out the post-weanling child when deprived of a high quality protein diet is not old enough to fend for himself in the family circle and is particularly subject to dietary taboos and prejudices (Campbell and Cuthbertson, 1963). Milder forms of dietary deficiency in respect of protein continue to occur among children and adolescents in low-income groups in developing countries and there the growth rates deviate sharply from the normal at the time of weaning and may continue at a low level throughout the entire period of growth, resulting in stunted adults (Jelliffe, 1955).

Apart from the effect on growth, mild or moderate protein deficiency renders infants and young children particularly susceptible to respiratory and gastrointestinal infections (FAO, 1965).

In contrast to infants and young children the adult is more likely to show marginal deficiency which is often seasonal or transient and does not cause obvious ill health unless it is complicated by acute or chronic disease.

Pregnant and lactating women form another vulnerable group in whom the effects of low protein intakes may manifest themselves in different ways.

The effect of shortages of calories and protein, especially animal protein, is to make populations more vulnerable to certain deficiency diseases and to reduce vitality and energy that are necessary for natural development. The Foreign Agricultural Service of the U.S.A. (1961) applied nutritional reference standards based on those developed by FAO, and these ranged from a needed daily consumption of 2300 kcal per person in the Far East to 2500 kcal in Latin America. The protein reference standard was set at 65 g total protein per head per day, and of this it was reckoned that at least a tenth, or 7 g should be animal protein, and the total animal and pulse protein should equal at least 17 g. The most critical shortages of animal protein are in the Far East, followed by Communist Asia and by Africa. Nearly all the pulse protein is located in Africa and the Far East.

Table 1 indicates how constant is the proportion of energy in the diet derivable from protein whereas that contributed by fat and carbohydrate moves in an opposite direction. In the temperate zones on mixed diets

Starling's dictum still tends to hold. "Take care of the calories and the protein will take care of itself." But this collapses where the staple foodstuff—that is, main energy source—is one with low and imperfect proteins in terms of amino acid requirements, such as cassava and yams. On the whole, it is remarkable how constant this proportion of protein (expressed as calories) is generally at all levels of energy expenditure, pregnancy, lactation and hard work. People eat more or less of the same pattern of foodstuffs but with upward change in economic status deficiencies are generally made good. Thus in Japan and Taiwan, animal protein intakes are now double the low levels prevailing in 1948–50, and still show an upward trend. There is, too, an upward trend in fish production in many countries, the outstanding feature being the spectacular advance in fish meal production in Peru. The safety and suitability of new sources of protein-rich food needs study and, on occasions, special investigation. There is an increasing demand for animal protein throughout the developing countries. Let us consider its production through farm livestock.

TABLE 1
Percentages of Calorie Supplies from Carbohydrates, Protein, and Fat in Average Regional Diets.

Region	Percentage of calories from			
	Carbohydrates	Proteins	Fats	Total
Far East	78	10	12	100
Near East	74	11	15	100
Africa	71	9	20	100
Latin America ..	69	10	21	100
Europe	65	10	25	100
North America ..	51	10	39	100
Oceania	54	10	36	100
World	71	10	19	100

From Sukhatme (1961).

DOMESTICATED LIVESTOCK PRODUCTION

There are about 1000 million cattle and buffaloes in the world and their numbers during the past decade have increased by some 1·5 per cent annually until recently when they have slowed down (FAO, 1964). Data for many countries are very uncertain. Buffaloes number about 102 millions. Assessments of the sheep population place them at around the 1000 millions with one-third native in type, and goats numbering 350 millions. World poultry meat production was assessed as 8·1 million tons for 1961 (FAO, 1952) and has been growing since. There are some 550 million pigs in the world. Llamas and vicunas in the Andean region of South America which stretches from

Colombia to Northern Chile, and with greatest concentration in Peru, number some 5 millions (FAO, 1962).

Cattle. Of the world's cattle population a very large proportion are unproductive or at best have a very low level of productivity. The more productive are in the more climatically suited regions, namely North America, Europe and Oceania. There, the demands for both meat and milk are high. On the other hand, Asia is essentially a meat-deficient area in spite of an estimated population of cattle and buffaloes approaching 335 millions (FAO, 1964). There is a strong religious bar to the eating of meat in parts of that continent and the main area for increased consumption of animal protein in these parts lies in the increased production and use of milk and milk products.

Improved husbandry, nutrition, breeding and extensive control of disease, parasites and vectors could bring about an immense increase in meat production throughout the African continent where the population is very deficient in animal protein.

In South America consumption of beef is high in Brazil, Argentina and Uruguay, but in the rest of the continent it is low, notwithstanding an overall population of 166 million cattle which is some 50 per cent greater than in Africa (FAO, 1962). While Brazil and Venezuela have increased their numbers, Argentina and Uruguay have shown a decline.

In general, throughout the world, much more beef and veal are produced from dairy cattle and from working oxen and buffaloes than from beef breeds. By crossing with beef breeds the average quality, conversion rate and meat yield could undoubtedly be increased.

Valuable sources of protein, minerals and vitamins may be lost when the wide range of useful and valuable animal by-products such as blood, bones and offal are not used for conversion to meat and bone or carcass meals for return to animals.

The high price of cow's milk in relation to man's income is a major obstacle in its consumption and the economic advantage of selling it rather than consuming it by the needy is, at first sight, obvious. Milk production, though growing, fails to keep pace in the developing countries with their increase in population. It has increased by 35 per cent from 1948–52 to 1963 (FAO, 1964) and this has mainly been as cow's milk.

FAO (1964) reports that in the Far East milk production is estimated to have increased by some 23 per cent from the 1948–52 level but because of population growth the *per caput* supply has not altered in the last few years. The increase in Africa was rather greater but production *per caput* declined slightly. In Latin America milk production rose by about 50 per cent, and here the output *per caput* rose similarly. With the exception of countries such as Argentina and Uruguay with favourable livestock conditions, the level of consumption still remains low in the majority of countries in that area.

Mexico has, however, shown a remarkable increase *per caput* and Uruguay a considerable one.

Sheep. It is precisely in the regions where low-productivity animals are found that the population depends most upon sheep for the production of meat and milk. In many of these countries the only meat consumed may be mutton or goat. In many areas these flocks have been maintained under desert conditions with long dry periods of semi-starvation separated by short spells of pasturing.

In the Americas, Australia, New Zealand and most of Northern Europe, sheep are kept exclusively for meat and wool. On the other hand, in North Africa, the Near East, Southern Europe and Southern Asiatic U.S.S.R., almost 75 million ewes are maintained primarily for milk production. In these regions ewe's milk accounts for over one-third of the total milk production (FAO, 1962). There has been little systematic attempt at increasing the output of sheep's milk. There seems little doubt but that sheep will continue to be the main source of milk in less-developed communities which have not adequate pasture for productive dairy cattle.

FAO (1962) reports evidence of increasing consumption of mutton and lamb in the Mediterranean countries, excluding Italy. In Afghanistan, Iraq, Iran, Jordan, Pakistan and Saudi Arabia, sheep supply 75 per cent of the meat consumed by the people. The demand is constant but there the economic status is low and severely restricts the consumption of this and other animal proteins. Although the demand for coarse wools for carpets is firm there has been little expansion in these areas.

The leading developed countries possess flocks which produce annually an average of 3·5 to 5·5 kg of wool per sheep, but in the less-developed flocks of Asia, North Africa, the Near East and Andean South America it is only 1·4 kg yearly. The lack of adequate husbandry practices and ineffective disease control mean that production of meat per breeding ewe is only about one-fifth that of sheep used for meat production in the advanced countries. Lack of winter grazing reduces the efficient use of summer pastures. There is, in many regions, an increasing shortage of experienced shepherds.

Goats. By proper management methods the capacity of the goat as a destructive grazer can be turned to account. It is an invaluable source of meat and milk in certain regions of Africa, Asia and South America.

Poultry. In some areas production has increased so rapidly as to have outstripped demand, and several of the more developed countries have discovered that they have an excess of poultry meat and find a surfeit of eggs an economic embarrassment. One great advantage of this industry is the smallness of the capital investment required and the ease in acquiring the necessary skills to raise poultry under village conditions as they adapt better than any other stock to variations of climate and geographical conditions. Rapidity in rate of growth and in reaching a productive stage of development has meant that improved strains of laying and meat-producing birds have been selected and propagated widely. The importance of small poultry units should not be overlooked but as countries become more developed there is a trend to larger and still larger units of the industry if it is to be viable. Given satisfactory nutrition and management and with disease under control the annual production of eggs should reach 300 eggs per hen.

If these conditions can be achieved for poultry meat production then, if fast-growing strains and properly balanced feedstuffs are available, less than 2 kg of feed per kg of liveweight will be required. Feed conversion rates (kg feed to produce 1 kg liveweight) around 2·5 are readily obtained in broiler units. In less developed units 3 to 4 kg of feed are required for the same weight increase. Because of the ravages of contagious diseases, developing countries need assistance to produce or acquire vaccines and to conduct vaccination programmes.

Pigs. The greatest increases in pig populations have occurred in Latin America, the Far East and Europe, while small increases have been recorded in Oceania. In North America and Europe there have been cyclical movements in pig numbers, while in Africa and the Near East numbers have remained fairly stationary (FAO, 1962).

Where religious objections against the pig applies to only a minority of the population the pig is nearly as promising as the bird as a potential source of acceptable protein for human consumption. Pork can be an inexpensive meat of nutritional importance. Improvement of the quality and quantity of pork by even a small fraction can significantly enrich the diet and raise the standard of living of many peoples.

For the small farmer in a developing country the production of meat and fat for family use and modest marketing can be achieved through pigs and the developmental plan can be on much the same lines as for poultry.

Where selected pigs are used and high concentrate grade feed is available and with means for disease control, conversion ratios of 3 can be achieved. The rates may be double and more in developing countries. This is due to the traditional method of keeping pigs on by-products and waste materials rather than on carefully balanced feeds.

The very heavy lard type of pig is wasteful when the market for the excess fat is declining. The general trend is to leaner pigs for higher meat and lower fat production.

Buffaloes. For many thousands of small farmers from Egypt to the Philippines the buffalo is their main support. They depend on it primarily as a draft animal. It has been estimated that there are some 80 millions and just because it is found mainly in the less developed countries there are considerable gaps in technical development and there is much scope for improvements in productivity. The buffalo has a great capacity to thrive on pastures where other cattle fail. Buffalo milk is richer than that of the dairy cow and can be used either whole or as a constituent of a "toned milk" made up of buffalo milk and dried skim milk. The obvious practical steps to improve health and husbandry practices must be taken to enhance the living standards of the populations concerned.

Llamas and vicunas. Amongst the llamas we can include the alpacas which are long woolly-haired llamas and the guanacos which are wild llamas with reddish brown wool. Llamas are widely used as beasts of burden and both they and the camel-like vicunas provide hair, hides, meat and milk, thereby playing a significant role in the agricultural production of the indigenous mountain people who herd them. Attempts to improve them have been small and uncoordinated. Adaptability to high altitudes must be taken into account in any breeding programme.

Elephants. Mature elephants are of value both as working animals and as a means of opening tracks through jungles for others to follow with appropriate road-making equipment.

Contribution of other wild herbivores to human welfare in developed lands. The economic utility of game biomasses can no longer be disregarded. Wildlife is increasingly coming to be regarded as a crop and to be harvested accordingly for there are increasing demands for animal protein to offset deficits in so many human diets. The biological communities in dynamic association have been evolved in adaptation with the edaphic, climatic and biological factors and their productivities are determined by the quantity and quality of the feed supplies. Man has, too often, in his cultivation of marginal zones with consequent over-population of the game habitats with his domestic livestock, caused these areas to degenerate into deserts. But the economic incentives for such continued wild-life conservation must be demonstrated before its future can be assured. No longer is a man satisfied with the products of his hunt.

There is no doubt that grazing by a single species in a varied vegetational habitat does not produce such yields as when animals of different grazing and browsing habits consume the natural flora. For example, browsing goats and grazing cattle are mutually complementary and together produce better results than when each operates alone in bush country: likewise the existing range of game species are equally supplementary and extract the maximum quantity of nutrients per hectare by consuming materials from each vegetative layer. Wild ungulates, with their differential, non-duplicating preferred diets, more flexible water requirements with disease tolerance, thus make more efficient use of available forage and water. In addition, if properly processed, most game meat is of high quality (Talbot *et al.*, 1965).

Wild-life is not restrained nor hampered by restrictive herding in the daytime nor by penning during the night—practices which curtail the efficiency of domestic livestock. On the other hand the economics of harvesting annual game take-offs of varied character requires critical assessment and must be set against the destruction of marginal areas by domestic livestock merely to obtain traditional meats with a low degree of efficiency. French (1965) has stated:

> Not only is organized and regulated cropping financially rewarding but it helps to conserve the vegetative cover. In fact, because of the small capital expenditures involved, compared with those required for establishing herds of domestic stock, such wild-life farming enterprises can contribute importantly to future animal production activities.
>
> The recorded meat outputs from wild-life populations, though sometimes exaggerated, can easily out-yield those attainable from livestock under extensive husbandry systems in poor environments . . . a case exists for the collection of reliable representative data on which future large scale and rational harvesting can be based. Much study is still needed on the ecological relationships of wild-life, often many factors affecting their well-being and maximum growth rates and of the economic factors involved in wild-life ventures such as the costs of collecting, handling, marketing game carcasses and meats.

On the whole, game units containing high biomasses normally contain the larger sized but slower growing game species, whereas the lower biomass units often exhibit a high proportion of smaller but more rapidly growing and sexually maturing species. The moisture contents of material in the digestive tracts of many game are lower than for cattle and this may be related to their lower water requirements and, in turn, this difference in moisture levels may explain the lower proportional weights of some game animals. Carcasses of game animals contain significantly less fat than those of cattle.

Hides and skins. The supply and demand for hides and skins generally depend on world factors outside the control of the country of origin. Increase in supply may follow a demand for meat but may also result from emergency slaughter in time of prolonged drought. Sometimes the hide is the only salvageable part of a diseased or drought-stricken animal. Such "fallen" hides are produced in large numbers in countries such as India where the slaughter of cows is prohibited and all cattle die from natural causes. With little additional effort, hides and skins of good quality can be produced for a remunerative market. Success depends upon the treatment of parasitic skin conditions and the control of other diseases, improved management and handling of livestock and better methods of removing, flaying and preparing the hides.

The development of local tanning industry will obviate the wasteful procedure of exporting raw hides and skins and re-importing as finished leather. Hides and skins can be important commodities in world trade.

CONSUMPTION LEVELS

The consumption of livestock products, as sources of protective foods, is extremely low in most of the less developed countries, and their scarcity results in large measure from the inadequate feeding of the flocks and herds which provide them. It is in respect of the disparity in consumption of livestock products that the welfare of the more developed and less developed countries is most clearly demarcated. FAO (1962) have recently estimated that, whereas total food consumption as measured in calories in the less developed regions as a whole averages some 70 per cent of the average for more developed regions, the corresponding figure for total protein is about 65 per cent and for animal protein only 20 per cent. Of the main sources of animal protein, the consumption of meat is estimated at about 20 per cent of the average in the more developed regions, of milk about 14 per cent, and of eggs about 12 per cent. For fish at 70 per cent the disparity appears to be much smaller, but the consumption of fish tends to be concentrated in coastal and riverine areas. Actually hundreds of millions of people consume only tiny quantities of livestock products.

Poverty is the main explanation for low production and uneven distribution in many of these countries. At low income levels the whole sum of money available for food is spent on the cheaper varieties, such as cereals and potatoes or other starchy roots, to satisfy hunger. It is only when these basic needs are satisfied that any appreciable expenditure on livestock products becomes possible. Since the market is low a high level of production seems hardly worthwhile.

Environment and Development

Where, however, there is not too much pressure on the land and the area is climatically suited for livestock production the situation may be modified as production costs are low and livestock products enter more largely into the diet. Thus we find in some Latin-American countries a high meat consumption and amongst the pastoral Fulani of West Africa a high milk intake. The natural advantages of the Argentine and Uruguay have allowed these countries not only to build up a large export business but also to permit one of the highest levels of meat consumption.

Other Factors influencing Expansion

The accelerating growth of population, mainly as a result of the control of disease, is producing a fitter population but a higher rate of population increase. As incomes rise the demand for livestock products is likely to rise in the same proportion. In Japan from 1955 to 1960 average expenditure of urban households rose by 35 per cent and expenditure on food by 19 per cent. During the same period expenditures on meat, milk and eggs rose by no less than 68 per cent. In detail (FAO, 1962): beef and veal rose by 19 per cent, pork by 41, eggs by 45, milk products by 80 and fluid milk by over 100 per cent. Greece with its developing tourist industry and Venezuela with its natural resources have also shown rapid rates of increase.

For the world as a whole the expected growth rate is about 2 per cent annually but in some less developed countries reaching 3 per cent. It has been calculated that the total effective demand for livestock products in less developed countries could be expected to increase by something like 5 per cent annually since in the less developed countries an increase of 1 per cent in *per caput* incomes normally leads to a higher percentage increase in livestock products. At that rate the effective demand for livestock products would increase by nearly 50 per cent by 1970 and by well over 100 per cent by 1980. Even so, consumption per head in the less developed countries would have risen by a little more than 25 per cent by 1980 which is much less than what is considered an essential and normal diet by millions of people in the more developed economies. It is true to say that if the cost of livestock products to consumers could be reduced by improvement in production and marketing, a larger increase in the volume of consumption would be possible for a given rise in income. Special efforts are thus required to raise the output of livestock products roughly in proportion to the expected combined rise in population and *per caput* income; that is to say, by some 5 per cent annually. It is imperative that increased production should be aimed first at the home market where the need there is greatest; markets must be carefully explored before production for export is initiated. It would appear that surpluses of milk products have been depressing world markets for some years, and prospects for international trade in meat have recently become less promising, but for some of the less developed countries favourable opportunities for exporting livestock products may occur.

Possibilities in Livestock Production

The growth target of some 5 per cent annually in livestock production is considerably greater than the rate during the last 10 years. Current FAO (1962) estimates suggest that world production rose by 2·6 per cent annually during the 1950's; the extremes being Western Europe with 3·8 and Africa with only 1·5 per cent. There has frequently been a tendency for prices of meat and other livestock products to rise faster than the general run of food prices.

Livestock in the Balance of Agriculture

It must be recognized that not only do livestock provide high-quality protein, as well as draft power and wool, hides and skins but they can also contribute to the maintenance and progress of crop production through increased soil fertility. Much of the earth's surface can only be utilized for the production of food for human consumption by grazing it with livestock. Range lands and moorlands which are unsuitable for cereal and other crops can be best used through livestock. Further, it is only through livestock—mainly ruminants—that crop residues and various waste products and by-products can be converted into human food.

Intensified systems of livestock husbandry are springing up with population increase and some of these are based on imported feedstuffs.

Overstocking

This is an age-long problem in some of the less developed countries. In India where the cow is sacred to the Hindu and may not be slaughtered the cattle and buffaloes number some 200 millions. Further, in Africa where numbers of cattle have prestige and currency value there can also be uncontrolled expansion. It is so obvious that a reduction in numbers of these unproductive and underfed cattle would with better management lead to greater supplies of milk for a human population lacking animal protein, and also seriously undernourished generally.

Over stocking may also lead to erosion through the destruction of most of the vegetation, particularly around water-holes and communal areas of grazing land. It has been calculated that by using better systems of management double the present stock could be carried.

The vicious cycle of uncontrolled breeding of low quality stock, overstocking, and undernutrition of livestock through lack of food and water must be broken if deterioration is to be stopped. Adequate markets may provide the best stimulus.

Range in Productive Efficiency

In addition to the foregoing restrictions there are also the obsolete methods of land tenure, lack of production credit at reasonable rates of interest and price instability: all retard the development of livestock production. Increase in productivity of flocks and herds rather than increase in numbers should be the keynote of policy. The ratios between the different classes of livestock do not vary greatly throughout the main regions of the world in respect of cattle and sheep. World numbers in relation to cattle, sheep, goats and pigs per thousand human inhabitants are not so very different from those in Western Europe though the numbers of goats in the Near East and Africa are greater than in any other region and the numbers of pigs least.

A very wide divergence exists between the levels of efficiency in animal productivity in the different regions. It has been pointed out (FAO, 1962) that in many parts of the tropics the yield of twenty milk cows is barely equal to that of one cow in such countries as Belgium, the Netherlands, New Zealand, Switzerland, and the United Kingdom. Poor livestock and inadequately managed pastures are the two chief reasons for low milk and meat production in the less developed countries. Overstocking with such poor quality material exists over vast areas of the Far East, Near East, Africa and Latin America.

The output per head of the cattle population in Europe is about ten times greater than in the Far East, seven times greater than in Africa and four times greater than in the Near East and Latin America. This is not to reason that livestock production in Europe cannot be improved. There is still a wide spectrum of achievement amongst farmers. With poor management goes failure to prevent, treat or control the diseases which reduce production. The indivisible facets of improving production are better nutrition, improved breeding and control of disease.

Means of Increasing Animal Protein Supply in Developing Countries

Initially there must be control of disease but there must go with this elimination, improved management, nutrition, genetics and breeding. With this must also go an enhanced appreciation of reproductive and environmental physiology in the habitats selected for expansion. With this must go the study of the handling of meat and other animal products.

Where advisory services do not exist they must be created and where they exist in these less developed countries they must be strengthened and the knowledge more widely disseminated among producers. A socially favourable climate to the contribution of agriculture must be created. Adequate credit must be created and improved conditions of land tenure established.

Animal Nutrition

To enhance production more use should be made of grass and forage and/or cereal grains and legumes, combined with the proper management of livestock and the best utilization of these feeds. There is a need for new varieties of pasture legumes which can survive in the tropics and sub-tropics, including the arid areas. The supply of nutrients must match the latent growth potential if optimal production is to be achieved. Obviously any mineral deficiencies in the improved pastures must be corrected.

Lack of water is the greatest limiting factor for plant growth in the arid tropics, including dry forest areas, steppes, savannas and scrub forests and, unless proper stocking and animal management are maintained, the land may become waste. The common practice of burning bush to control it is common in most of the savannah areas of Africa and if uncontrolled may lead to harm, even destruction, of grazing land and endanger forest areas. It may lead to soil erosion. Bush burning carried out at longish intervals (up to 4 years) should be adequate to control the bush. More research is needed on drought-resistant grasses, the forage of which would be of great value to ruminants. Currently great interest is being again centred on the use of non-protein nitrogenous supplements for ruminants under different conditions of feed, in particular as to carbohydrate source and minerals, including salt.

The humid tropics and temporarily flooded areas with their capacity for prolific growth of tropical grasses and legumes offer great possibilities for expansion of livestock production. Permanent pastures ensure an even and high production of forage and also, by covering the soil, protect it against erosion. One of the chief impediments to satisfactory nutrition amongst ruminants is the poor nutritional value of the forage.

Where the pressure of population is great, cereal grains will be used largely for human consumption. With increase in economic prosperity there will be an increased demand for a greater variety in the diet and this will free more cereals to improve animal production. Where climatic conditions are more favourable for pasture production, the by-products of milk and meat production can be used to support pig and poultry production. The administration of a small proportion of these may greatly increase the efficiency of utilization of other nutrients in feed materials suitable only for animal consumption.

It is maybe necessary in the first instance to estimate the value of local feeds on the same basis as the values established for similar kinds of feeds in the more highly developed countries. Refinements can follow later as analytical data become available.

While the story of trace element deficiency states has been a dramatic one and their cure spectacular, nevertheless the major minerals such as calcium, phosphorus and magnesium are as important though the body reserves are much greater. It must also be remembered that these major elements also act in trace amounts in enzyme systems.

The appropriate investigations to secure the best returns will demand research: in genetics to assess ability to produce, and in efficiency of feed utilization, work potential, adaptation to environment, quality and yield of wool; a study of the relative proportions of meat, fat and bone; and in reproductive efficiency and resistance to disease and other stresses. The heavy losses caused by periodic droughts can be eliminated or alleviated by abandoning the complete reliance of the livestock raiser upon pasture in favour of a combination of stock raising and feed production.

Forage Conservation and Grazing Management

Even in the more temperate climates where seasonal droughts are less acute, hay, silage and other methods of conservation play an important part in preventing reduction in the efficiency of the marketing system and so reduce costs. It is very necessary therefore to secure a more even distribution of animal feed throughout the year. This can only be done by adapting animal numbers to forage resources and complementing the rations during periods of shortage with concentrates or conserved fodder.

Greater efficiency in range management, the cultivation of fodder crops, the manufacture of silage, the development of ley farming systems, the utilization of grain cereal crops, the use of crop by-products and concentrates all have their parts to play in increasing forage supply. Fencing with controlled grazing increases carrying capacity but obviously the feasibility depends on costs in relation to productivity of the land thus managed.

Introduction of Exotic Breeds of Livestock

There is no doubt that the introduction of carefully selected strains of exotic breeds can be very successful, witness the spread of Friesian cattle in the United Arab Republic. Quarantine practices are not invariably such that stock can be guaranteed free from infectious disease. The possibilities for the transfer of major epizootics into free areas are very real. Artificial insemination is one of the most useful instruments for improvement of livestock quality, but unless properly applied can be the medium for widespread dissemination of both undesirable genetic factors and of disease.

Fuller Use of Indigenous Stock

Early consideration should be given first to improvement of the indigenous stock. Failure to cull unproductive lines is partially compensated by survival of animals able to thrive within the available feed and management resources and unlike the exotic stock, they have a marked resistance to the diseases common in the area. In general the less developed countries have not been able to maintain indigenous stock under conditions which will ensure the exhibition of their full genetic potential. It is essential to ensure that there is proper utilization and conservation of food if livestock improvement is to be achieved. Stock improvement can best be achieved more effectively by large units such as co-operative enterprises. A breeding policy should be capable of showing its results within the framework of what is economically feasible in animal husbandry practice.

The extensive feeding of livestock has to be done at a distance from the consumer centres and this raises transport problems particularly with perishable commodities. Where animals have to trek long distances suitable watering and feeding facilities should be provided, otherwise condition will be lost. The construction of roads and railways cannot be as a rule justified on the basis of livestock and livestock marketing requirements alone. In Western Europe and parts of Latin America the greater availability of refrigerated transport is having an important influence on the marketing channel and the location of processing facilities. By carrying only the most valuable parts of the carcass transport costs can be reduced. Movement of livestock by boat is cheap where conditions make this feasible.

In some areas access to a profitable market is only feasible for livestock products if they can be processed into a form less susceptible to deterioration in hot or hot/humid climates during the long time needed to reach the consumer. In many parts of Africa and in North-east Brazil the drying, smoking or salting of meat has become common practice. Canning is the obvious next stage in marketing. It limits risk of transmitting disease.

Schemes for the conversion of seasonal surpluses of milk into butter, ghee or cheese can be most easily put into effect if there are wholesaling and processing facilities and distribution through a canalized organization. This system also allows use to be made of the milk protein at reduced prices in "toned", "filled" or "reconstituted" milks.

Toning the high fat content of local cow or buffalo milk to 3 per cent or less by adding water and skim milk powder has permitted its sale in Bombay at retail prices less than that of whole milk. The reconstitution of skim milk with edible vegetable fats and oils and added vitamins has achieved rapid acceptance in the Phillippines. For milk, payment on a solids-not-fat basis should be introduced since protein is more important. If refrigeration throughout marketing of eggs is not economically feasible, then they should be collected and distributed very rapidly with effective price premiums for performance at each stage.

ANIMAL PRODUCTION AND DISEASE

Although it is not the purpose to include in this volume a monograph on diseases of domesticated animals nevertheless in passing we must note that we cannot reap the harvest of our preventive medicine if water or fodder is insufficient to maintain life and production. As Stableforth (1961) has pointed out:

> It is equally frustrating to improve fodder supplies, and potential production of meat, milk, eggs, wool, if the effort is to be crippled by sudden or insidious inroads of disease. In all our efforts to improve living conditions, it is self-evident that those concerned with water, soil, crop production, plant and animal genetic potential, and with disease, must work side by side, and also with those concerned with human health. It is obvious that with the advances against famine and pestilence the increased number of survivors can produce extra food. Control of enzootic and epizootic diseases such as trypanosomiasis, rinderpest, foot-and-mouth, parasitic diseases, anaerobic diseases, to say nothing of tuberculosis, brucellosis, and mastitis, is necessary. The extension of international organizations concerned in work of this kind is to be

encouraged. What has been demonstrated so clearly in rinderpest control in several areas of the world, and in foot-and-mouth control in Europe (a disease which costs nearly £150 million) may become the policy of every nation.

THE WORLD FOOD PROGRAMME OF FAO

The origins of the World Food Programme lie in the many efforts undertaken during the past decade to find constructive uses for the agricultural surpluses which have been accumulating in some countries in consequence of the application of advanced technology. Parallel to this development there has been a continuing scarcity of food in the much larger developing areas of the world. This prompted research into how surpluses may best be harnessed to combat worldwide hunger and malnutrition. At the same time it was recognised that the abundant output of the more fortunately placed producers must not be utilized in a manner which might injure the interests of producers and traders in other countries.

The General Regulations specify that the Programme may, on request from governments, provide food aid (i) to meet emergency food needs and emergencies inherent in chronic malnutrition; (ii) to assist in pre-school and school feeding; and (iii) to assist in implementing pilot projects using food as an aid to economic and social development.

Through the generosity of many donor countries in all parts of the world and the enlightened guidance of the Intergovernmental Committee and the two sponsoring organizations, the United Nations and FAO, the World Food Programme thus has come into being. Other agencies of the United Nations family and certain non-governmental organizations have played a part. But the challenge of the future is much greater than any the Programme has yet faced. By 1 November 1964 68 countries had pledged cash and commodities, valued at $91·9 millions. Nearly five-sixths of $53·3 millions worth of commodities already delivered or committed to projects and emergencies consisted of the traditional surplus, i.e. cereals, certain dairy products and vegetable oils but the short and shrinking supplies of high protein foods, together with the shortage of pledged cash, limit the balancing of diets in projects still to be adopted. On the livestock production side simultaneous efforts towards internal production of the feed supplies initially provided by the Programme, and a systematic plan for reducing economic numbers and improving productivity of herds were set out.

World Food Programme consultants agree that, irrespective of changes in agricultural policy, food surpluses will accrue at an increasing rate and it is not certain that sufficient scope for their effective use can be found even after adopting both the project and programme approaches in food aid. Integration with development planning and policy is important. While previous assessments of the possible encroachment on potential export markets by food aid overlook the greater injury to international food prices that is avoided by the "umbrella effect" of these aid programmes, great attention needs to be paid to the impact upon the domestic agriculture of beneficiary countries. The net effect in the latter respect is held by FAO to be positive if care be taken to avoid depressing food prices and to concentrate an appropriate proportion of the aid on projects which increase investment and improve productivity in domestic agriculture. Multilateral operation of the project approach has made practicable the mobilization of food aid from many countries not easy to obtain on a bilateral basis, and the resultant inclusion of non-surplus foods has permitted the provision of a more nutritionally adequate diet.

EDUCATION

Education is essential for industrialization, economic development and agricultural improvement, and to reduce social inertia and remove social barriers, and lastly for effective population control policies. How to break the vicious cycle; how to break through the custom, social and economic inertia of a purely agrarian policy, to acquire capital and get the economy on an accelerating industrial basis, is the problem of most developing countries. This has been described as the 'take-off' problem. There is a pressing need for a corps of general practitioners in animal nutrition and management and this will also hold for agronomy and soils.

EPILOGUE

Total calories, total proteins and particularly animal proteins are on the whole very low for populations of Western Asia, Africa, parts of the Far East, Mainland China, and large parts of Latin America. Although food supplies have improved in the worse-fed parts of the world, this improvement has been mainly in crop products because the primary need is for energy. The total supplies of animal products in parts of the Far East will require to be increased six times their present size in order to ensure a reasonably adequate level of nutrition for the people by the year 2000 and even for the world as a whole they will require to be trebled by the turn of the century. We recognize that the general trend is that with increase in economic circumstance there is an increase in consumption of animal protein in so far as religious tenets will permit. By 1980 a doubling of the world's production will be needed. We have 13 years in which to accomplish this.

REFERENCES

CAMPBELL, R. A. and CUTHBERTSON, D. P. (1963) In *Progress in Nutrition and Allied Sciences*, p. 395. Oliver & Boyd, Edinburgh.

CUTHBERTSON, D. P. (1962*a*) *World Food Forum*, p. 111. Proceedings Commemorating Centennial U.S. Department of Agriculture, 1862–1962. May 15–17. Washington, D.C.

CUTHBERTSON, D. P. (1962*b*) *Animal Nutrition and Management*. Proceedings of the United Nations Conference on the Application of Science and Technology for the Benefit of Less Developed Areas. Oct. 8–16. Geneva.

Food and Agricultural Organization of the United Nations (1962) *The Livestock Industry in Less Developing Countries*. Rome.

Food and Agricultural Organization of the United Nations (1964) *Production Year Book*, Vol. 18. Rome.

Food and Agricultural Organization of the United Nations (1965) *Protein Requirements*, Report of a Joint FAO/WHO Expert Group. Rome.

FRENCH, M. H. (1965) Contribution to *International Biological Programme Symposium on Herbivores*. Sept. 19–25, Aberdeen and Cambridge.

JELLIFFE, D. B. (1955) *Infant Nutrition in the Subtropics and Tropics*, World Health Organization. Monograph Series No. 29. Geneva.

STABLEFORTH, A. W. (1961) *Vet. Rec. 73*, 957.

SUKHATME, P. J. (1961) *J. Roy. Stat. Soc. 124*, 463.

TALBOT, L. M., PAYNE, W. J. A., LEDGER, H. P., VERDCOURT, L. D. and TALBOT, M. H. (1965) *The Meat Production Potential of Wild Animals in Africa*. Tech. Communication No. 16 of Commonwealth Bureau of Animal Breeding and Genetics. Edinburgh.

U.S. Department of Agriculture (1961) *The World Food Deficit. A First Approximation. Foreign Agricultural Service*, p. 15.

Author Index

Subject Index